Lecture Notes in Civil Engineering

Volume 46

Lecture Notes in Civil Engineering (LNCE) publishes the latest developments in Civil Engineering - quickly, informally and in top quality. Though original research reported in proceedings and post-proceedings represents the core of LNCE, edited volumes of exceptionally high quality and interest may also be considered for publication. Volumes published in LNCE embrace all aspects and subfields of, as well as new challenges in, Civil Engineering. Topics in the series include:

- Construction and Structural Mechanics
- Building Materials
- Concrete, Steel and Timber Structures
- Geotechnical Engineering
- Earthquake Engineering
- Coastal Engineering
- Ocean and Offshore Engineering; Ships and Floating Structures
- Hydraulics, Hydrology and Water Resources Engineering
- Environmental Engineering and Sustainability
- Structural Health and Monitoring
- Surveying and Geographical Information Systems
- Indoor Environments
- Transportation and Traffic
- Risk Analysis
- Safety and Security

To submit a proposal or request further information, please contact the appropriate Springer Editor:

- Mr. Pierpaolo Riva at pierpaolo.riva@springer.com (Europe and Americas);
- Ms. Swati Meherishi at swati.meherishi@springer.com (Asia—except China—and Australia/NZ);
- Ms. Li Shen at li.shen@springer.com (China).

Indexed by Scopus

More information about this series at http://www.springer.com/series/15087

Kaustubh Dasgupta · A. S. Sajith ·
G. Unni Kartha · Asha Joseph ·
P. E. Kavitha · K. I. Praseeda
Editors

Proceedings of SECON'19

Structural Engineering and Construction Management

Springer

Editors
Kaustubh Dasgupta
Indian Institute of Technology Guwahati
Guwahati, Assam, India

A. S. Sajith
NIT Calicut
Kozhikode, Kerala, India

G. Unni Kartha
Department of Civil Engineering
Federal Institute of Science
And Technology (FISAT)
Angamaly, Kerala, India

Asha Joseph
Department of Civil Engineering
Federal Institute of Science
And Technology (FISAT)
Angamaly, Kerala, India

P. E. Kavitha
Department of Civil Engineering
Federal Institute of Science
And Technology (FISAT)
Angamaly, Kerala, India

K. I. Praseeda
Department of Civil Engineering
NSS College of Engineering
Akathethara, Kerala, India

ISSN 2366-2557 ISSN 2366-2565 (electronic)
Lecture Notes in Civil Engineering
ISBN 978-3-030-26364-5 ISBN 978-3-030-26365-2 (eBook)
https://doi.org/10.1007/978-3-030-26365-2

This Springer imprint is published by the registered company Springer Nature Switzerland AG
The registered company address is: Gewerbestrasse 11, 6330 Cham, Switzerland

Preface

Sustainable construction practices affect every aspect of civil engineering activities with its vast and extensive scope. SECON'19, Third National Conference on Structural Engineering and Construction Management organised by Department of Civil Engineering, Federal Institute of Science and Technology, with its theme "Sustainable Construction and Design for Future" addressed the various challenges in sustainable construction and modern technologies to subdue it.

The two-day national conference held on 15 and 16 May 2019 at Federal Institute of Science of Technology intended to foster research relations and exchange new ideas between the researchers, academicians and industrial professionals in the fields of structural engineering and construction management. The subthemes of the conference included sustainable construction, numerical modelling and simulation in structural engineering, earthquake engineering, building information modelling and modern civil engineering practices.

The conference could witness a number of eminent speakers from reputed institutes across the country. Dr. Radhakrishna G. Pillai, Associate Professor, IIT Madras, inaugurated the conference and delivered the first keynote on "Challenges and Career Prospects in Civil Engineering in India". Er. C. Prathapmohan Nair, Associate Director, Pricewaterhouse Coopers Pvt. Ltd., delivered the second keynote on the topic "Resilient Cities", which touched upon the many aspects of sustainability and the future designs of cities.

There were also keynote addresses by Dr. Sudheesh T. K., Assistant Professor, IIT Palakkad, on the topic "Geotechnical Aspects of Foundation Design and Construction", Dr. Sunil Kumar N. Principal, Cochin University College of Engineering, Kuttanad, on the topic "A Glimpse on Flood Resilient Construction Practices and Analysis Tools Beyond the Finite Element Method", Dr. Job Thomas, Professor, Cochin University of Science and Technology, Kochi, on the topic "Multi-Criteria Decision-Making Models" and by Dr. Sajith A. S., Associate Professor, National Institute of Technology, Calicut, on the topic "Vibration Mitigation of Civil Engineering Structures—Techniques and Strategies" spread across the two days.

All papers presented in the conference were peer reviewed at two stages. It was reviewed after the abstract submission and double-blind reviewed after the submission of full paper. The review committee comprised of 45 reviewers from 19 institutions across the country. A total number of 239 abstracts were received, of which 119 were accepted for full paper submission. After an extensive double review, 86 papers were selected for publication in these proceedings, which is published by Springer in the SCOPUS indexed publication "Lecture Notes in Civil Engineering".

Angamaly, India — Dr. G. Unni Kartha
Angamaly, India — Dr. Asha Joseph
Angamaly, India — Dr. P. E. Kavitha
Guwahati, India — Dr. Kaustubh Dasgupta
Kozhikode, India — Dr. A. S. Sajith
Palakkad, India — Dr. K. I. Praseeda

Acknowledgements

The two-day national conference on Structural Engineering and Construction Management (SECON'19) held on 15 and 16 May 2019 at Federal Institute of Science and Technology (FISAT) could meet its objectives by becoming a platform for fostering research relations and exchange new ideas between the researchers, academicians and industrial professionals in the fields of structural engineering and construction management. The organising committee is indebted to the relentless support received from many.

I am indeed very happy to write this acknowledgement note on behalf of the Organising Committee of SECON'19.

SECON'19 is the third edition of the conference and the prime mover was the Governing body, FISAT, headed by the Chairman Mr. Paul Mundadan. The management has amply supported by financial and institutional support for all three conferences. The top administrative team, which consists of Dr. George Issac, Principal, Dr. K. S. M. Panicker, Director (Academics), Dr. C. Sheela, Vice Principal, and Dr. Sunny Kuriakose, Dean (Student Affairs), offered their complete support for organising the conference.

It was an honour for the Department of Civil Engineering to host the keynote speakers and chairs from very reputed institutions across the country during the two-day event. The reviewer team has been exceptionally supportive in helping the organising committee to meet the deadlines for reviewing the manuscripts and giving suggestions.

The backing provided by Springer for publishing the conference proceedings in "Lecture notes in Civil Engineering" gave good visibility and helped in making the conference appealing for the community. The endorsement from Springer is greatly acknowledged. The contributions of editors, Dr. Kaustubh Dasgupta, IIT Guwahati, Dr. A. S. Sajith, NIT Calicut, and Dr. K. I. Praseeda, NSS College of Engineering, Palakkad, are also acknowledged.

The support provided by Indian Concrete Institute—Kochi chapter, Institution of Engineers (India)—Kochi Chapter, and Indian Society for Technical Education—FISAT chapter for associating with our conference is greatly acknowledged.

The response of the conference was truly remarkable, and the organising committee wish to acknowledge the authors who contributed their papers for the conference.

The conference could materialise only because of the concerted efforts put in by the department team headed by Mr. Unni Kartha G. The untiring efforts put by Dr. Kavitha P. E. (Co-Convener), Ms. Rinu J Achison (Treasurer), and Ms. Lidiya P. M. (Department Editorial Coordinator) are highly commendable. Special note of acknowledgement to all staff members of the Department of Civil Engineering and first-year M.Tech. students who performed the numerous tasks involved in the organisation of the conference.

Dr. Asha Joseph
Convener
SECON'19

Contents

Evaluation of Structural Performance of Concrete with Ambient-Cured Alkali-Activated Binders

Kruthi Kiran Ramagiri, Darshan Chauhan, Shashank Gupta, Arkamitra Kar and Dibyendu Adak

Abstract Enormous global CO_2 emissions associated with cement production necessitates the use of sustainable cementitious alternatives. Alkali-activated binder (AAB) which utilizes industrial wastes as precursors is a promising substitute for cement. To promote the practical use of AAB concrete, this paper presents an investigation on the mechanical and microstructural properties of ambient-cured AAB concrete. Fly ash/slag ratio is varied and the optimum mix is proposed based on compressive strength test results. Pull-out test is performed to evaluate the bond strength of ambient-cured reinforced AAB concrete. The specimen-level tests are supplemented with results from X-ray diffraction (XRD), Fourier transform infrared (FTIR) spectroscopy and scanning electron microscopy (SEM) along with energy-dispersive spectroscopic (EDS) analysis of the AAB paste samples. This is done to corroborate the microstructural characteristics with the mechanical properties at specimen-level. Fly ash: slag ratio of 70:30 is recommended as the optimum proportion considering both strength and economical aspects. Incorporation of slag results in the formation of the additional reaction products, refining the pore structure and

K. K. Ramagiri (✉) · A. Kar
Department of Civil Engineering, BITS-Pilani Hyderabad Campus, Hyderabad, India
e-mail: p20170008@hyderabad.bits-pilani.ac.in

A. Kar
e-mail: arkamitra.kar@hyderabad.bits-pilani.ac.in

D. Chauhan
Department of Civil and Environmental Engineering,
Portland State University, Portland, OR, USA
e-mail: drc1807@gmail.com

S. Gupta
Civil, Environmental and Land Management Engineering, Politecnico di
Milano, Milan, Italy
e-mail: shashankgupta196@gmail.com

D. Adak
Civil Engineering, NIT Meghalaya, Shillong, India
e-mail: dadak@nitm.ac.in

K. Dasgupta et al. (eds.), *Proceedings of SECON'19*,
Lecture Notes in Civil Engineering 46,
https://doi.org/10.1007/978-3-030-26365-2_1

enhancing strength. The AAB mix with fly ash: slag ratio of 50:50 exhibits the highest compressive strength and bond strength.

Keywords Alkali-activated concrete · AAB · Ambient-cured · Bond strength · Microstructure

1 Introduction

Manufacturing of portland cement (PC) is both raw material and energy intensive. Production of one tonne of PC consumes nearly 1.5 tonnes of raw materials, 65–141 kWh of electrical energy along with 2.93–6.23 GJ of thermal energy [1]. Thereby, production of PC contributes to 5–7% of global anthropogenic CO_2 emissions [2]. Research on alternative binder for PC suggests alkali-activated binder (AAB) as a potential replacement [3]. AAB concrete exhibits enhanced mechanical strength [4] and durability characteristics [5, 6]. Different reaction products are formed on alkaline activation of low-calcium and high calcium AAB systems, with coexistence of reaction products in the consequent blended systems. Additionally, secondary reaction products increases the complexities in microstructural characteristics of these binder systems. This demands microstructural analyses using XRD, FTIR, and SEM-EDS to further understand the underlying reaction mechanism. Hence, this study presents an extensive microstructural analysis of ambient-cured AAB with varied proportions of precursors.

The composite behaviour of a reinforced concrete specimen is significantly governed by its bond strength. Few studies conducted on AAB concrete using pull-out test validates its superior bond behaviour compared to PC concrete [7, 8]. The greater tensile strength of AAB concrete and its characteristic compact interfacial transition zone (ITZ) contributes to this superior bond behaviour [7, 8]. Generally, the bond strength of AAB concrete is governed by many factors, including the compressive strength of concrete and the cover provided to the rebar and its geometry [7–9]. Curing conditions govern the bond behaviour of class F fly ash based AAB concrete and improved with the duration of heat curing [10]. The nano-silica modified AAB concrete displayed superior bond performance in contrast to PC concrete, with both mild and deformed steel bars [11]. However, these studies evaluated the bond behaviour between embedded rebar and heat-cured AAB concrete, with curing at temperatures of 60–150 °C for 24 h [7, 8].

Hence, the present study aims to promote the commercial use of AAB as a binder and hence investigates the ambient-cured AAB concrete mixes. The current study investigates the mechanical performance of AAB concrete with varying precursor proportions through compressive and bond strength tests.

Table 1 Details of AAB mix

Materials	FA (kg/m^3)	Slag (kg/m^3)	NaOH (kg/m^3)	Na_2SiO_3 (kg/m^3)	C.A. (kg/m^3)	F.A. (kg/m^3)	Water (kg/m^3)	SP (l/m^3)
FS 0	400	0	10.57	129.43	1209	651	67.65	0
FS 30	280	120	10.57	129.43	1209	651	67.65	3.14
FS 50	200	200	10.57	129.43	1209	651	67.65	4

2 Materials and Methods

2.1 *Materials and Mix Proportions*

Class F fly ash (FA) was procured from National Thermal Power Corporation (NTPC) Ramagundam. Ground granulated blast furnace slag (slag) was obtained from JSW Cement limited. For the alkali-activating solution, rayon grade NaOH flakes and industrial grade Na_2SiO_3 solution with 29.5% SiO_2 and 14.7% Na_2O by weight supplied by Hychem Chemicals were utilized. Aggregates with nominal maximum size of 10 mm and 4.75 mm graded river sand were used as coarse (C.A.) and fine (F.A.) aggregates respectively. Polycarboxylic ether (PCE) superplasticizer (SP) was supplied by BASF Chemicals. Deformed steel bar with a nominal diameter of 12 mm was used to study the bond behaviour of the specimens.

Three different mixes with fly ash: slag ratios varied as 100:0 (FS 0), 70:30 (FS 30), and 50:50 (FS 50) were employed in the present study (Table 1). The water-to-solids (w/s) ratio for AAB concrete was maintained at 0.3. The preparation of specimens commenced with the blending of dry ingredients, followed by addition of activating solution, water and SP, if required, in the sequence mentioned. 100 mm cubes without and with rebar embedded at the centre were cast for compressive and bond strength test respectively. Specimens were moist-cured for 7 days and then preserved in lab until tests were commenced. All the specimens were tested at the age of 28 days.

2.2 *Methods*

2.2.1 Microstructural Studies

XRD analysis of powdered samples was performed through Rigaku Ultima IV X-ray Diffractometer. The CuKα X-rays were operated at 40 kV and 30 mA. The range of 2θ was 5–100°, with a step width of 0.02° and scan speed of 1°/min. Fourier transform infrared (FTIR) spectroscopy was performed in the wavenumber range of 4000–400 cm^{-1} using Jasco FT/IR-4200 and a resolution of 4 cm^{-1} with 64 scans per spectrum. Scanning Electron Microscopy (SEM) and Energy Dispersive X-ray

Spectroscopy (EDS) analysis of the samples was performed using FEI Apreo SEM. The samples were sputtered with 10 nm layer of gold-palladium. The micrographs were captured at 2500× and an operating voltage of 20 kV. The working distance of 10 mm was maintained. A simultaneous EDS analysis was performed for each sample after recording the micrograph.

2.2.2 Compressive Strength

The compressive strength test of concrete was performed in accordance with specifications of ASTM C39/C39M [12] using HEICO compression testing machine of 2000 kN capacity.

2.2.3 Bond Strength

The bond strength of concrete was evaluated through HEICO automated universal testing machine (UTM) of 1000-ton capacity and a specially fabricated test frame. The test complies with the Indian standard IS 2770-1 [13]. A dial gauge is used to measure the slip corresponding to the load recorded. The specimens were subjected to load until they failed either by yielding of rebar or by splitting of concrete.

3 Results

3.1 XRD

XRD pattern of AAB concrete (Fig. 1) reveals its highly amorphous nature due to N-A-S-H and C-A-S-H [14]. Quartz (SiO_2–Q) and mullite ($2Al_2O_3 \cdot SiO_2$—M) were the major crystalline phases and the presence of slag contributed to minor peaks of gehlenite ($Ca_2Al_2SiO_7$—G). Owing to alkali-activation of precursors, minor crystalline phases of albite (Al) and analcime ($NaAlSi_2O_6 \cdot H_2O$—Ac) were also formed. Zeolites are known to form when NaOH is used as activator [15] and it is majorly governed by the chemical composition when slag is added to the system. Minor crystalline zeolitic phases of faujasite (F) and hydroxy sodalite (S) are hence formed in FS 0 and FS 30 with lower Ca/Si ratio. The broad humps indicating the amorphous phase increased in proportion with slag content, due to formation of additional C-A-S-H matrix [14].

Fig. 1 XRD pattern of AAB concrete

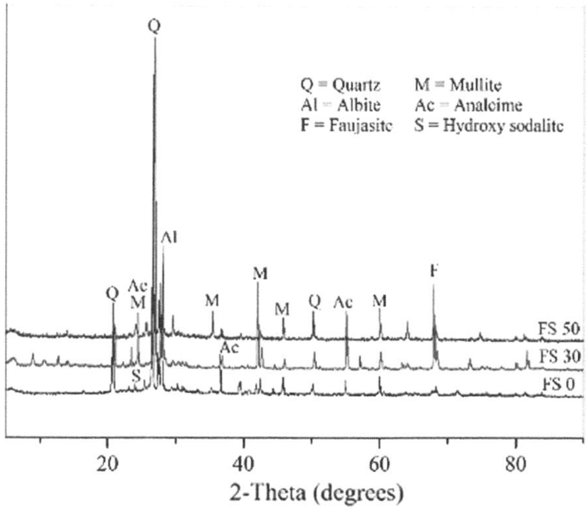

3.2 FTIR

In AAB concrete, the primary bands in the spectra are associated with T–O bonds (Fig. 2). The bands in the range of 3600–3200 and 1640 cm^{-1} are assigned to stretching and bending vibrations of O–H bonds of water in the system. Si–O stretching of SiO_4 tetrahedra in N-A-S-H matrix is ascribed to the band at 1007 cm^{-1}. Relative to FS 0, the chemical shift of these bands is due to increasing Ca content in FS 30 and FS 50 [14]. The band at 777 cm^{-1} is due to Si–O stretching of quartz, while at 446 cm^{-1} represents T–O (T=Al, Si) bending of (Al, Si)O_4 tetrahedra of analcime. In

Fig. 2 FTIR spectra of AAB concrete

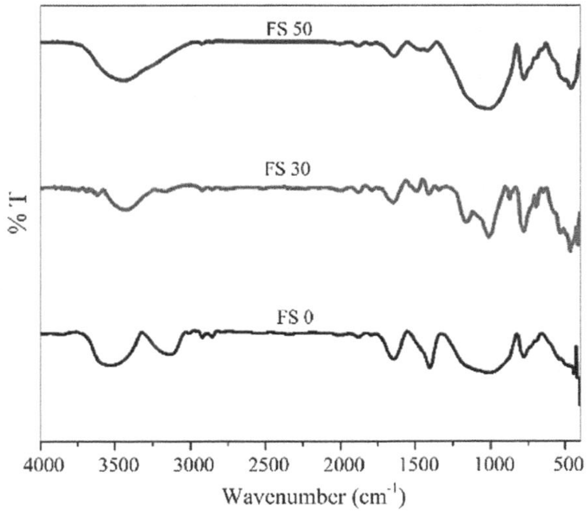

FS 30 and FS 50, the band at 460 cm^{-1} is ascribed to the in-plane bending vibrations of Si–O–Si.

3.3 SEM-EDS

SEM micrographs (Fig. 3) are examined to study the morphologies of different AAB concrete samples. The micrograph for FS 0 shows the maximum proportion of unreacted fly ash particles (10–200 μ in size) as identified by the largest concentration of the cavities surrounding them (Fig. 3a). This is due to the slow rate of reaction at ambient conditions. For FS 30, the concentration of these unreacted particles is considerably lesser. The inclusion of slag in AAB concrete results in the formation of multiple reaction products. With the increase in the proportion of slag in the mixes, the difference in the precipitation of reaction products is evident from the micrographs of FS 30 and FS 50. The microstructure of FS 30 is relatively less compact than FS 50, but the extent of reaction is evident by absence of fly ash spheres (Fig. 3c).

Table 2 presents the average atomic proportions (%) of the primary elements of the reaction products along with their respective ratios. A total of 18 points were analysed for a sample. Si, Al, Na and Ca form the major elements of the reaction products. The principal reaction product in fly ash-dominant systems is N-A-S-H matrix and as the availability of Ca increases with the addition of slag, C-A-S-H matrix begins to develop as well [14]. The decreasing Al content with slag is an indication of formation of additional reaction product, C-A-S-H [14]. Ca/Si ratio increases in proportion to the amount of slag, with FS 50 giving the highest value. FS 0 exhibits maximum Al/Ca ratio. Na/Al ratio for FS 0 is 1.28, which corresponds to

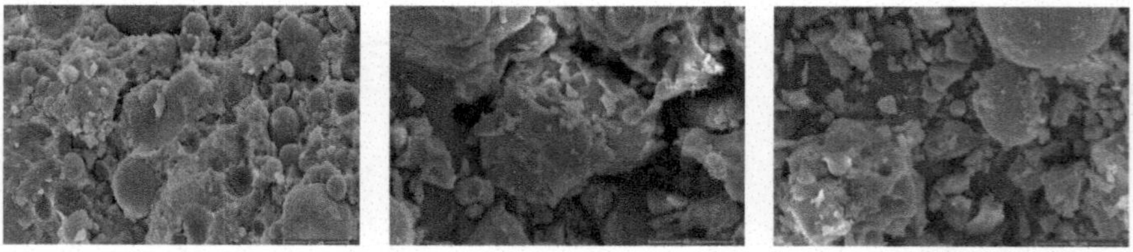

(a) FS 0 (b) FS 30 (c) FS 50

Fig. 3 SEM micrographs and EDS spectra of AAB concrete

Table 2 Results of EDS analysis

Mix	Si	Al	Na	Ca	Si/Al	Al/Ca	Ca/Si
FS 0	11.2	3.5	4.5	0.4	3.22	9.53	0.03
FS 30	15.9	5.7	1.3	2.4	2.79	2.41	0.15
FS 50	13.6	4.8	2.1	3.8	2.83	1.26	0.28

N-A-S-H [14, 16]. In the case of FS 0, Al content of 3.5% corresponds to that of N-A-S-H matrix. But in blended systems, it is difficult to ascertain the atomic percentages to a reaction product, as EDS analysis cannot completely give an indication of the extent of co-existence of these matrices.

3.4 Compressive Strength Test

The mean compressive strength of three ambient-cured AAB concrete at an age of 28 days is presented in Fig. 4. AAB concrete exhibits superior compressive strength with an exception of FS 0 (Fig. 4). The activation energy of fly ash is higher compared to slag and hence inevitably requires heat-curing to overcome the activation barrier and achieve comparable mechanical performance [17]. Enhanced compressive strength with increasing slag content is attributed to the formation of additional C-(A)-S-H matrix [18], as evident from the above mineralogical and microstructural analyses. FS 50 with the highest proportion of slag exhibits maximum compressive strength of all the mixes, with strength 114% higher in contrast to PC concrete. The substantially higher price of slag relative to fly ash and its positive effect of ameliorating the mechanical performance of AAB concrete emphasizes the necessity of arriving at optimum fly ash/slag proportions, in order to promote the use of AAB concrete on the field. The production cost of 1 m^3 for different AAB concrete mixes and their respective compressive strength are presented in Fig. 5. From the results of the present study, FS 30 can be recommended for practical use, subject to the requirements of the end-user.

Fig. 4 Compressive strength test results

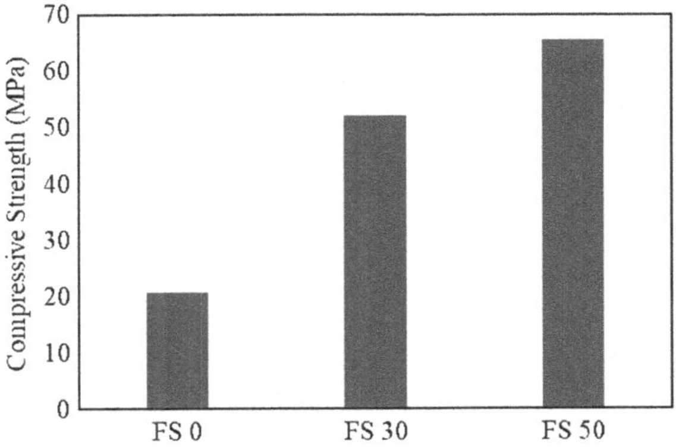

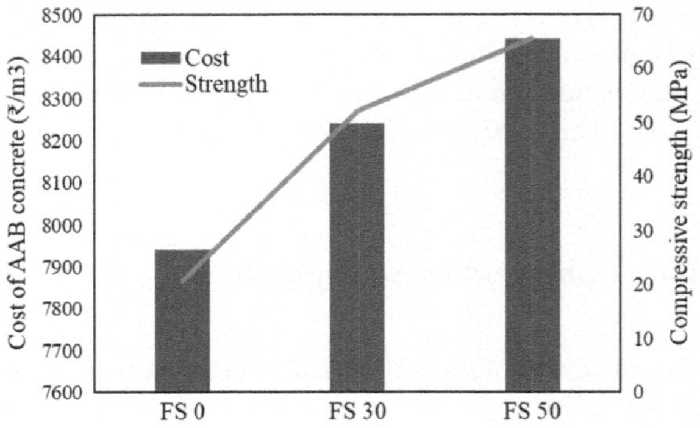

Fig. 5 Production cost versus compressive strength

3.5 *Bond Strength Test*

Figure 6 presents the variation in the bond strength of AAB concrete with different fly ash/slag ratios. Bond strength results follow the same trend as compressive strength. On performing the pull-out test, the failure of FS 0 specimen is due to the splitting of concrete. FS 30 and FS 50 specimens exhibit failure due to slippage of rebar. Rebar embedded in AAB concrete exhibit a greater resistance to pull-out with increasing proportion of slag in the mix. The poor bond behaviour of FS 0 can be explained by the slow reaction of fly ash when cured under ambient conditions. This results in the deficient interfacial transition zone (ITZ) between embedded rebar and concrete, contributing to its lower bond strength. Alkali-activation of slag forms additional reaction products resulting in a compact microstructure. The denser ITZ in AAB concrete with slag contributes to its superior bond behaviour. Based on these observations, the following recommendations are made for the practical use of AAB concrete.

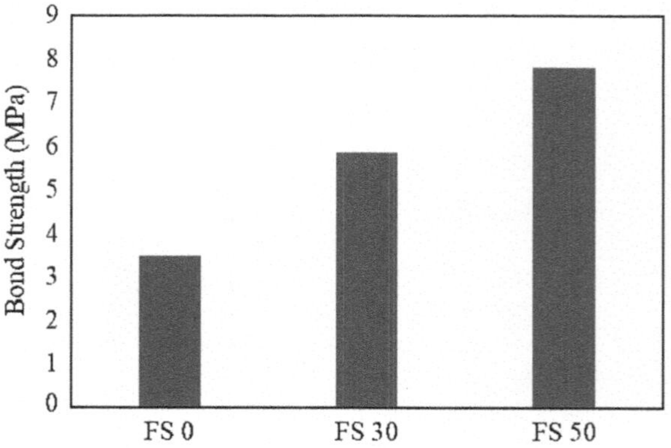

Fig. 6 Bond strength test results

4 Summary and Conclusions

- The compressive strength of ambient-cured AAB concrete with slag as one of the precursors exhibits higher strength as compared to other mixes, with strength increasing in proportion to the slag content. Increase in the proportion of broad peaks with slag content is attributed to the formation of amorphous C-(A)-S-H matrix. Relatively high amount of unreacted fly ash is observed in SEM micrograph of FS 0. SEM micrograph of FS 50 show a denser matrix. Thus, slag addition can compensate for the loss in strength due to the slow reaction of fly ash at ambient conditions.
- Bond strength results exhibit a similar trend as compressive strength. FS 0 exhibits lowest mechanical strength properties among all the mixes due to the splitting of concrete and slippage of the embedded rebars. FS 50 exhibits highest bond strength as the incorporation of slag improves the microstructure due to the formation of additional reaction products.
- From economic considerations, FS50 is the most expensive as fly ash is available free of cost, but slag is relatively expensive. Despite the highest compressive and bond strengths exhibited by FS50, this study recommends the implementation of FS30 for practical use to obtain a balance between expenses and the mechanical properties.

References

1. Stafford FN, Raupp-Pereira F, Labrincha JA, Hotza D (2016) Life cycle assessment of the production of cement: a Brazilian case study. J Cleaner Prod 137:1293–1299
2. Turner LK, Collins FG (2013) Carbon dioxide equivalent (CO_2-e) emissions: a comparison between geopolymer and OPC cement concrete. Constr Build Mater 43:125–130
3. Ke X, Bernal SA, Provis JL (2017) Uptake of chloride and carbonate by Mg-Al and Ca-Al layered double hydroxides in simulated pore solutions of alkali-activated slag cement. Cem Concr Res 100:1–3
4. Gebregziabiher BS, Thomas R, Peethamparan S (2015) Very early-age reaction kinetics and microstructural development in alkali-activated slag. Cem Concr Compo 55:91–102
5. Fernández-Jiménez A, García-Lodeiro I, Palomo A (2007) Durability of alkali-activated fly ash cementitious materials. J Mater Sci 42(9):3055–3065
6. García-Lodeiro I, Palomo A, Fernández-Jiménez A (2007) Alkali-aggregate reaction in activated fly ash systems. Cem Concr Res 37(2):175–183
7. Fernandez-Jimenez AM, Palomo A, Lopez-Hombrados C (2006) Engineering properties of alkali-activated fly ash concrete. ACI Mater J 103(2):106–112
8. Sarker PK (2011) Bond strength of reinforcing steel embedded in fly ash-based geopolymer concrete. Mater Struct 44(5):1021–1030
9. Yang KH, Song JK (2012) Empirical equations for mechanical properties of $Ca(OH)_2$-based alkali-activated slag concrete. ACI Mater J 109(4):431–440
10. Castel A, Foster SJ (2015) Bond strength between blended slag and Class F fly ash geopolymer concrete with steel reinforcement. Cem Concr Res 72:48–53
11. Adak D, Sarkar M, Mandal S (2017) Structural performance of nano-silica modified fly-ash based geopolymer concrete. Constr Build Mater 135:430–439

12. ASTM (2011) Standard test method for compressive strength of cylindrical concrete specimens. ASTM C39/C39M11, West Conshohocken, PA
13. Indian Standards (1967) Methods of testing bond in reinforced concrete, Part 1: Pull-out test. IS 2770-1, Bureau of Indian Standards, New Delhi, India
14. Criado M, Aperador W, Sobrados I (2016) Microstructural and mechanical properties of alkali activated Colombian raw materials. Materials 9(3):158(Multidisciplinary Digital Publishing Institute (MDPI))
15. Ma Y (2013) Microstructure and engineering properties of alkali activated fly ash-as an environment friendly alternative to Portland cement. Ph.D. dissertation, TU Delft, Netherlands
16. Walkley B, Rees GJ, San Nicolas R, van Deventer JS, Hanna JV, Provis JL (2018) New structural model of hydrous sodium aluminosilicate gels and the role of charge-balancing extra-framework Al. J Phys Chem C 122(10):5673–5685
17. Bakharev T (2005) Geopolymeric materials prepared using Class F fly ash and elevated temperature curing. Cem Concr Res 35(6):1224–1232
18. Kar A, Halabe UB, Ray I, Unnikrishnan A (2013) Nondestructive characterizations of alkali activated fly ash and/or slag concrete. Eur Sci J ESJ 9(24). http://dx.doi.org/10.19044/esj.2013.v9n24p%25p

Light Weight Foamed Concrete as a Substitute for Bricks in Framed Structure

K. Azmil Bani and Jiji Antony

Abstract Construction industries in India are extensively using various materials such as concrete block, bricks, hollow blocks, etc. for infill walls. These blocks are bulk in weight and have transportation problems. This study aims at the feasibility of cellular lightweight concrete using fly ash, cement and synthetic based foaming agent with density 800 kg/m^3. Mix design was prepared with cement to fly ash ratio as 375:375, 250:500, 500:250 kg/m^3, followed by water curing. Experimental investigation was conducted on optimized mix with respect to density, compressive strength, water absorption, fire resistance, in addition to extreme environmental conditions. Micro structural studies including SEM/XRD are also performed. The results indicated that foamed concrete block and panels can be used for infill purpose with improved performance and other characteristics.

Keywords Foamed concrete · Foaming agent · Compressive strength · Durability · Fire resistance · Thermal conductivity · Micro structure

1 Introduction

1.1 General

Construction sector is growing day by day with the invention of new building materials like solid block, hollow blocks, interlocking block etc. As in the case of multistoreyed structures, elements like partition walls and parapet are non-load bearing. These blocks impart a huge dead load on foundation and large number of vertical

K. Azmil Bani (✉) · J. Antony
Department of Civil Engineering, Federal Institute of Science and Technology, Angamaly, Kochi 683577, India
e-mail: azmilbani01@gmail.com

J. Antony
e-mail: jijiantony13@gmail.com

K. Dasgupta et al. (eds.), *Proceedings of SECON'19*,
Lecture Notes in Civil Engineering 46,
https://doi.org/10.1007/978-3-030-26365-2_2

joints in them results in the propagation of cracks through that joints. Foamed concrete is considered as an economical and easy method in fabrication of lightweight construction materials and components such as structural members, partitions, and filling material, due to its easy and economical in production from manufacturing process, transportation and final applications [1–4]. The strength properties of light weight concrete are low because of its low density. So, less weight, lowers transportation and handling costs which results in economical construction of structure. Several studies were concentrated on foamed concrete with its ability to flow, low cement content, low or no aggregate usage and its thermal insulation properties. A compressive strength more than 25 MPa can be achieved with foamed concrete under special curing conditions has been reported together by varying porosity, density and compressive strength [1–4].

Method of foam preparation, distribution of air voids, selection of materials, mix design etc. will determine the production of a good and stable foamed concrete. According to the studies, foamed concrete is commonly manufactured by two different methods. First method known as preforming method, is done by mixing a preformed foam or mix-foaming agents into the cement and water slurry. As the concrete hardens, the bubbles disintegrate, leaving air voids of uniform size [5–7]. Second method known as mixed foaming method [5, 6]. Autoclaved aerated concrete (AAC) most commonly used mixed foaming method. It is done by mixing lime, sand or fly ash, cement mix with expansion agent such as aluminium powder. The reaction between the cement and expansion agent causes hydrogen bubbles to form, results the expansion of concrete into five times than original volume. After the reaction process over, aerated concrete is cut into suitable sizes and steam-cured in an autoclave chamber. But it is preferred only for large scale production with high investment. This research aims to develop a new building material which consider strength, durability and by using fly ash and cement we can reduce weight without not much effecting cost. thermal properties of foamed concrete. Here in this project we selected pre foaming method due to easy availability.

2 Material and Methods

The materials used for this research work are OPC (53 grade) confirming to IS:12269 2013, class F fly ash confirming to IS:3812 part 1 2003, synthetic based foaming agent meet requirements of 9 of IS 9103 1999 and potable water used. The properties of cement, fly ash and foaming agent are Shown in Tables 1, 2 and 3 respectively.

Table 1 Physical properties of cement

Specific gravity	3.231
Fineness of cement	8%
Consistency of cement	30.5%
Initial setting time	90 min

Table 2 Details about fly ash

Class	F
Specific gravity	2.71
Specific surface area	606.4 m^2/kg

Table 3 Details of foaming agent

Type of foaming agent	Synthetic
Physical state	Liquid
Packaging size	50 kg barrel
Dilution ratio	1–50 L

3 Mix Design

Initially cement, fly ash and water are mixed to obtain a good workable mix (base mix). Trial mixes were initially prepared with varying water to binder ratios such as 3.0, 3.2, 3.5. The optimum water to binder ratio is selected as 0.35 based on workability and consistency. Mixes were cast by using cement, fly ash and foaming agent and are designated as FC1, FC2, FC3. Foam is prepared using a foam generator and it was mixed with base mix and again the mixing was continued three minutes. Figure 1 shows the picture of foam with base mix.

The mixing was done at a rate of 30 rpm. The density of batter was also verified in accordance with the procedure explained in [5]. After 28 days of curing, cubes were tested for compression. Mix design is tabulated in Table 4.

The results of compressive strength are tabulated in Table 5.

The compressive strength test results of FC2 mix are within the permissible limit according to IS 2185 part 4 2008. There for mix FC2 is selected for further studies. The 90 day compressive strength test was conducted on selected FC2 mix. There was a increase in strength with increase in age.

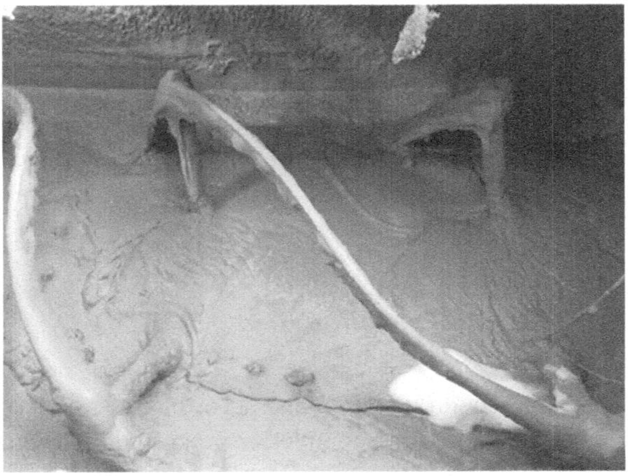

Fig. 1 Mixing foam with base mix

Table 4 Mix proportions of specimen

Sample name	Weight of cement (kg/m^3)	Weight of fly ash (kg/m^3)	Water/solid
FC1	375	375	0.35
FC2	250	500	0.35
FC3	500	250	0.35

Table 5 Compressive strength results of mix design

Mix	Compressive strength
FC1 (1.5:1.5)	2.1
FC2 (1:2)	2.22
FC3 (2:1)	1.22

4 Tests on Foamed Concrete Specimen

The selected samples were subjected to compressive strength, water absorption, aggressive environment, thermal conductivity, scanning electron microscopy, x-ray diffraction and elevated temperature studies.

4.1 SEM/XRD Test

SEM is a precise and good method to determine the pore size distribution of materials. The micrographs obtained from SEM gives a detail about porosity. It is a useful aid in determining the structure of compounds formed after the hydration process. The microstructure of powdered foamed concrete sample of FC2 mix at 28 day and 90 day are determined [8].

X-Ray Diffraction (XRD) analysis, can be used to obtain more information about crystallinity of foamed concrete, its compounds and peaks. The chemical compounds that formed on foamed concrete due to hydration reaction with age can be analysed. It is also useful tool in identifying morphology and details about the crystalline phases too. By conducting XRD analysis chemical modification that takes place in hydration reaction with respect to age can be analysed. XRD test of FC2 sample at 90 day were determined.

4.2 Water Absorption Test

Water absorption test are done according to IS 2185 part 4 2008. Specimen after curing were taken and completely immersed in water for 24 h. Then test specimens

Fig. 2 Water absorption test

shall be removed from water, allowed to drain, surfaces were cleaned by a cloth and weighed (A kg). In order to take the dry weight (B kg) all specimens shall be dried in an oven for 24 h at 100 °C to 115 °C. Figure 2 shows water absorption test.

$$\text{Water absorption, percent} = \frac{A - B}{B} \times 100$$

4.3 Aggressive Environment Test

The samples were immersed in sulphuric acid, sodium sulphate, hydrochloric acid and sodium chloride solutions to study the effects in aggressive. All the solutions were prepared with 3% of each substance. The studies were conducted on cured specimen for 28 days on solutions and compared with water immersed in water for same duration [9–11]

4.4 Elevated Temperature Test

Behaviour of foamed concrete under elevated temperatures is evaluated on 7 × 7 × 7 cm specimens after curing. The specimens were exposed to an elevated temperature of 200, 400, 600, and 800 °C in a muffle furnace for 2 h. Later the specimens were cooled in normal temperature and weight loss, crack pattern and compressive strength were checked out [12]. Fire resistance is done as per code ASTM 2748-11-2017.

4.5 Thermal Conductivity Test

Lee's disc test was conducted to determine the thermal conductivity of foamed concrete. It is done as per code ASTM D7340-07-2018. Lee's apparatus consists of a metallic disc provided with a hole for inserting thermo meter. A disc made of foamed concrete specimen is placed above steam chamber. Through the hole in the solid base of the chamber, a thermometer T_1 is inserted. A sensitive thermometer T_2 is inserted in the metallic disc. FC2 mix with density 800 kg/m^3, with specimen diameter same as that of disc and thickness about 10 mm is used. Figure 3 shows the photograph of thermal conductivity test.

The thermal conductivity is calculated using the equation as follows.

$$\lambda = \frac{MC * \frac{DT}{dt}}{\pi r^2 (T1 - T2)} * \frac{(r + 2h)d}{2(r + h)}$$

Fig. 3 Photograph of thermal conductivity measurement setup

Fig. 4 Compressive strength test on blocks and panels

4.6 Application of Foamed Concrete in Block and Panel

Blocks of sizes 60 × 20 × 20 cm and panels of size 70 × 70 × 10 cm casted with mix of FC2 (250:500) of density 800–900 kg/m^3. The size of block were determined according to IS 2185 part 4 2008. Wooden moulds are used for casting, after 28 days of water curing compressive strength is checked out. In foamed concrete panels, a chicken mesh of size 12–25 mm is put inside in order to avoid any temperature and shrinkage cracks. The bottom layer foamed concrete layer is first laid, then mesh is placed over, then top layer is laid. In both blocks and panels foamed concrete should be placed as soon as possible with a single stretch of pouring, in order to avoid a separation of layers. The compressive strength testing of blocks shown in Fig. 4a and that of panels are shown in Fig. 4b.

5 Results and Discussions

The results are discussed below.

5.1 SEM/XRD Results

The compressive strength of foamed concrete shows a continuous increase with age. The rate of strength development was greater at initial stage and decreased at later stage. However, a comparison study of strengths shows that, at 28 days foamed concrete developed almost 80–85% of the 90-day compressive strength. This clearly reveals that the foamed concrete with fly ash can gain long-term strength, 90 day results are 10–15% higher than that of 28 day strength of same density. The reason for

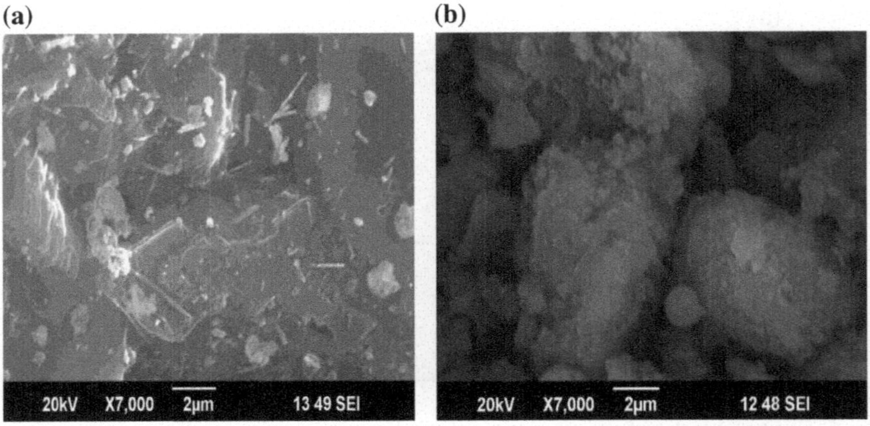

Fig. 5 SEM images of 28 day and 90 day

strength gains of foamed concrete with respect to age can be clearly identified from SEM/XRD results. SEM results of 28 and 90 day specimens are shown in Fig. 5a, b respectively.

In the earlier stage water from curing get diverted to cement for its hydration and strength gain. Later water finds its way to fly ash results formation of C–S–H gels Foamed concrete usually attains strength beyond 28–90 days. During 28 day, ettringite, large prismatic crystals of calcium hydroxide and very small fibrous crystals of calcium silicate hydrates begin to form. C–S–H have very high surface area, is not a well-defined compound and exact structure is not known. The SEM results of 90 day are more. So, more C–S–H is formed with in 90 day than 28 day.

Figure 6 shows XRD results of FC2 specimen in 90 days reveals higher proportion of C–S–H phase a from 33° to 59° two-theta positions. Higher peaks of silicon oxide between 26° and 27° two-theta positions are present. Moganite is also present throughout, which is an oxide mineral with chemical formula SiO_2.

5.2 *Water Absorption Test*

Water absorption is inversely proportional to density. The water absorption test was done with target density of 800 kg/m^3. The average value obtained 10.2% is less than 12.5% as prescribed by the code IS 2185 part 4 2008. It was investigated that the increase in fly ash/cement ratio in the mix proportionally increased the water vapor permeability especially at the lower densities.

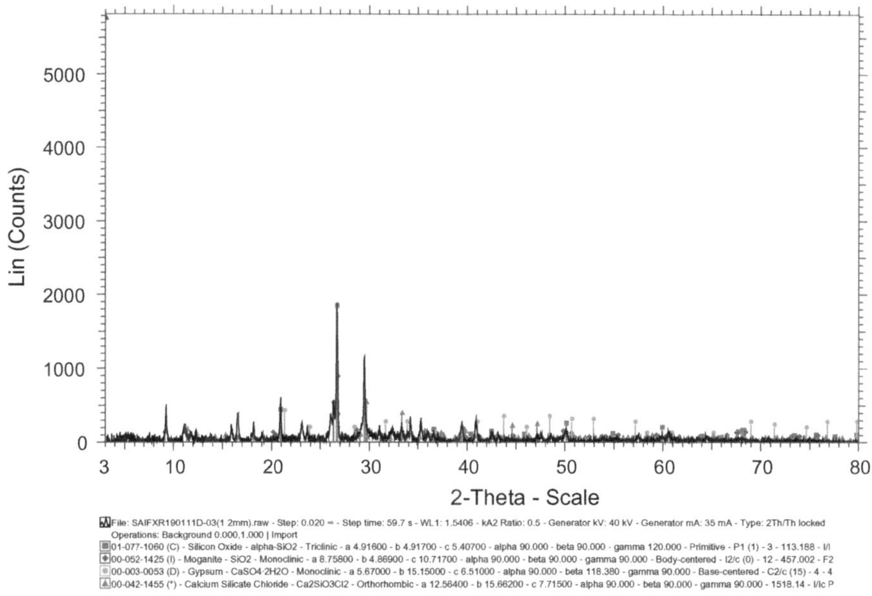

Fig. 6 XRD image of 90 day

5.3 *Aggressive Environment Test Results*

Visually specimen immersed in sulphate solution expanded and cracks formed throughout the body. Whereas the surfaces of the specimen immersed in chloride solution peeled off with surface deteriorations Weight gain occurred for specimens in sulphate environment and weight loss occurred for specimens in chloride environment. An average of 8.34% weight gain occurred and around 35% loss of compressive strength occurred for foamed concrete in sulphuric acid. In sodium sulphate environment an average of 4.3% of weight gain and 12.5% loss of compressive strength occurred.

Expansion of specimen in sulphate environment is due to formation of calcium sulphate and calcium sulpha aluminate which is 227% more volume than, thus increased volume. At initial stage foamed concrete less dense, but at later gypsum and ettringite fill the existing voids As the immersion period proceeds, more $Ca(OH)_2$ reacts with sulphates and expansion continues, results the formation of cracks [2, 9]. Figure 7 shows the weight loss/gain of specimen on different solutions. Figure 8 shows the loss of compressive strength on solutions.

In hydrochloric acid environment, an average weight loss about 7.9% and compressive strength loss about 13.16% occurred. In sodium chloride environment 3.7% of weight loss and 2.6% loss in compressive strength occurred. The main reason for

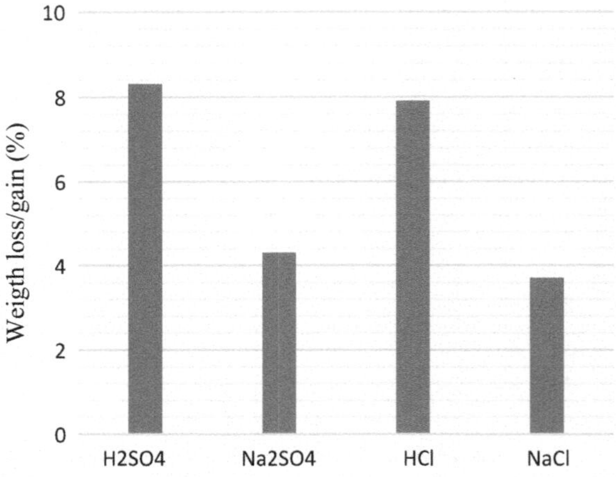

Fig. 7 weight loss/gain of specimen on different solutions

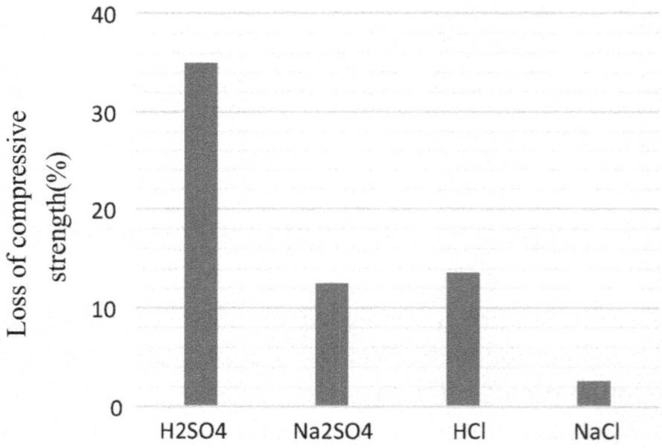

Fig. 8 Compressive strength loss on different solutions

mass loss is $Ca(OH)_2$ Formed during reaction react with HCl to form calcium chloride. The mass loss increases with time, formed compounds are soluble in external solution [10, 11].

5.4 Elevated Temperature Test

Visually when specimen exposed to elevated temperatures propagation of cracks increases as with increase in temperature. A decrease in weight is also observed. During the initial stage of heating foamed concrete lost its free water (evaporable water), absorbed water and then chemically bound water. Loss water in foamed concrete would induce microcrack and resulting in reduction in strength. Up to 200 °C loss of strength is less. Between 200 and 400 °C loss of strength increased, it may be due to decomposition of C–S–H gel and sulpha luminate phases and caused

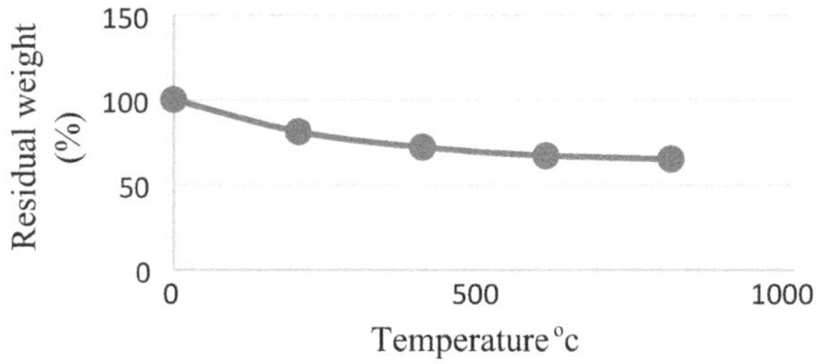

Fig. 9 Residual weight versus temperature

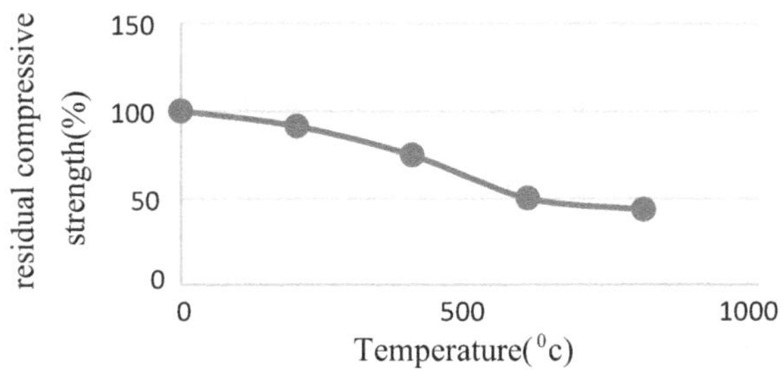

Fig. 10 Residual compressive strength versus temperature

increased cracks in specimen. After 600°–800° strength again decreased and concrete retained only about 40–50% of the original strength [7, 13]. Figures 9 and 10 shows relation between residual weight, residual compressive strength versus temperature.

5.5 *Thermal Conductivity Test*

Thermal conductivity values of foamed concrete varies directly with density. The average thermal conductivity value for obtained is 0.27 w/m/k is with in permissible limit according to IS 2185 part 4 2008 Table 1. The permissible value is up to 0.32 W/m/k for minimum density of 800 kg/m^3. The utilization of low thermal conductivity building materials like foamed concrete are important to decrease heat gain through the envelope into the building in hot climate country like India.

5.6 *Test Results on Blocks and Panels*

The average compressive strength of blocks obtained was 2.26 MPa, which is with in permissible limits according to IS 2185-part 4 2008 Table 1 (Minimum required is 2.2 MPa). Plates are placed above and below the specimen for uniform distribution of load through the surface. The average compressive strength of panel obtained was

2 MPa. Its compressive strength is less than permissible limits. It is because entire area not take part while loading on UTM, and cracks are formed at the top part only.

6 Conclusions

(1) The 28-day average compressive strength value obtained is 2.22 N/mm^2, satisfies the requirements as per IS 2185 part 4 2008.
(2) The water/solids ratio adopted for the mix was found to be 0.35, met the requirement as per the code IS 2185 part 4 2008.
(3) Foamed concrete has excellent thermal insulating properties, the average thermal conductivity value obtained is 0.27 w/m/k, which is with in permissible limit according to Table 1 of IS 2185 part 4 2008.
(4) Foamed concrete has good ability to with stand extreme temperature, it retained more than 45% of its compressive strength after keeping for 800 °C for 2 h.
(5) It performed a good result in resistance against extreme environmental conditions such as water absorption, sulfates and chloride environments, maximum loss of compressive strength in sulphuric acid environment with a value of 35% in 28 days.
(6) From the above research on foamed concrete, it can be recommended as infill block material.

References

1. Ramamurthy K, Nambiar EKK, Indu Siva Ranjani GA (2009) A classification of studies on properties of foam concrete. Cement Concr Compos 31:388–396
2. Amran M, Farzadnia N, Abang Ali AA (2015) Properties and applications of foamed concrete a review. Constr Build Mater 101:990–1005
3. Sattainathan Sharma A, Srisudharssan S, Rathna Priya P (2017) Experimental investigation on cellular lightweight concrete. IJSRD 5
4. Mastali M, Kinnunen P, Isomoisio H, Karhu M, Illikainen M (2017) Mechanical and acoustic properties of fiber-reinforced alkali-activated slag foam concretes containing lightweight structural aggregates. Constr Build Mater 187:371–381
5. Bing C, Zhen W, Ning L (2012) Experimental research on properties of high strength foamed concrete. ASCE 24:113–118
6. Harith IK (2018) Study on polyurethane foamed concrete for use in structural applications. Case Stud Constr Mater 2018(8):79–86
7. Vinith Kumar N, Arunkumar C, Srinivasa Senthil S (2018) Experimental study on mechanical and thermal behavior of foamed concrete. Proc Mater Today 5:8753–8760
8. Just M (2009) Microstructure of high-strength foam concrete. J Mater Mater Charact 60:741–748
9. Izzat AM et al (2014) Sulfuric acid attack on ordinary Portland cement and geopolymer material. Res Gate
10. Medine M et al (2018) Durability properties of five years aged lightweight concretes containing rubber aggregates. Periodica Polytech Civil Eng 62:386–397

11. Anandh S et al (2018) Behaviour of foamed concrete under high temperature. Int J Pure Appl Math 118
12. Ramadhansyah P et al (2012) Properties of concrete containing rice husk ash under sodium chloride subjected to wetting and drying. Proc Eng 50:305–313
13. Wang YC et al (2012) Mechanical properties of foamed concrete exposed to high temperatures. Constr Build Mater 26:638–654

Experimental Studies on Fly Ash Based Basalt Fibre Reinforced Concrete

T. Swathi and K. N. Resmi

Abstract The use of supplementary cementitious materials by replacing cement makes the concrete more economical and the use of industrial wastes like fly ash for this makes the disposal method of these wastes easy and ecofriendly. Basalt fibre is an inorganic fibre manufactured from the extrusion of melted basalt rock. Basalt fibres are characterized by high corrosion and thermal resistance, lightweight and high strength. The fibres act as a proactive reinforcement that provides immediate tensile load carrying capacity when micro-cracks are developed in concrete. It has excellent mechanical and physical properties and an environmental friendly manufacturing process. The advantages are fire resistance, good resistance to chemically active environments, vibration and acoustic insulation capacity. The optimum of fly ash in concrete is found to be 20% by weight of cement. The optimum percentage of basalt fibre is 0.5% by volume of concrete. Basalt fibre addition may not be an effective method to increase concrete compressive strength. By adding basalt fibre in concrete, the split tensile strength and flexural strength can be increased up to 37.73% and 36.71% respectively. Tests on the concrete are conducted according to IS: 516-1959.

Keywords Supplementary cementitious materials · Fly ash · Basalt fibre

T. Swathi (✉) · K. N. Resmi
Department of Civil Engineering, Federal Institute of Science and Technology, Angamaly, Kochi 683577, India
e-mail: swathitharol@gmail.com

K. N. Resmi
e-mail: reshmikn@gmail.com

K. Dasgupta et al. (eds.), *Proceedings of SECON'19*,
Lecture Notes in Civil Engineering 46,
https://doi.org/10.1007/978-3-030-26365-2_3

1 Introduction

1.1 General

Concrete is the most widely used construction material all over the world due to its good compressive strength and durability. The popularity of concrete is also due to the fact that from the common ingredients, the properties of concrete are tailored to meet the demand of any particular application. In order to increase the properties of concrete, other materials can also be incorporated in it.

Fly ash is a coal combustion product that is composed of the particulates i.e. fine particles of burned fuel that are driven out of coal-fired boilers together with flue gases [1]. Depending upon the composition and source of the coal being burned, components of fly ash vary considerably, but all fly ash includes substantial amounts of silicon dioxide (SiO_2) both crystalline and amorphous, aluminium oxide (Al_2O_3) and calcium oxide (CaO), the main mineral compounds in coal-bearing rock strata. In the past, fly ash was generally released into the atmosphere, but now air pollution control standards require that it be captured prior to release by fitting pollution control equipment. In the United States, generally fly ash is stored at coal power plants or placed in landfills. About 43% is recycled and often used as a pozzolan to produce hydraulic cement or hydraulic plaster and a replacement or partial replacement for Portland cement in concrete production. The pozzolans ensure the setting of concrete and plaster and provide concrete with more protection from wet conditions and chemical attack. Owing to its pozzolanic properties, fly ash is used as a replacement for Portland cement in the concrete. As pozzolan greatly improves the strength and durability of concrete, the use of ash is a key factor in their preservation. Use of fly ash as a partial replacement for Portland cement is particularly suitable but is not limited to Class C fly ashes [2]. Class F fly ashes can have volatile effects on the entrained air content of concrete, causing reduced resistance to freeze/thaw damage. Fly ash often replaces up to 30% by mass of Portland cement, but they can be used in higher dosages in certain applications. In some cases, fly ash can add to the concrete's final strength and increase its durability and chemical resistance [3]. Due to the spherical shape of fly ash particles; it can increase the workability of cement while reducing water demand.

Various types of fibres can be used in cement and concrete composites, such as steel or organic fibres. Basalt fibre (BF) is a new kind of inorganic fibre extruded from melted basalt rock [4, 5] and is currently available commercially [6]. The manufacturing process of this kind of fibre is similar to that of the glass fibre, but with less energy consumed and no additives, which makes it cheaper than glass or carbon fibres [7]. Other advantages are high modulus, heat resistance, good resistance to chemical attack, excellent interfacial shear strength and currently commercial availability [8, 9]. Basalt fibre is a good alternative to glass, carbon or aramid fibre as a reinforcing material in concrete composite. Requirements of the moderate strengthening in the civil structures and high fire resistance can be met with basalt fibres while FRP strengthening can be considered for pure strengthening [10].

1.2 Objectives

- To find the optimum percentage of Basalt Fibre and Fly Ash in concrete to enhance the properties of normal concrete.
- To test the properties of fly ash based basalt fibre reinforced concrete.
- To compare these properties with normal concrete.

2 Materials Used

2.1 Cement

OPC of grade 53 conforming to IS 8112:1989 is used. Cement tests were conducted as per IS 4031:1988 and the results are analyzed using IS 8112:1989. The properties of cement used are given in Table 1.

2.2 Aggregates

The fine aggregate type used is manufactured sand which falls under zone II according to IS 383:1970. Crushed stone of maximum size 20 mm is used as coarse aggregate which is confirming to IS: 383-1970. Tests on coarse aggregate was done confirming to IS 2386 (Part 1 and 3): 1963 and the test results are analyzed as per IS: 383-1970. The properties of aggregates used are given in Table 1.

Table 1 Properties of cement and aggregates

	Properties	Obtained value
Cement	Specific gravity	3.14
	Standard consistency	30.5%
	Initial setting time	85 min
	Fineness	8.66%
Fine aggregate	Specific gravity	2.487
	Water absorption	2.4%
	Fineness modulus	2.71
Coarse aggregate	Specific gravity	2.649
	Water absorption	0%

Table 2 Properties of basalt fibre

Properties	Value
Diameter	20 μm
Length	12 mm
Aspect ratio	600
Moisture content	≤0.30%
Density	2.65 g/cm^3

Fig. 1 Basalt fibre

2.3 *Basalt Fibre*

Basalt fibre is an inorganic fibre which is made from melted basalt rock [11, 12] and which has a tensile strength 3 times that of steel fibres [13]. The properties of basalt fibre used are listed in Table 2 and it is shown in Fig. 1.

2.4 *Fly Ash*

Class F fly ash is used as a partial replacement of cement. The specific gravity of the fly ash used is 2.16 and the standard consistency was obtained as 29%. The initial setting time was found to be 95 min.

2.5 *Superplasticizer*

MasterRheobuild 1125 is composed of synthetic polymers specially designed to allow considerable reduction of mixing water while maintaining control on extend of set retardation. A dosage range of 600–1800 ml per 100 kg of cementitious material is normally recommended. It is a ready-to-use liquid which is dispensed into the concrete together with the mixing water. The properties of superplasticizer used are shown in Table 3.

3 Mix Design

The mix design for M30 concrete as well as for fly ash concrete was calculated. The replacement ratios of fly ash were 5, 10, 15, 20, 25, 30 and 35% by weight of cement. The mix design of concrete specimens is shown in Table 4.

The replacement ratios of basalt fibre were 0.25, 0.5, 0.75, 1, 1.25 and 1.5% by volume of concrete. Mix design of the specimens made by varying fly ash is given in Table 5.

Table 3 Properties of superplasticizer used

Properties	Values
Relative density	1.24 ± 0.02 at 25 °C
pH	>6
Chloride ion content	<0.2%
Dosage	0.6% by weight of cement

Table 4 Mix design of specimen with varying percentage of fly ash

Notation	Cement (kg/m^3)	Fly ash (kg/m^3)	% replacement	Fine aggregate (kg/m^3)	Coarse aggregate (kg/m^3)	SP (l/m^3)	Water (l/m^3)
CM	370	0	0	692.73	1203.86	2.22	147.75
FA05	351.5	18.5	5	692.73	1203.86	2.22	147.75
FA10	333	37.5	10	692.73	1203.86	2.22	147.75
FA15	314.5	55.5	15	692.73	1203.86	2.22	147.75
FA20	296	74	20	692.73	1203.86	2.22	147.75
FA25	277.5	92.5	25	692.73	1203.86	2.22	147.75
FA30	259	111	30	692.73	1203.86	2.22	147.75
FA35	240.5	129.2	35	692.73	1203.86	2.22	147.75

Table 5 Mix design of specimen with varying percentage of basalt fibre

Notation	Cement (kg/m^3)	Basalt fibre (kg/m^3)	% addition	Fine aggregate (kg/m^3)	Coarse aggregate (kg/m^3)	SP (l/m^3)	Water (l/m^3)
CM	370	0	0	692.73	1203.86	2.22	147.75
BF1	370	6.625	0.25	692.73	1203.86	2.22	147.75
BF2	370	13.25	0.5	692.73	1203.86	2.22	147.75
BF3	370	19.875	0.75	692.73	1203.86	2.22	147.75
BF4	370	26.5	1	692.73	1203.86	2.22	147.75
BF5	370	33.125	1.25	692.73	1203.86	2.22	147.75
BF6	370	39.75	1.5	692.73	1203.86	2.22	147.75

Table 6 Mix design of specimen with optimum percentage of fly ash and basalt fibre

Notation	Cement (kg/m^3)	Fly ash (kg/m^3)	Fine aggregate (kg/m^3)	Coarse aggregate (kg/m^3)	Basalt fibre (kg/m^3)	SP (l/m^3)	Water (l/m^3)
CM	370	0	692.73	1203.86	0	2.22	147.75
FABF1	296	74	692.73	1203.86	13.25	2.22	147.75
FABF2	296	74	692.73	1203.86	19.875	2.22	147.75

The mix design for the specimens with optimum fly ash and basalt fibre are shown in Table 6. The details of the mixes would be discussed later.

4 Results and Discussions

4.1 Slump Test

Slump test was carried out to determine the workability of concrete.

4.1.1 Slump Results on Addition of Fly Ash

The slump was checked for different percentages of fly ash. The results obtained are shown in Table 7. With the addition of fly ash, the slump value also increased. The usage of fly ash in concrete increases the workability of concrete.

Table 7 Slump test of concrete with fly ash

Notation	Slump (mm)
CM	90
FA05	90
FA10	92
FA15	94
FA20	98
FA25	99
FA30	100
FA35	102

Table 8 Slump test of concrete with basalt fibre

Notation	Slump (mm)
CM	90
BF1	88
BF2	85
BF3	81
BF4	77
BF5	74
BF6	72

4.1.2 Slump Results on Addition of Basalt Fibre

The slump was checked for different percentages of basalt fibre. The results obtained are shown in Table 8. With the addition of basalt fibre, the slump value decreased. Addition of basalt fibre results in reduced workability. This is due to the greater surface area of fibres which requires higher amount of cement paste to wrap around it.

4.1.3 Slump Results on Fly Ash Based Basalt Fibre Reinforced Concrete

Table 9 shows the slump test results of control mix and the mix with fly ash and basalt fibre.

Table 9 Slump test of concrete with basalt fibre and fly ash

Notation	Slump (mm)
CM	90
FABF1	88
FABF2	85

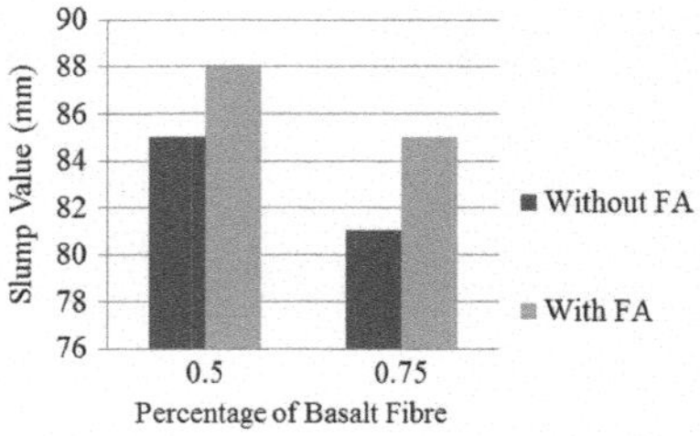

Fig. 2 Slump test of basalt fibre concrete with and without fly ash

Table 10 Compressive strength of concrete with varying fly ash

Notation	Compressive strength (N/mm^2)
CM	38.26
FA05	38.29
FA10	38.31
FA15	38.36
FA20	38.5
FA25	36.5
FA30	35.33
FA35	31.22

From Fig. 2 it is clear that the addition of fly ash in concrete with fibre increases the workability of concrete. When 0.5% by volume of fibre is added in concrete, the slump was found to be 85 mm and when 20% fly ash is added in it by replacing cement, the slump increased to 88 mm.

4.2 Compressive Strength

4.2.1 Compressive Strength of Concrete with Varying Percentage of Fly Ash

The compressive strength of concrete cubes by varying the percentage of fly ash is given in Table 10 and Fig. 3. The optimum of fly ash is found to be 20%. When the fly ash content is increased further, the compressive strength decreases.

4.2.2 Compressive Strength of Concrete with Varying Percentage of Basalt Fibre

The compressive strength of concrete cubes made with and without basalt fibre is tested after 7 and 28 days. The effects of fibre addition in concrete on the compressive strength are shown in Table 11 and Fig. 4.

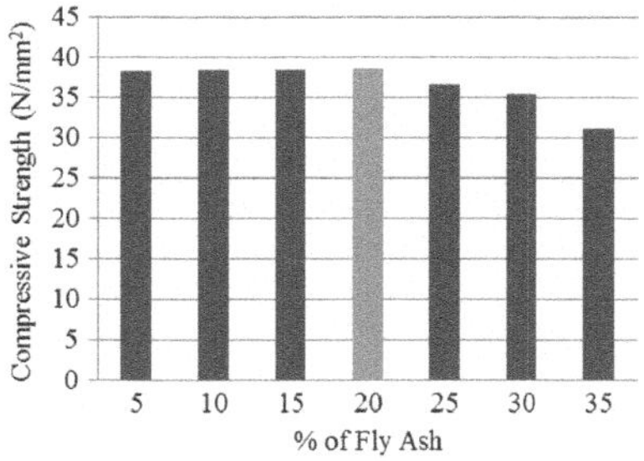

Fig. 3 Variation of compressive strength with percentage of fly ash

Table 11 Compressive strength of concrete with varying basalt fibre

Notation	Compressive strength (N/mm^2)	
	7 Days	28 Days
CM	24.78	38.26
BF1	24.73	33.43
BF2	24.69	36.13
BF3	22.58	30.61
BF4	20.3	25.20
BF5	19.5	20.21
BF6	18.6	18.85

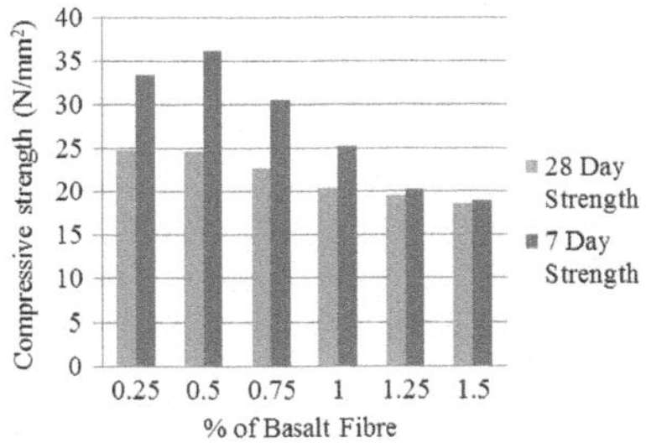

Fig. 4 Compressive strength of concrete with varying basalt fibre

Concrete with 0.5% basalt fibre gives the maximum compressive strength. After this when the fibre content is increased, the compressive strength is decreased. Basalt fibre addition cannot be adopted as a method to improve the compressive strength of concrete. It has little effect on the compressive strength. The decrease in compressive strength is mainly due to the non-homogeneity of the mix.

Table 12 Split tensile strength of concrete with varying basalt fibre

Notation	Split tensile strength (N/mm^2)	
	7 day strength	28 day strength
CM	2.2	2.56
BF1	2.33	2.58
BF2	2.43	3.43
BF3	2.48	3.62
BF4	2.41	3.45
BF5	2.4	3.42
BF6	2.3	3.36

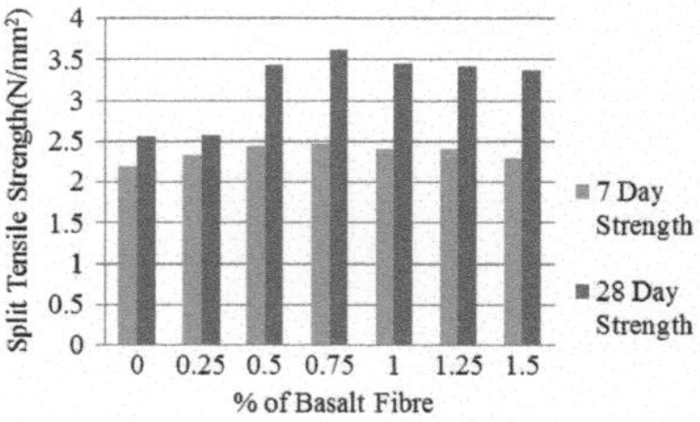

Fig. 5 Split tensile strength of concrete with varying basalt fibre

4.3 *Split Tensile Strength*

4.3.1 Split Tensile Strength of Concrete with Varying Percentage of Basalt Fibre

The tensile strength of concrete cylinders made with and without basalt fibre was tested after 7 and 28 days. The effects of fibre addition in concrete on the tensile strength are shown in Table 12 and Fig. 5. Concrete with 0.75% basalt fibre gives the maximum tensile strength. The tensile strength increased up to 37.73%. Unlike the control mix, the specimen with basalt fibre did not break after first cracking.

4.4 *Flexural Strength*

4.4.1 Flexural Strength of Concrete with Varying Percentage of Basalt Fibre

The flexural strength of concrete beams made with and without basalt fibre was tested after 7 and 28 days. The effects of fibre addition in concrete on the flexural strength are shown in Table 13 and Fig. 6. Concrete with 0.75% basalt fibre by volume gives the maximum flexural strength. The flexural strength increased up to 36.71%. Using basalt fibre for increasing the flexural strength is effective and the optimum may be

Table 13 Flexural strength of concrete with varying basalt fibre

Notation	Flexural strength (N/mm^2)
	28 day strength
	Average
CM	5.91
BF1	6.33
BF2	7.58
BF3	8.08
BF4	5.91
BF5	5.91
BF6	4.75

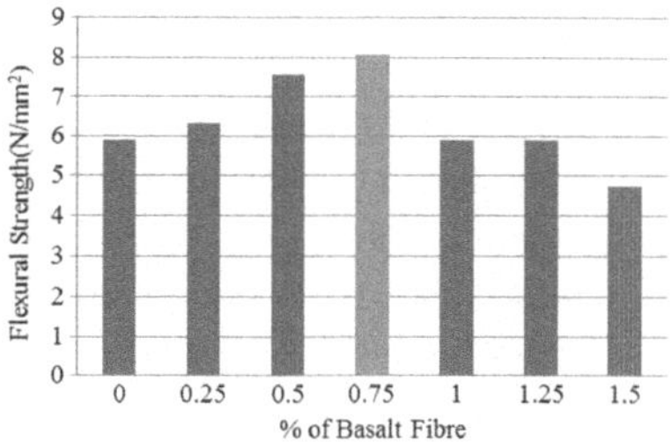

Fig. 6 Flexural strength of concrete with varying basalt fibre

adopted as 0.75% by volume of concrete. The flexural strength increased from 5.91 to 8.08 N/mm^2 and then decreased.

4.5 Density

The variation of density of concrete with the addition of basalt fibre is as shown in Table 14 and Fig. 7. It may be concluded from the table that the density of concrete decreases with the increase in percentage of basalt fibre.

Table 14 Density of concrete with varying basalt fibre

Notation	Density (kg/m^3)
CM	2498.33
BF1	2456.97
BF2	2448.90
BF3	2445.06
BF4	2386.52
BF5	2372.83
BF6	2312.17

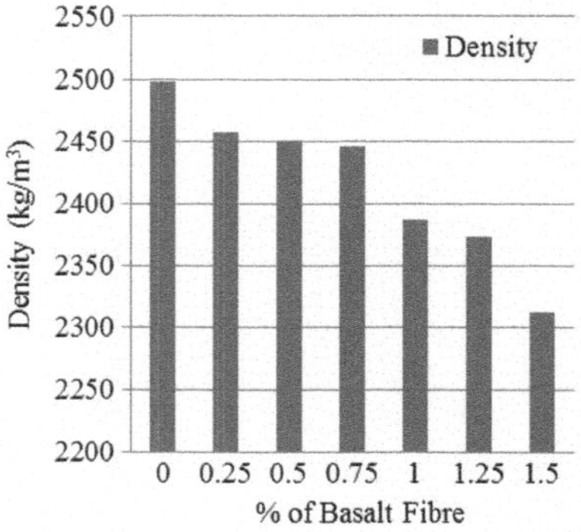

Fig. 7 Density of concrete with varying fibre content

4.6 Water Absorption

The difference in percentage of water absorption of mixes with increasing the fibre content is shown in Table 15 and Fig. 8. It is clear that as the fibre content is increased, the water absorption is increased slightly. The percentage of water absorption varies from 4.02 to 4.301 for concrete with fibre content 0 to 1.5%.

The addition of basalt fibre in concrete results in a non-homogeneous mix which incorporates voids in the concrete. The presence of these voids in concrete results in the increase in water absorption.

Table 15 Water absorption of concrete with varying basalt fibre

Notation	Water absorption (%)
CM	4.02
BF1	4.11
BF2	4.16
BF3	4.19
BF4	4.21
BF5	4.21
BF6	4.22

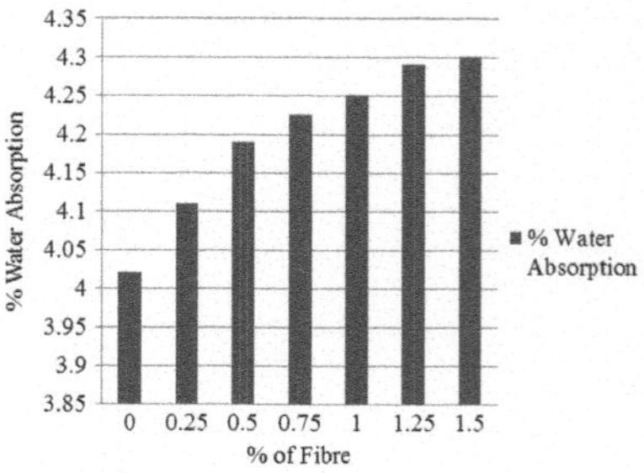

Fig. 8 Water absorption of concrete with varying basalt fibre

Table 16 Properties of concrete with optimum fly ash and basalt fibre

Properties	CM	0.5% BF		0.75% BF	
		Without FA	With FA	Without FA	With FA
7 day compressive strength (N/mm^2)	24.78	24.69	24.6	22.58	22.8
28 day compressive strength (N/mm^2)	38.26	36.13	37.21	30.61	32.09
7 day split tensile strength (N/mm^2)	2.2	2.433	2.35	2.48	2.51
28 day split tensile strength (N/mm^2)	2.56	3.43	3.47	3.616	3.72
28 day flexural strength (N/mm^2)	5.91	7.58	7.51	8.08	8.1
Density (kg/m^3)	2498.33	2448.90	2451.63	2445.06	2445.44
Water absorption (%)	4.02	4.16	4.098	4.198	4.15

4.7 Properties of Fly Ash Based Basalt Fibre Reinforced Concrete

The properties of concrete with optimum percentage of fly ash and basalt fibre are shown in Table 16 the optimum of fly ash was found to be 20% and for basalt fibre both 0.5 and 0.75% were checked.

5 Conclusions

- The properties such as split tensile strength and flexural strength of conventional concrete can be improved by using fly ash and basalt fibre.
- 20% replacement of cement with class F fly ash gives the maximum strength for concrete.
- Use of basalt fibre may be adopted for increasing both split tensile as well as the flexural strength of concrete. But the use of basalt fibre may not increase the compressive strength of concrete. The decrease in compressive strength is due to the non-homogeneity of the ix results as the addition of basalt fibre.
- Using basalt fibre alone in concrete may result in a decrease in workability as well as the compressive strength and this may be overcome by using fly ash in it. Fly ash improves both workability and compressive strength of concrete.
- The workability of the concrete is decreased with addition of basalt fibre and it is due to the greater surface area of fibres which requires more cement paste to wrap around it.

- It is observed that the compressive strength of concrete made using fly ash at replacement level of 20% and basalt fibre at 0.5% increases. Beyond that it decreases because of non-homogeneity of the mix.
- By using basalt fibre, the split tensile strength of concrete can be improved up to 37.73% and the flexural strength can be improved up to 36.71%.
- Similarly the flexural and tensile strength of the concrete made with basalt fibre up to 0.75% increases.
- The density of concrete decreases as the fibre dosage is increased. The density of concrete without fibre was found to be 2498.33 kg/m^3 and it reduces with the increases in dosage of fibre. The density of concrete with 1.5% fibre was 2312.17 kg/m^3.
- The density of the concrete with both basalt fibre and fly ash is greater than that with basalt fibre only. The smaller fly ash particle makes the concrete denser by filling up the voids and gaps.
- Water absorption of concrete increases with increase in fibre content. The addition of basalt fibre in concrete results in a non-homogeneous mix which incorporates voids in the concrete. The presence of these voids in concrete results in the increase in water absorption. The addition of fly ash in basalt fibre concrete leads to filling up of the voids and thereby decreases the water absorption.

References

1. George S, Sofi A (2017) Enhancement of fly ash concrete by hydrated lime and steel fibres. 4:807–811
2. Wang W, Lu C, Li Y, Yuan G, Li Q (2017) Flexural behavior of basalt fibre reinforced concrete beams with recycled concrete coarse aggregates. Constr Build Mater 169:165–178
3. Sujjavanich S, Suwanvitaya P, Chaysuwan D, Heness G (2017) Synergistic effect of metakaolin and fly ash on properties of concrete. Constr Build Mater 155:830–837
4. High C, Seliem HM, El-Safty A, Rizkalla SH (2015) Use of basalt fibres for concrete structures. Constr Build Mater 96:37–46
5. Alnahhal W, Aljidda O (2018) Flexural behavior of basalt fibre reinforced concrete beams with recycled concrete coarse aggregates. Constr Build Mater 169:165–178
6. Zhang H, Wang B, Xie A, Qi Y (2017) Experimental study on dynamic mechanical properties and constitutive model of basalt fibre reinforced concrete. Constr Build Mater 152:154–167
7. Jiang C, Fan K, Wu F, Chen D (2014) Experimental study on the mechanical properties and microstructure of chopped basalt fibre reinforced concrete. Mater Des 53:187–193
8. Afroz M, Patnaikuni I, Venkatesan S (2017) Chemical durability and performance of modified basalt fibre in concrete medium. Constr Build Mater 154:191–203
9. Niaki MH, Fereidoon A, Ahangari MG (2018) Experimental study on the mechanical and thermal properties of basalt fibre and nanoclay reinforced polymer concrete. Compos Struct 191:231–238
10. Wlodarczyka M, Jedrzejewskia I (2016) Concrete slabs strengthened with basalt fibres – experimental tests results. 163:460–470
11. Kabay N (2014) Abrasion resistance and fracture energy of concretes with basalt fibre. Constr Build Mater 50:95–101
12. Branston J, Das S, Kenno SY, Taylor C (2016) Mechanical behaviour of basalt fibre reinforced concrete. Constr Build Mater 124:878–886

13. Arslan ME (2016) Effects of basalt and glass chopped fibres addition on fracture energy and mechanical properties of ordinary concrete: CMOD measurement. Constr Build Mater 114:383–391

Experimental Studies on Brick Powder Replaced Concrete Exposed to Elevated Temperature

Ann Maria Santhosh and Abin Thomas

Abstract Industrial waste management constitutes one of the greatest global problems of our time, especially for large cities lacking landfill. Recycling non biodegradable wastes is a challenging task, and brick wastes are classified as this kind of waste with a long decomposition period. Since brick powder has high content of silica, it is pozzolanic in nature. So it can be used as an effective replacement for cement. The concrete structures are most likely to experience elevated temperatures during their service period due to fire. Therefore, the relative properties of concrete after exposure to fire are to be studied for the serviceability of buildings and the safety of inhabitants. In this study, experimental programs are to be performed regarding the use of brick powder for replacement of cement at elevated temperature. The different specimen samples are casted. The properties of the concrete specimens including compressive strength, SEM and XRD of the concrete after being exposed to the temperatures of 200, 400 and 600 °C and prior to experiencing heat are to be investigated. The obtained results are to be compared and the changes occurred are to be studied further.

Keywords Brick powder · Elevated temperature

1 Introduction

Concrete is the most widely used material in construction industry because of high structural strength and ability. The concrete is heterogeneous mix of cement, aggregate and water in which about 65% of the volume of the concrete is occupied by the volume of the aggregates and the remaining is occupied by cement and water. The majority of cementitious binder used in concrete is based on Portland Cement clinker

A. M. Santhosh (✉) · A. Thomas
Department of Civil Engineering, Federal Institute of Science and Technology, Angamaly, Kochi 683577, India
e-mail: annmariasanthosh@gmail.com

A. Thomas
e-mail: abinabraham80@gmail.com

K. Dasgupta et al. (eds.), *Proceedings of SECON'19*,
Lecture Notes in Civil Engineering 46,
https://doi.org/10.1007/978-3-030-26365-2_4

which is an energy intensive process. One ton of cement production is responsible for one ton of CO_2 emission [1]. A problem from the construction activity is the improper way of managing the construction waste. Year by year, the amount of construction wastes that are dumped at the land filling area is increasing. One of the ways is by recycling the wastes for other usage. The main idea is to substitute or reduce the materials usage for conventional construction in construction industry. The brick waste is a potential material to be used as an alternative for cement in concrete production [2–4]. Brick powder is very easily available at very low cost [5–7]. It may be recycled from destroyed buildings, bridges, and any other destroyed structures. The concrete materials in structures are most likely to experience elevated temperatures during their service period due to fire. Therefore, the relative properties of concrete after exposure to fire are of particular interest in the serviceability of buildings and the safety of inhabitants. In this study, experimental programs were performed regarding the use of brick powder replaced concrete at elevated temperature [8].

Concrete temperatures up to 95 °C have very little effect on the strength and other properties of concrete [9, 10]. Above that, the threshold cement paste undergoes shrinkage due to dehydration and aggregates expand due to temperature rise which results in overall expansion of concrete and reduction in its strength [9–11]. The rise in temperature causes a decrease in the strength for concrete. However, the rate at which the strength decrease depends on the rate of increase in the temperature of the fire and the insulating properties of concrete [11–13]. Concrete structures are frequently exposed to fire. It is not only the maximum temperature that is important but also the rate at which temperature rises. Compressive strength is the most studied property of concrete in fire.

2 Objectives

- To study the behavior of brick powder replaced concrete at elevated temperature
- To study the effectiveness of brick powder in imparting strength to the concrete at an elevated temperature
- To compare the characteristics of normal concrete and brick powder replaced concrete after being exposed to higher temperature.

3 Materials Used

3.1 Brick Powder

Bricks obtained are powdered and sieved through 90μ sieve and then analysed for its properties and are found to have similar cementatious property with cement (Table 1; Fig. 1).

Table 1 Chemical composition of brick powder

Constituents	Percentage of composition
CaO	0.52
SiO_2	63.21
Al_2O_3	16.41
Fe_2O_3	6.05
Na_2O	1.19
K_2O	2.83
MgO	1.11

Fig. 1 Brick powder

3.2 *Cement*

The cement used for this study is Ordinary Portland Cement conforming to IS 12269-1987 of grade 53 with a specific gravity of 3.124 (Table 2).

3.3 *Aggregates*

The Coarse Aggregate having specific gravity of 2.572 were used. M-sand with a specific gravity of 2.513 and fineness 2.51 is used as fine aggregate and it is sieved through sieves of different sizes (4.75–0.15 mm). It conforms to Zone II of the grading zone. The aggregates are conforming to IS 383-1970.

Table 2 Properties of cement and brick powder

Properties	Cement	Brick powder
Specific gravity	3.124	2.08
Fineness	7.67%	6.53%
Consistency	31%	–

3.4 *Super Plasticizer*

MasterRheobuild 1125 with a specific gravity of 1.24 ± 0.02 at 25 °C is used as the superplasticizer.

3.5 *Water*

Water is added according to the water cement ratio. The water quality conforms to IS456-2000.

4 Mix Design

See Table 3.

5 Results

5.1 *Compressive Strength*

The cubes of size 15 cm × 15 cm × 15 cm of control mix and brick powder replaced samples are casted and tested for the compressive strength. Cement is replaced by 10% brick powder which was obtained as the optimum among different percentages of 5, 10, 15 and 20. The compressive strength of control specimens and brick powder replaced specimens exposed to room temperature, 200, 400 and 600 °C for a duration of 30 min are shown in the table. The samples are named according to the mix (control mix concrete and brick powder concrete) and with their varying temperature. CM000 stands for control mix at room temperature and BP000 stands for brick powder concrete at room temperature and so on. The strength increases at 200 °C due to

Table 3 Mix design of specimens

Materials	CM	BP10
Cement (kg/m^3)	370	333
Fine aggregate (kg/m^3)	692.73	692.73
Coarse aggregate (kg/m^3)	1203.86	1203.86
Water (l/m^3)	147.75	147.75
Super plasticizer (l/m^3)	2.22	2.22
Brick powder (kg/m^3)	0	37.5

Table 4 Compressive strength of cubes

Property	CM000 (MPa)	CM200 (MPa)	CM400 (MPa)	CM600 (MPa)	BP000 (MPa)	BP200 (MPa)	BP400 (MPa)	BP600 (MPa)
7 day	21.27	27.87	25.13	4.54	21.6	24.98	30.7	9.29
14 day	29.78	30.66	28.56	5.13	30.495	32.36	31.095	7.23
28 day	37.775	40.28	36.45	11.125	39.315	41.455	39.795	14.9

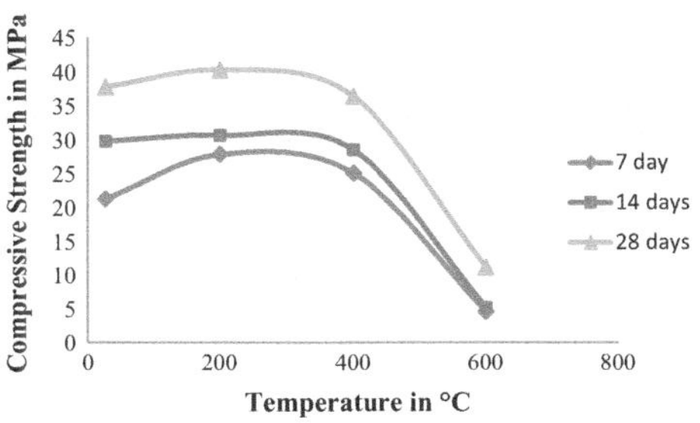

Fig. 2 Variation of compressive strength with temperature for normal concrete samples

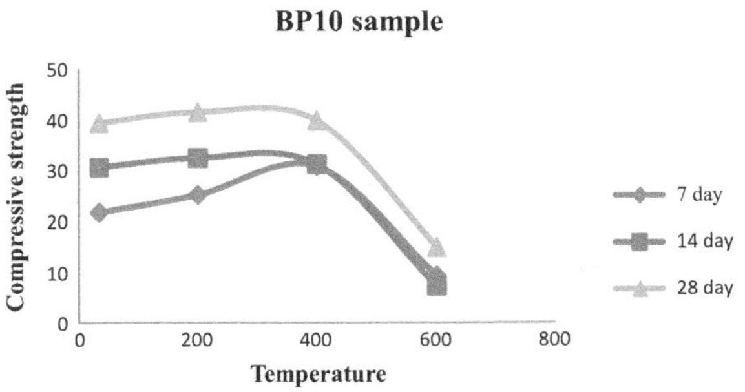

Fig. 3 Variation of compressive strength with temperature for brick powder replaced samples

the sudden hydration taking place when the sample is exposed to temperature. The decrease in strength at 400 and 600 °C is by losing the bond between the particles in the sample when exposed to higher temperature (Table 4; Figs. 2 and 3).

5.2 *Scanning Electron Microscope*

The SEM results show the microstructural changes happening to the samples while it is exposed to elevated temperatures [14]. As per the obtained SEM images, the presence of CSH enhances the strength and the presence of CH initiates hydration reaction when it is suddenly exposed to a higher temperature. With an increase in temperature, the binding between the particles get reduced which results in voids thus resulting in reduction of strength (Figs. 4, 5, 6, 7, 8 and 9).

Fig. 4 SEM image of CM000

Fig. 5 SEM image of CM200

Fig. 6 SEM image of CM600

Fig. 7 SEM image of BP000

Fig. 8 SEM image of BP200

Fig. 9 SEM image of BP600

5.3 X-Ray Diffraction

The albite content in normal concrete mix imparts strength to the sample. Albite which has a content of silica is more at 200 °C which results in its increased strength and is less at 600 °C. For brick powder replaced samples a compound named labradorite is also present along with albite which helps in filling the voids and imparting strength to the concrete (Figs. 10, 11, 12, 13, 14 and 15).

5.4 Visual Examinations

At 200 °C, there is no noticeable effect except for the colour change. At 400 °C, along with the colour change cracks started appearing on the surface. At 600 °C, the cracks became intense and the colour almost turned to slight red. The visual changes are almost same for normal concrete and brick powder replaced concrete except that by the addition of brick powder, the intensity of formation of cracks reduced (Fig. 16).

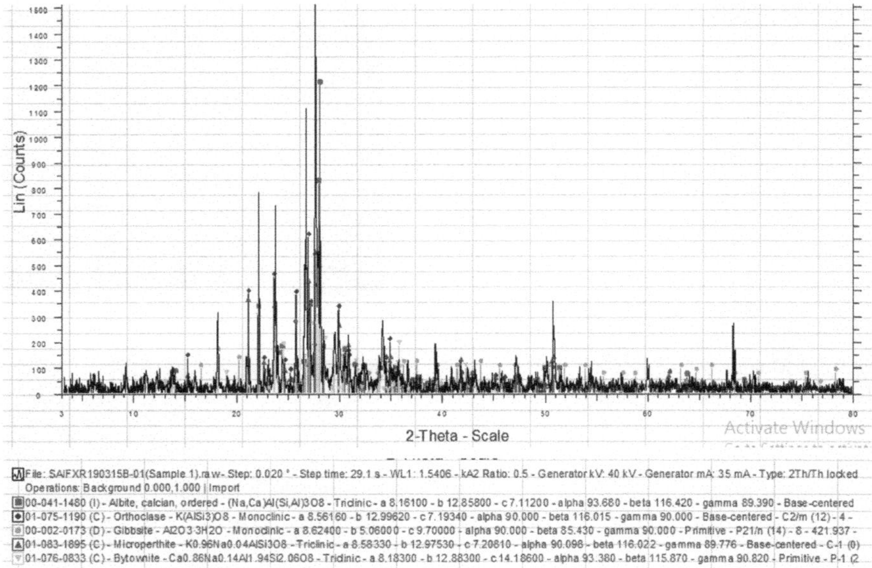

Fig. 10 XRD image of CM000

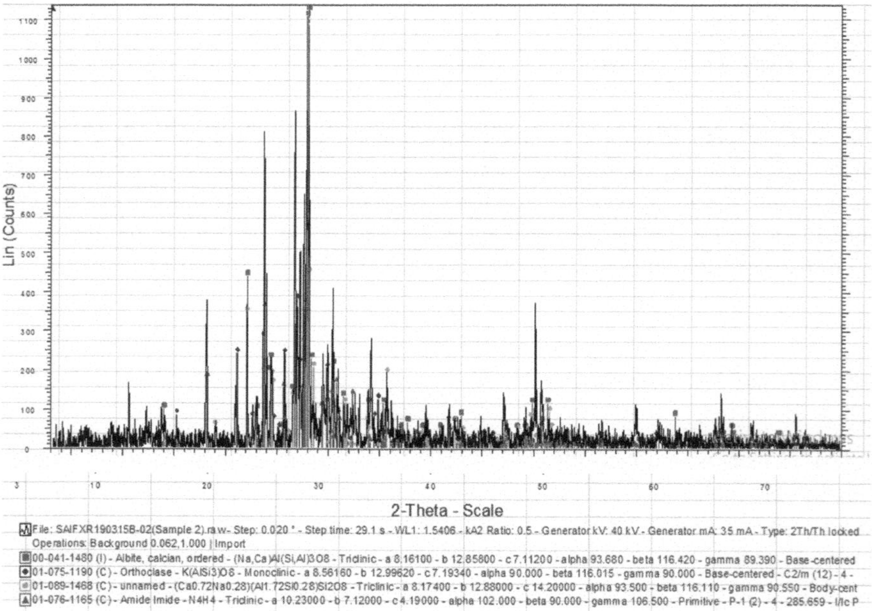

Fig. 11 XRD image of CM200

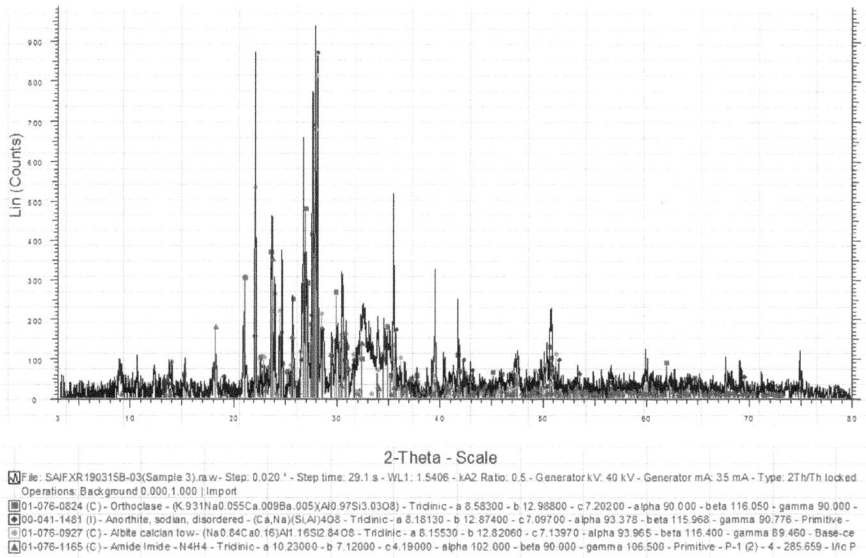

Fig. 12 XRD image of CM600

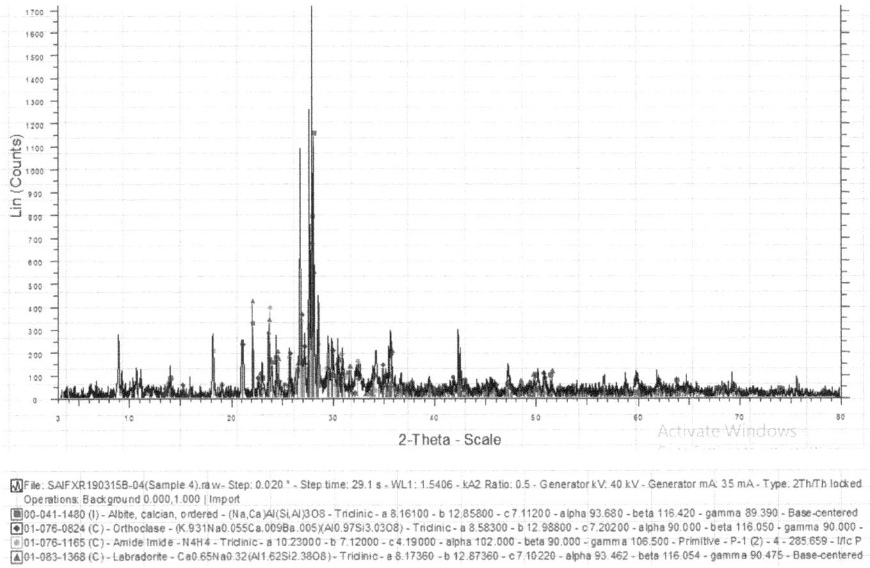

Fig. 13 XRD image of BP000

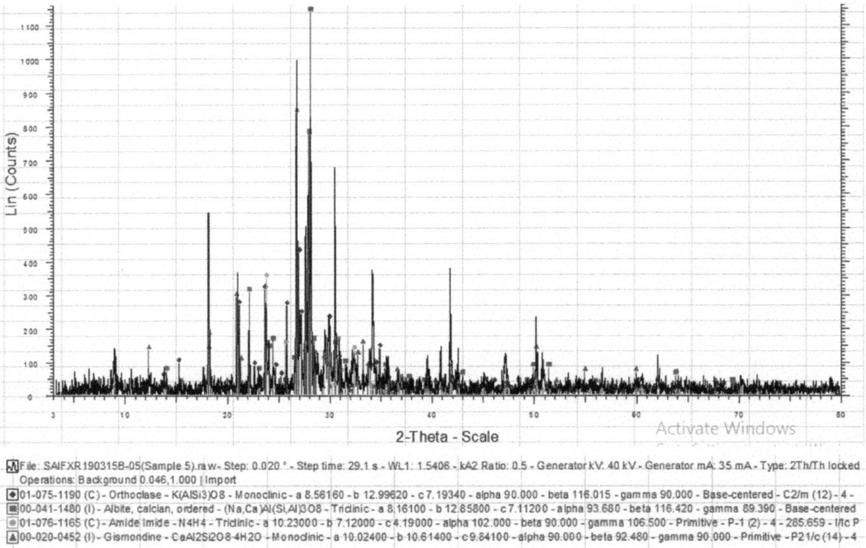

Fig. 14 XRD image of BP200

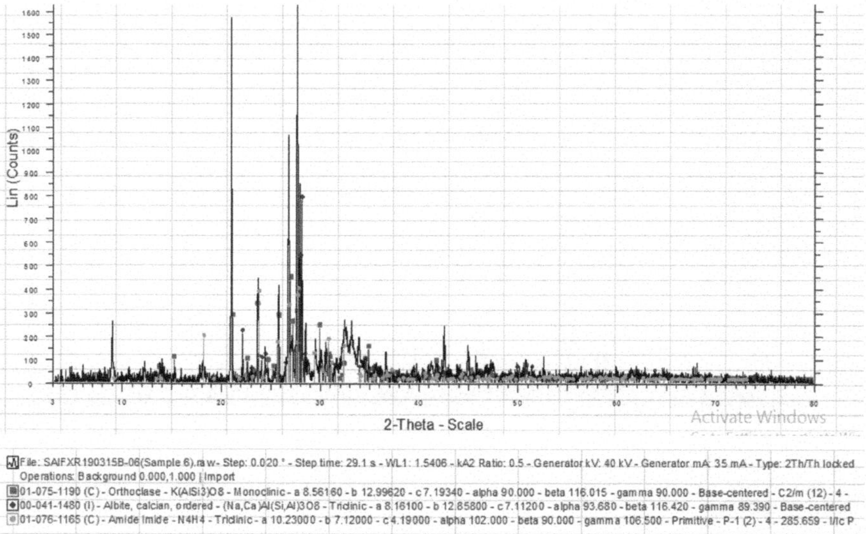

Fig. 15 XRD image of BP600

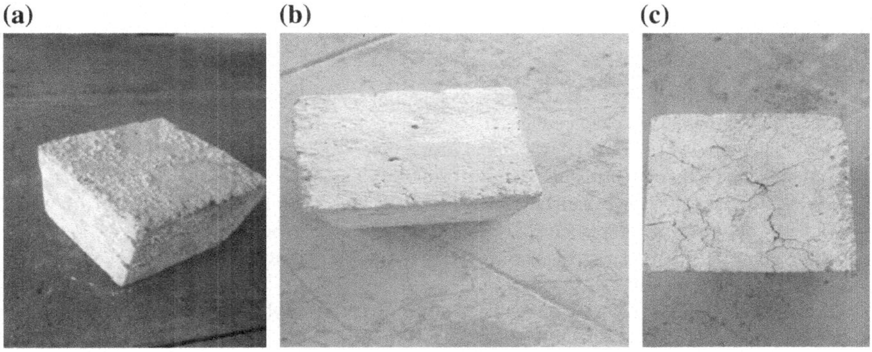

Fig. 16 **a** At 200 °C. **b** At 400 °C. **c** At 600 °C

6 Conclusions

- The properties of conventional concrete on exposure to elevated temperature can be improved by using brick powder.
- 10% replacement of cement with brick powder gives the maximum strength for concrete.
- The compressive strength of normal concrete at 200 °C has increased by 6.63% when compared to the concrete exposed to room temperature and at 400 °C it has decreased by 9.5% when compared with the concrete at 200 °C and there is a drastic decrease of 69.47% for the sample at 600 °C when compared with the concrete at 400 °C.
- The compressive strength of brick powder replaced concrete at 200 °C has increased by 5.44% when compared with the sample exposed to room temperature and at 400 °C there is a decrease of 4% when compared with the sample exposed to 200 °C. At 600 °C it has decreased by 62.55% when compared with the sample exposed to 400 °C.
- Thus, we can see that by the use of brick powder the percentage of reduction of strength has been reduced.
- While comparing the brick powder replaced samples and the samples of normal concrete at room temperature, 200, 400 and 600 °C, we obtain an increase of 4.076%, 2.91%, 9.17% and 33.93% respectively.
- Thus it proves that use of brick powder enhances the strength when exposed to elevated temperature.

References

1. Naceri A et al (2009) Use of waste brick as a partial replacement of cement in mortar. Waste Manage 29:2378–2384 (Science Direct)
2. Martina Jenifer J et al (2016) Mechanical properties of concrete with partial replacement of Portland cement by clay brick powder. Int J Eng Res Technol 5(02)
3. Letelier V et al (2017) Mechanical properties of concrete with recycled aggregate and waste brick powder as cement replacement. Proc Eng 171:627–632 (Science Direct)
4. Ramesh J et al (2017) Experimental analysis on partial replacement of cement with brick powder in concrete. IJIRT 4(3)
5. Pavan Kumar B et al (2017) Feasibility studies of partial replacement of cement by brick powder & sand by quarry dust. Int J Innov Res Sci Technol 4(1)
6. Antony J et al (2016) Roof tile powder as a partial replacement to cement in masonry mortar. Civil Eng Urban Plan: Int J 3(2)
7. Aliabdo AA et al (2014) Utilization of crushed clay brick in concrete industry. Alexandria Eng J 53:151–168
8. Nematzadeh M et al (2017) Residual properties of concrete containing recycled refractory brick aggregate at elevated temperatures. Am Soc Civil Eng
9. Nasiri AB et al (2017) The effect of elevated temperatures on the mechanical properties of concrete with fine recycled refractory brick aggregate and aluminate cement. Constr Build Mater 147:865–875 (Science Direct)
10. Husem M (2006) The effects of high temperature on compressive and flexural strengths of ordinary and high-performance concrete. Fire Safe J 41:155–163 (Science Direct)
11. Kodur V (2014) Properties of concrete at elevated temperatures. ISRN Civil Eng
12. Nimlyat PS et al (2013) Performance of concrete at elevated temperatures: utilizing a blended ordinary Portland cement (Opc)/saw dust ash (Sda) as binder. IOSR J Mech Civil Eng 9(3)
13. Arioz O (2007) Effects of elevated temperatures on properties of concrete. Fire Safe J 42(2007):516–522 (Science Direct)
14. Georgali B et al (2005) Microstructure of fire-damaged concrete. A case study. Cement Concr Compos 27:255–259 (Science Direct)

Experimental Investigation on Strength and Durability of Fly Ash and GGBS Based Geopolymer Concrete

G. Asha and Jiji Antony

Abstract Geopolymer is an aluminosilicate binding material synthesized by thermal activation of solid aluminosilicate base material such as GGBS and fly ash. Geopolymer concrete is produced due to alkalization of materials rich in aluminum and silicon with alkali solution, an aqueous solution of sodium hydroxide and sodium silicate. Alkali solution was varied in different molar concentration with sodium hydroxide to sodium silicate ratio of 2.5. The effect of age on geopolymer binder cured at ambient temperature and microstructure of binder were studied. The material used in this study consists of fly ash, ground granulated blast furnace slag, alkaline liquid, fine aggregate and coarse aggregate and sisal fibre. The mechanical properties of geopolymer concrete and binder with or without the sisal fibre were determined. The durability tests such as acid attack test and chloride attack for various mix samples were also conducted.

Keywords Geopolymer concrete · Geopolymer binder · Alkaline liquid · Fly ash · GGBS

1 Introduction

Construction industry is facing a number of challenges related to depletion of resources, accessibility to affordable materials, disposal of wastes and generation of greenhouse gases. Emission of polluting gases from construction industry leads to global warming. Cement is considered as binding material used in construction industry. Cement manufacture is an highly energy intensive and cause environmental pollution. Nearly 820 kg of carbon dioxide is released during the manufacturing of one tonne of cement [1].

G. Asha (✉) · J. Antony
Department of Civil Engineering, Federal Institute of Science and Technology, Angamaly, Ernakulam, India
e-mail: ashag628@gmail.com

J. Antony
e-mail: jijiantony13@gmail.com

K. Dasgupta et al. (eds.), *Proceedings of SECON'19*,
Lecture Notes in Civil Engineering 46,
https://doi.org/10.1007/978-3-030-26365-2_5

In 1978, professor joseph Davidovits introduced the development of mineral binders with an amorphous structure, named geopolymers. Geopolymer concrete is an innovative and eco-friendly construction material and an alternative to Portland cement concrete. The demand of Portland cement which is responsible for high CO_2 emission can be reduced by the use of geopolymer. Geo polymer concrete can be produced by the chemical action of inorganic molecules. Otherwise geopolymer is an inorganic alumino-hydroxide polymer synthesized from predominantly silicon and aluminum materials of geological origin or byproduct materials such as flyash.

The term geopolymer was introduced to represent the mineral polymers resulting from geochemistry. The process involves a chemical reaction under highly alkaline condition on Si–Al minerals, yielding polymeric Si–O–Al–O bonds in amorphous form. Due to its high mechanical properties combined with substantial chemical resistance, low shrinkage and creep and environment friendly nature [2]. It is a better construction material for future. Geopolymers were made by using aluminosilicate waste as base material. Hence substituting OPC with aluminosilicate material reduced cost and CO_2 emission. Geopolymers comprise of silicon and aluminum atoms bonded via oxygen atoms to form a polymer network [3]. They were prepared by dissolution and polycondensation reactions between a reactive aluminosilicate material and an alkaline solution, such as a mixture of an alkali metal silicate and metal hydroxide. Under highly alkaline conditions, polymerisation takes place when reactive aluminosilicates are rapidly dissolved and free [SiO_4] and [AlO_4] tetrahedral units are released in solution. The tetrahedral units are alternatively linked to polymeric precursor by sharing oxygen atom, thus forming polymeric Si–O–Al–O bonds. The following reactions occur during geopolymerisation [3].

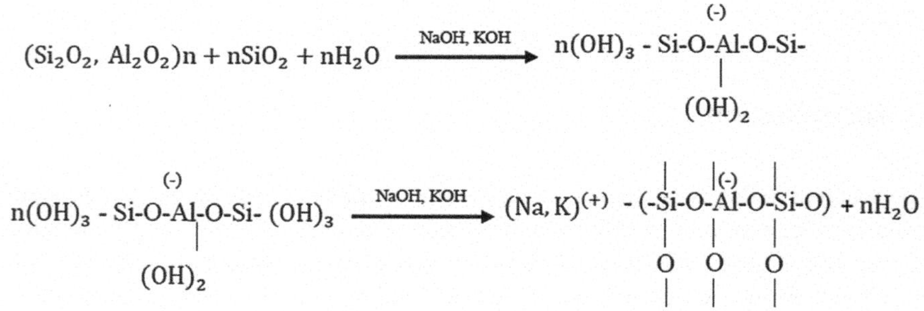

2 Experimental Investigation

Materials used are fly ash, ground granulated blast furnace slag (GGBS), sodium silicate solution, sodium hydroxide (pellet) fine aggregate, coarse aggregate and portable water.

Class F Fly ash obtained from Tripunithura at Ernakulam district of Kerala was used as the main source of aluminosilicate material for making geopolymer concrete

Table 1 Chemical composition of fly ash and GGBS

Sample	Fly ash (%)	GGBS (%)
SiO_2	50	32.46
Al_2O_3	28.25	14.3
Fe_2O_3	13.5	0.61
CaO	1.79	43.1
MgO	0.89	3.94
Na_2O	0.32	3.94
K_2O	0.46	0.33
SO_3	0.38	4.58
$P2O_5$	0.98	0.02
TiO_2	1.54	0.55

and mortar. Commercially available ground granulated blast furnace slag (GGBS) was used as an additive. The chemical compositions of fly ash and GGBS are shown in Table 1.

Coarse aggregate with a maximum size 20 mm, Manufactured-sand used as fine aggregate its specific gravity about 2.9. A mixture of sodium hydroxide (NaOH) and sodium silicate (Na_2SiO_3) solutions was used as the activator solution. Sodium hydroxide solution of desired concentration was prepared by mixing 97–98% pure NaOH pellets with tap water. The concentration of sodium hydroxide solution was kept constant (12 M) for all mixtures. Sodium silicate solution was collected from a local commercial produce.

2.1 Mix Design

Mix design was adopted based on the previous literatures [1–4] geopolymer mortar and concrete were prepared with the different mix proportion. Mix design is prepared by referring to B. V. Rangan's paper (Tables 2 and 3).

2.2 Mixing, Sample Preparation and Curing

The sodium hydroxide and sodium silicate solutions of desired quantity were mixed together about 30 min before mixing with other ingredients to enhance reactivity of the solution. Coarse and fine aggregates, prepared in saturated surface dry condition, and the binders (fly ash and slag) were dry mixed thoroughly in the mixing pan for 2 min. The premixed alkaline activator solution was then added gradually in the mixer. Mixing was continued for further 3–5 min to achieve a uniform mixing. Cubes

Table 2 Details of geopolymer mortar mix proportions

Mix No.	Curing temperature (23 °C)	Molarity	SS/SH ratio	A/B Ratio	Fly ash (g)	GGBS (g)	M-sand (g)	Na_2Sio_3 (g)	NaOH Pellet (g)	Water (ml)
50% flyash 50 % GGBS	23	12	2.5	0.5	100	100	600	71.429	7.577	20.992
	23	12	2.5	0.6	100	100	600	20.992	9.092	25.187
	23	12	2.5	0.7	100	100	600	100	10.60	29.39
	23	12	2.5	0.8	100	100	600	114.28	12.12	33.58
	23	12	2.5	0.9	100	100	600	128.572	13.64	37.78
60% flyash 40% GGBS	23	12	2.5	0.5	120	80	600	71.429	7.577	20.99
	23	12	2.5	0.6	120	80	600	20.992	9.092	25.18
	23	12	2.5	0.7	120	80	600	100	10.60	29.39
	23	12	2.5	0.8	120	80	600	114.28	13.64	37.78
	23	12	2.5	0.9	120	80	600	128.572	13.64	37.78
40% fly ash 60% GGBS	23	12	2.5	0.5	80	120	600	71.429	7.577	20.99
	23	12	2.5	0.6	80	120	600	20.992	9.092	25.18
	23	12	2.5	0.7	80	120	600	100	10.60	29.39
	23	12	2.5	0.8	80	120	600	114.286	12.12	33.58
	23	12	2.5	0.9	80	120	600	128.572	13.64	37.787

Table 3 Details of geopolymer concrete mix proportions

Mix No.	Coarse aggregate (kg/m^3)	Fine aggregate (kg/m^3)	Fly ash (kg/m^3)	GGBS (kg/m^3)	Water (kg/m^3)	Mass of alkaline solution (kg/m^3)
50% flyash 50% GGBS	1293.6	554	172.5	172.5	43.45	207
60% fly ash 40% GGBS	1293.6	554	207	138	43.45	207
40% fly ash 60% GGBS	1293.6	554	138	207	43.45	207

and Cylindrical concrete moulds, were filled with geopolymer concrete in two layers and compacted on a vibrating table.

2.3 *Tests on Geopolymer Mortar*

Geopolymer specimen subjected to compressive strength test and water absorption.

2.3.1 Compressive Strength Test

Compressive strength test for mortar cube specimens were carried out as per IS: 4031 (part 6)-1988. Test was done after 28 days of curing. The specimen was mounted on the testing platform of the CTM. Load was applied uniformly till breaking of specimen occurred. The maximum load applied was recorded (Fig. 1).

2.3.2 Water Absorption Test

Water absorption test for the block were conducted with three samples as per IS: 2185 (Part 1)-2005. Concrete cubes were immersed in water for 24 h and wet mass of block was recorded. The specimen were dried in oven at 100–115 °C for not less than 24 h and dry mass of the mortar cubes were noted. Water absorption of the blocks was calculated using the formula;

$$\begin{aligned}&\%\text{ water absorption of the mortar cubes}\\&=\frac{(\text{wet mass of cube} - \text{dry mass of cube}) \times 100}{\text{dry mass of block}}\end{aligned}$$

Fig. 1 Compression test setup of geopolymer concrete and mortar cubes

2.3.3 X-Ray Diffraction (XRD) and Scanning Electron Microscopy (SEM) of Geopolymer Mortar

XRD and SEM analysis were done for microstructural analysis on samples cured at room temperature for 28 days and ground to fine powder.

2.4 Test on Geopolymer Concrete

Geopolymer concrete cubes of size 150 × 150 × 150 mm are subjected to compressive strength and water absorption test. Split tensile strength test is conducted on cylinder specimen of size 150 × 300 mm.

2.4.1 Compressive Strength Test

Compressive strength test was conducted as per 516-1969 on three cubes using a compressive testing machine. The compressive strength of the specimen is calculated by dividing the maximum load attained by the cross sectional area of the specimen.

2.4.2 Split Tensile Strength

IS 5816-1999 explained about procedure for determining the split tensile strength of concrete. Three specimens were tested for each combination. Test was done after 28 days of curing. The specimen was mounted on the testing platform of the CTM. Load was applied uniformly till breaking of specimen occurred (Fig. 2).

Fig. 2 Split tensile strength test

2.4.3 Water Absorption Test

IS: 2185 (Part 1)-2005 Concrete cubes were immersed in water for 24 h and wet mass of block was recorded. The specimen were dried in oven at 100–115 °C for not less than 24 h and dry mass of the cubes were noted. Water absorption of the blocks was calculated using the formula;

$$\%\ \text{water absorption of the cube} = \frac{(\text{wet mass of cube} - \text{dry mass of cube}) \times 100}{\text{dry mass of cube}}$$

3 Results and Discussions

Tests results are discussed below.

3.1 Compressive Strength

Maximum compressive strength is observed with A/B ratio 0.8. The compressive strength of paste obviously increased with increasing GBFS content for all sodium hydroxide, sodium silicate solution and combination of NaOH and sodium silicate solution. Geopolymerization of fly ash is extremely slow which improve the mechanical activation by the addition of GGBS. In addition, the reaction of GGBS and alkali solution was an exothermal process and generated heat which promoted the geopolymerization process. Increase the GGBS contend leads to high compressive strength (Figs. 3, 4, 5 and 6; Table 4).

Compressive strength of GPC mix binder 0.6 (60% GGBS and 40% flyash) shows the high strength value of 40 N/mm^2. Compressive strength generally increased with the increase of Si/Al ratio in the constituents of the mixtures. The Si/Al ratio can also be altered by varying the composition of alkaline solution. The molar ratio of Si/Al, increase with the increase of sodium silicate in the activator solution. The presence

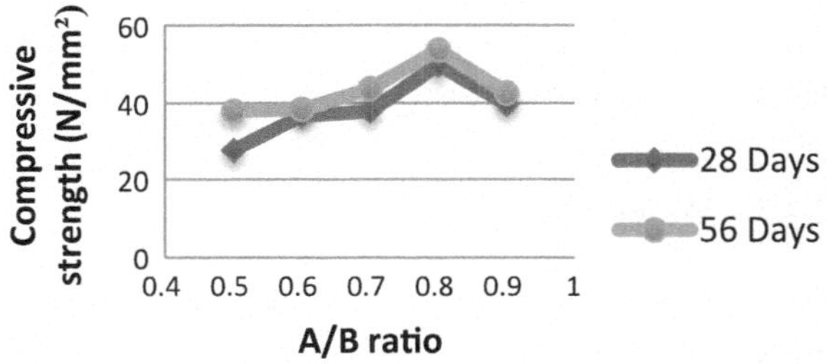

Fig. 3 Graphical variation of compressive strength of mortar cubes (50% fly ash and 50% GGBS)

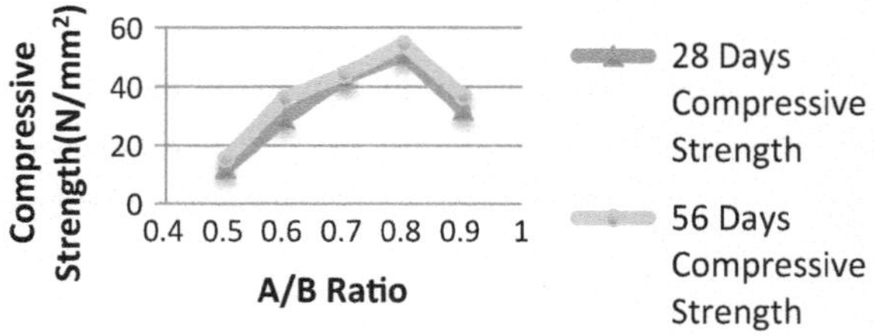

Fig. 4 Graphical variation of compressive strength of mortar cubes (60% fly ash and 40% GGBS)

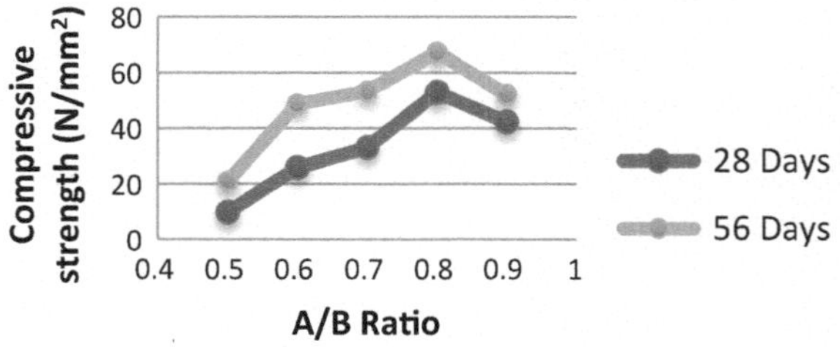

Fig. 5 Graphical variation of compressive strength of mortar cubes (40% fly ash and 60% GGBS)

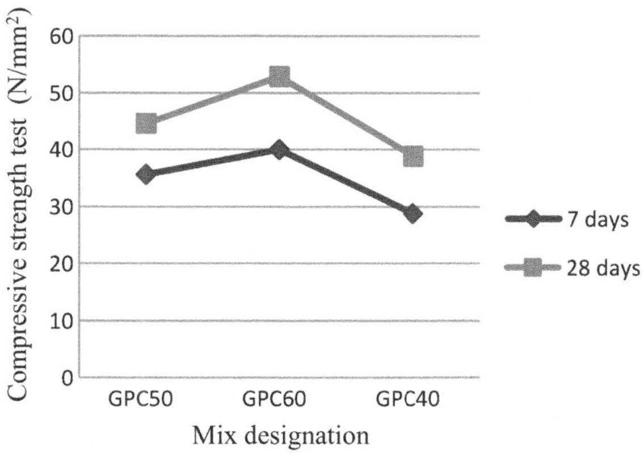

Fig. 6 Graphical representation of compressive strength of GPC

Table 4 Compressive strength of geopolymer concrete

S No.	Mix ID	A/B Ratio	7th day average compressive strength	28th day average compressive strength
1	50% fly ash 50% GGBS	0.6	28.79	38.36
2	60% fly ash 40% GGBS	0.6	20.72	32.72
3	40% fly ash 60% GGBS	0.6	29.05	39.82

of soluble silica enhance the condensation process and introduces more Si in the polymeric chain. This enhances the strength properties.

3.2 Split Tensile Strength

The tensile strength of fly ash and GGBS based geopolymer concrete was measured by performing the cylinder splitting test on 150 × 300 mm concrete cylinders. The test results are given in Table 5 (Fig. 7).

From the obtained results, split tensile strength of geopolymer concrete with 60 percentage GGBS and 40 percentage fly ash.

3.3 SEM Analysis

The geopolymerization reaction of SS/SH ratio 2.5. the maximum compressive strength is observed with A/B = 0.8 The activator content being optimum, it allows continual dissolution of raw materials. Hence the polycondensation rate during the

Table 5 Split tensile strength of geopolymer concrete

Description	Load (KN)	Split tensile strength (N/mm^2)
50% fly ash 50% GGBS	250	3.536
60% fly ash 40% GGBS	182	2.574
40% fly ash 60% GGBS	355	5.022

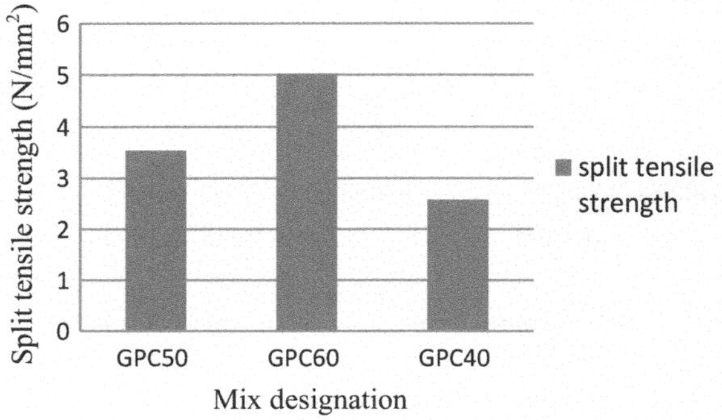

Fig. 7 Graphical variation of split tensile strength of GPC

geopolymer synthesis is not hindered and allowed the maximum geopolymeric gel formation. The increase in A/B ratio above 0.8 leads to excess activator solution in the mix. Extra voids are formed in the structure due to the escape of this excess water. Additionally the excess OH concentration left in the system weakens the structure of geopolymer and cause a decline strength (Figs. 8 and 9).

Fig. 8 SEM image with SS/SH ratio

Fig. 9 SEM image with high content of fly ash

3.4 *XRD Analysis*

From XRD analysis replacement of fly ash the geopolymeric gels were clearly found to co-exist with C–S–H gel. Calcium and silicon resulting from the dissolution of GGBS and react the C–S–H. Formation of C–S–H gel along with silica-aluminates

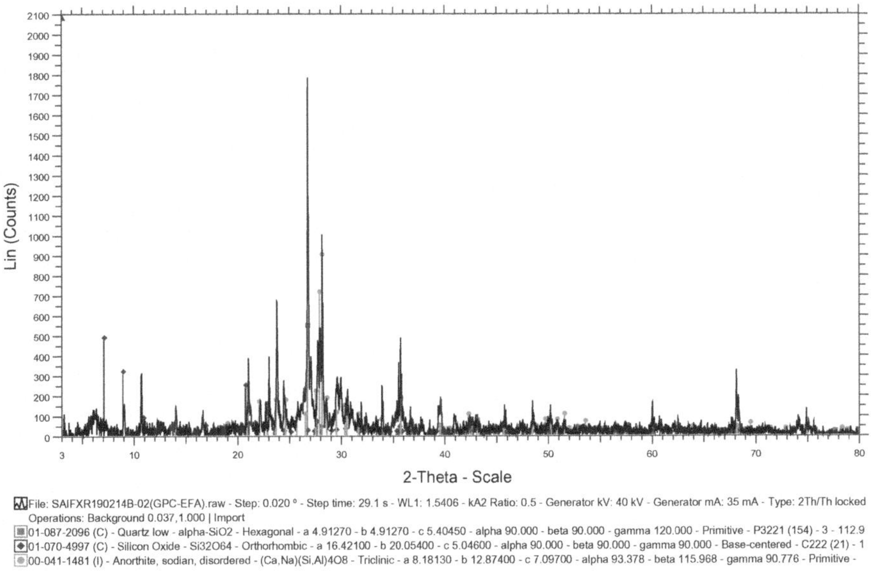

Fig. 10 XRD pattern of ambient temperature of geopolymer mortar

structure and it will gives the high strength. The specimen with higher dosage of GGBS have denser structure as a results high strength is observed (Fig. 10).

4 Conclusion

- The compressive strength of geopolymer mortar increases as the GGBS content is increased up to 60% due to densification of matrix. Further increase in GGBS reduces the strength as it act as filler material.
- Maximum compressive strength of mortar cube is observed with A/B ratio 0.8.
- Geopolymerization of fly ash is extremely slow, addition of GGBS results in an exothermic reaction and the heat generated enhances the geopolymerization process.
- Geopolymer binder was optimized with the parameter such as molarity of NaOH 12M, sodium silicate to sodium hydroxide solution ratio 2.5, alkali activator to binder ratio is 0.8.
- GGBS and fly ash enhances the strength characteristics of geopolymer concrete.
- Superior performance in terms on durability can be attributed to the dense microstructure of geopolymer concrete.

References

1. Nath P, Sarker PK (2014) Effect of GGBFS on setting, workability and early strength properties of fly ash geopolymer concrete cured in ambient condition. Constr Build Mater 66:163–171
2. Duan P, Yan C (2016) Fresh properties, mechanical strength and microstructure of fly ash geopolymer paste reinforced with sawdust. Constr Build Mater 111:600–610
3. Mane S, Jadhav HS (2012) Investigation of geopolymer mortar and concrete under high temperature. Int J Civ Eng Technol 2:384–390
4. Cheah CB, Samsudin MH (2018) The use of high calcium wood ash in the preparation of ground granulated blast furnace slag and pulverized fly ash geopolymers a complete microstructural and mechanical characterization. Constr Build Mater 17:1–30

Load Settlement Behavior of Ceramic Columns

Aaron Rodrigues, V. Gopika, Anupriya Saji, Safna Salam and A. A. Abishek Kumar

Abstract Indian ceramic production is 100 Million ton per year in the ceramic industry, about 15–30% waste is generated from the total production. Dumping of ceramic waste leads to serious environmental and dust pollution and occupy large area. To avoid these conditions, it is desirable to reduce the stockpiles of ceramic waste by recycling or other alternative methods. Stone columns are one of the versatile techniques for engineering the ground. It is the most effective solution for improving the strength of soil and thereby protecting against liquefaction with less installation time. Our aim is to investigate the feasibility of using ceramic waste as aggregates in stone columns there by reducing the waste disposal, to provide a replacement to aggregates in columns and to minimize the cost of soil stabilization. **Method/analysis**: Experimental study on the load settlement behavior (IS 1888–1982) in test model of Ceramic stone column, Ceramic stone columns with geotextile casing, Ceramic stone columns with horizontal arrangement of geotextile, Ceramic stone columns with both horizontal layering and casing of geotextile and to compare with the ordinary stone column. Result shows that improvement in the load carrying capacity is found less in ceramic as compared to stone column but by the introduction of geotextile casing and layering, found an increase in the load carrying capacity.

Keywords Stone column · Geotextile · Load settlement · Ceramic column · Reinforcement

1 Introduction

Marine clays possess higher water content and they have low strength and high compressibility. Due to low supporting strength, these soils pose serious problems to construction of structures . Marine clays are normally present in seashore areas where ports are built for import and export operations. These operations involve stacking of materials at ports. If proper ground improvement is not done in soft marine clays

A. Rodrigues · V. Gopika · A. Saji · S. Salam · A. A. Abishek Kumar (✉)
Adi Shankara Institute of Engineering and Technology, Kalady, India
e-mail: abishek.ce@adishankara.ac.in

K. Dasgupta et al. (eds.), *Proceedings of SECON'19*,
Lecture Notes in Civil Engineering 46,
https://doi.org/10.1007/978-3-030-26365-2_6

and materials are excessively stacked, shear movements occur in subsoil affecting the neighboring structures. Constructions on such soil are being managed by stone columns.

Ceramic wastes are produced every day. Their main contribution are from tile manufacturing companies, houses etc. Stoke piling is the traditional way of disposal of these waste which may cause several environmental problems and dust pollution which also may occupy large useful land.

So in order to avoid such problems we are trying to cooperate ceramic wastes as aggregates in stone columns.

1.1 Materials Used

The marine clay material is collected from the place near to the MULT Project of the Port Trust. Crushed aggregates was collected from Metro Crushers, Kalady. They are having a size of between 6 and 8 mm. Ceramic wastes are collected from Surabhi tiles, Angamaly. The geotextile used is Non-woven polypropylene having a GSM OF 250 and thickness 1 mm which is collected from Shri Raghavendra Geotextiles, Chennai (Fig. 1).

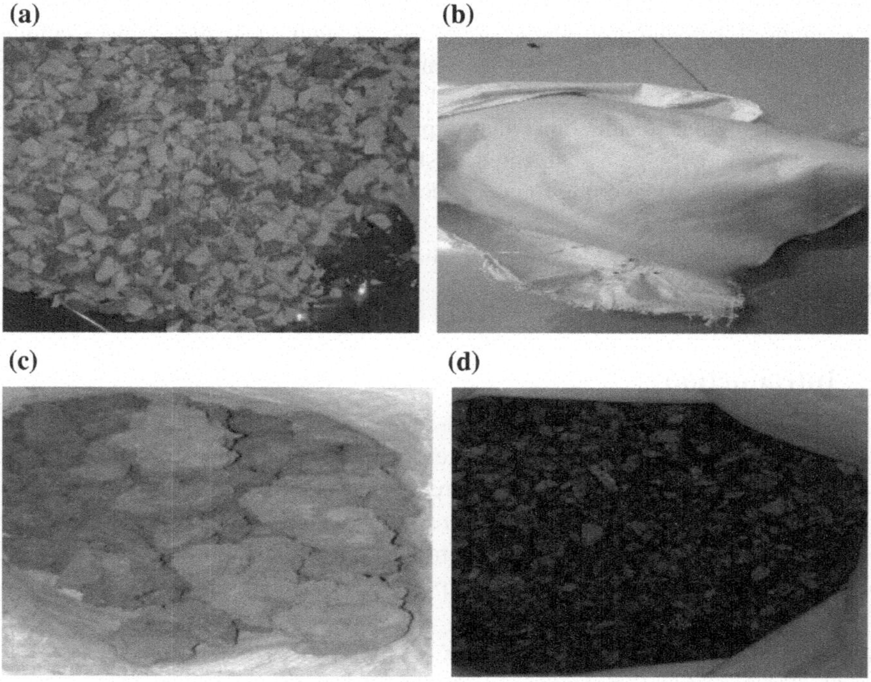

Fig. 1 **a** Ceramic chips. **b** Non-woven polypropylene geotextile. **c** Marine clay. **d** Aggregate

2 Experimental Study

Index and engineering properties of marine clay are given in Table 1.

2.1 Experimental Set up and Loading Condition

Tests were conducted on a test tank of dimension 150 mm diameter and 175 mm height. A circular steel plate of 14 mm diameter and 2 mm thickness is placed on the surface of the test specimen to provide even distribution of loading. The load was applied on the plate at a rate of 1.25 mm/min.

2.2 Preparation of Soft Clay Bed

The air dried clay sample was mixed with required quantity of water. The optimum moisture content of 20% was determined by conducting standard proctor test. After adding the required water to the clay, it was thoroughly mixed to get a consistent paste. This soil mix was then filled in the test tank in 3 layers by hand compaction such that no air voids are left in the soil. Before filling soil in the tank, the inner surface of the tank wall was first coated with grease to minimize the friction between soil and the tank wall. For each load test, a fresh clay was prepared in the test tank and stone columns were installed in it. Tests were conducted on stone columns formed in a clay bed of 150 mm diameter and 175 mm height.

Table 1 Index properties of soil

Properties	Values
Specific gravity	2.48
Liquid limit	34.5%
Plastic limit	15.62%
Plasticity index	18.78%
Optimum moisture content	20%
Maximum dry density	1.768 gm/cc
Consolidation	3.778×10^{-4} cm^2/s
CBR	0.0058%

2.3 *Construction of Stone Column*

After preparing the clay bed, a PVC pipe of internal diameter 40 mm and 1 mm thick was inserted into the clay bed manually. Later the pipe was pulled out by rotating slightly so that the soil is removed. Then the stone column was casted by filling the stone chips by four layers by compaction. Each layer was given a light compacted of 25 blows.

2.4 *Construction of Ceramic Columns*

After preparing the clay bed and the hole, ceramic tile chips of size 6–8 mm was filled into the hole as same as that of stone column preparation procedure.

2.4.1 Construction of Ceramic Column with Geotextile Casing

In the prepared clay bed, hole was created by using PVC pipe and it is provided with geotextile casing. Then ceramic tile of required size is filled as the above procedure.

2.4.2 Construction of Ceramic Column with Geotextile Layering

Soil bed was prepared and hole was created. At the bottom of the hole circular geotextile of diameter 40 mm was placed and ceramic tile chips were filled as layer by layer. The circular geotextiles were introduced between the layers by pushing the circular geotextiles to the desired position by the means of PVC pipe.

2.4.3 Construction of Ceramic Column with Casing and Layering of Geotextiles

In the prepared clay bed, circular hole was created. The circular hole was encased with geotextile and was filled ceramic tile pieces, layer by layer. In between each layers geotextile was placed by using PVC pipe (Fig. 2).

3 Result and Discussions

3.1 *Load Settlement Behaviour of Plain Clay Bed*

Figure 3 shows the load settlement behavior of clay bed from the test conducted.

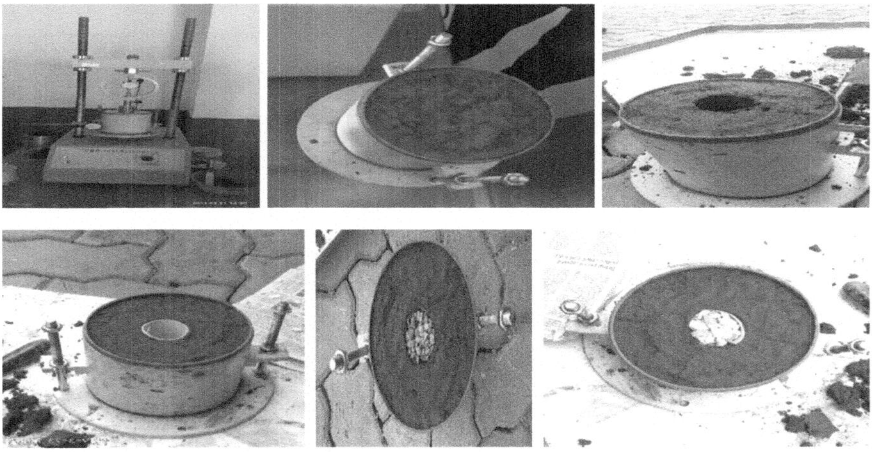

Fig. 2 Experimental set up

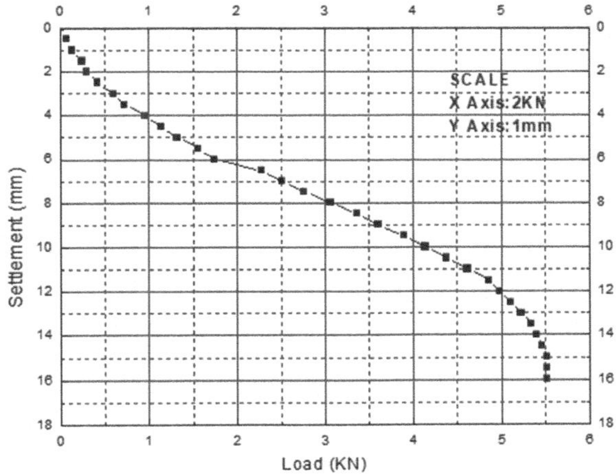

Fig. 3 Settlement curve of clay bed

The ultimate bearing capacity of clay bed found to be 5.52 kN for the corresponding settlement of 16.5 mm.

3.2 Load Settlement Behavior of Clay Bed with Stone and Ceramic Column

From Fig. 4 it is inferred that load carrying capacity of clay bed with stone column gives a better result when compared to clay bed. This result found to be true with in the case of clay bed with ceramic column.

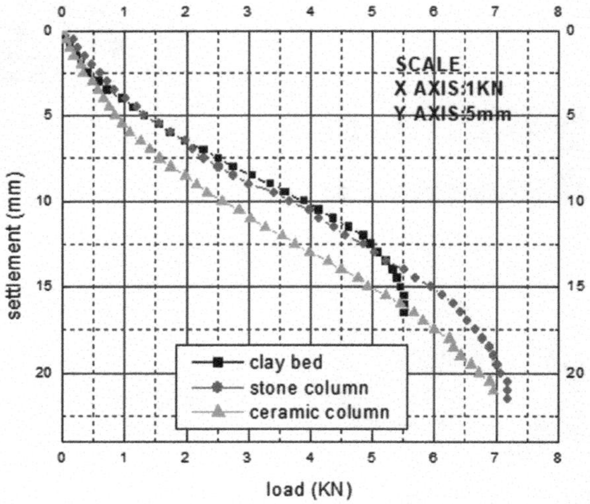

Fig. 4 Settlement curve of-clay bed, clay bed with stone and ceramic column

The load capacity corresponding to the settlement value of 21 mm is 7.2 kN for soil with stone column. Whereas the value obtained for soil with ceramic column is 6.96 kN.

3.3 Load Settlement Behavior of Clay Bed with Reinforced Ceramic Column (Casing, Layering and Combination of Both)

From Fig. 5 it is inferred that the load bearing capacity of ceramic column with geotextile layering is less than other two cases and the highest value is found for the ceramic column with both casing and layering. The ceramic column with layering can bear a loading value of 6.84 kN. The loading value for the settlement of 22 mm is 7.92 kN for ceramic column with both casing and layering and 7.17 for ceramic column with casing only.

4 Conclusions

The conclusions obtained from the test conducted is listed below:

- The load carrying capacity of clay bed is seems to be very low.
- By the installation of stone column the load carrying capacity of the clay bed is increased by 30%.

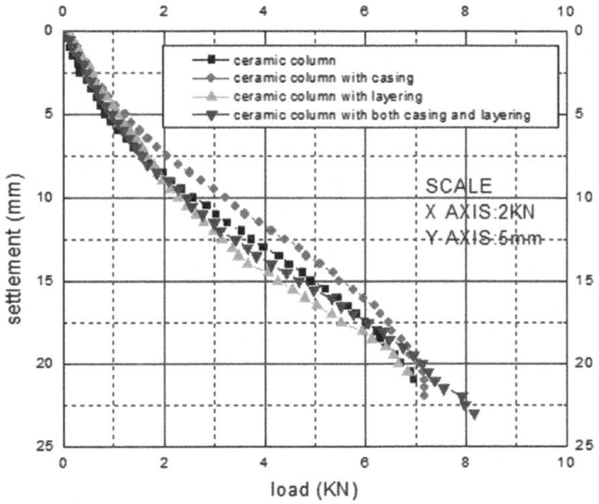

Fig. 5 Settlement curve of different cases of ceramic column

- In case of ceramic column 26% increase in load carrying capacity is found when compared to clay bed but it shows 3.3% decrement in loading capacity as compared to stone column.
- There is no significant change in load carrying capacity of clay bed with geotextile layered ceramic column when compared with the clay bed with stone column.
- Ceramic column with geotextile casing gives 0.5% decrement in load bearing capacity as compared to clay bed with stone column.
- The load carrying capacity of clay bed can be increased by 10% than stone column by the introduction of ceramic column with both geotextile casing and layering by the extra densification of the soil.
- By the completion of this project it can be concluded that the replacement of stone column can be done by ceramic column with the provision of geotextile casing and layering.

References

1. Nazari Afshar J et a l (2012) A simple analytical method for calculation of bearing capacity of stone column. Int J Civil Eng 12(1)
2. Andreas A et al (2009) Ground improvement of clayey soil formation using stone column: a case study from Greece. Int J Geotech Eng 3:493–498
3. Sarvaiya KH (2015) An experimental study on load capacity of floating stone column in soft soil. Int J Adv Eng Technol 8(6):965–975
4. Siva Gowri S (2016) Improvement of soft soil performance using stone columns improved with circular geogrid discs. Indian J Sci Technol 9(30)
5. Kiran Kumar TK (2017) Ground improvement and load carrying capacity using stone column in clay soils. Int J Innov Res Sci Eng Technol 6(2):2892–2897

Experimental Investigation on Partial Replacement of Cement with Fly Ash and Glass Powder

K. Devu and S. Sreerath

Abstract Use of glass in concrete is an interesting possibility for economy on wastage disposals. It is an amorphous material with high silica content, thus making it potentially pozzolanic when particle size is less than 75 μm [1, 2]. The amorphous silica (SiO_2) in the glass powder reacts with the portlandite [$Ca(OH)_2$] generated during cement hydration and convert to gels of calcium silicate hydrate (C–S–H) [3]. A major concern regarding the use of glass in concrete is the chemical reaction that takes place between the silica rich glass particle and the alkali in pore solution of concrete, which is called Alkali Silica Reaction. It can be very detrimental to the stability of concrete, unless appropriate precautions are taken to minimize its effects. The inclusion of fly ash in glass powder concrete has shown reduction in the alkali silica reaction and improvement in the workability and durability properties of concrete [4, 5]. This study summarizes information on the mechanical properties of concrete containing fly ash and glass powder. The main objective of this work is to study the suitability of these materials as partial replacement of cement. In this study, cement is partially replaced with fly ash and glass powder in different proportions by weight and the combined optimum percentage is determined. The properties of concrete containing optimum percentage of these materials are compared with that of control concrete. A better understanding of the performance of these materials could lead to its increased usage, consequently contributing to sustainability.

Keywords Glass powder · Fly ash

1 Introduction

Cement manufacturing industry is one of the major carbon dioxide (CO_2) emitting sources. Among the greenhouse gases, carbon dioxide contributes about 65% of global warming. The global cement industry contributes about 7% of greenhouse gas

K. Devu (✉) · S. Sreerath
Department of Civil Engineering, Federal Institute of Science and Technology, Angamaly, Kochi 683577, India
e-mail: devukish94@gmail.com

K. Dasgupta et al. (eds.), *Proceedings of SECON'19*,
Lecture Notes in Civil Engineering 46,
https://doi.org/10.1007/978-3-030-26365-2_7

emission to the earth's atmosphere [6]. Thus, there is a need to economise the use of cement. One of the practical solutions to economise cement is to replace cement with supplementary cementitious materials. Consequently, extensive research is ongoing in this topic, regarding the replacement of cement with industrial by-products and various other substitutes. Also, million tons of waste glass as well as fly ash is being generated annually all over the world. The effective utilization of these materials in concrete making is attracting serious considerations.

Glass is an amorphous material with high silica content, thus making it potentially pozzolanic when particle size is less than 75 μm. Use of ground glass in concrete as partial replacement of cement could be an important step toward development of sustainable infrastructure systems.

Alkali Silica Reaction was identified as the greatest challenge to the acceptance of glass powder as a common SCM since it is not possible to predict the reaction product with any refined level of control. It can be prevented or reduced by adding mineral admixtures in the concrete mixture [7]. Design professionals often specify prescriptive requirements such as quantities of fly ash, slag, etc., to avoid alkali silica reaction-related distress in structures [8]. Efforts have been made in the concrete industry to use glass powder and fly ash as partial replacement of cement. In this work, the use of finely powdered glass, fly ash and their combination as a partial replacement of cement in concrete will be studied and compared with conventional concrete.

2 Material Used

2.1 Cement

Portland cement of 53 grade conforming to IS 12269-1987 was used in this study.

2.2 Fly Ash

Class F Fly ash was used in this study. These are generally low-calcium fly ashes with carbon contents less than 5% but sometimes as high as 10%. Class F ashes are produced from bituminous or anthracite coals. This fly ash is pozzolanic in nature, and contains less than 20% lime (CaO) [9].

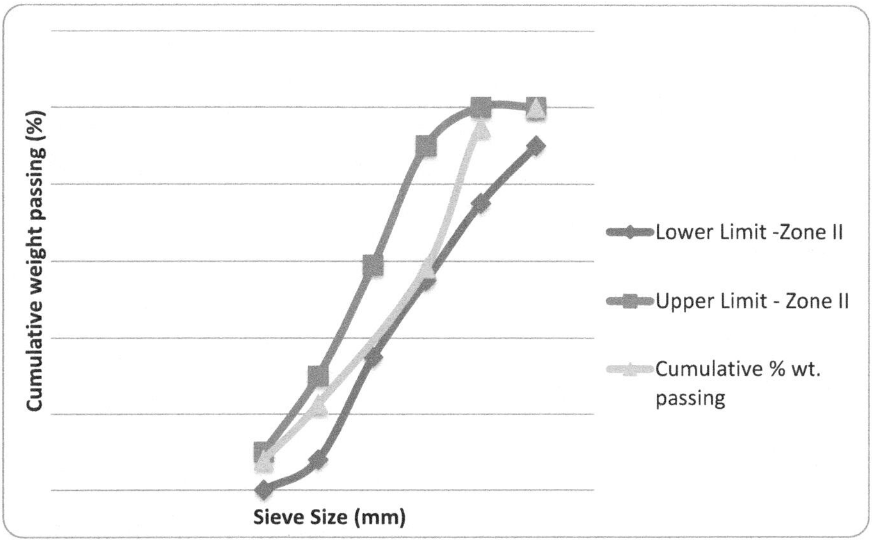

Fig. 1 Particle size distribution curve of fine aggregate used

2.3 Glass Powder

Glass is principally composed of silica. When glass is milled down to micro size particles, it is expected to undergo pozzolanic reactions with cement hydrates, forming secondary Calcium Silicate Hydrate (C–S–H). Clear glass powder with particle size less than 75 μ was used in this study.

2.4 Aggregates

Coarse aggregate conforming to IS 383-1970 was used in this work. Fine aggregate conforming to Zone II of IS 383-1970 was used in this work. The particle size distribution curve of fine aggregate is shown in Fig. 1.

2.5 Super Plasticizer

The super plasticizer used in this study was Master Rheobuild 1125 at a dosage rate of 0.6% by weight of cement. Master Rheobuild 1125 is composed of synthetic polymers specially designed to allow reduction of mixing water while maintaining control on extend of set retardation.

Table 1 Details of control mix

Details of mix	Grade M30
Mix proportion	1:1.86:3.25
Cement content	370 kg/m^3
Fine aggregate	692.71 kg/m^3
Coarse aggregate	1202.5 kg/m^3
Superplasticizer	2.22 l
Water cement ratio	0.4
Amount of water	147.75 l

3 Methodology

3.1 *Mix Design*

Mix proportioning was done as per IS 10262:2009, IS 383:1970 and IS 456:2000. Control concrete was proportioned using OPC alone as binder for target strength of 38.25 MPa at the age of 28 days and a slump of 100 mm. Final mix proportions were arrived on the basis of trial castings. Details of the control mix is shown in Table 1.

3.2 *Preparation of Specimen*

Casting of cubes, beams and cylinders were done as per IS 10086:1982. Mortar cubes of size 75 × 75 × 75 mm, concrete cubes of size 150 × 150 × 150 mm, beams of size 100 × 100 × 500 mm and cylinder of size 300 × 150 mm were cast to conduct test for compressive strength, flexural strength and splitting tensile strength of mixes.

4 Test Results and Discussion

4.1 *Slump Test*

The workability of concrete containing glass powder and fly ash as partial replacement of cement was tested by conducting slump test. The slump values obtained are shown in Table 2.

The slump values increased for both fly ash and glass powder till 20% replacement. Beyond 20%, the slump test results showed a decreasing trend. The slump obtained for concrete containing fly ash was slightly higher than that containing glass powder (Fig. 2).

Table 2 Slump test results on concrete containing GP and FA as cement replacement

Percentage replacement of cement (%)	Slump obtained for fly ash (mm)	Slump obtained for glass powder (mm)
0	90	90
5	115	100
10	120	115
15	128	125
20	136	130
25	122	114
30	99	95
35	84	80
40	76	72

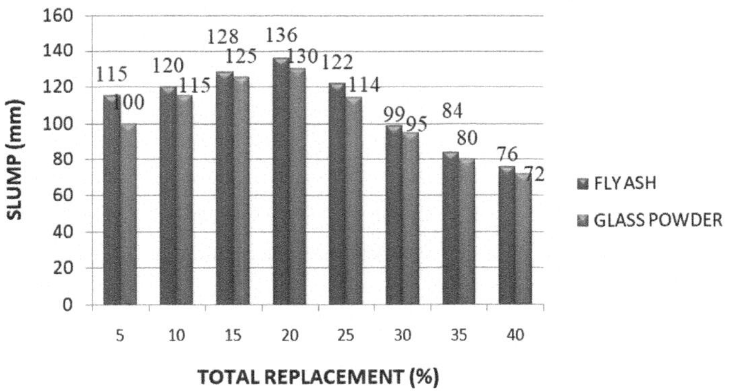

Fig. 2 Bar chart showing variation in slump values of concrete containing fly ash and glass powder for different percentage replacement

4.2 Compressive Strength of Mortar Cubes

The average 28th day compressive strength of all mortar cube specimens are shown in Table 3 (Fig. 3).

The combined optimum percentage for fly ash and glass powder was obtained as 25%. When more than 5% fly ash was incorporated to glass powder-based cement mortar, an increase in compressive strength was observed.

Table 3 28th day compressive strength of mortar cubes

Total percentage replacement	Specimen	Description	Average 28th day compressive strength (MPa)
0%	Control mix	0% replacement	53.75
10%	F0-G10	0% fly ash + 10% glass powder	50.43
	F10-G0	10% fly ash + 0% glass powder	52.51
	F5-G5	5% fly ash + 5% glass powder	52.93
15%	F0-G15	0% fly ash + 15% glass powder	51.59
	F15-G0	15% fly ash + 0% glass powder	53.03
	F5-G10	5% fly ash + 10% glass powder	53.24
	F10-G5	10% fly ash + 5% glass powder	53.92
20%	F0-G20	0% fly ash + 20% glass powder	54.43
	F20-G0	20% fly ash + 0% glass powder	55.34
	F10-G10	10% fly ash + 10% glass powder	56.98
	F5-G15	15% fly ash + 5% glass powder	55.93
	F15-G5	5% fly ash + 15% glass powder	56.07
25%	F0-G25	0% fly ash + 25% glass powder	48.21
	F25-G0	25% fly ash + 0% glass powder	49.95
	F10-G15	10% fly ash + 15% glass powder	58.93
	F15-G10	15% fly ash + 10% glass powder	57.72
	F5-G20	5% fly ash + 20% glass powder	55.48
	F20-G5	20% fly ash + 5% glass powder	56.55
30%	F0-G30	0% fly ash + 30% glass powder	47.36
	F30-G0	30% fly ash + 0% glass powder	48.71
	F5-G25	5% fly ash + 25% glass powder	48.91
	F25-G5	25% fly ash + 5% glass powder	49.09
	F15-G15	15% fly ash + 15% glass powder	51.23
	F10-G20	10% fly ash + 20% glass powder	50.47
	F20-G10	20% fly ash + 10% glass powder	49.98
40%	F0-G40	0% fly ash + 40% glass powder	45.39
	F40-G0	40% fly ash + 0% glass powder	46.22
	F5-G35	5% fly ash + 35% glass powder	46.81
	F35-G5	35% fly ash + 5% glass powder	47.4
	F10-G30	10% Fly Ash + 30% Glass Powder	47.78
	F30-G10	30% fly ash + 10% glass powder	47.01
	F15-G25	15% fly ash + 25% glass powder	48.37
	F25-G15	25% fly ash + 15% glass powder	47.91
	F20-G20	20% fly ash + 20% glass powder	49.19

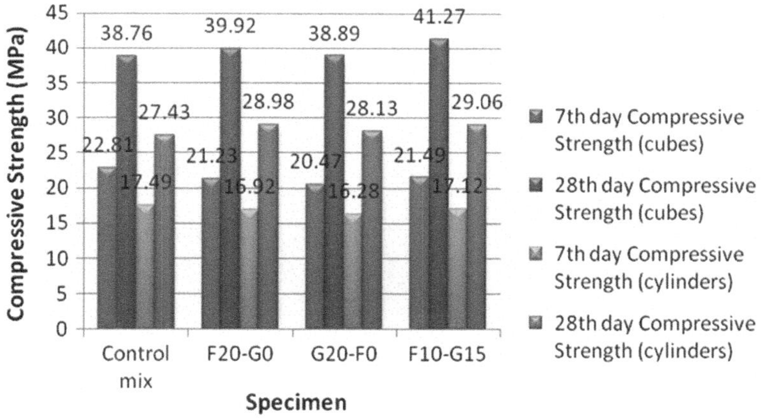

Fig. 3 Highest Compressive Strength obtained for mortar cubes which contain fly ash, glass powder and fly ash + glass powder for each percentage replacements

Table 4 Mechanical properties of concrete specimens

Property (MPa)	Control mix	F20-G0	G20-F0	F10-G15
7th day compressive strength (cubes)	22.81	21.23	20.47	21.49
28th day compressive strength (cubes)	38.76	39.92	38.89	41.27
7th day compressive strength (cylinders)	17.49	16.92	16.28	17.12
28th day compressive strength (cylinders)	27.43	28.98	28.13	29.06
7th day split tensile strength	2.52	2.27	2.23	2.49
28th day split tensile strength	3.34	3.41	3.39	3.46
7th day flexural strength	4.57	4.35	4.28	4.61
28th day flexural strength	6.31	6.59	6.42	6.74

4.3 Tests on Hardened Concrete

The mechanical properties (7th day and 28th day) of control mix, mix with 20% fly ash as replacement, mix with 20% glass powder as replacement and mix with overall replacement of 25% (F10-G15) is shown in Table 4 (Figs. 4, 5 and 6).

5 Observations

- The individual optimum percentage of fly ash and glass powder as replacement was obtained as 20%.
- The combined optimum percentage of fly ash and glass powder was obtained as 25% with 10% fly ash and 15% glass powder.

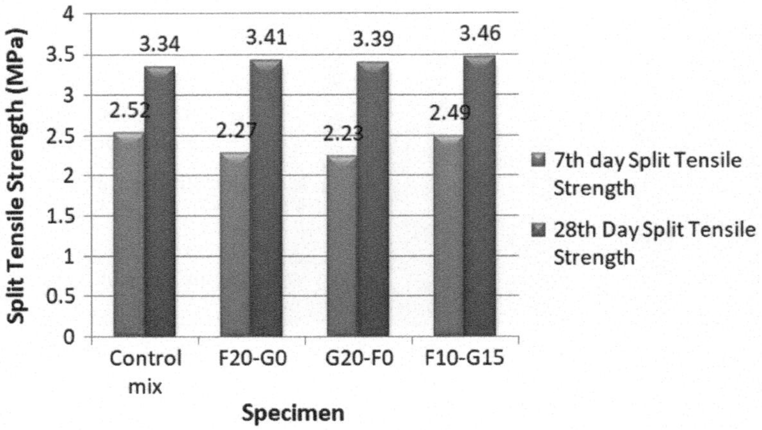

Fig. 4 Comparison of 7th day and 28th day compressive strength of cubes and cylinders

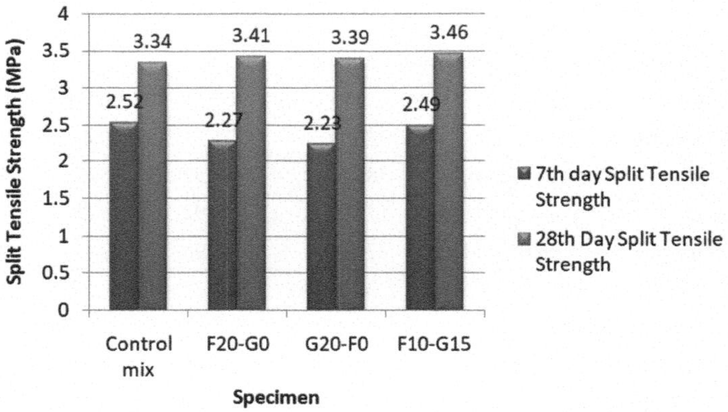

Fig. 5 Comparison of 7th day and 28th day split tensile strength

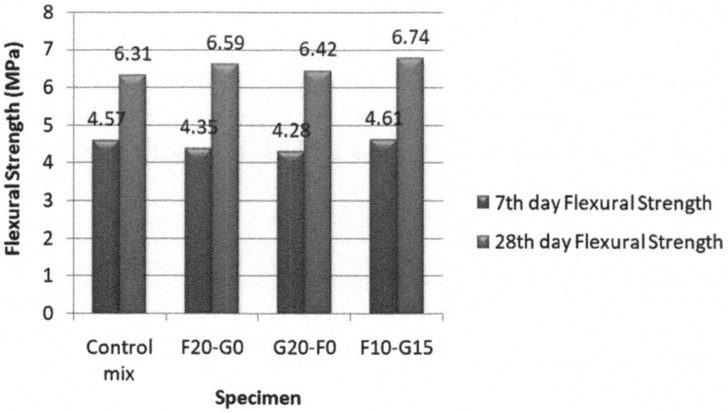

Fig. 6 Comparison of 7th day and 28th day flexural strength

- Beyond 25%, the compressive strength decreased for the specimens with combined replacement.
- When more than 5% fly ash was incorporated to glass powder-based cement mortar, an increase in compressive strength was observed.
- The properties of specimens with combined replacement were higher than that of specimens with individual replacement.
- The 7th day compressive strength of cubes and cylinders with replacement was lower than that of the cubes with no replacement.
- The 28th day compressive strength of cubes and cylinders were higher than that of control mix.
- The compressive strength of cylinders was approximately 0.7 times that of cubes.
- The 7th day split tensile strength of samples with replacement was slightly lower than that of samples with no replacement.
- The 28th day split tensile strength of specimens with fly ash and glass powder as replacement was slightly higher than that of control mix specimens.
- The 7th day flexural strength of specimens with replacement showed a slight decrease compared to the control mix.
- The 28th day flexural strength of samples with replacement was slightly higher than that of control mix.
- The strength values of glass powder concrete were lower than that of fly ash concrete.
- When fly ash was incorporated to glass powder concrete, there was an improvement in the mechanical properties.

6 Conclusions

The replacement of cement with materials such as fly ash and glass powder will add to the engineering as well as economic and ecological benefits. A major concern regarding the use of glass in concrete is the chemical reaction that takes place between the silica-rich glass particle and the alkali in pore solution of concrete (ASR). The inclusion of fly ash in glass concrete reduces this reaction and improves the workability and durability properties of concrete.

In this study, the individual optimum percentage and combined optimum percentage of fly ash and glass powder as cement replacement was determined. It was noted that when glass powder was used in concrete, the mechanical properties had no improvement. The strength properties observed for glass powder concrete was lower than that of fly ash concrete. But, when more than 5% fly ash was incorporated to glass powder concrete, the strength was enhanced. The mechanical properties of specimens which contain both fly ash and glass powder was much more than that of control concrete.

References

1. Omran A et al (2016) Performance of glass-powder concrete in field application. ScienceDirect
2. Aliabdo AA et al (2016) Utilization of waste glass powder in the production of cement and concrete. ScienceDirect
3. Bignozzi MC et al (2015) Glass waste as supplementary cementing materials: the effects of glass chemical composition. ScienceDirect
4. Paris JM et al (2016) A review of waste products utilized as supplements to Portland cement in concrete
5. Jain JA et al (2009) Chloride transport in fly ash and glass powder modified concretes—influence of test methods on microstructure. ScienceDirect
6. Goud V et al (2016) Partial replacement of cement with fly ash in concrete and its effects. IOSR J Eng
7. Lee H et al (2017) Performance evaluation of concrete incorporating glass powder and glass sludge wastes as supplementary cementing material. ScienceDirect
8. Elaqra H et al (2017) Effect of using glass powder as cement replacement on rheological and mechanical properties of cement paste. ScienceDirect
9. Saha AK et al (2017) Effect of class F fly ash on the durability properties of concrete. ScienceDirect

Environmental and Economic Impact Assessment of Flooring Materials

P. A. Ajusree and Jose Jenson

Abstract Many materials are used during construction which goes through different manufacturing and transporting process, which may adversely affect the environment. Choosing an environmentally friendly material will reduce the inimical environmental effect caused due to construction material. But for the proper selection of sustainable building material, it must be balanced against the economic performance also. So proper decision-making criteria have to be developed. This study analyses and compare five flooring materials (terrazzo, linoleum, ceramic, and marble and vinyl composition tiles) on their environmental and economic performance. The environmental impact analysis was done using the Building for Environmental and Economic Sustainability (BEES) software. The Life cycle cost of the materials was found with the help of data from various journals and from National Building estimator and Indian context. The result obtained was then analyzed using VIKOR (VIseKriterijumska Optimizacija I Kompromisno Resenje, that means: Multicriteria Optimization and Compromise Solution) to develop a combined performance score for the selection of a best suitable material. Then the Contingent Valuation Method (CVM) is utilized to estimate the environmental cost associated with the selected alternative flooring materials. The method will provide improved decision-making criteria for the selection of suitable flooring material.

Keywords Life cycle assessment · Life cycle cost assessment · MCDM · VIKOR

1 Introduction

Nowadays the impact caused by human activities on the environment is increasing drastically and a number of environmental problems have emerged at local, regional and global levels. According to Environmental Performance Index (EPI) 2018, developed by Yale University and Columbia University in collaboration with the World Economic Forum and the Joint Research Centre of the European Commission, India

P. A. Ajusree (✉) · J. Jenson
Department of Civil Engineering, SJCET, Palai, India
e-mail: ajusree.aju@gmail.com

K. Dasgupta et al. (eds.), *Proceedings of SECON'19*,
Lecture Notes in Civil Engineering 46,
https://doi.org/10.1007/978-3-030-26365-2_8

has been ranked 177 among 180 countries. Construction industry which is a major developing sector also contributes to the increasing environmental pollution. Construction industry consists of various building materials which undergo different manufacturing and transportation process. This may result in the release of emission of various greenhouse gases and other harmful gases into the air, soil, and water, which may lead to global warming, eutrophication, acidification, and various other environmental impacts. As a result of this, there comes a need for the assessment of environmental soundness of products. So there is a demand for improving the product from an environmental point of view both for internal use and marketing purpose. Life cycle assessment (LCA) is becoming an increasingly important method for making product related environmental assessments in India, as the housing sector accounts for a large part of the primary energy use [7]. The life cycle of building materials include the manufacture, operation, maintenance, and end of life [2]. Flooring materials which is an important component of a building, is any a permanent covering of a floor. It cover about 50% of the building surface. The manufacture and transportation of flooring materials utilizes a large amount of natural resources leading to environmental impact. So a proper decision criteria has to be developed for choosing a suitable flooring material [1].

But while choosing a flooring material its environmental preference cannot be considered as a selection criteria. Selection of a material depends on economical factor also. While some prioritize the environmental impact of a product the others priorities the economic factor. There comes the need to integrate the life cycle cost with life cycle assessment. Life cycle cost assessment incudes the calculation of total cost of a product starting from its initial cost to the disposal cost [3]. In this paper a comparative life cycle assessment and life cycle cost assessment of five flooring materials is done to assess the environmental end economic impacts [8]. The materials include composite marble, ceramic tile, terrazzo tile, linoleum flooring and vinyl composition tile. The building for environmental and economic sustainability (BEES) and VIKOR (VIseKriterijumska Optimizacija I Kompromisno Resenje) that means: Multicriteria Optimization and Compromise Solution approach of Multi Attribute Decision Making (MCDM) where used in this study.

2 Input for Environmental Impact Calculations

Input is provided from data obtained through survey. An expert survey was conducted through email to get the weight out of 100 for each environmental effect considered for the investigation. A sum of 26 replays was acquired from experts having experience of around 3–20 years. Among all the environmental impact factors global warming, fossil fuel depletion, and human health scores the maximum. The information from SimaPro LCA database and United States Life Cycle Inventory Database were utilized in the BEES for the analysis of various flooring materials utilized in this examination.

3 Output

3.1 Environmental Impact Result

From the environmental impact result from analysis using BEES it is found that the composite marble possess the maximum global warming potential of about 4141 g CO_2 and Vinyl composition tile contribute the least of about 1042 g CO_2. Composite marble emits a large amount of carbon dioxide and methane, which is a main source of increasing global warming. Since linoleum flooring is manufactured using all natural elements the emission from this flooring is the minimum which can contribute less to global warming. In a similar manner the environmental impact caused by each flooring materials are calculated.

The algorithm used in BEES to calculate the impact result is as follows [4, 5]:

$$IA_{jk} = \sum_{i=1}^{n} I_{ij} * \text{IA factor}_i \tag{1}$$

where,

n number of inventory flows in impact category k;
I_{ij} inventory flow quantity for alternative j with respect to inventory flow i,
IA factor_i impact assessment characterization factor for inventory flow i.

Normalizing impact values

The effect esteems acquired from BEES is in various units for example global warming potential in CO_2 proportionate, fossil fuel depletion in MJ and so on. So for getting a typical positioning for each environmental impact these effect esteems are to be changed over into standardized score. The standardized score is determined utilizing the US Normalization factor, since it is a universally acknowledged standardization esteems. The algorithm used to ascertain the standardized score is given beneath [5]:

$$NS_i = \frac{IA_{jk}}{NF_i}, \tag{2}$$

where,

NS_i normalized score of impact category i
IA_{jk} environmental impact result for impact category i
NF_i normalization factor for impact category i.

Figure 1 shows the overall normalized score for all the environmental impact for each flooring material used in the study.

From the outcome the score acquired by composite marble (0.20401634) is more noteworthy than different materials and the least score is acquired by Vinyl Composition Tile (0.034164662), which implies that VCT is more environmental friendly

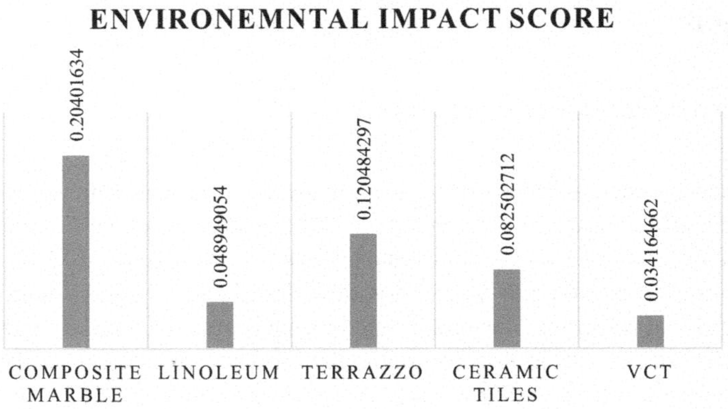

Fig. 1 Total environmental impact score

than composite marble since the vitality used the destructive gases radiated amid the Life cycle of more when contrasted and VCT.

4 Computation of Economic Impact

The life cycle cost during the use phase of the material is calculated using the life cycle cost analysis. For LCC calculation, the discount rate is taken as 6.25%, which is the discount rate in India. The annual maintenance cost is assumed to be 3% of initial cost and the life cycle cost is calculated for per ft^2 of flooring material. For example, the LCC computation of composite marble, the total initial cost is evaluated to be Rs. 185/ft^2 (Purchase cost + Installation cost), where purchase cost is Rs. 150/ft^2 and installation cost is Rs. 35/ft^2 Concerning the way that it is a floor, it was expected that maintenance would be done once at regular intervals, considered for 5 year interval. The present value is given by PV = F × SPV, where F is the future value and SPV $= (1 + d)^{-n}$ is the single present value at time t, with the rebate rate of d.

Table 1 outlines the estimation for the maintenance and replacement costs for composite marble. The value Rs. 14.9182 is the PV of the re-pair/maintenance cost decided at regular intervals of the investigation time frame at a discount rate of 6.25 and 3% repair rate. Since the life expectancy of composite is 75 years, which is more than the investigation time frame, no replacement is considered for composite marble.

The LCC is given by

$$\text{Total life cycle cost} = I_c + M_{PV} + REPL_{PV} + R_{PV}$$

where, I_c = initial installation cost; M_{pv} = present value of the maintenance cost; $Repl_{pv}$ = present value of the replacement cost; and R_{pv} = PV of the residual value.

Table 1 Weightage for impact factors

Impact factors	Average weightage
Global warming	25
Acidification	5.5
Eutrophication	5
Fossil fuel depletion	13
Indoor air quality	9
Habitat alteration	6
Water intake	2
Criteria air pollutants	3
Smog	2
Eco toxicity	4
Ozone depletion	6.1
Human health	19.4
Sum	100

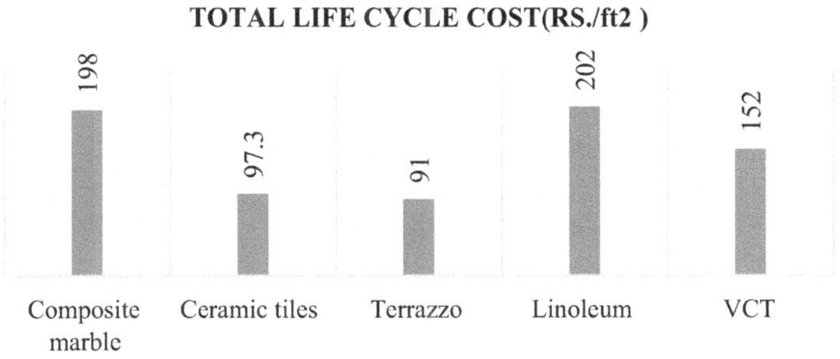

Fig. 2 Total life cycle cost

The value of LCC can therefore be calculated as LCC = (185 + 14.9182 + 1.978), where Rs. 1.9878 is the salvage value of composite material. So the total LCC of composite marble is calculated to be Rs. 198/ft^2 Similarly the LCC of all the other four flooring materials are for 50 year study period. The graph 2 shows the calculated life cycle cost of all the flooring materials (Fig. 2).

5 Developing Overall Performance Score

Selecting a suitable flooring materials based on one criteria is cannot be considered as a proper method. So a combined performance score has to be developed considering both environmental end economic impact [6]. For this VIKOR method is used, which is a multi-criteria decision making (MCDM) or multi-criteria decision

Table 2 Maintenance and replacement cost for composite marble

Maintenance cycle (years)	Discount factor	Present value (Rs./ft^2)	Replacement year
5	0.7385	4.0987	–
10	0.5454	3.0270	–
15	0.4028	2.2355	–
20	0.2975	1.6511	–
25	0.2197	1.2194	–
30	0.1622	0.900	–
35	0.1198	0.6649	–
40	0.0885	0.4911	–
45	0.0653	0.3624	–
50	0.0483	0.2687	–
Total		14.9182	

Table 3 Ranking of flooring materials

	S_i	R_i	Q_i	Rank
Composite marble	0.126659	0.047843	0.000	5
Linoleum	0.142767	0.738498	0.851	2
Terrazzo	1.51449	0.580716	0.768	4
Ceramic	1.607484	0.544095	0.770	3
VCT	1.847305	0.778482	1.000	1

analysis method. The steps involved in VIKOR method is given below. (Table 2) provide the result for combined performance score calculation.

Step 1 Establish the decision matrix.
Step 2 Determine the normalized decision matrix.
Step 3 Determine the ideal and negative ideal solutions.
Step 4 Calculate the utility measure (S_i) and the regret measure (R_i).
Step 5 Calculate the VIKOR index (Q_i).
Step 6 The alternative solution with the highest VIKOR index is the best solution (Table 3).

6 Conclusions

The LCA and LCC investigations are two incredible ways to deal with measure the environmental and economic impact of materials. MADM is a procedure for analyzing choices with qualities in disproportionate units. This investigation utilized the NIST's BEES model to analyze the environmental impact and cost based on Indian context. It is found that among the selected flooring materials Vinyl Composition tile is the best solution when considering both environmental and economic. This means that when balancing environmental against economic performance Vinyl Composition Tile is the most effective materials than others. Also the results shows that Composite marble is the lowest scoring material based on environmental and economic impact. The application of this method improved the bias in choosing the best flooring material that is merely based on environmental and economic performance.

References

1. Athawale VM, Chakraborty S (2012) Material selection using multi- criteria decision-making methods: a comparative study. Proc Inst Mech Eng Part L J Mater Des Appl 226(4):266–285
2. Cabeza LF, Rincón L, Vilariño V, Pérez G, Castell A (2014) Life cycle assessment (LCA) and life cycle energy analysis (LCEA) of buildings and the building sector: a review. Renew Sustain Energy Rev 29:394–416
3. Hunkeler D, Rebitzer G (2005) The future of life cycle assessment. Int J Life Cycle Assess 10(5):305–308
4. Lippiatt BC (2000) Building for environmental and economic sustainability technical manual and user guide. U.S. Environmental Protection Agency Office of Pollution Prevention and Toxics, Washington, DC
5. Lippiatt BC (2017) BEES 3.0 building for environmental and economic sustainability: technical manual and user guide. NIST Interagency/Internal Rep. (NISTIR)-6916. NIST, Gaithersburg, MD
6. Makutėnienė D, Šostak OR, Maceika A (2016) Decision support system for roof installation. Stud Inf Control 25(2):164
7. Senitková I, Bednárová P (2015) Life cycle assessment. JP J Heat Mass Transf 11(1):29
8. Sim J, Sim J, Park C (2016) The air emission assessment of a South Korean apartment building's life cycle, along with environmental impact. Build Environ 95:104–115
9. Weidema B (2014) Has ISO 14040/44 failed its role as a standard for life cycle assessment? J Ind Ecol 18(3):324–326

Step-Wise Multiple Linear Regression Model Development for Shrinkage Strain Prediction of Alkali Activated Binder Concrete

Sriman Pankaj Boindala, Kruthi Kiran Ramagiri, Anju Alex and Arkamitra Kar

Abstract Among the recent discoveries for alternatives of portland cement (PC) concrete, Alkali-activated binder (AAB) concrete is prolifically being considered as the most eco-friendly and sustainable alternative. The present study evaluates the shrinkage behaviour for three different AAB mixtures containing fly ash and/or slag at different proportions which are activated by sodium hydroxide and sodium silicate. Multiple linear regression models are developed to predict shrinkage strains of ambient-cured AAB concrete as a function of age and percentage of fly ash in the precursor. The aim of this work is to come up with a generalized equation that can predict the shrinkage of various binary blended AAB mixes cured at room temperature. The predicted models are ranked based on RMSE and then compared with the experimental data. The correlations were found to be quite satisfactory ($R^2 = 0.937$) and can be used to estimate the shrinkage for similar AAB mixtures. It is observed that the proposed model agrees more closely with the experimental results from the present study.

Keywords Alkali-activated concrete · AAB · Ambient-cured · Shrinkage strain · Regression model

1 Introduction

Alkali-activated binders (AAB) are one of the most popular *ecolonomical* (Ecological + Economical) alternatives for portland cement (PC). It has been reported by several researchers that AAB has better mechanical properties than PC [1–3]. This is due to polymerization reaction of AAB upon the activation of industrial wastes such

S. P. Boindala · K. K. Ramagiri · A. Alex · A. Kar (✉)
BITS-Pilani Hyderabad Campus, Hyderabad, Telangana 500078, India
e-mail: arkamitra.kar@hyderabad.bits-pilani.ac.in

S. P. Boindala
e-mail: srimanpanakaj@gmail.com

K. K. Ramagiri
e-mail: p20170008@hyderabad.bits-pilani.ac.in

K. Dasgupta et al. (eds.), *Proceedings of SECON'19*,
Lecture Notes in Civil Engineering 46,
https://doi.org/10.1007/978-3-030-26365-2_9

as fly ash and slag with alkaline solution containing sodium silicate and sodium hydroxide. The polymerization leads to the formation of a stronger aluminosilicate hydrate matrix compared to the calcium silicate hydrate formed by portland cement hydration. The alkali-activation of fly ash takes place at a slow rate under ambient conditions and consequently, thermal curing is preferred. Hence, to make this kind of binders more practical and more energy efficient ambient cured AAB mixes are prepared by partial replacement of fly ash with slag. This is due to the higher reactivity of slag compared to fly ash under ambient conditions. Though the slag-based ambient-cured AAB mixes have shown promising results in improving the mechanical and fire resistance properties compared to PC mixes, the durability properties like shrinkage and creep have not been well established. To overcome this issue, binary and tertiary blended AAB mixes are investigated.

Most of the research on AAB shrinkage is reported either using mortar or paste samples [3–6]. Moreover, the majority of these studies use thermal-cured mixes [7]. Limited information is available on the shrinkage characteristics of AAB concrete cured under ambient conditions. There are various PC shrinkage models namely EN 1992-1-1, B3, ACI 209, Model Code 2010, GL 2000. Most of these models are complex and they take into consideration various parameters like slump of the concrete, relative humidity, compressive strength value and various coefficients which have to be determined based on the type of the mix. Existing models to predict shrinkage for PC are not suitable for AAB concrete [8–11]. 10–20% mass replacement of fly ash with slag resulted in a decrease in shrinkage with increase in slag content for ambient cured AAB mixes over a period of 180 days [1]. Another study carried on binary blended AAB reported that shrinkage increases with increasing slag content and decreases with increasing the Ms modulus (i.e. SiO_2/equivalent Na_2O in the activating solution) [12, 13]. Evidently, there is no conclusive report on the variation of shrinkage with increasing slag content. Hence, the present study focuses on binary blended ambient-cured AAB mix and its shrinkage properties. The objective is to use multiple linear regression to generate a relatively simple model to predict the shrinkage strain of AAB concrete by using age of the specimen and percentage of fly ash in the mix as independent variables (IN) and shrinkage strain as the dependent variable (DV). This would not only make the model less complex but also unique for all binary blended AAB mixes. The materials and methods used in this study are presented in the next section.

2 Materials and Methods

2.1 Materials and Mix Proportions

The materials used for this study are Class F fly ash (FA) and Ground Granulated Blast Furnace Slag (GGBS). These materials comply with the guidelines of ASTM C618 [14] and ASTM C989 [15] respectively. FA is obtained from a thermal power plant in

Ramagundam in the state of Telangana, India. GGBS is obtained from JSW cements, Telangana. Alkali-activating solution is prepared by adding sodium hydroxide pellets to sodium silicate solution. Sodium hydroxide used conforms to food grade with 99% purity and the sodium silicate solution is of industrial grade containing 29.5% SiO_2 and 14.7% Na_2O by weight. Both of these chemicals are obtained from Hychem Chemicals limited. For all the mixes the activating solution was prepared a day before casting. Granite aggregates of nominal size 10 mm that are obtained from local quarries in Hyderabad and 4.75 mm passing local river sand are used for the preparation of concrete. Both comply with the ASTM C33/C33M-18 [16] standard. Chemical admixture used for this work is commercially available polycarboxylate ether based high range water reducer supplied by BASF chemicals according to the guidelines of ASTM C494 [17]. Normal tap water is used for the additional water required in the mix.

2.2 Method

Three different mixes of ambient cured AAB concrete with varying FA to slag ratios, FS0 (100:0), FS30 (70:30) and FS50 (50:50) are used in this study as shown in Table 1. The water to solids ratio (w/s) is maintained at 0.3 for each mix and Ms modulus is taken to be 1.4. Addition of water and superplasticiser vary according to the workability required for a slump of 150 mm. For measuring AAB concrete shrinkage, prisms of dimension 280 × 25 × 25 mm tested in accordance with the specifications of ASTM C157/157M [18] the width of the specimen is taken as 25 mm as the maximum size of aggregate in the concrete is 10 mm. Shrinkage strains (SS) were then calculated as the ratio of change in length to the initial length. The length change of the specimen at any age is calculated using the following equation.

$$\begin{aligned} \Delta \mathrm{L} &= (\mathrm{CRD} - \mathrm{initialCRD}) \\ SS &= \Delta \mathrm{L}/\mathrm{G} \end{aligned} \quad (1)$$

Table 1 Details of AAB Concrete mix

Materials	FS 0	FS 30	FS 50
FA/slag	100/0	70/30	50/50
FA (kg/m^3)	400	280	200
Slag (kg/m^3)	0	120	200
Activator solution (kg/m^3)	140	140	140
Coarse aggregate (kg/m^3)	1209	1209	1209
Fine aggregate (kg/m^3)	651	651	651
Water (kg/m^3)	67.5	67.65	67.65
SP (l/m^3)	0	3.14	4

where

ΔL = length change of specimen at any age,

CRD = difference between the comparator reading of the specimen and the reference bar at any age, and

G = the gauge length [280 mm].

The regression analysis and model formation are performed using the R 3.1.3 statistical software package. The procedure followed for arriving at a model equation is explained as follows [19]:

- Data Assimilation: Complete data available is tabulated in one file which can be accessible to the user.
- Data Separation: The complete dataset is randomly divided into training and testing set in a ratio of 70:30 by using "caTools" library in R.
- Model assumption and Validation: Since there are two independent variables and one dependent variable, a linear relationship is assumed initially i.e. $DV = a_0 + a_1*IN_1 + a_2*IN_2$ and the coefficients a_0, a_1 and a_2 are obtained using 'lm' function in R. After finding the relationship between the variables, the statistical significance of each variable is calculated using 't test' which can be obtained by using 'summary' function in R. The insignificant variables are eliminated and the regression analysis is performed again. Subsequently, the statistically significant variables and their relationship with the DV are attained. The most efficient model is selected based on the least root mean square error (RMSE) value. For large RMSE values, then the proposed model form is modified by increasing the polynomial power i.e. $DV = a_0 + a_1*IN_1 + a_2*IN_2 + a_3*IN_1^2 + a_4*IN^2 + a_5*IN_1*IN_2$. Statistically insignificant terms are dropped from the model and the RMSE is computed again to arrive at an optimal model form.

3 Results

A combined set of data containing shrinkage strains, age and percentage of FA readings of all the three mixes is taken as a dataset. The data is divided into training and testing set with 70:30 ratio as explained above and the regression analysis is performed. Five different models are established. These model performances are evaluated and ranked based on the RMSE value obtained from the testing set data. The details of the models with the RMSE value are presented in Table 2. The observed and predicted value of maximum shrinkage strain using the best model is tabulated in Table 3.

Here '*t*' represents the age in days and F represents percentage of FA content in the mix and SS represents shrinkage strain.

Model 5 has the least RMSE value of 58.51 and R^2 of 0.937 and is used for the prediction of shrinkage strains of AAB concrete.

The relationships between the predicted and experimental shrinkage of AAB concrete are presented in Fig. 1. The experimental values reveal that the strains

Table 2 Model equations

Name	Equation	RMSE	R^2
Model 1	$SS = 1.885 * t + 4.84 * F$	215.6532	0.91
Model 2	$SS = 5.59 * 10^2 + 6.64 * t - 1.73 * 10^1 * F - 1.4 * 10^{-2} * t^2 + 1.437 * 10^{-1} * F^2$	157.1211	0.801
Model 3	$SS = 13.15 * t - 6.199 * 10^{-2} * t^2 + 9.043 * 10^{-5} * t^3 - 1.096 * 10^{-1} * F^2 + 1.201 * 10^{-3} * F^3$	123.842	0.9771
Model 4	$SS = 7.64 * t - 0.013 * t^2 + 0.0209 * F^2 - 79.166 * t/F$	101.046	0.961
Model 5	$SS = -35.04 + 23.43 * t - 0.06061 * t^2 + 8.875 * 10^{-5} * t^3 - 1333 * t/F + 38810 * t/F^2$	58.5119	0.937

Table 3 Observed and predicted readings of shrinkage strain observed at 250 days

Mix	Observed maximum strain at 250 days (micro strain)	Predicted maximum strain at 250 days (micro strain)
FS 0	1046.205	1036.554
FS 30	685.524	618.4517
FS 50	684.821	615.5538

increase at a relatively swift rate over the first 90 days and then gradually become constant with time up to an age of 365 days. The graphs plotted are for model 5, since the RMSE value is the least for that model. It is evident from the results that the proposed model closely follows the experimental data for all the three AAB mixes and satisfactorily predicts its shrinkage. Hence, unlike previous models containing several IN, the proposed model adequately predicted the shrinkage strains of AAB concrete using a significant mix proportion parameter, i.e. FA proportion in the mix.

4 Conclusions

In the present study, three different AAB concrete mixes are examined for its shrinkage characteristics. Of various multiple linear regression models developed, the model with the least RMSE is used to predict the shrinkage of AAB concrete. The conclusions from the present study are as follows:

- Compared to other models, model 5 has the least RMSE value and is proposed to predict the shrinkage strains of binary blended ambient cured AAB concrete mixes.
- The RMSE and R^2 value of model 5 are 58.51 and 0.937 respectively.
- The shrinkage strains of AAB concrete increase at a rapid rate until 90 days. After 90 days, the increase in shrinkage strain continue at a slower rate.

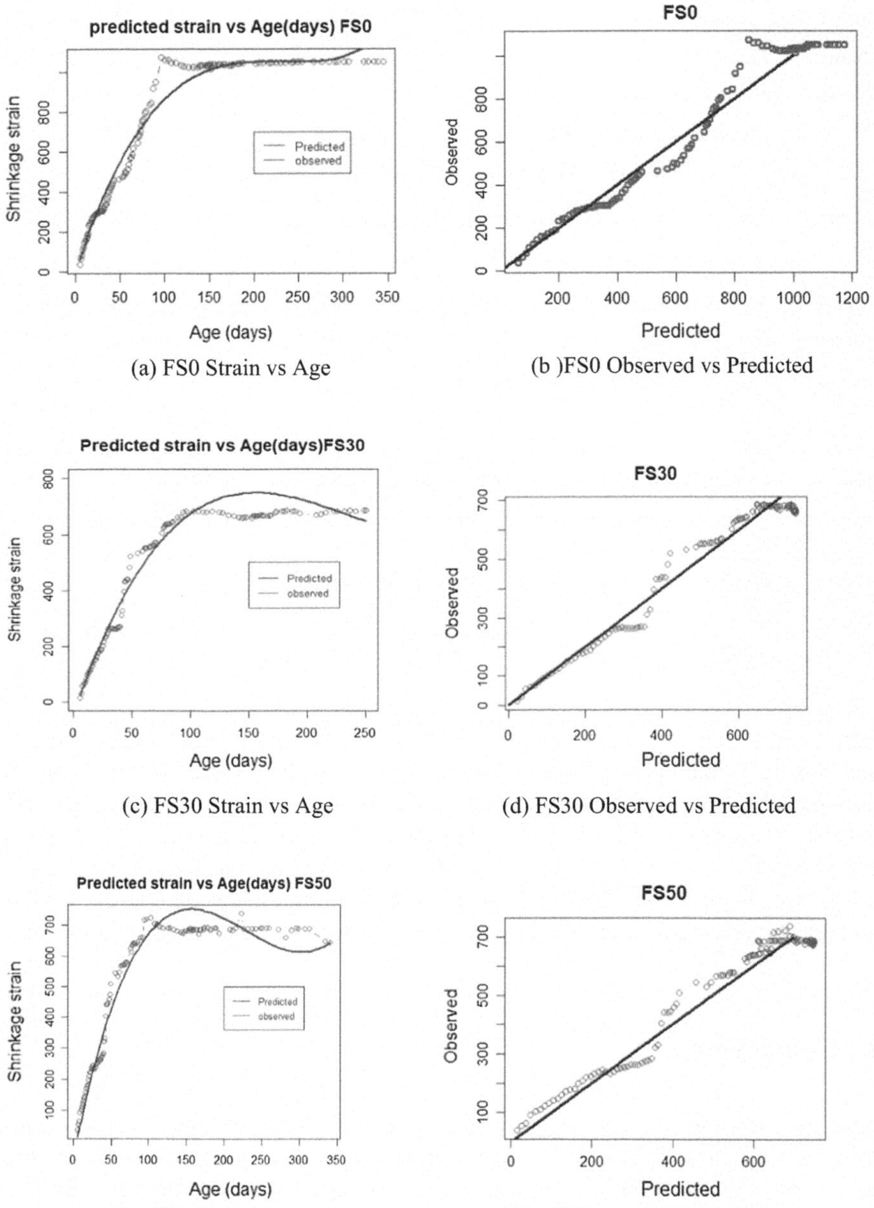

(a) FS0 Strain vs Age

(b)FS0 Observed vs Predicted

(c) FS30 Strain vs Age

(d) FS30 Observed vs Predicted

(e) FS50 Strain vs Age

(f) FS50 Observed vs Predicted

Fig. 1 Experimental and predicted shrinkage of AAB concrete

- It is observed that shrinkage of AAB concrete decreased with increasing slag content. FS 0 exhibits higher shrinkage, whereas FS 30 and FS 50 exhibit almost similar shrinkage strains.
- The predicted shrinkage of AAB concrete using model 5 follow the same trend as experimental data for mixes with varying fly ash/slag proportion.
- The next phase of this study will compare the shrinkage strains of concrete mixes made from AAB and PC having similar compressive strength, to encourage the practical use of AAB concrete.

References

1. Deb PS, Nath P, Sarker PK (2014) The effects of ground granulated blast-furnace slag blending with fly ash and activator content on the workability and strength properties of geopolymer concrete cured at ambient temperature. Mater Des 62:32–39
2. Jia Z, Yang Y, Yang L, Zhang Y, Sun Z (2018) Hydration products, internal relative humidity and drying shrinkage of alkali activated slag mortar with expansion agents. Constr Build Mater 158:198–207
3. Lee NK, Lee HK (2013) Setting and mechanical properties of alkali-activated fly ash/slag concrete manufactured at room temperature. Constr Build Mater 47:1201–1209
4. Ma Y, Ye G (2015) The shrinkage of alkali activated fly ash. Cem Concr Res 68:75–82
5. Nath P, Sarker PK (2014) Effect of GGBFS on setting, workability and early strength properties of fly ash geopolymer concrete cured in ambient condition. Constr Build Mater 66:163–171
6. Thomas JJ, Allena J, Jennings HM (2012) Density and water content of nano-scale solid C-S-H formed in alkali-activated slag (AAS) paste and implications for chemical shrinkage. Cem Concr Res 42(2):377–383
7. Castel A, Foster SJ, Ng T, Sanjayan JG, Gilbert RI (2016) Creep and drying shrinkage of a blended slag and low calcium fly ash geopolymer Concrete. Mater Struct 49(5):1619–1628
8. Gilbert RI (2002) Creep and shrinkage models for high strength concrete–proposals for inclusion in AS3600. Aust J Struct Eng 4(2):95–106
9. Rashad AM (2013) Properties of alkali-activated fly ash concrete blended with slag. Iran J Mater Sci Eng 10(1):57–64
10. Wallah S, Rangan BV (2006) Low-calcium fly ash-based geopolymer concrete: long-term properties, Research Report GC 2, Faculty of Engineering, Curtin University of Technology, Perth, Australia
11. Ye H, Radlińska A (2016) Shrinkage mechanisms of alkali-activated slag. Cem Concr Res 88:126–135
12. Fang G, Bahrami H, Zhang M (2018) Mechanisms of autogenous shrinkage of alkali-activated fly ash-slag pastes cured at ambient temperature within 24 h. Constr Build Mater 171:377–387
13. Gao X, Yu QL, Brouwers HJH (2015) Reaction kinetics, gel character and strength of ambient temperature cured alkali activated slag–fly ash blends. Constr Build Mater 80:105–115
14. American Society for Testing and Materials International, ASTM C618-17a (2017) Standard Specification for Coal Fly Ash and Raw or Calcined Natural Pozzolan for Use in Concrete, West Conshohocken, PA
15. American Society for Testing and Materials International, ASTM C989/C989M-17 (2017) Standard Specification for Slag Cement for Use in Concrete and Mortars, ASTM International, West Conshohocken, PA
16. American Society for Testing and Materials International, ASTM C33/C33M–16e1 (2016) Standard Specification for Concrete Aggregates, West Conshohocken, PA

17. American Society for Testing and Materials International, ASTM C494/C494M-17 (2017) Standard Specification for Chemical Admixtures for Concrete, West Conshohocken, PA
18. American Society for Testing and Materials International, ASTM C157/C157M-17 (2017) Standard Test Method for Length Change of Hardened Hydraulic-Cement Mortar and Concrete, West Conshohocken, PA
19. Thomas RJ, Peethamparan S (2017) Stepwise regression modeling for compressive strength of alkali-activated concrete. Constr Build Mater 141:315–324

Experimental Study on Self Compacting Self Curing Concrete Using Copper Slag as Partial Replacement of Fine Aggregate

Anisha Mariya Paul and Elba Helen George

Abstract Self-Compacting Concrete has gained wide use for placement in congested reinforced concrete structures with difficult casting conditions while in Self Curing Concrete, water is restricted by means of some chemical compounds to move out from the concrete body. The main problem due to acute shortage and high price of river sand led to the enormous usage of M sand in construction. Use of copper slag as a fine aggregate is a good alternative to M sand and a better remedy to the disposal of copper slag. This investigation is aimed at utilizing the benefits of both self-compacting and self-curing concrete incorporating Copper Slag. In this study, the fresh and hardened properties of M50 equivalent Self-Compacting Concrete is obtained by incorporating 15% of Fly Ash as constant. Self-curing agent Polyethylene Glycol is added in 0, 0.5,1 and 1.5% to Self-compacting Concrete and its optimum is evaluated based on strength parameters. The optimum dosage of Copper Slag by the replacement of fine aggregate in different percentages of 10, 20, 30, 40 and 50% in Self-Compacting Self-Curing Concrete is obtained and their strength parameters are compared with normal Self Compacting Self-Curing Concrete. The utilization Fly Ash and Copper Slag which indirectly facilitate waste reduction helps in maintaining the ecological balance thus reduces the consumption of cement and aggregates.

Keywords Polyethylene Glycol · Self-compacting concrete · Self-curing concrete · Fly Ash · Copper slag

1 Introduction

Concrete is a very strong and versatile construction material. It consists of cement, sand and aggregate (e.g., gravel or crushed rock) mixed with water. Since the time that concrete has been acknowledged as an asset for development of building, researchers

A. M. Paul (✉) · E. H. George
Toc H Institute of Science and Technology, Ernakulam, India
e-mail: anishamariya.paul95@gmail.com

E. H. George
e-mail: elbaheleng@gmail.com

K. Dasgupta et al. (eds.), *Proceedings of SECON'19*,
Lecture Notes in Civil Engineering 46,
https://doi.org/10.1007/978-3-030-26365-2_10

have been attempting to help its strength and enhance its performance. Current advances in construction industry demand better quality strength of structures. There is a helpful modification in the design of concrete from strength-oriented idea to a performance-oriented design. Present-days there is a vast prominence on performance aspect of concrete. One such thought has prompted the evolution of Self Compacting Self Curing Concrete which is reflected as "the most innovative development in concrete construction" [1].

A very limited work is reported from this area having the benefits of both self-curing as well as self-compaction. The future for this type of concrete is very bright due to scarcity of skilled man power, non-mechanization of construction industry, abundant availability of construction materials available at very low cost. The properties of this type of concrete, if found satisfactory would be a great step in concrete technology compiling the advantages of both internal curing as well as self-consolidation [2].

Copper slag is a waste material of matte smelting and refining of copper, which could have a promising future in construction industry as a partial or complete substitute of aggregates. Each ton of copper generates approximately 2.5 tons of copper slag. The research on using copper slag as a partial or complete replacement of fine aggregate is becoming prominent in the present decade because of its sustainable approach [3–5].

1.1 Self-compacting Concrete

Self-compacting concrete (SCC) represents one of the most outstanding advancement in concrete technology during the last decade. SCC is another sort of concrete with huge deformability and segregation resistance. SCC was first developed in 1988 by professor Okamura intended to improve the durability properties of concrete structures. SCC is a flowing concrete mixture which is able to consolidate under its own weight. The highly fluid nature of SCC makes it suitable for placing in difficult conditions and in sections with congested reinforcement [6].

The method for achieving self-compactibility involves not only high deformability of paste or mortar, but also resistance to segregation between coarse aggregate and mortar when the concrete flows through the confined zone of reinforcing bars. Okamura and Ozawa have employed the methods to achieve self-compactability such as limited aggregate content, low water-powder ratio and use of super plasticizer [6].

1.2 Self-curing Concrete

A self-curing concrete is provided to absorb water from atmosphere to achieve better hydration of cement in concrete. It solves the problem that the degree of cement hydration is lowered due to no curing or improper curing and thus unsatisfactory properties of concrete [7]. The self-curing concrete means that no curing is required for concrete, or even there is no external supplied water is required after placing. The properties of this self-cured concrete of this invention are at least comparable to and even better than those of concrete with traditional curing. The internal water is maintained by incorporating the self-curing agent which reduces the evaporation of water from the concrete, thereby increasing the water retention capacity of concrete [8]. In past decades, the effect of Self Curing Concrete possesses improved properties while comparing to identically cured controls. It was found that, initial surface absorption, chloride ingress, carbonation, corrosion potential and freeze and thaw resistance characteristics were comparatively better by self-cure concrete than the water cured concrete [9].

2 Objectives of the Study

The main objective of this investigation is:

- To establish M50 grade equivalent Self-Compacting Concrete (SCC) incorporating 15% of Fly Ash as constant.
- To obtain M50 grade equivalent Self-Compacting Self-Curing Concrete (SC SCC) using Polyethylene Glycol (PEG 400) in 0, 0.5, 1 and 1.5% as self-curing agent and obtain its optimum percentage.
- To investigate strength parameters of SCC SC with Copper Slag (CS) in 10, 20, 30, 40 and 50% as partial replacement of fine aggregate and obtain its optimum percentage.
- To analyze and compare the results obtained with normal SCC and SC SCC and interpret the results.

3 Experimental Methodology

The experimental program is designed to investigate the strength of self-curing self-compacting concrete by adding poly ethylene glycol PEG400 at 0, 0.5, 1 and 1.5% by weight of binder to the concrete. The experimental program is aimed to study the workability and strength parameters. The slump flow test and J ring test were conducted for all mixes to know the fresh property of concrete. Compressive strength, Flexural strength and Split tensile strength test was conducted at 7 and 28 days. In this investigation the maximum dosage of self-curing agent was restricted to 1.5% and minimum dosage to 0.5%.

Table 1 Material properties

Materials	Properties	Test results	Reference code
Cement	Specific gravity	3.15	IS 4031-1988 Part-IV (Reaffirmed 2009)
	Fineness	6%	IS 4031-1988 Part-IV (Reaffirmed 2009)
	Consistency	32%	IS 4031-1988 Part-XI (Reaffirmed 2009)
	Initial setting time	50 min	IS 4031-1988 Part-V (Reaffirmed in 2009)
Fine aggregate	Specific gravity	2.75	IS 2386-1963 Part-III (Reaffirmed 2016)
	Water absorption	2.3%	IS 2386-1963 Part-III (Reaffirmed 2016)
Coarse aggregate	Specific gravity	2.82	IS 2386-1963 Part-III (Reaffirmed 2016)
	Water absorption	0.35%	IS 2386-1963 Part-III (Reaffirmed 2016)
Fly Ash	Specific gravity	2.3	
Copper slag	Specific gravity	3.5	

3.1 *Materials and Properties*

The different materials used in this investigation are listed below and their physical properties are illustrated in the Table 1 given:

1. Cement
2. Fine aggregate
3. Coarse aggregate.
4. Polyethylene Glycol (PEG-400)
5. Polycarboxylate Ether (superplasticizer)
6. Fly Ash
7. Water
8. Copper Slag.

3.2 *Mix Design*

There is no standard method for SCC mix design and many academic institutions, admixture, ready-mixed, pre cast and contracting companies have developed their own mix proportioning methods. Several methods exist for the mix design of SCC. The mix designs were carried out for concrete grade 50 MPa based on European

Table 2 Mix design

Material	Cement	Fine aggregate	Coarse aggregate	Water	Fly Ash
Quantity (kg/m^3)	501.5	865	785	200	88.5
Ratio	1	1.72	1.57	0.40	0.18

Federation for Specialist Construction Chemicals and Concrete Systems (EFNARC) guidelines [10]. The details of mixes are given in Table 2. All the ingredients were first mixed in dry condition [11]. Then 70% of calculated amount of water was added to the dry mix and mixed thoroughly. Then 30% of water was mixed with the super plasticizer and added in the mix. Then the mix was discharged for SCC workability tests [12, 13].

4 Results and Discussion

4.1 Fresh Properties

The main characteristics of Self-Consolidating Concrete are the properties in the fresh state. The mix design is focused on the ability to flow under its own weight without vibration, the ability to flow through heavily congested reinforcement under its own weight, and the ability to retain homogeneity without segregation [14]. The workability of Self-Consolidating Concrete is higher than "very high" degree of workability mentioned in IS 456:2000. A concrete mix can only be classified as Self Consolidating Concrete if it has the following characteristics.

- Filling ability
- Passing ability
- Segregation resistance [15]

Several test methods have been developed in attempts to characterize the properties of Self Consolidating Concrete. Figure 1 shows the J Ring test. In this study three Properties are used to evaluate the fresh properties of Self Consolidating Concrete. The flow values of different mix proportions are listed in Table 3 in which SCC is the normal Self Compacting concrete, SCC SC –0.5, SCC SC –1 and SCC SC –1.5 represents the Self Compacting Self Curing Concrete containing self-curing agent PEG-400 of 0.5%, 1% and 1.5% respectively.

4.2 Hardened Properties

The properties of hardened SCC SC were measured in terms of Compressive Strength, Split Tensile Strength test and Flexural Strength Test confirming to IS 516:1959

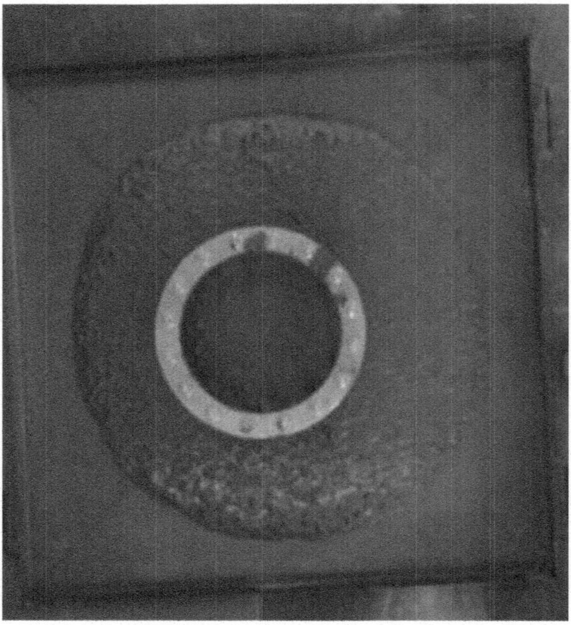

Fig. 1 J Ring test

Table 3 Workability results

Sl. No.	Mixes	Slump flow (mm)	J-Ring (mm)	Passing ability (mm)	Remarks (as per ASTM 1621/C 1621M)
1	SCC SC -0	678	669	9	No visible blocking since passing ability values are between 0 and 25 mm
2	SCC SC -0.5	680	672	8	
3	SCC SC -1	678	669	9	
4	SCC SC -1.5	696	686	10	
5	SCC SC -1 C10	688	680	8	
6	SCC SC -1 C20	691	682	9	
7	SCC SC -1 C30	699	690	9	
8	SCC SC -1 C40	708	698	10	
9	SCC SC -1 C50	716	708	8	

Fig. 2 Compressive strength test

(Reaffirmed 2004), IS: 5816: 1999 (Reaffirmed 2004) and IS 516: 1959 (reaffirmed 2004) respectively. Compression Test is conducted on 54 cubes of 150 × 150 × 150 mm size, which are casted in the laboratory. The test specimens are marked and removed from the moulds. 6 cubes were immediately submerged in clean fresh water and kept there for normal curing. 48 cubes were cured with Polyethylene Glycol of molecular weight 400 (PEG-400) in which fine aggregate was replaced with Copper Slag in different percentages in 30 cubes. 2000 KN capacity Compression Testing Machine (CTM) is used to conduct the test. The specimens are place between the steel plates of the CTM and the failure load in KN will be observed from the load indicator of the CTM. From each type of curing 7 days and 28 days compressive Strength is obtained by using the relation Compressive strength = Load/Area (MPa). Figure 2 shows the compressive strength test and the results of variation in compressive strength with and without external curing are as shown in Table 4.

After 28 days of curing the SCC SC having 1% of PEG-400 showed better result than the normally cured SCC.SCC SC having 40% of Copper Slag as partial replacement of fine aggregate results in higher compressive strength. From the test results, it is evident that the reduction in strength resulting from increasing copper slag is due to increased voids in the concrete, since copper slag possesses fewer fine particles than fine aggregate. It could also be due to the increase of the free water because copper slag absorbs less water than the fine aggregate. The compressive strength of concrete at 7 and 28 days increased gradually up to 40% fine aggregate replacement and then decreased with increase in percentage of replacement. From the 7th and

Table 4 Compression test results

Sl. No.	Mix	7-day compressive strength (N/mm^2)	28-day compressive strength (N/mm^2)
1	SCC SC -0	39.80	58.90
2	SCC SC -0.5	38.85	57.62
3	SCC SC -1	41.70	59.50
4	SCC SC -1.5	39.20	56.30
5	SCC SC -1 C10	39.02	57.25
6	SCC SC -1 C20	39.61	58.58
7	SCC SC -1 C30	41.10	58.81
8	SCC SC -1 C40	42.25	59.92
9	SCC SC -1 C50	39.80	57.46

28th day compressive strength test results, the optimum percentage of copper slag as a partial replacement for fine aggregate in M50 grade concrete was found to be 40%.

Split tensile Test is conducted on 18 cylinders of 150 × 300 mm size, which are casted in the laboratory. Split tensile test was done only on normal SCC, the obtained optimum percentage of SCC SC having 40% of fine aggregate replaced with Copper Slag. The test specimens are marked and removed from the moulds. 6 cylinders were immediately submerged in clean fresh water and kept there for normal curing. 12 cylinders were cured with Polyethylene Glycol of molecular weight 400 (PEG-400). 2000 KN capacity Compression Testing Machine (CTM) is used to conduct the test. The specimens are place between the steel plates of the CTM and the failure load in KN will be observed from the load indicator of the CTM. From each type of curing 7 days and 28 days split tensile strength is obtained by using the relation $\sigma SP = 2P/\pi DL = 0.637P/DL$ (MPa). Figure 3 shows the split tensile strength test and the results of variation in split tensile strength with and without external curing are as shown in Table 5.

From the 7th and 28th day split tensile strength. The maximum tensile strength was found to be at 40% fine aggregate replacement of about 4.53 and 5.56 N/mm^2. The split tensile strength of cylinder showed a similar behaviour to the compressive strength of the cube for all mixtures.

Flexural Strength Test is conducted on 18 beams of 700 × 150 × 150 mm size, which are casted in the laboratory. Flexural strength test was done only on normal SCC, the obtained optimum percentage of SCC SC having 40% of fine aggregate replaced with Copper Slag. The test specimens are marked and removed from the moulds. 6 beams were immediately submerged in clean fresh water and kept there for normal curing. 12 beams were cured with Polyethylene Glycol of molecular weight 400 (PEG-400). The flexural strength of concrete beams was found using Universal Testing Machine after 7 days and 28 days of curing. From each type of curing 7 days and 28 days flexural strength is obtained by using the relation $= Pl/bd^2$ (MPa). The

Fig. 3 Split tensile strength test

Table 5 Split tensile strength results

Sl. No.	Mix	7-day split tensile strength (N/mm^2)	28-day split tensile strength (N/mm^2)
1	SCC SC -0	3.81	4.83
2	SCC SC -1	3.92	4.90
3	SCC SC -1 C40	4.53	5.56

Fig. 4 shows the flexural strength test and the results of variation in flexural strength with and without external curing are as shown in Table 6.

After 28 days of curing, Flexural Strength for mixes SCC SC -0, SCC SC -1 and SCC SC -1 C40 was reported as 4.83 N/mm^2, 4.90 N/mm^2 and 22.8 N/mm^2 respectively. The flexural strength of the control specimen is lower than that of the optimum copper slag contain specimen.

Fig. 4 flexural strength test

Table 6 Flexural strength results

Sl. No.	Mix	7-day flexural strength (N/mm^2)	28-day flexural strength (N/mm^2)
1	SCC -0	3.25	6.04
2	SCC SC -1	3.59	6.23
3	SCC SC -1 C40	3.92	6.79

5 Conclusions

This research work has presented the results of an experimental study on the mechanical of Self-compacting Self-curing concrete prepared with partial replacement of fine aggregate with copper slag. From the studies, it is concluded that optimum percentage of copper slag as a partial replacement was found to be 40% in M50 grade concrete. Based on the experimental investigation conducted in this work, the following conclusions were drawn.

- SCC can be effectively used where compaction is very difficult due to the presence of heavy reinforcements like beams, columns and for structural members with typical architectural requirements [16].
- Self-curing concrete is the answer to many problems faced due to lack of proper curing. Self-curing concrete is an alternative to conventional concrete in desert regions where scarcity of water is a major problem.
- The slump test was conducted on fresh concrete for various percentage replacement of fine aggregate with copper slag. The workability increased substantially with the increase of copper slag content in the concrete mixture due to the low water absorption and glassy surface of copper slag compared with fine aggregate in concrete.

- Strength properties like compressive strength, split tensile strength and flexural strength properties were evaluated. Compressive strength of self-cured concrete for dosage of 1% of PEG was higher than water cured conventional concrete. At the places of water scarcity, these types of agents will give a better result.
- The compressive strength of concrete at 7th and 28th days increased gradually up to 40% fine aggregate replacement with copper slag and then decreased with increase in percentage of replacement. Therefore the optimum percentage of copper slag is 40%. The maximum compressive strength obtained at 7 and 28 day are 42.25 and 59.92 N/mm^2 for 40% partial replacement of fine aggregate with copper slag.
- From the 7th and 28th day split tensile strength, the maximum tensile strength was found to be at 40% fine aggregate replacement of about 4.53 and 5.56 N/mm^2. The split tensile strength of copper slag added concrete was gradually increased up to 40% replacement and then decreased with further fine aggregate replacement.
- The use of copper slag as a replacement for fine aggregate is environmentally helpful due to the utilization of the waste produced from the copper manufacturing process. Copper slag is an alternative material for replacing fine aggregate for manufacturing of concrete.

References

1. Chandar SP, P Sandeep, Raj J (2016) Experimental investigation on self-compacting and self-curing concrete with various admixtures for M30 grade concrete. Int J Innovative Res Sci, Eng Technol 5(2):2458–2465
2. Jadhav DB, Ghate R (2017) A study on self-curing and self-compacting concrete using polyethylene. Int Res J Eng Technol 4(2):1014–1019
3. Wu W, Zhang W, Ma G (2010) Optimum content of copper slag as a fine aggregate in high strength concrete. Constr Build Mater 31(6):933–938
4. Sharma R, Khan RA (2017) Sustainable use of copper slag in self-compacting concrete containing supplementary cementitious materials. J Cleaner Produc 151:179–192
5. Al-Jabri KS, Al-Saidy AH, Taha R (2011) Effect of copper slag as a fine aggregate on the properties of cement mortars and concrete. Constr Buil Mat. 25(2):933–938
6. Kumar P, Reena K (2015) Experimental study on self compacting self curing concrete by replacing river sand by quarry dust for M50 grade concrete. Int J Eng Manage Res 5(5):180–186
7. Tyagi S (2015) An experimental investigation of self-curing concrete incorporated with polyethylene glycol as self-curing agent. Int Res J Eng Technol 2(6): 129–132
8. Mohanraj A, Rajendran M, Ramesh AS, Mahalakshmi M, Prabhakar MS (2014) An experimental investigation of eco-friendly self-curing concrete incorporated with polyethylene glycol. Int Adv Res J Sci, Eng Technol 1(2):85–89
9. Mousa MI, Mahdy MG, Abdel-Reheem AH, Yehia AZ (2015) Physical properties of self-curing concrete. Housing Build Nat Res Center 11(2):167–175
10. EFNARC (2002) Specification and guidelines for self-compacting concrete. www.efnarce.org
11. Sun Z, Young C (2014) Bleeding of SCC pastes with fly ash and GGBFS replacement. J Sustain Cement-Based Mat 220–229
12. Karamloo M, Mazloom M, Payganeh G (2016) Effects of maximum aggregate size on fracture behaviors of self-compacting lightweight concrete. Constr Build Mater 123:508–515

13. Bharali B (2015) Experimental study on self-compacting concrete using GGBS and fly ash. Int J Core Eng Manage 2(6):1–11
14. Olafusi OS, Adewuyi AP, Otunla AI, Babalola AO (2015) Evaluation of fresh and hardened property of self-compacting concrete. Open J Civ Eng 5:1–7
15. Raman JVM, Krishnan VM (2017) Partial replacement of cement with GGBS in self-compacting concrete for sustainable construction. SSRG Int J Civ Eng 4(3):22–25
16. Adari MP, Rao EVR, Sateesh D (2015) An experimental development of M40 grade self-compacted concrete and comparison in behaviour with M40 conventional concrete. Int J Eng Sci Res Technol 4(9):305–319

Experimental Study on the Properties of Bendable Concrete

S. Aishwarya, M. Shenoy Kavya, Roy Reffin, S. Veena and R. Vasudev

Abstract Bendable concrete also known as Engineered Cementitious Composites abbreviated as ECC is a class of ultra-ductile fibre reinforced cementitious composites, characterized by high ductility and tight crack width control. The bendable concrete has very good flexibility. Ductile property of normal concrete can be improved by using natural and artificial fibers like jute fibre and PVA fibre. In this project the strength characteristics of M30 bendable concrete is evaluated by incorporating different percentage of jute and PVA fibres and by partial replacement of cement by fly ash. The PVA and jute fibers will be added to the concrete by 0, 0.5, 1, 1.5 and 2% by weight of binders. Suppressing brittle fracture prevents the formation of wider cracks and it prevents water and other aggressive agents to penetrate easily into concrete structures thus leading to structural safety and stability.

Keywords ECC · Strain capacity · Polyvinyl alcohol fibers · Jute fibers

1 Introduction

Conventional concretes are almost unbendable and have a strain capacity of only 0.01% making them highly brittle and rigid. This lack of bendability is a major cause of failure under strain and has been a pushing factor in the development of an

S. Aishwarya · M. S. Kavya (✉) · R. Reffin · S. Veena · R. Vasudev
Department of Civil Engineering, Toc H Institute of Science and Technology, Ernakulam, India
e-mail: kavyamshenoy1997@gmail.com

S. Aishwarya
e-mail: aishu.santhosh09@gmail.com

R. Reffin
e-mail: reff9090979@gmail.com

S. Veena
e-mail: 123veenas@gmail.com

R. Vasudev
e-mail: vasudevr@tistcochin.edu.in

K. Dasgupta et al. (eds.), *Proceedings of SECON'19*,
Lecture Notes in Civil Engineering 46,
https://doi.org/10.1007/978-3-030-26365-2_11

elegant material namely, bendable concrete also known as Engineered Cementitious Composites abbreviated as ECC. ECC acts more like a ductile metal than a brittle glass which then leads to a wide variety of application. It has enhanced flexibility [1].

A bendable concrete is reinforced with micro mechanically designed polymer fibers. It has a strain capacity of more than 3% whereas conventional concrete have a strain capacity of only about 0.01% and hence are almost unbendable and brittle. When ordinary concrete is designed to resist cracking, ECC is designed to crack only in a carefully controlled manner. It is used for various construction applications all around the world because of the unique ability of ECC to bend while maintain its physical properties [2].

Cementitious materials, such as fly ash, silica fume, blast furnace slag, silica fume etc may be used in addition to cement to increase the paste content. The fibers used in the project are Poly vinyl Alcohol fibres and Jute fibres. Polyvinyl Alcohol fibers are known for their low modulus of elasticity, ductility, tensile strength and bonding strength [1].

Polymer modified jute fibers have been decided to be used as reinforcing element in cement concrete in which polymer will chemically bridge jute in one side and cement on the other side. Polymer modified jute fiber is expected to act as a flexible reinforcing agent in cement concrete enabling it to transmit both static and dynamic stresses to its surrounding bulk as well as absorb a portion of the stress by virtue of its flexible nature [3, 4].

The advantages of natural fibers over the conventional reinforcing fibers like glass, synthetic, carbon, steel etc., are: abundant availability, low cost, less abrasiveness, ability to absorb mechanical impact, easy to handle and process and environmental friendliness [5].

2 Experimental Programme

A. Materials

For the study, Ordinary Portland Cement 53 Grade was used. Locally available fine and coarse aggregate was used with specific gravity of 2.72 and 2.73 respectively. The maximum size of aggregate was 10 mm. Poly Vinyl Alcohol fibers and Jute fibers were used for the study.

B. Mix Design

The mix design was prepared as per IS 10262:2009. The following mix proportion was obtained from the mix design. The water cement ratio for the mix is 0.38 (Table 1).

Table 1 Mix proportion

Mix designation	Cement (kg/m^3)	Fly Ash (kg/m^3)	Fine aggregate (kg/m^3) (blended)	Coarse aggregate (kg/m^3) (blended)
M 30	362	91	800	1036

C. **Casting and Curing**

The concrete was prepared and the fibers were added according to their percentages then the concrete was placed in the moulds. The following mould were used to cast the specimen.

1. Cube mould of size 150 × 150 × 150 mm
2. Cylinder mould of diameter 150 mm and height 300 mm
3. Prism mould of size 100 × 100 × 500 mm.

After casting the specimen was left for 24 h after which the specimen was demoulded. They were then put in water of curing till the end of testing.

3 Strength Tests on Concrete

3.1 Tests on Fresh Concrete Samples

A. **Slump Test**

Slump test is used conveniently as a control test and it gives an indication about the uniformity of concrete. Concrete is said to be workable if it can be easily mixed, placed, compacted and finished. Workable concrete should not show any segregation or bleeding. Slump test is a field test conducted to determine the workability of fresh concrete. The test is carried out using a mould known as a slump cone or Abrams cone. The cone is placed on a hard non-absorbent surface. This cone is filled with fresh concrete in three stages. Each time, each layer is tamped 25 times with a rod of standard dimensions. At the end of the third stage, concrete is struck off flush to the top of the mould. The mould is carefully lifted vertically upwards with twisting motion, so as not to disturb the concrete cone. Concrete subsides in the mould subsides. This subsidence is termed as slump, and is measured to the nearest 5 mm if the slump is <100 mm and measured to the nearest 10 mm if the slump is >100 mm.

3.2 *Tests on Hardened Concrete Samples*

A. **Compressive Strength**

Compression test is the most common test conducted on hardened concrete, partly because it is an easy test perform, and partly because most of the desirable characteristics properties of concrete are qualitatively related to its compressive strength. To find the compressive strength 150 × 150 × 150 mm cube specimens were prepared. The compressive strength of concrete was found by compression testing machine of 2000 kN capacity. The specifications of compressive and flexural strength is given in IS 516:1959 [4–9].

$$\text{Compressive Strength} = \text{P/A} \tag{1}$$

where, P is the Compressive load. A is the Cross sectional area.

B. **Split Tensile Strength**

The test is carried out by placing cylindrical specimen horizontally between the loading surface of a universal testing machine and the load is applied until failure of the cylinder along the vertical diameter. To find split tensile strength of concrete cylinder specimen of 150 × 300 mm prepared. The split tensile strength of concrete was found by Universal Testing Machine of 400 kN capacity [4, 6, 8].

$$\text{Split Tension} = 2\text{P}/\pi\text{LD} \tag{2}$$

where, P is the compressive load on the cylinder. L is the length of the cylinder. D is the diameter of the cylinder.

C. **Modulus of Rupture**

Concrete is relatively strong in compression and weak in tension. To find Modulus of rupture of concrete, 100 × 100 × 500 mm size prism specimens were prepared. The Modulus of rupture was found by Universal Testing Machine of 400 kN capacity. The Modulus of rupture of the specimen is expressed as the modulus of rupture fb which if “a” equals the distance between the line of fracture and the nearer support, measured on the center line of tensile side of the specimen in cm, is calculated to the nearest 0.05 MPa as follows [4, 10].

$$f_b = \text{Pl/bd}^2 \tag{3}$$

4 Result and Discussion

A. Slump Test on Concrete Samples

The strength of concrete is inversely proportional to the workability. It is because when the water from the concrete dries up, it leaves behind voids during the setting of concrete. The increase in number of voids decreases the compressive strength. One of the methods used for determining workability is slump cone test. This method is suitable for concretes of high workability. It was observed that the concrete had medium workability (Table 2).

B. Compressive Strength of Concrete Cubes

The addition both the fibers shall be done carefully as it causes the reduction in workability and also strength in many cases. In order to understand the influence of fibers in concrete the specimens were cast into basic properties of 0.5% individual that is 100% of each fibers. The results of compressive strength on cube specimen and split tensile strength on cylinder specimen is illustrated in Tables 3 and 4. The specifications for compressive strength is given in IS 516-1959.

The compressive strength and split tensile strength of cubes and cylinders incorporated with 0.5% Jute is more than that of cubes and cylinders incorporated with 0.5% PVA respectively. Thus in hybrid incorporation of fibers, the proportion of jute fibers is marginally increased to retain the strength. Hence the ratio of Jute: PVA is maintained is 60:40. By maintaining this ratio, both jute and PVA fibers were added in combination in varying proportions of 0.5, 1, 1.5 and 2% and cubes were casted.

Table 2 Slump values

Trial No	Slump value (mm)		Range (mm)	Remarks
1	100	97.66	50–100	Medium workable mix
2	98			
3	100			

Table 3 Compressive strength results of concrete cubes with 0.5% jute and 0.5% PVA separately

Specimen designation	Sample	Load (kN)	7th day compressive strength (N/mm^2)	Load (kN)	28th day compressive strength (N/mm^2)
0.5% PVA	1	708.1	31.47	860	38.4
	2	707		865	
	3	709.6		868	
0.5% Jute	1	836.9	36.9	936.6	41.4
	2	833		933	
	3	827		925	

Table 4 Split tensile strength results of cylinders with 0.5% jute and 0.5% PVA separately

Specimen designation	Sample	Load (kN)	28th day split tensile strength (N/mm^2)
0.5% PVA	1	61.85	3.5
	2	60.08	
	3	63.62	
0.5% Jute	1	67.15	3.9
	2	68.92	
	3	70.68	

The 7th day and 28th day compressive strength was tested in compressive testing machine. In this experimental study, the concrete is incorporated with Jute and PVA fibers in combination in various proportions of 0.5, 1, 1.5 and 2% of the binder material and form MECC-1, MECC-2, MECC3 and MECC-4 respectively. The Jute and PVA fibers are combined together at a ratio of 60:40.

The results obtained are as shown (Table 5).

The cubes casted with 0.5, 1, 1.5 and 2% of fibers were tested for its compressive strength on the 7th day of casting. The results thus obtained are shown in Fig. 1. When the Jute and PVA fibers were added at different percentages together at a fixed ratio of 60:40, it was found that the compressive strength for the bendable concrete was maximum at 0.5% for both 7th day and 28th day compressive strength test. The strength obtained was 25.71 N/mm^2 for 7th day and 31.5 N/mm^2 for 28th day (Fig. 2).

Table 5 Compressive strength results of bendable concrete cubes

Specimen designation	Sample	7th day		28th day	
		Load (kN)	Compressive strength (N/mm^2)	Load (kN)	Compressive strength (N/mm^2)
MECC 1	1	580.1	25.71	709.3	31.5
	2	575.5		708.1	
	3	580		709	
MECC 2	1	514	23.09	546.7	24.2
	2	525		544	
	3	520		542.8	
MECC 3	1	410.7	18.3	468	20.8
	2	412		470	
	3	413		469.1	
MECC 4	1	371.4	16.49	435	19.3
	2	372		438	
	3	370		440	

Fig. 1 Slump observed

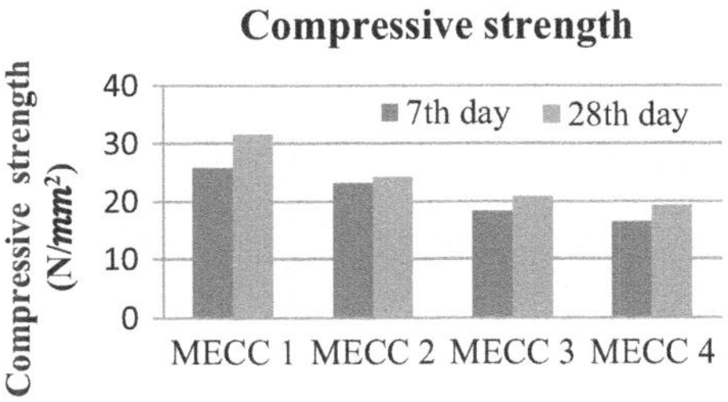

Fig. 2 Compressive strength of bendable concrete samples

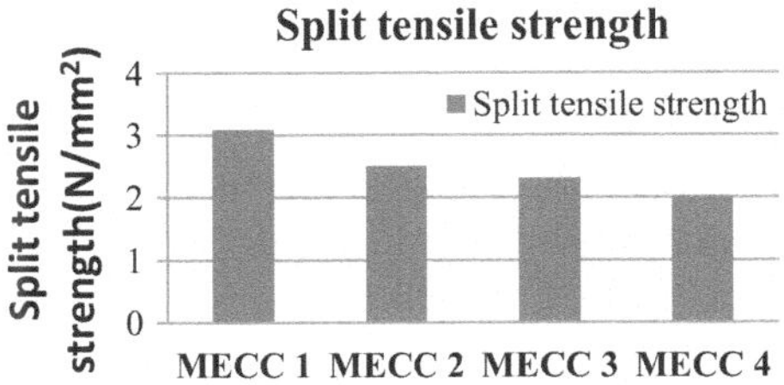

Fig. 3 Split tensile strength of bendable concrete samples

C. **Split Tensile Strength of Concrete Cylinders**

When the Jute and PVA fibers were added at different percentages together at a fixed ratio of 3:2, it was found that the split tensile strength for the bendable concrete was maximum at 0.5%. The strength obtained was 3.08 N/mm^2 for 28th day (Fig. 3; Table 6).

Table 6 Split tensile strength results

Specimen designation	Samples	Load (kN)	28th day split tensile strength (N/mm^2)
MECC-1	1	220.8	3.08
	2	215	
	3	218	
MECC-2	1	177.9	2.5
	2	180	
	3	178	
MECC-3	1	164.3	2.3
	2	162	
	3	163	
MECC-4	1	144.3	2
	2	140	
	3	142	

Table 7 Flexural tensile results

Specimen designation	Samples	Load (kN)	28th day flexural strength (N/mm^2)
Conventional concrete	1	9	4.833
	2	10	
	3	10	
MECC1	1	17	8.66
	2	17	
	3	18	

D. **Flexural Strength of Concrete Beams**

The cubes and cylinders incorporated with 0.5% fibers (Jute and PVA) yield maximum compressive strength and split tensile strength respectively. Beams with 0.5% fibers (MECC-1) and Conventional beams were also casted. Flexural strength of Engineered Cementitious Composite was conducted at 28 days. The flexural strength of beams incorporated with optimum value of fibres is more than that of conventional beams (Table 7) [5, 6, 8, 9].

E. **Ductility Index**

Ductility of a material is defined as its ability to absorb energy without critical failure. It is the amount of inelastic deformation that a material or structure experience before undergoing complete failure. The deformation thus occurred is measured in terms of displacement, strain or curvature. Ductile behavior allows a structure to undergo large plastic deformations with little decrease in strength and hence prevents brittle failure.

$$\mu = \frac{\Delta_f}{\Delta_{cr}}$$

where 'μ' is the deflection ductility index of the member, Δ_f is the deflection at failure of the member and Δ_{cr} is the deflection of the member at first crack load (Figs. 4 and 5).

Ductility index is defined as the ratio between the mid span deflection at the beam ultimate load and the mid span deflection at the yield load. Ductility Index increased when 0.5% of jute and PVA fibers were added together (Table 8).

The ductility index of 0.5% reinforced bendable concrete is 23.35% more than that of conventional concrete.

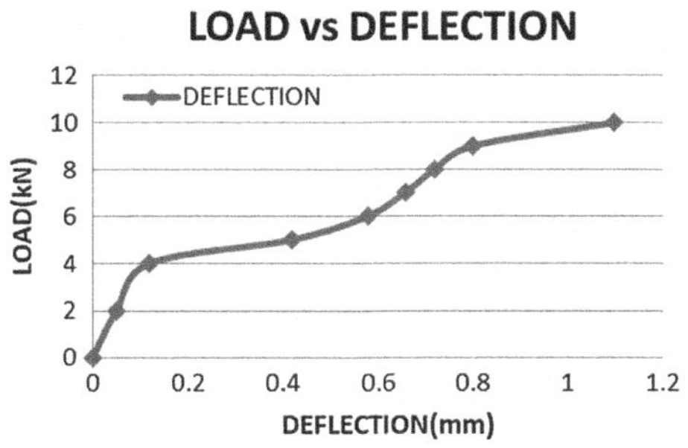

Fig. 4 Load deflection curve for conventional beams

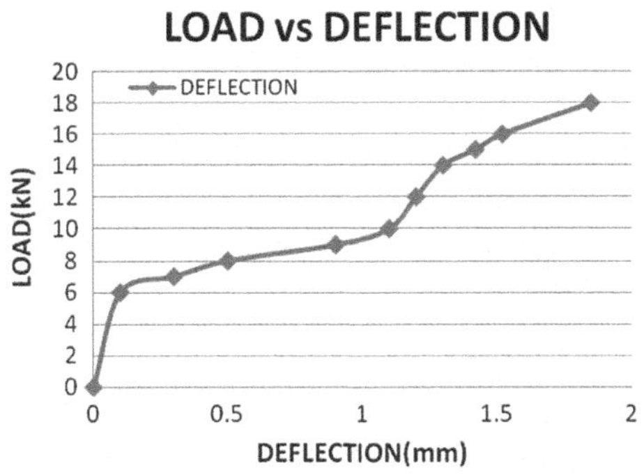

Fig. 5 Load deflection curve for beams with optimum value of fibres

Table 8 Ductility index of specimen

Specimen designation	Fiber content	Deflection at first crack load (mm) Δ_{cr}	Deflection at failure load (mm) Δ_f	Ductility index μ
Conventional concrete	0%	0.66	1.1	1.67
MECC 1	0.5%	0.9	1.85	2.06

5 Conclusion

- The 28th day compressive and split tensile strength of the samples incorporated with 0.5% jute alone was found to be more than that of the samples incorporated with 0.5% of PVA fibers.
- When jute fibres and PVA fibres were added as hybrid fibres in 0.5, 1, 1.5 and 2% at a fixed ratio of 3:2, the optimum value was obtained at the ratio 0.5% and the compressive strength are 25.71 N/mm^2 for 7th day and 31.5 N/mm^2 for 28th day respectively. It was observed that on increasing the percentage of fibres the tensile strength decreased.
- The 28th day tensile strength of M30 concrete when the jute fibers and PVA fibres are added at different percentage, the optimum percentage was found as 0.5% for the 28th day split tensile test. The strength obtained was 31.5 N/mm^2.
- The flexural strength of conventional concrete was 4.83 N/mm^2 and that of 0.5% fiber reinforced bendable concrete was 8.66 N/mm^2. The flexural strength of beams incorporated with 0.5% of fibres was 79.3% more than that of conventional beams.
- The ductility index of bendable concrete with 0.5% fibers is 23.35% more than that of conventional concrete. Hence bendable concrete with 0.5% fibers is more ductile than conventional concrete.
- Hence it can be concluded that the ECC with a combination of Jute and PVA fibers at 0.5% proportion has better flexural behaviour and ductility than conventional beams.

From the results, it is seen that the maximum compressive strength is obtained at 0.5% of fibers. Upon increasing the proportion of fibers incorporated in bendable concrete, the compressive strength reduces. The increased content of fibers cause balling of fibers and as a result, the desired performance is not achieved and there are greater chances of developing cracks in them. The cracking behaviour was observed visually and it was found that no micro cracks occurred in beams casted with 0.5% of fibers.

6 Future Scope

The study can be extended to the properties such as creep, shrinkage and self-healing properties of bendable concrete structure. Carbon dioxide emission of ECC (bendable concrete) specimens can be studied. The effect of ECC on thermal expansion in concrete, variation of the thermal expansion with normal concrete can be studied. The effect of bendable concrete specimens on application of dynamic loading can be studied. The endeavor of this project is to explore the possibilities of flexible concrete as a solution to cracks in a cost-effective manner. This can be achieved by using smaller size coarse aggregates, partially replacing cement by fly ash, incorporating PVA fibers (Artificial fibers) and Jute fibers (Natural fibers). The partial replacement of cement by fly ash also favors sustainability. Sleek structures with less cracks can be obtained by using flexible concrete.

References

1. Brinila Bright BN, Beer MS (2018) Experimental investigation on bendable concrete using natural and artificial fibers (Jute and Nylon). Int Res J Eng Technol 5(4)
2. Bindu Madhavi K, Venugopal M (2016) Experimental study on bendable concrete. Int J Eng Res Technol 5(10)
3. Raval G, Kansagra M (2017) Effects of Jute fibers on fiber reinforced concrete. Int J Innovative Emeg Res Eng 4(8)
4. Devi M, Kannan L, Ganesh kumar M, Venkatachalam TS (2017) Flexural behavior of polyvinyl alcohol fiber reinforced concret. Int J of Civil Eng 4(6):26–30
5. Ghodke GS, Daphal NS (2017) Experimental study of bendable concrete by using admixture and fiber. Int J Tech Res Eng 4(9)
6. Sharmila S (2016) Behaviour of hybrid fiber engineered cementitious composites. Int J Sci Environ Technol 5(5)
7. Zakaria M, Ahmed M, Hoque MM, Islam S (2017) Scope of using jute fiber for the reinforcement of concrete material. Text Clothing Sustain 2(1)
8. Kumar M, Khadwal A (2014) Strength evaluation of steel-nylon hybrid fibre reinforced concrete. Int J eng Res Appl 4(7):32–36
9. Mithanthaya IR, Shriram M, Sandeep VS (2017) Flexural behaviour of fiber reinforced concrete using PVA-ECC. IJCRT 5(4)
10. Satheesh VS, Yuvaraja N (2017) Experimental study on flexural behaviour of bendable concrete. Int J Sci Eng Appl Sci 3(3)

Effect of Alkaline Solution Content on Strength and Chloride Induced Corrosion of Steel in Geopolymer Concrete Made from Fly Ash

Sathishraj Mani and Bulu Pradhan

Abstract In this paper, the effect of alkaline solution content on compressive strength and corrosion behavior of steel reinforcement in geopolymer concrete (GPC) has been evaluated. For the study, fly ash was used as the source material for preparation of geopolymer concrete. The mixture of NaOH solution (10 M and 14 M) and Na_2SiO_3 solution was used as the alkaline solution. Alkaline solution contents of 190 and 210 kg/m^3 were used in the preparation of geopolymer concrete mixes. Sodium chloride was added at the time of preparation of GPC mixes and its concentrations were 0% and 3% by mass of geopolymer solids content. From GPC mixes, cube specimens and prismatic reinforced concrete specimens were prepared. Half-cell potential and corrosion current density by linear polarization resistance (LPR) measurement were carried out on prismatic reinforced specimens. From obtained results, it is observed that there was no systematic variation in 28 day compressive strength of GPC with alkaline solution content and the compressive strength decreased in the presence of chloride ions. Further, GPC made with higher alkaline solution content showed higher probability of occurrence of reinforcing steel corrosion and higher corrosion current density as compared to that made with lower alkaline solution content.

Keywords Geopolymer concrete · Fly ash · Alkaline solution · Admixed chloride · Steel reinforcement · Corrosion

1 Introduction

Portland cement is one of the main constituents of concrete, which is manufactured in large quantities all over the world. However, the manufacture of Portland cement is associated with environmental related problems due to emission of CO_2 to the atmosphere. The annual global production of Portland cement contributes nearly

S. Mani (✉) · B. Pradhan
Department of Civil Engineering, Indian Institute of Technology Guwahati,
Guwahati 781039, India
e-mail: sathishraj@iitg.ac.in

K. Dasgupta et al. (eds.), *Proceedings of SECON'19*,
Lecture Notes in Civil Engineering 46,
https://doi.org/10.1007/978-3-030-26365-2_12

1.35 billion tons of greenhouse gas emissions [1]. For the purpose of reducing, CO_2 emission due to the manufacture of Portland cement, several alternative materials are now being used to replace partly or completely Portland cement in the production of concrete. In this context, geopolymer concrete is nowadays receiving more attention for the purpose of lowering CO_2 emission.

Geopolymer is an inorganic alumino-silicate polymer that is synthesized from geological origin materials or industrial by-products such as fly ash, which is rich in aluminum and silicon [2]. Geopolymers are similar to zeolite materials with amorphous microstructures [3]. The geopolymer binder comprises mainly of alumino-silicate source materials and alkaline solutions. During geopolymerization process, the rapid reaction of alumino-silicate source materials in the presence of alkaline solution results in the formation of a three-dimensional (3D) polymeric chain and a structure consisting of Si–O–Al–O bonds [3, 4]. The various parameters that affect the geopolymerization process are properties of alumino-silicate source materials such as the Si/Al ratio, calcium content, Si and Al speciation, presence of impurities, particle size distribution, curing temperature and period, and type, proportion and quantity of alkaline solution [5]. The constituents used in the preparation of geopolymer concrete are fly ash and/or slag, alkaline solutions and aggregates. In case of low calcium fly ash geopolymer concrete, the sodium alumino-silicate hydrate (N–A–S–H) gel is the final product that governs its properties [4].

From view point of durability, encouraging results about the resistance of geopolymer to sulfate attack and alkali-silica reactions have been obtained in addition to its high stability in fire or freeze-thaw cycles and also excellent adhesion to steel reinforcement [6]. In the literature, very few studies have been reported on durability aspects such as corrosion of reinforcing steel in geopolymer concrete in the presence of internal chloride. In the present study, the effects of alkaline solution content on workability, compressive strength and corrosion of steel reinforcement in geopolymer concrete in the presence of internal chloride have been evaluated. For this purpose, fly ash based geopolymer concrete was prepared with different quantities of alkaline solution and admixed with sodium chloride (for internal chloride). Tests for workability, compressive strength and corrosion parameters were conducted on geopolymer concrete and the obtained results from different tests were analyzed and discussed.

2 Experimental Program

2.1 Materials

In this paper, fly ash passing through 150 μm sieve was used as the source material for the production of geopolymer concrete. For the preparation of geopolymer concrete, crushed stone particles of size of 20 mm MSA (Maximum size of aggregates) and 10 mm MSA were used as coarse aggregates. Locally available natural river sand

Table 1 Mix quantities of fly ash based geopolymer concrete

Molarity of NaOH solution	Alkaline solution (kg/m^3)	Fly ash (kg/m^3)	NaOH solution (kg/m^3)	Na_2SiO_3 solution (kg/m^3)	Coarse aggregate (kg/m^3)	Fine aggregate (kg/m^3)
10 M and 14 M	190	425	69.09	120.91	1155.0	622.0
	210		76.36	133.64		

with maximum grain size of 4.75 mm was used as fine aggregate. Tempcore TMT (Thermomechanically treated) steel bars of 12 mm diameter and 330 mm length were used as steel reinforcement.

The combined mixture of sodium hydroxide (NaOH) and sodium silicate (Na_2SiO_3) solutions was used as the alkaline solution. The NaOH solutions of 10 M and 14 M were prepared by mixing required quantities of sodium hydroxide pellets with laboratory tap water. Commercially available sodium silicate solution (weight % of Na_2O: 7.5–8.5%, and that of SiO_2: 25.0–28.0%) was used in the preparation of alkaline solution. The NaOH solution was mixed with Na_2SiO_3 solution 24 h prior to casting of geopolymer concrete (GPC) mixtures.

2.2 *Mix Quantities of Fly Ash Based Geopolymer Concrete (GPC) Mixtures*

A fly ash content of 425 kg/m^3 was used in the preparation of GPC mixtures. The total aggregate content was kept at 74% by weight of the total mix. In the total aggregate content, the proportion by mass of fine and coarse aggregates was 35% and 65% respectively. Similarly, in the total coarse aggregate content, the proportion by mass of 20 mm MSA and 10 mm MSA coarse aggregates was 62% and 38% respectively. The quantities of alkaline solution used in the preparation of GPC mixtures were 190 and 210 kg/m^3. The ratio of sodium silicate solution to sodium hydroxide solution by mass was fixed at 1.75 for the GPC mixtures. The mix quantities of fly ash based geopolymer concrete are shown in Table 1.

2.3 *Mixing, Casting and Curing of GPC*

Cubes specimens of size 150 mm and prismatic specimens of size 72 mm × 72 mm × 300 mm with a centrally embedded steel bar were prepared from GPC mixtures for determining compressive strength and corrosion parameters respectively. The mixing of geopolymer concrete was carried out in a laboratory drum mixer. Initially, the dry materials (coarse aggregates, fine aggregates and fly ash) were mixed for 3 min in the mixer. After that the alkaline solution was added to the dry mixture and the wet mixing was continued for another 4 min. To investigate the steel reinforcement corrosion in GPC in the presence of internal chloride, different dosages of sodium chloride (NaCl) such as 0% and 3% by mass of geopolymer solids were dissolved in the alkaline solution before mixing of geopolymer concrete. The geopolymer solids is the sum of the mass of fly ash, NaOH solids and Na_2SiO_3 solids. Immediately after mixing, slump test was conducted on fresh GPC mixture to evaluate its workability. Subsequently, the fresh GPC mixtures were cast in steel cube moulds and wooden prismatic moulds. From each GPC mixture, three replicate cube specimens and three replicate prismatic specimens were prepared. Before embedding into the prismatic specimen, the steel bar was cleaned with wire brush. For preventing crevice corrosion, the steel bar was applied with insulating tape followed by epoxy coating over its surface at the locations of its discontinuity with the surrounding concrete. The concrete cover to the embedded steel bar at bottom and on all sides in the prismatic specimen was 30 mm. The schematic diagram of the prismatic GPC specimen with the embedded steel bar is shown in Fig. 1.

After casting, all the cube and prismatic GPC specimens were kept in ambient laboratory condition for 48 h. After 48 h, all the GPC specimens were kept in an oven at a temperature of 80 °C for another 48 h. After 48 h of oven curing, the GPC specimens were removed from oven and were kept in ambient laboratory condition till testing.

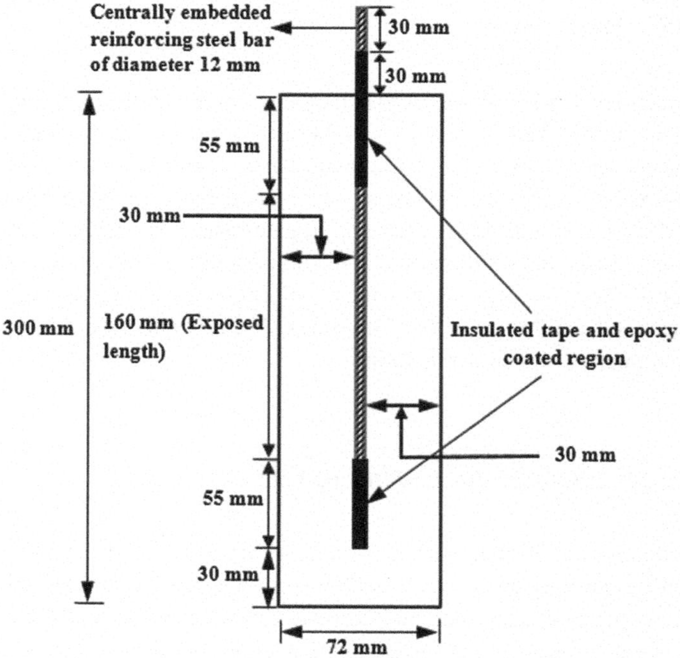

Fig. 1 Schematic diagram of prismatic GPC specimen of size 72 mm × 72 mm × 300 mm

2.4 *Tests Conducted on GPC Specimens*

2.4.1 Compressive Strength Test

The geopolymer concrete cube specimens were tested in a compression testing machine at the age of 28 days from the day of casting for determining the compressive strength. Three replicate cube specimens from each GPC mix were tested and the mean compressive strength of three specimens was reported as the compressive strength of that mix.

2.4.2 Electrochemical Measurements

In this study, to assess the durability performance of geopolymer concrete, the electrochemical measurements such as corrosion potential and linear polarization resistance (LPR) measurements were carried out on prismatic reinforced geopolymer concrete specimens at the ages of 180 and 360 days from the day of casting using the corrosion monitoring instrument: Make ACM, Gill AC serial no. 1542.

The experimental set up shown in Fig. 2 for electrochemical measurement consists of the potentiostat (corrosion monitoring instrument), a plastic container filled with the test solution, partially immersed prismatic specimen with embedded steel bar as working electrode, saturated calomel electrode as reference electrode, and a pair of stainless steel plates as auxiliary electrode. The concentration of sodium chloride

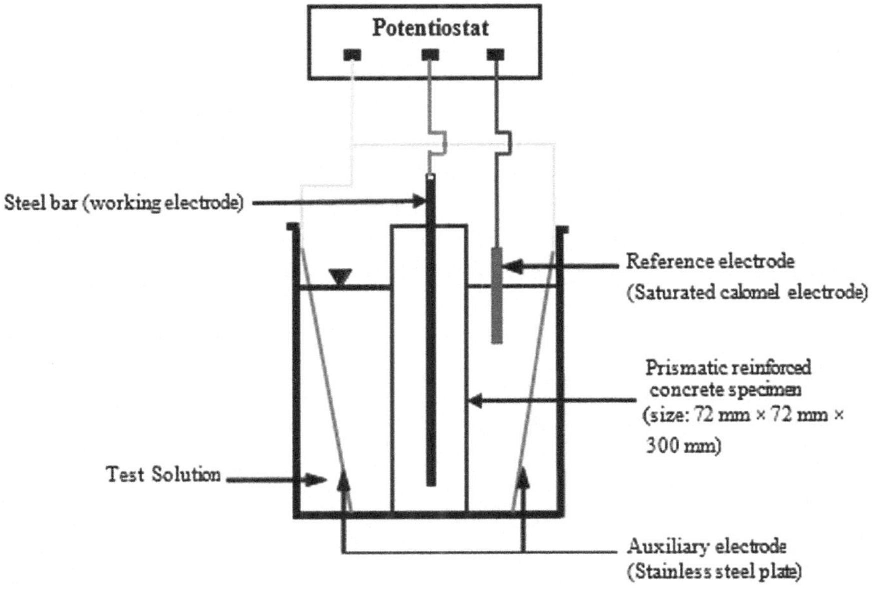

Fig. 2 Schematic diagram of test setup for electrochemical measurements

(NaCl) in the test solution was same as that admixed during the preparation of GPC mixtures. The corrosion potential (half-cell potential) of the embedded reinforcing steel bar was measured with reference to saturated calomel electrode (SCE). During LPR measurement, the reinforcing steel bar embedded in the prismatic specimen was polarized to ±20 mV from equilibrium potential at a scan rate of 6 mV/min.

The corrosion current density of steel reinforcement was determined using the Stern-Geary equation [7], which is as follows;

$$I_{corr} = \frac{B}{R_p} \quad (1)$$

where, I_{corr}= corrosion current density

B = Stern-Geary constant

R_p = polarization resistance of steel.

The expression for Stern-Geary constant B is as follows;

$$B = \frac{\beta_a \times \beta_c}{2.3(\beta_a + \beta_c)} \quad (2)$$

where, β_a *and* β_c are anodic and cathodic Tafel constants respectively. In this study, the value of B was taken as 26 mV, considering the reinforcing steel bar in active condition [7].

2.4.3 FESEM Analysis of Fly Ash Based GPC

To study the morphology of GPC, the field emission scanning electron microscope (FESEM) analysis was carried out. After completion compressive strength test at the age of 28 days, the broken GPC cube specimens were further crushed and the crushed material was then passed through 75 μm sieve. The sieved material was then stored in air tight plastic bags. The FESEM analysis was conducted on the powder sample in Zeiss Gemini field emission scanning electron microscope. The sample was placed on an aluminum stub with carbon tape. The powder sample was then coated with a thin layer of gold by sputtering method. The powder sample was observed through in-lens mode.

3 Results and Discussion

3.1 Workability

To evaluate the workability, the slump test was conducted on fresh GPC mixtures (stated earlier). The measured slump values of GPC mixtures are shown in Figs. 3

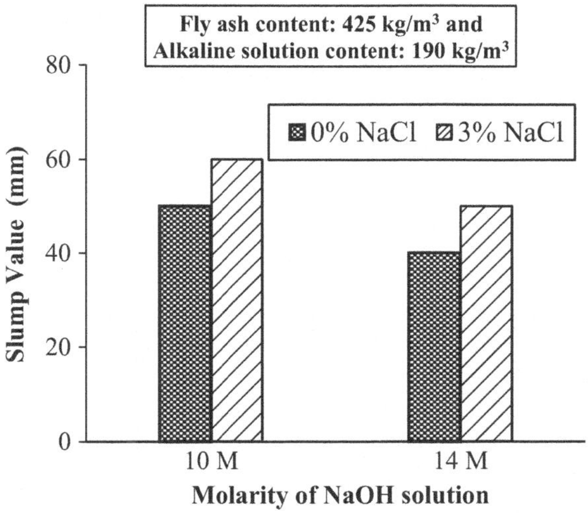

Fig. 3 Slump values of geopolymer concrete made with alkaline solution content of 190 kg/m^3 and admixed with different concentrations of sodium chloride

and 4 for alkaline solution contents of 190 kg/m^3 and 210 kg/m^3 respectively. From these figures, it is observed that slump values of the GPC mixtures decreased with increase in the molarity of NaOH solution, which may be due to the presence of higher solids content of NaOH solution at higher molarity. Further, the slump values of GPC increased in the presence of admixed sodium chloride as observed from Figs. 3 and 4. This may be attributed to the increase in particle mobility in the presence of sodium chloride in GPC mixes.

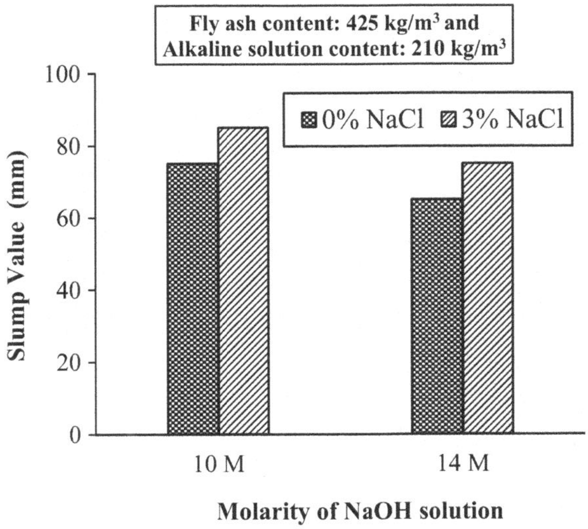

Fig. 4 Slump values of geopolymer concrete made with alkaline solution content of 210 kg/m^3 and admixed with different concentrations of sodium chloride

From Figs. 3 and 4, it is noted that the slump values of GPC mixture increased with increase in alkaline solution content. This is ascribed to reduction in particle-to-particle interaction at higher alkaline solution content in the fresh GPC mixture [8].

3.2 Compressive Strength

The compressive strength values of geopolymer concrete obtained at the age of 28 days are shown in Figs. 5 and 6 for the alkaline solution contents of 190 kg/m^3 and 210 kg/m^3 respectively.

From Figs. 5 and 6, it is observed that the compressive strength increased with increase in molarity of NaOH solution. This is due to dissolution of comparatively higher amount of silica and alumina from fly ash at higher molarity of NaOH solution that resulted in improved geopolymerization process [9]. Further, the compressive strength of GPC decreased in the presence of admixed sodium chloride, which may be attributed to the crystallization of sodium chloride in the alumino-silicate gel. While analyzing the effect of alkaline solution content, it is observed that there is no systematic variation in compressive strength of GPC with respect to alkaline solution content.

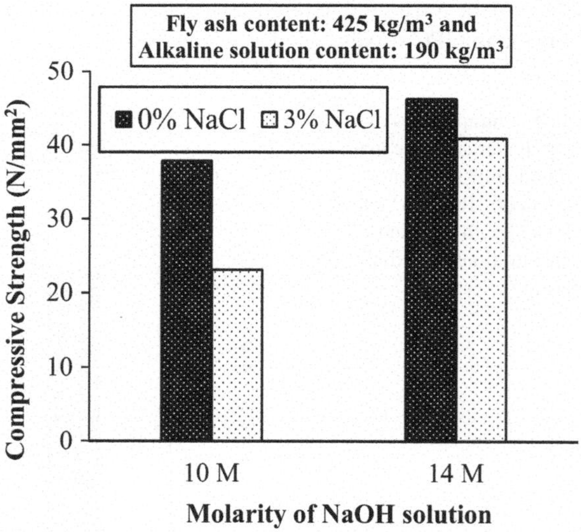

Fig. 5 Compressive strength of geopolymer concrete made with alkaline solution content of 190 kg/m^3 and admixed with different concentrations of sodium chloride

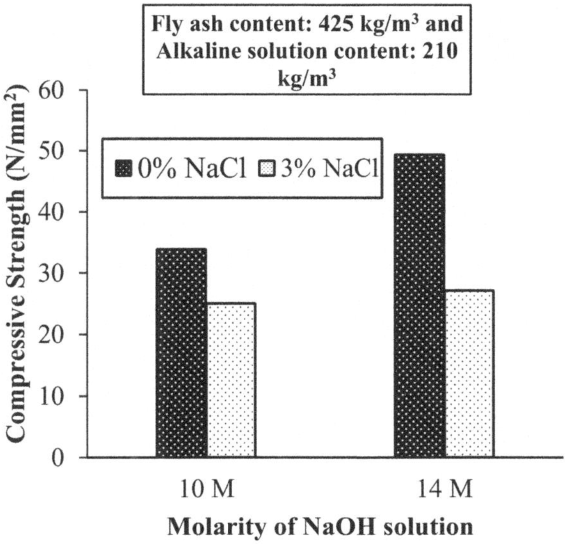

Fig. 6 Compressive strength of geopolymer concrete made with alkaline solution content of 210 kg/m^3 and admixed with different concentrations of sodium chloride

3.3 Corrosion Parameters

As mentioned earlier, the corrosion parameters such as corrosion potential and corrosion current density of reinforcing steel bar in prismatic GPC specimens were measured at different testing ages. The obtained results are reported as the average value of three replicate prismatic specimens.

3.3.1 Corrosion Potential

The measured corrosion potential values of reinforcing steel bar in prismatic GPC specimens at testing ages of 180 and 360 days are shown in Figs. 7, 8, 9 and 10.

From Figs. 7, 8, 9 and 10, it is observed that the corrosion potential values of reinforcing steel bar in GPC specimens made without admixed NaCl were less negative than −270 mV (SCE)/−350 mV (Cu/$CuSO_4$ electrode) whereas in GPC specimens made with admixed NaCl, the corrosion potential values were more negative than −270 mV (SCE)/−350 mV (Cu/$CuSO_4$ electrode) at both testing ages, indicating greater than 90% probability of occurrence of reinforcing steel corrosion in the presence of chloride ions. Further, it is observed that there is no systematic variation in corrosion potential of reinforcing steel bar with molarity of NaOH solution at different testing ages. In GPC specimens made with and without admixed NaCl, the corrosion potential of reinforcing steel bar became more negative with increase in alkaline solution content for both molarity of NaOH solution and at both testing ages as observed from Figs. 7, 8, 9 and 10.

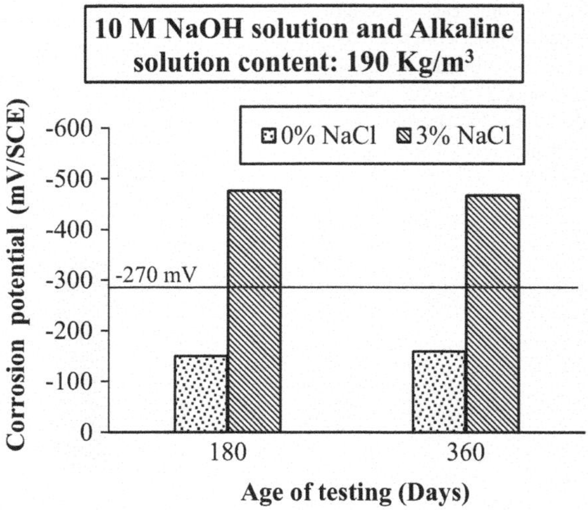

Fig. 7 Corrosion potential values of reinforcing steel bar in GPC specimens made from 10 M NaOH solution with alkaline solution content of 190 kg/m^3

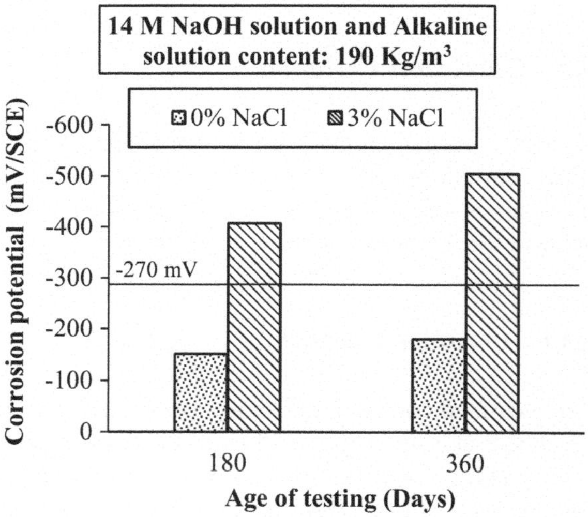

Fig. 8 Corrosion potential values of reinforcing steel bar in GPC specimens made from 14 M NaOH solution with alkaline solution content of 190 kg/m^3

3.3.2 Corrosion Current Density

The measured corrosion current density of reinforcing steel bar in prismatic GPC specimens at testing ages of 180 and 360 days are shown in Figs. 11, 12, 13 and 14. From these figures, it is inferred that the corrosion current density values of reinforcing steel bar in NaCl admixed GPC specimens were significantly higher than those in GPC specimens admixed without NaCl. This is due to increase in conductivity of GPC in the presence of chloride ions.

From Figs. 11, 12, 13 and 14, it is noted that in the GPC specimens made without admixed NaCl, the corrosion current density decreased with increase in testing age

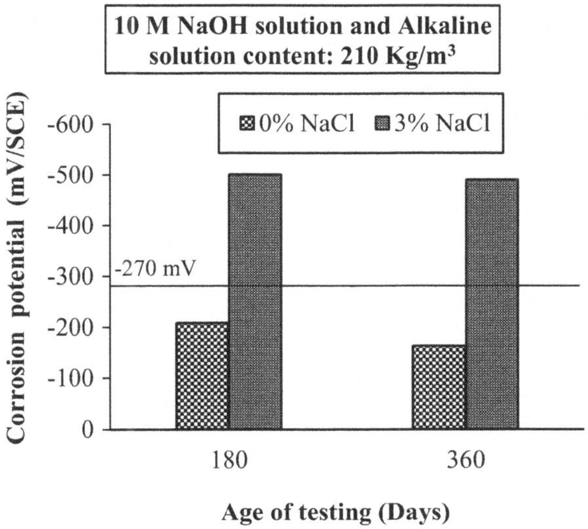

Fig. 9 Corrosion potential values of reinforcing steel bar in GPC specimens made from 10 M NaOH solution with alkaline solution content of 210 kg/m^3

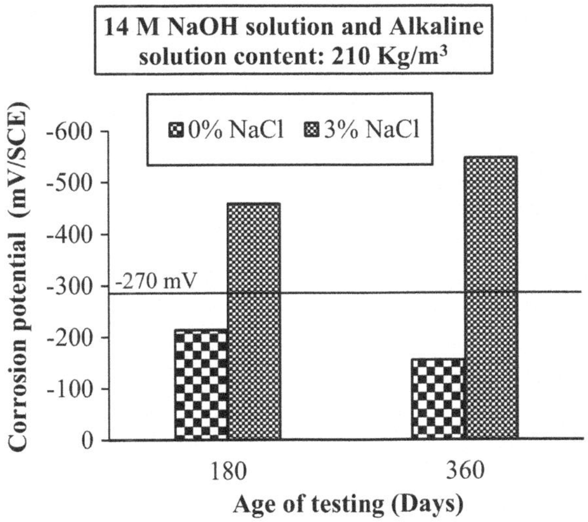

Fig. 10 Corrosion potential values of reinforcing steel bar in GPC specimens made from 14 M NaOH solution with alkaline solution content of 210 kg/m^3

whereas in the NaCl admixed GPC specimens the corrosion current density mostly increased with increase in testing age from 180 days to 360 days. These variations in corrosion current density with testing age may be attributed to the effect of chloride ions altering the passivity of reinforcing steel bar in the GPC specimens. Further, the corrosion current density of reinforcing steel bar in GPC specimens made without admixed NaCl decreased with increase in molarity of NaOH solution. However, there was no systematic variation in corrosion current density with increase in molarity of NaOH solution in NaCl admixed GPC specimens.

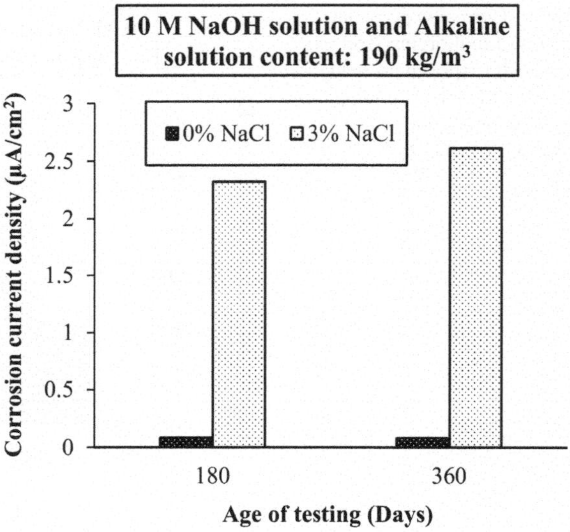

Fig. 11 Corrosion current density of reinforcing steel bar in GPC specimens made from 10 M NaOH solution with alkaline solution content of 190 kg/m^3

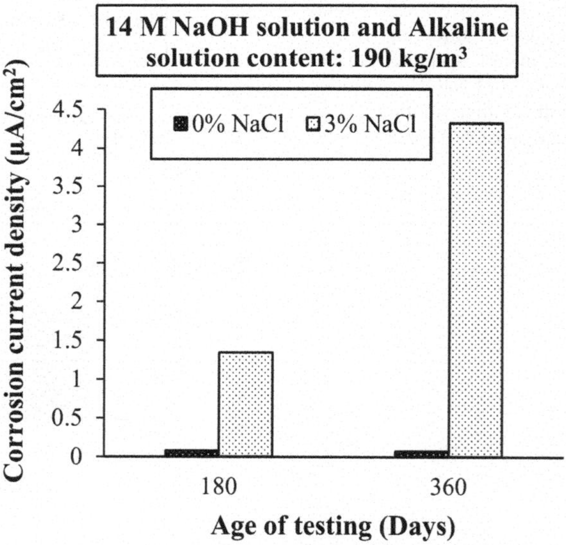

Fig. 12 Corrosion current density of reinforcing steel bar in GPC specimens made from 14 M NaOH solution with alkaline solution content of 190 kg/m^3

While analyzing the effect of alkaline solution content, it is inferred that the corrosion current density increased with increase in alkaline solution content in all the GPC specimens (both with and without admixed NaCl) for both molarity of NaOH solution and at both testing ages as evident from Figs. 11, 12, 13 and 14. This may be attributed to increase in conductivity of GPC due to higher alkaline solution content.

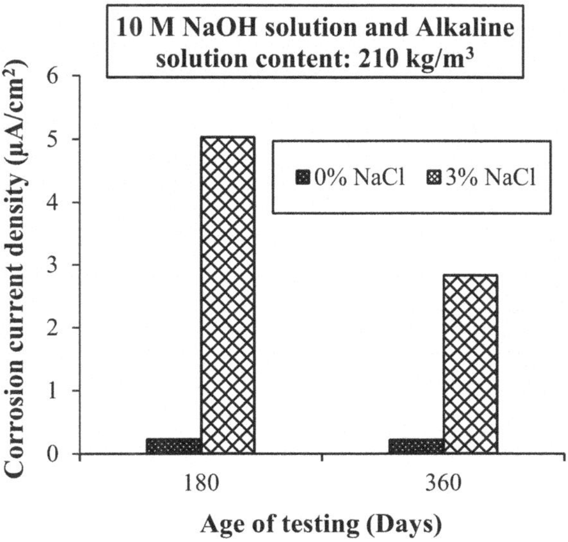

Fig. 13 Corrosion current density of reinforcing steel bar in GPC specimens made from 10 M NaOH solution with alkaline solution content of 210 kg/m^3

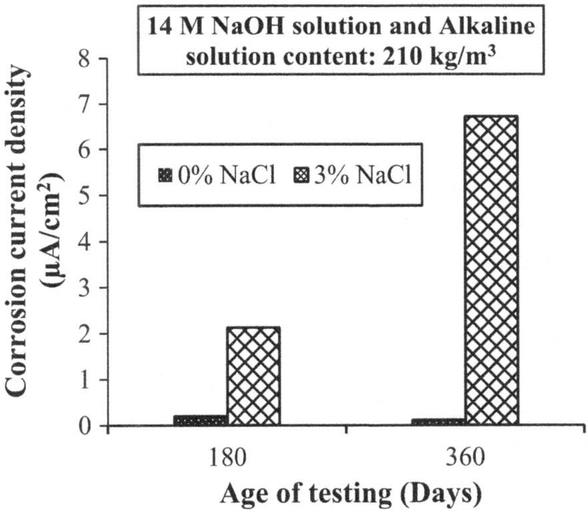

Fig. 14 Corrosion current density of reinforcing steel bar in GPC specimens made from 14 M NaOH solution with alkaline solution content of 210 kg/m^3

3.4 *FESEM Analysis*

Typical FESEM images for GPC mixtures made without admixed NaCl are shown in Fig. 15a, b.

From these micrographs, the sodium alumino-silicate hydrate (N–A–S–H) gel is appeared as cotton-shaped materials [10]. Further, the GPC made with NaOH solution of 14 M showed a denser microstructure as compared to that made with NaOH solution of 10 M, as observed from Fig. 15a, b. This also corroborates with

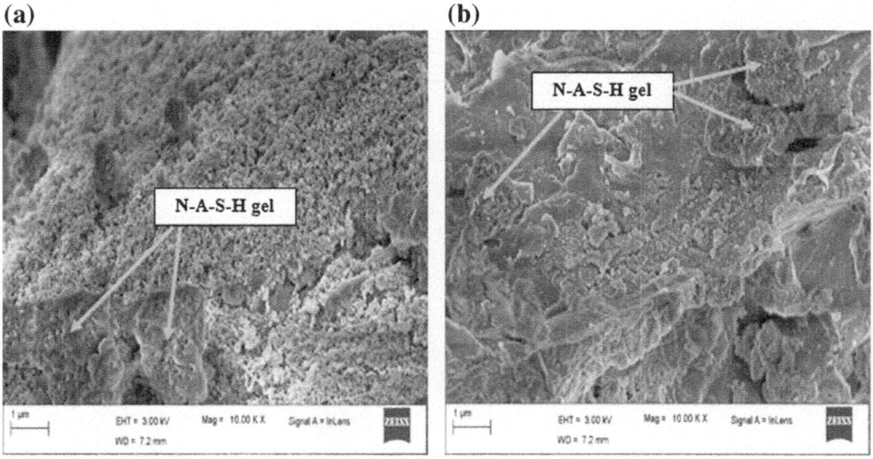

Fig. 15 Micrograph of GPC made without admixed NaCl for **a** 10 M NaOH solution with alkaline solution content of 190 kg/m^3, at the age of 28 days. **b** 14 M NaOH solution with alkaline solution content of 190 kg/m^3, at the age of 28 days

higher compressive strength of GPC made with 14 M NaOH solution as compared to that made with 10 M NaOH solution.

4 Conclusions

From this investigation, the following conclusions were obtained.

- The slump of GPC increased in the presence of chloride ions and with increase in alkaline solution content.
- The 28-days compressive strength of the GPC decreased in the presence of admixed sodium chloride. However, there was no systematic variation in compressive strength of GPC with alkaline solution content.
- There was greater than 90% probability of occurrence of reinforcing steel corrosion in GPC specimens admixed with chloride ions. Further, the corrosion potential became more negative with increase in alkaline solution content.
- The GPC specimens made with admixed NaCl showed significantly higher corrosion current density as compared to those made without admixed NaCl. Further, the corrosion current density of reinforcing steel bar increased with increase in alkaline solution content.

References

1. Demie S, Nuruddin MF, Shafiq N (2013) Effects of micro-structure characteristics of interfacial transition zone on the compressive strength of self-compacting geopolymer concrete. Constr Build Mater 41:91–98
2. Okoye FN, Durgaprasad J, Singh NB (2015) Mechanical properties of alkali activated fly-ash/Kaolin based geopolymer concrete. Constr Build Mater 98:685–691
3. Kupawade-Patil K, Allouche EN (2013) Impact of alkali silica reaction on fly ash-based geopolymer concrete. J Mater Civ Eng 25:131–139
4. Wardhono A, Gunasekara C, Law DW, Setunge S (2017) Comparison of long term performance between alkali activated slag and fly ash geopolymer concretes. Constr Build Mater 143:272–279
5. Malkawi AB, Nuruddin MF, Fauzi A, Almattarneh H, Mohammed BS (2016) Effects of alkaline solution on properties of the HCFA geopolymer mortars. Proc Eng 148:710–717
6. Montecelli C, Natali ME, Balbo A, Chiavari C, Zanotto F, Manzi S, Bignozzi MC (2016) Corrosion behavior of steel in alkali-activated fly ash mortars in the light of their microstructural, mechanical and chemical characterization. Cem Concr Res 80:60–68
7. Gonzalez JA, Algaba S, Andrade C (1980) Corrosion of reinforcing bars in carbonated concrete. Br Corros J 15:135–139
8. Sathonsaowaphak A, Chindaprasirt P, Pimraksa K (2009) Workability and strength of lignite bottom ash geopolymer mortar. J Hazard Mater 168:44–50
9. Hanjitsuwan S, Hunpratub S, Thongbai P, Maensiri S, Sata V, Chindaprasirt P (2014) Effects of NaOH concentrations on physical and electrical properties of high calcium fly ash geopolymer paste. Cement Concr Compos 45:9–14
10. Cui Y, Wang D, Wang Y, Sun R, Rui Y (2019) Effects of the n(H_2O: Na_2Oeq) ratio on the geopolymerization process and microstructures of fly ash-based geopolymers. J Non-Cryst Solids 511:19–28

Relationship Between Flexural and Compressive Strength of Concrete Made of Alkali Activated Binder

Lashhanth Dhevaraju, E. Aakash Reddy, Naga Dheeraj Dogiparthy and Arkamitra Kar

Abstract The cement industry contributes to 5% of global CO_2 emissions and is thus one of the biggest contributors to global warming. Alkali-Activated binders offer a more sustainable solution to the problem, as they are produced from industrial wastes such as fly ash and slag. Their production has a lower environmental impact in terms of greenhouse gas emissions. Hence, AABs are sustainable and can be considered as green construction materials. Most research into the field of alkali-activated binders indicates the superiority of AABs in terms of mechanical properties and durability as compared to portland cement. This study aims to find the correlation between the flexural strength and compressive strength of alkali activated binder concrete. This is achieved by formulating a regression model which correlates the flexural and compressive strengths of AAB concrete by incorporating grade of concrete as a factor. The results of this study will promote the practical use of AABs as an alternative to portland cement.

Keywords Alkali-activated binders · Flexural strength · Compressive strength · Regression modelling

1 Introduction

Portland cement (PC) is one of the most widely used materials in the construction industry. Manufacturing of PC results in a considerable amount of carbon dioxide

---

L. Dhevaraju · E. A. Reddy · N. D. Dogiparthy · A. Kar (✉)
Department of Civil Engineering, BITS-Pilani Hyderabad Campus, Telangana 500078, India
e-mail: arkamitra.kar@hyderabad.bits-pilani.ac.in

L. Dhevaraju
e-mail: lash.dhevaraju@gmail.com

E. A. Reddy
e-mail: eaakash998@gmail.com

N. D. Dogiparthy
e-mail: nagadheerajdogiparthy@gmail.com

K. Dasgupta et al. (eds.), *Proceedings of SECON'19*,
Lecture Notes in Civil Engineering 46,
https://doi.org/10.1007/978-3-030-26365-2_13

emission. The production of 1 tonne of cement produces roughly 0.99 tonnes of carbon dioxide. The construction industry is continuously growing at 3.6% annually with an expected rise of 2.7% on the existing value by the end of 2019 [1]. PC production contributes to roughly 7% of the world's carbon dioxide emission [2]. PC production depletes the Earth's natural limestone reserves and a large amount of energy is consumed during the clinkering process. As a result, possible alternatives to PC are investigated and alkali-activated binders (AAB) can be one of the viable options [3].

AAB are produced by the reaction between a dry precursor blend containing industrial wastes and by-products such as fly ash and slag and an alkaline activating solution containing sodium silicate and sodium hydroxide [4]. In order to establish AAB as a suitable alternative to PC, the mechanical properties of these binders have to determined. One of the most important characteristics for any binder used in construction practices is the flexural strength of the concrete prepared with it. It is well established that concrete can withstand compression loads and it is prone to crack in tension and flexure. Existing literature shows limited reports on the flexural behavior of AAB concrete and even fewer studies on the development of models to predict the flexural strength of AAB concrete as a function of its compressive strength [2, 5–8].

A study on the effects of aggregate content on strength properties of AAB concrete made with fly ash reported 4.95 MPa as the flexural strength [9]. The mixture was cured at 100 °C for 24 h, and tested at 28 days. Another study reported the compressive and flexural tensile strength of AAB concrete as 60 and 6 MPa respectively [10]. A study on the effect of alkali activator solution on AAB mortar cured at 65 and 85 °C showed 22 and 8 MPa as the compressive and flexural tensile strength respectively. The mixture was prepared with 6 M NaOH solutions and cured at temperature of 85 °C, for 24 h heat curing duration. It was observed from their results that the strength property increased with increase in curing periods [11].

A study on the engineering properties of fly ash/slag-based AAB was conducted according to Australian Standards by using three different sources of Class F fly ash [12]. The results show that the experimental splitting tensile strength and flexural strength of AAB are higher than the values predicted by the models presented by the Australian Standard (AS 3600) for PC concrete. Another study provides the following Eq. (1) for expressing the flexural strength ($f_{ct.f}$) of AAB concrete as a function of its compressive strength (f_{cm}) [13].

$$f_{ct.f} = 0.69\sqrt{f_{cm}} \tag{1}$$

An investigation was carried out into the properties of ambient-cured AAB to encourage the use of AAB in cast in situ applications [14]. The results showed that for similar compressive strengths, ambient cured AAB concrete exhibits higher flexural strengths than its PC counterpart. Additionally, the following Eq. (2) is provided to find the relationship between the flexural and compressive strength values of AAB concrete using regression analysis.

$$f_{ct.f} = 0.93\sqrt{f_{cm}} \tag{2}$$

Another study reports the variation in mechanical properties of AAB by varying the composition of slag added to the mix design [15]. The compressive strength was determined using British Standards (BS 1881: Part 115) while the splitting tensile strength and flexural strengths were determined according to ASTM C496 and ASTM C293 respectively. Experimental results showed that the compressive, splitting tensile, and flexural strengths increased as the percentage of fly ash replaced by slag increased.

Another existing study has presented the following Eq. (3) correlating the flexural strength and compressive strength [16].

$$f_{ct.f} = 0.639\sqrt{f_{cm}} \tag{3}$$

IS 456 [17], AS-3600 [18], and ACI 318-14 [19] recommend Eqs. (4–6) respectively for the correlation of flexural strength and compressive strength of portland cement concrete.

$$f_{ct.f} = 0.7\sqrt{f_{cm}} \tag{4}$$

$$f_{ct.f} = 0.6\sqrt{f_{cm}} \tag{5}$$

$$f_{ct.f} = 0.62\sqrt{f_{cm}} \tag{6}$$

Based on the findings from previous studies, it is essential to develop a unified model to predict the flexural strength of AAB concrete as a function of its compressive strength. Consequently, the primary objective of this study is to experimentally generate data on compressive and flexural strength values of AAB concrete and combine them with existing data to formulate a regression model. The materials and detailed methodology used in this study are provided in the next section.

2 Materials and Methodology

2.1 Materials

Locally available Class F Fly ash complying with ASTM C618 acquired from NTPC Ramagundam is used in this study. Slag conforming to ASTM C989 is used for partial replacement of fly ash in some of the mixes. Locally available aggregates and river sand complying with ASTM C33M—13 are used. The alkali activator used is a mixture of sodium hydroxide (99% rayon grade), sodium silicate (55.9% water, 29.4% SiO_2, and 14.7% Na_2O), and additional water to maintain a slump height of

Table 1 Mix design for various compositions

	50:50	70:30	100:0
Fly ash (kg/m^3)	200	280	400
Slag (kg/m^3)	200	120	0
Coarse aggregate (kg/m^3)	1209	1209	1209
Fine aggregate (kg/m^3)	651	651	651
Sodium silicate (kg/m^3)	129.43	129.43	129.43
Sodium hydroxide (kg/m^3)	10.57	10.57	10.57
Water (kg/m^3)	67.65	67.65	67.65
Admixture (ml/m^3)	4	3.14	0

150 mm. The molarity of the NaOH used in this activating solution is 3.9. High range water reducing admixture (HRWRA) composed of a polycarboxylate ether in water complying with ASTM C494 Type F is used.

2.2 *Experimental Methodology*

Beam specimens of dimensions 150 mm × 150 mm × 1220 mm were prepared using AAB concrete for evaluation of flexural strength at an age of 28 days. Fly ash:slag ratios were varied as 100:0, 70:30, 50:50 to prepare these specimens. Triplicate specimens were cast for each mix proportion to ensure the reliability of the results. They were cured for three days using ambient conditions. The beams were tested for flexural strength according to ASTM C78/C78M—18 and the compressive strength data was obtained according to. ASTM C39 C39M—18. Table 1 shows the composition of AAB concrete by weight.

In order to obtain the compressive strengths of AAB concrete, cube specimen of 150 mm × 150 mm × 150 mm were cast, cured for 3 days and tested for 28-day strengths in compliance with ASTM C39, for various mix proportions. The data thus obtained was used to derive normalized values of compressive strengths for corresponding cylindrical specimens with 150 mm diameter and 300 mm height.

For regression-based analysis of flexural and compressive strength, data from 5 external sources was collected [12–16]. This data was then compiled together to obtain a relationship between flexural and compressive strength. The relevant plot is shown in Fig. 1. Coefficient of correlation and the root mean square errors of various models were compared to identify the model that represented the correlation with the best accuracy.

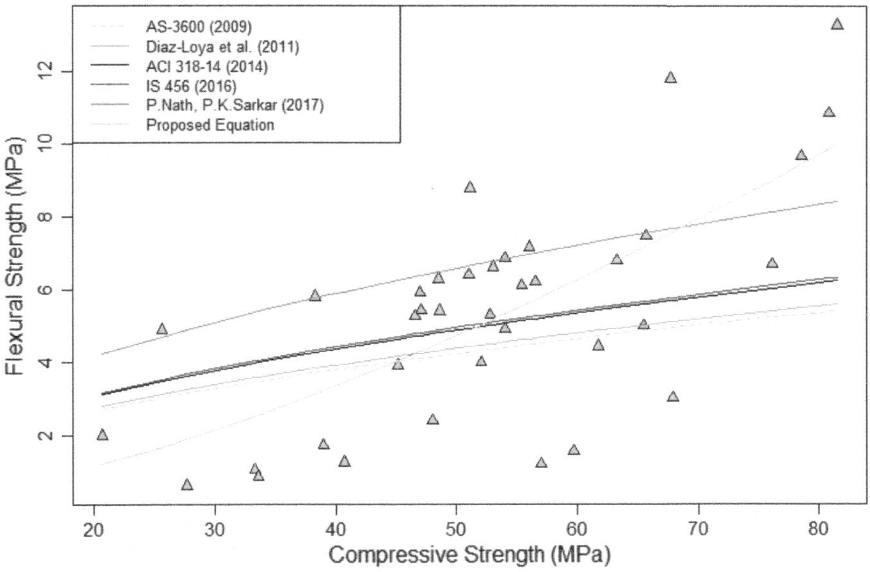

Fig. 1 Comparison of proposed equation with existing models and experimental data

3 Results

Using the statistical software package R, a nonlinear regression analysis was performed on the data points collected from sources previously mentioned. An approximate relationship (7) between compressive strength and flexural strength is proposed.

$$f_{ct.f} = 0.012 * f_{cm}^{1.530} \tag{7}$$

Figure 1 compares the predicted values of flexural strength of AAB concrete obtained from the different equations mentioned in this paper with experimentally determined values from this study. The metrics used for these comparisons are Root Mean Square Error (RMSE) and Mean Absolute Error (MAE).

Table 2 shows that the RMSE and MAE values of the proposed equation are lower than that of the already existing equations while comparing the predicated and experimental strength values. This shows that the proposed equation can be recommended in practice to predict the flexural strength of AAB concrete as a function of its compressive strength.

Table 2 RMSE and MAE

	RMSE	MAE
Proposed equation	2.1906	1.7599
Diaz-Loya et al. [13]	2.6242	2.0953
Nath and Sarkar [14]	2.8358	2.1988
Kar [16]	2.7246	2.1926
IS 456 [17]	2.6103	2.0784
AS-3600 [18]	2.8319	2.2818
ACI 318-14 [19]	2.7738	2.2323

4 Conclusion

- Existing models show considerable deviation from the experimental flexural strength data, especially for ambient-cured concrete.
- The proposed equation considers data from various sources and is observed to be in agreement with the experimental results on the flexural strength of AAB concrete.
- The proposed equation can be further refined by conducting more experiments at various locations with AAB concrete made from fly ash collected from different sources. This can help in the development of a standard equation to predict the flexural strength of AAB concrete.

References

1. Global Construction Market 2018; Expected to drive a galloping growth to US$12.7 trillion by 2022. (n.d.). Retrieved from https://www.reuters.com/brandfeatures/venture-capital/article?id=48295
2. Sarker PK (2008) Analysis of geopolymer concrete columns. Mater Struct 42(6):715–724
3. Mindess S, Young FJ, Darwin D (2003) Concrete, 2nd edn. Technical Documents
4. Ke X, Bernal SA, Provis JL (2017) Uptake of chloride and carbonate by Mg-Al and Ca-Al layered double hydroxides in simulated pore solutions of alkali-activated slag cement. Cem Concr Res 100:1–3
5. Fernandez-Jimenez AM, Palomo A, Lopez-Hombrados C (2006) Engineering properties of alkali-activated fly ash concrete. ACI Mater J 103(2):106
6. Pan Z, Sanjayan JG (2010) Stress–strain behaviour and abrupt loss of stiffness of geopolymer at elevated temperatures. Cem Concr Compos 32(9):657–664
7. Ganesan N, Abraham R, Raj SD, Sasi D (2014) Stress–strain behaviour of confined geopolymer concrete. Constr Build Mater 73:326–331
8. Thomas RJ, Peethamparan S (2015) Alkali-activated concrete: Engineering properties and stress–strain behavior. Constr Build Mater 93:49–56
9. Joseph B, Mathew G (2012) Influence of aggregate content on the behavior of fly ash based geopolymer concrete. Sci Iranica 19(5):1188–1194
10. Luo X, Xu J, Bai E, Li W (2012) Systematic study on the basic characteristics of alkali-activated slag-fly ash cementitious material system. Constr Build Mater 29:482–486

11. Görhan G, Kürklü G (2014) The influence of the NaOH solution on the properties of the fly ash-based geopolymer mortar cured at different temperatures. Compos B Eng 58:371–377
12. Sofi M, Deventer JV, Mendis P, Lukey G (2007) Engineering properties of inorganic polymer concretes (IPCs). Cem Concr Res 37(2):251–257
13. Diaz-Loya EI, Allouche EN, Vaidya S (2011) Mechanical properties of fly-ash-based geopolymer concrete. ACI Mater J 108(3):300
14. Nath P, Sarker PK (2017) Flexural strength and elastic modulus of ambient-cured blended low-calcium fly ash geopolymer concrete. Constr Build Mater 130:22–31
15. Rashad Alaa M (2013) Properties of alkali-activated fly ash concrete blended with slag. Iran J Mater Sci Eng 10(1):57–64
16. Kar A (2013) Characterizations of concretes with alkali-activated binder and correlating their properties from micro-to specimen level. Doctoral dissertation, West Virginia University
17. IS 456:2000 (2016) Indian Standard PLAIN AND REINFORCED CONCRETE - CODE OF PRACTICE, 4th Amendment, Bureau of Indian Standards
18. AS 3600—2009 (Incorporating Amendment Nos 1 and 2), Australian Standard® Concrete structures, Standards Australia
19. Building Code Requirements for Structural Concrete (ACI 318-14), An ACI Standard, Reported by ACI Committee 318, American Concrete Institute

Study on the Effectiveness of Shrinkage Reducing Admixtures on Plastic Shrinkage of Concrete

E. K. Arya, Jerison Scariah James and Elson John

Abstract Concrete is plastic for a very short time from being cast to final setting time. But even in this early age, the concrete shrinks and induces cracking. Aiming to mitigate the impact of shrinkage in concrete, shrinkage reducing admixtures can be used in concrete. But the effect of shrinkage reducing admixture on the plastic period of concrete is not well defined, so there is a need for study in this area. This paper quantifies the effectiveness of shrinkage reducing admixture on cracking of plastic concrete. This is achieved by measuring the slump, compressive strength and plastic shrinkage cracks of concrete mixes. ASTM C1579 based mould is used to promote the plastic shrinkage crack in concrete. To assess the performance of shrinkage reducing admixture and to determine its optimum amount 0, 0.5, 1, 2, 3 and 4% dosage were analysed. The optimum percentage of shrinkage reducing admixture was found to be 2%.

Keywords Shrinkage reducing admixture · Plastic concrete · Cracking · Shrinkage

1 Introduction

Shrinkage of concrete, along with the cracking has been a major concern in the concrete construction industry on maintaining durable structures. Structures that complied with code requirements still fail due to excessive cracking and shrinkage is primarily responsible for such failures. Shrinkage in concrete can be broadly classified into two different stages, namely early age and later age. There are difficulties in measuring the properties of early age concrete as the concrete is plastic and moist

E. K. Arya (✉) · J. S. James · E. John
Department of Civil Engineering, Mar Athanasius College of Engineering, Kothamangalam, Ernakulam, Kerala, India
e-mail: aryapramodek@gmail.com

J. S. James
e-mail: jerissss@gmail.com

E. John
e-mail: elson@mace.ac.in

K. Dasgupta et al. (eds.), *Proceedings of SECON'19*,
Lecture Notes in Civil Engineering 46,
https://doi.org/10.1007/978-3-030-26365-2_14

at this age [1]. The comprehensive physical testing of shrinkage in plastic concrete is hindered due to these difficulties. Most of the research conducted on the shrinkage study of concrete mainly focuses on the later age autogenous and drying shrinkage of concrete [2]. But even in the plastic period, the concrete shrinks and over time, this shrinkage occurring in plastic concrete induces cracking which can severely decrease concrete life expectancy. During the plastic period, three dimensional volume reductions occur in the concrete and due to this, stresses will develop within the concrete which may lead to cracking.

In the past years, many shrinkage reducing or mitigating agents have been used in the industry. Shrinkage reducing admixtures is popular among them. Shrinkage reducing admixtures functions by reducing the surface tension of the pore fluid in the concrete [3]. Shrinkage reducing admixtures are non-ionic organic surfactants that can reduce the surface tension of the capillary pores in concrete [4]. When the surface tension of the pores in concrete is reduced, it changes the shape of drying profile of concrete by forming a concentrated later of admixture on the surface of concrete, thus reducing the plastic shrinkage [5]. Shrinkage reducing admixture can potentially reduce the stresses that arise in concrete. In spite of all the advantages of shrinkage reducing admixture, it has some drawbacks too. It reduces the rate of cement hydration as the water is replaced in the concrete mix and thus reduces the development of mechanical stresses [6]. This paper only studies the cracking that occur in the plastic period, specifically the period up to the final setting of concrete, due to the importance of this period with regard to the durability and service life of structure.

2 Experimental Framework

2.1 *Materials*

Ordinary Portland Cement (OPC) of 53 grade, conforming to IS 12269-2013 was used [7]. Specific gravity of cement was found to be 3.15. Initial and final setting time of the cement used was 44 min and 348 min respectively. 31% is the standard consistency of the cement used and fineness of cement by dry sieving was found to be 7%. Test for cement were conducted as per IS 4031-1996 [8]. Manufactured sand and 20 mm sized aggregates were used as fine and coarse aggregate. The fine aggregate used in concrete is chemically inert, clean and well graded. The concrete mix proportioning was done as per IS 10262-1982 and the control mix of characteristic compressive strength 30 N/mm^2 was prepared [9].

Fine aggregate and coarse aggregate of mass 667 kg/m^3 and 1107 kg/m^3 respectively was used in all the concrete mixes. Shrinkage reducing admixture used is of the brand Master Life SRA200 with relative density 0.98 and pH of 7.5. The constituents, proportions and properties of concrete mixes used for testing the plastic shrinkage of concrete are shown in Table 1.

Table 1 Constituents, proportions and properties of concrete mixes used for the study

Mix abbreviation	Cement	Water	SRA	Slump (mm)	Average 28-day compressive strength (MPa)
M_0	438	195	–	108	40.17
$M_{0.5}$	438	192.81	2.19	104	39.74
M_1	438	190.62	4.38	95	37.97
M_2	438	186.24	8.76	88	36.33
M_3	438	181.86	13.14	76	31.57
M_4	438	177.48	17.52	70	30.21

2.2 *Plastic Shrinkage Test*

Plastic shrinkage is measured using a steel mould of 600 × 200 × 100 mm size with three stress risers. The central riser is with a height of 63.5 mm which is used to provide maximum stress concentration on the central area and promote cracking. The other two risers of height 32 mm each are use to provide restrain in concrete. The mould is made out of cast iron and transparent acrylic sheet is provided on one side, this is to make the crack visible which is formed along the depth of the mould (Fig. 1).

During testing the formation of the full length plastic shrinkage crack was observed with naked eyes. The crack was formed above the central riser throughout the depth and across the width of the mould. The time of occurrence of the initial centre line crack and propagation time of the crack on the concrete surface was recorded.

Fig. 1 ASTM C1579 based mould

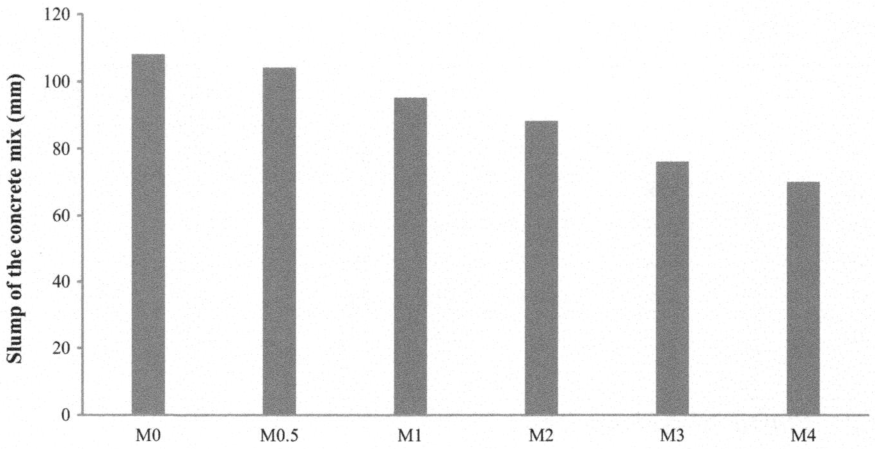

Fig. 2 Workability of different concrete mixes

3 Results and Discussion

3.1 Workability

The property of concrete which determines the amount of useful internal work necessary to produce full compaction is known as workability. The workability of fresh concrete depends mainly on the material, mix proportion and environmental conditions. The workability of concrete is measured using slump test. Figure 2 shows that as the percentage of SRA used in the mix increased, the slump decreased. The decrease in workability can be due to the reduced fluidity arising from the decrease in the mixing water as higher amount of SRA is added.

3.2 Compressive Strength

The compressive strength tests were conducted on concrete cube specimens of size 150 mm. The cubes are to be tested after curing periods of 7 days and 28 days. Concrete with shrinkage reducing admixture has weaker compressive strength than that of the normal concrete mixture (Fig. 3).

When SRA was added to the concrete mix the compressive strength decreased. The compressive strength is observed to decrease as the dosage of SRA is increased. When shrinkage reducing admixture is added to concrete mix, the rate of cement hydration decreases and this affects the mechanical property of the concrete. To assess the performance of shrinkage reducing admixture and to optimise its dosage, the effects on conventional concrete were analysed at varying percentages of shrinkage reducing admixture. For 0.5, 1, 2, 3 and 4% SRA dosage, the compressive strength is equivalent

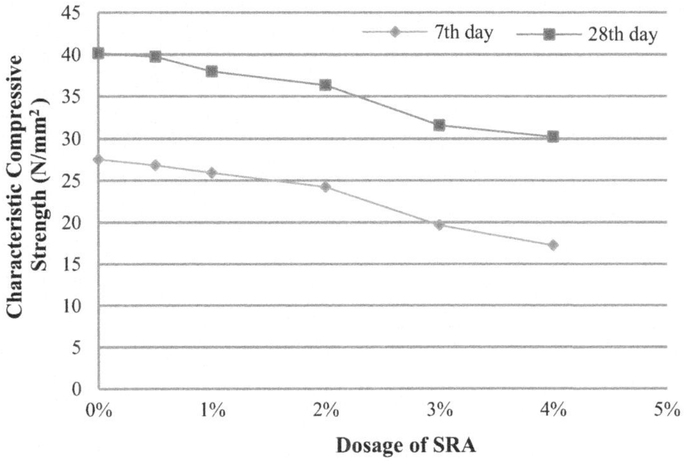

Fig. 3 Compressive strength of concrete with varying dosage of shrinkage reducing admixtures

to 98.93, 94.52, 90.44, 78.59 and 75.2% that of the conventional concrete mix. A drastic drop in strength can be observed when the percentage of SRA varied from 2 to 3%.

3.3 *Plastic Shrinkage Behaviour*

The fresh concrete was placed in the fabricated ASTM C1579 based mould [10]. After placing the concrete in the mould, the time of occurrence of hair line cracks, initial centre line crack and centre line full length crack was recorded. The centre line initial and full length crack is shown in Figs. 4, 5 and 6.

The addition of shrinkage reducing admixture to the normal concrete mix delayed the occurrence of crack in the concrete. As the dosage of shrinkage reducing admixture in concrete increases, the time of occurrence of cracks delayed (Table 2).

The time of occurrence of cracks for the 2, 3 and 4% shrinkage reducing admixture dosage is nearly the same. But there observed a drastic drop in compressive strength when the dosage varied from 2 to 3%. So 2% shrinkage reducing admixture dosage can be considered as the optimum dosage of shrinkage reducing admixture. Figure 7 shows that the effect of different dosage of shrinkage reducing admixture on the time for crack propagation in concrete.

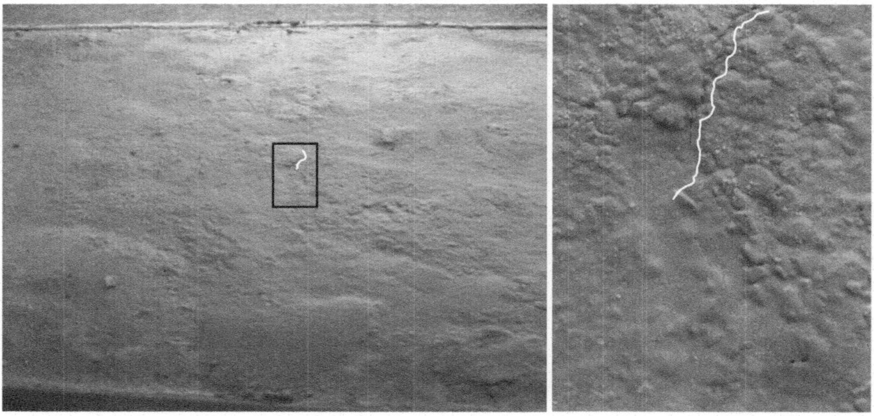

Fig. 4 Initial centre line crack evolution of specimen in ASTM C1579 based mould

Fig. 5 Full length crack evolution of specimen in ASTM C1579 based mould

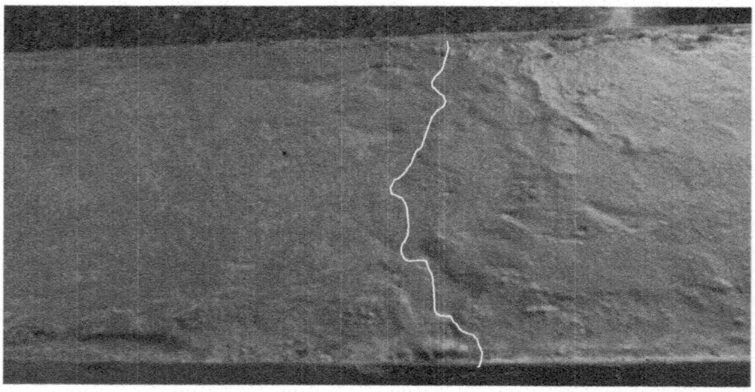

Fig. 6 Plastic shrinkage crack observed from the side of ASTM C1579 based mould

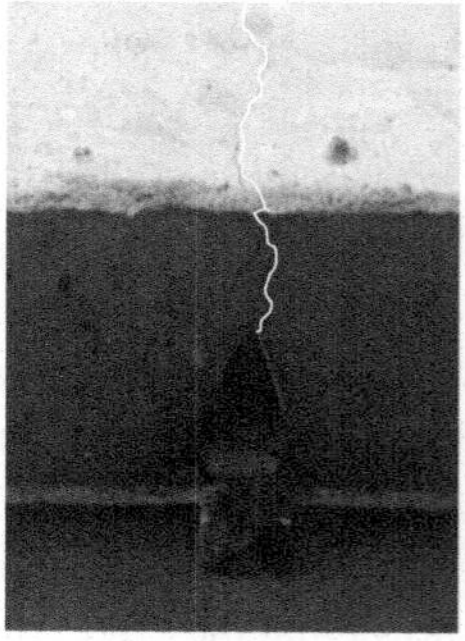

4 Conclusions

This paper discusses the effect of shrinkage reducing admixtures on the early age shrinkage behaviour of conventional concrete. The paper also optimises the percentage of shrinkage reducing admixture to be used in concrete to sufficiently reduce the

Table 2 Time of occurrence of centre line initial crack and time for crack propagation of mixes with varying percentage of shrinkage reducing admixture

Concrete mix	Centre line initial crack (min)	Time for crack propagation
M_0	25	130
$M_{0.5}$	35	135
M_1	50	130
M_2	190	45
M_3	200	30
M_4	220	30

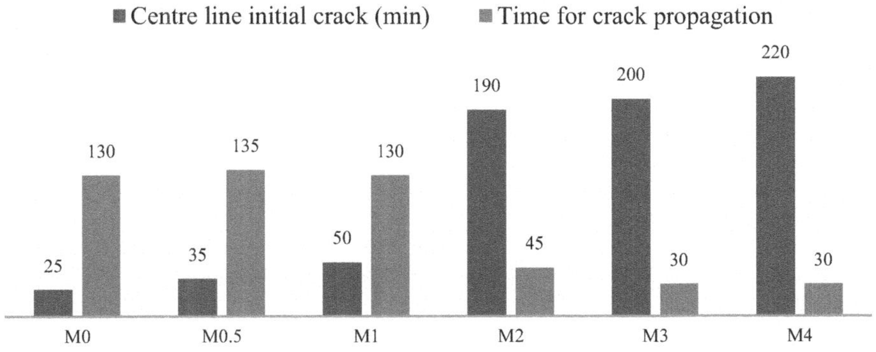

Fig. 7 Comparison of plastic shrinkage behaviour of different concrete mixes

shrinkage performance of concrete without much affecting the compressive strength of the concrete. The following are the significant conclusions drawn from this paper:

- As the dosage of shrinkage reducing admixture increases, the workability decreased. This is due to the reduced fluidity arising from the decrease in the mixing water as higher amount of shrinkage reducing admixture is added.
- As the percentage of shrinkage reducing admixture increased, the compressive strength reduced for the concrete mixes. This can be due to the reduction in rate of cement hydration in concrete as the mixing water is replaced by shrinkage reducing admixture.
- Adding shrinkage reducing admixture to the control mix delayed the occurrence of crack in the concrete.
- Considering the compressive strength pattern and the time of occurrence of the cracks in concrete with varied dosage of shrinkage reducing admixture, the optimum percentage of shrinkage reducing admixture was observed as 2%.

References

1. Holt E (2005) Contribution of mixture design to chemical and autogenous shrinkage of concrete at early ages. Cem Concr Res 35(3):464–472. https://doi.org/10.1016/j.cemconres.2004.05.009
2. Zhang J, Gao Y, Han Y, Wang J (2017) Evaluation of shrinkage induced cracking in early age concrete: from ring test to circular column. Int J Damage Mech 26(5):771–797. https://doi.org/10.1177/1056789515618531
3. Bentz DP (2005) Curing with admixtures beyond drying shrinkage reduction. Concr Int 27(10):55–60
4. Rajabipour F, Sant G, Weiss J (2008) Interactions between shrinkage reducing admixtures (SRA) and cement paste's pore solution. Cem Concr Res 38:606–615. https://doi.org/10.1016/j.cemconres.2007.12.005
5. Wittmann F (1968) Surface tension skrinkage and strength of hardened cement paste. Matériauxet Constr 1(6):547–552. https://doi.org/10.1007/BF02473643
6. Valipour M, Khayat KH (2018) Coupled effect of shrinkage-mitigating admixtures and saturated lightweight sand on shrinkage of UHPC for overlay applications. Constr Build Materials 184:320–329. https://doi.org/10.1016/j.conbuildmat.2018.06.191
7. IS 8112 (March 2013) Ordinary Portland Cement, 43 Grade—Specification, Bureau of Indian Standards, New Delhi
8. IS 4031 (March 1996) Method Of Physical Tests For Hydraulic Cement—Specification, Bureau of Indian Standards, New Delhi
9. IS 10262-1982 Vxeivesvabaya HC, Cj. D., A E D, S. T., Shr, S., DUGGAL Al, N. C., Shri, mute V, Gupta, K., R Kuckarnx, SS (1999). Indian standard recommended guidelines for concrete mix design cement and Concrete Sectional Committee, BDC 2 Engineering Research Laboratories, Hyderabad (1982)
10. ASTMC1579 (2013) Standard test method for evaluating plastic shrinkage cracking of restrained fiber reinforced concrete (Using a Steel Form Insert). ASTM international. Am Soc Test Mater. 1–7. https://doi.org/10.1520/C1579-13.2

Effect of Silica Fume in the Mechanical Properties of Ambient Cured GGBS Based Geopolymer Concrete

Basma Basheer and Gouri Antherjanam

Abstract Geopolymer concrete (GPC) is an emerging sustainable construction material alternative to the conventional Ordinary Portland Cement Concrete (OPCC), which utilizes the cementitious properties of industrial or agricultural by-products such as fly ash, Ground Granulated Blast furnace Slag (GGBS), silica fume, red mud, rice husk ash (RHA) etc. activated by an alkaline solution and thereby reduce the demand for cement and the related environmental issues. In the present work, Ground Granulated Blast furnace Slag (GGBS) is blended with silica fume in various proportions at ambient temperature condition to produce the geopolymer concrete and the mechanical properties are investigated and compared with OPCC of M30 grade. Sodium hydroxide and sodium silicate solution are used as alkali activators throughout this work. The compressive strength test results show that the GGBS based geopolymer concrete (GGPC) gives high compressive strength concrete at ambient temperature conditions and the inclusion of silica fume at various percentages (10, 20, 30, 40, 50 and 60%) significantly reduced the compressive strength. However, a considerable increase in workability was observed with increasing percentage of silica fume in GGPC. The test results of GGPC blended with silica fume at 40% are comparable to OPCC of M30 grade.

Keywords Geopolymer concrete · Ground granulated blast furnace slag · Silica fume

1 Introduction

Infrastructure plays a significant role in the development of any country. The increase in infrastructural developments causes an increase in demand for concrete. A thousand tonnes of concrete is been produced every year in the construction sector. The

B. Basheer · G. Antherjanam (✉)
Sree Buddha College of Engineering, Pattoor, Alappuzha, India
e-mail: gantherjanam@gmail.com

B. Basheer
e-mail: getbasma@gmail.com

K. Dasgupta et al. (eds.), *Proceedings of SECON'19*,
Lecture Notes in Civil Engineering 46,
https://doi.org/10.1007/978-3-030-26365-2_15

versatility of this material has contributed largely to its widespread use and will be continuing its application far into the future.

Ordinary Portland cement (OPC) is indispensable for any construction activity as it is used as the binding material in concrete. However, the production of cement is associated with two major problems. First is that in the manufacturing of OPC, raw materials such as limestone, clay and other natural resources are required. To yield 1 ton of cement about 1.6 tons of raw materials are required and the time taken in the formation of these natural raw materials is far lengthier than the rate at which humans use it. On the other hand, the demand for concrete is increasing day by day. It is reported that India's cement demand is expected to reach about 550–600 million tons per annum by 2025 with a shortfall of 230 million tones i.e. 58% shortage. Second is that in the production of OPC huge amount of carbon dioxide (CO_2) is released. It is reported that the production of 1 ton OPC produces about 1 ton of CO_2 to the atmosphere. The cement industry contributes about 5–6% of total global carbon dioxide emissions. This tremendous amount of CO_2 released into the environment causes an alarming rise in the CO_2 level, which further leads to global warming.

On the other side, each year million tons of industrial and agricultural wastes are produced in our country. The management of these wastes and other related issues is a big bottleneck for development. Moreover, the increasing concern about the ecological significances of waste disposal has led researchers to investigate the consumption of the wastes as potential construction materials. The abundance and availability of waste materials such as fly ash (FA), ground granulated blast furnace slag (GGBS), silica fume (SF), red mud and rice husk ash (RHA) worldwide create an opportunity to make use of these by-products of different industries as partial or complete replacement for OPC in concrete. By wisely incorporating these waste materials in concrete, we can not only minimize the problems associated with the production of cement but also solve the major problem of waste disposal. In view of this, there is a need to develop sustainable alternatives to the conventional binder (cement) utilizing the cementitious properties of industrial and agricultural by-products.

Many intense types of research and experimental works were conducted to find a suitable replacement material for cement. Studies on partial replacement and high volume replacement of OPC with materials having binding property were carried out. In 1978, Davidovits proposed that binders can be produced by the polymeric reaction of alkaline liquid with alumino-silicate materials such as fly ash, silica fume, blast furnace slag, etc. and termed as Geopolymers. Thus the two main constituents of geopolymer are the source material and the alkaline liquids. The source material for geopolymer is basically an alumina-silicate that is rich in silicon (Si) and aluminum (Al). These could include natural minerals such as kaolinite, clays, etc. Alternatively, byproduct materials such as fly ash, silica fume, slag, rice-husk ash, red mud, etc. could also be used as source material. The choice of the source material for making geopolymers depends on factors such as availability, feasibility, type of application, and particular demand of the end users. The alkaline liquids are derived from soluble alkali metals that are usually sodium or potassium based. The most commonly used alkaline liquid in geopolymerization is a combination of sodium hydroxide or potassium hydroxide and sodium silicate or potassium silicate. In contrast to OPC, principle binder in geopolymer concrete (GPC) is not calcium silicate hydrates (CSH) gel, the function of the binder is presumed by an alumino-silicate polymeric gel, formed by the tetrahedrally bonded silicon and aluminum with oxygen atoms shared in between. The main factors thus influencing geopolymerization are thus identified as the type and properties of raw materials, alkaline activators, and curing conditions. Geopolymers can be hence considered as a sustainable alternative for cement, since it relies on minimally processed natural materials or industrial byproducts, thus dropping down its carbon footprint. Geopolymers have also gained considerable attention for their rapid strength attainment, corrosion resistance, and chemical stability, low rate of shrinkage and freeze-thaw resistance [1].

Ground granulated blast furnace slag (GGBS) is an industrial by-product resulting from rapid water cooling of molten steel whereas Silica fume or micro silica is a by-product resulting from the silicon and ferrosilicon alloy production collected as ultrafine powder and consists of particles size less than GGBS. Ground granulated blast furnace slag and silica fume are thus industrial byproducts which are rich in silica and alumina and can be therefore effectively used as source materials in geopolymer concrete which otherwise causes major disposal problems. Moreover, the works of literature on the GGBS or silica fume based geopolymer concrete at ambient curing conditions are very limited. This is because the geopolymerization reaction of GGBS based GPC is fast at ambient temperature, so it starts setting after 8–10 min of mixing and also GGBS based GPC mix is highly cohesive and stiff which makes less workable. On the other hand, the geopolymerization reaction of silica fume based GPC is very slow at ambient temperature. This is a major limitation from a practical point of view. This study focuses to provide a sustainable novel solution to the problems related to ambient cured GGBS based geopolymer concrete by incorporating silica fume.

2 Objectives

1. To obtain an optimum mix proportion for GGBS based Geopolymer concrete in terms of workability.
2. To investigate and compare the mechanical characteristics of GGBS based geopolymer concrete under various percentages of silica fume at ambient curing condition with conventional concrete of M30 grade.
3. To estimate the optimum percentage of GGBS and Silica fume in geopolymer concrete with regard to strength and workability.

3 Methodology

- A detailed literature survey was carried out on the problems associated with the production of cement and waste disposal problems of industrial byproducts, the requirement of use of alternative binder material in concrete, geopolymer concrete using various industrial byproducts, generation and availability of GGBS and silica fume, and their potential to be used in concrete.
- Detailed tests were conducted in the laboratory to evaluate the required properties of the individual materials. Properties of the constituent materials were tested as per the methods prescribed by the relevant IS codes.
- The basic mix proportion of GGBS based geopolymer concrete was adopted from "*Modified guidelines for geopolymer concrete mix design using Indian standard guidelines*" by Sreevidya et al. [2] and the optimum mix proportion was chosen using trial and error method in terms of workability.
- Standard specimens were cast for OPCC of M30 grade; GGBS based geopolymer concrete with varying percentages (0, 10, 20, 30, 40, 50 and 60%) of silica fume.
- Specimens were tested for fresh properties at the time of casting and tested for hardened properties on 7th day and 28th day of casting.
- The test results for fresh and hardened properties of GGBS based geopolymer concrete under the influence of silica fume at various proportions were studied and compared to that of OPCC of M30 grade. The optimum percentage of GGBS and silica fume in geopolymer concrete with regard to strength and workability comparable to OPCC of M30 grade was chosen.

4 Materials

A. *Cement*

Ordinary Portland cement of 53 grade conforming to IS 12269-2003 was used in this study to produce conventional concrete of M30 grade.

B. *Fine aggregate (FA)*

Manufactured sand conforming to zone II was used as fine aggregate having fineness modulus 2.93 and specific gravity 2.72. Tests are conformed to IS: 383-2016 [3].

C. *Coarse aggregate (CA)*

Crushed stone aggregate of size between 20 and 4.75 mm and specific gravity 2.74 and fineness modulus 7.16 was used as coarse aggregate. Tests are conformed to IS: 383-2016 [3].

D. *Ground Granulated Blast furnace Slag (GGBS)*

Ground granulated blast furnace slag (GGBS) is an industrial by-product resulting from rapid water cooling of molten steel and is obtained from Ultratech Pvt. Ltd. Tests are conformed to IS: 12089:1987.

E. *Silica fume*

Silica fume (SF), also referred to as micro-silica or condensed silica fume, is a byproduct as a result of the reduction of high-purity quartz with coal in an electric arc furnace in the manufacture of silicon or ferrosilicon alloy and is obtained from BSS Pvt. Ltd. Tests are conformed to IS: 15388: 2003.

F. *Alkaline solution*

The alkaline liquid used in this study was a combination of Sodium Hydroxide (NaOH) and Sodium Silicate (Na_2SiO_3). The sodium hydroxide and sodium silicate of laboratory grade were purchased from Hazeena Chemicals, Trivandrum. Sodium hydroxide is available in the form of flakes and sodium silicate is available in the form of liquid. Sodium hydroxide solution was prepared by dissolving flakes in distilled water. 10 M concentration of sodium hydroxide was used for the entire study.

G. *Super plasticizer (SP)*

The super plasticizer used was Conplast SP430 supplied by M/s FOSROC CHEMICAL (INDIA) (P) LTD. The super plasticizer was taken as 1.1% by amount of binding material for all the mixes.

H. *Water*

Potable water is generally considered as being acceptable for casting the test specimens. Clean drinking water available in the college water supply system was used for mixing and preparing alkaline liquid.

5 Mix Design

Since no standard mix design approaches are available for geopolymer concrete so far, the mix design procedure suggested by Sreevidya et al. [2] in "*Modified guidelines for geopolymer concrete mix design using Indian standard guidelines*" was used for determining the basic mix proportion for M30 geopolymer concrete. Based on the results from the trial mixes a better mix proportion was chosen to produce GGBS based geopolymer concrete based on workability. The optimized mix proportion for GGPC from trial and error method and the mix proportion obtained for OPCC of M30 grade are given in Table 1.

To study the mechanical properties of GGBS based GPC under the influence of silica fume, standard specimens were cast for 7 different proportions each with different percentages of silica fume by weight i.e., GGBS from the mix proportions

Table 1 Mix proportion of GGPC and OPC for M30 grade

Constituents (kg/m^3)	GGPC	OPCC
GGBS	420	–
Fine aggregate	608.1	716
Coarse aggregate	1203.322	1114
Sodium silicate	171.2	–
Sodium hydroxide	68.5	–
Extra water	25.2	197
Cement	–	438

was replaced by silica fume in percentages of 0, 10, 20, 30, 40, 50, and 60%. Standard specimens for Ordinary Portland Cement Concrete of M30 Grade were also cast to compare the test results.

6 Mixing Procedure

The manufacture of geopolymer concrete is carried out using the usual concrete technology methods. Mixing is done in machine for 3–4 min until it becomes homogeneous and uniform in appearance. As the setting time of GGBS based geopolymer concrete (GGPC) concrete was much less than normal concrete, care was taken to place this mix in moulds as quickly as possible. T he mixing effort was reduced and the setting time was visibly improved with increasing percentage of silica fume.

7 Test for Workability

Concrete was tested for workability as soon as possible after mixing. Slump test and compaction factor test was carried out for each mix as per IS: 1199-1959 [4]. After the concrete tested for workability, the concrete is mixed again uniformly and placed into the moulds. Care must be taken to place concrete in the moulds as quickly as possible after testing. The test results for slump and compaction factor test are given in Table 2.

8 Placing of Concrete

The uniformly mixed concrete was placed into the moulds in three layers. After placing the first layer a tamping rod was used to give 25 blows.

Table 2 Slump and compaction factor test results

S. No.	Mix	Ref. Id	Slump (mm)	Compaction factor
1	OPCC M30	OPCC	73	0.93
2	100% GGBS	GGPC	66	0.74
3	90% GGBS + 10% SF	M1	69	0.77
4	80% GGBS + 20% SF	M2	72.5	0.80
5	70% GGBS + 30% SF	M3	75	0.84
6	60% GGBS + 40% SF	M4	80.5	0.88
7	50% GGBS + 50% SF	M5	87	0.93
8	40% GGBS + 60% SF	M6	95	0.96

9 Curing

After demoulding the concrete specimens were kept for curing. As there will be no heat of hydration in GPC the specimens were not water cured. There are two types of curing for GPC, namely ambient curing and heat curing. In ambient curing the specimens are cured at ambient temperature. In heat curing the specimens are cured at a specified temperature in oven by covering the specimens with polythene covers in order to restrict the humidity. In this work ambient curing was done in practical point of view. Specimens were cured for 7 and 28 days at ambient temperature.

10 Tests on Specimens

The specimens were tested for compressive strength, split tensile strength, and flexural strength for 7 and 28 days as per IS: 516-1959 [5]. In the present study, compression tests were carried out on 100 mm cube specimens. Cylinder specimens having diameter 150 mm and height 300 mm were cast to perform split tensile strength test. Flexural strength test was conducted on standard beam specimens of size 500 $\times$ 100 $\times$ 100 mm for this investigation. Two-point loading was applied and breaking load was noted at 28th day. The reported strength values are average of three test results. The test results for the hardened properties of GPC are noted in Table 3.

11 Results and Discussions

From the works of literatures, a less workable mix was expected for GGPC due to the fast geopolymerization reaction of GGBS when activated by an alkaline solution at ambient temperature and the test results from Table 2 prove that a comparatively low workability value was observed for GGPC. It was also observed there that is a

Table 3 Hardened properties of GGPC incorporating silica fumeat various proportions

S. No.	Mix	Compressive strength (MPa)		Split tensile strength (MPa)	Flexural strength (MPa)
		7 day	28 day	28 day	28 day
1	OPCC	23.9	42.5	4.03	4.28
2	GGPC	55.55	58.34	3.98	4.15
3	M1	51.89	54.56	3.67	4.03
4	M2	47.12	52.11	3.43	3.52
5	M3	43.12	49.56	3.19	3.17
6	M4	39.41	44.77	2.93	2.72
7	M5	32.01	37.41	2.12	2.4
8	M6	29.35	33.08	1.64	2.24

significant improvement in workability of GGPC as the percentage of silica fume was increased. The improvement in workability may be accounted for the fineness of silica fume than GGBS and the presence of increased water content.

In GPC, an average of about 80–90% of 28th day strength was observed within 7 days at ambient temperature curing condition. Unlike ordinary concrete, the rate of strength development of geopolymer concrete beyond 28th day is not significant. Because the chemical reaction of the geopolymer gel is due to substantially fast geopolymerization process, the compressive strength does not vary with the age of concrete. This observation is in contrast to the well known behaviour of OPC concrete, which undergoes hydration process and hence gains strength over the time.

From the test results for hardened properties of GGBS based geopolymer concrete as noted in Table 3, it is clear that the compressive strength of GGBS based geopolymer concrete obtained was of high strength concrete. Though a significant improvement in workability was observed for GGPC with increase in silica fume, the mechanical properties such as compressive strength, split tensile strength, flexural strength of GPC decreased with increase in percentage of silica fume Fig. 1 as seen in Fig. 1. However, the reduction in compressive strength is continuous and significant from 0 to 60% of replacement of GGBS by silica fume as can be observed in Fig. 1a is suitable for various structural applications with regard to compressive strength. In comparing the test results of GGPC incorporating silica fume with that of OPCC of M30 grade, it was observed that the target compressive strength for M30 was achieved by the mix M4 with 40% replacement of GGBS by silica fume. But the tensile and flexural properties were observed well below the required target strength as observed in Fig. 1c.

The reduction in strength can be due to the presence of oxides like Al_2O_3, CaO and MgO in GGBS, which may absorb more water for the reactions with oxides. Whereas Silica fume contains 90% of SiO_2 and the other oxides like Al_2O_3, FeO, and CaO in minute quantities it may not absorb more water for the reactions with oxides. The presence of more water in the mix may leave minute pores in the hardened concrete which may cause reduction in the concrete strength.

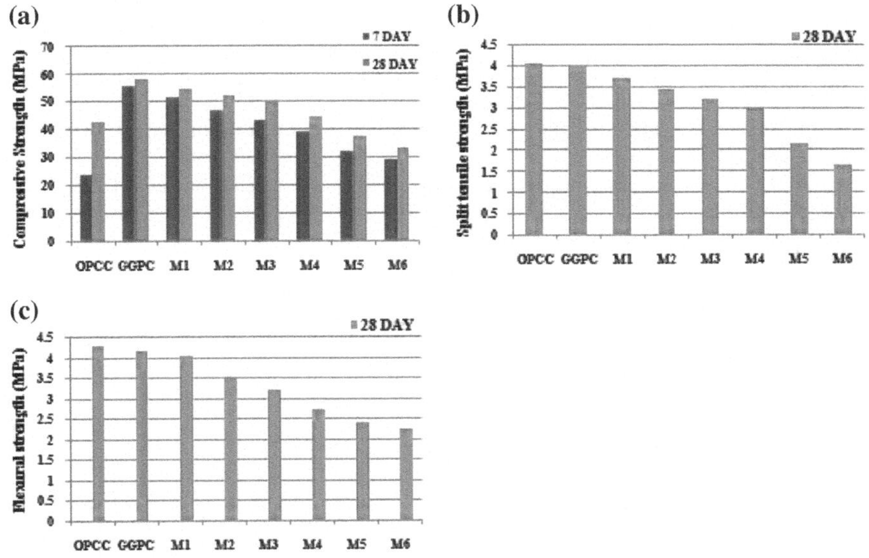

Fig. 1 **a** Variation in Compressive strength for varying CGBS-Silica fume binder ratio. **b** Variation in Split Tensile strength for varying CGBS-Silica fume binder ratio. **c** Variation in Flexural strength for varying CGBS-Silica fume binder ratio

12 Conclusion

- GGBS based geopolymer concrete gives high compressive strength concrete and an increase of 37.27% was observed when that compared to OPCC of M30 grade.
- Workability of the GGBS based geopolymer concrete is significantly improved by the addition of silica fume.
- In GPC, as an average about 80–90% of 28th day strength was observed within 7 days at ambient temperature curing condition.
- The mechanical properties such as compressive strength, split tensile strength, flexural strength of GPC decreased with increase in percentage of silica fume. However, the reduction in compressive strength is continuous and significant from 0 to 60% of replacement of GGBS by silica fume which make it suitable for various structural applications with regard to compressive strength.
- The target compressive strength for OPCC of M30 grade was achieved by mix 4 of GPC with 40% replacement of GGBS by silica fume.

References

1. Roy DM, Jiang W, Silsbee M (2000) Chloride diffusion in ordinary, blended, and alkali-activated cement pastes and its relation to other properties. Cem Concr Res 30(12):1879–1884

2. Anuradha R, Sreevidya V, Venkatasubraman R, Rangan BV (2011) Modified guidelines for geopolymer concrete mix design using Indian standard. Asian J Civ Eng 13:353–364
3. IS:383-1970, Specifications for coarse and fine aggregate from natural sources for concrete. Bureau of Indian Standards, New Delhi
4. IS:1199-1959 (1959) Indian standard methods of sampling and analysis of concrete. Bureau of Indian Standards, New-Delhi
5. IS:516-1959, Indian standard methods of test for strength of concrete, Bureau of Indian Standards, New Delhi

Effect of GGBS on Self-compacting Recycled Aggregate Concrete

Gopika Krishnan and Greeshma Subhash

Abstract Self compacting concrete (SCC) is an extremely fluid mix which does not require tamping or vibration and gains its fluid property from high proportion of fine aggregate, super plasticizers and viscosity enhancing admixtures. One of the major draw back of SCC is its cost due to the utilization of high amounts of cement and chemical admixtures. Also large scale cement production causes the discharge of high amount of CO_2 into the atmosphere causing global warming and ozone layer depletion. So Ground Granulated Blast Furnace Slag (GGBS) can be used as a better substitute to cement which possessing cementitious property. Shortage of natural coarse aggregate in future can be solved by replacing it with recycled coarse aggregate from demolition waste itself. This study aim to determine the utilization of GGBS and recycled aggregate in SCC as an alternate material to cement and natural coarse aggregate by determining its workability and mechanical properties. Cement is replaced with GGBS at 20, 30, 40 and 50% by weight of cement in SCC and 40% of natural coarse aggregate is replaced with recycled aggregate (Vivek and Dhinakaran in Eng Sci Technol Int J 20:1173–1179, 2017 [2]). Optimum percentage of GGBS is evaluated by compressive, split tensile and flexural strength tests. From the experiment, the optimum percentage of GGBS obtained is 30%.

Keywords Superplasticisers · GGBS · Viscosity

1 Introduction

Self-compacting concrete (SCC) is an extremely fluid concrete mix which does not require any tamping or vibration. It gains its fluid property from high proportion of fine aggregate, super plasticizers and viscosity enhancing admixtures. High deformability, low yield stress, good segregation resistance and moderate viscosity are some

G. Krishnan · G. Subhash (✉)
Sree Buddha College of Engineering, Pattoor, Alappuzha, India
e-mail: subhashgreeshma93@gmail.com

G. Krishnan
e-mail: gopika.krishnan888@gmail.com

K. Dasgupta et al. (eds.), *Proceedings of SECON'19*,
Lecture Notes in Civil Engineering 46,
https://doi.org/10.1007/978-3-030-26365-2_16

of the advantages of SCC. Setting, curing and strength of SCC are similar to conventional concrete. Supplementary Cementitious materials such as fly ash and silica fume increase volume of paste, enhancing deformability, cohesiveness and stability. SCC is characterised by its filling ability, passing ability and segregation resistance. One of the major draw back of SCC is its cost due to the utilization of high amounts of cement and chemical admixtures. To solve these issues, the cement powder in SCC is partially replaced with supplementary cementitious materials to produce an ecofriendly concrete. The introduction of Ground Granulated Blast Furnace Slag (GGBS) as a partial replacement of cement reduces the heat of hydration, which is of particular importance for massive concrete elements, decrease the construction cost as it is a by-product of iron and steel manufacturing process reduces the utilization of high amount of cement whose production has globally became a sole reason for the emission of green house gases. Huge demands of aggregates leads to massive exploration of natural resources. By collecting the demolition waste of old building and breaking it up, recycled aggregates are obtained. The use of recycled aggregate in concrete is relatively a new technique. The main reason behind this is to make concrete green, environment friendly and reduces the future shortage of natural aggregate.

This study aim to determine the utilization of recycled coarse aggregate and GGBS in self-compacting concrete as an alternative material to cement and natural coarse aggregate by determining its workability mechanical properties. Mix design was carried out according to the suggestions of EFNARC. Here cement is replaced with GGBS at 20, 30, 40 and 50% by weight of cement in SCC keeping 40% recycled aggregate constant [1] in all mixes by replacing natural coarse aggregate. Optimum percentage of GGBS is evaluated by compressive, split tensile and flexural strength tests.

2 Objectives

1. To investigate the mechanical behaviour of self compacting recycled aggregate concrete (SCRC) using GGBS.
2. To determine optimum percentage of GGBS in SCC recycled aggregate concrete.
3. To compare the strength properties of GGBS incorporated SCRC with that of normal SCC and SCC with recycled aggregate.

3 Literature Review

Olafusi et al. made a comparative study on the rheological properties and compressive strengths of SCC and conventional concrete. The fresh properties of SCC were tested by slump test, L-box and U-box test. Compressive strength of both normal concrete and SCC was also tested. The rheological properties of SCC are incomparable with those of the conventional concrete due to their diverse testing methods and characteristics of individual flow. The compressive strength results of hardened concrete showed that SCC gained strength slowly compared to the conventional cement concrete due to the presence of admixtures and its 28 days.

Manzi et al. [1] investigated the shrinkage and creep of self-compacting concrete prepared with coarse and fine recycled aggregate. Physical properties and porosity measurements are studied and related to the mechanical properties. Results highlights that the self-compacting characters are maintained when the recycled aggregates are used and their qualities promotes high mechanical properties. Mechanical properties are maximum for 40% replacement of coarse aggregate with recycled one.

Vivek et al. [2] studied the effect of different mineral admixtures such as GGBS, silica fume and metakaolin as a partial substitute to cement in self-compacting concrete. Fresh and hardened properties of all the three types of SCC were investigated. For all the mixes the obtained value of fresh property tests were within the specification of EFNARC guidelines. From the results, 50% GGBS, 10% Silica Fume and 20% Metakaolin were found to be the optimum values as partial substitute to cement in SCC.

4 Research Significance

Sustainability in concrete production can be achieved by substitutions of materials used. The demand of aggregates in the construction industry has consequently increased, resulting in future shortage and an increase in price. From the literature review, it is evident that self-compacting characteristics are maintained when recycled aggregates are utilized. Mechanical property is higher for recycled aggregate up to 40 vol%. Utilization of Supplementary Cementitious materials enhances compressive strength. There in no previous study conducted on self compacting recycled aggregate concrete using GGBS.

This study aim to investigate the utilization of recycled coarse aggregate and GGBS in self-compacting concrete as an alternative material to cement and natural coarse aggregate by determining its workability and mechanical properties. The optimum percentage of replacement was identified on the basis of strength and workability of the specimen.

5 Methodology

- Literature Review: A detailed literature survey was carried out on self compacting concrete, Properties of GGBS in concrete.
- Collection and testing of raw materials: The materials used for the study are ingredients of normal concrete with a mineral admixture as fly ash, super plasticizer as

Conplast SP 430 and cement replaced with GGBS. Laboratory tests were carried out on the raw materials used for this study. Physical properties of materials were tested and found that all properties are conforming to IS standards.
- Using Trial and error method and fixing the mix proportion of SCC by EFNARC guidelines. The mix proportion obtained is 1:2.4:1.9.
- The fresh properties tests such as workability were done as per the guidelines of European Federation of National Associations Representing for Concrete (EFNARC), 2002. The workability of SCC can be characterised by the properties like Filling ability, Passing ability and Segregation resistance.
- Casting of cubes, beams and cylinders for control mix of self compacting concrete with 35% fly ash as an addition of cement, different percentage of cement replaced with GGBS (20, 30, 40, 50%).
- Testing of specimens for compressive strength, flexural strength and split tensile strength after 28 days of normal water curing.
- Identify the optimum percentage of GGBS from the test results.
- Presenting the test results in tables and graphs and comparing the test results with normal SCC and SCRC. Conclusions and recommendations are finally made based on the findings and observations.

6 Materials and Properties

A. *Cement*

Ordinary Portland cement of 53 grade conforming to IS 12269-1987 was used in the study. Physical properties of cement were tested using Le-Chatlier apparatus and Vicat apparatus. The standard consistency of cement was obtained as 30.25%. The initial setting time and final setting time was 90 min and 300 min respectively. The specific gravity and fineness of cement was obtained as 3.06% and 4% respectively.

B. *Fine aggregate*

Fine aggregate are soil particles passing through 4.75 mm IS sieve. Generally river sand, crushed stone, crushed gravel, M sand etc. are used as fine aggregate. In this study, M sand conforming to Zone II was used.

C. *Coarse Aggregate*

Coarse aggregates passing through 16 mm IS sieve and retained on 4.75 mm IS sieve was used for the study. Aggregates used should be clean and free from deleterious matter and confirm to the IS: 383-1970. Aggregates consist of naturally occurring stones, gravel. They should be hard, strong, dense and durable. Properties of coarse aggregate are given in Table 1.

Table 1 Properties of fine, coarse and recycled aggregate

Properties	Fine aggregate	Coarse aggregate	Recycled aggregate
Specific gravity	2.72	2.74	2.74
Fineness modulus	2.96	5.22	7.476
Effective size	150 μ	16 mm	16 mm
Void ratio	0.34	0.816	0.79
Bulk density (g/cc)	1.847	1.557	1.66

D. *Recycled Coarse Aggregate*

Recycled aggregates are collected by breaking the casted cubes from the laboratory. Aggregates are then crushed into the size of 16 mm. Due to the presence of adhered mortar, the water absorption of recycled aggregate is higher than that of natural aggregate. 40% recycled aggregate is used in the work by replacing natural coarse aggregate. Laboratory test were conducted to determine the different physical properties. Properties of recycled coarse aggregate are given in Table 1.

E. *Fly Ash*

Fly ash is a byproduct of the thermal power plants. Fly ash normally produced from burning anthracite or bituminous coal. Class F fly ash was used and it exhibit Pozzolonic properties. Specific gravity of fly ash is 2.2 as per Specific gravity Test, IS: 2386 Part III, 1963.

F. *GGBS*

Ground Granulated Blast furnace Slag (GGBS), a co-product produced simultaneously with iron, molten blast furnace slag is cooled instantaneously by quenching in large volumes of cold water, known as granulation, to produce Granulated Blast furnace Slag. Specific gravity of GGBS is 2.94 as per Specific gravity Test, IS: 2386 Part III, 1963.

G. *Super plasticizer (SP)*

Super plasticizer is a chemical used as an admixture to improve the properties of concrete. High Range Water reducing admixture (HRWRA) or Super plasticizer is an important constituent in the production of SCC to achieve self-compatibility and flow ability. The HRWRA helps in achieving excellent flow at low water content. Conplast SP 430 is used for the work as super plasticizer.

H. *Water*

Casting and curing of specimens is done with the portable water which is available. The correct quantity of water in the concrete is critical. It helps to form the strength giving cement gel. The quantity and quality of water required to be observed very carefully.

Table 2 Mix proportion of SCC

Cement (kg/m^3)	Fly ash (%)	Fine aggregate (kg/m^3)	Coarse aggregate (kg/m^3)	Super plasticizer (L/m^3)	Water (L/m^3)
400	35	951	763	1.6	185

7 Experimental Investigation

The experimental investigation is done on the mechanical properties of the GGBS in self-compacting recycled aggregate concrete by varying proportions of GGBS. The study includes industrial by-products like GGBS and demolition waste recycled aggregate.

A. Mix Design

The mix design procedure for SCC is based on trial mixes. Based on the results from the trial mixes, a better mix proportion was chosen. Mix design was carried out according to the suggestions of EFNARC (European Federation of National Associations Representing for Concrete) and the obtained mix proportion is given in Table 2.

B. Casting and Testing of Specimens

To study the influence of GGBS on the mechanical properties of Self-compacting recycled aggregate concrete, different proportions are prepared with different percentages of GGBS (20, 30, 40 and 50%) by weight of cement and 40% natural coarse aggregate is replaced with recycled aggregate [1]. For each mix at least 3 specimens are tested for compressive strength, split tensile strength, flexural strength at 7 days and 28 days. The compressive strength test is done by casting concrete cube of size 150 mm × 150 mm × 150 mm. Concrete cylinder of height 300 mm and diameter 150 mm is used to test the split tensile strength and flexural test is conducted by casting plain concrete beam of size 500 mm × 100 mm × 100 mm. The concrete is poured in the mould with no compaction. After 24 h these moulds are removed and test specimens are put in water for curing. The test is conducted on 7th day and 28th day.

8 Results and Discussions

A. *Workability Test*

Fresh properties of self compacting concrete such as Slump Flow test, L-box test and V-funnel were conducted and results are tabulated in Table 3.

Table 3 Fresh properties of SCC

Tests	SCC	SCRC	SCRC 20%	SCRC 30%	SCRC 40%	SCRC 50%
Slump flow (mm)	690	670	700	720	725	730
T50 cm slump flow (s)	4.2	4.1	3.5	3.2	3.15	3.0
L-Box	0.85	0.86	0.88	0.94	0.951	0.962
V-Funnel (s)	9	9	8.5	8	7.9	7.82

B. *Hardened properties*

The strength of concrete is obtained when it gets hardened. Compressive strength test, split tensile strength test and flexural tensile strength test are mainly used to measure the strength of concrete. All the specimens were tested after 7 days and 28 days of water curing. For each mix, three specimens were cast and tested to obtain the average value. From the result, it is clear that the compressive strength increases with addition of GGBS percentage due to its pozzolanic property and higher content of CaO up to 30% replacement. The decrease in compressive strength above 30% cement replacement is due to decrease in calcium hydroxide content in concrete. On the replacement of 40% natural aggregate with recycled aggregate, slight decrease in compressive strength is observed and this decrease was compensated on the replacement of cement with GGBS. With 30% replacement of cement with GGBS, the compressive strength is increased by 10% compared to SCC with recycled aggregate and 5.05% increase compared to normal SCC. The result showed that the split tensile strength decrease on the replacement of natural coarse aggregate with recycled one but it start increase on the addition of GGBS up to 30% and then decreases. Flexural strength continued to increase with the increase in GGBS percentage at 28 days with 30% replacement. Hardened properties of self compacting recycled aggregate concrete with GGBS is shown in Table 4. Graph showing variation of compressive

Table 4 Hardened properties of SCC

Mixes	Compressive strength (N/mm^2)		Split tensile strength (N/mm^2)	Flexural strength (N/mm^2)
	7 day	28 day	28 day	28 day
SCC	32.22	47	4.59	5.2
SCRC	30.5	45	3.91	4.29
SCRC 20%	33.6	48.3	4.25	4.34
SCRC 30%	34.5	49.5	4.4	4.39
SCRC 40%	31	44.69	4.22	4.1
SCRC 50%	28.7	42	4.13	3.8

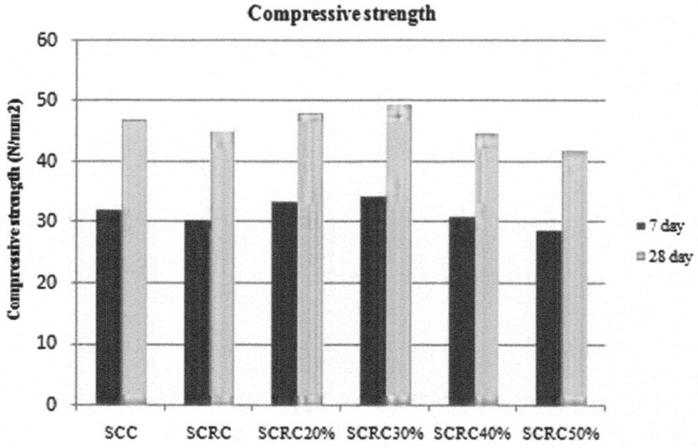

Fig. 1 Compressive strength

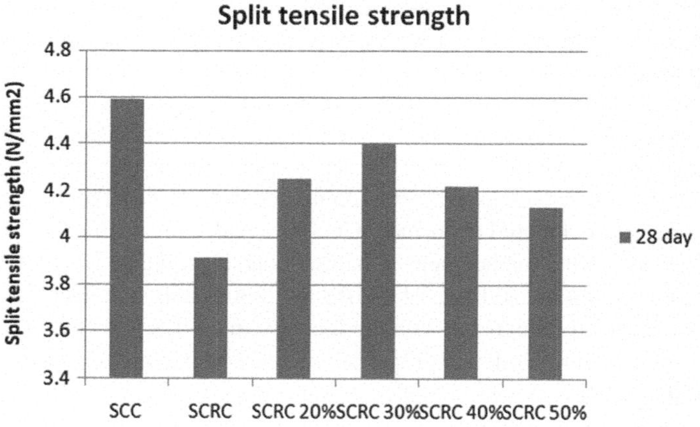

Fig. 2 Split tensile strength

strength of 7 and 28 day is shown in Fig. 1, split tensile strength in Fig. 2 and flexural strength in Fig. 3.

9 Conclusions

- From the experiment, it is found that replacement of cement with GGBS improve the compressive strength of Self compacting recycled aggregate concrete.
- Split tensile strength and flexural strength is decreased on the replacement of 40% coarse aggregate with recycled one.

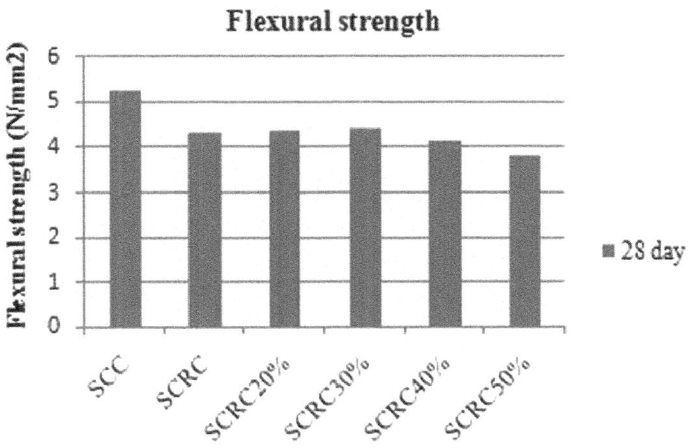

Fig. 3 Flexural strength

- The sample shows increase in compressive strength, split tensile strength and flexural strength on the replacement of cement with GGBS.
- The mechanical property is higher for 30% cement replacement. The decrease in mechanical property above 30% replacement is due to decrease of CaOH content in concrete.
- Percentage increase in compressive, split tensile and flexural strength for 30% replacement in normal SCC is 5.05%, 10% and 10.34% respectively and that of SCC with recycled aggregate is 10%, 19.6% and 20.58% respectively.
- Addition of GGBS also improves the workability of concrete and these waste materials reduces the cost of construction and make the concrete ecofriendly.

References

1. Manzi S et al (2017) Self-compacting concrete with recycled concrete aggregate: Study of the long-term properties. Constr Build Mater 157:582–590
2. Vivek SS, Dhinakaran G (2017) Fresh and hardened properties of binary blend high strength self compacting concrete. Eng Sci Technol Int J 20(2017):1173–1179

Strength Comparison of Cement Mortar and Geopolymer Mortar

V. Revathy and Gouri Antherjanam

Abstract In construction industry large number of buildings are repaired and retrofitted due to the decrease in the load bearing capacity of RC beams within the life span of the building. Repair materials with reduced carbon footprint and environmental sustainability have been in great demand by the construction industry worldwide. Cement is the main constituent of the mortar used for the repair and retrofitting techniques. The production of cement is energy-intensive and releases large amount of carbon dioxide (CO_2) to the atmosphere that significantly contributes to greenhouse gas emissions. It is estimated that one ton of CO_2 is released into the atmosphere for every ton of ordinary Portland cement produced. Environmentally sustainable repair materials with reduced carbon footprint have been in great demand by the construction industry worldwide. Cement-free geopolymer mortars prepared from waste materials with high content of silicate aluminium and alkaline activator solution are emerging as prominent sustainable repair materials. Industrial waste Fly ash is used in this study as the replacement of cement. The current study focuses on strength comparison of cement mortar cubes and geopolymer mortar cubes in ambient curing conditions. 10M, 12M, 15M geopolymer mortar cubes are prepared for the study. Useage of geopolymer mortar leads to the reduction in cement quantity in construction industry and its use should be promoted for better performance and environmental sustainability.

Keywords Geopolymer mortar · Sustainable material

1 Introduction

Cement is most popularly used as a binder material in concrete. At present, the production of cement is increasing significantly to cope with demand of construction

V. Revathy · G. Antherjanam (✉)
Sree Buddha College of Engineering, Pattoor, Alappuzha, India
e-mail: gantherjanam@gmail.com

V. Revathy
e-mail: revathyv481996@gmail.com

K. Dasgupta et al. (eds.), *Proceedings of SECON'19*,
Lecture Notes in Civil Engineering 46,
https://doi.org/10.1007/978-3-030-26365-2_17

industries. The cement manufacturing industry has a large contribution to rising level of carbon dioxide emissions in the environment [1]. Therefore, it is the time requirement to develop an eco-friendly and economic binder material that can totally replace cement as a binder in mortar and concrete. In context of this, materials prepared by geopolymerisation process can be effectively used as binders, instead of cement binder in mortar and concrete. Geopolymerisation is a process that can transform alumina and silica rich waste materials into valuable building/binding materials, having the same properties as of cement based construction materials [2]. The mortar prepared with geopolymerisation process is known as geopolymer mortar. Geopolymer mortar is developed by alkali activating alumina and silica rich waste materials such as fly ash, rice husk ash, GGBS etc.

The utilization of fly ash as source material in geopolymer mortar and concrete not only helps in the reduction of disposing problem of fly ash but also eliminates the adverse impact causes to the environment during cement production by totally replacing cement as a binding material in mortar and concrete. Fly ash based geopolymer mortar is prepared by activating fly ash with alkaline solutions. Sodium and potassium based alkaline activators are most commonly used as alkaline activators. From the previous studies it has been proved that Class F fly ash is a good source for geopolymer mortar and sodium based alkaline activators are more efficient than potassium based activators for the activation of the fly ash.

Geopolymer binders are preferred because the abundance and availability of class F flyash (FA) worldwide create opportunity to utilize these by-products and also they generate 70–80% less carbon dioxide with remarkably lesser greenhouse gas emissions than ordinary Portland cement. Geopolymer mortar is prepared with alkali activator solution and is composed of sodium silicate and sodium hydroxide. It is recommended that the alkaline liquid is prepared at least 24 h prior to use [3]. The NaOH solids must be dissolved in water to make a solution with the required concentration. The concentration of sodium hydroxide solution can vary in the range between 8 Molar (8M) and 16 Molar (16M).

2 Objectives

1. To make a mortar without using cement (i.e. Geopolymer mortar).
2. To prepare the 10M, 12M, 15M geopolymer mortar cubes.
3. To investigate the mechanical behaviour of geopolymer mortar cubes.
4. To compare the strength properties of conventional mortar and geopolymer mortar.

3 Methodology

- A detailed literature survey was carried out on Studies on geopolymer mortar.
- Physical properties of materials were tested and found that all properties are conforming to IS standards.
- Cubes of cement mortar are cast (1:3).
- 10M, 12M, 15M geopolymer mortar cubes are cast.
- Testing of specimens for compressive strength on 7th day and 28th day.
- Comparing the strength of cement mortar and geopolymer mortar.
- Conclusions and recommendations are finally made based on the findings and observations.

4 Materials

A. *Cement*

Ordinary Portland cement of 53 grade conforming to IS 12269-2013 was used in this study to produce conventional concrete of M30 grade.

B. *Fine aggregate (FA)*

Manufactured sand conforming to zone II was used as fine aggregate having fineness modulus 2.93 and specific gravity 2.72. Tests are conformed to IS: 383-2016.

C. *Coarse aggregate (CA)*

Crushed stone aggregate of size between 20 and 4.75 mm and specific gravity 2.74 and fineness modulus 7.16 was used as coarse aggregate. Tests are conformed to IS: 383-2016.

D. *Fly Ash (FA)*

Fly ash is a by-product from burning pulverized coal in electric power generating plants, and is widely used as an ingredient in hydraulic-cement concrete. Because it improves many desirable properties of concrete, it is introduced either as a separately batched material or as a component of blended cement. Class F fly ash is used and it exhibit Pozzolonic properties. Tests is conducted on fly ash as per IS 2386 (Part III)-1963.

E. *Alkaline solution*

The alkaline liquid used in this study was a combination of Sodium Hydroxide (NaOH) and Sodium Silicate (Na_2SiO_3). The sodium hydroxide and sodium silicate of laboratory grade were purchased from Hazeena Chemicals, Trivandrum. Sodium hydroxide is available in the form of flakes and sodium silicate is available in the form of liquid. Sodium hydroxide solution was prepared by dissolving flakes in distilled water. 10M concentration of sodium hydroxide was used for the entire study.

F. *Water*

Potable water is generally considered as being acceptable for casting the test specimens. Clean drinking water available in the college water supply system was used for mixing and preparing alkaline liquid.

5 Experimental Investigation

The experimental investigation is done on the cement mortar and geopolymer mortar. Cement mortar cubes of 100 × 100 × 100 mm where cast and water curing is provided for 28 days. Geopolymer mortar cubes of 100 × 100 × 100 mm where casted and ambient curing is provided for 28 days.

Mix design of geopolymer mortar is given in Table 1.

A. *Specimen Details*

Mechanical properties include compressive strength. The compressive strength test is done by casting mortar cube of size 100 mm × 100 mm × 100 mm. Figure 1 shows the cement mortar cubes. The cement mortar is poured in the mould with compaction. 10M, 12M, 15M geopolymer mortar is prepared and is poured into the mould with compaction. Figure 2 shows the geopolymer mortar cubes. After 24 h these moulds are removed and cement mortar specimens are put in water for curing and geopolymer mortar cubes are cured in ambient temperature. The compression test is conducted on 7th day and 28th day. Figures 3 and 4 shows the curing of mortar cubes.

Table 1 Mix design of geopolymer mortar

Materials	Molarity	Mix 1:3 (kg/m^3)
Fly ash	10M	495
	12M	501
	15M	509
Sand	10M	1484
	12M	1519
	15M	1525
Sodium hydroxide	10M	110
	12M	111.5
	15M	141
Sodium silicate	10M	134.5
	12M	140
	15M	141.2

Fig. 1 Casting of mortar cubes

Fig. 2 Casting of mortar cubes

Fig. 3 Casting of mortar cubes

Fig. 4 Casting of mortar cubes

Table 2 Compressive strength of mortar cubes

S. No.	Specification	Average seven day compressive strength (N/mm^2)	Average 28 day compressive strength (N/mm^2)
1	GPM—10M	15	35
2	GPM—12M	16	37
3	GPM—15M	16	38
4	Cement mortar	20	37

6 Results and Discussions

A. Strength tests on hardened mortar

Compressive strength test is used to measure the strength of mortar cubes. The specimens for compression test are tested on 7th day and 28th day. For each mix, six specimens are cast and cement mortar cubes are water cured and geopolymer mortar cubes are ambient cured. Table 2 shows the compressive strength of mortar cubes. Figure 5 shows the compression test and Fig. 6 shows compressive strength variation.

7 Conclusion

Geopolymer mortar or cement less mortar is a promising sustainable solution to many of the environmental problems in the near future. Flyash based geopolymer mortar is used for this study and ambient curing conditions are provided for the mortar cubes.

Fig. 5 Testing of mortar cube

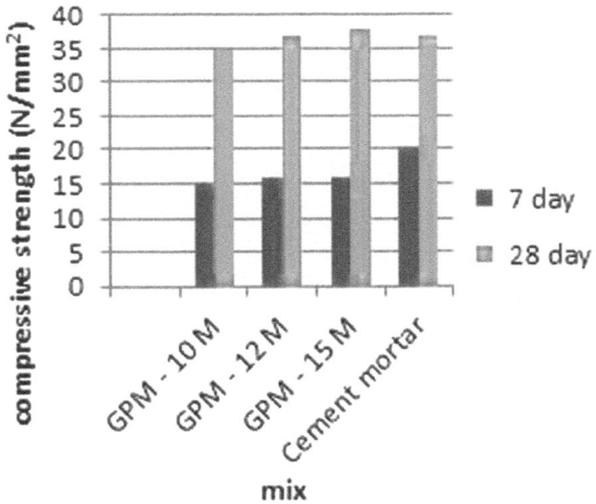

Fig. 6 Compressive strength of mortar cubes

The geopolymerisation reaction is slow in ambient curing conditions so the 7th day strength of geopolymer mortar is less than cement mortar. The 28th day strength of geopolymer mortar is comparable with cement mortar. 12M and 15M cubes having approximately same compressive strength but workability of 15M mortar is very less. 12M geopolymer mortar is suitable for every constructional purpose where cement mortar is used.

References

1. Ortega I et al (2018) An experimental study on RC columns repaired on all four sides with cementitious mortars. Constr Build Mater 161:53–62 (Science Direct)
2. Karuppuchamy K et al (2018) Effect of alkaline solution with varying mix proportion on geopolymer mortar. IOP Conf Ser Mater Sci Eng 310:1–9
3. Mahajan LS et al (2017) Experimental study of conventional and geopolymer mortar. Int J Sci 5(2):2277–7261

Experimental Investigation on Performance of Internally Cured Self Compacting Concrete Using Sintered Fly Ash Aggregates

Muthulakshmi Ajay and B. Aswathy Lal

Abstract In recent decades, Self compacting concrete (SCC) plays a vital role in building construction due to its unique ability to fill formwork of complex steel reinforcements by its own weight with no mechanical vibrations. Due to severe environmental impact caused by over usage of natural aggregates and water in concrete production, a novel approach in producing a sustainable SCC is by replacing natural aggregates by light weight aggregate. The lightweight aggregate used plays a vital role in enhancing the fresh and hardened properties of SCC due to its better internal curing property. This experimental investigation was conducted to study the fresh and hardened properties of SCC incorporated with sintered fly ash aggregates (Sintag). In this experimental study the coarse aggregate is replaced with various percentages replacement of Sintag such as 5, 10, 15, 20, 25 and 30% respectively. The fresh properties of SCC were studied in accordance with EFNARC guidelines. Hardened properties include compressive strength, split tensile strength, flexural strength and hardened density testing. The optimum percentage replacement was obtained as 15% replacement. The presence of internal moisture content helps to obtain hardened cement paste compared to conventional SCC. The results showed that compressive strength, split tensile strength and flexural strength of internally cured SCC has increased significantly.

Keywords Self compacting concrete · Internal curing · Sintered fly ash aggregates · Fresh properties · Hardened properties

M. Ajay · B. Aswathy Lal (✉)
Sree Buddha College of Engineering, Pattoor, Padanilam, India
e-mail: lalaswathy89@gmail.com

M. Ajay
e-mail: lakshmimuth08@gmail.com

K. Dasgupta et al. (eds.), *Proceedings of SECON'19*,
Lecture Notes in Civil Engineering 46,
https://doi.org/10.1007/978-3-030-26365-2_18

1 Introduction

The construction industry of every country is an imperative indicator of the development as it creates venture opportunities across various related sectors. The worldwide ever increasing usage of natural resources and water in the construction sector leads to severe natural resource depletion. Also, the generation and disposal of abundant industrial waste materials has led to set environmental consciousness. This scenario will form a basis of sustainable development in the construction industry with greater prominence on the implementation of ingenious industrial by-products as a substitute of the natural resources used in concrete including recycling and reuse of waste materials. The present scenario of frequent failures of tall buildings necessitates the use of a sustainable concrete that is ductile and durable.

Self-compacting concrete (SCC) is able to flow and compact under its own weight and can occupy all the spaces in the form without any vibration effect. Trial and error method is used for finding the correct mix. The higher volume of cement is one of the reasons for increasing the cost of self compacting concrete. It can be reduced by replacing the cement with any cementitious material. Silica fume (SF) can be used to replace cement.

On site concrete failures are associated to several reasons; right from mix design, materials used, mixing, placing, compaction and curing procedures. Improper curing is considered as one of the significant reasons for concrete failures observed in the form of cracks. Unfortunately, adequate curing is not given much importance at most of the sites leading to reduction in the durability of the structure. Internal curing is hastily emerging as an effective way to advance curing of concrete. It holds promise for producing concrete with increased compressive strength, reduced permeability and cracking; thereby lengthen the service life of infrastructure.

Lightweight aggregate is used for internal curing because of its increased absorption, which range from less than 10% to more than 30% depending on the type of aggregate and how thoroughly it has been prewetted. The moisture is not released from the aggregate until after the concrete has set and the pore size in the paste becomes smaller than the pores in the lightweight aggregate. The absorbed water does not affect the water-cement ratio. The addition of lightweight aggregate to a mix for internal curing will also reduce the density of the concrete. However, for precast concrete, weight reduction is an important benefit.

The possibility of developing SCC incorporating sintered fly ash aggregates (Sintag) was a novel approach to combine the advantages of both SCC and self curing concrete. This seemed to be a promising technology in enhancing both the strength and durability parameters of concrete. Very few studies have been carried out so far on internal curing of self compacting concrete using sintered fly ash aggregate (SCCS).the main objectives of the investigation were to evaluate the strength parameters of SCC and SCCS.

2 Objectives

1. To obtain the mix proportion of self compacting concrete.
2. To obtain the optimum percentage of Sintag as partial replacement of coarse aggregate.
3. To compare the fresh and hardened properties of SCC and SCCS.

3 Methodology

- A detailed literature survey was carried out on Studies on self compacting concrete using various materials in SCC.
- Material collection.
- Material testing.
- Fixing the mix proportion of SCC by trial and error method as per the EFNARC guidelines.
- The mix proportion adopted is 1:1.96:1.67.
- Cubes, beams and cylinders with control mix self compacting concrete using natural aggregates and with 5, 10, 15, 20, 25 and 30% of coarse aggregates replaced with sintered fly ash aggregates were casted.
- Testing of specimens done after 7 days of water curing and thereafter left exposed to open atmosphere for 28 day testing.
- Determining the optimum percentage of sintered fly ash aggregates from the test results.
- Strength of SCC specimens and SCCS specimens were compared.
- Conclusions and recommendations are finally made based on the findings and observations.

4 Materials

4.1 Cement

Ordinary Portland cement of 53 grade conforming to IS 12269-2013 is used in the study.

4.2 Fine Aggregate (FA)

Manufactured sand conforming to zone II was used as fine aggregate having fineness modulus 2.96 and specific gravity 2.72. Tests were conformed to IS: 383-2016.

4.3 Coarse Aggregate (CA)

Crushed stone aggregate of size between 16 and 4.75 mm and specific gravity 2.71 and fineness modulus 5.4 was used as coarse aggregate. Tests are conformed to IS: 383-2016.

4.4 Fly Ash (FLA)

Fly ash used in this study is low calcium (ASTM Class F) Fly ash collected from Q-Crete ready-mix plant. Fly ash is used as fines along with fine aggregate. 20% of fine aggregate was replaced with Fly ash.

4.5 Silica Fume (SF)

The mineral additive silica fume obtained from Bison shelters limited, Kerala was used. It is a white coloured, odourless, micronized powder with a specific gravity 2.63.

4.6 Sintered Fly Ash Aggregate (Sintag/FAA)

Sintag can be used as a lightweight aggregate which is up to 50% lighter than natural aggregate. The sintered fly ash aggregate of size 16 mm and below were used in this study. The material was collected from Jintal Steel power plant, Orissa.

4.7 Water

Casting and curing of specimens were done with the portable water which is available. The correct quantity of water in the concrete is critical.

4.8 Super Plasticizer (SP)

High Range Water reducing admixture (HRWRA) or Superplasticizer is an important constituent in the production of SCC to achieve self-compatibility and flow ability. Conplast SP 430 was used for the work as super plasticizer.

5 Experimental Investigation

5.1 Mix Proportion

For experimental study mix proportion selected was 1:1.96:1.67 with a w/c 0.38. Coarse aggregate in normal SCC was replaced with Sintag at 5, 10, 15, 20, 25 and 30% of volume of fine aggregate. SCC is the control mix; SCCS5, SCCS10, SCCS15, SCCS20, SCCS25 and SCCS30 are the mixes with equivalent volume replacement of coarse aggregate with sintered fly ash aggregate at 5, 10, 15, 20, 25 and 30% respectively. Final mix proportions for the different combinations of the mixes are shown in Table 1.

5.2 Casting and Curing Conditions

Rotating drum mixer was used for mixing the concrete to make the test specimens. The concrete specimens for compressive strength test, split tensile strength test and flexural strength test were cured for 7 days in water and followed curing in room condition till the day of testing. 7 days of external curing is adopted as the external top surface of concrete may exposed to direct sunlight which may results in crack formation.

Table 1 Mix proportions for various mixes

Mix	Cement (kg/m^3)	SF (kg/m^3)	CA (kg/m^3)	FAA (kg/m^3)	FA (kg/m^3)	FLA (kg/m^3)	W/C ratio	SP (L)
SCC	485	15	808.038	–	761.87	190.46	0.38	8
SCCS5	485	15	767.63	23.38	761.87	190.46	0.38	8
SCCS10	485	15	727.23	46.76	761.87	190.46	0.38	8
SCCS15	485	15	686.83	70.14	761.87	190.46	0.38	8
SCCS20	485	15	646.43	93.52	761.87	190.46	0.38	8
SCCS25	485	15	606.03	116.89	761.87	190.46	0.38	8
SCCS30	485	15	565.63	140.27	761.87	190.46	0.38	8

6 Results and Discussions

6.1 Fresh Properties

Fresh properties were used to determine the self compatibility of the concrete. All tests were carried out as per EFNARC 2002. Slump flow test, V-funnel test and L-box test were conducted and results are tabulated in Table 2.

6.2 Hardened properties

The strength of concrete is obtained when it gets hardened. Compressive strength test, Split tensile strength test and Flexural strength test is mainly used to measure the strength of concrete. All the mixes were tested after 7 days of water curing to know the effect of internal curing.

- *Compressive Strength Test*

Compressive strength test was carried out in cube specimens of size 150 mm after 7 days water and followed by air curing till the day of testing. Additional percentage replacements of SCC mixes were casted in such a way the coarse aggregate volume was replaced by Sintag of (SCCS12.5 and SCCS17.5) 12.5 and 17.5%. The reason for decrease in compressive strength with 30% sintag as coarse aggregate replacement was due to reduction in density of specimens. Table 3 shows the compressive strength of concrete with various percentages of Sintag and Fig. 1 shows the graphical representation of compressive strength.

Table 2 Fresh properties of different mix

Tests	Unit	Results						
		SCC	SCCS5	SCCS10	SCCS15	SCCS20	SCCS25	SCCS30
Slump flow	mm	700	708	712	720	720	694	680
T_{50} cm slump	s	4	4	4.2	4.3	4.5	4.1	3.9
V-funnel	s	10	10	10.2	10.5	10.5	10.7	11
V-funnel at T_5 min	s	+2	+2.05	+2.18	+2.2	+2.26	+2.2	+2.52
L-box (H_2/H_1)	–	0.82	0.82	0.85	0.87	0.9	0.93	0.95

Table 3 Compressive strength results

Mix	7 day (N/mm^2)	28 day (N/mm^2)	56 day (N/mm^2)
SCC	36	43.11	62.23
SCCS5	38.32	47.74	63.16
SCCS10	40.35	50.81	64.54
SCCS12.5	41.92	53.26	65.97
SCCS15	44.29	57.85	67.78
SCCS17.5	42.36	55.34	67.14
SCCS20	45.48	51.89	65.17
SCCS25	40.82	45.75	64.27
SCCS30	33	42.95	61.57

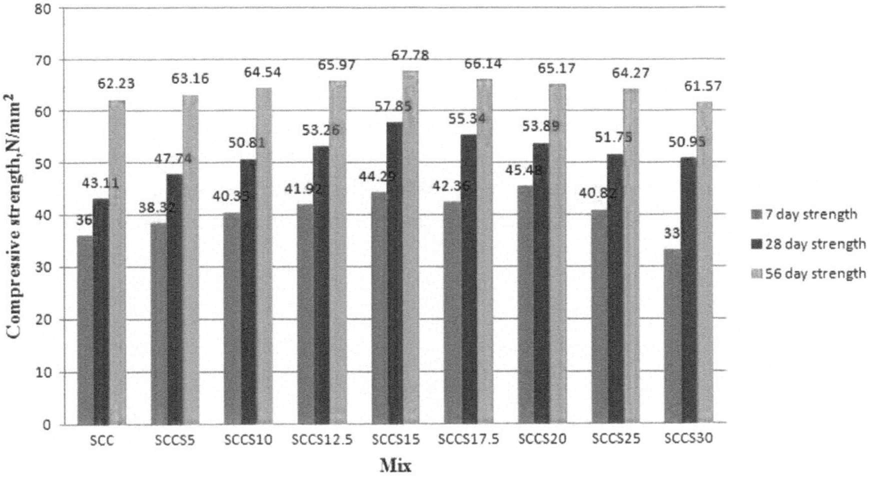

Fig. 1 Compressive strength of different mixes of SCC

- *Split Tensile Strength Test*

The low tensile strength and brittle nature of concrete doesn't allow resisting direct tension. Cylindrical specimens of length 300 and 150 mm diameter was used to find the split tensile strength after 7 days of water curing. The results were shown in Table 4 and the graphical representation is shown in Fig. 2.

Table 4 Split tensile strength test result

MIX	28 day split tensile strength
SCC	4.28
SCC15	4.39

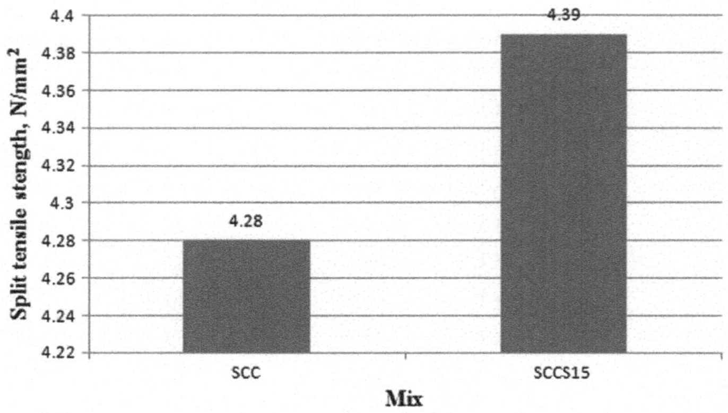

Fig. 2 Split tensile strength of SCC

- *Flexural Strength Test*

Plain cement concrete beams and reinforced cement concrete beams having dimension 100 mm × 100 mm × 500 mm were used to study the flexural strength. Test was carried out after 28 days with 7 days of water curing only and three specimens were casted for each mix to take the average. Table 5 shows the flexural strength of SCC and SCCS15 mix and Fig. 3 shows the graphical representation of flexural strength.

Table 5 Flexural strength test result

Mix	28 day strength P.C.C. (N/mm^2)	Increase in strength (%)	28 day strength R.C.C. (N/mm^2)	Increase in strength (%)
SCC	6.8	–	22.4	–
SCCS15	7.6	11.76	25.76	15.4

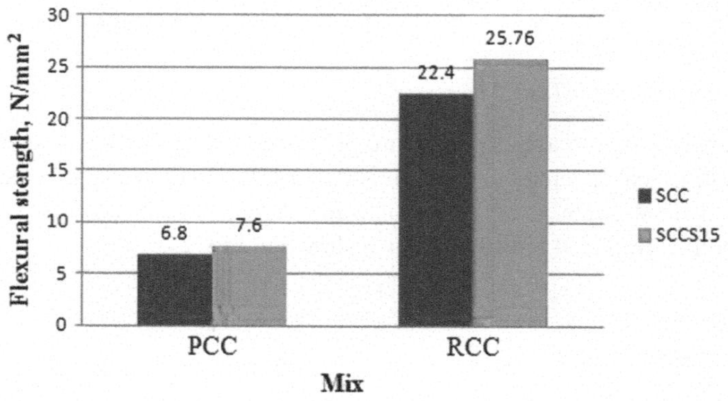

Fig. 3 Flexural strength of SCC

7 Conclusion

From the test results the replacement of coarse aggregate in normal SCC with sintered fly ash aggregate is found effective. The following are the conclusions derived from the experimental results.

- From the 28 days test results, compressive strength of internally cured SCC with 15% replacement of Sintag is 34% higher compared to conventional SCC and is taken as the optimum replacement percentage.
- Self Compacting Concrete with Sintag 25% has attained the 28 day compressive strength of 45.75 N/mm^2 thereby promising to replace coarse aggregates up to 25% by volume of coarse aggregate.
- The split tensile strength of SCC mix with 15% replacement of Sintag at 28 days of test is 4.39 N/mm^2. The mix with 15% replacement of Sintag is 2.6% higher compared to control SCC mix.
- The flexural strength of SCC mix with 15% replacement of sintag is greater than control mix in both P.C.C and R.C.C. The SCCS15 P.C.C beam has increased 11.76% compared to normal P.C.C beam (SCC) whereas the R.C.C beam with 15% replacement (SCCS15) has increased 15.4% compared to normal R.C.C beam (SCC).

Investigation on Strength Characteristics of Silica Fume Incorporated Foamed Concrete

P. S. Anusree and R. Gopakumar

Abstract An experimental investigation was conducted to study the strength characteristics of silica fume incorporated foam concrete. Light Weight Foamed concrete (LFC) has become most commercial material in construction industry for non-structural applications owing to its lesser density, stability and high thermal insulation properties. The aim of this study is to develop structural foamed concrete by using silica fume and fly ash. Constant cement: fly ash: sand ratio of 1:0.5:1.5 and W/C ratio of 0.65 was used for all mixes. The study presents the use of fly ash for partially replacing sand to produce foamed concrete. Fine silica fume is used to improve properties of foamed concrete. Four variations of foam (2, 3, 4 and 5%) were added and cement was partially replaced with silica fume of various percentages (0, 5, 10, 15 and 20%). The results indicate that foamed concrete with a density of 600–1450 kg/m^3 and compressive strength of 6–13 N/mm^2 can be made by using silica fume. The silica fume improved the compressive strength of foamed concrete.

Keywords Foamed concrete · Fly ash · Silica fume · Compressive strength · Wet density

1 Introduction

With the increase in demand for structures which are light in weight, researchers attention in foamed concrete to be used in structural applications is steadily increasing [1, 2]. Foamed concrete, also known as lightweight concrete is a type of concrete which contains cement, fine aggregate, fly ash, water and voids, but coarse aggregate. Foamed concrete can be produced with dry density of 400–1600 kg/m^3, and 7-day strength of approximate 1–10 N/mm^2.

P. S. Anusree · R. Gopakumar (✉)
Sree Buddha College of Engineering, Pattoor P.O., Alappuzha 690529, Kerala, India
e-mail: r.gopakumar@outlook.com

P. S. Anusree
e-mail: anujeevan000@gmail.com

K. Dasgupta et al. (eds.), *Proceedings of SECON'19*,
Lecture Notes in Civil Engineering 46,
https://doi.org/10.1007/978-3-030-26365-2_19

Foamed concrete is a light weight, low cost and easy to manufacture material with good workability and excellent performance in thermal insulation, acoustic insulation, corrosion resistance, fire resistance, and shock absorption [1]. Due to its unique properties and relative cost effectiveness.

Foamed concrete has many engineering applications. Usage of this concrete reduces dead loads on the structure and foundation, contributes to energy conservation, and lowers labor cost during construction. It also reduces the cost of production and transportation of building components compared to normal concrete. Some of the applications of foam concrete include; production of lightweight blocks and precast panels, fire insulation, thermal and acoustic insulation, road sub-base construction, and trench re-instatement. High density foam concrete has the potential of being used as a structural material.

Stable foamed concrete production depends on many factors such as; type of foaming agent, method of preparation of foaming agent to initiate a homogeneous distribution of air voids, and design calculation of the mixture [2]. Compressive strength is considered as the primary function of the desirable design density, and as a main consideration for this concrete to be finally used to fabricate structural, non-structural or semi-structural components. Meanwhile, durability is another property of foamed concrete that needs to be at a level which can effectively allow it to resist the aggressive environments [1, 2].

In this study, the strength characteristics of foamed concrete incorporated with silica fume are evaluated by compressive strength tests. To effect economy and sustainability in construction, the effects of partial replacement of cement with silica fume, and replacement of sand with fly ash are also studied.

2 Scope and Objectives of the Study

The applications of foamed concrete are presently limited due to lack of know-how on production of this material and its structural performance. This study was aimed to examine the potential to improve the strength of foamed concrete by including silica fume, and to make it a more cost effective and sustainable construction material by partially replacing sand with fly ash. The main objective of the research was to develop structural fibre reinforced foamed concrete.

The objectives of the completed first phase of the study were:

- To obtain the mix proportion for the ordinary foamed concrete.
- To obtain the optimum percentage of silica fume as partial replacement of cement.
- To investigate the strength characteristics of foamed concrete using silica fume.
- To compare the strength properties of normal foamed concrete and silica fume incorporated foamed concrete.

3 Methodology

A detailed literature survey was carried out on studies on foamed concrete using various materials

- Physical properties of materials were tested and found that all properties are conforming to IS standards.
- Fixing the mix proportion of foamed concrete by trial and error method with four proportions of foam (2, 3, 4 and 5%) based on target density.
- For the optimum proportion of foam found from above samples, 6 numbers of cubes of size 150 mm are cast by partially replacing cement with silica fume in varying percentages (5, 10, 15, and 20%).
- Testing of specimens for compressive strength on 7th day and 28th day.
- Determining the optimum percentage of silica fume from the test results.
- Conclusions and recommendations are finally made based on the findings and observations.

4 Materials

The constituent materials used to produce foamed concrete [3] in the study were:

A. *Cement*

Ordinary Portland cement of 53 grade conforming to IS 12269-1987 was used. Physical properties of cement were tested using Le-Chatlier apparatus and Vicat apparatus. The standard consistency of cement was obtained as 32%. The initial setting time and final setting time were 90 min and 360 min respectively. The specific gravity and fineness of cement are obtained as 3.15 and 7.33% respectively.

B. *Fine aggregate*

A fine aggregate passing through 4.75 mm IS sieve was used. Generally river sand, crushed stone, crushed gravel, M sand etc. are used as fine aggregate. In this study, Manufactured sand (M sand) conforming to Zone II of IS 2386 (part I)-1963 were used.

C. *Fly Ash (FA)*

Fly ash is a by-product from burning pulverized coal in electric power generating plants, and is widely used as an ingredient in hydraulic-cement concrete. Because it improves many desirable properties of concrete, it is introduced either as a separately batched material or as a component of blended cement. Class F fly ash is used and it exhibit Pozzolanic properties. Tests is conducted on fly ash as per IS 2386 (Part III)-1963.

D. *Silica Fume (SF)*

Silica Fume is a by-product of silicon and ferrosilicon alloy production and consists of spherical particles with an average particle diameter of 150 nm, which makes it approximately 100 times smaller than the average cement particle. It is a white colored, odorless, micronized powder with a specific gravity 2.63.

E. *Foaming agent*

Foaming agents are mainly used to control the density of foamed concrete. When foaming agents are mixed with water, it will produce disconnected bubbles and through mixing will produce stable foam which becomes incorporated in the cement paste. For this study, Sodium Lauryl Ether Sulphate (SLES) is used as a foaming agent, which is an anionic detergent as well as surfactant found in many personal products (e.g. soap, shampoos, toothpaste etc.) that behave similar to soap.

F. *Water*

As per IS standards, the water free from adverse amount of soils, organic and inorganic impurities was used for this work. Casting and curing of specimens were done with the potable water. The correct quantity of water in the concrete is critical. It helps to form the strength giving cement gel. The quantity and quality of water required are observed very carefully.

5 Experimental Investigation

The experimental investigation is done on the fresh concrete (slump flow, wet density test) and hardened concrete (compressive strength) of foamed concrete using silica fume.

A. *Mix Design*

Presently, there are no standard guidelines for the mix proportioning of foamed concrete, and the mix design procedure is based on trial mixes. Based on the results obtained from the trial mixes, a better mix proportion was chosen. When foamed concrete is considered, not only the strength, but also the density is an important factor [4]. Since the compressive strength of foamed concrete is a function of its density, strength can be altered by changing density but this does not give any indication about the water requirement in the mix. It is not an easy task to attain a precise measurement of the density of foamed concrete on site, because the hardened density depends on the saturation concentration in its pores. So, the mix proportion was obtained based on trial and error.

Table 1 Composition of foam concrete

Trial mix	Proportion			w/c ratio	Foaming agent (% by weight of cement)	Wet density (kg/m^3)	Compressive strength (N/mm^2)	
	Cement	Fly ash	Sand				7 day	28 day
Mix 1	1	0.5	1.5	0.65	5	600	3	4
Mix 2	1	0.5	1.5	0.65	4	1000	3.4	4
Mix 3	1	0.5	1.5	0.65	3	1200	4	4.5
Mix 4	1	0.5	1.5	0.65	2	1450	4.2	5

B. *Mix proportion*

The present experiments were to make lightweight foamed concrete without coarse aggregate. For that, fixing a mix proportion of 1:0.5:1.5 by weight of cement, fly ash and sand, and a water cement ratio of 0.65 is used. The optimized mix of foamed concrete is found by trial and error the details of which are given in Table 1. Four variations of foam (2, 3, 4 and 5%) were added and cement was partially replaced with silica fume of various percentages (0, 5, 10, 15 and 20%).

C. *Test Procedure*

For assessing the stability of foam concrete, the fresh density of the foam concrete was measured by filling a standard container of known volume and the density ratio was calculated by comparing it with the designed density as proposed in [1]. The water required for obtaining a density ratio was determined by trials. The mix proportion of foam concrete was obtained by trial and error method based on target density and details are shown in Table 1. The appropriate quantity of foam required was generated and added immediately to the base mix and mixed for a minimum duration until there was no physical sign of the foam on the surface and all the foam was uniformly distributed and Incorporated into the mix. Figure 1 shows the foam generation. For the optimum proportion of foam found from the strength tests on above samples, 6 numbers of cubes of size 150 mm are cast by partially replacing cement with silica fume in varying percentages (5, 10, 15, and 20%) (Fig. 2).

6 Results and Discussions

The optimized mix of foamed concrete is found by trial and error the details of which are given in Table 1.

A. Workability

The inverted slump test was conducted according to ASTM C995 by using a slump cone and flat base. The test is conducted in order to know the spread ability of foam

Fig. 1 Generated foam

Fig. 2 Wet density test

concrete. Since foam is free flowing, self-leveling material, the slump obtained is in a collapsed form and diameter of spread is 320 mm which is within the prescribed limit. Maximum limit of spread should be 500 mm (Table 2).

B. Compressive Strength Tests

By foam addition, the foamed concrete shows decrease in compressive strength. By increasing amount of foam, both the density and compressive strength gets reduced. When foam addition is about 5% compressive strength is very small. Therefore 2% is considered as optimum quantity in terms of density and compressive strength. Mix 4 is adopted as the base mix of normal foam concrete based on the target wet density of 1450 kg/m^3. Figure 3 shows compressive strength test of foam concrete. Table 3 shows the 7 day and 28th day compressive strength of normal foam concrete. Compressive strengths of foamed concrete incorporating silica fume at 7 and 28 days are given in Table 3 (Figs. 4, 5, 6, 7 and 8).

C. Effect on age

The compressive strength of foamed concrete in almost all mixes showed a continuous increase with age. The rate of strength development was greater initially and

Table 2 Optimized foamed concrete mix proportion

Cement (kg/m^3)	Fly ash (kg/m^3)	Fine aggregate (kg/m^3)	Foaming agent %	w/c (%)
483	242	725	9.66	0.65

Fig. 3 Compressive strength test of foam concrete

Table 3 Compressive strength of foamed concrete incorporating silica fume

S. No.	Mixes	Density	Compressive strength (N/mm^2)	
			7 day	28 day
1	FC-0% SF	1450	4.2	5
2	FC-5% SF	1456	6	7.5
3	FC-10% SF	1462	6	7
4	FC-15% SF	1468	10	13
5	FC-20% SF	1470	8	9.5

Fig. 4 Casting of foam concrete specimen

decreased as age increased. However, a comparison of 7-day and 28-day strengths revealed that concrete with no silica fume developed small percentages of the 28-day strength, while those containing silica fumes developed greater variation of the corresponding 28-day strength.

D. Effect of Density and Foam Volume

The strength of foamed concrete appeared to increase exponentially with an increase in concrete density, or with a decrease in foam volume. For foamed concrete without silica fume, when the foam volume is 2% by weight of cement, strength attained is 5.5 MPa. This shows that while the target wet density of foamed concrete is

Fig. 5 Curing of foam concrete specimen

Fig. 6 Casted foam concrete cubes

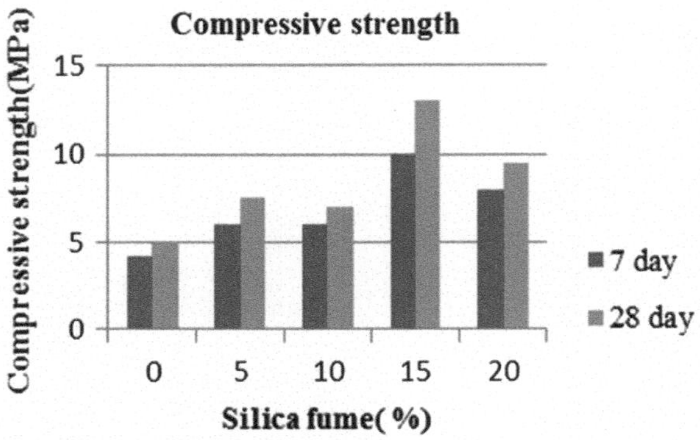

Fig. 7 Compressive strength variation of foamed concrete with varying percentages of silica fume

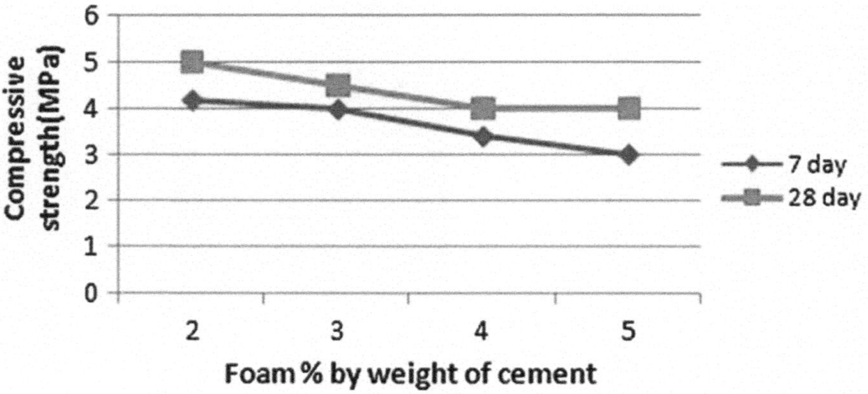

Fig. 8 Compressive strength with various percentages of foam

1450 kg/m^3, its compressive strength reaches to 5.5 MPa. For foamed concrete with silica fume, when its foam volume is 2%, compressive strength can reach up to 13 MPa.

E. Effect of Silica Fume

With the same volume content of foam, fine silica fume can significantly increase the compressive strength of foamed concrete. By 15% replacement of cement with silica fume, 28 day strength of 13 MPa can be attained.

7 Conclusion

- Compared to other foamed concrete mixes, the concrete mix with 2% foaming agent gave rise to desired density.

- The foamed concrete shows reduction in compressive strength and density with increase in foam addition. Compressive strength is very low when foam addition is about 5%. Thus, when density and compressive strength are considered, the optimum dosage is adopted as 2%.
- Fly ash which is a waste product from coal industry replaces sand which makes the foamed concrete much less denser and stronger. 25% replacement of sand with fly ash gives much larger compressive strength. Fly ash which is freely and readily available can be used to make foamed concrete cost effective.
- Effect of silica fume can significantly increase the compressive strength of foam concrete. The largest increase acquired is 13 MPa at 28 days by adding foaming agent 2% by weight of cement, and by replacement of cement with 15% silica fume.

References

1. Bing C et al (2012) Experimental research on properties of high-strength foamed concrete. J Mater Civ Eng (ASCE) 0899–1561
2. Kunhanandan Nambiar EK et al (2008) Fresh state characteristics of foam concrete. J Mater Civ Eng (ASCE) 0899–1561
3. Mugahed Amran YH et al (2015) Properties and applications of foamed concrete: a review. Constr Build Mater 101:990–1005
4. Rommel E et al (2017) Characteristics of foam concrete with usage of foam agent which varies. Int J Sci Eng Res 8

Mud Bricks Using Oyster Shells

Jibin Idiculla Thomas, Josin Jose, Namitha Rose Mathew and Resma Reji

Abstract Sustainable construction practice is essential for the safeguarding and manageability of our earth just as the wellbeing of ourselves and what is to come. Even if brick is one of the most common construction material, scarcity of natural materials have led to the incorporation of different types of waste into the bricks and have been investigated [1]. In the growing concern of sustainability, compressed earth bricks are more energy efficient, economic and environment friendly. The aim of this project is to study the suitability of clayey soil and crushed Oyster Shell (OS) powder using Rice Husk Ash (RHA) as binder for unfired rammed earth brick. OS is an industrial waste, disposed of in open dumps at coastal areas in huge quantities. In order to reduce the environmental problems created by the waste OS, the soil in the bricks were replaced by Oyster shell powder by varying the percentage from 5 to 20 at an interval of 5%. To evaluate the performance, tests such as compressive strength, efflorescence and total water absorption were conducted on samples of size 190 mm × 90 mm × 90 mm. Compared to conventional unfired bricks the samples showed a better compressive strength for 10% replacement.

Keywords Oyster shells · Sustainable building material · Unfired rammed earth bricks · Compressive strength · RHA

1 Introduction

Brick is one of the most important materials for construction. Compressive strength is the important parameter when considering the design of bricks. Currently there is increase in interest in unfired clay brick that would meet strength requirement. In the manufacturing process of fired brick, several gases, including CO_2, SO_2, etc., are released from the brick kilns while a large amount of land resources are consumed [2]. These emissions and consumptions are becoming a major environmental concern

J. I. Thomas · J. Jose · N. R. Mathew (✉) · R. Reji
Department of Civil Engineering, Amal Jyothi College of Engineering,
Kanjirapally, Kerala, India
e-mail: namitharosemathew12@gmail.com

K. Dasgupta et al. (eds.), *Proceedings of SECON'19*,
Lecture Notes in Civil Engineering 46,
https://doi.org/10.1007/978-3-030-26365-2_20

in many countries. The use of waste and recycled materials has many economic and environmental benefits for our society. In this study, oyster shells (OS), the by-products of ostreaculture, is used as alternative in mud bricks.

Extensive research has been conducted on oyster shell properties in concrete. When crushed oyster shells are used as an aggregate, compressive strength decreased as replacement level increased [3]. Researches were done on partial replacement for coarse aggregate by seashell and cement by lime in concrete [4] and on mud brick with different sand/mud ratio varying from 0.05 to 1 and keeping the rice husk ratio constant [5]. The feasibility of Oyster Shell Foamed Brick (OSFB) as a replacement for traditional alkaline chemical agents to neutralize recycled rainwater was investigated [6]. Some researchers have conducted comparative studies like that of ordinary Portland cement concrete (OPCC) and OPCC with sea shell [7]. Crushed shell pervious concrete strength is lower than that of pervious concrete due to their high permeability values when compared to control pervious concrete [8].

From the literature studies we learned that the $CaCO_3$ in OS can be converted to CaO by calcining and can be used as a substitute for aggregate in concrete. In our experimental analysis it is clear that the properties of OS are similar to that of soil; hence we used OS shell as a partial replacement for soil in the manufacture of mud bricks using RHA as the binder.

2 Materials and Methodology

2.1 Materials

The materials used for this study are soil containing adequate amount of sand, silt and clay, Oyster Shell powder for replacing the soil, Rice Husk Ash as the binder and water.

2.1.1 Soil

About 50 kg of sand and mud was collected from Cherpunkal, Pala. Various material property tests such as Atterberg's Limit (liquid limit, plastic limit and shrinkage limit), Sieve Analysis, Hydrometer Analysis and Specific Gravity were conducted for the soil sample and the result obtained are shown in Table 1.

2.1.2 Oyster Shells

Oyster shell was collected from the oyster farm, Kasaragod. About 30 kg oyster shell was collected which was then dried. The dried oyster shells were calcined at 550 °C for 1 h [9]. Then it was powdered using a rammer. The portion of oyster shell powder

Table 1 Soil properties

Property	Test procedure	Results	Remarks
Specific gravity	IS 2386 part III	2.62	Within the range of 2.6–2.8
Plastic limit	IS 2720 part V 1965	20.2%	Medium plastic
Liquid limit	IS 2720 1985 part V	31%	Clayey soil (lies above A-line in plasticity Chart)
Shrinkage limit	IS 2270 part VI	19.56%	Less shrinkage
% clay	IS 2720 part IV 1985	50%	30–70%
% silt	–	31%	–
% sand	–	19%	–
Fineness modulus	IS 2720 part IV 1980	4.66	–

Table 2 Oyster shell properties

Property	Test procedure	Results	Remarks
Specific gravity	IS 2386 part III	2.72	Within the range of 2.6–2.8
Fineness modulus	IS 2720 part IV 1980	4.15	–

passing 2.36 mm IS sieve was taken for the experiment. Specific gravity test and sieve analysis was conducted to determine the oyster shell properties. The results are shown in Table 2.

2.1.3 Rice Husk Ash

Rice Husk Ash was used as the binding material for the mud brick. Rice Husk was collected from a paddy field in Pala and then it was fired to make ash. About 1 kg of RHA passing through 425 μ was made.

2.2 *Methodology*

2.2.1 Sample Preparation

The soil collected was dried and crushed manually. Oyster shell was dried in the furnace at 550 °C for 60 min and then crushed manually. Varying proportions of oyster shell powder by weight of dry soil are 5, 10, 15 and 20%. The sand, oyster shell and rice husk ash were weighted and mixed according to the desired ratio in dry state to procure a homogeneous mix. 2% of RHA [10] by weight of soil was used as shown in Table 3.

Table 3 Proportioning of materials

Group	% of soil	% OS	% RHA
A	98	0	2
B	93	5	2
C	88	10	2
D	83	15	2
E	78	20	2

Fig. 1 Preparation of bricks

Water (OMC) was added to this homogeneous mix and further mixed thoroughly until obtain a uniform consistency. Lubrication with oil was provided on the inner surface of the mould (190 mm × 90 mm × 90 mm) to ensure prevention of damage by demoulding. The moistened soil was compacted in three separate layers, each covering about one third of the height of the mould after being compacted using the hydraulic punching machine (Fig. 1). A total of 44 bricks were prepared, 11 of each category and tested. The bricks were demoulded from the mould of the punching machine after applying suitable compression and were kept for drying in shade at room temperature for 12 h followed by sun drying.

2.2.2 Test Methods

Tests were conducted on the samples in order to evaluate their performance such as compressive strength (7, 14, 28 days), total water absorption and efflorescence on which the durability of bricks depends. Tests were carried out as per the recommendations in Indian Standards (IS 3495 (Part 1-3): 1992).

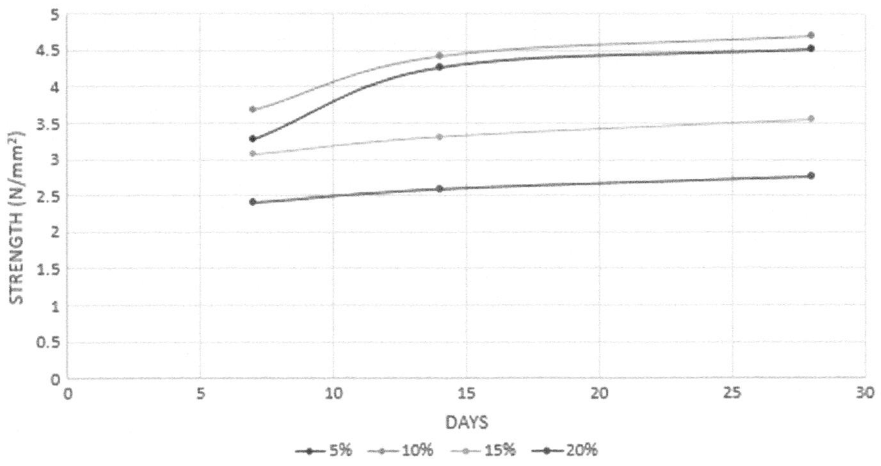

Fig. 2 Compressive strength for various percentages

3 Results and Discussions

3.1 Compressive Strength

As per IS 3495 part 1: 1992 compression strength test was performed. Nine bricks of each proportion was tested for 7, 14 and 28 days compressive strength. The dimensions of the bricks were noted initially and load was applied continuously till failure. The load failure was noted and was used to compute the compressive strength. Compressive strength of different proportions are shown in Fig. 2.

Maximum compressive strength was obtained for Group C samples. The compressive strength of mud bricks have was observed to have increased by 56.6% after adding 10% Oyster shell and 2% RHA (group C). This is because RHA contains 85–95% by weight of amorphous silica. It act as a pozzolanic reactive material. It can be used to improve the surface area of transition zone between the microscopic structures.

The compressive strength of mud brick increased with increase in the OS up to 10% OS. But at 20% OS replacement (group E) the compressive strength was found to decrease by 8%. The average 28 day compressive strength of group A, B, C, D and E are 3 N/mm^2, 4.51 N/mm^2, 4.7 N/mm^2, 3.54 N/mm^2 and 2.76 N/mm^2 respectively.

3.2 Water Absorption

As per IS 3495 part 2: 1992 water absorption test was performed. When these mud bricks were tested for water absorption: blocks got dissolved in water within 45 min.

At the point when these blocks are utilized for brick work development they are to be safeguarded from downpour.

3.3 *Efflorescence*

As per IS 3495 part 3: 1992 efflorescence test was performed. When these mud bricks were tested for efflorescence; blocks got dissolved in water.

4 Conclusions

The use of oyster shell and RHA as a partial replacement for soil in the manufacture of mud bricks is a relief on waste disposal and also a sustainable approach. The use of waste and recycled materials has many economic and environmental benefits for our society. It helps to reduce the cost of disposal in landfill sites and to maintain a sound environment.

Compressive strength of mud bricks increased with increase in the percentage of oyster shell powder up to 10% replacement of soil (4.7 N/mm^2). It is greater than minimum compressive strength requirement for sundried clay bricks.

Since these bricks are unfired they show high water absorption. When these mud bricks were tested for water absorption and efflorescence; blocks got dissolved in water within 30 min. At the point when these blocks are utilized for brick work development they are to be safeguarded from downpour. This problem can be rectified by burning the bricks.

The minimum compressive strength requirement for bricks to be used for construction purpose is 3.5 N/mm^2. Therefore these bricks can be used for construction purpose and Oyster shells prove to be a partial replacement for soil.

Acknowledgements The authors would like to thank the management and Faculties of Amal Jyothi College of Engineering, Kanjirapally and all those who have provided valuable help for this study.

References

1. Al-Fakih A et al (2018) Incorporation of waste materials in the manufacture of masonry bricks: an update review. J Build Eng, 1–60. https://doi.org/10.1016/j.jobe.2018.09.023
2. Li G et al (2015) Properties of cement-based bricks with oyster-shells ash. J Clean Prod, 279–287. http://dx.doi.org/10.1016/j.jclepro.2014.12.023
3. Moa KH et al (2018) Recycling of sea shell waste. Constr Build Mater, 751–764. https://doi.org/10.1016/j.conbuildmat.2017.12.009
4. Sasi Kumar P et al (2016) A partial replacement for coarse aggregate by sea shell and cement by lime in concrete. Imperial J Interdisc Res (IJIR) 2(5):1131–1136

5. Al-Ajmi F et al (2016) strength behavior of mud brick in building construction. Open J Civ Eng, 482–494. http://dx.doi.org/10.4236/ojce.2016.63041
6. Chiou et al (2014) Using oyster-shell foamed bricks to neutralize the acidity of recycled rainwater, 480–487. http://dx.doi.org/10.1016/j.conbuildmat.2014.04.101
7. Olivia M et al (2015) Mechanical properties of seashell concrete. Proc Eng, 760–764. https://doi.org/10.1016/j.proeng.2015.11.127, http://creativecommons.org/licenses/by-nc-nd/4.0/
8. Nguyen DH et al (2017) Durability of pervious concrete using crushed seashells. Constr Build Mater, 137–150. http://dx.doi.org/10.1016/j.conbuildmat.2016.12.219
9. Nordin N et al (2015) Effect of temperature in calcination process of seashells. Malays J Anal Sci 19(1):65–70
10. Suta J et al (2012) Effect of rice husk and rice husk ash to properties of bricks. Proc Eng, 1061–1067. https://doi.org/10.1016/j.proeng.2012.02.055

Experimental Study on Hybrid Fibre Reinforced Geopolymer Concrete

Ann Sabu and Lathi Karthi

Abstract Annual production of cement, the most important ingredient in concrete is increased by 3%. It is also a fact that the process of cement production is highly energy intensive and releases large volume of greenhouse gases like CO_2 to the atmosphere. The environmental effects associated with production of Portland cement, created an urgent need to develop an alternative binder to make concrete [1]. In this respect, Davidovits introduced geopolymer technology which provides an alternative low emission binding agent to conventional cement concrete. Many research works are being carried out to replace cement either fully or partially from concrete. Geopolymer is an innovative construction material produced by polymeric reaction of alkaline liquid with alumino-silicate materials such as fly ash, blast furnace slag, rice husk ash etc., making it a promising material to replace ordinary Portland cement as a green binder for concrete production [2]. This paper presents the results based on experimental investigation on the mechanical strength of hybrid fibre (1% of polypropylene fibre from which 10% of polypropylene fibre is replaced by glass fibre) reinforced geopolymer concrete. The addition of hybrid fibre shows an increase in strength parameters with respect to geopolymer concrete without fibres.

Keywords Portland cement concrete · Ordinary Portland cement · Geopolymer concrete · Polymeric reaction · Flyash · Blast furnace slag · Rice husk ash · Green binder · Hybrid fibre · Alkali activated

1 Introduction

Concrete is the most common, and versatile construction material used in the world. Use of concrete is increasing due to infrastructure growth and developments in the field of construction. The production of cement consumes considerable energy and at the same time contributes a large volume of carbon dioxide to the atmosphere. The calcination of $CaCO_3$ to make 1 ton of Portland cement releases 0.53 tons of

A. Sabu (✉) · L. Karthi
Department of Civil Engineering, Toc H Institute of Science and Technology, Arakkunnam, India
e-mail: annsabu95@gmail.com

K. Dasgupta et al. (eds.), *Proceedings of SECON'19*,
Lecture Notes in Civil Engineering 46,
https://doi.org/10.1007/978-3-030-26365-2_21

CO_2 into the atmosphere, and if carbon fuels used in the production of Portland cement is then another 0.45 tons of CO_2 are generated. Therefore, the production of 1 ton of PC produces approximately 1 ton of CO_2 in the atmosphere. An effort with this regard was the introduction of Geopolymer concrete (GPC). Geopolymers are inorganic polymers synthesized via a chemical reaction between a highly alkaline solution and the Si–Al minerals present in the fly ash [2, 3]. This process is termed geopolymerization.

This results in a 3-D polymeric network consisting of Si–O–Al–O bonds with the formula of $[Mn–(SiO_2)z–AlO_2]·wH_2O$ where M is an alkaline element, n is the degree of polymerization, z is a value between 1 and 32, and w is the hydration extent, which is a function of the type and amount of the alkaline solution used [4]. Common aluminosilicate material used for producing geopolymer is fly ash and slag, which are both industrial byproducts and both of these materials have much lower carbon dioxide emission factor compared to cement [5]. The alkaline liquids are usually sodium or potassium based. A combination of sodium hydroxide (NaOH) or potassium hydroxide (KOH) and sodium silicate or potassium silicate is the most common alkaline liquids used in geopolymerization [6].

Geopolymer binder is a low CO_2 cementitious material. It do not rely on the calcination of limestone that generates CO_2. In general, GPC emit less green-house gases due to their lower calcium carbonate-based raw materials and production temperature. This technology could bring down the overall carbon dioxide emission by up to 64% in comparison with the use of cement [7, 8]. The use of geopolymer technology not only helps in the reduction of CO_2 emissions by the cement industries, but also utilises the industrial wastes and by-products of alumino-silicate composition to produce added-value construction materials [7].

The use of geopolymers as binders in concrete is gaining attention in the construction industry due to their environmental benefits, and also their improvement of concrete strength. The properties of geopolymer concrete includes high early strength, low shrinkage, sulphate resistance, freeze-thaw resistance and corrosion resistance [9]. Geopolymers exhibit high chemical and thermal stability, and have excellent adhesive behaviour, mechanical strength, and long-term durability [10]. In addition, geopolymer concrete exhibits similar strength properties as in the conventional concrete. However, similar to that of plain concrete, geopolymer concrete is brittle and has low flexural strength. It is a well-known practice that fibre reinforcements are provided to overcome these shortcomings. Many research studies indicate that the incorporation of fibres in concrete improve several properties like cracking resistance, ductility and fatigue resistance, impact and wear resistance.

1.1 Fibre Reinforced Geopolymer Concrete

Fibre reinforcing of concrete is the process by which fibrous materials are added in order to strengthen the matrix resulting in a product with increased performance specifications. Fibre reinforced geopolymer concrete is relatively a new composite material where fibres are added in the matrix as micro reinforcement so as to improve the strength properties. Development of cracks due to plastic shrinkage, drying shrinkage and other reasons of changes in volume of concrete, can be minimized by introducing fibres in concrete [11]. The fibre mechanism is that the strong fibres pick up the load when the first crack occurs, thereby acting as the crack arresters. Fibres, due to their lower price and excellent characteristics they impart to the concrete, makes them popular for its increased use in concrete in the recent years. The fibres which are usually used in geopolymer concrete are metallic fibres, synthetic fibres and natural fibres. They include glass, polypropylene, steel, basalt, carbon etc. The amount of fibres added to geopolymer mix is expressed as a percentage of the total volume of the composite which is same as in normal cement concrete.

1.2 Hybrid Fibre Reinforced Geopolymer Concrete

If two or more types of fibers are combined in a common matrix to produce a composite that derives benefits from each of the individual fibers, then the composite can be termed as hybrid. It has been observed that the concept of hybridization of two different fibres included in a common cement matrix can offer more attractive engineering properties because the presence of one fiber allows the more effective utilization of the potential properties for the other fiber [12]. Fibers of different materials and geometric properties are used in construction applications which can be divided into two major categories viz., with high modulus (metallic) and low modulus (non-metallic). Each category enhances particular properties of the matrix. Generally, incorporation of metallic fiber results in flexural strength enhancement due to their higher stiffness while non-metallic fibers control the plastic shrinkage of the matrix since they have a higher aspect ratio and surface contact area. Glass fiber and polypropylene fiber can be used as high modulus and low modulus fiber reinforcement respectively in concrete.

The hybridized fibres exhibits improvement in the compressive strength due to the better mechanical bond between the fibres and binding matrix which delays micro crack formation [13]. The hybridization of fibres may play key roles in arresting cracks and there by helps in gaining high performance of concrete [14]. The major factors to be considered upon selecting a type of fibre comprises fibre volume, fibre geometry, fibre orientation and fibre distribution. However, hybridization can proffer more attractive engineering properties as the presence of one fibre permits the more efficient utilization of the potential properties of the other fibre. The mixing of hybrid fibres makes the concrete homogeneous and isotropic and therefore it is transformed from brittle to more ductile material. The applications of hybrid fibre reinforced geopolymer concrete in various civil engineering include precast concrete pipe, highway pavement, airport runway, industrial flooring, etc.

2 Experimental Programme

2.1 *Materials*

Materials used for making fibre reinforced geopolymer concrete for this study are fly ash, sodium silicate, sodium hydroxide, fine aggregate (M-sand), coarse aggregate, polypropylene fibre, glass fibre and super plasticizer.

Low calcium fly ash (ASTM class F) was used as the source material to make geopolymer concrete in the laboratory. Class F Fly Ash is produced from burning anthracite or bituminous coal that meets the applicable requirements. This class of Fly Ash has pozzolanic properties and will have a minimum silicon dioxide plus aluminium oxide plus iron oxide of 70%. Manufactured sand (M-sand) is used as fine aggregate in this study. Fine aggregate having a specific gravity of 2.71, and fineness modulus of 2.81 was used. Coarse aggregates having a fineness modulus of 3.17 and specific gravity of 2.79 were used.

In this investigation, a combination of Sodium hydroxide solution and sodium silicate solution was used as alkaline activators for geopolymerisation. Sodium hydroxide is available commercially in flakes or pellets form. For the present study, sodium hydroxide flakes with 98% purity were used for the preparation of alkaline solution. Sodium silicate is available commercially in solution form and hence it can be used as such. The chemical composition of sodium silicate is: Na_2O −14.7%, SiO_2 −29.4% and water −55.9% by mass. Water conforming to the requirements of water for concreting was used throughout.

In this work alkali resistant glass fibers of 12 mm length and 0.014 mm nominal diameter having a tensile strength of 1700 MPa and polypropylene fibre of 12 mm length and 0.014 mm nominal diameter having a tensile strength of 400–600 MPa were used. MasterGlenium SKY 8233 is used as superplasticizers.

2.2 *Mix Design*

The mix designation adopted in this study is as follows:

GC-Geopolymer concrete.

HGC-1% PP fibre and glass fibre in 10% of PP fibre.

The calculations were based on the density of concrete as per the design given by Lloyd and Rangan [15]. The density of geopolymer concrete was assumed as 2400 kg/m^3, the combined total volume occupied by the coarse and fine aggregate was 76%. The alkaline liquid to binder ratio was taken as 0.4 target strength of 30 MPa was fixed considering as a regular strength concrete. Fibre proportion is calculated by multiplying fibre density and dosage of fibre (Table 1).

Table 1 Mix proportions

ID mix	Fly ash (kg/m^3)	Fine aggregate (kg/m^3)	Coarse aggregate (kg/m^3)	NaOH (kg/m^3)	Na_2SiO_3 (kg/m^3)	Polypropylene fibre (kg/m^3)	Glass fibre (kg/m^3)
GC	412	601.92	1222	47	117	–	–
HGC	412	601.92	1222	47	117	8.28	0.92

2.3 *Preparation of Geopolymer Concrete*

To prepare 8 molarity concentration of sodium hydroxide solution 320 g of sodium hydroxide pellets was dissolved in water and made up to one litre. The sodium hydroxide solution was prepared 24 h prior to use because after dissolving pellets of NaOH in water the temperature of solution goes up. Therefore, it is required to cool it at room temperature. The sodium hydroxide solution thus prepared was mixed together with sodium silicate solution to get desired alkaline solution. Then the other ingredients used for preparing geopolymer concrete are flyash, fine and coarse aggregate and were dry mixed for about 2 min. For fibre reinforced geopolymer concrete, hybrid fibres (polypropylene + glass fibre) in relevant proportion is added and mixing was continued for 2 more minutes. After this the alkaline solution and superplasticizers were added to the dry materials and mixed for 4 to 5 min. After mixing slump of fresh concrete were determined. After that the concrete was placed in steel mould by giving proper compaction. Cubes were oven cured (60 °C for 8 h) after demoulding and after 8 h they were allowed to cure in the room temperature till the required day of testing. The specimens were tested at 7 and 28 days from the day of casting (Figs. 1 and 2).

2.4 *Results and Discussions*

The compressive strength test was conducted as per IS 516-2004. Concrete cubes are tested in compression testing machine of 200 tonne capacity. The load was increased continuously at a rate of approximately 140 $kg/cm^2/min$. Compressive strength test is a mechanical test measuring the maximum compressive load a material can sustain before failure. The split tensile strength test was conducted as per IS 516-2004. The split tensile strength was carried out after 7th and 28th day curing by placing the cylindrical specimen horizontally between the loading surfaces compression testing machine of 200 tonne capacity. The load was applied until cylinder failure along vertical diameter.

The compressive strength result obtained in this study on 7th day and 28th days are 35.65 and 38.65 MPa for geopolymer concrete and 37.72 and 39.90 MPa for hybrid fibre reinforced geopolymer concrete. The split tensile strength result obtained in

Fig. 1 Compressive strength test on fiber reinforced GPC

Fig. 2 Split tensile strength test on fiber reinforced GPC

Table 2 Comparison of compressive strength available in literatures on GPC and HGC

Author	Compressive strength (MPa)			
	Geopolymer concrete		Hybrid fibre reinforced GPC	
	7th day	28th day	7th day	28th day
Ann, Lathi	35.65	38.65	37.72	39.90
Aswani and Karthi [16]	36.08	38.47	37.61	39.73
Eliza and Vanreyk [17]	33.44	38.85	37.63	39.93

Table 3 Comparison of split tensile strength available in literatures on GPC and HGC

Author	Split tensile strength (MPa)			
	Geopolymer concrete		Hybrid fibre reinforced GPC	
	7th day	28th day	7th day	28th day
Ann, Lathi	3.31	3.45	3.69	3.73
Aswani and Karthi [16]	3.36	3.47	3.52	3.61
Eliza and Vanreyk [17]	3.09	3.30	3.42	3.56

this study on 7th day and 28th days are 3.31 and 3.45 MPa for geopolymer concrete and 3.69 and 3.73 MPa for hybrid fibre reinforced geopolymer concrete.

The above results are being compared with previous works done, available in literatures. The comparison of results show that the compressive strength and split tensile strength are comparable (Tables 2 and 3).

3 Conclusions

In this study, an experimental investigation on the strength parameters of geopolymer concrete and hybrid fibre reinforced GPC was conducted. The following conclusions were made,

- The addition of hybrid fibre in GPC shows an increase in strength parameters with respect to GPC without fibres.
- The average increase in compressive strength of oven cured GPC from 7 days to 28 days is 7.7% and that of hybrid fiber reinforced GPC is 5.46%. The average increase in split tensile strength of oven cured GPC samples with and without fibre at 7 days is 11.4% and at 28 days is 8.11%. This indicates that GPC will gain the target strength faster provided conclusive atmosphere is given to augment the polymerization process.

References

1. Rajamane NP, Nataraja MC, Lakshmanan N (2016) An introduction to geopolymer concrete. Ind Concr J (Research Gate publication)
2. Davidovits J, Comrie DC, Paterson JH, Ritcey DJ (1990) Geopolymeric concretes for environmental protection. Concr Int: Des Constr 7:1230–1240
3. Patil S, Patil A (2015) Properties of polypropylene fiber reinforced geopolymer concrete. Int J Curr Eng Technol 5:2909–2912
4. Aswani E, Karthi L (2017) A literature review on fiber reinforced geopolymer concrete. Int J Sci Eng Res 8(2):408–411
5. Ganesan N, Deepa Raj S, Abraham R, Sasi D (2014) Stress–strain behaviour of confined Geopolymer concrete. Constr Build Mater 73:326–331
6. Krishnan L, Karthikeyan S, Nathiya S, Suganya K (2014) Geopolymer concrete, an eco-friendly construction material. Int J Res Eng Technol 3:164–167
7. Malhotra VM (2002) Introduction: sustainable development and concrete technology. ACI Concr Int 7:2422
8. McLellan BC, Williams RP, Lay J, Van Riessen A, Corder GD (2011) Costs and carbon emissions for geopolymer pastes in comparison to ordinary Portland cement. J Cleaner Prod 19:1080–1090
9. Venugopal V, Arthanareswaran R (2015) Experimental Study on strength properties of polypropylene fibre reinforced geopolymer concrete. Int J Innov Res Sci Eng Technol 4:32–40
10. Alomayri T (2017) Effect of glass microfibre addition on the mechanical performances of fly ash-based geopolymer composites. J Asian Ceram Soc 5:334–340
11. Vijai K, Kumuthaa R, Vishnuram BG (2012) Properties of glass fibre reinforced geopolymer concrete composites. Asian J Civ Eng (Build Hous) 13(4):511–520
12. Bentur A, Mindess S (1990) Fiber reinforced cementitious composites. Elsevier Applied Science, London
13. Dawood ET, Ramli M (2011) Contribution of hybrid fibres on the properties of high strength concrete having high workability. Proc Eng 14:814–820
14. Subbiah Ilamvazhuthi S, Gopalakrishna GVT (2013) Performance of geopolymer concrete with polypropylene fibers. Int J Innovations Eng Technol (IJIET), 148–156
15. Lloyd NA, Rangan BV (2010) Geopolymer concrete with fly ash. In: 2nd international conference on sustainable construction materials and technologies, Italy
16. Aswani E, Karthi L (2017) Experimental investigations on fiber reinforced geopolymer concrete, Thesis report, APJ Abdul Kalam Technological University (2017)
17. Edison E, Vanreyk AJ (2018) Strength properties of fibre and hybrid fibre reinforced geopolymer concrete made with low calcium flyash. Int J Develop Res 08(05):20245–20252
18. Thaarini D, Venkatasubramani R (2015) Strength studies on geopolymer concrete using steel and polypyopylenefibres. Int J Appl Eng Res 10:14088–14092

Effect of Chemically Activated Fly Ash on Concrete

Akshay Ajith and K. Gokul Raveendran

Abstract Fly ash is most commonly used as a replacement material for cement in concrete. The fly ash based concrete mixes exhibits low initial strength development due to low pozzolanic reactivity of fly ash particles. It limits the usage of fly ash up to a certain percentage in case of cement manufacturing process and also reduces the application of high-volume fly ash structures in construction field. A number of studies had been conducted on the chemical activation of fly ash using various sulphate sources. Phosphogypsum (PG), a particular type of industrial waste material which is an another form of calcium sulphate, has not been used for this purpose. This paper presents an exploratory study on the effectiveness of phosphogypsum on activating fly ash cement system. The initial strength development of chemically activated fly ash mixes is mainly due to the formation of ettringite (AFt) and tobermorite crystals during the early age of hydration. The fly ash cement concrete mixes were prepared by replacing cement with fly ash by 15, 25, 35, 45% and these mixes were treated with 0, 6, 8, 10, 12% of phosphogypsum. The activation effect happens mainly during the first 3–7 days. The strength parameters and durability nature of chemically activated fly ash mixes were compared with that of the normal fly ash mixes. Strength tests such as compressive, split tensile and flexural strengths were conducted at age of 3, 7 and 28 days on various fly ash mixes. Durability nature of activated and non-activated fly ash mixes were studied by acid and base durability tests at age of 90 days. The obtained results indicate that the initial strength development is optimum at replacement of 35% fly ash and 8% PG. Based on the test results the chemically activated fly ash mix shows improved strength and durability nature compared to normal fly ash mixes.

Keywords Chemical activation method · Durability · Ettringite formation · Fly ash · Initial strength development · Phosphogypsum · Tobermorite

A. Ajith (✉)
Vidya Academy of Science and Technology, Thrissur, India
e-mail: akshayajithaj@gmail.com

K. Gokul Raveendran
Department of Civil Engineering, Vidya Academy of Science and Technology, Thrissur, India
e-mail: gokul.r.k@vidyaacademy.ac.in

K. Dasgupta et al. (eds.), *Proceedings of SECON'19*,
Lecture Notes in Civil Engineering 46,
https://doi.org/10.1007/978-3-030-26365-2_22

1 Introduction

There is a growing interest in the development of new cementitious binders which enhance optimal utilization of industrial by-products such as fly ash, blast furnace slag, metakaolin, etc. Among all the industrial by-products, fly ash predominates as an alternative building material for construction activities. Fly ash is the by-product obtained from the combustion of pulverized coal and is collected by mechanical and electrostatic separators from the fuel gases of power plants where coal is used as a fuel. Traditionally, fly ash used in structural concrete is limited to 25–35% of total cement content. Any further increase in the amount of fly ash usage, especially in high volumes reduces the early strength of concrete significantly. Decrease in early strength of fly ash is attributed to the slow pozzolanic reactivity between fly ash and calcium hydroxide.

The early strength behaviour of concrete is a significant parameter, with the present-day emphasis on speedier construction. This invariably defines the minimum safe time for striking of forms and removal of precast units from moulds which facilitates an effective utilisation of the moulds in the precast industry. The transfer or application of prestress in pre-tensioning or post-tensioning also require a minimum compressive strength and the early age behaviour assumes considerable importance. Similarly, the very early age strength behaviour, within the first few hours, may also be significance in slip forming, ready mixed and pumped concrete application. Different approaches are used to accelerate the pozzolanic reaction of fly ash and therefore increase the early strength of the concrete containing fly ash. These approaches include (i) mechanical treatment (grinding), (ii) accelerated curing and autoclaving, and (iii) chemical activating. A number of studies had been carried out on the activation of fly ash using chemical activators. These studies involved using different activating methods including alkali activation and sulphate activation method. The sulphate activation method is based on the ability of a sulphate compound to react with aluminium oxide in the glass phase of fly ash to produce ettringite (AFt) that contributes to the strength at early ages. The strength parameters of fly ash mix is also improved by the formation of tobermorite crystals due to the reaction of silica content in fly ash with the sulphate source. In this study, a chemical known as Phosphogypsum (PG) used to provide a suitable atmosphere for the occurrence of sulphate activation process for the fly ash particles. Phosphogypsum is a by-product generated from the phosphorus fertilizer industry (Phosphoric acid production). It is chemically denoted as $CaSO_4{\cdot}2H_2O$. This paper presents an exploratory study on the activation of fly ash cement systems using phosphogypsum. A comparison is then made between the chemically activated fly ash mixes and normal fly ash mixes regarding their strength parameters and durability properties.

Table 1 Chemical composition of Class F fly ash

Compounds	Percentage
SiO_2	55
Al_2O_3	26
Fe_2O_3	7
CaO	9
MgO	2
SO_3	1

2 Experimental Details

2.1 Materials

2.1.1 Cement

The cementitious materials used were Ordinary Portland cement (53-grade). It conformed to the requirements of Indian Standard Specifications IS 12269-1987. Cement has specific gravity 3.15 and initial setting time 65 min.

2.1.2 Fly Ash

Class F fly ash is used for the experimental purpose. The fly ash has a specific gravity of 2.2 and fineness of 4%. Fly ash for use in Portland cement concrete shall conform to the requirements of ASTM C 618, standard specification for fly ash. Chemical composition of fly ash is given in Table 1.

2.1.3 Activator

A material known as Phosphogypsum (PG) used as an activator for the experimental purpose. It is chemically denoted as $CaSO_4 \cdot 2H_2O$. Phosphogypsum was tested according to IS: 12679-1989 and found satisfy the requirement. Phosphogypsum was collected from FACT (The Fertilizers and chemicals Travancore limited) in Ambalamugal, Ernakulum. The physical properties of the phosphogypsum is given in Table 2.

The chemical composition of Phosphogypsum is given in Table 3.

Table 2 Physical properties of phosphogypsum

Colour	Pale yellow
Particle size	Majority of the particles about 50–75% are finer than 0.075 mm
Dry bulk density	1470–1670 kg/m^3
Specific gravity	2.3–2.6
Moisture content	18–30%

Table 3 Chemical composition of phosphogypsum

Chemical constituents	Percentage (%)
$CaSO_4$	95.32
F	0.028
Fe_2O_3	0.260
K_2O	2.82
Na_2O	0.0016
P_2O_5	0.320
Total P_2O_5	1.04

2.1.4 Coarse Aggregate

Machine crushed 12.5 mm nominal size angular granite metal from local source confirming to IS 383:1970 is used as coarse aggregate. It is free from impurities such as dust, clay particles and organic matter, etc. The coarse aggregate has specific gravity 2.72.

2.1.5 Fine Aggregate

The locally available sand conforming to IS 383:1970 is used as fine aggregate in the present investigation. The sand is free from clay matter, silt and organic impurities. The sand has specific gravity 2.68 in accordance with IS 2386-1963, and fineness modulus 2.6.

2.1.6 Reinforcement Bar

Thermo mechanically treated (TMT) high yield strength deformed (HYSD) bars of characteristic tensile strength of 500 Mpa was used in this project. The bars used were manufactured by Kairali TMT. 8 mm diameter bars are used as main reinforcement and hanging bars. 6 mm diameter bars are used as stirrups for the construction of reinforced cement concrete beams.

2.1.7 Water

Ordinary potable water was used for mixing and curing operations.

2.2 Mixture Proportions

In this work, control mix was designed per Indian Standard specifications IS: 10262-1982 to have 28 day compressive strength of 32.55 MPa. The mix proportions of various mixes were prepared according to the percentage replacement of cement by fly ash. The mixes were prepared by replacing cement with fly ash by 15, 25, 35, 45% and these mixes were treated with 0, 6, 8, 10, 12% of phosphogypsum for the chemical activation purpose. In doing so, water to cementitious materials ratio was kept same to investigate the effect of chemically activated fly ash concrete.

2.3 Preparation and Casting of Test Specimens

Concrete cubes, 150 mm in size were cast for compressive strength and durability tests, 150 × 300 mm cylinders for splitting tensile strength and 3200 × 150 × 200 mm beams for the flexural strength. Beams were provided by 3–8 mm diameter bars as main reinforcement. All specimens were prepared in accordance with Indian Standard Specifications IS: 516-1959. After casting, test specimens were left at room temperature. Specimens were demoulded after 24 h and were cured in water until the time of the test.

3 Result and Discussion

3.1 Effect of Phosphogypsum on Hydration Products

In the case of normal fly ash concrete the strength formation depended mainly on pozzolana reaction between the fly ash particles. But this reaction progress very slowly, so the early time of the fly ash concrete is very slowly, In the case of activated fly ash the pozzolana reaction [1] is hastened significantly phosphogypsum. The product xCaOA–l_2O_3 (n + x) H_2O of the pozzolana reaction is substituted by AFt also. The SEM images activated fly ash concrete (Fig. 1a, b) indicate that the smooth surface of fly ash balls are eroded by Ca $(OH)_2$, but there are just a few hydration products surrounding the fly ash particles. Figure 1a, b indicates that a large quantity of hydrated products is propagated in the 28-day fly ash—Phosphogypsum binder, needle like crystals known as Ettringite have covered the surface of fly ash's spherical

Fig. 1 **a** SEM image of the normal fly ash particles surface. **b** SEM image of chemically activated fly ash particles surface

particles, the hydration process of activated fly ash concrete is much further than the normal concrete binder, so the strength of the former is much higher than the normal case. Further development in the strength parametric is due to the formation of tobermorite crystals. Ettringite is flaky crystals, so the frictional force between fly ash particle increases due to the roughness created on the surface and also contribute to the strength formation. The strength characteristics of the crystals formed is measured in terms of Mohr's scale. The SEM images show that some needle like AFt crystals have been formed, they can the conjunctures of the different solid particles in concrete.

Most of the AFt crystal formation was happened in the initial days after concreting. Figure 2a, b represents the microscopic pictures of tobermorite crystals that formed within the mix of fly ash particles due to the reaction of silica content present in the fly ash with the phosphogypsum. The activated fly ash is a kind of poriferous

Fig. 2 SEM image of the tobermorite crystal formation on the surface of fly ash particles

material, the hydration products make the particles integrate with each other, the formation of AFt always follows a slight expansivity, it can fill the small pores in the mix. So, strength of chemically activated fly ash mixes is much bigger than the normal fly ash mixes.

3.2 Compressive Strength

Compressive strength of concrete mixtures was determined at the ages of 3, 7, and 28 days. At 28 days, the control mix achieved compressive strength of 32.55 N/mm^2. The results are given in Fig. 3a–d.

The results shows that the initial strength of various fly ash mixes was improved significantly by the chemical activation process. The effectiveness of activation process is increases up to 8% addition of PG in each case, on further addition of PG beyond 8% the mixes shows a slight decrease in the strength characteristics. At the end of 3 days, on addition of 8% PG the initial strength of activated fly ash mixtures F15 (fly ash 15%), F25 (fly ash 25%), F35 (fly ash 35%), and F45 (fly ash 45%) was improved by 37.5%, 40%, 76%, and 45.5% respectively compared to the corresponding normal fly ash mixes. While at the end of 7 days the increase in strength development is 40.2%, 50.9%, 41%, and 40% respectively on the same case. The strength development was most predominant in first 3–7 days compared to the 28th day strength. The effectiveness of activation process is due to nature of change occurred in the process of hydration of fly ash particles and also due to it's improved pozzolanic reacting nature.

3.3 Tensile Strength

The splitting tensile strength of the mixtures was determined at the age of 28. Results are given in Fig. 4a–d. The variation in splitting tensile strength with fly ash and PG content was similar to that observed in case of compressive strength. At 28 days, splitting tensile strength control mixture was 2.97 N/mm^2. The maximum tensile strength is obtained for the activation 35% fly ash by 8% PG. The maximum tensile strength obtained was 3.87 N/mm^2.

The tensile strength of chemically activated fly ash mixes F15, F25, F35, and F45 was improved by 26.03%, 25.52%, 38.21%, and 39.53% respectively compared to the corresponding normal fly ash mixes. The nature of increase in tensile strength by activation process of fly ash was improved up to 8% addition of the activator. On any further addition in the PG content beyond 8% for the activation purpose resulted in the drastic reduction in tensile strength.

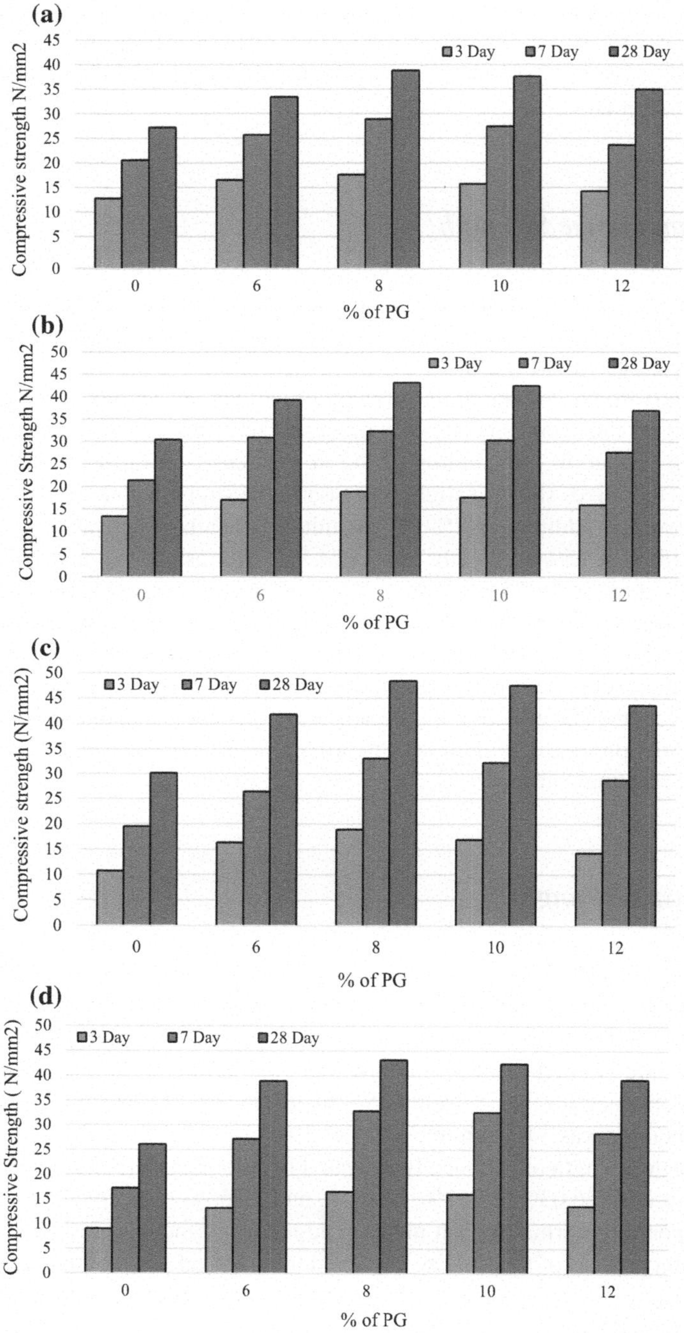

Fig. 3 **a** Compressive strength of F15 mixes. **b** Compressive strength of F25 mixes. **c** Compressive strength of F35 mixes. **d** Compressive strength of F45 mixes

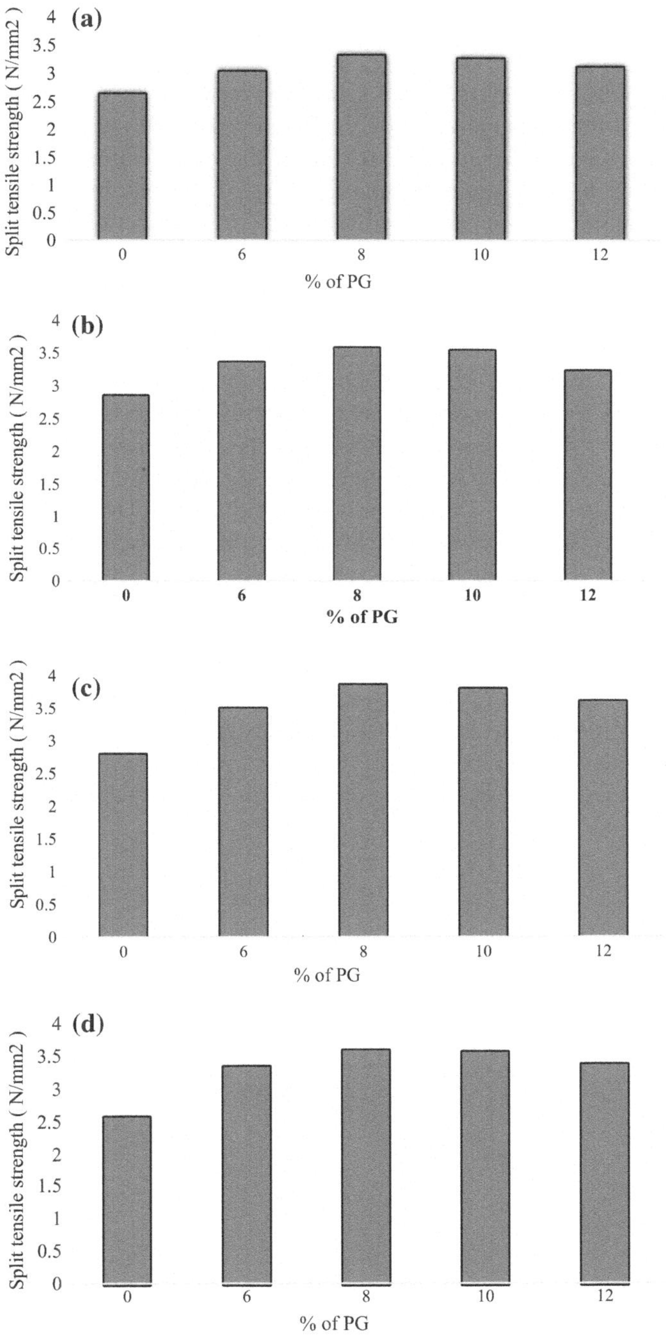

Fig. 4 **a** Split tensile strength of F15 mixes. **b** Split tensile strength of F25 mixes. **c** Split tensile strength of F35 mixes. **d** Split tensile strength of F45 mixes

3.4 Flexural Strength

Flexural strengths of the of the control mix, F35, and F35 mix activated by using 8% PG was determined at the ages of 7 and 28 days. The test specimen has a size of 3200 × 150 × 200 mm. Beams were provided with a main reinforcement of 3 no. of 8 mm bars and 2 no. of 8 mm bars as hanging bars. Two-point loading method was adopted for the flexural tests. The load versus deformation curve of test specimens for the 7th day test is given in Fig. 5.

Figure 5 shows load versus deflection behaviour of the beams F35 (fly ash 35%) and F35 + PG8 (fly ash 35 and 8% PG as activator) on 7th day flexural test. The normal fly ash beam shows slow strength development nature compared to that of the activated fly ash beam. The first crack on normal fly ash beam (F35) was occurred at a load of 0.55 ton and on chemically activated beam (F35 + PG8) at a load of 0.75 ton. Before the occurrence first flexural cracks, both beams displayed a steep linear elastic behaviour. The F35 beam exceeds the deflection failure criteria at a load of 1.15 ton and F35 + PG8 beam at a load of 1.7 ton. On further addition of load beyond these loads, the width of flexural cracks was increased. The failure of the beams occurred when the flexural cracks penetrated to the compression zone of the beam near the loading plate prior to yielding of longitudinal reinforcement. The width of the crack was 4 mm at the failure of F35 beam where in the case of F35 + PG8 beam it was 3 mm. The beam with activated fly ash shows highly improved flexural strength nature compared to the normal fly ash beams. The F35 + PG8 beam attains a strength which is close to that of the design strength of the control beam at the end of 7 day itself. From the test results, it understands that activated fly ash beam takes an additional ultimate flexural load of 37.93% than the normal fly ash beams.

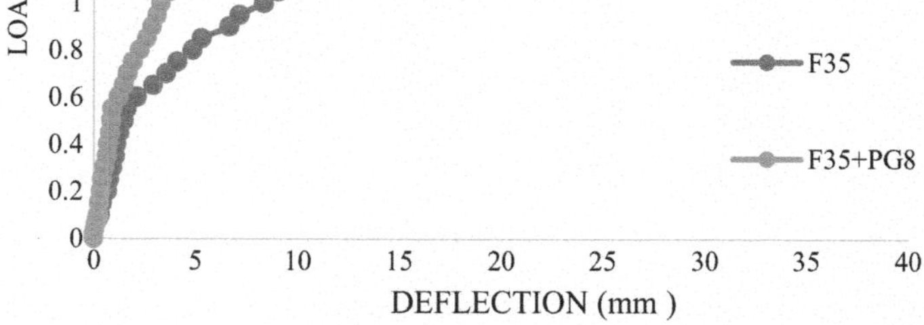

Fig. 5 Load versus Deflection graph of test specimens on 7th day flexural test

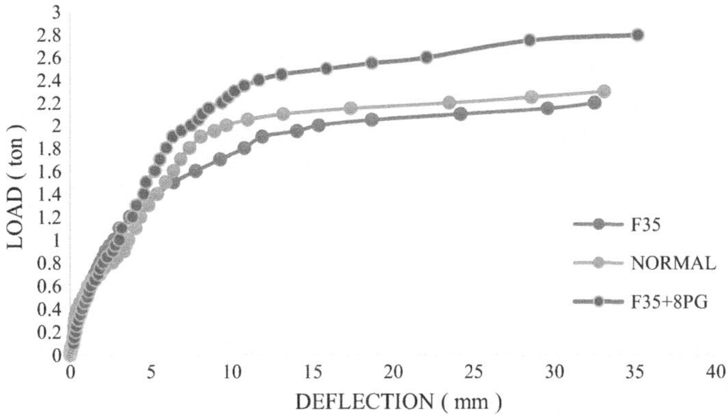

Fig. 6 Load versus Deflection graph of test specimens on 28th day flexural test

Figure 6 shows load versus deflection graph of 28th day flexural test of the Normal, F35, and F35 + PG8 beam specimens. All the beams exhibit linear elastic behaviour until the formation of flexural cracks. The chemically activated beam shows much improved flexural strength behaviour than the normal concrete beam and normal fly ash beam. The ultimate flexural strength F35 + PG8 beam was 27.27% higher than the flexural strength normal fly ash beam and 21.73% than that of the normal concrete beam. Flexural cracks were initiated in normal fly ash beam earlier than in the chemically activated fly ash beams. No shear cracks were developed on all beams until load of failure of the beams. Crack width in activated fly ash beam was smaller than that of normal fly ash beams.

3.5 Setting Time

The setting time of the fly ash concrete mixes was found out by adding 35% of fly ash with different proportions of PG. Consistency of fly ash—cement paste was estimated before testing the setting time. Setting time result is presented in Fig. 7. On addition of PG to the fly ash-concrete mix, results in the reduction of setting time of the paste. It happens due to the increase in the rate of reaction between the fly ash particles and CaOH. The chemical activation of fly ash using the addition of PG results in an improved pozzolanic reacting nature of fly ash particles and reduction in it's setting time to attain sufficient early strength. The setting time of fly ash (35%)—cement reduces up to the addition of 10% PG. On any further addition of PG beyond 10% results in a slight increase in the setting time of the mix.

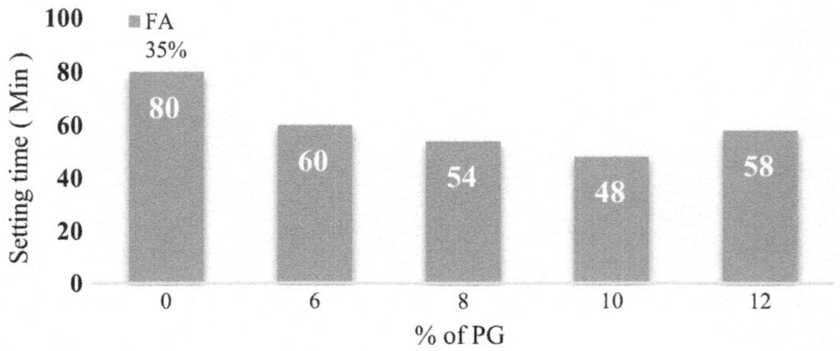

Fig. 7 Initial setting time of FA 35 concrete mixes

3.6 Durability Test

The durability nature of the concrete was studied by (i) Acid Durability Test and (ii) Base Durability Test. The tests were conducted based on IS: 456-2000. The procedures of acid and base durability tests are mentioned below. The concrete cube specimens of various concrete mixtures of size 150 mm were cast and after 28 days of water curing, the specimens were removed from the curing tank and allowed to dry for one day. The weights of concrete cube specimen were taken. The acid attack test on concrete cube was conducted by immersing the cubes in the acid water for 90 days after 28 days of curing. Sulphuric acid (H_2SO_4) with pH of about 2 at 5% weight of water was added to water in which the concrete cubes were stored. For determine the resistance of various concrete mixtures to alkaline attack, the residual compressive strength of concrete mixtures of cubes immersed in alkaline water having 5% sodium chloride (NaCl) of by weight of water was found. The pH was maintained throughout the period of 90 days. After 90 days of immersion, the concrete cubes were taken out of acid or alkaline water. Then the specimens were tested for compressive strength. The resistance of concrete to acid attack or alkaline attack was found by the % loss of weight of specimen and the % loss of compressive strength on immersing concrete cubes in acid or alkaline water.

The experimental results of acid attack test with 5% Sulphuric acid (H_2SO_4) of M25 grade concrete with different proportions of PG and 35% Fly ash is displayed in Table 4.

Each and every fly ash concrete specimen gets affected by acid and alkali attack. The outer portion of cubes gets bust and there is a maximum reduction of 2 mm at all sides for all fly ash concrete specimen. The results about the weight loss of cubes due to the acid attack are given in Fig. 8. The results about the % reduction in compressive strength of fly ash—concrete cubes due to the acid attack is given in Fig. 9

The experimental results of alkali attack test with 5% Sodium Chloride of M25 grade concrete with different proportions of PG and 35% Fly ash is displayed in Table 5.

Table 4 Effect of acid attack on weight and compressive strength of FA 35% concrete mixes

No.	Mix designation	Weight of cubes (kg)		% Loss in weight	Comp. strength (N/mm$^{2)}$		% Loss in Comp. strength
		Before acid attack	After 90 days of acid attack		Before acid attack	After 90 days of acid attack	
1	Control concrete	8.35	8.20	1.82	32.71	30.92	5.85
2	FA35% + PG0%	8.10	7.85	3.18	30.15	27.21	10.80
3	FA35% + PG6%	8.10	7.90	2.53	41.82	38.80	7.78
4	FA35% + PG8%	8.15	8.00	1.87	48.44	45.74	5.90
5	FA35% + PG10%	8.20	8.05	1.86	47.51	45.27	4.94
6	FA35% + PG12%	8.25	8.05	2.48	43.64	40.56	7.59

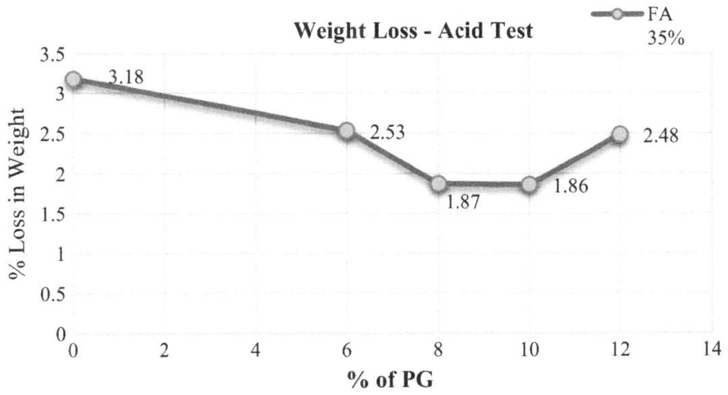

Fig. 8 % Loss in weight of fly ash concrete cubes due to acid attack

The results about the weight loss of cubes due to the alkali attack are given in Fig. 10.

Results about the % reduction in compressive strength of fly ash—concrete cubes due to the alkali attack is given in Fig. 11.

From the results it's clear that weight losses and reduction in compressive strength decrease up to 10% addition of the PG as activator. Any further addition of PG, weight loss and loss in compressive strength gradually increases in both acid durability test and base durability test.

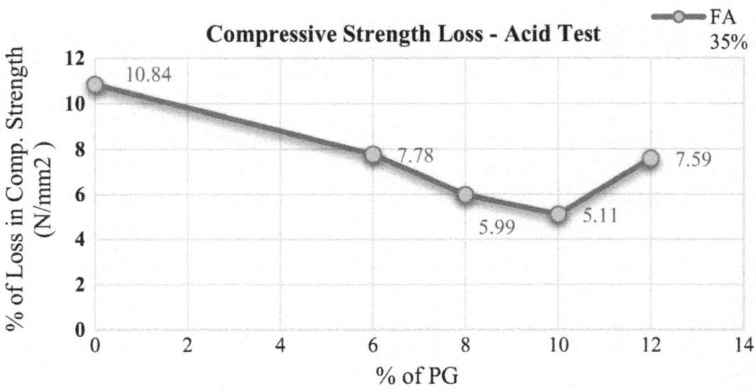

Fig. 9 % Loss in Comp. strength of fly ash concrete cubes due to acid attack

Table 5 Alkali attack on weight and compressive strength of FA 35% concrete mixes

No.	Mix designation	Weight of cubes (kg)		% Loss in weight	Comp. strength (N/mm^2)		% Loss in Comp. strength
		Before alkali attack	After 90 days of alkali attack		Before alkali attack	After 90 days of alkali attack	
1	Control concrete	8.35	8.15	2.45	32.71	30.10	8..67
2	FA35% + PG0%	8.15	7.80	4.48	30.15	26.90	12.08
3	FA35% + PG6%	8.20	7.95	3.14	41.82	37.60	11.22
4	FA35% + PG8%	8.20	8.00	2.50	48.44	44.80	8.12
5	FA35% + PG10%	8.30	8.10	2.46	47.51	44.30	7.24
6	FA35% + PG12%	8.35	8.10	3.08	43.64	39.75	9.78

4 Conclusions

The low initial strength development of fly ash becomes a major issue when it is used in high volumes for construction. Even if it has sufficient latter strength, low initial strength development causes some limitations in the usage of high-volume fly ash structures. The chemical activation process is highly effective to improve the strength characteristics of the fly ash. The addition of phosphogypsum to the fly ash accelerates the pozzolanic reaction and improves the early strength development nature due to the formation of AFt (ettringite) and tobermorite crystals. Most of the crystal formation occurred in the early age of hydration mainly before 3–7 days.

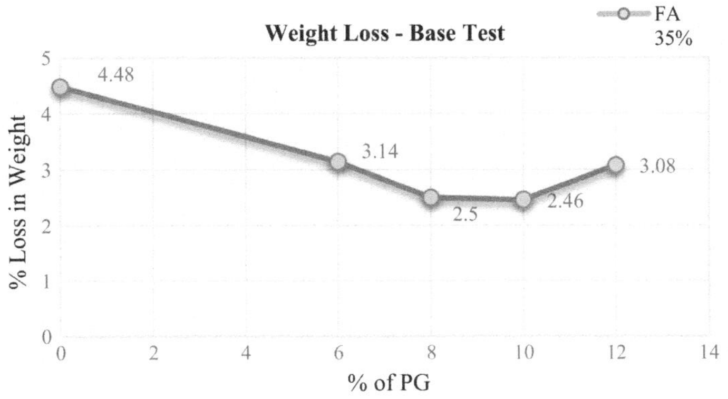

Fig. 10 % Loss in weight of fly ash concrete cubes due to alkali attack

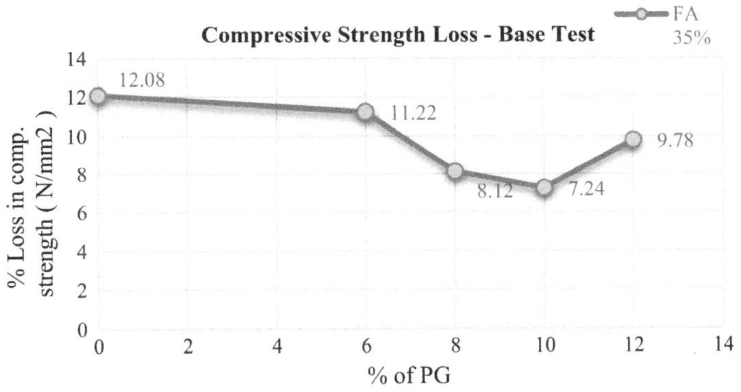

Fig. 11 % Loss in Comp. strength of fly ash concrete cubes due to alkali attack

The FA35% mix activated with 8% of PG shows higher initial strength compared to other mixes. The initial strength improvement shown by this mix at the end of 3 and 7 day was almost 76.4 and 69.7% higher than the normal fly ash mix. The chemically activated fly ash structures also exhibits good durability nature similar to that of the conventional concrete. The durability nature of fly ash mixes improves on chemical treatment by the addition of PG up to 10%. It helps to reduce the weight loss and reduction in compressive strength occurred in normal fly ash mix in both acid and alkali durability tests.

Reference

1. Shen W, Zhou M, Zhao Q (2007) Study on lime—fly ash—phosphogypsum binder. Constr Build Mater 21:1480–1485 (ScienceDirect)

Development of SCC Mix Using Jute and Coir as Additives

H. Neha, Ramees Hassan, Sonia Shaji, P. V. Sreelakshmi and C. A. Abin Thomas

Abstract Coir fibres and jute fibres are natural materials that are abundantly available in tropical regions of our country. Waste generated by the industrial and agricultural processes has created disposal and management problems, which causes serious challenges for environmental conservation. A considerable amount of coir fibres and jute fibres remain in the environment as waste, so the utilization of these materials for construction is an important step to improve sustainability and eco-friendly construction. The current study deals with the addition of coir and jute fibres to concrete such as attain sustainable construction material without sacrificing the strength of concrete. To check its suitability, test for compressive strength, split tensile strength, flexure strength and water absorption test were performed. To ensure the suitability of coir fibres and jute fibres as ingredients of concrete the results obtained from above were compared with conventional concrete. Also, a comparative study of fibre reinforced concrete and self-compacting fibre reinforced concrete was carried out. The work was carried out by conducting tests on the raw materials to determine their properties and suitability as engineering material. Concrete mix designs were prepared using the IS method for an M30 grade concrete. The specimens were cast with percentage addition of 0.7 jute fibre and 0.3 coir fibre, 0.3 Jute fibre and 0.7 coir fibre and 0.5 Jute fibre and 0.5 coir fibre.

Keywords Jute fiber · Coir fiber · Self compacting concrete (SCC) · Fibre reinforced concrete (FRC)

H. Neha (✉) · R. Hassan · S. Shaji · P. V. Sreelakshmi · C. A. Abin Thomas
Department of Civil Engineering, Federal Institute of Science and Technology, Angamaly, Kochi 683577, India
e-mail: nehaph97@gmail.com

C. A. Abin Thomas
e-mail: abinabraham80@gmail.com

K. Dasgupta et al. (eds.), *Proceedings of SECON'19*,
Lecture Notes in Civil Engineering 46,
https://doi.org/10.1007/978-3-030-26365-2_23

1 Introduction

The design of a durable and low cost fibre reinforced cement concrete for building construction is a technological challenge. Natural fibres have the potential to be used as reinforcement to overcome the inherent deficiencies in cementitious materials. Use of natural fibres in a relatively brittle cement matrix has achieved considerable strength, and toughness for the composite [1].

Investigations to overcome this brittle response and limiting post-yield energy absorption of concrete has led to the development of fibre reinforced concrete using discrete fibres within the concrete mass [2]. Agricultural wastes can be used to prepare fibre reinforced polymer composites for commercial use. Although glass and other synthetic fibre-reinforced plastics possess high specific strength, their fields of application are very limited because of their inherent higher cost of production [3]. Out of the commonly used fibres, easily available low cost natural fibres are renewable source materials such as jute and coir. This study deals with the addition of coir and jute fibres as combination to concrete in order to study the suitability of material based on its strength development [4].

This study will investigate the mechanical properties like compressive strength, split tensile strength, flexural strength and water absorption test of coir and jute fibre reinforced concrete. This study will also include the development of SCC for the optimum strength mix.

2 Experimental Program

2.1 Materials

Ordinary Portland cement of 53 grade conforming to IS 12269-1987. Manufactured sand (M sand) conforming to zone II of IS 2386 (part I)-1963 were used [5]. Crushed metal of 20 mm size from a local source will be used as coarse aggregate. It is tested for specific gravity as per IS 2386 Part III [6]. There are many general advantages of coconut fibres e.g. they are resistant to fungi and rot, provide excellent insulation against temperature and sound, not easily combustible, flame-retardant, unaffected by moisture and dampness, tough and durable, easy to clean [7]. Advantages of jute fibre include good insulating, low thermal conductivity and unaffected by dampness [8]. The properties were tested as shown in Tables 1, 2, 3, 4 and 5.

Specific gravity of Fly ash is 2.71.

Table 1 Physical properties of cement

Property	Value
Specific gravity	3.03
Standard consistency	30.5%
Initial setting time	90 min
Fineness	9.67
Compressive strength of mortar cube	51.5 N/mm^2

Table 2 Physical properties of fine aggregate

Specific gravity	2.44
Water absorption	1%

Table 3 Physical properties of coarse aggregate

Specific gravity	2.7
Water absorption	0.5%
Bulk density (compacted)	1.577 kg/l
Bulk density (non compacted)	1.493 kg/l

Table 4 Physical properties of coir fiber

Property	Value
Length	20 mm
Diameter	0.24 mm
Aspect ratio (L/d)	83

2.2 Experimental Study

Initially the optimum strength obtained when jute and coir fiber are added to concrete was studied by casting concrete specimens of M30 mix in accordance with IS 10262:2019 [6]. The percentage of jute and coir added were 0.7 coir fiber and 0.3 jute fiber; 0.5 coir fiber and 0.5 jute fiber and 0.3 coir fiber and 0.7 jute fiber to the weight of total mixtures. And further, the SCC mix was developed for the optimum strength (Table 6).

Table 5 Physical properties of jute fiber

Property	Value
Length	10 mm
Diameter	0.105 mm
Aspect ratio (L/d)	95

Table 6 Final mix of fibre reinforced SCC

Specimen	Quantity
Cement	508 kg
Water	399 kg
Fine aggregate	564 kg
Coarse aggregate	438 kg
Fly ash	353 kg
Chemical admixture	5.74 l

Mix of FRC using Jute and Coir Fibre as per IS 10262:2019 [6].

Cement = 394.32 kg/m^3
Water = 197.16 kg/m^3
Fine aggregate = 601.02 kg/m^3
Coarse aggregate = 1152.04 kg/m^3
Water-cement ratio = 0.5.

3 Results and Discussions

The aim of study was to develop jute and Coir reinforced fiber concrete along with development of SCC. The effect of variation in jute and coir fiber proportions was also studied.

3.1 Results and Discussions on Fresh Properties

3.1.1 Slump Test

See Table 7.

Table 7 Slump values

Mix designation	Slump (mm)
Control mix	88
0.7 CF + 0.3 JF	83
0.5 CF + 0.5 JF	84
0.3 CF + 0.7 JF	86

Table 8 Fresh properties of SCC

SCC sample	V-Funnel test (in sec)	L-Box test	Slump flow time	Slump flow diameter (in mm)
Typical range of values	6–12	0.8–1	2–5	550–800
0%	8.5	0.84	2.5	650
0.7 CF + 0.3 JF (%)	11	0.93	3.7	560

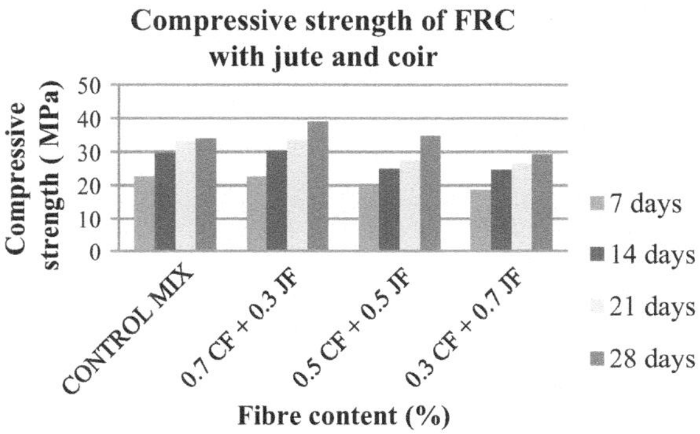

Fig. 1 Compressive strength of FRC

3.1.2 Fresh Properties of SCC

During the addition of jute and coir fiber, it is found that the slump flow decreases. Thus it affects the flow ability of concrete mix as shown in Table 7. Addition of jute and coir fiber also affects all the factors including the segregation resistance and filling ability in SCC as shown in Table 8.

3.2 Results and Discussions on Hardened Properties

3.2.1 Compressive Strength

Compressive Strength of Fibre Reinforced Concrete

Conventional concrete had a compressive strength of 33.98 MPa while the compressive strength of 0.7% of coir fibre and 0.3% of jute fibre was found to be 39.08 MPa on the 28th day of curing. It was found that the 7th day strength of 0.7% of coir fibre and 0.3% of jute fibre was less than plain concrete. Figure 1 shows the plot of compressive strength against fibre content of M30 grade concrete.

Table 9 Compressive strength of Control Mix SCC

Specimen	Compressive strength of control mix SCC			
	Sample 1 (MPa)	Sample 2 (MPa)	Sample 3 (MPa)	Average (MPa)
7 days	18.58	17.54	17.43	17.85
14 days	24.35	24.37	24.83	24.52
21 days	26.76	26.85	26.57	26.73
28 days	30.45	30.85	30.22	30.51

Table 10 Compressive strength of 0.7% Coir Fibre + 0.3% jute fibre SCC

Specimen	Compressive strength of 0.7% coir fibre + 0.3% jute fibre SCC			
	Sample 1 (MPa)	Sample 2 (MPa)	Sample 3 (MPa)	Average (MPa)
7 days	19.43	19.74	19.6	19.59
14 days	27.56	27.45	27.39	27.47
21 days	29.58	29.54	28.53	29.22
28 days	32.36	32.67	32.3	32.44

Compressive Strength of SCC

Tables 9 and 10 show the compressive strength of M30 grade of SCC concrete at the ages of 7, 14, 21 and 28 days. The compressive strength of SCC was found to be 30.51 Mpa for plain concrete and compressive strength of 0.7% of coir fibre and 0.3% of jute fibre was found to be 32.44 Mpa on the 28th day of curing.

It was found that the compressive strength of SCC was reduced compared to conventional form of compacted fibre reinforced concrete.

3.2.2 Tensile Strength

Split Tensile Strength of Fibre Reinforced Concrete

Conventional concrete had a split tensile strength of 3.29 Mpa while the split tensile strength of 0.7% of coir fibre and 0.3% of jute fibre was found to be 3.34 Mpa on the 28th day of curing. It was found that the 7 day strength of 0.7% of coir fibre and 0.3% of jute fibre was less than plain concrete. Figure 2 shows the plot of split tensile strength against fibre content of M30 grade concrete.

Split Tensile Strength of SCC

Tables 11 and 12 show the split tensile strength of M30 grade SCC concrete at the ages of 7, 14, 21 and 28 days. The split tensile strength of SCC was found to be

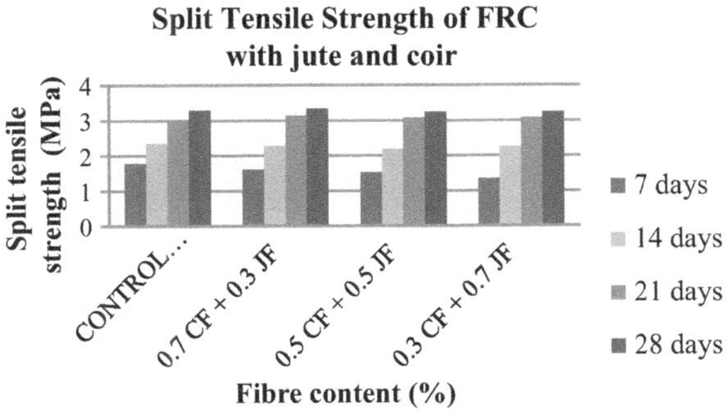

Fig. 2 Split tensile strength of FRC

Table 11 Split tensile strength of control mix SCC

Specimen	Split tensile strength of control mix SCC			
	Sample 1 (MPa)	Sample 2 (MPa)	Sample 3 (MPa)	Average (MPa)
7	1.32	1.27	1.22	1.27
14	2.23	2.28	2.33	2.28
21	2.96	2.96	2.93	2.95
28	3.21	3.22	3.23	3.22

Table 12 Split tensile strength of 0.7% coir fibre + 0.3% jute fibre SCC

Specimen	Split tensile strength of 0.7% coir fibre + 0.3% jute fibre SCC			
	Sample 1 (MPa)	Sample 2 (MPa)	Sample 3 (MPa)	Average (MPa)
7 days	1.45	1.34	1.64	1.48
14 days	2.73	2.65	2.76	2.71
21 days	3.12	3.15	3.14	3.14
28 days	3.42	3.47	3.35	3.41

3.22 Mpa for plain concrete and compressive strength of 0.7% of coir fibre and 0.3% of jute fibre was found to be 3.41 Mpa on the 28th day of curing.

It was found that the split tensile strength of SCC was reduced compared to conventional form of compacted fibre reinforced concrete.

3.2.3 Flexural Strength

Flexural Strength of Fibre Reinforced Concrete

Conventional concrete had a flexural strength of 8.68 Mpa while the split tensile strength of 0.7% of coir fibre and 0.3% of jute fibre was found to be 10.84 Mpa on the 28th day of curing. It was found that the 7 day strength of 0.7% of coir fibre and

Fig. 3 Flexural strength of FRC

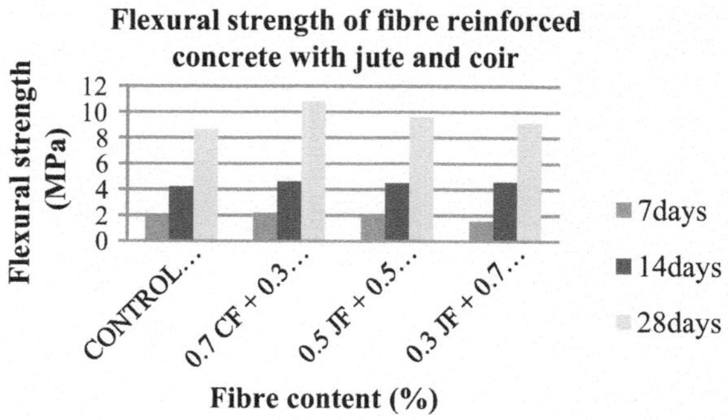

Table 13 Flexural strength of control mix SCC

Specimen	Flexural strength of control mix SCC			
	Sample 1 (MPa)	Sample 2 (MPa)	Sample 3 (MPa)	Average (MPa)
7 days	1.87	1.48	1.83	1.73
14 days	5.34	5.55	5.53	5.47
28 days	9.53	9.87	10.62	10.01

Table 14 Flexural strength of 0.7% coir fibre + 0.3% jute fibre SCC

Specimen	Flexural strength of 0.7% coir fibre + 0.3% jute fibre SCC			
	Sample 1 (MPa)	Sample 2 (MPa)	Sample 3 (MPa)	Average (MPa)
7 days	2.01	1.87	1.81	1.9
14 days	7.43	7.23	7.30	7.32
28 days	12.47	10.93	12.61	12.00

0.3% of jute fibre was more than that of plain concrete. Figure 3 shows the plot of flexural strength against fibre content of M30 grade concrete.

Flexural Strength of SCC

Tables 13 and 14 show the flexural strength of M30 grade of concrete at the age of 7, 14 and 28 days. The flexural strength of SCC was found to be 10.01 Mpa for plain concrete and compressive strength of 0.7 percentage of coir fibre and 0.3% of jute fibre was found to be 12 Mpa on the 28th day of curing. Figure 3 shows the plot of flexural strength against fibre content of M30 grade SCC concrete.

It was found that the flexural strength of SCC was greater compared to conventional form of compacted FRC.

Table 15 Water absorption test

Specimen	Water absorption
Control mix	1.25
0.7 JF + 0.3 CF	4.29
0.5 JF + 0.5 CF	5.85
0.3 CF + 0.7 JF	6.27

Table 16 Water absorption values of SCC

Specimen	Water absorption
Control mix	1.35
0.3 JF + 0.7 CF	2.25

3.2.4 Water Absorption Test

Water Absorption Test of FRC

Table 15 shows the variation in water absorption capacity of concrete with different fibre content. It was found that fibre reinforced concrete absorbs more water than normal concrete for M30 grade concrete.

Water Absorption Test of SCC

Table 16 shows the variation in water absorption capacity of concrete with different fibre content. It was found that fibre reinforced concrete absorbs more water than normal concrete for M30 grade concrete. It was found that control mix has a water absorption of 1.35 and 0.3% of jute fibre and 0.7% coir fibre has a water absorption of 2.25.

The SCC has lower water absorption capacity compared to conventional fibre reinforced concrete.

4 Conclusions

Following conclusions were made after testing the specimens:

- Addition of fibres increased the compressive strength, split tensile strength and flexural strength of M30 grade fibre reinforced and fibre reinforced SCC concrete of 0.3% JF + 0.7% CF combination. The compressive strength of 0.7% CF + 0.3% JF reinforced concrete was increased up to 15% when compared to conventional concrete while the SCC mix of the same fibre proportion had an increase of 6% strength when compared to the control SCC mix.

- The split tensile strength of 0.7% CF + 0.3% JF reinforced concrete was increased up to 9% when compared to conventional concrete while the SCC mix of the same fibre proportion had an increase of 6% when compared to the control SCC mix.
- The flexural strength of 0.7% CF + 0.3% JF reinforced concrete was increased up to 24% when compared to conventional concrete while the SCC mix of the same fibre proportion had an increase of 20% when compared to the control SCC mix.
- As the coir fibre content increases water absorption of concrete was also increased. The water absorption of fibre reinforced concrete (0.7% CF + 0.3% JF) was found to be higher than fibre reinforced SCC for the same proportion.
- Fibre reinforced concrete absorbed more water than normal concrete. The increasing water absorption was due to hydrophilic nature of coir and jute fibre and the greater interfacial area between the fibre and matrix.
- It was found that compressive strength and split tensile strength of fibre reinforced self compacting concrete was lesser than that of fibre reinforced concrete. While the flexural strength of fibre reinforced self compacting concrete was greater than that of fibre reinforced concrete.

References

1. Yan L, Su S, Chouw N (2015) Microstructure, flexural properties and durability of coir fibre reinforced concrete beams externally strengthened with flax FRP composites. Compos Part B 80:343–354
2. Ali M, Li X, Chouw N (2013) Experimental investigations on bond strength between coconut fibre and concrete. Mater Des 44:596–605
3. Ali M, Liu A, Sou H, Chouw N (2012) Mechanical and dynamic properties of coconut fibre reinforced concrete. Constr Build Mater 30:814–825
4. Girisha C, Gunti R (2012) Sisal/coconut coir natural fibres-epoxy composites: water absorption and mechanical properties. Int J Eng Innov Technol 2:166–170
5. Dinesh A, Harini S, Jasmine Jeba P, Jincy J, Javed S (2017) Experimental study on self compacting concrete. Int J Eng Sci Res Technol 25(6):42–50
6. IS 10262:2019. Concrete mix proportioning—guidelines
7. Sahaya Ruben J, Baskar G (2014) Experimental study of coir fiber as concrete reinforcement material increment based composites. Int J Eng Res Appl 4(1):128–131
8. Ramli M, Kwan WH, Abas NF (2013) Strength and durability of coconut-fiber-reinforced concrete in aggressive environments. Constr Build Mater 38:554–566

Investigating the Growth of Microbial Colonies in Cement Paste to Aid in Concrete Repair

Sk Rahaman, Jayati Ray Dutta, Arkamitra Kar and Mohna Bandyopadhyay

Abstract Synthetic polymers generally used for repair of concrete are harmful to the environment. Hence, the use of alternative repair techniques for concrete is being investigated. Calcium carbonate-precipitating bacteria can be used to plug the pores and impart a self-healing capacity to the concrete. However, these bacteria require specific conditions to survive. The environment inside concrete has a pH of 10.2–13.3 and an internal temperature that can go up to as high as 45 °C due to the exothermic nature of the hydration reactions. This study investigates the microbial growth as colony formation inside the hydrated cement paste to check the survival characteristics in these severe conditions. A direct technique of streaking to grow microorganism in a cement paste plate is used to check that, the cement paste has enough nutrients to support this growth. In this study, different species from same genus *Bacillus*, *Bacillus cereus*, *Bacillus licheniformis* bacteria are taken for comparative study along with *Bacillus subtilis*. From the current study, it was found that bacterial growth is clearly visible with 1% yeast extract. Also, from the SEM-EDS analysis calcium carbonate precipitation is evident. Further study can be performed to investigate the cell viability in cement paste as an extension of this work.

Keywords Bacterial concrete · Self-healing concrete · Repair

Sk Rahaman (✉) · A. Kar
Department of Civil Engineering, BITS-Pilani Hyderabad Campus, Hyderabad, India
e-mail: rahamansk58@gmail.com

A. Kar
e-mail: arkamitra.kar@hyderabad.bits-pilani.ac.in

J. R. Dutta
Department of Biological Science, BITS-Pilani Hyderabad Campus, Hyderabad, India
e-mail: jayati@hyderabad.bits-pilani.ac.in

M. Bandyopadhyay
Department of Biology, University of Pittsburgh, Pittsburgh, USA
e-mail: mohnab04@gmail.com

K. Dasgupta et al. (eds.), *Proceedings of SECON'19*,
Lecture Notes in Civil Engineering 46,
https://doi.org/10.1007/978-3-030-26365-2_24

1 Introduction

The modern techniques used for concrete repair and maintenance are both costly and harmful to the environment. Hence, there is a need for alternative repair and maintenance techniques like self-healing concrete. Usually, in concrete, calcium precipitating microorganisms are used for self-healing [1, 2]. Bacterial concrete is a relatively recent research domain. This concept can be used in concrete made with cementitious binder systems that can undergo automatic self-healing through the mechanism of bio-mineralization [3]. The idea of bacterial inclusion in concrete is to utilize the calcium precipitation produced by bacteria due to their metabolic activity. The presence of micro-cracks and voids in the hydrated cement paste can highly deteriorate concrete characteristics due to the ingress of moisture, chlorides, and other aggressive chemical agents. These micro-cracks can expand in the absence of due attention and result in weakening of the concrete strength and durability [4].

Based on existing literature, different types of microorganisms are used as self-healing agents in concrete. Several researchers found that *Bacillus subtilis* can be used as self-healing bacteria as they precipitate calcium carbonate as their metabolic product [5, 6]. Bio-mineralization of calcium carbonate has found application in strengthening and consolidation of sand, limestone monument repair, reduction of permeability of cement mortar, pores and cracks filling in concrete [7, 8]. In the presence of an alkaline environment rich in Ca^{2+} ions, most bacterial species can precipitate carbonates [9]. The crack healing mechanism in bacterial concrete is believed to occur due to the metabolic conversion of calcium lactate to calcium carbonate [10]. Research shows that the *Sporosarcina pasteurii* in fly ash-blended concrete escalate its compressive strength and also reduce its porosity and permeability [11]. However, there is no existing standard for the practical use of bacterial concrete. Moreover, there is no reported study regarding the utility of two or more strains of bacteria in consortia being used for self-healing purposes of concrete.

Based on the above findings, the present study aims to find out the growth of *Bacillus cereus* and *Bacillus licheniformis* along with *Bacillus subtilis* in the cementitious binder paste. Their ability to precipitate calcium carbonate is determined. Also, the growth of microorganisms based on the nutrient present in the cement paste is analyzed. The materials and methods used in this study are presented below.

2 Materials and Methods

PC conforming to ASTM C33/C33 M-18 is used in the present study to make the cement paste plate with a water-to-cement ratio (w/c) of 0.5. For the preparation of each cement paste plate, 20 gm of autoclaved cement is mixed with 10gm of autoclaved water. And also as a nitrogen source yeast extract is added in three different percentages 0.25, 0.5 and 1% of the weight of cement.

For each bacterial culture, four plates are prepared in a Laminar Air Flow (LAF) setup to counter any kind of contamination. Firstly, all the materials and equipment are sterilized by autoclaving at 120 °C for 1 h. After the autoclaving, all the materials and equipment are shifted to LAF and exposed to ultraviolet light for 15–20 min. Then, in an autoclaved plate, 20 gm of cement is measured and 10 gm of water is added and mixed properly to make a thick cement paste. In the same manner, other plates are also prepared along with varying percentages of yeast extract as mentioned. After preparing the plates they are kept overnight in the LAF to let them harden.

After hardening, the cement paste plate is streaked [12] with above mentioned three bacterial strains taken from three different full grown LB agar master plates. Each master plate containing 1×10^{11} cells per ml. uniformly streaked to all cement paste plates accordingly and kept in the incubator overnight. After incubation, the plates were removed from the incubator and observed carefully using the naked eye to check the growth of bacteria in the cement paste plate.

Additionally, using FEI Apreo scanning electron microscope (SEM) equipped with a secondary electron detector and energy dispersive x-ray spectroscopy (EDS) detector, the microstructural characteristics of the samples are studied. The elemental composition of the cement paste along with bacteria is analyzed using the EDS. The samples are dehydrated at a temperature of 100 °C for 2 h prior to SEM analysis. Following that, the samples were sputtered with 10 nm layer of silver coating using a Leica EM ACE200 sputter coater to make them electrically conductive. A working distance of 10 mm is maintained throughout the analysis. The micrographs were captured at magnifications of 2500× and an operating voltage of 20 kV. At the same time, EDS analysis is performed for each sample after recording the micrograph.

3 Results

The growth of bacteria in the cement paste plate is clearly visible on the completion of the incubation period. As mentioned earlier, bacterial growth depends on the nutrients present in the media and some physical parameters like temperature and pH. As seen from Figs. 1, 2, 3, 4, 5, 6, 7, 8, 9, 10, 11 and 12 cement paste is a suitable medium for bacteria. Although the temperature and the pH are suitable for bacterial growth, not all of the different nutrients required for bacterial growth are present in the cement paste. Hence, to attain the minimal nutrient content, yeast extract is added as a nitrogen source. From the different plates, it is evident that the growth of bacteria is negligible when there is no nitrogen source. With 1% yeast extract in cement paste the growth of bacteria was clearly visible, which proves sufficient utilization of nitrogen source by the bacteria (Figs. 4, 8 and 12).

From the SEM micrographs, it was evident that the calcium precipitation and bacterial colony formation modify the morphology of hydrated cement paste. EDS spectrum confirms the formation of precipitates through their elemental compositions. From Fig. 13 formations of the hydrated cementitious product such as calcium silicate hydrate (Spines), calcium hydroxide (hexagonal crystals <0.1 mm) and

Fig. 1 *B. subtilis* growth in cement paste

Fig. 2 *B. subtilis* growth in cement paste with 0.25% yeast extract

Fig. 3 *B. subtilis* growth in cement paste with 0.5% yeast extract

ettringite (long slender prismatic needles) are visible [13]. From Figs. 14, 15 and 16 it is visible that the precipitation of calcium carbonate [14] results from bacterial inclusions in the cementitious paste.

Fig. 4 *B. subtilis* growth in cement paste with 1% yeast extract

Fig. 5 *B. cereus* growth in cement paste

Fig. 6 *B. cereus* growth in cement paste with 0.25% yeast extract

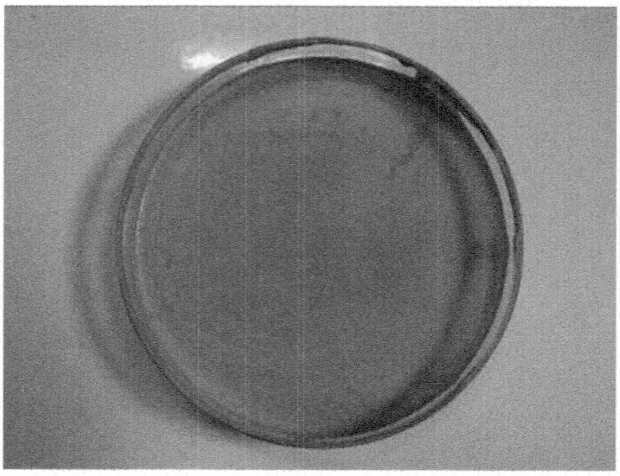

Fig. 7 *B. cereus* growth in cement paste with 0.5% yeast extract

Fig. 8 *B. cereus* growth in cement paste with 1% yeast extract

Fig. 9 *B. licheniformis* growth in cement paste

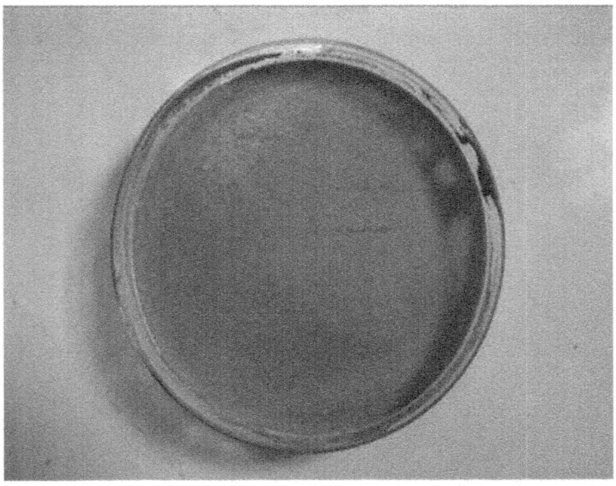

Fig. 10 *B. licheniformis* growth in cement paste with 0.25% yeast extract

Fig. 11 *B. licheniformis* growth in cement paste with 0.5% yeast extract

Fig. 12 *B. licheniformis* growth in cement paste with 1% yeast extract

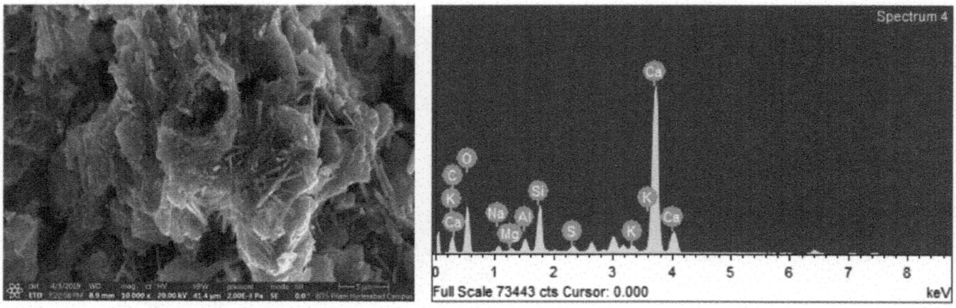

Fig. 13 SEM-EDS analysis of cement paste

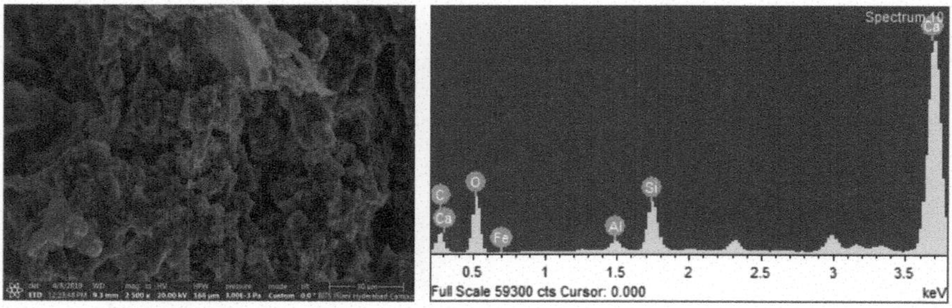

Fig. 14 SEM-EDS analysis of cement paste with *B. subtilis*

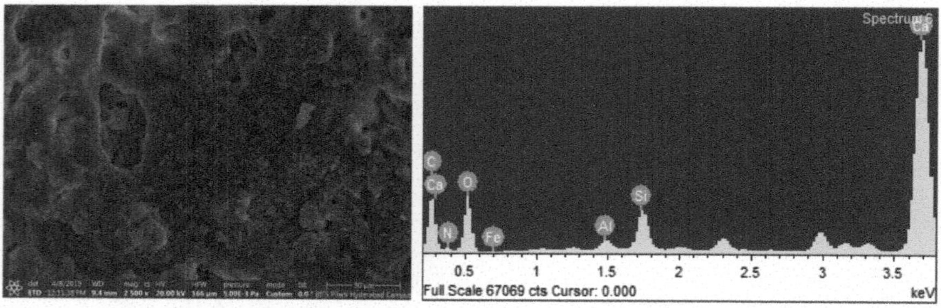

Fig. 15 SEM-EDS analysis of cement paste with *B. cereus*

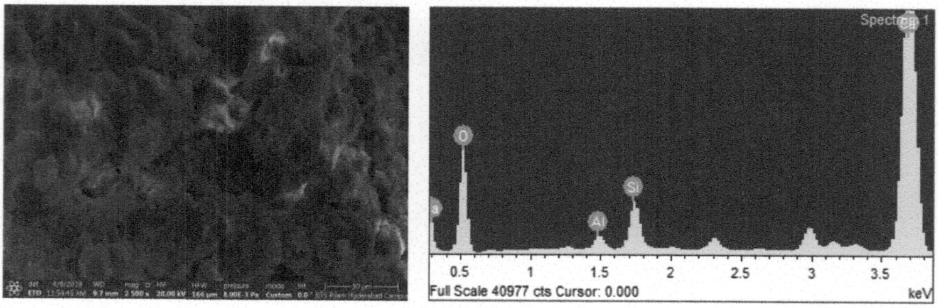

Fig. 16 SEM-EDS analysis of cement paste with *B. licheniformis*

4 Conclusions

Based on the observations from these study the following conclusions were made

- Bacterial colonies can grow inside hydrated cementitious paste with the addition of yeast extract as a nitrogen source.
- Increase in yeast concentration leads to an increase in bacterial growth inside the cementitious paste.
- The rate at which bacterial colonies can grow inside concrete is directly proportional to the amount of calcite precipitation which can consequently govern the extent of crack healing.
- As a continuation of these studies, bacterial cell viability can be checked using the spread plate technique through the determination of optimum temperature and pH for bacterial growth inside the concrete.

Acknowledgements The authors would like to show their sincere gratitude to the Council of Scientific and Industrial Research (CSIR) for providing sponsorship of this study. The authors would also like to acknowledge the central analytical laboratory facilities at BITS Pilani, Hyderabad campus for providing the necessary setup to conduct SEM-EDS analyses. The authors would also like to acknowledge Miss. J. Geethanjali, Lab technician of the Biology Department for her continuous assistance and guidance during experimental work.

References

1. Jonkers HM (2011) Bacteria-based self-healing concrete. Heron 56(1/2)
2. Wang JY, Soens H, Verstraete W, De Belie N (2014) Self-healing concrete by use of microencapsulated bacterial spores. Cem Concr Res 56:139–152
3. Başaran Bundur Z, Bae S, Kirisits MJ, Ferron RD (2017) Biomineralization in self-healing cement-based materials: investigating the temporal evolution of microbial metabolic state and material porosity. J Mater Civ Eng 29(8):04017079
4. Wiktor V, Jonkers HM (2011) Quantification of crack-healing in novel bacteria-based self-healing concrete. Cem Concr Compos 33(7):763–770
5. Vijay K, Murmu M, Deo SV (2017) Bacteria based self healing concrete—a review. Constr Build Mater 152:1008–1014
6. Rao MS, Reddy VS, Hafsa M, Veena P, Anusha P (2013) Bioengineered concrete—a sustainable self-healing construction material. Res J Eng Sci. ISSN: 2278-9472
7. Dhami NK, Reddy MS, Mukherjee A (2012) Improvement in strength properties of ash bricks by bacterial calcite. Ecol Eng 39:31–35
8. Dhami NK, Reddy SM, Mukherjee A (2012) Biofilm and microbial applications in biomineralized concrete. In: Advanced topics in biomineralization, IntechOpen
9. Stocks-Fischer S, Galinat JK, Bang SS (1999) Microbiological precipitation of $CaCO_3$. Soil Biol Biochem 31(11):1563–1571
10. Jonkers HM, Thijssen A, Muyzer G, Copuroglu O, Schlangen E (2010) Application of bacteria as self-healing agent for the development of sustainable concrete. Ecol Eng 36(2):230–235
11. Chahal N, Siddique R, Rajor A (2012) Influence of bacteria on the compressive strength, water absorption and rapid chloride permeability of fly ash concrete. Constr Build Mater 28(1):351–356

12. Maheshwari DK (2002) Practical microbiology. S. Chand Publishing, New Delhi
13. Mindess S, Young FJ, Darwin D (2003) Concrete 2nd Editio. Technical Documents
14. De Muynck W, Cox K, De Belie N, Verstraete W (2008) Bacterial carbonate precipitation as an alternative surface treatment for concrete. Constr Build Mater 22(5):875–885

Bitu Block—A Sustainable Building Block

P. V. Aparna, Kesiya Elizabeth Samuel, N. Bhagya, P. A. Abdul Majeed and C. J. Chithra

Abstract This paper describes an investigation on a new sustainable building block in cooperating waste materials namely, rice husk ash (RHA), fly ash, crushed glass, aggregates bound with bitumen as the binder. The bitublocks were prepared using four different mixes and its properties such as compressive strength, volume stability, the initial rate of suction (IRS), density, etc. were determined. It shows that all these properties are dependent on the type of aggregates and bitumen used as well as on compaction and the temperature at the time of mixing. In the UK similar experiments were done but the mixes used were different. Our results show that the properties of bitublock units are approximately similar to masonry building blocks available in India and it can be used in non-structural walls such as partition walls etc.

Keywords Bitublock · Bitumen · Binder · The building block

1 Introduction

India has the world's second largest road network after the USA. Annually an average of 4MMT (million metric tonnes) of bitumen is used in India for road construction since a large amount of waste bitumen is available in India. At the same time over 170 million tonnes of fly ash, crushed glass and 20 million tonnes of rice husk ash are been produced in India. Since the cost of disposal of these waste materials into landfill is really high and the impact of waste materials on environment is increasing daily the method needed is recycle and reuse. And in bitublock, instead of cement bitumen is used as a binder, hence eliminating the adverse effects caused by the production of cement. The masonry blocks available in India need a large amount of clay and it is prepared by temperature curing up to 20000C hence energy consumption is really high. But for bitublock only air cooling is needed and after 3 days it can be directly put into use.

P. V. Aparna (✉) · K. E. Samuel · N. Bhagya · P. A. Abdul Majeed · C. J. Chithra
Civil Engineering, Adi Shankara Institute of Engineering and Technology, Kalady, India
e-mail: aparnapvappu97@gmail.com

K. Dasgupta et al. (eds.), *Proceedings of SECON'19*,
Lecture Notes in Civil Engineering 46,
https://doi.org/10.1007/978-3-030-26365-2_25

Research in utilizing waste materials in the civil engineering field is been done already. In UK similar experiments were done by I. N. A. Thanaya, J. P. Forth but with different mixes. In this study, an innovative and sustainable topic is been investigated i.e. reuse of the waste materials using bitumen as a binder. So as to decrease the usage of aggregates and all wastes such as crushed glass and also the scrap aggregates available from the pavement is been used. Also, wastes such as fly ash and rice husk ash (RHA) is used as filler.

Crushed glass is the waste available from the glass factory. By the usage of crushed glass in building block, the usage of aggregates and clay in building blocks can be minimized by protecting the natural resource. It is proved that the usage of crushed glass as both fine and coarse aggregate improves the compressive strength.

Fly ash is the waste by-product available from the coal industry and RHA is a waste product available from rice factory. The usage of fly ash and RHA as filler in bitublock not only improves its properties but also reduce the impacts on the environment as it converts the wastes into an useful product. The properties such as compressive strength, shrinkage and the initial rate of suction of building block are tested. The objectives of this study are to evaluate the performance of bitublock and to produce the bitu block meeting the performance criteria of masonry blocks available in India.

2 Methodology

2.1 Material Used and Its Description

Bitumen:

In general, all type of bitumen can be used based on availability. In this study, the waste bitumen is been collected and it is having a penetration value of 82. So as it has a penetration value of 80/90 it is been classified as hard bitumen. The bitumen has a softening point of 49.7 °C. The ductility of bitumen obtained was above 100 and is having 0% stripping value and viscosity of 17.96 s. The range of bitumen varies from 8 to 12% for mixes using fly ash as filler. But for mixes using RHA as filler have more absorbing power hence 17–20% bitumen is been used (Fig. 1).

Aggregate type and mix:

So as to reduce the amount of bitumen used and ensure sufficient coating of bitumen waste coarse aggregates of size 14–10 mm and 10–5 mm is used and at the same time, the crushed glass of size 5–2.36 mm is also used as per the journal. This mix proportion is been taken so as to reduce the usage of aggregates. For mix 1 and 2 crushed glass is coarse and fine aggregate and flyash is the filler. Whereas in mix 3 and 4 aggregate is been used as coarse aggregate and crushed glass is used as fine aggregate and the filler is RHA. The mix consists of about 40% of coarse aggregate,

Fig. 1 Bitumen

50% of fine aggregate and 10% of filler since this grade consists of enough fine fraction and hence low compaction is required. The size of the sample chosen is 19 × 9 × 4.5 cm. This size was preferred for easy production and so as to compare it with the masonry building blocks available in India. The density of the materials used is given in Table 1 (Fig. 2; Table 2).

2.2 Experimental Details

During the initial stage one of the mixes from Table 1 is chosen and the aggregates are dry mixed and they are preheated before mixing with bitumen at a temperature of 180 °C and bitumen were preheated to a temperature of 160 °C. Then the calculated quantity of bitumen is added to the aggregate materials and mixed thoroughly until

Table 1 Density of materials used

Materials	Density (gm/cm^3)
Crushed glass	2.51
Aggregate	
12 mm	1.45
6 mm	1.466
3 mm	1.52
Fly ash	2.16
Rice husk ash	2.04

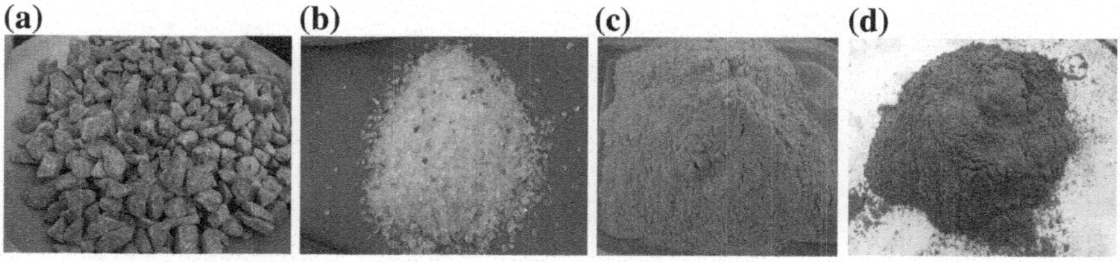

Fig. 2 Materials used **a** aggregate, **b** crushed glass, **c** fly ash, **d** rice husk ash

Table 2 Mix proportion

Mix No.	Coarse aggregate 40%		Fine aggregate 50%	Filler 10%
	5% (14–10 mm) 20% (10–5 mm)	15% (5–2.36 mm)		
1	Aggregate	Crushed glass	Crushed glass	Fly ash
2	Aggregate	Crushed glass	Crushed glass	RHA
3	Aggregate	Aggregate	Crushed glass	Fly ash
4	Aggregate	Aggregate	Crushed glass	RHA

a uniform mix is obtained. Then the loose mix is filled into the mould in layers by compacting each layer uniformly. The sample is then placed for air cooling for 3 days. To determine the properties of bitu block and to compare it with masonry blocks following tests are been conducted.

2.2.1 Compressive Strength

Compressive strength is the ability of material to carry the loads on its surface without any crack or deflection. Compressive strength formula is the load applied at the point of failure to the cross-section area of the face on which load was applied. These specimens are tested by compression testing machine. Load should be applied

Fig. 3 Compressive testing machine

gradually at the rate of 140 kg/cm^2 per min till the Specimens fails. Load at the failure divided by area of specimen gives the compressive strength of concrete (Fig. 3).

2.2.2 The Initial Rate of Suction

The IRS test is conducted by immersing the sample in 3 mm depth water for 60 s. The IRS value is obtained by calculating the weight of water absorbed by the sample and it is divided by the area in contact with water. IRS value gives us information about the effect of bitublock on the sand-cement mortar.

2.2.3 Volume Stability

For volume stability, the expansion of the four sides of the sample is been tested using Demec gauge. The volume stability of sample is been affected by water absorption from the environment and the humidity. The result gives us information about the stability of the sample affected by moisture absorption.

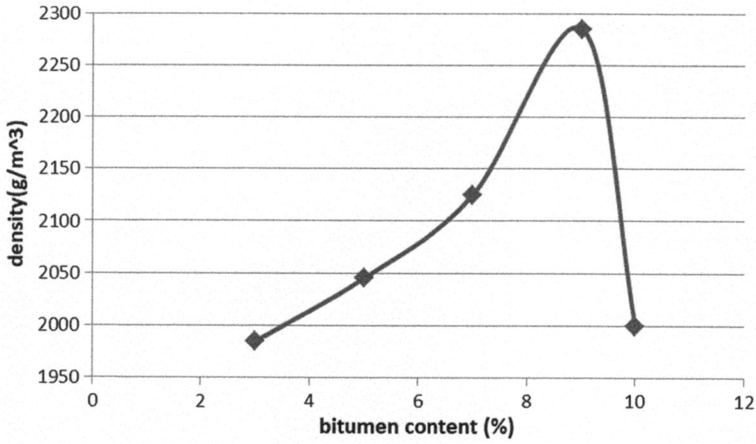

Graph 1 Density against bitumen content for mix 1

3 Results

3.1 Density

Density is the ratio of mass by volume. For mix 1 and 3 at first density increases with bitumen content up to 9%. After that, the density starts to decrease with increase in bitumen content. Hence for mixes 1 and 3 which are using fly ash as filler a 9% of bitumen is added as a binder. Similarly for mix 2 and 4 at first density varies proportionally with bitumen content up to 19% but after that, they started to behave inversely. Hence for mixes having RHA as a filler (mix 2 and 4) 19% of bitumen is used (Graphs 1, 2, 3 and 4; Tables 3, 4, 5 and 6).

3.2 Compressive Strength

For mix 1 and 3 which are having fly ash as filler, at 9% bitumen content a compressive strength of 7.07 and 4.2 MPa is obtained. For mix 2 and 4 which are having rice husk ash as filler, at 19% of bitumen content a compressive strength of 8.19 and 5 MPa is obtained (Graph 5 and Table 7). Mix 2 is having the highest compressive strength of 8.19 MPa.

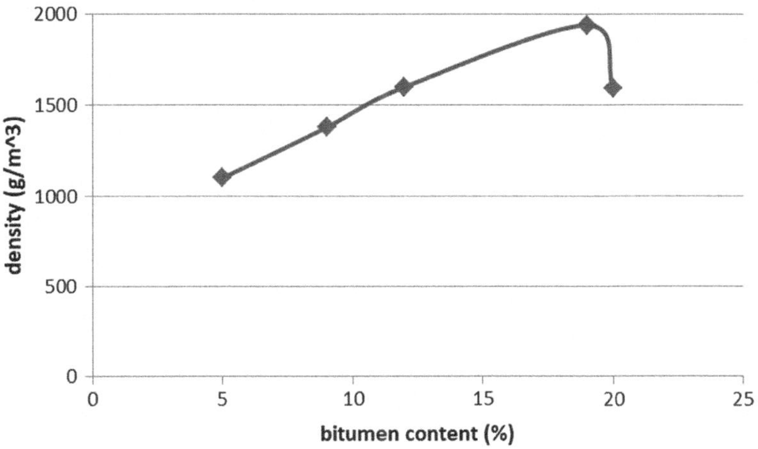

Graph 2 Density against bitumen content for mix 2

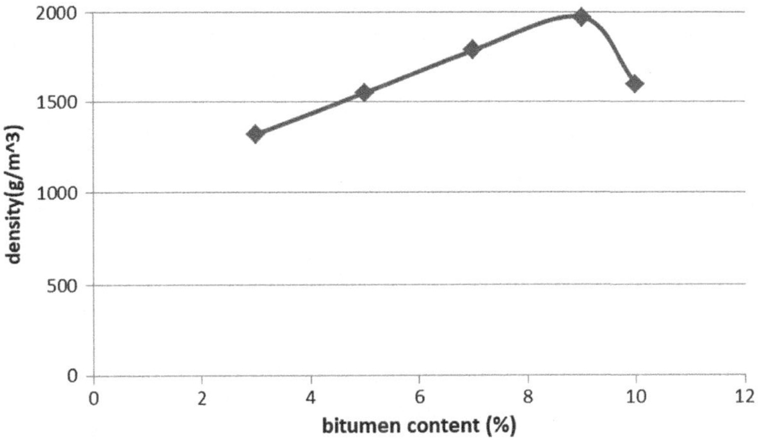

Graph 3 Density against bitumen content for mix 3

3.3 Initial Rate of Suction

The initial rate of suction of masonry bricks available in India is between 0.25 and 1.5 kg/m^2min. Lower IRS values indicates that bitumen is well coated with aggregate and makes the block impermeable. So if the bitumen content does not lie within the calculated range then there occurs a chance for the IRS value to decrease, hence bitublocks would require stiffer mortar. Since the obtained IRS value for bitublock is within the limit only adequate amount of mortar will be needed (Fig. 4; Table 8).

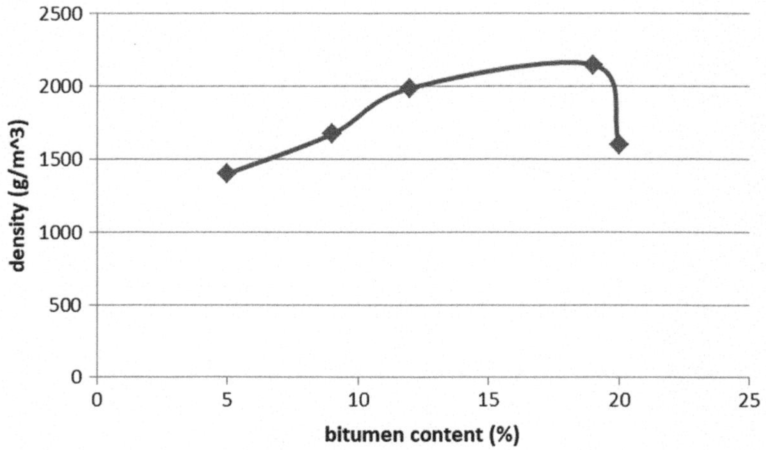

Graph 4 Density against bitumen content for mix 4

Table 3 Density of bitublocks for mix 1

Sl. No.	Bitumen content (%)	Density (g/m^3)
1	3	1985
2	5	2045
3	7	2125
4	9	2285
5	10	2000

Table 4 Density of bitublocks for mix 2

Sl. No.	Bitumen content (%)	Density (g/m^3)
1	5	1100
2	9	1380
3	12	1600
4	19	1938
5	20	1590

Table 5 Density of bitublocks for mix 3

Sl. No.	Bitumen content (%)	Density (g/m^3)
1	3	1320
2	5	1550
3	7	1789
4	9	1970
5	10	1600

Table 6 Density of bitublock for mix 4

Sl. No.	Bitumen content (%)	Density (g/m^3)
1	5	1400
2	9	1670
3	12	1985
4	19	2145
5	20	1600

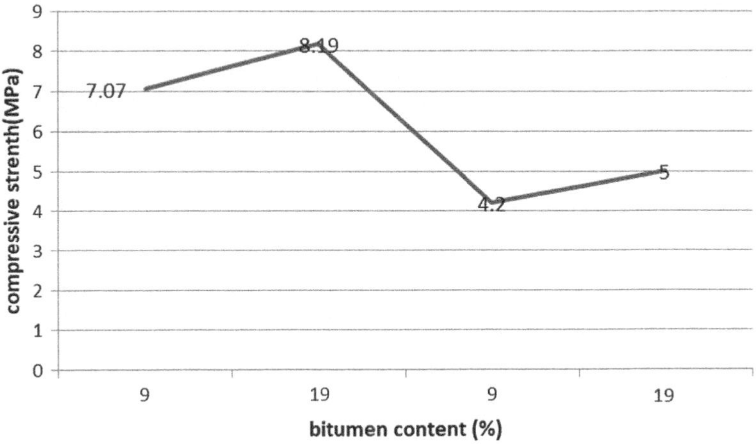

Graph 5 Compressive strength versus bitumen content

Table 7 Compressive strength of bitublock

Mix	Bitumen content (%)	Avg. compressive strength (MPa)
1	9	7.07
2	19	8.19
3	9	4.2
4	19	5

Fig. 4 Bitu block

Table 8 Initial rate of suction of bitublock

Mix	IRS (kg/m^{2}min)
1	0.52
2	0.52
3	0.779
4	0.779

4 Conclusion

The compressive strength of the existing masonry block in India is around 10 MPa. For the construction of non-structural wall such as partition wall, the obtained compressive strength of bitublock (7–8.2 MPa) is sufficient. The properties of bitublock can be improved by replacing the aggregates with steel slag. The further details respect to bitublock can be obtained by conducting tests such as creep test and volume stability test.

References

1. Thanaya INA (2010) Building blocks in-cooperating waste materials bound with bitumen. Civ Eng Dimension 12:36–43
2. Raju S, Kumar PR (2014) Effect of using glass powder in concrete. Int J Innov Res Sci Eng Technol 3(Special Issue 5)
3. Marthong C, Agarwal TP (2012) Effect of fly ash additive on concrete properties. Int J Eng Res Appl 2(4):1986–1991
4. Forth JP, Zoorob SE, Thanaya INA (2006) Development of bitumen-bound waste aggregate building blocks. Constr Mater 159(CM1):23–32
5. Habeeb GA, Mahmud HB (2010) Study on properties of rice husk ash and its use as a cement replacement material. Int J Innov Res Sci Eng Technol

Experimental Investigation on Geopolymer Masonry Units

K. R. Reema, Sethulakshmi S. Shajan, V. S. Soumya, Swetha Vinod and Vidya Jose

Abstract Masonry is the oldest method of construction and is composed of masonry units and mortar. The traditional masonry units like burnt bricks and conventional cement blocks are not considered as sustainable. Burnt brick consumes fossil fuel and top fertile soil. Hence there is a need to develop and use alternatives to traditional masonry units. Geopolymer technology is one among many alternatives in which complete elimination of cement is achieved without compromising the strength and durability. This technique makes use of supplementary cementitious materials like fly ash, Ground granulated blast furnace slag, etc. as binders [1]. The objective of this project is to study the effect of fly ash and alccofine on the physical and mechanical properties of geopolymer masonry bricks at different replacement levels of fly ash with 0, 10, 20, 30, 40% of alccofine and comparing the optimum results with fly ash brick [2, 3]. The polymerization is done with the help of alkaline activators. Sodium silicate and sodium hydroxide are the alkaline activators used. It is proposed to study the properties such as compressive strength, water absorption, density etc. and these properties are compared with that of that of fly ash bricks [4, 5].

Keywords Geopolymer · Masonry · Fly ash · Alccofine · GGBS · Compressive strength · Water absorption · Density

1 Introduction

Masonry is the oldest method of construction and is composed of masonry units and mortar. The traditional masonry units like burnt brick and regular cement blocks are not considered as sustainable. Burnt brick consumes fossil fuel and top fertile soil. Cement block needs traditional cement which produces carbon dioxide during manufacturing. Hence there is a need to develop and use alternatives to traditional masonry units.

K. R. Reema · S. S. Shajan (✉) · V. S. Soumya · S. Vinod · V. Jose
Department of Civil Engineering, Toc H Institute of Science & Technology, Arakkunam, Ernakulam, India
e-mail: sethu2111@gmail.com

K. Dasgupta et al. (eds.), *Proceedings of SECON'19*, Lecture Notes in Civil Engineering 46,
https://doi.org/10.1007/978-3-030-26365-2_26

Geopolymer technology is one among many alternatives in which complete elimination of cement is achieved without compromising the strength and durability. Geopolymer was the name given by Daidovits in 1978 to materials which are characterized by chains or networks or inorganic molecules. Geopolymers are inorganic, ceramic materials which form long chain covalent bond amorphous network. The quick reaction of the alkaline activated solution with silica alumina materials takes place and formation of a three dimensional polymeric chain and ring structure made of Si-O-Al-O bonds takes place [6].

Geopolymer concrete is made from utilization of waste materials such as fly ash and ultra-fine ground granulated blast furnace slag(GGBS) [1]. Fly ash is a coal combustion product. It is part of a set of products that makes up the most abundant waste materials worldwide. If not collected, this waste material is blown out with the flue gas in a coal fired power plant. Fly ash exists after combustion because ash adheres to coal, making up between 1 and 15% of its weight. About 90% of the ash is fly ash, while 10% is bottom ash. Fly ash is composed of tiny, airborne particles and is thus considered to be a type of particulate matter or particle pollution. The abundant availability of fly ash is creating problems in disposal operations and tremendous environmental concerns. The utilisation of fly ash will be beneficial when treated as valuable resources in the production of good quality geopolymer masonry units. Both fly ash and alccofine are processed by appropriate technology and used for concrete works in the form of geopolymer concrete. The use of this concrete helps to reduce the stock of wastes and also reduces carbon emission by reducing Portland cement demand.

2 Materials

2.1 Fly Ash

Class F fly ash is obtained from the burning of anthracite and bituminous coal. It is a coal combustion product that is composed of the fine particles of burned fuel that are driven out of coal-fired boilers together with the flue gases. Depending upon the source and composition of the coal being burned, the components of fly ash vary considerably, but all fly ash includes substantial amounts of silicon dioxide (SiO_2), aluminium oxide (Al_2O_3) and calcium oxide (CaO), the main mineral compounds in coal bearing rock strata. As the production of fly ash arises continuously and creates serious environmental pollution problems, fly ash should be reused as a raw material in new technology with good properties. The specific gravity of fly ash was determined by use of Le Chatelier's flask as per IS 2720-Part 3 (Fig. 1).

Two classes of fly ash are defined by ASTM C618: class F fly ash and class C fly ash. The chief difference between these classes is the amount of calcium, silica, alumina and iron content in the ash. The chemical properties of the fly ash are largely influenced by the chemical content of the coal burned (Table 1).

Fig. 1 Fly ash powder

Table 1 Chemical composition of class F fly ash

Constituents	Percentage
SiO	54.9
Al_2O_3	25.8
Fe_2O_3	6.9
CaO	8.7
MgO	1.8
SO_3	0.6
Na_2O and K_2O	0.6

2.2 *Alccofine*

Alccofine is a new generation, micro fine material of particle size and is much finer than other hydraulic materials like cement, fly ash, silica etc. being manufactured in India. Alccofine has unique characteristics to enhance performance of concrete in fresh and hardened stages due to its optimized particle size distribution. And it also enhance the workability as well as the strength properties.

There are two types of Alccofine:

Alccofine 1203: It is an alccofine with low calcium silicate. Alccofine 1200 series is of 1201, 1202, 1203 which represents fine, micro fine, ultrafine particle size respectively. Alccofine 1203 is a slag based SCM having ultra-fineness with optimized particle size distribution. Alccofine 1203 provides reduced water demand for a given workability, even up to 70% replacement level as per requirement of concrete performance.
Alccofine 1101: It is an Alccofine with high calcium silicate. It is a micro finer Cementitious grouting material for soil stabilization and rock anchoring. The performance of Alccofine is superior to all other admixtures used in India, due to high

Fig. 2 Alccofine 1203

Table 2 Chemical composition of alccofine

Constituents	Percentage
SiO_2	35.4
Al_2O_3	21.6
Fe_2O_3	1.2
CaO	34.0
SO_3	0.12
MgO	6.5

calcium oxide (CaO) content. The specific gravity of alccofine was determined as 2.9 (Fig. 2; Table 2).

2.3 *Fine Aggregate*

Fine aggregate or sand is an accumulation of grains of mineral matter derived from the disintegration of rocks. It is distinguished from gravel only by size of grain or particle, but is distinct from clays which contain organic minerals. Usually commercial sand is obtained from river beds or from sand dunes originally formed by the action of winds. The properties of fine aggregates was studied by conducting experiments like specific gravity, bulk density, bulking, Fineness modulus (Fig. 3).

2.4 *Sodium Hydroxide*

Sodium hydroxide is also known as caustic soda is a white solid ionic compound comprising of sodium cations Na^+ and hydroxide anions OH^-. Sodium hydroxide

Fig. 3 Fine aggregate

is caustic base and alkali that decays proteins at ambient temperatures. The specific gravity of sodium hydroxide was 2.13. Sodium hydroxide is also known as lye and caustic soda is an inorganic compound with the formula NaOH. It is a white solid ionic compound consisting of sodium cations and Na+ and hydroxide anions OH^-. Sodium hydroxide is a highly caustic base and alkali that decomposes proteins at ordinary ambient temperatures and may cause serve chemical burns. it is highly soluble in water and readily absorbs moisture and carbon dioxide from the air. it forms a series of hydrates NaOH . nH_2O (Fig. 4).

Fig. 4 Sodium hydroxide pellets

2.5 *Sodium Silicate*

Sodium silicate is also known as water glass or liquid glass, these materials are available in aqueous solution and in solid form. The pure compositions are colorless or white, but commercial samples are often greenish or blue owing to the presence of iron-containing impurities. They are used in cement, passive fire protection, textile and lumber processing, refractories, and automobiles. Sodium carbonate and silicon dioxide react when molten to form sodium silicate and carbon dioxide. Anhydrous sodium silicate contains a chain polymeric anion composed of corner-shared $\{SiO_4\}$ tetrahedral and not a discrete $SiO_3{}^{2-}$ ion. In addition to the anhydrous form, there are hydrates with the formula $Na_2SiO_3 \cdot nH_2O$ (where n = 5, 6, 8, 9), which contain the discrete, approximately tetrahedral anion $SiO_2(OH)_2^{2-}$ with water of hydration. For example, the commercially available sodium silicate pentahydrate $Na_2SiO_3 \cdot 5H_2O$ is formulated as $Na_2SiO_2(OH)_2 \cdot 4H_2O$, and the monohydrate $Na_2SiO_3 \cdot 9H_2O$ is formulated as $Na_2SiO_2(OH)_2 \cdot 8H_2O$. The specific gravity of sodium hydroxide was 1.39 (Fig. 5).

3 Preparation and Testing of Geopolymer Masonry Units

Geopolymer bricks and are hand moulded. Raw materials are mixed manually by varying the proportion of fly ash with alccofine. The mix proportion used for the preparation of bricks is shown in Table 3. Mould size was selected as 225 × 105 × 75 mm for bricks. The materials such as fly ash, M sand and alccofine were mixed together. Then alkaline activators are added to the dry materials and mixed

Fig. 5 Sodium silicate solution

thoroughly. The fresh mortar was then casted into the moulds immediately after mixing and well compacted and the top surface is finished (Fig. 6).

Table 3 Mix Proportion for the preparation of geopolymer masonry units

Sodium hydroxide:sodium silicate	0.4
Alkaline solution:Binder	0.5
Aggregate:Binder	3
Fly ash:Alccofine	100:0 90:10 80:20 70:30 60:40

Fig. 6 Mixing process of geopolymer brick

Fig. 7 Compressive strength test of geopolymer brick

4 Result and Discussion

4.1 Compressive Strength

The brick masonry units were placed in the compression testing machine in such a manner that the load shall be applied on opposite sides. The specimen was aligned centrally on the base plate of the machine. The movable portion was rotated gently by hand so that it touches the top surface of the specimen. The load was applied gradually without shock and continuously till the specimen fails. The maximum load was noted.

The compressive strength of geopolymer bricks was determined according to BS 6073 as shown in Figs. 7 and 8 (Table 4).

$$\text{Compressive strength} = \text{Maximum load at failure/Average area of bed face}$$

4.2 Water Absorption of Geopolymer Bricks

The water absorption was determined according to IS as shown in Fig. 9 and the results of water absorption with the increase in proportion of alccofine is shown in

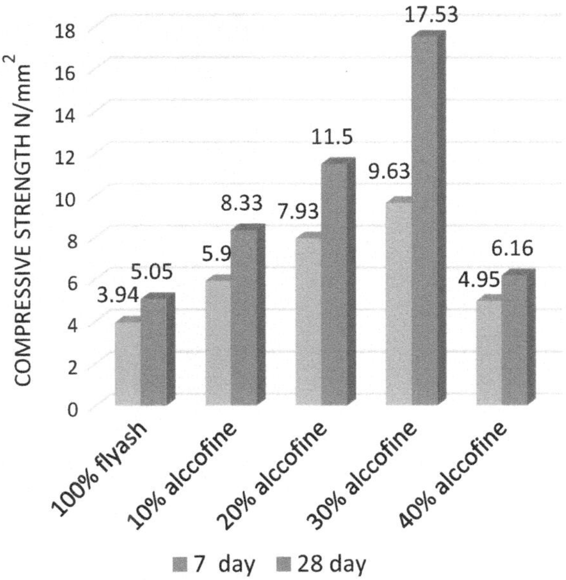

Fig. 8 Variation of 7 day and 28 day strength with increase in percentage of alccofine

Table 4 Compressive strength values for geopolymer bricks

Fly ash (%)	Alccofine (%)	Compressive strength (N/mm^3)	
		7 days	28 days
100	0	3.94	5.05
90	10	5.90	8.33
80	20	7.93	11.50
70	30	9.63	17.53
60	40	4.95	6.16

Table 5. More the water absorption the weaker will be the brick (Fig. 10).

$$\text{Water absorption} = (M_2 - M_1)/M_1$$

where

M1 = weight of dry specimen
M2 = weight of wet specimen

Fig. 9 Water absorption of bricks

Table 5 Water absorption test results

Fly ash (%)	Alccofine (%)	Water absorption (%)
100	0	12.2
90	10	8.9
80	20	6.62
70	30	5.86
60	40	6.41

4.3 Bulk Density of Geopolymer Bricks

Bulk density of geopolymer brick is defined as dry weight of brick per unit volume of brick. Table 6 shows the bulk density values of the geopolymer bricks (Fig. 11).

5 Conclusion

In this paper, the curing is done under ambient temperature. The optimum compressive strength was obtained by replacing 30 percentage fly ash with alccofine. Strength of 9.63 MPa was found to be the highest strength of geopolymer bricks produced when cured for 7 days and a strength of 17.53 MPa for 28 days. The lowest

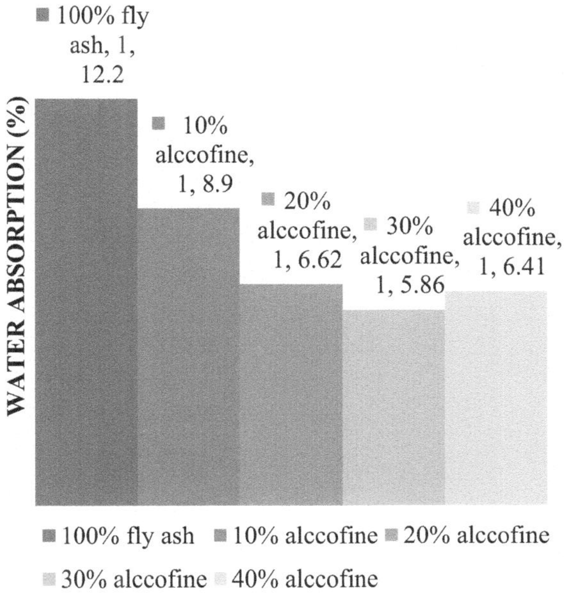

Fig. 10 Variation of water absorption with the increase in percentage of alccofine

Table 6 Bulk density of the tested geopolymer bricks

Fly ash (%)	Alccofine (%)	Bulk density (kg/m^3)
100	0	2084.23
90	10	2103.42
80	20	2141.80
70	30	2145.18
60	40	2157.60

percentage water absorption of geopolymer brick is 5.86% shown by bricks cured for 28 days. Highest value of bulk density was found to be 2145.18 kg/m^3 at 30% replacement of fly ash with alccofine.

The obtained results satisfy the relation between compressive strength, water absorption and bulk density, that is, compressive strength increases with decrease in water absorption and increase in bulk density. Good quality of geopolymer bricks can be produced from stipulated proportion of fly ash and alccofine. The manufacturing process of geopolymer brick is simple and the energy consumption will be much less as compared to the conventional burnt clay bricks as it is cured under ambient temperature.

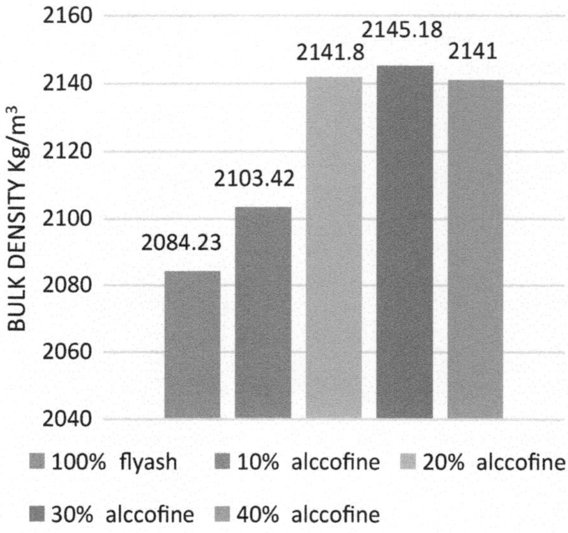

Fig. 11 Variation of bulk density with the increase in percentage of alccofine

References

1. Rajini B, Narasimha Rao AV (2014) Mechanical properties of geopolymer concrete with fly ash and GGBS as source materials. Int J Innov Res Sci Eng Technol
2. Abdullah MMA, Ibrahim WMW, Tahir MFM (2015) The properties and durability of fly ash based geopolymeric masonry bricks
3. Radhakrishna, Venugopal K (2015) Properties and application of geopolymer masonry units. SSRG Int J Civ Eng
4. Chitawadagi MV, Tejas O (2014) Experimental investigations on strength, durability, sustainability & economic characteristics of geopolymer concrete blocks. Int J Res Eng Technol 3(6):115–122
5. Radhakrishna, Venugopal K (2016) Structural behaviour of geopolymer masonry. Int J Sci Technol 9
6. Raut A, Bhasker T, Bhasker S, Sreelatha D, Ibrahim SM, Thriveni R (2018) Experimental investigation on geopolymer brick by using fly ash and quarry dust. Int J Adv Res Ideas Innov Technol 4(2)

Fly Ash Based Geopolymer Bricks: A Sustainable Construction Material

Niveditha Balakrishnan, S. Usha and Ponny K. Thomas

Abstract Considering the increasing demand for sustainable and environment friendly construction materials, the feasibility of fly ash based alkali activated geopolymer bricks, has been aimed in this project as an alternative to ordinary bricks. Geopolymers are amorphous to semi crystalline alumina silicate polymer originated by inorganic polycondensation reaction of solid alumina silicate with highly concentrated alkali hydroxide or silicate solution. Production of building materials, particularly bricks using fly ash is considered to be one of the solutions to the ever increasing fly ash disposal problems in the country. Fly ash based geopolymer with alkali liquid to binder ratio as 0.4 and sodium silicate to sodium hydroxide ratio as 2.5 was studied in this project. Geopolymer bricks of size 230 mm × 110 mm × 70 mm were optimized by varying the molarity of NaOH, percentage of extra water added, curing conditions. The compressive strengths, water absorption and durability of optimized fly ash based geopolymer bricks were compared with cement bricks and country burnt bricks. Considering the economical aspect, the mix proportions were changed with fly ash to fine aggregate ratio as 1:6 and molarity of NaOH as 5M. Fine aggregate as river sand, M sand and quarry dust were also studied to find the most affordable brick.

Keywords Geopolymer · Brick · Fly ash

N. Balakrishnan
College of Engineering, Trivandrum, India
e-mail: nivedithabalakrishnan9@gmail.com

S. Usha (✉)
Sree Narayana Gurukulam College of Engineering, Cochin, India
e-mail: ushaushus11@rediffmail.com

P. K. Thomas
Desai Homes, Cochin, India
e-mail: ponnykthomas@gmail.com

K. Dasgupta et al. (eds.), *Proceedings of SECON'19*,
Lecture Notes in Civil Engineering 46,
https://doi.org/10.1007/978-3-030-26365-2_27

1 Introduction

Worldwide, millions of tons of fly ash are generated as waste by the power plants in every year. Among this large quantity of fly ash, only a small quantity is utilized and disposal of fly ash become huge problems across the nation. Geopolymers are produced by the alkali activation of alumino-silicate materials at ambient or slightly elevated temperatures. Geopolymerization process is a green technology which can utilize waste materials. Geopolymers consume less energy during production and release less carbon dioxide. Several research works has been carried out to investigate the possibility of utilizing waste material in the production of geopolymer cement [1–4].

Mastura et al. studied the influence of curing time and curing temperature on properties of fly ash based geopolymer bricks. Based on the experimental results, the geopolymer bricks prepared with 12M NaOH concentration and 1:3 ratio of fly ash to sand give the optimum compressive strength when they were cured at elevated temperature [5]. According to Muduli et al., the reaction process of geopolymerization depends on the type of raw materials and on the concentration of alkaline activator [6]. Mustafa Al-Bakari et al. studied the ratio of alkaline activator to fly ash and optimized the alkaline activator/fly ash ratio as 0.4 [7]. Hardjito et al. observed that the compressive strength of fly ash based geopolymer concrete was optimum with sodium silicate to sodium hydroxide ratio 2.5 [8] According to Bakri et al., the maximum compressive strength of fly ash based geopolymer was obtained at 60 °C for 24 h [9]. Tejas and Manojkumar conducted acid resistance test with fly ash and GGBS based geopolymer concrete blocks by immersing the specimens in 3% solution of sulphuric acid and hydrochloric acid to prove its high resistance against acid [10]. This paper presents the possibilities of fly ash based geopolymer brick as a environmental friendly brick.

2 Experimental Methods

2.1 Materials

Materials used for the investigation are fly ash, cement, clay soil, river sand, M sand, quarry dust, sodium hydroxide, sodium silicate and water.

Fly ash

Fly ash waste was collected from Neptune RMC, Edayar, Kerala. Table 1 shows the chemical composition of fly ash obtained by X-Ray Fluorescence (XRF) analysis (Fig. 1).

Table 1 Chemical composition of fly ash

Compound	SiO_2	Al_2O_3	Fe_2O_3	TiO_2	CaO
%wt.	50.5	25	9.3	1.2	3.5

Fig. 1 Fly ash

Fine Aggregates

Locally obtained river sand, M sand and Quarry dusts were used as fine aggregates in this investigation. The river sand is obtained from Perumbavoor belongs Zone I as per Table 4, IS:388-1970. Fineness modulus of river sand obtained is 3.5. The quarry dust used has a specific gravity of 2.88 and was taken from Airapuram. The M sand with specific gravity 2.56 was used. Fineness of M sand is 3.35 and it belongs to zone II as per Table 4, IS:388-1970 (Fig. 2).

Alkali activator solution

In geopolymer fly ash bricks, sodium hydroxide pellets, sodium silicate solution and water were used as alkaline solution for geo polymerization.

97% purity sodium hydroxide pellets and 1.39 g/cm^3 density sodium silicate solution was used for alkali activator. Alkali activator to binder ratio 0.4 and sodium

Fig. 2 M sand, quarry dust and river sand

silicate to sodium hydroxide ratio 2.5 were fixed as per the literature survey. The activator solution was prepared 24 h prior to casting of geopolymer bricks (Figs. 3 and 4).

Cement

Ordinary Portland cement of 33 grade confining to IS 269:1989 was used throughout the work.

Water

Potable water was used for the investigation.

Clay

Non calcareous clay containing feldspar and iron oxides suitable for burnt clay bricks was used.

Fig. 3 Sodium hydroxide pellets

Fig. 4 Sodium silicate solution

Fig. 5 Fly ash based geopolymer brick

2.2 Preparation of Fly Ash Based Geopolymer Bricks

Keeping alkali activator to binder ratio 0.4 and sodium silicate to sodium hydroxide ratio 2.5, the alkali activator was prepared by dissolving NaOH pellets in portable water and mixed with sodium silicate solution at least one day prior to the casting of geopolymer specimens. The fly ash and fine aggregates in the ratio 1:3 were first dry mixed manually in a container. The alkali activator solution, required additional water for a workable mix were added to the dry mix and mixes thoroughly for 4 min to get the geopolymer mix. The mixture was then filled in the (230 mm × 110 mm × 70 mm) mould and compacted manually using wooden rod. The specimens were covered with aluminium foil during elevated temperature curing in order to retain moisture content for the chemical reaction (Fig. 5).

In the first phase of experiments geopolymer bricks were casted in different molarities of NaOH (5M and 11M) with fly ash to fine aggregate mix ratio 1:3 using fine aggregate as river sand keeping additional water for workable mix as 1.8 and 2.4% under elevated (60 °C for 24 h) temperature. The optimized geopolymer brick was casted and cured under ambient temperature. To study economical geopolymer brick, with fly ash to fine aggregate mix ratio 1:6 and 5M molarity, fine aggregate was varied with river sand, M sand and quarry dust under elevated temperature (60 °C for 24 h) (Fig. 6).

2.3 Preparation of Cement Bricks

Ordinary Portland cement 33 grade and river sand in the ratio 1:3 were dry mixed and water was added to get required workable mix. The bricks were casted using 230 mm × 110 mm × 70 mm size mould with proper compaction. The mould was removed after 24 h of casting and bricks were cured in water till testing (Fig. 7).

Fig. 6 Curing in hot air oven at 60 °C

Fig. 7 Cement brick

2.4 Preparation of Country Burnt Clay Bricks

Country burnt bricks were casted and burnt at Kalady. The clay soil suitable for burnt brick was dugged and cleaned. The clay was powdered and exposed to atmosphere. Required ingredients were added and blended properly. Water was added to soil and mixed well in pug mill. Ground moulded bricks (230 mm × 110 mm × 70 mm) were casted using the prepared clay. After the molding process, the bricks contain some amount of moisture in it. They are unmoulded and stacked on the ground with sufficient gaps between them to dry under sun for 10–15 days. The dried bricks were burned in kilns about 1100 °C (Fig. 8).

2.5 Different Tests Conducted on Bricks

The compressive strength test and water absorption test of the bricks were carried out as per standard specifications.

Fig. 8 Country burnt bricks

Acid resistance test was conducted on geopolymer, cement and country burnt bricks. Since there is no standard procedures for acid resistance test of bricks exist, test was carried out using 3% of diluted acid (HCl and H_2SO_4) solution. The evaluations were conducted after 28 and 56 days from the date of immersion.

2.6 *Results and Discussion*

Fly ash based geopolymer bricks with different molarities (5M and 11M) and different percentage of water content are tested and the results are shown in Figs. 9 and 10.

Figures 9 and 10 represents the 7th and 28th day compressive strength of fly ash based geopolymer bricks with different molarities respectively. The brick with 11M molarity and 1.8% additional water show optimum strength under 60 °C for 24 h curing.

The optimized proportion geopolymer brick cured under ambient temperature was tested and the strength of elevated temperature cured and ambient cured geopolymer bricks are compared and shown in Fig. 11. Ambient temperature cured geopolymer

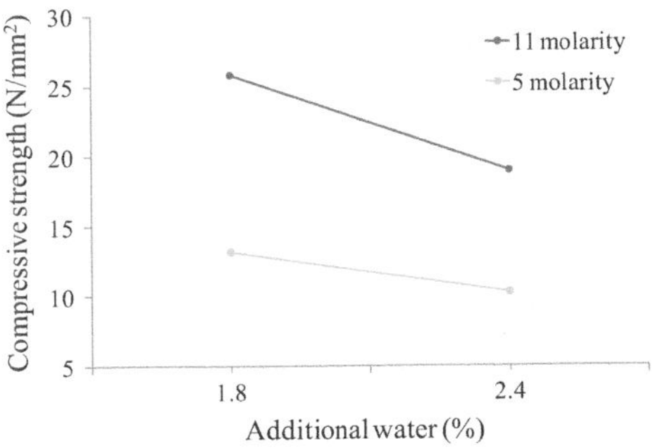

Fig. 9 7th day compressive strength of fly ash based geopolymer bricks with different molarities

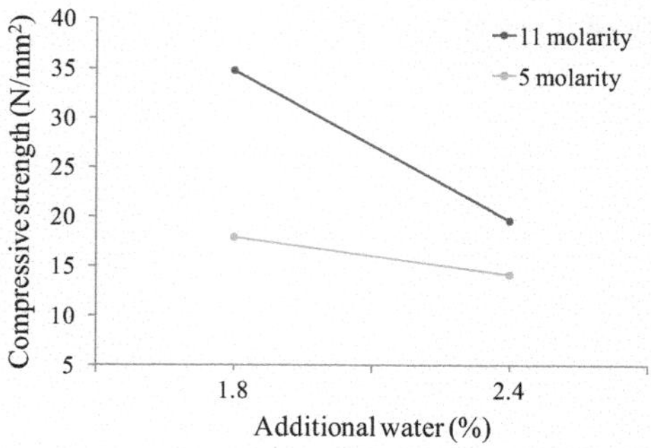

Fig. 10 28th day compressive strength of fly ash based geopolymer bricks with different molarities

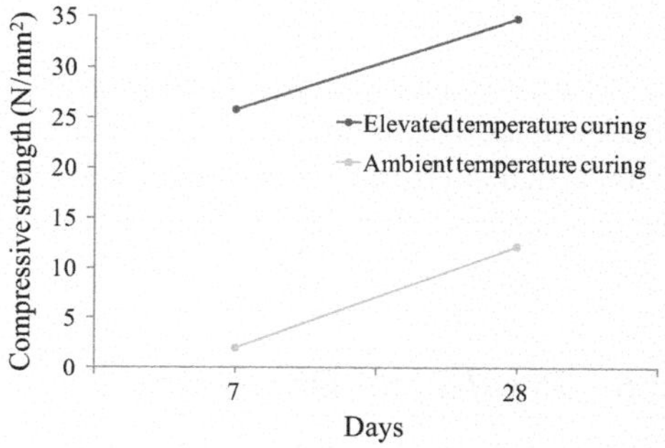

Fig. 11 Compressive strength of fly ash based geopolymer bricks with different curing temperatures

brick show negligible strength at 7th day whereas elevated temperature cured brick show reasonable strength.

In Fig. 12 28th day compressive strength of optimized fly ash based geopolymer brick, cement brick and burnt clay brick are compared. Among the three, geopolymer brick show highest strength.

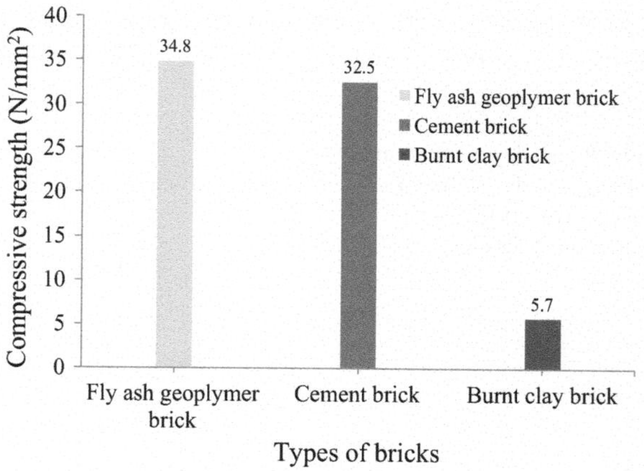

Fig. 12 Compressive strength (28th day) comparison of different bricks

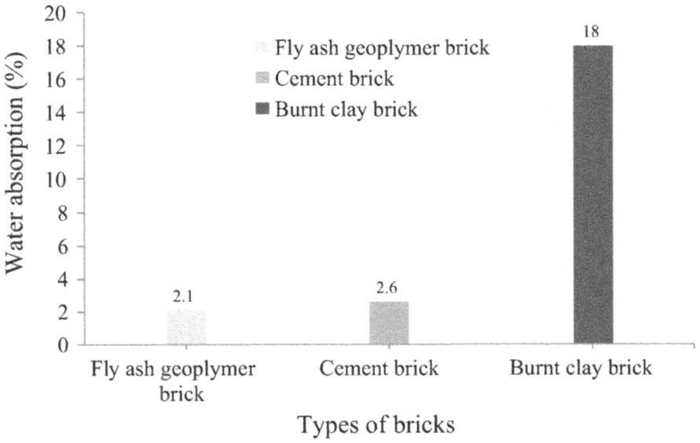

Fig. 13 Water absorption comparison of different bricks

The water absorption % of three types were compared in Fig. 13. In the above figure the fly ash based geopolymer has the lowest water absorption %.

The durability of these bricks was tested by immersing in acid solutions. The results are depicted in Figs. 14 and 15. The highest residual compressive strength in % is observed with fly ash based geopolymer brick with both the acids.

The geopolymer brick with compressive strength 5 N/mm^2 and more with reduced molarity (5M) and fly ash to fine aggregate 1:6 ratio is depicted in Fig. 16. Here geopolymer brick with 5 M cured under 60 °C for 24 h show more than 5 N/mm^2 after 3 days itself. After 28th day ambient cured geopolymer is also good for construction.

Fly ash based geopolymer bricks were then casted by replacing river sand by M sand and quarry dust for further economical mix. Bricks were cured in the oven at a temperature of 60 °C for 24 h and then at room temperature. Compressive strength was tested at 28th day and results are shown in Fig. 17. Geopolymer brick with M sand and quarry dust as fine aggregate are also suitable for construction.

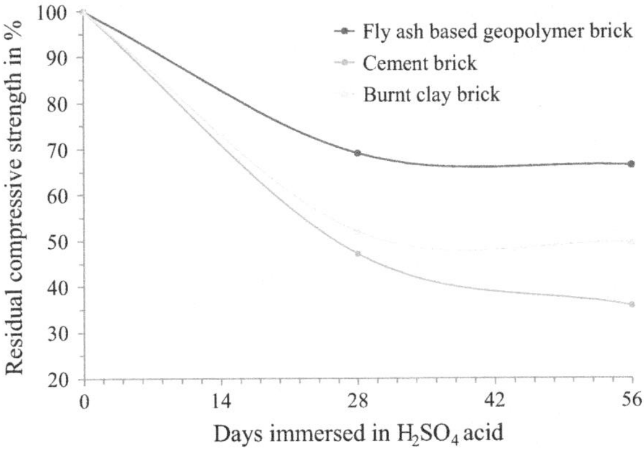

Fig. 14 Effect of sulphuric acid on compressive strength of the brick

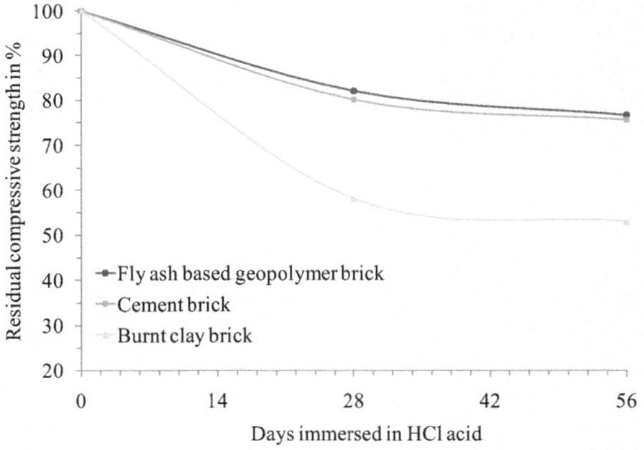

Fig. 15 Effect of hydrochloric acid on compressive strength of the brick

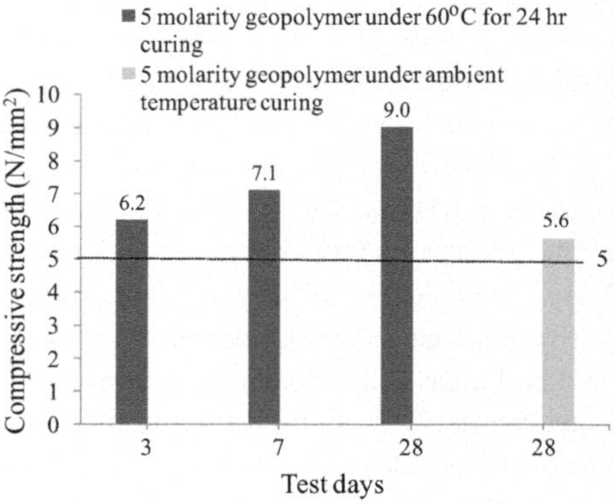

Fig. 16 Compressive strength of fly ash based geopolymer bricks with 5 molarity and 1:6 mix proportion with river sand as fine aggregate

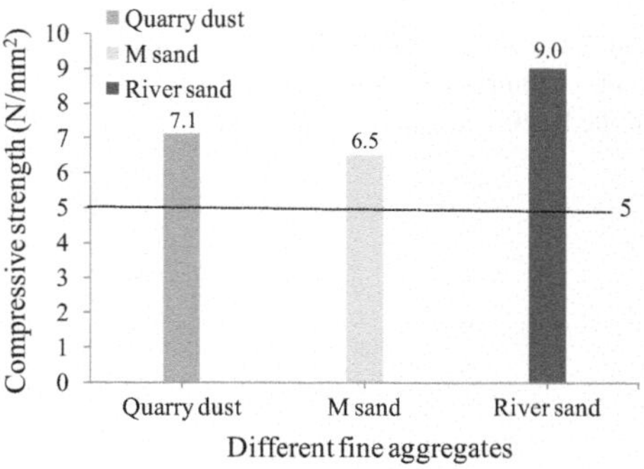

Fig. 17 28th day compressive strength of fly ash based geopolymer bricks with 5 molarity and 1:6 mix proportion with different fine aggregates

2.7 Cost Analysis

Table 2 display the cost comparison between different bricks. On comparing the cost of different geopolymer bricks with cement brick and burnt clay brick, the

Table 2 Cost comparison between different bricks

Type of brick	Molarity	Binder: fine aggregates	Curing	Type of aggregate	Cost per brick	28th day Compressive strength (N/mm^2)
Geopolymer brick	11M	1:3	60 °C for 24 h	River sand	Rs. 23.9	34.8
Geopolymer brick	5M	1:6	60 °C for 24 h	River sand	Rs. 13.5	9.0
Geopolymer brick	5M	1:6	60 °C for 24 h	Quarry dust	Rs. 5.9	7.1
Geopolymer brick	5M	1:6	60 °C for 24 h	M sand	Rs. 7.5	6.5
Geopolymer brick	5M	1:6	Ambient temperature	River sand	Rs. 11.9	5.6
Cement brick	–	1:3	Water curing	River sand	Rs. 41.4	32.5
Country burnt brick	–	–	–	Clay soil	Rs. 7.0	5.7

geopolymer brick with fly ash to fine aggregate ratio 1:6 and molarity of sodium hydroxide as 5M is found to be affordable.

3 Conclusion

- Fly ash: river sand (1:3) with molarity of NaOH-11M, ratio of sodium silicate to sodium hydroxide solution—2.5, additional water content 1.8%, under curing at 60 °C for 24 h provides optimum compressive strength for fly ash based geopolymer brick.
- On comparing the 28th day strength of three different types of bricks, fly ash based geopolymer brick shows the highest strength.
- Fly ash based geopolymer brick exhibits lowest water absorption % compared to cement and burnt clay bricks.
- Resistance to sulphuric acid and hydrochloric acid of geopolymer brick is better than cement brick and burnt clay brick.
- Since river sand is costly it can be replaced with quarry dust or M sand. Quarry dust is better than M sand for geopolymer bricks.
- The geopolymer bricks with fly ash to fine aggregate ratio 1:6 and molarity of NaOH 5M using quarry dust as fine aggregate shows lowest cost compared to all other bricks.

References

1. Pacheco-Torgal F, Castro-Gomes J, Jalali S (2008) Alkali-activated binders: a review part 1, historical background, terminology, reaction mechanisms and hydration products. Constr Build Mater 22:1305–1314
2. Radhakrishna VK, Sasalatti V, Venumadhav T (2015) Study on geopolymer masonry as sustainable building material. J Environ Res Dev 9(3):925–932
3. Puertas F, Martinez-Ramirez S, Alonso S, Vazquez T (2000) Alkali-activated fly ash/slag cement strength behaviour and hydration products. Cem Concr Res 30:1625–1632
4. Wan Mastura L, Kamarudin H, Khairul Nizar I, Mustafa Al Bakri AM, Mohammed H (2013) Effect of curing system on properties of fly ash based geopolymer bricks. Int Rev Mech Eng (IREME) 7(1):67–71
5. Muduli SD, Sadangi JK, Nayak BD, Sadangi JK, Mishra BK (2012) Effect of NaOH concentration in manufacture of geopolymer fly ash building brick. Greener J Phys Sci. ISSN:2276-7851
6. Mustafa Al-Bakari AM, Abdul Kareem OAKA, Myint S (2012) Optimization of alkaline activator/fly ash ratio on the compressive strength of fly ash-based geopolymer. Paper prepared for Conference of the University Malaysia Perlis
7. Bakharev T (2005) Rsistance of geopolymer materials to acid attack. Cem Concr Res 35:658–670
8. Hardjito D, Wallah SE, Sumajouw DMJ, Rangan BV (2005) Introducing fly ash-based geopolymer concrete: manufacture and engineering properties. In: 30th conference on our world in concrete & structures
9. Mustafa Al Bakri AM, Kamarudin H, Binhussain M, Khairul Nizar I, Zarina Y, Rafiza AR (2011) The effect of curing temperature on physical and chemical properties of geopolymers. Phys Procedia 22:286–291
10. Ostwal T, Chitawadagi MV (2014) Experimental investigations on strength, durability, sustainability and economic characteristics of geopolymer concrete blocks. Int J Res Eng Technol 3(6):115–122

Study on Geo-polymer Concrete Under Acidic and Saline Conditions

Vaibhav Kumar, Akash Kumar and A. Sofi

Abstract The prominent issue in the construction business is the ecological and environmental contamination. In the construction industry, for the most part the generation of Portland cement causes the discharge of pollutants resulting in environmental contamination. We can lessen such consequences for nature by expanding the utilization of modern alternatives in our infrastructure development process. Geo-polymer concrete is one such substitute and is in effect progressively utilized on-site. To create the geo-polymer, concrete the Portland cement is completely replaced with the F-type fly-ash and alkaline fluids are utilized for the binding of materials. The alkaline liquids utilized in this examination for the polymerization are the solution of Sodium Hydroxide (NaOH) and Sodium Silicate (Na_2SiO_3). The cube samples taken are of size 100 mm × 100 mm × 100 mm for compression test, sorptivity test, carbonation depth test and 100 mm diameter × 200 mm height of cylinder are casted for splitting tensile strength and kept under acidic and saline conditions. The results are henceforth compared with conventional concrete of the same grade.

Keywords Geopolymer concrete · Compression test · Sorptivity test and carbonation depth test · Spilt tensile strength

Nomenclature

P	Maximum load
L	Length of cylindrical specimen

This project was funded by School of Civil Engineering, VIT, Vellore.

V. Kumar · A. Kumar
School of Civil Engineering, VIT, Vellore, India

A. Sofi (✉)
Department of Structural and Geotechnical Engineering, School of Civil Engineering, VIT, Vellore, India
e-mail: asofi@vit.ac.in

K. Dasgupta et al. (eds.), *Proceedings of SECON'19*,
Lecture Notes in Civil Engineering 46,
https://doi.org/10.1007/978-3-030-26365-2_28

D Diameter of cylindrical specimen

Abbreviation

ASTM American Society for Testing and Materials
GGBS Ground Granulated Blast-Furnace Slag
GPC Geo Polymer Concrete
CCS Conventional Concrete Specimens
LOI Loss on Ignition

1 Introduction

In recent times the use of pre-casted concrete has elevated drastically as they are more efficient and significantly reduce the time of completion of project. Researchers are constantly working on finding an efficient alternative to the conventional concrete, thus brings the existence of geopolymer concrete. The presence of sulphate deteriorates the concrete in coastal and regions with high chances of acid rain [1]. Geopolymer concrete is an inventive and eco-friendly construction material and an alternative in contrast to Portland cement concrete. Utilization of geopolymer decreases the requirement of Portland cement which is responsible for high CO_2 emission.

This research highlights the aspects of sulphate deterioration on concrete as a comparative study. In coastal areas (severe zone) minimum grade of concrete required as per IS 456:2000 is 30 MPa. Hence M-30 grade of concrete was used. Moreover, same values of permeability can be obtained in mix proportions of 25 MPa and 30 MPa [2]. In this study, F-type Fly ash geopolymer concrete is studied. Class F fly ash additionally has a lower calcium content than Class C fly ash [3]. The geopolymer concrete is hardened by the process of polymerization. The hardening of the geopolymer is known to be because of the polycondensation of hydrolyzed aluminate and silicate components [4]. Oven curing is sought to be best method for curing of geopolymer samples [5]. The hardening occurs at controlled temperatures. Curing of geopolymer concrete is another major constraint in casting. It requires controlled high temperature curing 60°–80 °C, 60° being optimum temperature [6]. Geopolymer concrete gains better strength when the particle size is finer. Additionally, higher fineness demonstrates higher workability and quality with early time of heating. In this way, the fundamental accentuation is given on amount and fineness of fly ash in the improvement of mix proportioning system of geopolymer concrete [7]. Usage of fiber reinforcement also enhances the strength of the concrete considerably [8]. Alkali binding materials provide strength in the fly ash based concrete. The solution of sodium hydroxide and sodium silicate were mixed together one day prior to the use in preparing the geopolymer concrete [9]. The geopolymer concrete can be cured for up to 96 h for better strength. The longer the curing period, the better the strength up to 96 h. Although post 24 h there is no significant change in strength of concrete [10].

The sulphate exposure determines the durability properties of the geopolymer concrete under extreme conditions. For a better and clearer understanding Sodium sulphate and Sulphuric acid environment was used as they display significant changes and better values [11].

The ratio of sodium silicate to sodium hydroxide also determines the compressive strength of geopolymer concrete. A value more than 2.5 reduces the strength due to increase in alkalinity [12].

During the casting, the water seeps out of the mix and does not have any role in the strength of the mix [13]. The effect of carbonation is studied to determine the extent of corrosion of steel reinforcement. At higher water cement ratio, the carbonation depth is higher [14]. Carbonation reduces the alkalinity of concrete which results in the corrosion of the steel reinforcements [15]. The presence of salts in concrete reduces the sorptivity of the samples as it reduces the overall movement of liquids due to chloride binding effects [16] Sea water being 3.5% saline has significant effects on the surrounding concrete. It results in irregular trend in change in strength [17].

With increase in amount of acid and salts in water surrounding the concrete, better measures shall be taken to counter corrosion [18]. Compressive strength of concrete exposed to salts during curing and casting displays comparatively higher strength than fresh water concrete of same grade [19]. In general, the sorptivity of fly ash based concrete is higher than that of conventional concrete [20].

2 Materials

2.1 Cement

The concrete specimens utilized in this study were of M-30 grade and ordinary Portland 53 grade cement was used for the same. The specific gravity of cement was found to be 3.13.

2.2 Fly-Ash

Low calcium (class F) fly ash from S.R bricks industry were used in the geopolymer concrete. The specific gravity of fly-ash was found to be 2.17.

2.3 Aggregate

Aggregates of different sizes from 20 to 12.5 mm were utilized in this work. The specific gravity of the coarse aggregates was found to be 2.79, and the water absorption of the coarse aggregates was 1.38%. The specific gravity of the fine aggregates was 2.65 and its water absorption was 0.62% and these are found as per ASTM C 127-01 [21], ASTM C128-01 [22].

2.4 Alkaline Binders

Sodium hydroxide and Sodium silicate were used for the activation of fly ash based geopolymer concrete. Sodium hydroxide is available in the market in pellet form. 10 molar solution was used in this study. The ratio of sodium silicate to Sodium Hydroxide was fixed 2.5. The chemical solution to fly-ash ratio was fixed to be 0.35.

3 Experimental Program

3.1 Compressive Test

For compressive tests, specimens of size 100 mm × 100 mm × 100 mm cubes were casted with design mix of M-30 as per IS:10262 specifications. The cubes were then cured for a duration of 14 days in water maintained at room temperature. For geopolymer concrete cubes, the specimens were cured in oven at 60°–80 °C for 24 h. The loading rate for tests was set to be 2.5 KN/m^2. The measured compressive strength of the cubes is then determined by dividing the maximum load applied to the cubes during the test by the cross-sectional area, calculated from the mean dimensions of the section and shall be expressed to the nearest 0.5 N/mm. The compressive strength can be calculated as (Tables 1 and 2):

$$\text{Compressive strength} = \frac{\text{failure load(kN)}}{Area\left(\text{m}^2\right)}$$

Table 1 Compressive strength in acidic exposure (N/mm^2)

Duration of exposure (days)	Conventional concrete	Geopolymer concrete
7	34.93	36.9
14	33.9	35.77
28	31.47	34.7

Table 2 Compressive strength in saline exposure (N/mm^2)

Duration of exposure (Days)	Conventional concrete	Geopolymer concrete
7	35.37	37.47
14	36.43	37.9
28	37.43	39.1

Table 3 Splitting tensile strength in acidic exposure (N/mm^2)

Duration of exposure (days)	Conventional concrete	Geopolymer concrete
7	2.46	2.8
14	2.28	2.57
28	2.05	2.42

Table 4 Splitting tensile strength in saline exposure (N/mm^2)

Duration of exposure (days)	Conventional concrete	Geopolymer concrete
7	2.28	2.43
14	2.3	2.46
28	2.42	2.64

3.2 Splitting Tensile Test

The cylinder samples were casted of size 100 mm in diameter and 200 mm in height. The samples were made with a design strength of 30 KN/mm^2 as per IS: 10262 2009 specifications [23]. Curing process was kept as same as for cube samples The splitting tensile strength was then determined by the equation, $fct = 2P/\pi ld$; where 'P' is maximum load in Newtons applied to the specimen, 'l' is the length of the specimen, and 'd' is the cross sectional dimension of the specimen (Tables 3 and 4).

3.3 Sorptivity

The samples were prepared as per IS4031-Part 6. It can be calculated by using the below mentioned formula:

$$I = S \,.\, t^{0.5}$$

where

S = sorptivity (in mm)
t = elapsed time (in min)
$I = \Delta w/Ad$
Δw = change in weight = W2 − W1

W1 = Oven dry weight of cube specimens (in kg)
W2 = Weight of cube specimens after 60 min capillary suction of water (in kg)
A = surface area of the specimen through which water penetrated
d = density of water

3.4 Carbonation Test

Carbonation of the concrete, brought about via carbon dioxide in the air, has the impact of diminishing the pH. Carbonation depth is observed utilizing phenolphthalein indicator that changes to pink in contact with basic cement with pH values more than 9 and colorless at lower pH. The test is mostly completed by sprinkling the indicator on freshly uncovered surfaces of specimen broken from the structure or on split centers.

3.5 Change in Mass

For this test, the cube samples were casted as per IS4031-Part-6. The specimens were weighed before and after immersing them in the desired exposure and the difference in their masses was expressed in percentage.

3.6 Visual Appearance

It is a qualitative parameter which tells about the external appearance of the specimen taken out from the acidic and saline exposures.

4 Results and Discussion

4.1 Compressive Test

On observing the compressive strength in acidic environment from Fig. 1, it has been observed that the decrement in compressive strength of conventional concrete is comparatively more than geopolymer concrete.

Figure 2 shows that there is increase in the compressive strength of both CC and GPC when soaked in saline exposure.

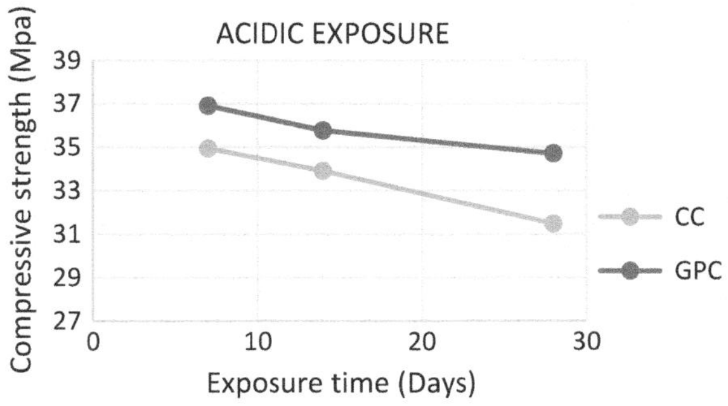

Fig. 1 Compressive strength under acidic exposure

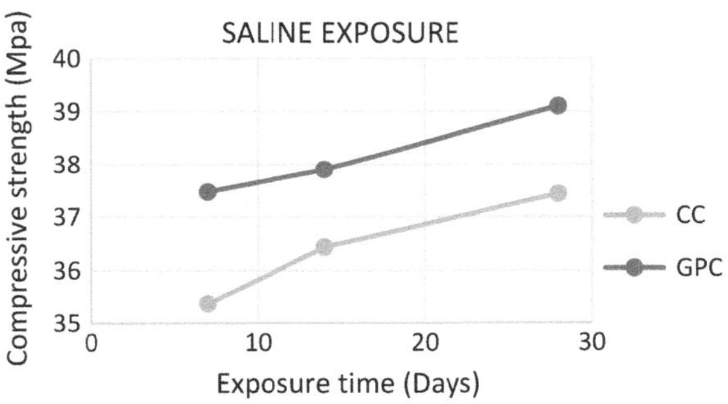

Fig. 2 Compressive strength under saline exposure

4.2 *Splitting Tensile Test*

On observing splitting tensile strength in acidic environment from Fig. 3 it can be seen that the decrement in conventional concrete is comparatively more than geopolymer concrete. It can be inferred that geopolymer concrete displayed less loss of strength in acidic environment.

Figure 4 shows no significant change in splitting tensile strength of both types of concrete in saline solution. This is due to that fact that the salt sodium contributes in enhancing the bonding of cement particles.

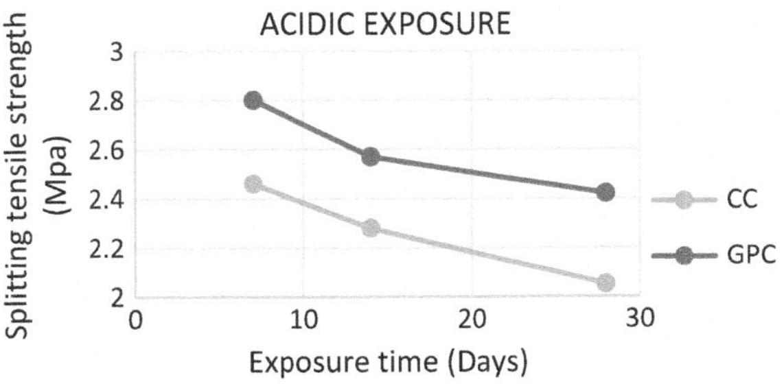

Fig. 3 Splitting tensile strength under acidic exposure

Fig. 4 Splitting tensile strength under saline exposure

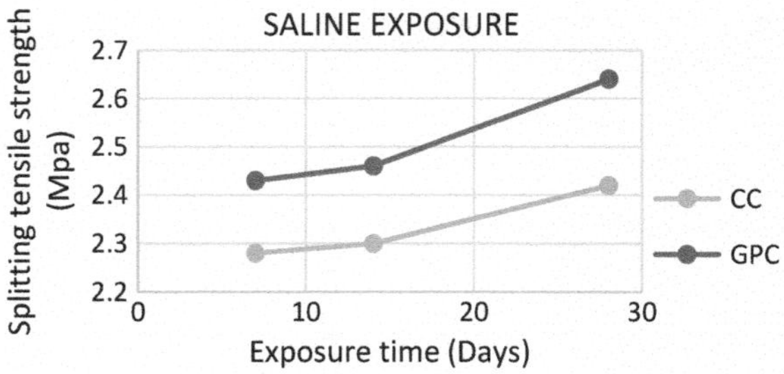

Table 5 Sorptivity ($mm/min^{0.5}$)

	Acidic	Saline
Conventional concrete	0.09	0.129
Geopolymer concrete	0.129	0.181

Table 6 % change in mass in acidic exposure (N/mm^2)

Duration of exposure (days)	Conventional concrete	Geopolymer concrete
7	0.073	0.4
14	1.085	0.306
28	2.621	1.743

4.3 Sorptivity

Table 5 shows that the sorptivity of conventional concrete in acidic exposure is 0.09 $mm/min^{0.5}$ and it increases to 0.129 $mm/min^{0.5}$ in saline exposure. Also, it follows the same trend in the geopolymer concrete as it increases from 0.129 to 0.181 $mm/min^{0.5}$.

It can be clearly seen that the sorptivity is higher in geopolymer concrete as compared to that of conventional concrete. It can also be inferred from Table 5 that sorptivity of specimens soaked in saline exposure have comparatively higher values than the acidic exposure (Table 6).

4.4 Carbonation

Figure 5 clearly shows that the external surface of the specimens kept in acidic exposure is carbonated as the surface did not change to pink after spraying phenolphthalein solution on the surface whereas inner surface is non-carbonated.

Figure 6 shows that both the outer and inner surface of the concrete specimens are non-carbonated as the surface turned to pink after spraying the phenolphthalein solution on the specimen.

Fig. 5 Carbonation on specimen soaked in acidic exposure

Fig. 6 Change in mass of samples exposed to saline conditions

The same trend has been observed in both conventional and geopolymer concrete.

5 Change in Mass

Figure 7 demonstrates linear incremental variation in % change in mass for both geopolymer and conventional concrete. Moreover, it can be observed that geopolymer concrete is less susceptible to deterioration in acidic environment.

It can be observed from Fig. 8 that it shows similar linear increment in percentage change of mass. However, the effect of degradation is almost equal in both conventional and geopolymer concrete.

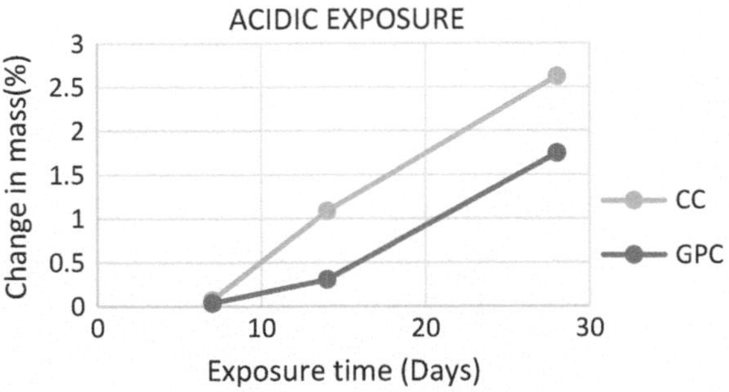

Fig. 7 Change in mass of samples exposed to acidic conditions

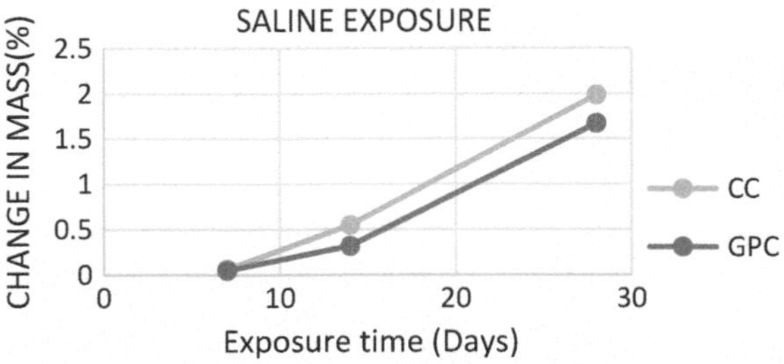

Fig. 8 Change in mass of samples exposed to saline conditions

5.1 *Visual Appearance*

There was no significant change in the external appearance of the geopolymer concrete in both the exposures while the conventional concrete specimens kept in acidic exposure got high erosion and it can be clearly seen in Fig. 9.

Fig. 9 Geopolymer concrete depicting lesser deterioration than conventional concrete

6 Conclusion

The following conclusions have been inferred from the results:

- The compressive strength of both types of concrete decreased in acidic exposure but it slightly increased in saline exposure.
- Since the strength decreases in the acidic environment, it is more severe as compared to the saline environment.
- Same trend as compressive strength has been observed in splitting tensile strength of conventional concrete while there is no significant change in GPC.
- Also, it has been seen that conventional concrete soaked in acidic exposure gets the maximum deterioration (change in mass) and GPC in saline exposure gets the least.
- The GPC has more capillary suction as compared to the CC i.e. higher sorptivity observed in GPC as that of CC.
- The carbonation of both the concrete is observed to be more in saline exposure as compared to the acidic exposure.
- The visual appearance of concrete specimens showed that there is more deterioration in acidic environment as compared to saline environment.

References

1. Stewart MG, Rosowsky DV (1998) Time-dependent reliability of deteriorating reinforced concrete bridge decks. Struct Saf 20(1):91–109. ISSN 0167-4730
2. Kumar G, Salkar A, Master A, Bhattad A, Venkataramana K, Prabhu KR (2015) Durability of concrete in coastal areas (severe zone)
3. American Society for Testing and Materials (ASTM), ASTM C618-94a (1994) Standard specification for coal fly ash and raw or calcined natural pozzolan for use as a mineral admixture in Portland cement concrete. Annual Book of ASTM Standards, vol 04.02, Philadelphia, Pennsylvania
4. Ryu GS, Lee YB, Koh KT, Chung YS (2013) The mechanical properties of fly ash-based geopolymer concrete with alkaline activators
5. Nurruddin MF, Haruna S, Mohammed BS, Galal Sha´aban I (2018) Methods of curing geopolymer concrete: a review
6. Abdul Aleem MI, Arumairaj PD (2012) Optimum mix for the geopolymer concrete. Indian J Sci Technol
7. Patankar SV, Jamkar SS, Ghugal YM (2012) Effect of sodium hydroxide on flow and strength of fly ash based geopolymer mortar. J Struct Eng 39(1):7–12
8. Sofi A, Phanikumar R (2016) Durability properties of fibre-reinforced pond ash-modified concrete. J Eng Sci Technol 11(10):1385–1402
9. Wallah SE, Hardjito D, Sumajouw DMJ, Rangan BV (2005) Performance of geopolymer concrete under sulfate exposure. Paper for Ed Nawy Symposium, American Concrete Institute, April 2005
10. Hardjito D, Rangan BV (2005) Development and properties of low-calcium fly ash-based geopolymer concrete

11. Nurhayat Degirmenci F (2017) Effect Of sodium silicate to sodium hydroxide ratios on durability of geopolymer mortars containing natural and artificial pozzolans. Balikesir University, Architecture Faculty, Department of Architecture, Cagis Campus, 10145, Balikesir, Turkey
12. Yahya Z, Mustafa Al Bakri AM, Hussin K, Ismail KN, Razak RA, Sandu AV (2015) Effect of solids-to-liquids, Na_2SiO_3-to-NaOH and curing temperature on the palm oil boiler ash (Si+ Ca) geopolymerisation system
13. Patankar SV, Ghugal YM, Jamkar SS (2015) Mix design of fly ash based geopolymer concrete
14. Singh N, Singh SP (2016) Reviewing the carbonation resistance of concrete. Department of Civil Engineering, Dr B R Ambedkar National Institute of Technology, Jalandhar, India
15. Zulu S, Allopi D (2014) Influence of high content fly ash on concrete durability
16. Ye H, Jin N, Jin X (2017) An experimental study on relationship among water sorptivity, pore characteristics, and salt concentration in concrete
17. Maniyal S, Patil A (2015) An experimental review of effect of sea water on compressive strength of concrete. IJETAE 5(3) (March)
18. Nagabhushana, Hebbal D, Akash N, Deepak S, Kumar M. Effect of salt water on compressive strength of concrete
19. Gawande S, Deshmukh Y, Bhagwat M, More S, Nirwal N, Phadatare A (2017) Comparative study of effect of salt water and fresh water on concrete
20. Pitroda J, Umrigar FS (2013) Evaluation of sorptivity and water absorption of concrete with partial replacement of cement by thermal industry waste (fly ash)
21. ASTM Designation C 127-01 Standard test method for density, relative density (specific gravity), and absorption of coarse aggregate. Annual Book of ASTM Standards; 2005
22. ASTM Designation C 128-01 Standard test method for density, relative density (specific gravity), and absorption of fine aggregate. Annual Book of ASTM Standards; 2005
23. IS 10262:2009 Indian Standard Recommended guidelines for concrete mix design

Study on Mechanical and Durability Properties of Recycled Coarse Aggregate in Concrete

Shubham Kumar, Ritesh Bharti, Prakeern Gupta and A. Sofi

Abstract The amount of construction waste has been dramatically increased in the last decade, and social and environmental concerns on the recycling of the waste have consequently been increased. Waste concrete is one of the crucial among the construction wastes. Recent technology has also improved the recycling process. In this modern industrialized world, recycling construction material plays an important role to preserve the natural resources. Recycled aggregates can be used for cost reduction of the constructional project. In this research work concrete waste from demolished structure has been collected and recycled coarse aggregate (RCA) of different percentage is used for preparing fresh concrete. This study aims to evaluate physical properties of concrete having different proportion of RCA. Mechanical properties such as compressive strength and split tensile strength was determined for various replacement percentages of recycled aggregates. The results obtained are compared with conventional concrete of the same grade.

Keywords Recycled aggregates · Water absorption · Compressive strength · Split tensile strength · Flexural strength etc.

Nomenclature

P	Maximum load
l	Length of cylindrical and beam specimen
d	Diameter of cylindrical specimen
b	Width of beam specimen
f_{ck}	Characteristics compressive strength of concrete

S. Kumar · R. Bharti · P. Gupta
School of Civil Engineering, VIT, Vellore, India

A. Sofi (✉)
Department of Structural and Geotechnical Engineering, School of Civil Engineering, VIT, Vellore, India
e-mail: asofi@vit.ac.in

K. Dasgupta et al. (eds.), *Proceedings of SECON'19*,
Lecture Notes in Civil Engineering 46,
https://doi.org/10.1007/978-3-030-26365-2_29

f_{ct} Characteristics flexural strength of concrete
f_{cts} Characteristics splitting strength of concrete

Abbreviation

CTM Compression testing machine
FTM Flexural testing machine
RCA Recycled coarse aggregate
NCA Normal coarse aggregate
CCS Conventional concrete

1 Introduction

In recent times concrete industry comprises approximately 30% of the total market for aggregates. Concrete is the most important and widely used construction material in all forms of engineering works, including low and high rise buildings, infrastructure, dams, bridges and other developments [1]. Various studies have been conducted in India to estimate the waste generated in the construction sites. It is estimated around 14.4 million tons/year [2]. Mainly the waste comprises of unused and demolished concrete, bricks, woods and other hazardous material [3]. Concrete constitutes approx. 40 of all the waste generated in construction and demolition work. Also the trends shows that in coming years the waste generated by the construction industry will only increase. So it becomes significant that we have to reduce the waste generation during construction [4]. It is now widely accepted that there is a potential to use recycled and reused concrete while construction to minimize the cost and also it will be environment friendly [5]. Researchers are constantly working on finding the efficient concrete mix design for replacing conventional concrete. Also the cost of recycled concrete aggregate will be 20–30% less than conventional concrete [6].

This research highlights the advantage and possible methods to make recycled concrete aggregate more efficient in terms of strength and durability [6]. The research was done using M-30 grade of concrete as it is most common and readily used in construction of buildings. As known the workability of concrete depends on the nature and strength of aggregate [7]. Acid treatment test on recycled coarse aggregate to remove the mortar, to make it more efficient [8]. Recycled concrete aggregates contain not only the original aggregates, but also hydrated cement paste. This paste reduces the specific gravity and increases the porosity compared to similar natural aggregates [9]. Higher porosity of recycled aggregates leads to a higher absorption [10]. To reduce the water absorption of recycled aggregates, chemical treatment in three different chemicals i.e. sulfuric acid, hydrochloric acid and acetic acid in 1.5, 2.0 and 2.5% concentration has been done. Suitable percentage for replacement for natural aggregate has been found. The percentage of replacement for natural aggregate taken were 20, 25 and 30%. For research, 7 and 28 days strength of concrete is important parameter to which gives us the idea of selection or rejection of concrete design [11]. So compared study of the compressive strength, splitting tensile and flexural strength of concrete cubes, cylinders and beams respectively for both conventional and recycled concrete aggregate.

2 Material

2.1 *Cement*

The concrete specimens utilized in this study were of M-30 grade and ordinary Portland 53 grade cement was used for the same. The specific gravity of cement was found to be 3.13.

2.2 *Untreated Recycled Coarse Aggregate*

The size of aggregate was kept in between 12 and 20 mm in the project. The specific gravity and water absorption of untreated RCA was found to be 2.514 and 3.78% respectively.

2.3 *Treated Recycled Coarse Aggregate*

Recycled coarse aggregate is treated for 3 days in the solution of H_2SO_4, HCl and CH_3COOH solution with different percentage of concentration. Specific gravity and water absorption are in Tables 1 and 2.

Table 1 Specific gravity of treated RCA

%	Chemicals		
	H_2SO_4	HCl	CH_3COOH
1.5	2.632	2.683	2.736
2	2.594	2.617	2.695
2.5	2.551	2.52	2.672

Table 2 Water absorption (%) of treated RCA

%	Chemicals		
	H_2SO_4	HCl	CH_3COOH
1.5	3.223	3.327	3.574
2	3.065	3.219	3.398
2.5	2.908	3.071	3.215

2.4 *Normal Coarse Aggregate*

The size of aggregate was kept in between 12 and 20 mm in the project. The specific gravity and water absorption of NCA was found to be 2.514 and 3.78% respectively.

2.5 *Fine Aggregate*

The size of aggregate was kept in between 12 and 20 mm in the project. The specific gravity and water absorption of FA was found to be 2.514 and 3.78% respectively.

3 Experimental Program

3.1 *Compressive Test*

For compressive tests, specimens of size 100 mm × 100 mm × 100 mm cubes were casted with design mix of M-30 as per IS: 10262 specifications. The cubes were then cured for a duration of 7 and 28 days in water maintained at room temperature. The loading rate for tests was set to be 2.5 KN/m^2. The measured compressive strength of the cubes is then determined by dividing the maximum load applied to the cubes during the test by the cross-sectional area of the cube i.e.

$$\text{Compressive strength} = \frac{Failure\, load(\text{N})}{Area\left(\text{mm}^2\right)}$$

For durability test, cube specimens are kept in water for curing for 28 days and after they were kept in acidic and basic solutions for 56 days (Tables 3 and 4).

Table 3 Compressive strength of various percentage of RCA concrete (MPa)

Sample	0%	20%	25%	30%
7 days	23.267	27.567	30.667	29
28 days	36.967	34.6	38.33	35.4

Table 4 Compressive strength of various percentage of RCA concrete after acidic and basic exposure (MPa)

Sample	0%	20%	25%	30%
H_2SO_4	30.55	26.5	30.45	28.95
Na_2SO_4	30.5	31.25	36.4	33.2

Table 5 Split tensile strength of various percentage of RCA concrete (MPa)

Sample	0%	20%	25%	30%
7 days	2.282	2.113	2.272	1.539
28 days	2.972	2.792	2.951	2.813

3.2 *Split Tensile Test*

For split tensile strength, the cylinders were casted of size 100 mm in diameter and 200 mm in height. The samples were made with a design strength of 30 KN/mm^2 as per IS: 10262 specifications. The cylinders were cured for a duration of 7 and 28 days in water at room temperature. The loading rate was set to be 2.5 KN/m^2. The split tensile strength was then determined by the equation, $f_{cts} = 2P/\pi ld$; where 'P' is maximum load in Newton's applied to the specimen, 'l' is the length of the specimen, and 'd' is the cross sectional dimension of the specimen (Table 5).

3.3 *Flexural Test*

For flexural strength, the beams were casted of size 100 mm in both depth and width and 500 mm in length. The samples were made with a design strength of 30 KN/mm^2 as per IS: 10262 specifications. The beams were cured for a duration of 7 and 28 days in water at room temperature. The beams are kept in the setup of 4 point bending. The flexural strength or modulus of rupture was then determined by the equation, $f_{ct} = 3F\,(L - L_i)/2bd^2$; where 'F' is the load (force) at the fracture point, 'L' is the length of the support span, 'b' is the width, 'd' is the thickness and 'L_i' is the length of the loading (inner) span (Table 6).

Table 6 Flexural strength value for various percentage of RCA (Mpa)

Sample	0%	20%	25%	30%
7 days	6.66	6.15	6.92	6.36
28 days	7.5	7.45	8.28	8.62

Table 7 Rebound index value for various percentage of RCA (MPa)

Sample	0%	20%	25%	30%
7 days	31.367	30.100	32.767	30.367
28 days	39.700	39.167	40.800	40.133

Table 8 Pulse velocity value for various percentage of RCA (km/s)

Sample	0%	20%	25%	30%
7 days	4.223	4.333	4.480	4.273
28 days	4.533	4.517	4.563	4.413

3.4 *Rebound Hammer Test*

As per IS: 13311(Part 1)-1992 specifications, rebound Hammer or Schmidt Hammer test is a Non-destructive testing method of concrete which gives a convenient and rapid indication of the compressive strength of the concrete for which we have used cube samples of dimensions 100 mm × 100 mm × 100 mm. The measured value is designated as Rebound Number or Rebound Index (Table 7).

3.5 *Ultrasonic Pulse Velocity Test*

As per IS: 13311(Part 1)-1992 specifications, ultrasonic testing of concrete or ultrasonic pulse velocity test on concrete is a non-destructive test to assess the homogeneity and integrity of concrete for which we have used cube samples of dimensions 100 mm × 100 mm × 100 mm (Table 8).

3.6 *Durability Test*

As per IS 456:2000 specifications, for durability of concrete, weight of concrete specimen are taken. Durability of concrete is very wide aspect while considering concrete properties. The specimen are kept in two chemicals, one in acidic and other in basic solution in specific concentration. The samples are kept in curing in each

Table 9 Percentage change in weight in both acidic and basic exposure (%)

Sample	0%	20%	25%	30%
H_2SO_4	5.461	5.131	4.912	4.917
Na_2SO_4	3.907	3.889	3.748	3.734

chemical for about 56 days. Then weight of each concrete specimen are noted down. The acidic chemical taken was sulfuric acid and basic chemical taken was sodium sulphate which was taken in a ratio of 5% by mass. Change in weight of the specimen was considered for concrete specimen (Table 9).

4 Result and Discussion

4.1 Compressive Test

After observing compressive strength of 7 days from Fig. 1, it has been seen that in 7 day test there was increase in compressive strength till 25% and after 25% there is decrease in the compressive strength.

And from observing Fig. 2, we came to know that the compressive strength decreases from 0 to 20% and then there is a sudden increase at 25% and then it decreases till 30%.

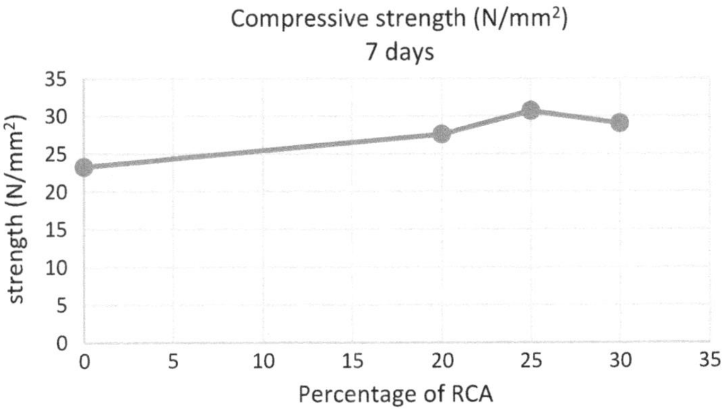

Fig. 1 Compressive strength of 7 days at different percent of RCA

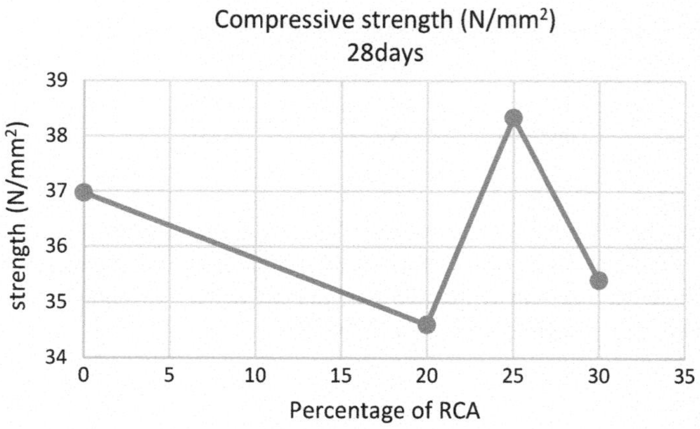

Fig. 2 Compressive strength of 28 days at different percent of RCA

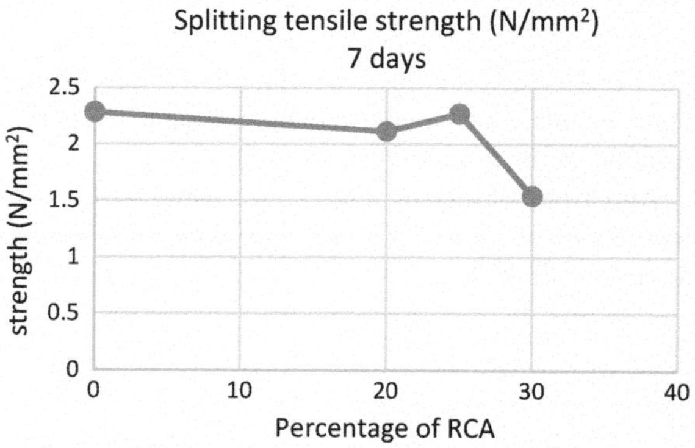

Fig. 3 Tensile strength of 7 days at different percent of RCA

4.2 Splitting Tensile Test

After observing splitting strength of 7 days from Fig. 3, it has been seen that splitting tensile strength at 7 days partially decreases from 0 to 20% and again increases at 25% and a sudden decrease happens after 25%.

And from observing Fig. 4, we came to know that the compressive strength decreases at a rapid rate from 0 to 20% and then there is a sudden increase at 25% and then it again suddenly decreases till 30%.

4.3 Flexural Test

After observing flexural strength of 7 days from Fig. 5, it has been seen that in 7 day test there was decreases at a gradually from 0 to 20% and then there is a sudden increase at 25% and then it again suddenly decreases till 30%.

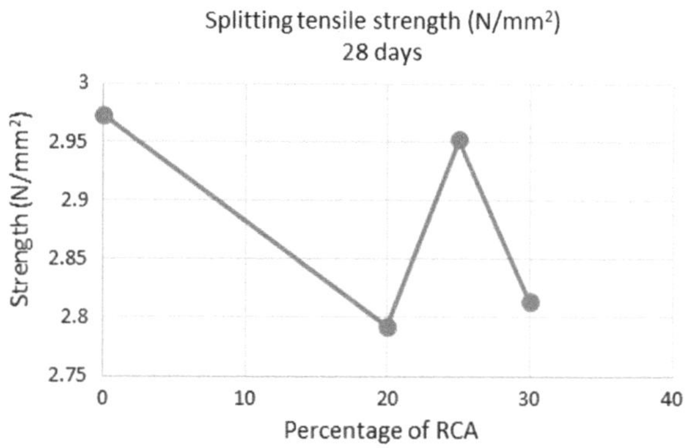

Fig. 4 Tensile strength of 28 days at different percent of RCA

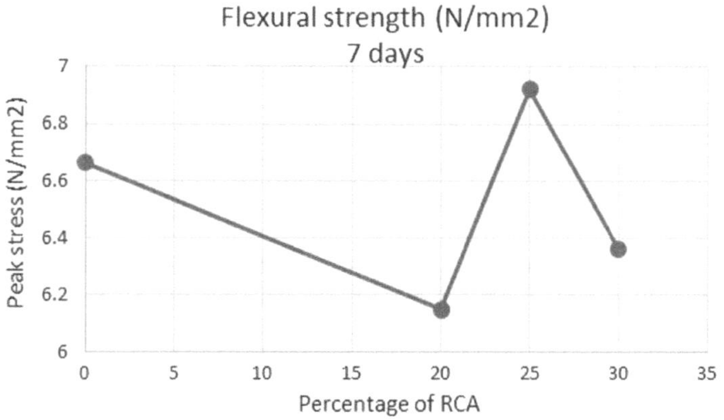

Fig. 5 Flexural strength of 7 days at different percent of RCA

And from observing Fig. 6, we came to know that the flexural strength is approximately constant from 0 to 20% and then there is gradual increase in the flexural value till 30%.

5 Conclusion

Mechanical and durability properties of recycled coarse aggregate concrete has been done in comparison to the conventional concrete. These concrete specimen was cured in water for 7 and 28 days for checking out the compressive, split tensile and flexural strengths. In addition we took some specimen of cubical shape and cured it in Sulphuric solution for a period of 56 days to check the durability properties. Various properties that has been observed are as follows.

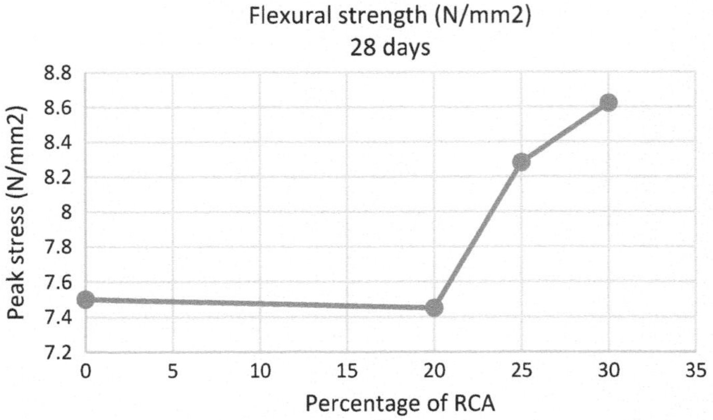

Fig. 6 Flexural strength of 28 days at different percent of RCA

- The slump value of concrete mix decreases with increase in the percentage of RCA.
- The compressive strength for 7 days increases slightly from 0% till 25% and decreases afterwards. Whereas compressive strength for 28 days decreases from 0 to 20% and then suddenly increases from 20 to 25% and then decreases rapidly till 30%.
- The split tensile strength for 7 days decreases slightly from 0 to 20% and increases slightly from 20 to 25% and then decreases suddenly till 30%. Whereas split tensile strength for 28 days decreases from 0 to 20% and then suddenly increases from 20 to 25% and then decreases rapidly till 30%.
- The flexural strength for 7 days decreases exponentially from 0 to 20% and then suddenly increases from 20 to 25% and then decreases rapidly till 30%. Whereas flexural strength for 28 days decreases slightly from 0 to 20% and then exponentially increases from 20 to 30%.
- From durability tests we got that average percentage weight change is maximum in the concrete specimen with no RCA and the least change occurs in the specimen having more RCA percent for both acidic and basic sulphate curing.
- Compressive strengths of the sulphate cured specimen showed slightly weak strength of the specimen in acidic sulphate curing and showed a good strength of the specimen in basic sulphate curing.
- The NDT showed a good average for all the specimen after 7 and 28 days of curing.
- Cost effectiveness was shown for different percentage replacement of RCA's.

References

1. Sharma P (2015) An introduction to recycled aggregate concrete: production and applications. Int J Res 296–303

2. Patil SP, Ingle GS, Sathe PD (2013) Recycled coarse aggregates. Int J Adv Technol Civ Eng 2(1):27–33
3. Sorato R (2016) Recycled aggregate concrete; an overview, p 28
4. Tam VW, Tam C, Le K (2016) Removal of cement mortar remains from recycled aggregate using pre-soaking approaches. Elsevier Ltd., pp. 82–101
5. Safiuddin M, Alengaram UJ, Rahman MM, Salam MA, Jumaat MZ (2016) Use of recycled concrete aggregate in concrete: a review. J Civ Eng Manage 1–35
6. Yehia S, Helal K, Zaher A, Abusharkh A, Istaitiyeh H (2015) Strength and durability evaluation of recycled aggregate concrete. Int J Concr Struct Mater 9(2):219–239
7. Vázquez E (2016) Recycled aggregates for concrete: problems and possible solutions. Int J Earth Environ Sci 1(122)
8. Saravanakumar P, Abhiram K, Manoj B (2016) Properties of treated recycled aggregates and its influence on concrete strength characteristics, vol 111. Elsevier Ltd., pp. 611–617
9. Sharma J, Singla S (2014) Study of recycled concrete aggregates. Int J Eng Trends Technol (IJETT) 13(3):123–125
10. Tiwari PK, Nateriya DR (2016) Replacement of recycled coarse aggregates with natural coarse aggregates in concrete. Int J Sci Eng Appl Sci (IJSEAS) 2(7):174–183
11. Shetty MS (2005) Concrete Technology. College of Military Engineering Pune

Inplane Lateral Load Behaviour of Masonry Walls

Anusree Sivadas, K. N. Kavya, Praseetha Prakash, Seethal Sharma and Jacob Alex Kollerathu

Abstract Masonry is one of the commonly used construction technology both in urban and rural areas. In this paper the in-plane behaviour of masonry walls is analytically studied considering existing closed form equations. Previous studies have proven that the lateral load behaviour mainly depends on the aspect ratios (h/L) as well as the axial loads. From this analysis the governing failure is determined and the lateral load versus lateral deflection curve is plotted for various percentages of axial loads. This graph gives the ductility of the wall. This concept is further applied to a simple masonry structure and the push over curve is plotted.

Keywords Unreinforced masonry · Storey-shear mechanism · Pushover analysis

1 Introduction

Masonry has been used, in a wide variety of forms, as a basic construction material specially to make a load bearing walls (structural walls) and infill walls (non-structural walls) for public and residential buildings (Fig. 1).

Seismic response of masonry structures is typically characterised based on two mechanisms:

1. In-plane mechanism
2. Out of plane mechanism

URM has low resistance under in-plane and out-of-plane seismic demand as the shear capacity and displacement ductility is lower. Hence URM is vulnerable to even low-moderate seismic loads. A lot of research has been done to explore ways

A. Sivadas · K. N. Kavya · P. Prakash · S. Sharma
Adi Shankara Institute of Engineering and Technology, Ernakulam, India

J. A. Kollerathu (✉)
Department of Civil Engineering, Faculty of Engineering, Christ (Deemed to be) University, Bengaluru, India
e-mail: jalexutd@gmail.com

K. Dasgupta et al. (eds.), *Proceedings of SECON'19*,
Lecture Notes in Civil Engineering 46,
https://doi.org/10.1007/978-3-030-26365-2_30

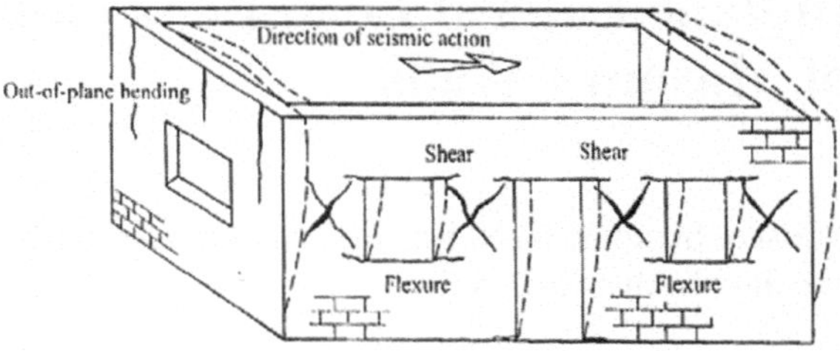

Fig. 1 Global and local failure mechanisms in masonry buildings (adapted from Alex and Menon [2])

to increase the shear capacity and displacement ductility of URM. Observations from research investigating the seismic response of URM structures such as [1] indicate that based on geometrical parameters and the level of axial load, the failure mechanisms of masonry piers subjected to seismic actions can be summarized as follows:

1. Flexural rocking
2. Diagonal tension
3. Sliding shear.

2 Lateral Load Behaviour of URM Walls

Generally, masonry walls are subjected to a combination of lateral and vertical loading. There are three simplified criteria used in the study of in-plane strengths of masonry walls [3].

- Maximum principal tensile stress criteria
- Coulomb-type criteria
- Flexural compression criteria

Here, l and t are length and thickness of the masonry wall respectively, σ_o and N is the Axial stress and axial load on the wall respectively, V is the Shear force on the wall, f_t is the tensile strength of masonry, b is the Aspect ratio of the wall. α_V is the shear span which is given as Ho/l, where H_o is the effective height of the wall, c is the Cohesive strength, μ is the Coefficient of friction, σ is the axial stress on the bed joints (Table 1).

An empirical criterion to evaluate b is [4]:

- $b = 1.5$ when $h/l \geq 1.5$ (slender walls),
- $b = 1.0$ when $h/l \leq 1.0$,
- $b = h/l$ for $1.0 < b < 1$.

Average compressive stress on the section, $\sigma_0 = N/lt$

Table 1 Failure criteria equations

Failure criteria	Failure mode	Equation
Maximum principal tensile stress criterion	Diagonal cracking	$V_u = f_t l t \sqrt{\left(1 + \frac{\sigma_O}{f_t}\right)}$
Coulomb-type criterion	Sliding shear	$V_u = lt\left(\frac{1.5c+\mu\sigma}{1+\left(\frac{3c\alpha}{\sigma}\right)}\right)$
Flexural compression criterion	Flexural rocking	$V_u = \frac{\sigma_O H}{2}\left(1 - \frac{\sigma_O}{kf_m}\right)$

3 Unreinforced Masonry Wall Analysis

3.1 Shear Capacity Versus Axial Load

The input data taken for this analysis are Length of the range 1000–4500 mm at an interval of 500 mm, thickness of 230 mm, height of 3000 mm, axial Load of 0–Nu at an interval of 25 kN, cohesion c and Poisson's ratio μ is taken as 0.3 and 0.4 respectively, Compressive strength of masonry (f_m) = 10 N/mm^2 and Tensile strength of masonry (f_t) = f_m/10, Shear block factor (κ) = 0.85. Using the above data, a graph is plotted between shear capacity (Vu) and axial load (N) with varying axial loads and aspect ratios. From these graphs, governing shear is determined and it is observed that governing shear changes with change in aspect ratio.

Figure 2 represents the shear capacity versus axial load graph for L = 1000 mm from the graph it was observed that the governing mode of failure is flexure alone. Figure 3 represents the graph for L = 4500 mm, in which the governing mode of failure is a combination of flexure and sliding shear as the axial load changes. It can be inferred that as the aspect ratio of the wall varies there is change in the behavior of the wall as well. Further, it was observed that from L = 1000 mm to 2500 mm the governing mode of failure is flexure. This observation is represented further in a Table 2.

Fig. 2 Vu versus N for L = 1000 mm

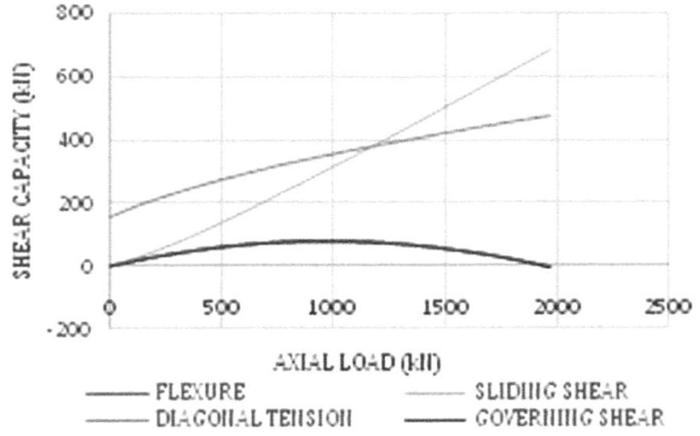

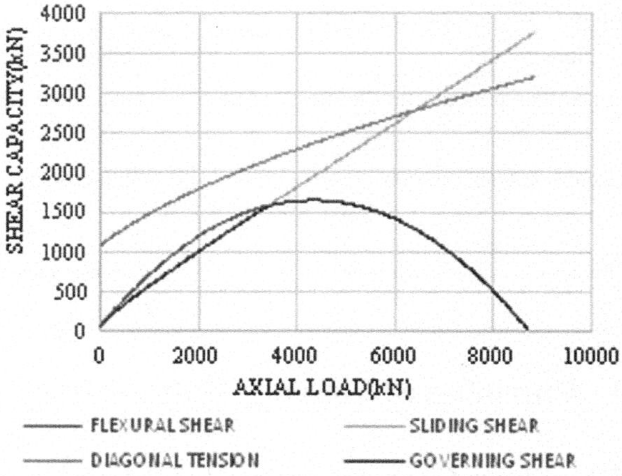

Fig. 3 Vu versus N for L = 4500 mm

Table 2 Behavior of walls for various aspect ratio

Aspect ratio	Failure mode
$\left(\frac{h}{l}\right) > 1.2$	Flexural rocking
$1 \geq \left(\frac{h}{l}\right) \geq .67$	Sliding shear/flexural rocking

3.2 *Development of Non-linear Force Displacement Curve*

In this case a graph between shear force and deflection is plotted. Here the wall is assumed to be one end fixed and other end hinged and loading is taken as the combination of axial as well as lateral load. The graph is plotted for different walls of varying lengths ranging from 1000 to 4500 mm, for each case, a certain percentage of ultimate loads are assumed as the axial load for determining the ductility.

The input values taken are:

$$\delta e = \mathrm{V/(TS)}$$

where TS = $K_f + K_s$, K_f is the flexural stiffness, K_s is the shear stiffness and TS is the Total Stiffness, shear block factor k varies from 0.85 to 1

$$\delta \mathrm{u} = \begin{cases} 0.4 \text{ to } 0.8\mathrm{h} \text{ for flexure} \\ 0.8 \text{ to } 1.2\mathrm{h} \text{ for shear or diagonal tension} \end{cases}$$

The graph between shear force (V) and deflection (δ) is obtained as follows:

From the graph, it is observed that the graph varies linearly up to a certain axial load i.e. ultimate load (V_U). It shows the elastic behav- iour of the structure. The structure tends to regain its original shape even after deflection up to ultimate load. After a certain value of de- flection (δ_b) the structure behaves linearly with no increase

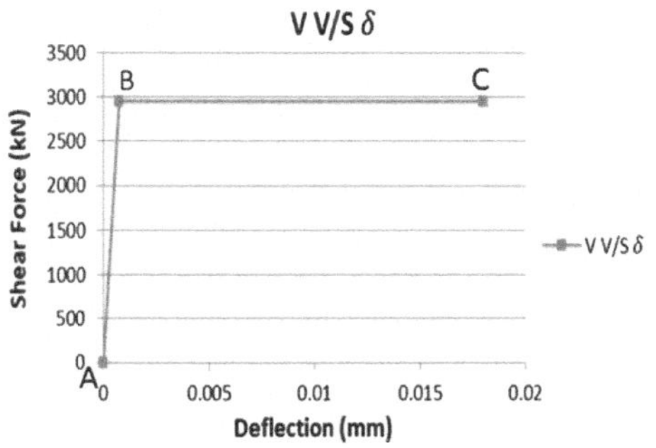

Fig. 4 Shear force versus displacement curve

Table 3 Ductility variation for different percentage of ultimate load

L	Nu	N/N_u	δ	L	Nu	N	δ
1000	1966.14	.20Nu	30.41	3000	5898.43	.20Nu	16.37
		.40Nu	15.20			.40Nu	8.188
		.60Nu	10.13			.60Nu	5.45
		.80Nu	7.60			.80Nu	4.09
4000	7864.57	.20Nu	24.37	4500	8847.64	.20Nu	23.29
		.40Nu	7.31			.40Nu	7.01
		.60Nu	4.87			.60Nu	4.67
		.80Nu	3.65			.80Nu	3.50

in the shear capacity constant (up to δu) until failure. After ultimate load, the structure fails (Fig. 4).

$$\text{Ductility} = \frac{\delta_b}{\delta_u}$$

Table 3 represents the ductility of walls subject to increasing axial loads. Similarly when the value of shear force decreases the value of ductility increases for different values of length. Results indicate that as the axial load demand on a wall increases there is a reduction in the ductility of the wall. Hence for ideal seismic behvaiour it is desired that the axial load demand on the wall be kept minimum.

4 Pushover Analysis

After determining the failure pattern and ductility behavior of a single wall, the analysis is further extended to a simple single storeyed building. A simple building plan in Fig. 5 having 4 walls is selected. The minimum value of shear force resisted

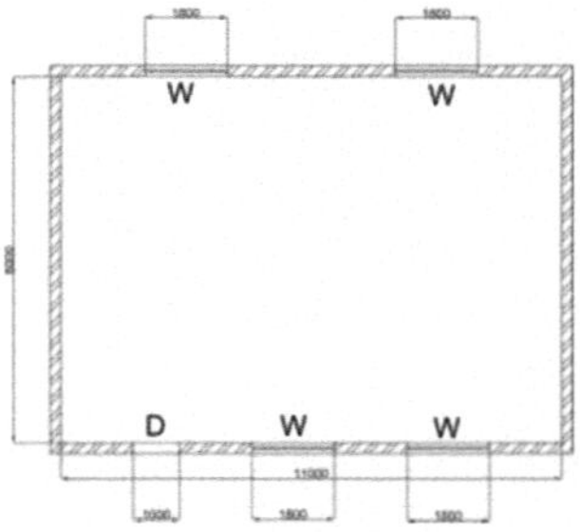

Fig. 5 Plan of a single storied structure

by these three behaviors is considered as the governing shear which represents the failure pattern of the unreinforced masonry structure. After determining the mode of failure the elastic displacement produced by the structure is determined using a ductility curve corresponding to the four walls.

$$\text{Elastic displacement,}\quad \delta_e = \frac{V}{T_s}$$

where

V = Shear for acting on the wall
TS = total stiffness of the wall.

After obtaining the values of displacement and shear forces of the individual piers, the displacement and shear force corresponding to the entire wall is obtained by taking the summation of all individual displacements and shear forces of all the piers. From the graph, it is observed that the graph varies linearly up to a certain displacement i.e. elastic displacement (δ_e). The wall tends to regain its original shape even after deflection up to ultimate load. After a certain value of deflection (δ_e) the shear force is constant until failure or ultimate displacement (δ_u).

Figures 6 and 7 represent the pushover curve is plotted in x and y directions by taking the summation of shear forces and displacements of walls prevailing in the corresponding directions.

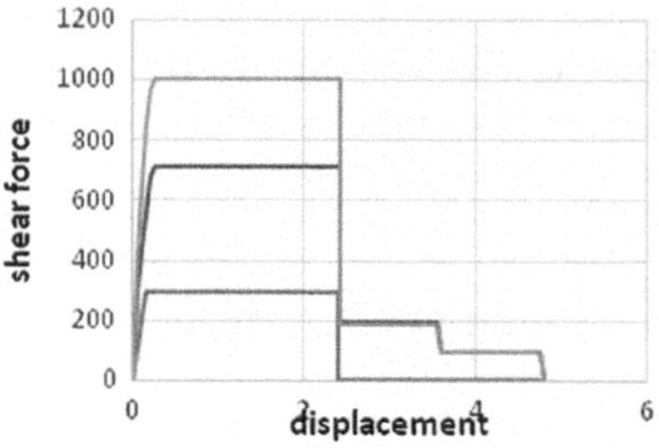

Fig. 6 Capacity in X direction

Fig. 7 Pushover in Y direction

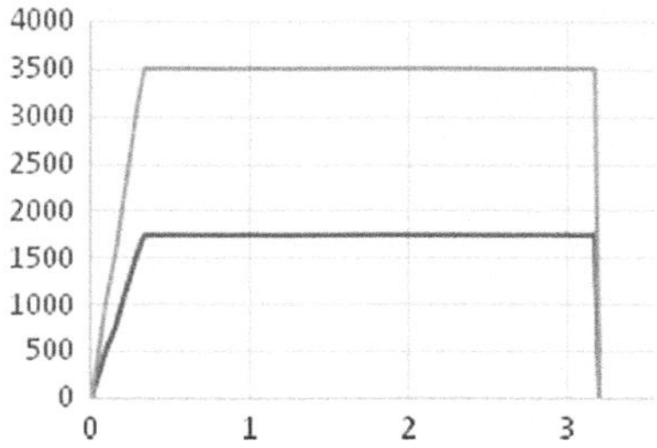

5 Observations

From the analysis it is observed that as the aspect ratio changes the governing mode of failure of the masonry walls also changes. For aspect ratio (h/l) > 1.2, the governing failure is flexural rocking and if it is between 1 and 0.67 the governing failure observed is sliding shear and flexure rocking.

From the non-linear force displacement curve it can be easily noted that, as the axial load increases there is a considerable reduction in the ductility for the observed masonry walls.

References

1. Gandhi R, Menon A (2014) Lateral load performance of bed joint reinforced masonry walls. In: International masonry conference, Portugal
2. Kollerathu JA, Menon A (2017) Role of diaphragm flexibility modelling in seismic analysis of existing masonry structures. Struct Res J Inst Struct Eng 11:22–39
3. Magenes G, Calvi GM (1997) In plane seismic response of brick masonry walls. Earthq Eng Struct Dynam 26:1091–1112
4. Tomaževič M (1996) Recent advances in earthquake resistant design of masonry buildings: European perspective. In: Proceedings of the 11th world conference on earthquake engineering, Acapulco

Numerical Analysis of Windowed Steel Tube Embedded in Concrete

Amrutha Anilkumar and P. E. Kavitha

Abstract Concrete-Filled Steel Tubes (CFSTs) are composite members consisting of steel tube filled with concrete. One of the main advantages of CFST is that the interaction between steel tube and concrete will increase load carrying capacity of the specimen. The local buckling of steel tube is delayed by restraint of concrete and strength of concrete is increased by confining effect provided by steel tube. Steel tubes in CFSTs are exposed to environmental deterioration. In the present study, this disadvantage is overcome by modifying CFST as steel tube embedded in concrete. The modified specimen showed better compressive strength but the bond strength was reduced. Further modification was made by providing holes in steel tube and embedding it in concrete. This improved the bond strength of specimen due to better bonding between inner and outer concrete and the compressive strength reduced due to reduction of steel section. To improve strength the specimen was modified as windowed steel tube embedded in concrete. The resulting specimen had better compressive strength and bond strength. In this paper standard size specimens are modelled and analyzed for compressive strength and bond strength by varying steel tube thickness and an optimum configuration of windowed steel section is obtained.

Keywords CFST · Bond strength · Compressive strength

1 Introduction

Columns are compression members that are provided on buildings to support the load coming from above and distribute them evenly downwards. Strength of a column and he manner in which it will fail is greatly dependent on its effective length. Columns that has slenderness ratio less than 12 are categorized under short columns.

A. Anilkumar (✉) · P. E. Kavitha
Department of Civil Engineering, Federal Institute of Science and Technology, Angamaly, India
e-mail: amrutha.ak94@gmail.com

P. E. Kavitha
e-mail: pe_kavitha@yahoo.com

K. Dasgupta et al. (eds.), *Proceedings of SECON'19*,
Lecture Notes in Civil Engineering 46,
https://doi.org/10.1007/978-3-030-26365-2_31

These are compression members and fails due to crushing. Since it is meant to take up only compression, reinforcements only take part in distribution of loads evenly throughout the column and to the adjacent stories. Hence, there is least participation of reinforcement in attaining strength as it is a tensile member. RCC and steel frames have been the most common frame systems for long times whereas composite frame system has recently been introduced for high rise buildings, bridge piers etc. The use of concrete filled steel tubes (CFST) in building construction has seen renaissance in recent years due to their numerous advantages, apart from its superior structural performance making a typical composite frame structure. CFST columns are composite structural members formed by encasing concrete into a hollow steel tube. The CFST section resists applied load through the composite action of concrete and steel due to the confinement effect on concrete by steel tube. Hence it has become popular in recent days in different countries worldwide and is being used in structures such as bridges, electricity towers, buildings etc. Extensive works have been carried out on CFST in past years and indicated that the CFST sections possess high ductility, strength and stiffness properties. The major drawback of using a CFST column is that the outer steel tube is exposed and is subjected to deterioration and eventually causes failure. Also, core concrete and surrounding steel tube are two different elements and they lack proper bonding. Apart from environmental deterioration of steel tube there can also be a chance of separation of steel tube from concrete. In this paper, all these conditions are considered to derive a suitable configuration of CFST.

2 Literature Review

Several studies were conducted by researchers on compressive strength and bond strength of CFST short columns.

Research on the bond strength in circular CFST was conducted [1]. Virdi and Dowling et al. performed push-out tests on stocky CHS filled with concrete of various grades and examined the effect of various parameters on the bond strength, such as age, strength, compaction and curing conditions of concrete, interface length, tube size and various surface treatments following which they concluded that the most significant factor contributing to the bond strength is the mechanical keying between steel and concrete.

Push out tests using static loading (SL) and variable repeated loading (VRL) on concrete filled steel tubular (CFT) circular stub columns were performed [2]. The bond strength obtained for CFT circular columns under static load conditions varied between 0.41 and 0.85 MPa. The incremental collapse low bound load limit was found to be 70% of the static collapse push out load. The influence of debonding on circular CFST stub columns was investigated experimentally and numerically [3]. The failure mode of debonding specimens is nearly the same as that of the specimens without debonding. However, the local buckling phenomenon in steel tube of debonding specimens is more serious than that of the specimens without debonding.

Relation between compressive strength and water cement ratio was experimentally tested [4]. It was proved that the strength increased with lower water cement ratio and using high strength concrete. Research were done on development of family of CFST and drew a research framework on CFST members [5]. Internal rings and shear studs were welded onto inner surface of steel tube and the bond strength was examined [6]. Axial loading was done on CFST short columns with shear and have a great effect on load distribution between the concrete and steel tubes and the enhancement was in the range of 13–15% [7].

Bond strength between the steel tube and core concrete was experimentally studied in full-scale concrete-filled steel tubes (CFST) through push-out [8]. It was found that welding internal rings onto the inner surface of the steel tube is the most effective method in enhancing the bond strength in CFST specimens.

Push-out tests were performed on concrete-filled stainless steel circular hollow section (CHS) tubes with different values of height-to-diameter ratio, diameter-to-thickness ratio and concrete strengths. It was concluded that 70% of the shear resistance of the bonding strength is taken by the interface friction force, while the remaining 30% of the shear resistance of the bonding strength is sustained by the chemical adhesive force and the mechanical interlock force [9].

Nonlinear analysis on CFST columns with and without stiffeners were done using ABAQUS 6.10-1 and verified with experimental test. The hollow and composite columns are extensively developed by considering different configuration of steel stiffeners [10].

In this study some modification are done to overcome the limitations of CFST. Apart from RC column (MR) and simple CFST columns (MC), three other configurations were analyzed in ANSYS to arrive at the required configuration.

3 Test Programme

In order to converge at the required objective finite element analysis was done on various modifications of CFST using ANSYS 16.0 software to check the strength and bonding between steel and concrete.

3.1 Material Properties

The specimens used in this project are composite in nature as it has both steel and concrete. S235 steel was chosen for the analysis. Length of steel tube being 300 mm throughout. Concrete used was of 28 days compressive strength equals to 56.2 MPa. Table 1 shows summarized material properties used for modelling in ANSYS 16.0.

Table 1 Material properties used for analysis

Material	Density (kg/m^3)	Elastic modulus (MPa)	Poison's ratio	f_{ck} (MPa)	f_y (MPa)
M56 concrete	2400	37,488	0.15	56.2	–
Steel bars	7850	210,000	0.3	–	415
Steel plates	7850	210,000	0.3	–	235

Table 2 Bond strength (MPa) results from ANSYS 16.0 for 3.31 mm thick steel tube

ID	20 mm × 20 mm	25 mm × 25 mm	30 mm × 30 mm
MC-WH	12.62	16.48	19.18

3.2 *Finite Element Analysis*

Software validation for ANSYS 16.0 was done with an experimental study done by Talha Ekmekyapar for a cylindrical specimen of 300 mm height and 114.3 mm diameter using mesh sizes 25, 20 and 10 mm. Better convergence of analytical results with experimental results were for those having mesh size 10 mm. Hence, uniform meshing of 10 mm was chosen for all models in further study. Standard cylindrical specimens of 300 mm long and 150 mm diameter with different configurations of steel plates were modelled in ANSYS 16.0. Initially all the engineering data of concrete and steel were fed as input values and geometry was created. Under static structural, load and support were created and assigned in terms of displacement. Force reaction, equivalent stress and total deformation were chosen as output variables based on which performance of specimens were compared in terms of bond strength and compressive strength. Loading and support conditions varied based on which analysis was performed.

3.3 *Methodology*

Existing literatures and Indian standard codes were referred and a gap in research was set for further research work. 5 types of column configurations were chosen each being an upgradation to the previous one (MR, MC, MC-WOH, MC-WH, and MC-WIN). In order to set a parameter to do further research, preliminary analysis were done in ANSYS 16.0 for compressive strength and bond strength on all the models by varying the thickness of steel tube and size of holes in steel plates respectively and a common pattern was identified in which the results varied. Load and support were given in terms of displacement at top and bottom respectively.

For compressive strength test, axial load was applied on the top face and support was given at the bottom face. End results will give the maximum axial load it can carry. For bond strength test, load was applied on concrete at the top and support was placed on steel tube at the bottom. The end results showed the maximum load at which the concrete slips out.

The major parameter set was volume of steel tube, on which the compressive strength mainly depend on and secondary parameter being size of holes in steel plates.

Results were compared with specimens MR and MC and obtained the best configuration in terms of maximum compressive strength and minimum cost.

3.4 Modelling and Analysis

All models are of 300 mm length and 150 mm diameter. Variations were made on steel tube thickness and holes in steel tube. The element types used for the analysis are SOLID186, CONTA174, TARGE170.

Modelling of irregular meshes are done using element type SOLID186 homogenous structural solid element. The contact between 3D target surfaces in case of general contact are represented by element type CONTA174. The various target surfaces associated with contact elements are represented as TARGE170. In the present study, TARGE170 is the target surface associated with contact element CONTA174.

3.4.1 RC Column (MR)

These are simple reinforced short columns whose reinforcement was designed based on IS456-2000. Column was designed for minimum number of longitudinal and transverse reinforcement and modelled in ANSYS 16.0 as shown in Fig. 1a. The minimum number of longitudinal bars provided in a column shall be six in circular column and the bars shall not be less than 12 mm in diameter. Therefore, 6 numbers of 12 mm bars are provided. Pitch of transverse reinforcement shall not be more than the least of following distances: {least lateral dimension, 16 times least diameter of longitudinal bars, 300 mm). Spacing of stirrups were chosen to be 150 mm. Total volume of steel used is 380×10^3 mm^3.

3.4.2 CFST Columns (MC)

CFST (MC) column of 300 mm height and 150 mm diameter with volume of steel tube being the same as that of reinforcements in RC column was considered. Thickness of steel tube chosen was 2.74 mm. ANSYS 16.0 model of MC is as shown in Fig. 1b.

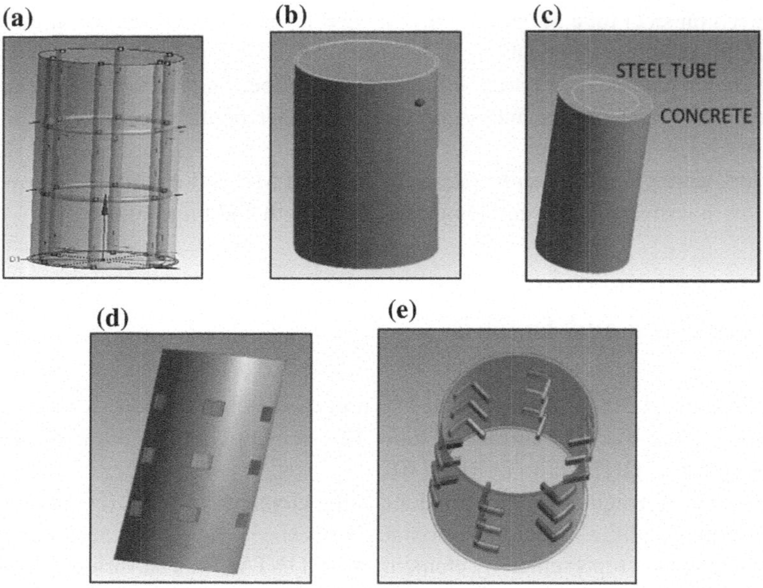

Fig. 1 Model of **a** MR, **b** MC, **c** MC-WOH, **d** Steel tube with holes in MC-WH, **e** windowed steel tube in MC-WIN

3.4.3 Steel Tube Embedded in Concrete Column (MC-WOH)

While considering CFSTs, the outer steel tube is subjected to extreme weather conditions. In India being a tropical region life of CFSTs might not be satisfying and will require excessive maintenance frequently over time. A solution to this problem was to embed same volume of steel tube inside the concrete leaving a concrete cover around steel tube and hence the modified CFST as shown in Fig. 1c.

3.4.4 Steel Tube with Holes Embedded in Concrete Column (MC-WH)

In case of MC-WOH profile of the cross section is a steel tube sandwiched between concrete. Steel and concrete are two different materials and their bonding will be poor. As a result there can be irregular deformation of steel and concrete. This was studied through bond strength test in which axial load was applied on concrete alone keeping the support only on steel. There was large amount of slip of concrete from the steel tube. As a solution to this problem the steel tube was provided with holes around it in a pattern similar to that of reinforcement in MR. 6 numbers of 20 mm^2 holes were given in an equidistant manner in 3 layers as shown in Fig. 1d and it was embedded in concrete.

Table 3 Compressive strength (MPa) of specimens by varying thickness of steel tube (mm)

ID	3.31	2.9	2.74	2.5	2	1.75
MC	86.37	81.43	79.9	77.96	73.54	70.68
MC-WHO	83.03	80.59	77.68	76.82	72.07	44.86
MC-WH	76.86	74.64	73.63	72.18	68.51	69.78
MC-WIN	79.89	76.47	74.64	77.48	69.90	70.54

3.4.5 Windowed Steel Tube Embedded in Concrete Column (MC-WIN)

In the previous model, holes were provided to improve the bonding between steel and concrete. But at the same time the compressive strength reduced since some amount of steel was removed while providing holes. So instead of wasting steel, an opening like a window inwards was provided as shown in Fig. 1e.

4 Results and Discussions

4.1 Preliminary Analysis

Compressive strength of all specimens by varying the thickness of steel tube are tabulated in Table 3. The variations showed that when steel tube was placed inside the specimen instead of providing it as outer cover, the volume of steel reduced with the change in tube diameter. Compressive strength reduced when the steel tube is provided with a reduced diameter and concrete cover. Hence it is evident that steel played a vital role in strength characteristics of the specimen. Another analysis was done to optimize the size of holes in MC-WOH for which it was analysed for bond strength and the results are shown in Table 2. Holes with 20, 25 and 30 mm^2 were chosen for comparison. The results showed that with the increase in hole size bonding also increased between steel and concrete as shown in Table 2. 25 mm^2 hole was chosen as the best option considering the maximum size of aggregates in concrete and the minimum spacing between reinforcements in RC specimens shall not be less than maximum of (a) diameter of bar if the diameters are equal, (b) diameter of larger bar if the diameters are unequal and (c) 5 m more than nominal maximum size of aggregate as suggested by IS 456 cl. 26.3.2.

4.2 Parametric Analysis

From the preliminary analysis volume of steel shall be kept as constant by varying tube thickness while changing from MC to MC-WOH. Size of holes in the rest of

Table 4 Compressive strength (MPa) and bond strength (MPa) results and comparison

ID	Thickness (mm) of steel plate	Volume (mm^3) of steel	Bond strength (MPa)	Compressive strength (MPa)	% increase in compressive strength
MR	–	377×10^3	–	58.90	–
MC	2.74	380×10^3	0.01	79.90	36.00
MC-WHO	3.31	380×10^3	3.25	83.03	40.90
MC-WH	3.31	380×10^3	12.62	76.83	30.40
MC-WIN	3.31	380×10^3	15.55	79.89	36.00

the models were set as 25 mm × 25 mm. Table 4 shows the results of parametric analysis of all configurations. Deformation of the specimens in bond strength test and compressive strength test are shown in Figs. 2 and 3 respectively.

5 Conclusions

RC columns are the most conventional type of columns used now a days. As discussed earlier while considering a short column, reinforcements are simply a means to transfer load and take up only tensile loads. Compressive loads are always taken up by concrete alone. In CFST columns both steel and concrete contribute towards carrying compressive loads. Also spalling of concrete is avoided by the confinement effect of concrete by steel. Use of CFST has been common in countries worldwide for building piers of bridges, columns of high rise buildings, arches etc. India being a tropical country CFST columns are a great risk economically as it will require frequent maintenance of outer steel tube due to prolonged exposure to extreme weather conditions.

Present study is intended to find a solutions for this problem through analysis of various models in ANSYS 16.0 and to suggest a better configuration that can be suggested instead of reinforced concrete (MR) and CFST columns (MC). The modified columns suggested are the columns with steel tubes embedded in concrete (MC-WOH), steel tube with holes embedded in concrete (MC-WH) and windowed steel tubes embedded in concrete (MC-WIN). There was 36% increase in compressive strength of MC compared to MR but lacked bond strength. Providing equal volume of steel tube inside the specimen MC-WOH as a protection for steel tube and also increased the compressive strength by 40.9% than MR. But as a composite structure there is always a chance of unequal load distribution and hence splitting of steel and concrete cover. To avoid that square holes of 25 mm × 25 mm were introduced in steel tube before embedding them in concrete (MC-WH) which improved bonding by 4 times but strength was 10% lesser than the result obtained for MC-WOH. The end configuration MC- WIN showed similar strength characteristics as that of MC

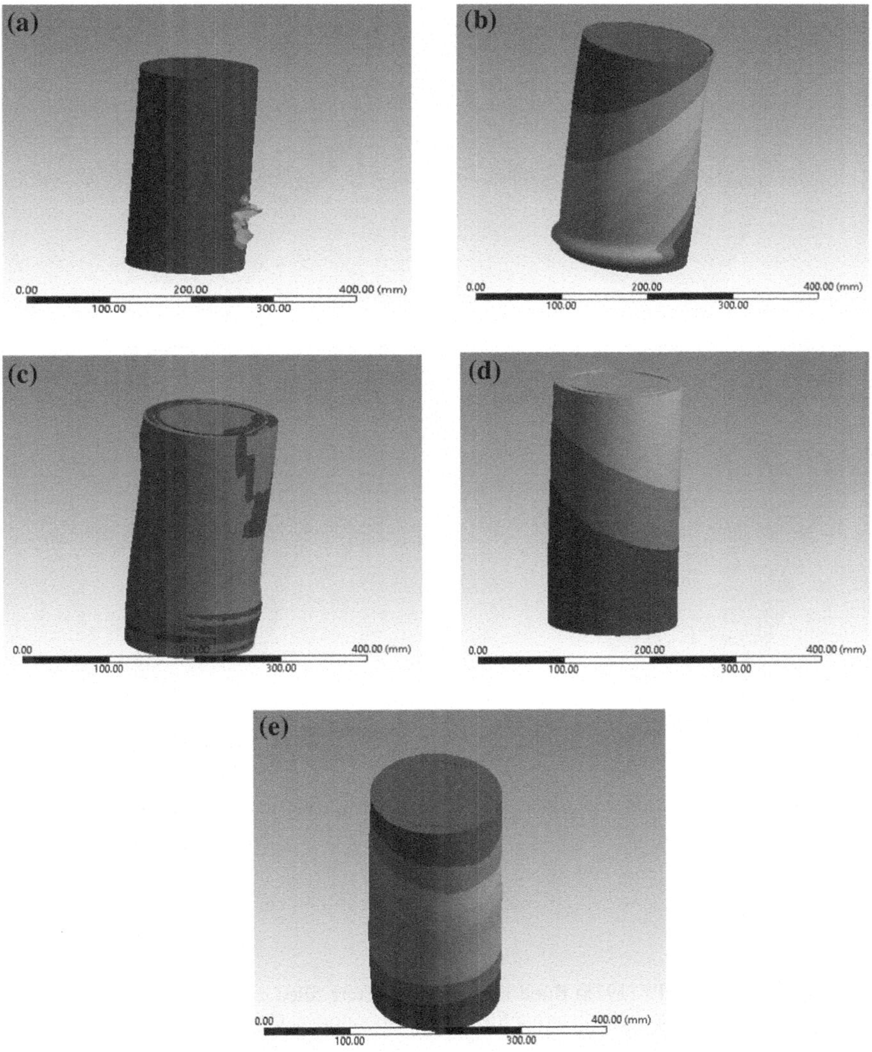

Fig. 2 Deformation of **a** MR, **b** MC, **c** MC-WOH, **d** MC-WH, **e** MC-WIN due to compression

with an additional advantage of steel being safe from environmental degradation and also being monolithic with concrete.

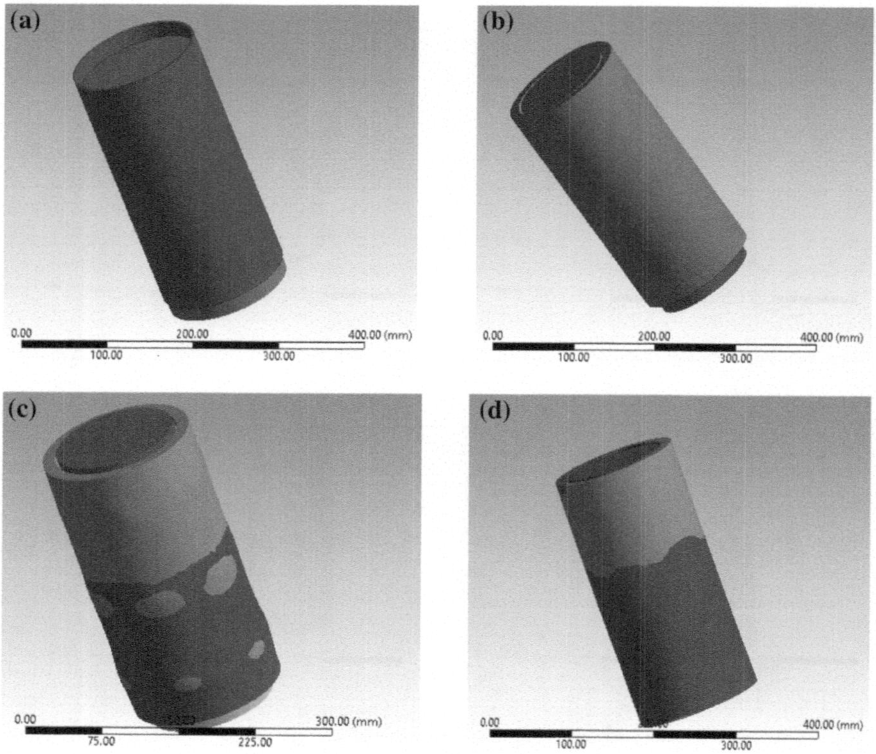

Fig. 3 Slip of concrete from steel **a** MC, **b** MC-WOH, **c** MC-WH and **d** MC-WIN

References

1. Virdi KS, Dowling PJ (1975) Bond strength in concrete filled circular steel tubes. CESLIC Report. CC11. London: Engineering Structures Laboratories, Civil Engineering Department, Imperial College, London
2. Aly T, Elchalakani M (2010) Incremental collapse threshold for pushout resistance of circular concrete filled steel tubular columns. J Constr Steel Res 66:11–18
3. Xue JQ, Brisagella B (2012) Effects of debonding on circular CFST stub columns. J Constr Steel Res 69:64–76
4. Khodaie N (2013) Effect of concrete strength on concrete steel bond in concrete filled steel tube. J Persian Gulf (Mar Sci) 4(11):9–16
5. Han LH, Li W (2014) Developments and advanced applications of concrete filled steel tubular structures. J Constr Steel Res 100:211–228
6. Tao Z, Song TY (2016) Bond behavior in concrete filled steel tubes. J Constr Steel Res 120:81–93
7. Younes SM, Ramadan HM (2016) Stiffening of short small size circular composite steel concrete columns with shear connectors. J Adv Res 7:525–538
8. Song TY, Tao Z (2017) Bond behavior of concrete filled steel tubes at elevated temperatures. ASCE 2017

9. Chen Y, Feng R (2017) Bond-slip behaviour of concrete-filled stainless steel circular hollow section tubes. J Constr Steel Res 130:248–263
10. Khanr AU, Kumar NS (2015) Nonlinear analysis of with and without stiffeners of hollow and concrete filled steel tube column by using ABAQUS 6.10-1. IJRET. eISSN: 2319-1163, pISSN: 2321-7308

Cyclic Response of Reduced Beam Section

Lakshmi Priya and Gayathri Krishna Kumar

Abstract Recent earthquakes have shown that steel moment frame (SMF) with weld connections are so brittle. According to the studies conducted, great damages are due to the cracking of the weld between the beam flange and the column face and inducing concentrated stresses in this area. A natural way to solve the problem is to reduce the ductility demand on the welded areas by reducing the stress concentration level. A useful approach to reduce the stress concentration at the panel zone could be the use of reduced beam section (RBS). A portions of the beam flanges are trimmed away in the region adjacent to the beam-to-column connection (Han et al. in J Constr Steel Res 70:256–263, 2012) [1]. The RBS can be viewed as a ductile fuse that forces the formation of the plastic hinge away from the joint so that much of ductility demand on beams may result from the RBS instead of the welded beam-to-column interface. RBS connection is widely investigated and used in US, Japan and Europe. In this study moment connections with different shape of reducing beam flange have been modeled using ANSYS and compared with each other during cyclic behavior. The sloped moment connection with best configuration is analyzed to find the cyclic response of the connection (Shen et al. in Eng Struct 22:968–983, 2000) [2].

Keywords Steel moment frame · Reduced beam section · Ductility demand · Cyclic behavior

1 Introduction

Recent earthquakes have shown that steel moment frame (SMF) with weld connections are so brittle. According to the studies conducted, great damages are due to the cracking of the weld between the beam flange and the column face and inducing

L. Priya · G. Krishna Kumar (✉)
Department of Civil Engineering, Federal Institute of Science and Technology, Angamaly, India
e-mail: gaya3krishnakumar92@gmail.com

L. Priya
e-mail: lakshmiumidhun95@gmail.com

K. Dasgupta et al. (eds.), *Proceedings of SECON'19*, Lecture Notes in Civil Engineering 46,
https://doi.org/10.1007/978-3-030-26365-2_32

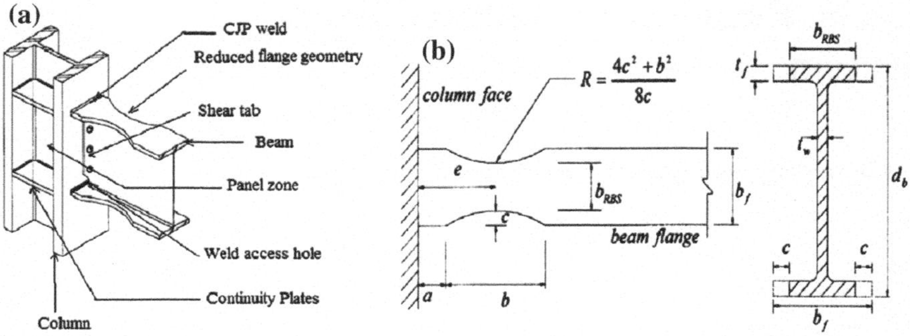

Fig. 1 RBS connection and geometry details. *Source* Swati and Gaurang [4]

concentrated stresses in this area. A useful approach to reduce the stress concentration at the panel zone could be the use of reduced beam section (RBS) Fig. 1. The reduced beam section is also called dog bone connection. In this we are reducing a portion of beam flange at a short distance from column's face so that the plastic hinge formation can be shifted to the beam regions protecting the connection from any type of failure. In general various shape cutouts are possible such as constant cut, rectangular cut, trapezoidal, holes etc. Extensive studies demonstrating that the radius cut behaves with highest rotational capacity. The advantages of the RBS are listed below [3]:

- RBS connection exhibits very good cyclic performance due to the hinge forms at RBS area.
- No weld fracture in RBS connection and connection region is elastic.
- Connection is full strength and offers satisfactory ductility, and can be used in special moment resisting frames.
- The energy dissipation capacity can be significantly improved by optimization compared with the normal beam with uniform flange width.
- RBS can be viewed as ductile fuse, force yielding occur in RBS area and can sustain large inelastic strains while at same time limiting stress in less ductile region.

Swati and Gaurang [4] conducted a study to learn the advantages and usefulness of RBS connection against connection without RBS for Indian profiles. They tested two connections under cyclic loading. The objective of the study was to investigate experimentally the cyclic behaviour of welded moment connections with and without RBS. Two external joint specimens were tested to compare and observe connection behaviour. Nonlinear finite element analysis of the connection models performed using the computer programme ANSYS. The results of the theoretical model created with the finite element are compared with those obtained from the experimental study. The analysis observed that specimen without RBS performed poorly due to cracks started at the bottom flange weld. The specimen with RBS reached rotation capacity of 0.02 radians without damage in the welds. This study concluded that Both the experimental and numerical results observed that cyclic performance of the RBS moment connection was much superior to the connection without RBS. No weld fracture was observed in RBS connection while there was a crack observed near beam bottom flange weld for connection without RBS. A reduction in material and labour cost is possible due to elimination of continuity for RBS moment connection. Numbers of tests conducted in above study are quite limited and more extensive testing is recommended to understand the behavior of RBS for Indian profiles [5].

An experimental program was performed by the Pachoumis et al. [6] in order to evaluate the proposed values of the geometrical characteristics of the RBS, and the results are presented in the present paper. Investigates the cyclic performance of the steel moment resisting connection with reduced beam section. Two full-scale subassemblages were tested under cyclic loading and the results are compared with those obtained from the theoretical model, using the finite element method. Both the experimental and numerical specimens exhibited satisfactory levels of connection ductility required of special moment-resisting frames. Although the values of the geometrical parameters were not according to the recommendations proposed, the plastic rotation exceeded the acceptable 0.03 rad without any weld fracture or any sign of distress at the face of the column. The cyclic performance of the RBS moment-resisting connection is excellent when the plastic hinge is formed at the RBS area. The RBS can be viewed as a ductile 'fuse' that forces yielding to occur within the reduced section of the beam, an area that can sustain large inelastic strains while at the same time limiting the stress in the less ductile region near the face of the column [7].

2 Test Programme

In order to converge at the required objective finite element analysis was done on various configurations of RBS using ANSYS 16.0 software to study the response of the reduced beam section under cyclic loading.

2.1 Material Properties

An element SOLID 186 from ANSYS element library was used for the 3-D finite element modelling of the RBS moment connection. The fundamental assumptions made to idealize steel mechanical properties are including: Young's modulus of 2×10^5 MPa, Poisson's ratio of 0.3. Multi-linear stress strain curve are input directly as element material property for cyclic analyses. Yield strength for column and beam will be 334 MPa and 330 MPa respectively [8]. The details of the specimen chosen for the study is shown in the Table 1. The column was assumed as pin connected at both the ends and at the joint beam to column element connection is configured as fully restrained. The table shows the different parameters used for modeling in ANSYS 16.0.

Table 1 Details of the specimen

Member	Beam	Column
Depth, d (mm)	200	162
Web thickness, t_w (mm)	5.6	8
Flange width, b_f (mm)	100	154
Flange thickness, t_f (mm)	8.5	11.5

2.2 Methodology

The methodology adopted to converge to the above mentioned objective is as follows, The geometric model is created by using the I beam cross sections. Different RBS configuration models such as RBS with circular cut, trapezoidal cut, rectangular cut and with holes on each flange of the beam are created. Simple weld connections are provided as the beam column connection and the column ends are pin connected at both the ends. Each subassemblies loaded at the beam free end in the displacement control. Analyze the connection to find out the seismic behavior by applying the cyclic loading. Comparison of the results with the conventional beam column connection to obtain the results.

2.3 Loading and Boundary Conditions

The column is pin connected at both the ends thereby restricting the translations in x, y and z directions and at the joint beam to column element connection is fully restrained. To determine the seismic behavior of the different configurations, cyclic loading was applied at the beam tip. This was applied in accordance of the loading sequence as shown in the Fig. 2. The loading is applied as cyclic vertical displacement at beam tip. The details of the loading sequence is presented in Table 2.

2.4 Modelling and Analysis

Different RBS configurations were modelled in ANSYS 16.0. Initially all the engineering data of concrete and steel were fed as input values and geometry was created. Medium size meshing was done for all models. Under static structural, load and support were created and assigned in terms of displacement. Force reaction, equivalent stress and total deformation were chosen as output variables.

The study aims to develop a better steel moment connection with the reduced beam sections. Different configurations are tested to evaluate the behavior. The three dimensional modeling of this is done by using the finite element software ANSYS.

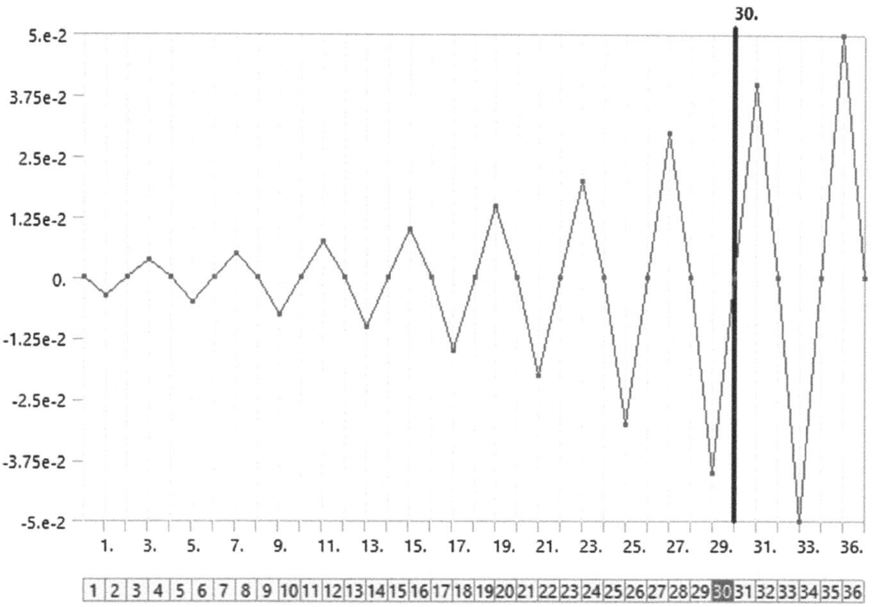

Fig. 2 Loading protocol

The modeling part include the modeling of beam, column, reduced sections on beam. The geometry is defined for each configurations and suitable properties are assigned to them [9].

2.4.1 RBS with Circular Cut

In this configuration a circular portion of the flanges of the beam is reduced near the column face. Beam web and flanges are connected directly by simple welded connection. The geometrical details of the RBS with circular cut connection is shown in the Fig. 3a. The detailed dimensions of the model is as per shown in the Table 3.

2.4.2 RBS with Rectangular Cut

In this configuration a rectangular portion of the flanges of the beam is reduced near the column face. Area of reducing the section for different configuration is made constant for the comparison of results. Beam web and flanges are connected directly by simple welded connection. The detailed dimensions of the model is as per shown in the Table 4. The FE model of the connection is shown in Fig. 3b.

Table 2 Loading schedule

Step	1	2	3	4	5	6	7	8	9
Beam tip displacement (mm)	±3.75	±5	±7.5	±10	±15	±20	±30	±40	±50

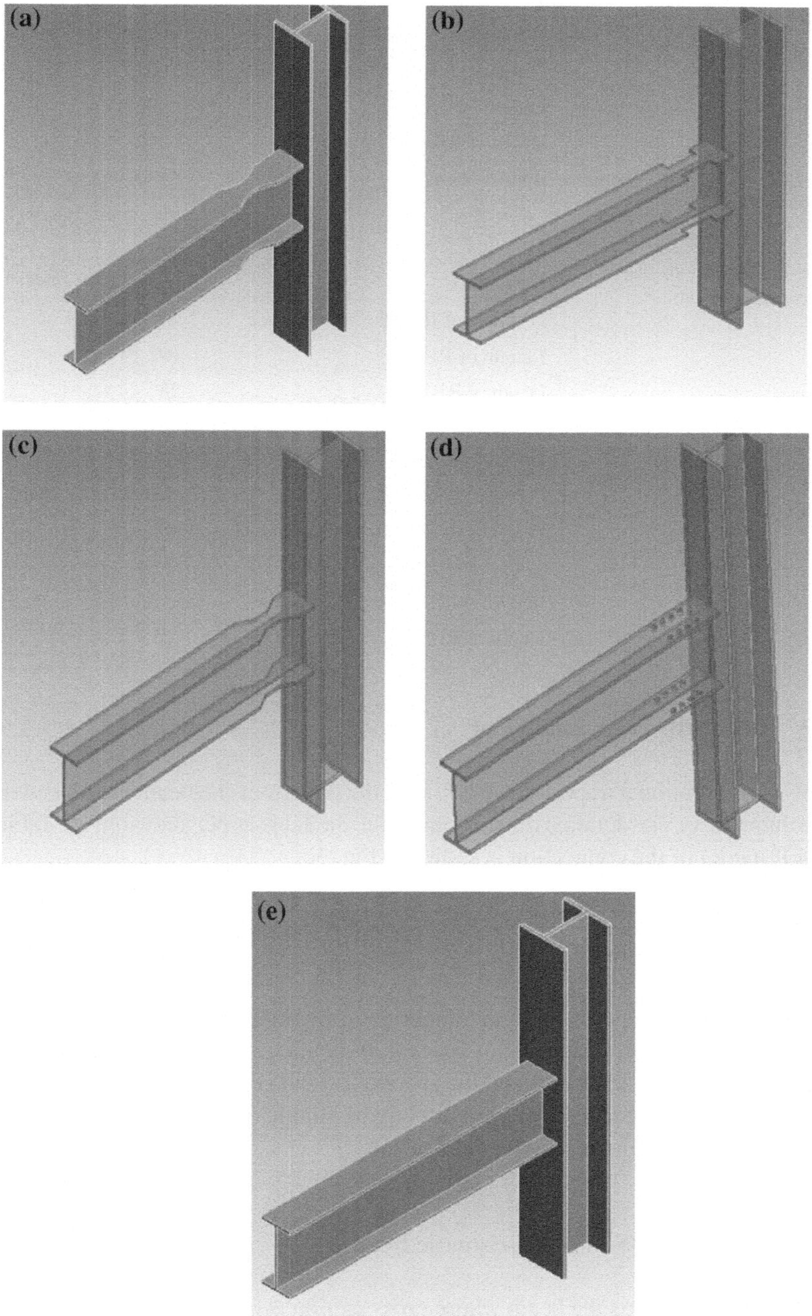

Fig. 3 Model of RBS with **a** circular cut, **b** rectangular cut, **c** trapezoidal cut, **d** holes and **e** without RBS

Table 3 Dimensions of the RBS with circular cut

Particulars	Dimension (mm)
Distance of RBS from column face	60
Length of RBS	160
Depth of RBS	25
Radius of cut	140.5

Table 4 Dimensions of RBS with rectangular cut

Particulars	Dimensions (mm)
Distance of RBS fro column face	60
Length of RBS	102.19
Depth of RBS	25

Table 5 Dimensions of RBS with trapezoidal cut

Particulars	Dimensions (mm)
Distance of RBS from column face	60
Length of RBS, b1	150
Length of RBS, b2	54.376
Depth of RBS	25

2.4.3 RBS with Trapezoidal Cut

In this configuration a trapezoidal portion of the flanges of the beam is reduced near the column face. The detailed dimensions of the model is as per shown in the Table 5. The FE model of the connection is shown in Fig. 3c.

2.4.4 RBS with Holes

In this configuration holes of equal diameter of the flanges of the beam is reduced near the column face. 8 number of holes with diameter of 14.258 mm is reduced from each flanges of the beam and the spacing is made constant between the holes. Beam web and flanges are connected directly by simple welded connection. The FE model of the connection is shown in Fig. 3d.

2.4.5 Effect of RBS in Non-prismatic Beams

Non-Prismatic beam is the beam whose cross section is not constant along its entire length. The non-prismatic beam is also known as tapered beam. Commonly the non-prismatic beams are generated by varying their tapered ratios. Commonly used tapered ratios are 1.5, 2 and 2.5. And the non-prismatic beams can be classified

Table 6 Details of the non-prismatic beams

Tapered ratios	Section constant		Weight constant	
	Size (mm)	Weight (kg)	Size (mm)	Weight (kg)
1.5	200 × 133.33	17.65	240 × 160	18.99
2	200 × 100	16.696	250 × 125	18.99
2.5	200 × 80	16.572	285 × 114	18.99

into two categories; section constant non-prismatic beams and weight constant non-prismatic beams.

Table 6 shows the details of the non-prismatic beams used for the analysis.

3 Results and Discussions

The behavior of different configurations of connection have been studied. The stress concentration and strain are studied. The location of plastic hinge is identified in each configuration. According to the capacity design concept, the formation of plastic hinge should be away from the column face. The seismic performance of all the configuration are studied by applying a cyclic loading at the free end of the beam.

3.1 Stress Distribution

The structure is analyzed after the application of the cyclic loading at the free end of the beam. The stress distribution for each configuration is noted. The location of inelastic action is identified in each case. According to the capacity design concept, the location of the plastic hinge should be away from the column face. The stress distribution for each RBS configuration is shown in the following Fig. 4.

The following Table 7 shows the results from the cyclic analysis. From this table considering the number of cycles, deformation, load and stress on column the reduced beam section with circular result gives the good result.

In the Table 7 stress of column in each configuration is measured from beam column connection face and for beam is measured from the RBS area excluding the beam without RBS. In the case of conventional connection highest stress will be located at the beam column interface. The following Fig. 5 shows the load deflection curves for the different RBS configurations.

The reduced beam sections are modeled with non-prismatic beams. The various non-prismatic beams are modeled by varying the tapered ratios. These different models are analyzed under cyclic loading. These models are compared by considering the number of cycles, load and stress on columns. The load deflection curves for non-prismatic beam is shown in the Fig. 6 (Table 8).

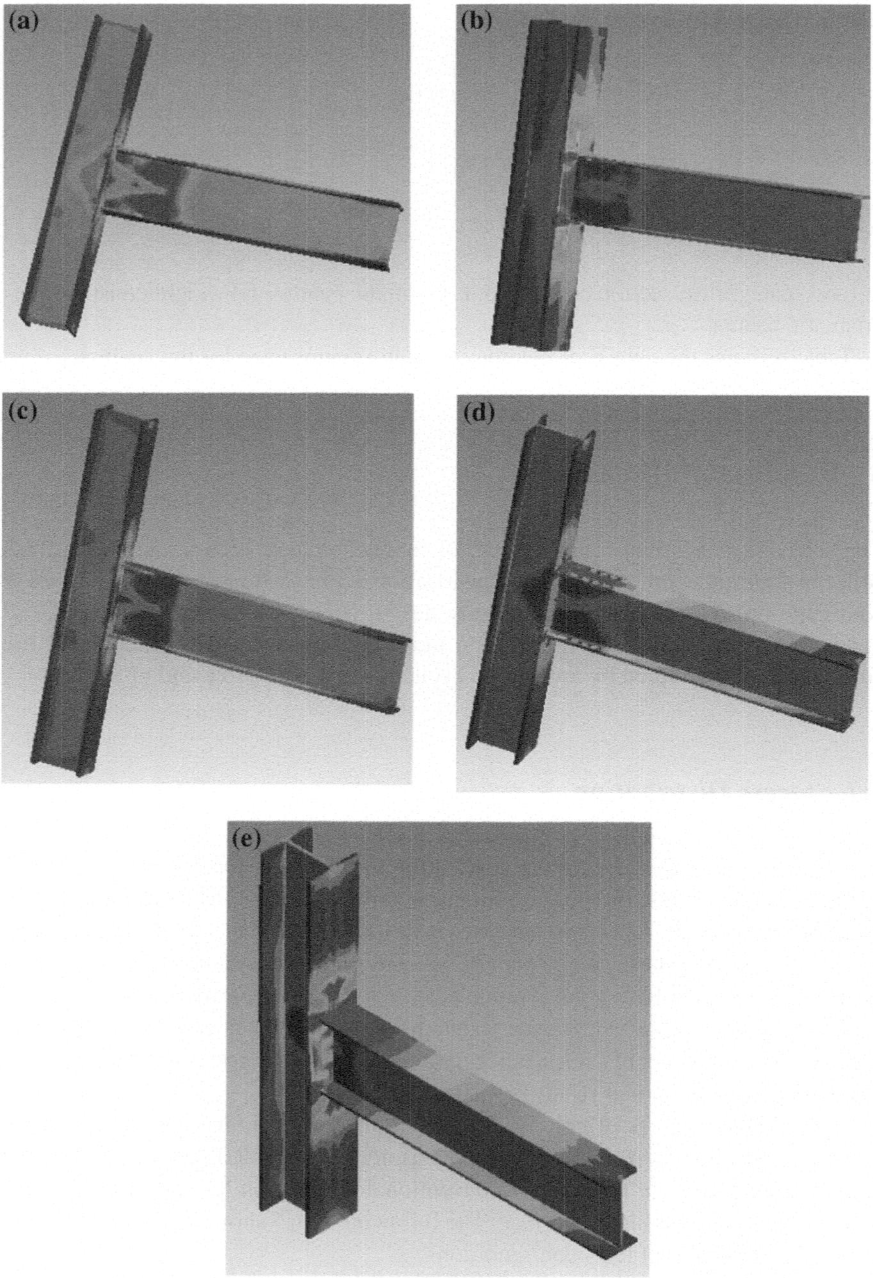

Fig. 4 Stress distribution of RBS with **a** circular cut, **b** rectangular cut, **c** trapezoidal cut, **d** holes and **e** without RBS

Table 7 Results of cyclic analysis for RBS with different configurations

	Cycles	Deformation (mm)	Load (kN)	Stress column (MPa)	Stress beam (MPa)	Energy dissipation (kJ)
WRBS	6.25	−30	−80.1	498.58	353.27	30.29
	6.75	30	80.1	470.17	342.84	
Circular RBS	6.25	−30	−61.5	372.22	371.92	15.19
	5.75	20	61.5	371.82	360.52	
Rectangular RBS	4.25	−15	−56.2	356.56	387.9	7.58
	4.675	10.5	50.8	376.25	350.15	
Trapezoidal RBS	5.25	−20	−56.2	345.58	383.68	14.15
	5.75	20	56.2	346.74	347.8	
RBS with holes	5.25	−20	−68.9	440.07	384.22	14.78
	5.75	20	69.1	464.47	428.1	

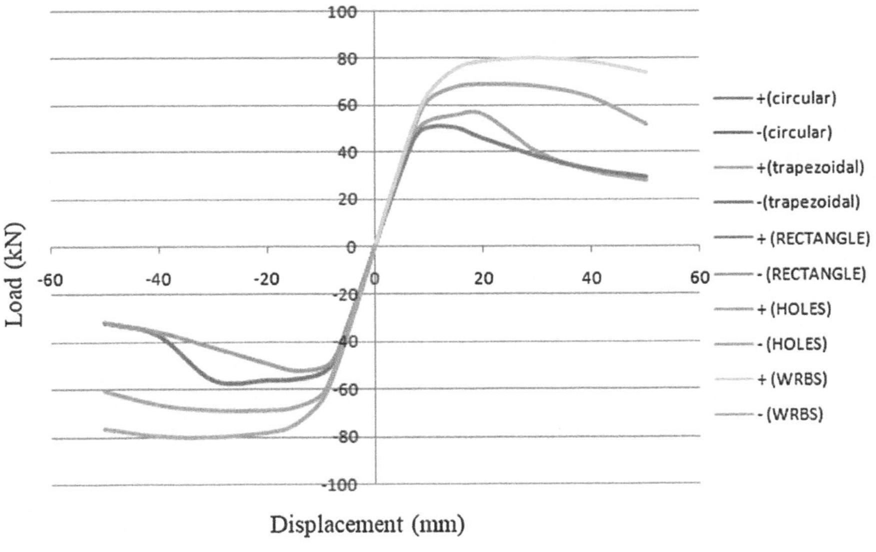

Fig. 5 Load deflection curves for different RBS configurations

4 Conclusions

The conventional steel beam moment connection will show the brittle failure due to the seismic actions. The high stress concentration critical factors causing such failures. A natural way to solve the problem is to reduce the ductility demand on the welded areas and alleviate the stress concentration level. One of these is the reduced beam section (RBS) configuration. Portions of the beam flanges are trimmed away

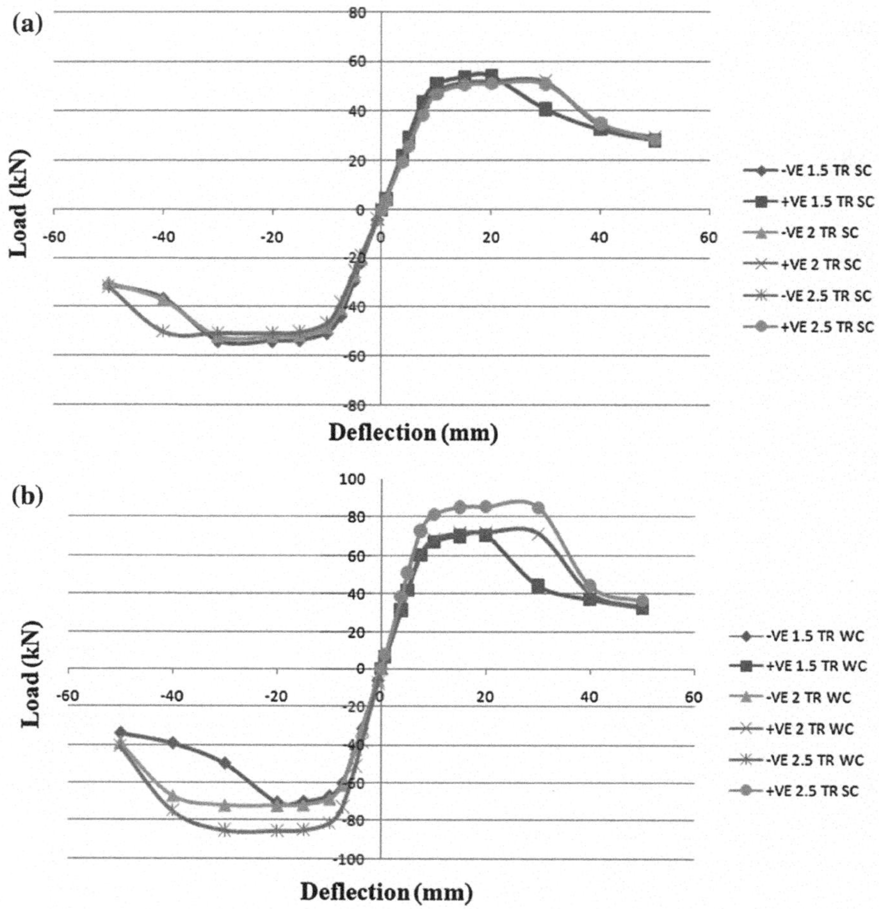

Fig. 6 Load deflection curves of non-prismatic beams **a** section constant and **b** weight constant

in the region adjacent to the beam-to-column connection. Portion of beam flanges near beam-to-column joint is cut to form a plastic hinge away from the joint so that much of ductility demand on beams may result from the RBS instead of the welded beam-to-column interface.

The seismic performance of the different configurations were examined under cyclic loading. The different configurations used for the study the study are reduced beam section with circular cut, rectangular cut, trapezoidal cut and with same sized holes. By adopting the RBS the plastic hinge location is shifted from the column face to the RBS region. Considering the stress concentration of the column the circular configuration is found to be the better model. The stress was reduced by 23.195% on comparing with the conventional moment connection and also it has a load carrying capacity of 61.55 kN.

Table 8 Results of cyclic analysis for non-prismatic beams

Tapered ratio	Cycles	Deformation (mm)	Load (kN)	Stress column (MPa)	Stress beam (MPa)
Section constant					
1.5	6.175	−21	−53.965	328.97	399.48
	5.75	20	54.025	335.19	399.2
2	6.175	−21	−51.929	406.15	377.25
	5.75	20	52.009	408.29	376.8
2.5	6.175	−21	−50.673	464.97	364.32
	5.75	20	50.767	462.75	364.48
Weight constant					
1.5	5.25	−20	−70.591	329.69	393.89
	5.72	18.2	70.633	329.29	354.68
2	5.25	−20	−72.102	402.21	407.23
	5.75	20	72.209	408.95	397.64
2.5	5.25	−20	−85.568	361.18	397.91
	5.75	20	85.788	373.16	393.6

Swati and Gaurang [4] conducted both the experimental and numerical analysis to find out the cyclic performance of the RBS moment connection and observed that was much superior to the connection without RBS. No weld fracture was observed in RBS connection while there was a crack observed near beam bottom flange weld for connection without RBS. Load carrying capacity for the circular RBS moment connection was 60.1 kN.

The cyclic analysis is also conducted for the non-prismatic beams having different tapered ratios. The tapered ratio having 1.5 will shows the lowest stress value of 332.08 N/mm^2 for section constant tapered beams compared to the others. And for the weight constant non-prismatic beams, the tapered ratio with 2.5 will gives the lowest stress value of 367.17 N/mm^2 for the higher load carrying capacity of 85.675 kN.

References

1. Han SW, Moon K-H, Hwang S-H, Stojadinovic B (2012) Rotation capacities of reduced section with bolted web (RSB-B) connections. J Constr Steel Res 70:256–263
2. Shen J, Kittjasateanphun T, Srivanich W (2000) Seismic performance of steel moment frames with reduced beam sections. Eng Struct 22:968–983
3. Ohsaki M, Tagawa H, Pan P (2009) Shape optimization of reduced beam section under cyclic loads. J Constr Steel Res 65:1511–1519
4. Swati AK, Gaurang V (2014) Study of steel moment connection with and without reduced beam section. J Struct Eng 1:26–31

5. Rahnavard R, Hassanipour A, Siahpolo N (2015) Analytical study on new types of reduced beam section moment connection affecting cyclic behavior. J Struct Eng 3:33–51
6. Pachoumis DT, Galoussis EG, Kalfas CN, Efthimiou IZ (2010) Cyclic performance of steel moment-resisting connections with reduced beam sections—experimental analysis and finite element model simulation. Eng Struct 33:2683–2692
7. Jin J, El-Tawil S (2005) Seismic performance of steel frames with reduced beam section connections. J Constr Steel Res 61:453–471
8. Naghipour M, Javadi N, Naghipour A (2011) Investigation of RBS connection ductility in eccentrically braced frames. Eng Struct 14:743–752
9. Huang Y, Yi W, Zhang R, Xu M (2014) Behavior and design modification of RBS moment connections with composite beams. Eng Struct 59:39–48

Time Optimization of Parallel Dynamic Analysis Using Greedy Algorithm in FEA

M. Chandana, G. Unni Kartha and C. Mahesh

Abstract The Finite Element Method (FEM) is the most widely used numerical technique to predict the approximate response of a structure under various loading conditions. Predicting the response of a structure to seismic loading using FEM can be computationally intensive and time-consuming. Parallel FEM is one solution to such situations where the computation is distributed efficiently among multiple cores available in modern supercomputers. In order to utilise the advantage of parallel computing in FEM, Pacific Earthquake Engineering Research Centre (PEER), has developed the open source software, OpenSees, with advanced capabilities for performing parallel FEM specifically for carrying out earthquake engineering simulations. In this paper, a new methodology is proposed to improve the efficiency of parallel computation using greedy algorithm in OpenSees for the time history analysis of framed structures for multiple earthquakes. Greedy algorithm finds an optimal solution in a number of steps by effective scheduling and proper load balancing. This method is verified by studying the time required for analysis of arbitrary framed structures using a high performance computing machine with a 32-core CPU, 62-GB RAM and 256-GB memory. A percentage increase of 16.35 is observed in the speedup factor for a two dimensional model studied.

Keywords Parallel finite element method · OpenSees · Time optimization · Greedy algorithm

M. Chandana (✉)
Department of Civil Engineering, Vedavyasa Institute of Technology,
Karadparamba, Malappuram 673632, India
e-mail: chandumanat@gmail.com

G. Unni Kartha
Department of Civil Engineering, Federal Institute of Science and Technology,
Angamaly, Kochi 683577, India
e-mail: unnikartha@gmail.com

C. Mahesh
Department of Computer Science and Engineering, Federal Institute of Science
and Technology, Angamaly, Kochi 683577, India
e-mail: mahesh.fisat@gmail.com

K. Dasgupta et al. (eds.), *Proceedings of SECON'19*,
Lecture Notes in Civil Engineering 46,
https://doi.org/10.1007/978-3-030-26365-2_33

1 Introduction

Performance-based methodologies in structural engineering require the need to study the response of structures to different seismic loads. Dynamic analysis for simple structures can be carried out manually, but for complex structures, they are typically performed using the nonlinear numerical methods, such as the Finite Element Method. Many applications in science and engineering for which the finite element method is used, requires the creation and subsequent computation on very large data sets. Hence, predicting the response of a structure to seismic loading is computationally very intensive and time consuming. The introduction of parallel computing in this stage is able to give the most significant result in terms of saving time, analysing and designing. The efficiency of parallel computing depends on its performance, which in turn depends on the effective scheduling and proper load balancing among the process available in the system. In this study, greedy algorithm is used to optimize the parallel analysis by which it sorts the jobs and finds a solution in a number of optimal steps.

2 Literature Survey

A variety of specializations in various fields of engineering is being carried out in finite element analysis. Nowadays, parallel processing of finite element method is a fast growing field of research. McKenna [1] in 1997 studied about object oriented finite element programming and a more flexible design was developed by breaking the various classes of finite element analysis into several component classes. In 2000, Jimack and Touheed [2] introduced the programming model, Message Passing Interface (MPI), for communicating messages among the processors for finite element parallel programming. Inorder to achieve the goals of flexibility, extensibility and portability of finite element software, Frank McKenna et al. [3] introduced in his paper a new object- oriented architecture in 2010. The main drawback in the object oriented structural analysis was the computational expense of dynamic memory management. To overcome this, OpenSees framework was developed which has various classes for creating finite element models and for the solution of governing equations.

McKenna and Fenves [4], carried out a study on the parallel capabilities of the OpenSees interpreters for conducting earthquake simulations, and explains about the communication mechanism between the processors. The procedure for building and running the interpreters in various operating systems is also described. Heister et al. [5] in his paper explains that computation time must scale linearly with respect to the number of processors and the problem size. Gavali and Shah [6], also implemented OpenSees on high performance computing system and evaluated its performance. The result showed that the simulation time reduces with increasing number of processors. In 2013, Gavali et al. [7] studied also on the time performance and the effectiveness of parallel OpenSees, to handle large size simulations.

The speedup of parallel computation depends on the allocation of tasks or activities to the processes in high performance computing machines. Malik et al. [8] in 2013, presented a study on greedy algorithm, used to solve various optimization problems. There is actually no general method to apply greedy algorithm to a given problem. Hence, the efficiency depends on the algorithm method chosen. Singh et al. [9] proposed a scheduling algorithm with load balancing for cloud computing in 2014. The important factors that improve the performance of a computing environment are job scheduling, load balancing and resource allocation. Greedy algorithm is very easy to implement and is fast. They can be used for optimizing parallel distribution.

3 Parallel Dynamic Analysis in OpenSees

The interpreter OpenSeesMP has the ability to perform parallelization based on the Process ID (PID) on which it runs and the Number of Processes (NP) involved in the parallel simulation. The initial procedure used for the present study for parallel finite element analysis is as shown in the Fig. 1.

The program runs on a user specified number of processes and each process will use the same input file. The very first step in the parallel analysis is to set the unique process id and the number of processes in the system. The command getNP returns the total number of processes and the command getPID returns the unique process number, through the value returned in [expr [getNP] − 1]. Each line in the file which holds the ground acceleration time history of several earthquakes is split and distributed among all the NP − 1 processes for parallel analysis. Each process, when it starts will determine its unique id and the number of process involved in the computation. They will each open the time history file and run through the list of earthquakes, only processing the earthquake assigned to that process id. Thus, each record is handled only once and each process handles roughly the same number of records. To study the computation time for analysis, a set of 2D models were created and analysed for different earthquakes. The models used for the study are:

1. A 2D frame model of 3 m beam span and six storeys of fiber section properties.
2. A 2D frame model of 4 m beam span and six storeys of fiber section properties.

The models are analysed for 100 earthquakes from the NGA West 2 PEER ground motion database for the load case Dead Load + Earthquake Load. The list of ground motion records to be run in the program for analysis was saved in text files. The minimum time duration of the ground motions considered is 761.159 s and the maximum is of 5527.317 s. The computation time taken for the analysis, when using different number of processes is recorded using an in-built function. The displacement at the top-most corner node and also the storey drift is recorded for all the 100 earthquake motions.

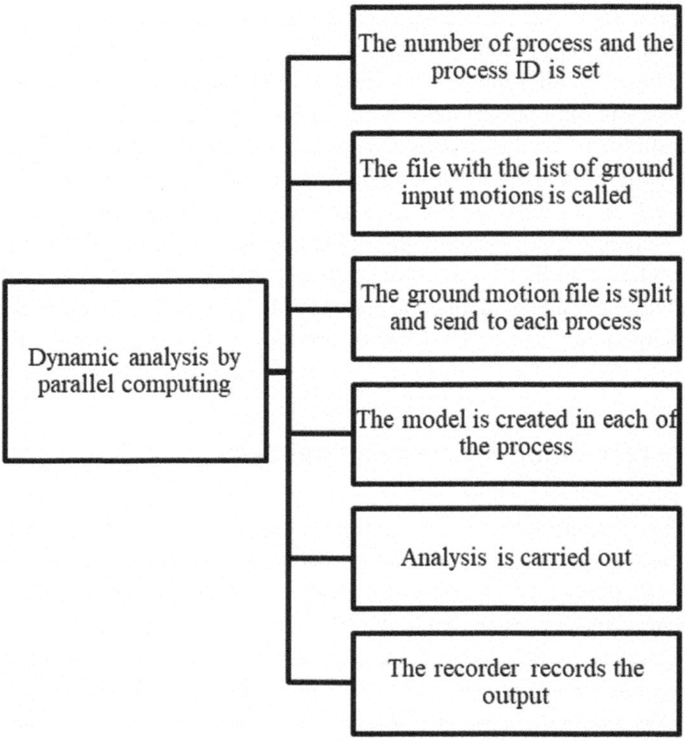

Fig. 1 General procedure of analysis

The analysis is carried out in a High Performance Computing machine with a 32-core CPU, 62 GB RAM and 256 GB memory. The machine runs on Ubuntu 14.04.5 operating system. The software OpenSeesMP of version 2.5.0 is built in the machine from the source code downloaded from OpenSees website. OpenMPI is used for message passing. The Number of Processes is varied from 1, 8, 16, 24, and 32 and the computation time for the analysis using each of the mentioned NP is recorded.

The speed of parallel computation is usually measured using a parameter speedup factor $S(p)$. It is defined as the ratio of execution time using one processor (t_s) in a best sequential algorithm to the execution time using a multiprocessor with p number of processors (t_p) as in the equation below.

$$S(p) = \frac{t_s}{t_p}$$

This speedup factor value is used to determine the parallelization efficiency, which in turn is used to calculate the maximum theoretical speedup $S(n)$. It is to be noted that the maximum speedup factor that can be achieved is limited by Amdahl's law for parallelization, which states that if P is the proportion of a program that can be made parallel and $1 - P$ is the proportion that remains serial, then the maximum speedup factor that can be achieved using n number of process is given by the parallelization equation below.

$$S(n) = \frac{1}{1 - P + \frac{P}{n}}$$

4 Parameter for Speeding up Parallel Execution

The time taken for parallel dynamic analysis depends many factors, but, if the model is same, the time taken will depend on the time history of the earthquake motion. Normally, the earthquake with largest duration will take more time. To check this hypothesis, analysis were carried out for two 2D six storey frame models, one of 3 m and other of 4 m beam span of fiber section, for 100 different earthquakes with different Number of Points (NPTS) in acceleration data. Figure 2 shows the variation of time taken in seconds for analysis when the number of points in the acceleration data or the duration of earthquake increases. It is observed that when the duration of earthquake is more, the time taken for analysis is also more.

The total time taken for completing all analysis will depend on the number of earthquakes being considered. During parallel analysis, when the distribution of analysis cases is in the order in the text file in which the earthquake cases are saved,

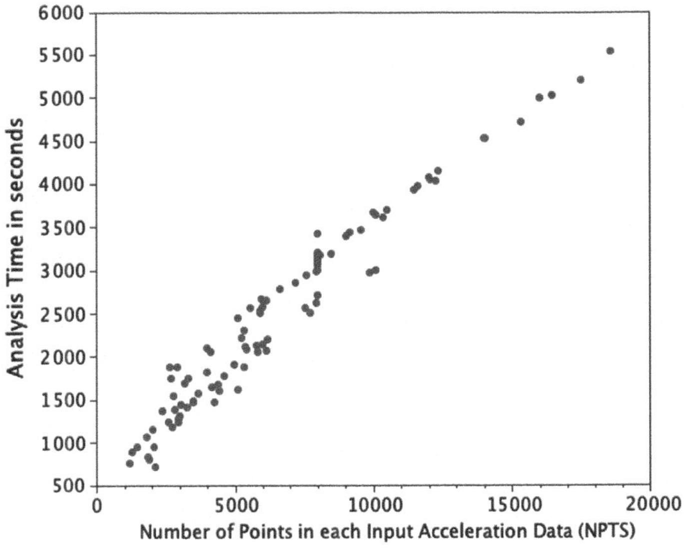

Fig. 2 Analysis time versus number of points in the acceleration data

the total computation time depends on the process that handles the file with the longest duration. While executing so, one process may receive two or more files with large duration, another may receive files with small durations, or both. Hence, if in a worst case scenario, if all the time history files having long duration goes to the same process, then there is a chance that, the process handling files with small duration completes the analysis and stays idle, even when the process handling the long duration files haven't completed the analysis of even the first file. The processes are allotted tasks unequally or inefficiently here.

An efficient distribution of tasks, to further reduce the time for analysis can be achieved by allotting tasks to each of the process equally. For this, the process should be loaded with equal time history durations.

5 Optimization Using Greedy Algorithm

Greedy algorithm is an algorithm that finds a solution in a number of optimal steps. Scheduling algorithm will make a sequence of process such that the efficiency of time and the output is increased and load balancing algorithm divide the load properly between all the available processes. In greedy algorithm, for a list of activities, the activity whose duration is least or highest among the remaining activities is selected. The activities are sorted according to their duration so that, the activities always have the minimum finishing time.

The program first calls the text file having the list of the 100 input ground motions' name and reads the value of NPTS of each of them. The list is then sorted based on the value of the NPTS in a descending order. A new file is created to save the sorted list.

The unique PID and NP in the system is set. Each line in the sorted file which holds the ground acceleration time history of several earthquakes till the NP'th ground motion is split and distributed among all the $NP - 1$ processes for parallel analysis. Initially, only NP number of files are split and distributed among the NP number of processes. They will each open the time history file and run through the list of earthquakes, only processing the earthquake assigned to that process id. The value of the NPTS of the time history file assigned to each process is saved as a list.

The time history file after the NP'th number in the list is sent to that process which had the file with minimum NPTS value. The sum of NPTS value of first file and newly allotted files is calculated and retained as the new NPTS value. Now, the next file is allotted to the process with minimum NPTS value. This is repeated for the rest of the input motion files in the list till all input motions are allotted. This methodology uses the idea of greedy algorithm where an optimal strategy is chosen at every step. Figure 3 shows the algorithm developed.

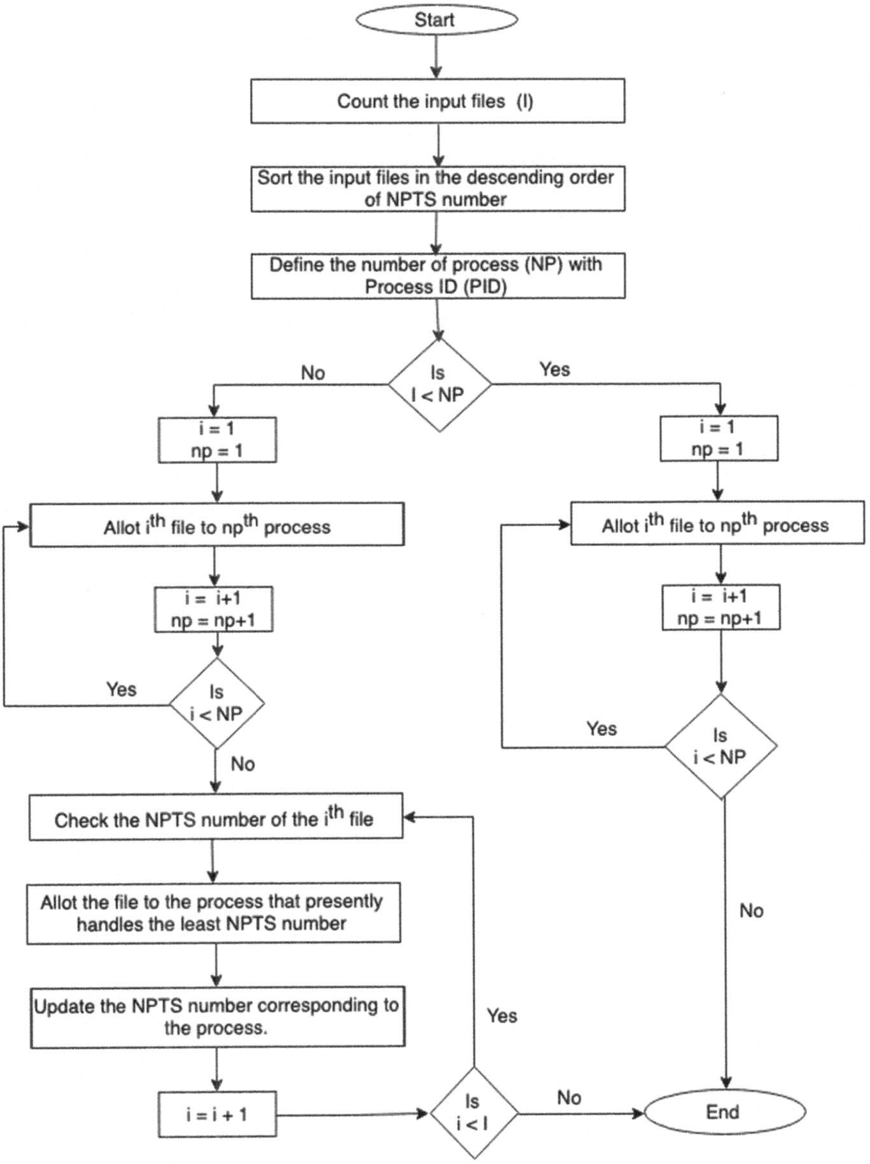

Fig. 3 Algorithm developed for optimization of parallel computing

6 Results

The responses of a structure, subjected to different earthquakes with respect time required for parallel analysis with different number of processors and processes is studied. It is observed that the time taken for analysis reduces from 1837.693 to 164.009 s for model 1 and from 1533.494 to 127.127 s for model 2 when NP is increased from 1 to 32.

The total computation time is reduced further when greedy algorithm is used for parallel analysis. The change in analysis time for all the models with the increase in number of processes before and after implementing greedy algorithm are plotted in the Fig. 4. The change in the speedup factor for the analysis of all models for different number of processes before and after implementing greedy algorithm are plotted in the Fig. 5.

The graph shows that the speedup factor increases with increasing number of processes used for analysis and it also has a higher value when greedy algorithm is implemented for both models. There is a decrease of 1.65 and 2.45% in time of analysis and an increase of 3.55 and 16.35% in speed factor for the model 1 and model 2 respectively, when the analysis is done parallelly in 32 NP using greedy algorithm. Also, the analysis time increases when the model becomes more complex.

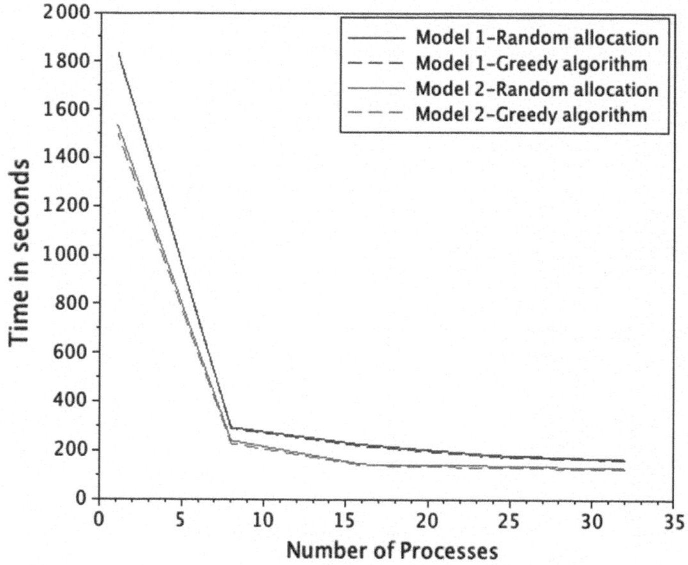

Fig. 4 Analysis time versus number of processes for all the models

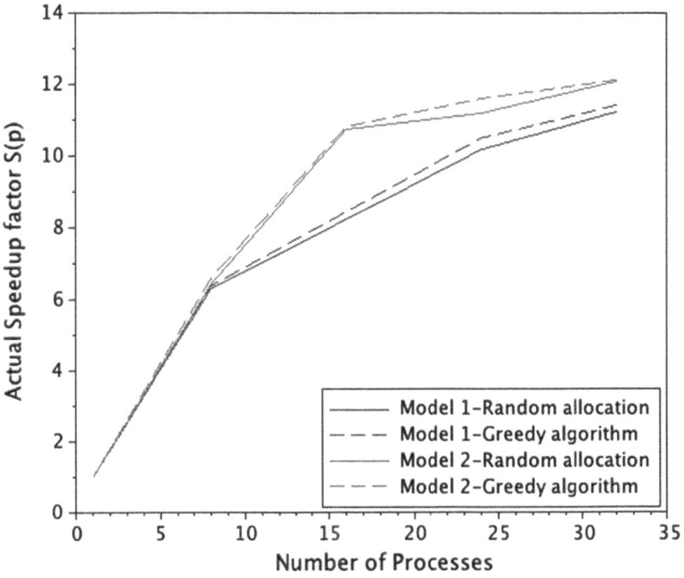

Fig. 5 Actual speedup factor versus number of processes

7 Conclusion

Time history analysis of two 2D models of different spans subjected to 100 earthquakes from the NGA West2 PEER ground motion database are performed and it is observed that the computation time reduces with increase in number of processes for all the analysis. But, there is always a maximum speedup that can be achieved for a particular problem in a particular machine. Hence, there will be no much decrease in time even if the total number of processes used for the analysis is more than the parallelizable section in a program.

Optimization of parallel computing is achieved by implementing greedy algorithm in the allocation of analysis jobs. The parallel efficiency, actual speedup and Amdahl's Law speedup shows improvement when greedy algorithm is implemented for parallel distribution of tasks. Even though the improvement presented here are not very significant, it can be easily concluded that significant improvement can be obtained where the time for individual processes are larger i.e., in the case of bigger and more complex 3D models and when there are earthquake data with large variation in duration.

References

1. McKenna FT (1997) Object-oriented finite element programming: frameworks for analysis, algorithms and parallel computing. Ph.D., dissertation, University of California, Berkeley
2. Jimack PK, Touheed N (2000) Developing parallel finite element software using MPI. High Perform Comput Comput Mech 15–38
3. McKenna F, Scott MH, Fenves GL (2010) Nonlinear finite-element analysis software architecture using object composition. J Comput Civ Eng (ASCE)
4. McKenna F, Fenves GL (2007) Using the OpenSees interpreter on parallel computers, NEESit, TN-2007-16
5. Heister T, Kronbichler M, Bangert W (2010) Massively parallel finite element programming. EuroMPI, LNCS, vol 6305. Springer, Heidelberg, pp 122–131
6. Gavali PG, Shah MS (2008) Earthquake simulations of large scale structures using Opensees software on grid and high performance computing in India. In: The 14th world conference on earthquake engineering, Beijing, China, 12–17 Oct
7. Gavali P, Shah MS, Kadam G, Meher K (2013) Seismic response and simulations of reinforced concrete bridge using OpenSees on high performance computing. CSIT 1:215–220
8. Malik A, Sharma A, Saroha V (2013) Greedy algorithm. Int J Sci Res Publ 3(8):1
9. Singh AK, Sahu S, Tiwari MN, Katare RK (2014) Scheduling algorithm with load balancing in cloud computing. Int J Sci Eng Res 2(1):38–43

Performance Evaluation of Polyurethane Cement Composite as a Retrofit Against Seismic Loading

Sheba Susan Abraham and Asha Joseph

Abstract Polyurethane Cement Composite (PUC) is an emerging material, versatile of all polymers. The stress-strain curve of PUC shows that it is equally strong in tension and compression and is an elastic material. It can be used for retrofitting purposes due to their excellent strength properties, light weight and ease of application. PUC also exhibit great ductility and durability properties due to the formation of dense microstructure. It has the combined properties of polyurethane and Portland cement. There is no need for adhesives as it excellent bonding characteristics. The only practical application was in the strengthening of T-Beam Bridge in Harbin, China (2016). The possibility of PUC as a retrofitting technique has not been explored yet. In this paper the static and dynamic response of the structure retrofitted using PUC, due to seismic loading is analyzed using finite element method. The finite element model is created in ANSYS software. The response of RC frame and RC frame strengthened using PUC composite and CFRP are also compared.

Keywords Polyurethane cement composite · Finite element model · Time history analysis · Reinforced concrete frame

1 Introduction

To improve the working ability of structures, many techniques have been used in strengthening. The most common methods for strengthening beams have been the use of Carbon Fiber Reinforced Polymer (CFRP), steel plate bonding, external prestressing reinforcement and others, these methods are widely used at present. CFRP materials have good structural performance, high strength and light weight. CFRP can be easily installed, as they can be attached to a curved profile. However, these

S. S. Abraham (✉) · A. Joseph
Department of Civil Engineering, Federal Institute of Science and Technology, Angamaly, India
e-mail: shebasusanabraham@gmail.com

A. Joseph
e-mail: ashameledath@gmail.com

K. Dasgupta et al. (eds.), *Proceedings of SECON'19*,
Lecture Notes in Civil Engineering 46,
https://doi.org/10.1007/978-3-030-26365-2_34

materials have their own shortcomings. The major drawback of CFRP is the high cost. Bonding steel plates have the disadvantages of weakened bonding caused by steel corrosion, increased dead load weight and difficulties in adapting to the concrete surface profile. Stress concentration can be caused at the end zone of the beam for the external pre-stressing reinforcement method, which unfavorably influences the beam [1].

The introduction of new construction materials such as polyurethane-cement composite (PUC) to civil engineering can provide a potential solution. PUC is a kind of composite material composed of polyurethane raw materials mixed with cement. Polyurethane is a high-performance polymer elastic material mainly based on the chemical compounds of poly isocyanate 51–52% and polyether polyol 48%, water 0–1% and silicon oil 1%. The harden range of PU is from 10 to 100, with good abrasion resistance performance, corrosion resistance, toughness and cohesiveness. PUC has the advantages of light weight, significant strength in compression and bending [2]. PUC has excellent bonding with concrete and additional adhesives are not required for reinforcing [3]. Addition of polyurethane to Portland cement covers for all shortcomings of concrete such as brittle failure, rigidity and poor ductility [4]. PU has been selected as the best potential polymer with seismic and blast resistance because of its high energy absorption and less fragmentation properties [5, 6].

The aim of this study is to evaluate the role of PUC in the enhancement of performance of RC structures so that they have a better resistance to earthquakes. A comparison is done between CFRP retrofitted frame and PUC attached frame under static loading and seismic loading by carrying out static structural analysis, modal analysis and time history analysis.

2 Finite Element Formulation

In order to converge at the required objective finite element analysis was done on RC frame and RC frame retrofitted using PUC and CFRP using ANSYS 19.0 to study the effect of PUC on the seismic performance of the frame.

2.1 Geometric Description

The RC frame consisting of beam and column members is of M25 grade concrete. The size of beam is 300 mm × 300 mm × 2700 mm and column size 300 mm × 300 mm × 3300 mm [7] as shown in Fig. 1. The beam consists of reinforcement 3 longitudinal bars of 16 mm diameter and 6 mm diameter stirrups at 125 mm spacing. Columns are reinforced with 4 longitudinal bars of 16 mm diameter and lateral ties of 6 mm diameter at 300 mm spacing. CFRP is attached to the frame as a surface element. The thickness of CFRP is taken as 1 mm which is the commercially available form. Only one layer is provided so that the structure remains ductile and not stiff.

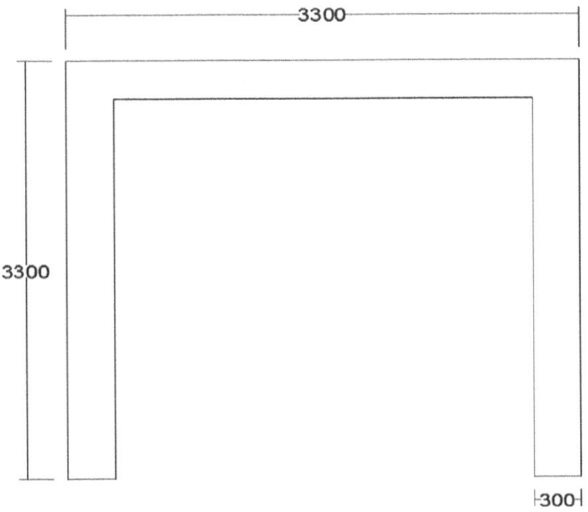

Fig. 1 Dimensions of RC frame in mm

Also CFRP is costlier and one layer of CFRP keeps the cost of PUC and CFRP comparable. CFRP is wrapped around the columns and on the underside of the beam [8].

PUC is attached to the frame as a solid body with bonded connection. PUC is attached same as that of CFRP with 25 mm thickness [7, 9].

2.2 *Material Properties*

Isoparametric linear elastic material properties and multilinear plasticity properties are used for PUC. The stress strain curve of PUC is given in Fig. 2. Steel is considered as a Bilinear elasto-plastic material with yield strength 200GPa and tangent modulus as 5000 Pa as given in Fig. 3. Concrete material has elastic and concrete material definitions. In Elastic definition, the modulus of elasticity and Poisson's ratio are necessary. For Concrete definition, the compressive and tensile strength are given. Since there is no issue of debonding bond stiffness value may not be considered for concrete or PUC. The material properties of concrete PUC and steel are taken according Haleem et al. (2013). Since the failure of CFRP is mainly by debonding the bond stiffness is considered as 50 N/m^2 [8]. Material yielding was set to be predicted using Von Mises yield criterion. Table 1 gives the properties of all the materials considered for the present study [10].

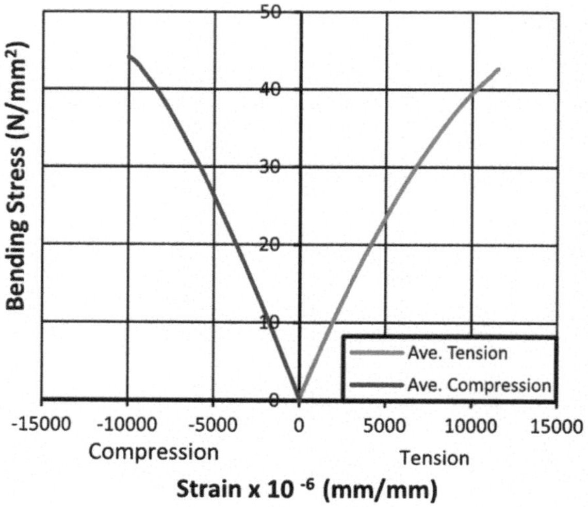

Fig. 2 Stress-Strain curve for PUC

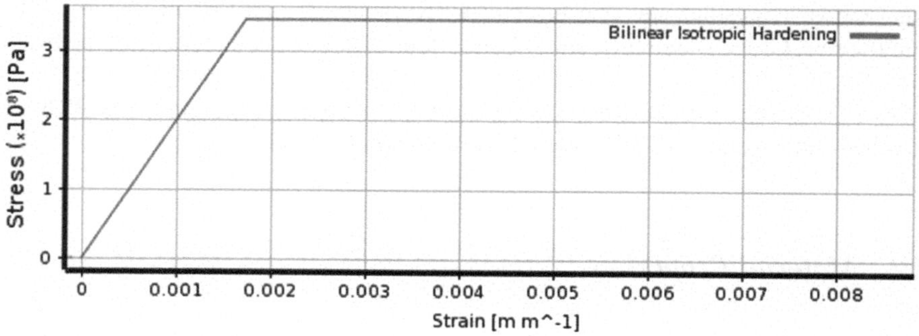

Fig. 3 Stress-Strain curve for steel

Table 1 Material properties used in modeling

Material	Concrete	Steel	PUC	CFRP
Yield strength (MPa)	–	200,000	–	3800
Modulus of elasticity (MPa)	25,000	200,000	4540	227,000
Compressive strength (MPa)	30.70	–	60.61	–
Density (kg/m^3)	2350	7800	1648	1800
Tensile strength (MPa)	2.52	–	42.70	–
Poison's ratio	0.20	0.3	0.27	0.3

2.3 Finite Element Modelling

Modeling in ANSYS includes providing appropriate elements, defining geometry and assigning suitable material models. Quadrilateral mesh is used for meshing all elements. The model consists of the column, beams, reinforcing steel, PUC coating and CFRP. Linear elastic material properties and multilinear plasticity properties were assigned for PUC. Concrete and PUC is modeled using Solid 185 element type available in ANSYS element library. The element consists of 20 nodes and has three translation degrees of freedom at each node. It can be used to model concrete with or without rebars. Rebars are smeared within the element and can be defined in three different axes. Full adherence between concrete and rebar and is assumed and is modeled using Beam 188 element which is a linear 2 noded element having 6 degrees of freedom. Bilinear elasto-plastic behaviour was assigned for the material property of steel. PUC can also be modeled same as concrete with Solid 185 element. CFRP is modeled as a surface structure and Shell 181 element is used. It is a four noded element with 6 degrees of freedom at each node. Bonded connection is provided between all the contact surfaces. Fixed support condition was provided at the column base. Material yielding was set to be predicted using Von Mises yield criterion. Force reaction, equivalent stress and total deformation were chosen as output variables for determining the performance of the structures.

3 Response of Retrofitted Structure to Static Loading

To determine the static performance a load of 100 kN was uniformly distributed over the beam surface. Self weight was also considered to include inertial effect. Boundary conditions and loading is depicted in Fig. 4.

Static analysis was carried out in RC frame and frame retrofitted with CFRP and PUC and their respective deformation characteristics are shown in Fig. 5.

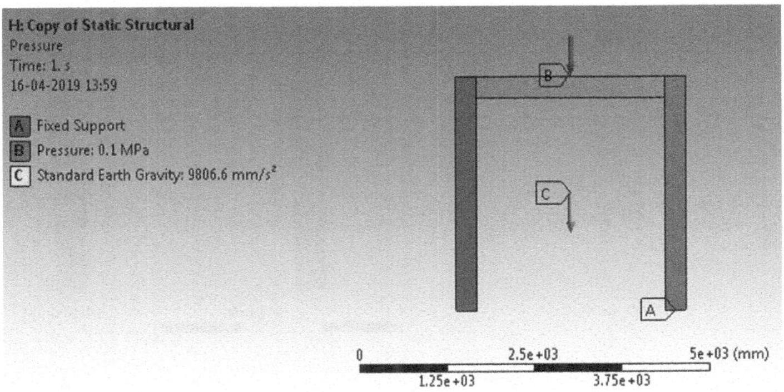

Fig. 4 Loading and support condition for static analysis

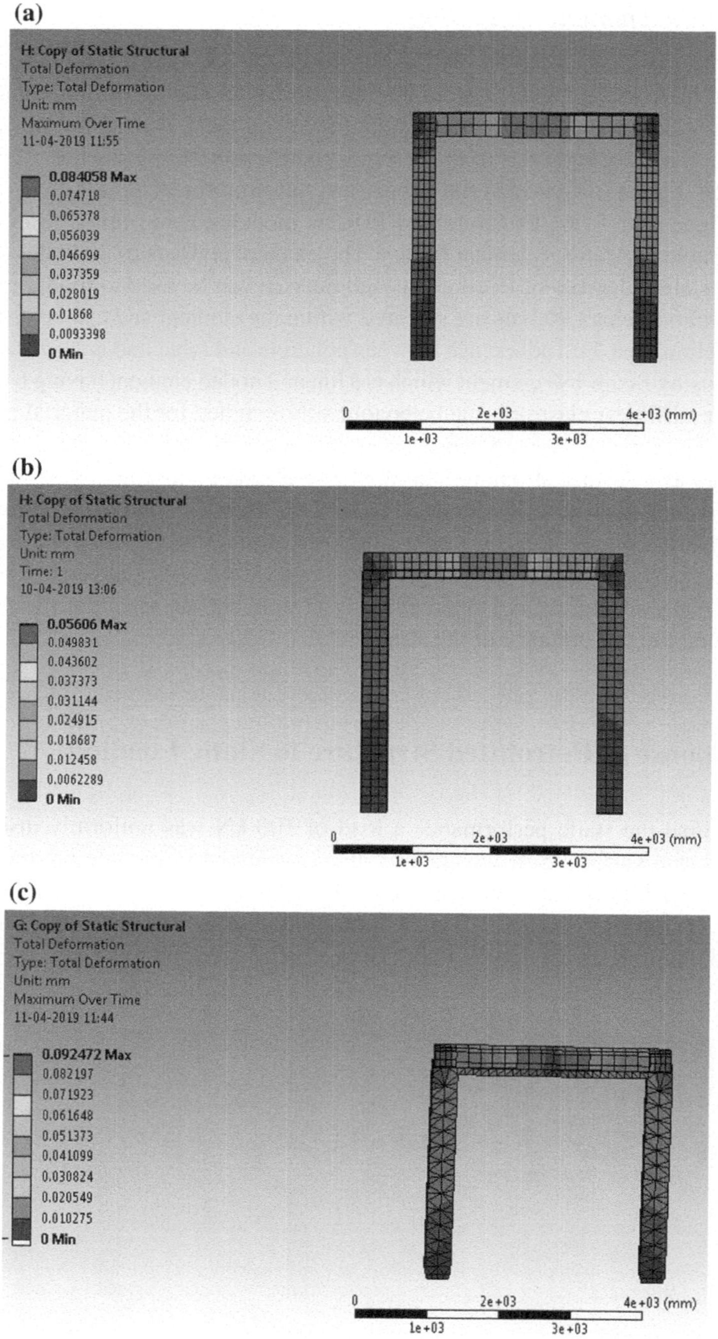

Fig. 5 Deformation contours of **a** RC frame, **b** CFRP frame, **c** PUC frame

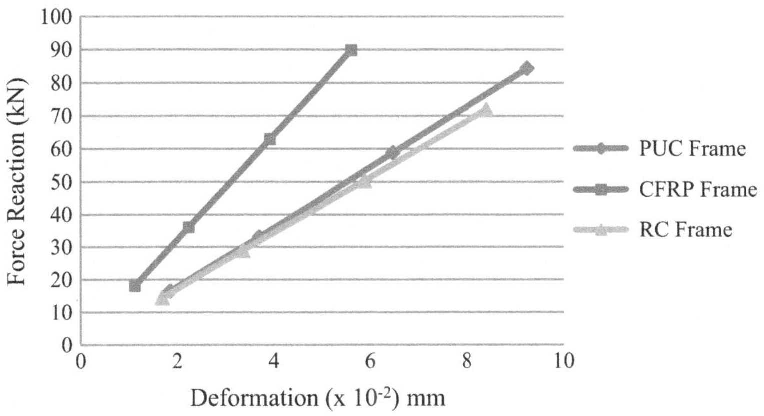

Fig. 6 Force-Deformation graph

Table 2 Maximum Stress and its Location

	Maximum stress (MPa)	Location
PUC	0.39	Beam
CFRP	0.68	CFRP
RC frame	0.44	Beam column joint

The load deformation plot (Fig. 6) shows that the load carrying capacity is more for CFRP retrofitted frame but the deformation is very less i.e., 0.058 mm. But in PUC retrofitted frame even though the load carrying capacity is less than CFRP it shows a deflection of 0.098 mm which is about 60% greater than CFRP retrofitted frame.

The major stress and its occurrence is given in Table 2 shows that in RC frame stress is concentrated at the beam column junction indicating that the plastic hinge is formed at the column which is not good under seismic conditions. But in the retrofitted structure the position of maximum stress changes to reinforcement in PUC and to CFRP.

4 Free Vibration Responses

Modal analysis uses the overall mass and stiffness of a structure to find the various periods at which it will naturally resonate. These periods of vibration are very important to note in earthquake engineering, as it is imperative that a building's natural frequency does not match the frequency of expected earthquakes in the region in which the building is to be constructed. If a structure's natural frequency matches an earthquake's frequency, the structure may continue to resonate and experience structural damage.

Table 3 Comparison of natural frequency and time period

	RC frame	PUC	CFRP
Natural frequency (Hz)	15.14	12.64	18.44

Only self weight was considered for the modal analysis. The fundamental mode of vibration was the translation along the X-axis. The corresponding frequency was taken as the natural frequency of all the 3 structures as given in Table 3.

From the analysis we can see that PUC attached frame has a longer time period than CFRP attached frame and the frequency of vibration is reduced in the case of PUC attached frame.

5 Response of Retrofitted Structure to Seismic Excitation

Time History analysis is a nonlinear dynamic analysis which utilizes the combination of ground motion records with a detailed structural model, therefore is capable of producing results with relatively low uncertainty. In nonlinear dynamic analyses, the detailed structural model subjected to a ground-motion record produces estimates of component deformations for each degree of freedom in the model and the modal responses are combined using schemes such as the square-root-sum-of-squares.

In non-linear dynamic analysis, the non-linear properties of the structure are considered as part of a time domain analysis. However, the calculated response can be very sensitive to the characteristics of the individual ground motion used as seismic input; therefore, several analyses are required using different ground motion records to achieve a reliable estimation of the probabilistic distribution of structural response. Since the properties of the seismic response depend on the intensity, or severity, of the seismic shaking, a comprehensive assessment calls for numerous nonlinear dynamic analyses at various levels of intensity to represent different possible earthquake scenarios [11].

To evaluate the seismic performance the time history plot of El Centro Earthquake of duration 54 s was given the base excitation. The acceleration time history is given in Fig. 7.

The results of the analysis are summarized in Table 4 and the deformation time plot of all the 3 models given in Fig. 8 gives a picture of the time of maximum displacement and its magnitude.

The results show that PUC attached frame undergoes the maximum displacement. The stress concentration is less in PUC attached frame than CFRP attached frame. But compared to RC frame the equivalent stress is slightly higher in PUC frame.

When we consider the time maximum deformation, in CFRP attached frame it is at 10.74 s while for other two models it is around 5 s. This is an indication of debonding of CFRP from the frame as shown in Fig. 9. So PUC have shown a better performance towards seismic loading.

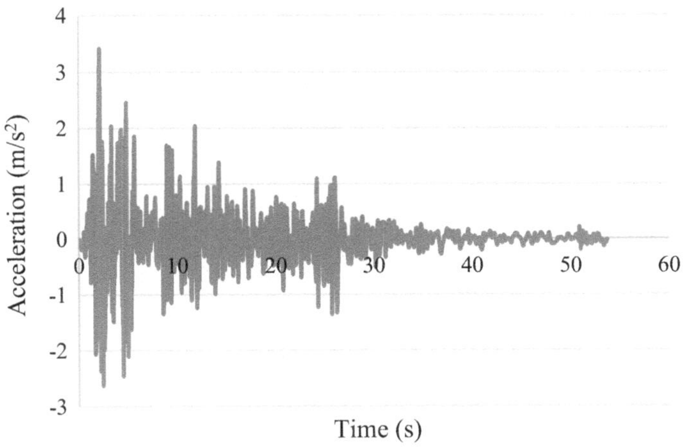

Fig. 7 Time-history of El Centro earthquake

Table 4 Result of time history analysis

	RC frame	PUC frame	CFRP frame
Max stress (MPa)	11.209	11.47	21.57
Max. deformation (mm)	5.1633	7.46	3.58
Location	Column	Beam	CFRP
Time (s)	4.88	4.48	10.74

6 Conclusions

To evaluate the performance of PUC static analysis, free vibration analysis and time history analysis was conducted and the results have been compared with RC frame and CFRP retrofitted frame.

The static analysis of the structures shows that

- The load carrying capacity is more in CFRP attached frame. This is because CFRP has greater modulus of elasticity and tensile strength. Also maximum stress occurs on CFRP which means CFRP takes up the entire load.
- In case of PUC retrofitted frame the maximum stress at the beam column joint is shifted to the reinforcement which means that by allowing the frame to deform more, PUC enhances the load carrying capacity of the RC frame. Further the increase in load carrying capacity may be attributed to Portland cement and reduction in stress to polyurethane components in the composite.
- If we consider the energy absorption of the structure, area under the load - deformation curve is more for PUC attached frame and therefore has the maximum energy absorption capacity.

It can be inferred from the modal analysis that

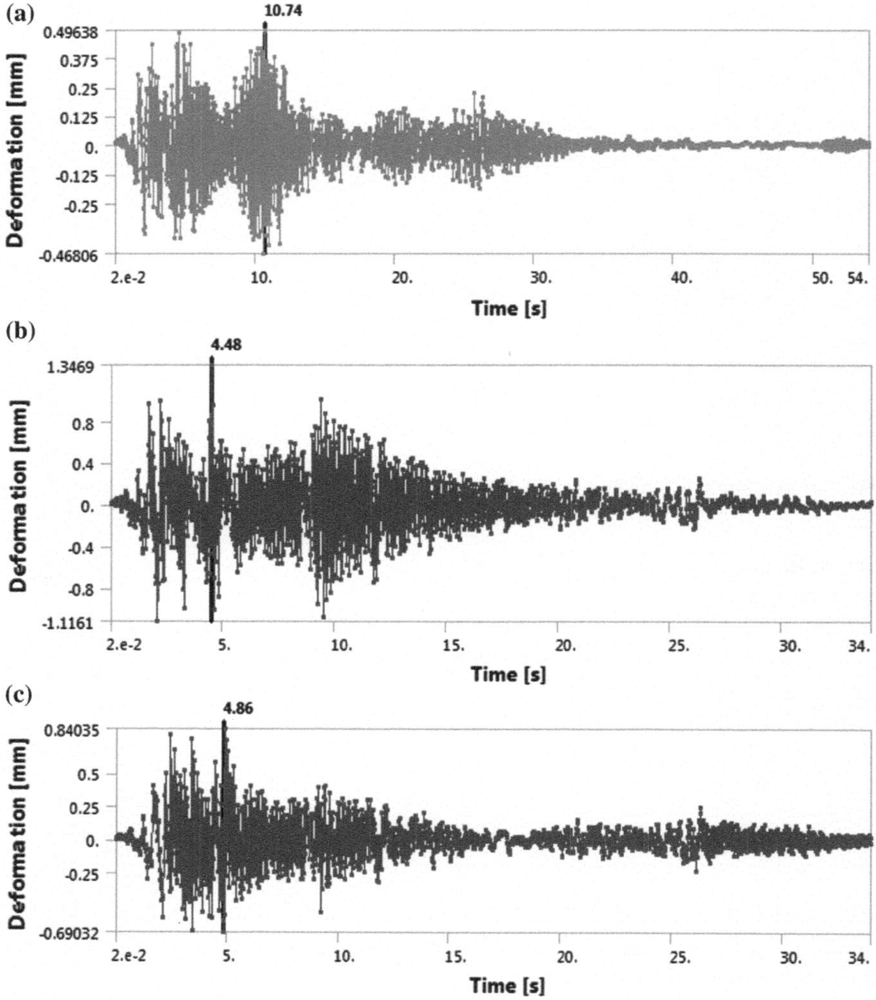

Fig. 8 Time history plot of deformation of **a** CFRP frame, **b** PUC frame, **c** RC frame

- PUC attached frame has a longer time period because the attachment of PUC has reduced the stiffness and made the structure more ductile.
- Whereas wrapping of CFRP made the structure stiffer and reduced the time period.

From the time history analysis we can see that

- The stress concentration in PUC attached frame is less than CFRP attached frame but slightly greater than RC frame. This can be attributed to the increase in volume of the structure due to the attachment of PUC.
- PUC attached frame underwent maximum deformation without debonding.
- CFRP underwent debonding failure.

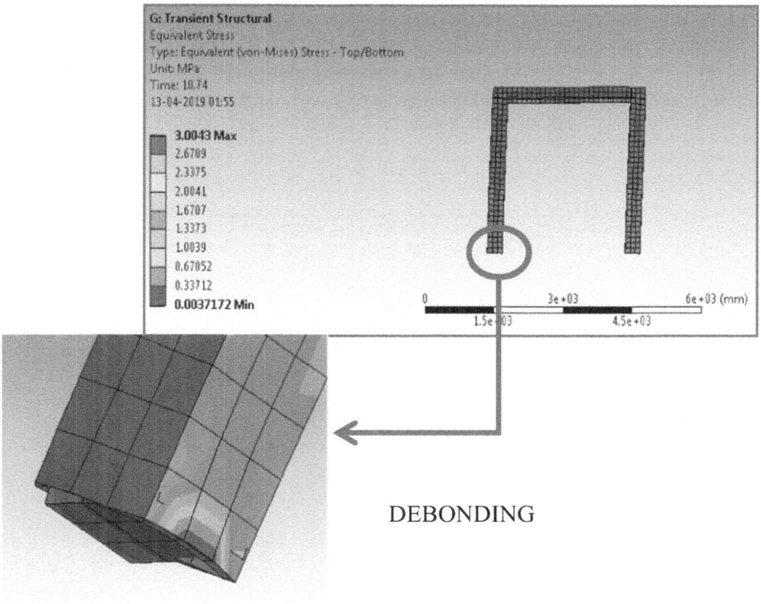

Fig. 9 Debonding of CFRP

So we can conclude that in the seismic point of view PUC makes the structure more ductile and is better retrofit than CFRP.

References

1. Buchan PA, Chen JF (2007) Blast resistance of FRP composites and polymer strengthened concrete and masonry structures—a state-of-the-art review. Compos Part B 38:509–522
2. Gbongbon W, Mounanga P, Poullain PT (2008) Proportioning and characterization of lightweight concrete mixtures made with rigid polyurethane foam wastes. Cem Concr Compos 30:806–814
3. Haleem KH, Lian Z, Gui WL (2012) Study of concrete strain for T-Beams retrofitting by poly-urethane-cement material (PUC). Res J Appl Sci Eng Technol 5(7):2354–2359
4. Melo DMA, Martinelli AE, Lima FM, Bezerra UT, Marinho EP, Henrique DM (2002) Addition of polyurethane to Portland cement. Composites 1805–1812
5. Davidson JS, Fisher JW, Hammons MI, Porter JR, Dinan RJ (2005) Failure mechanisms of polymer reinforced concrete masonry walls subjected to blast. J Struct Eng 131(8):1194–1205
6. Ju-Hyung H, Na-Hyun Y, Jong-Kwon C, Jang-Ho JK (2011) Experimental study on hybrid CFRP-PU strengthening effect on RC panels under blast loading. Compos Struct 93:2070–2082
7. Xavier M, Binol V (2018) Strength analysis-building frame with polyurethane cement composite (PUC). Int Res J Eng Technol (IRJET) 5
8. Zhao M, Farhad A (2004) Bond properties of FRP fabrics and concrete joints. Adv Struct Eng 3 (Springer)
9. Zhang K, Sun Q (2016) Strengthening of a reinforced concrete bridge with polyurethane cement composite (PUC). Open Civ Eng J 10:768–781

10. Yong YW, Haleem KH, Gui WL (2014) Experimental study to investigate mechanical properties of new material polyurethane–cement composite (PUC). Constr Build Mater 50:200–205
11. Chang TL, Ye X, Li K (2008) Analysis of seismic energy response and distribution of RC frame structures. In: The 14th world conference on earthquake engineering, China, 12–17 Oct

A Study on Effect of Shear Connectors in the Structural Performance of Steel-Concrete-Steel Sandwich Shear Walls

Reenu Eldhose and Reshma Prasad

Abstract Steel-concrete-steel (SCS) shear wall consist of two steel face plates and a sandwiched concrete core which are bonded together by mechanical shear connectors to form an integral unit to resist external loads. The steel face plates were connected together using tie rods and to the infill concrete using headed studs respectively. The advantages of the SCS Sandwich structure over the traditional reinforced concrete are that the external steel skin plates act as the permanent formwork and flexural reinforcement and offer water proofing protections. This paper presents a robust finite element model developed in ANSYS Workbench for Nonlinear cyclic analysis of the SCS walls. Rectangular SCS shear walls with an aspect ratio of 1.0 were modeled and analysed under displacement-controlled, in-plane cyclic loading. A parametric study was carried out to investigate the effects of different shear connector configurations and push over analysis was performed to determine the best connector configuration. A new structural topology has been introduced for better performance.

Keywords Steel-concrete-steel (SCS) sandwich shear wall · Shear connectors · Nonlinear cyclic analysis · Push over analysis

1 Introduction

A Steel-concrete-steel (SCS) sandwich shear wall to be made consists of steel face plates, infill concrete, and shear connectors to anchor the steel face plates together and to the infill concrete. SCS construction have been widely used as an effective lateral load resisting construction due to the reason including significant strength, ductility, reduced cost, high initial stiffness, lighter weight, less invasive construction; faster erection, increased speed of construction, better acoustical and

R. Eldhose (✉) · R. Prasad
Department of Civil Engineering, Federal Institute of Science and Technology, Angamaly, Kochi 683577, India
e-mail: reenu20795@gmail.com

R. Prasad
e-mail: reshma.prasad1@gmail.com

K. Dasgupta et al. (eds.), *Proceedings of SECON'19*,
Lecture Notes in Civil Engineering 46,
https://doi.org/10.1007/978-3-030-26365-2_35

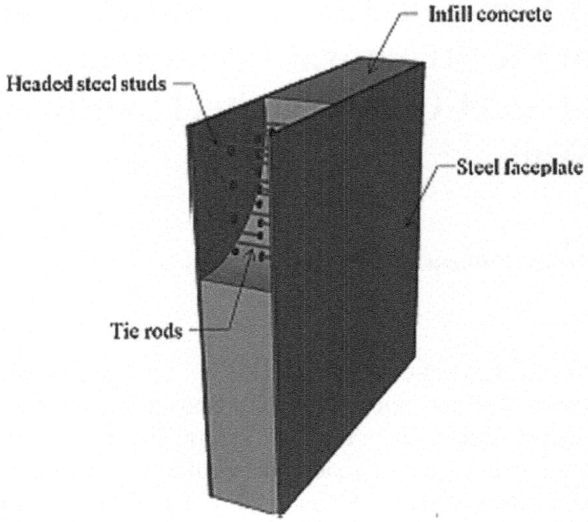

Fig. 1 Steel-concrete-steel sandwich shear wall

thermal insulation [1]. SCS wall shells can be fabricated offsite, assembled on-site, and filled on-site with concrete to create a monolithic structure which enhances the flexural rigidity of the structure without adding substantial weight as shown in Fig. 1 [2].

In SCS wall, the shear connectors lead to enhanced composite action, prevent local buckling of the steel plates, provide transverse shear resistance, and also improve the bond between the steel and the concrete structure without vertical separation [3]. Therefore, performance of SCS sandwich system depends upon the efficiency of the shear connector and study regarding this is very important. In this paper, the lateral load resistance and energy absorption of SCS walls are investigated by conducting cyclic and push-over analysis and the results have been compared between different shear connector configurations. 3D non-linear finite element analysis software, ANSYS 16.1 has been chosen to conduct the investigation.

2 Objectives

This study will mainly be focused towards the Maximum load and Energy absorption of SCS wall. The main objectives include:

- To study the cyclic response of shear-critical SCS walls subjected to in-plane loadings.
- To investigate the effect of various shear connector configurations in the lateral load resistance of SCS shear walls.
- To compare and discuss the energy absorption behaviour and maximum load of SCS walls using parametric study.

3 Structural Model and Material Properties

Steel concrete steel shear wall with shear connectors of different configuration such as Tie rods and headed studs, L-shaped, T-shaped, rectangular plate, Plate inserted with studs and tie rods (PST), Staggered Plate inserted with studs and tie rods (SPST), Single truss reinforcement (ST), Double truss reinforcement (DT) were modeled [4]. The cyclic response of SCS sandwich shear wall with tie rods and headed studs are studied on a rectangular wall of aspect ratio of 1.0 and analysed with finite element software ANSYS 16.1. Table 1 summarizes the geometry of the SCS wall [5].

The models were meshed and then fixed support condition were given at the bottom face of SCS wall and free at the top face as shown in Fig. 2a. Boundary conditions were provided according to the experimental setup [5]. Material properties are illustrated in Tables 2 and 3. Displacement load is applied horizontally on the top face of the wall and direction of displacement is as shown in Fig. 2b.

Table 1 Geometry of SCS wall

Specimen	Wall dimensions H × L× T (mm)	Stud spacing (mm)	Tie rod spacing (mm)	Thickness of steel plate (mm)	Diameter of stud and tie rod (mm)
SC1	1524 × 1524 × 305	102	305	4.8	9.8

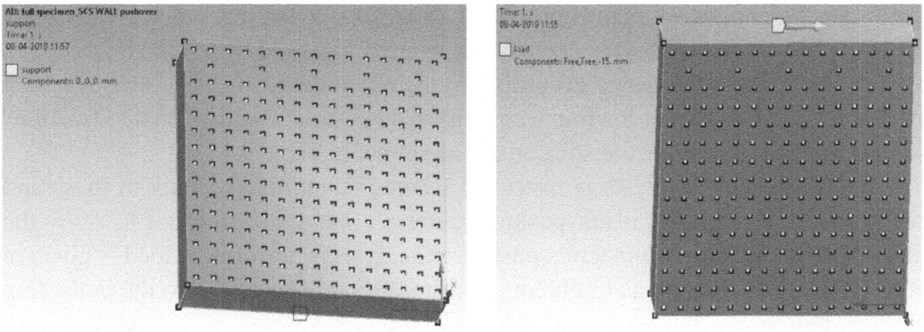

(a) Fixed support (b) Loading diagram

Fig. 2 Boundary conditions and loading diagram

Table 2 Concrete material properties

Young's modulus (MPa)	Poisson's ratio	Uniaxial compressive strength (MPa)	Uniaxial tensile strength (MPa)
20,685	0.18	31	2.4

Table 3 Steel material properties

Element	Young's modulus (MPa)	Poisson's ratio	Yield strength (MPa)	Ultimate strength (MPa)
Face plate	200,000	0.30	262	380
Connectors	200,000	0.30	345	517

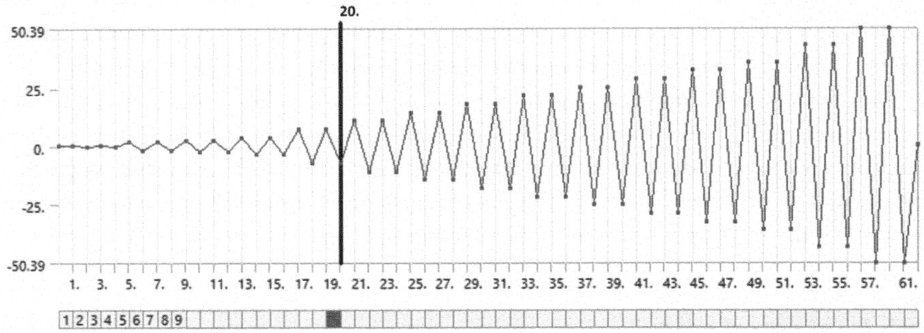

Fig. 3 Quasi-static cyclic displacement loading graph

3.1 Loading Program

In order to investigate the behaviour of SCS sandwich shear wall and to simulate earth quake load, quasi-static cyclic load were applied. Quasi-static cyclic loading were performed under displacement controlled loading and the loading graph is illustrated in Fig. 3. In each loading cycle, a push was exerted followed by a pull [6].

SCS shear wall was loaded according to the recommendations of ACI 374.1-05 (ACI 2005). Two cycles of loading were imposed at displacements equal to fractions and multiples of a reference displacement as shown in Table 4 [7].

The use of linear analysis is restricted when discontinuity occurs in mass and stiffness, hence for better result pushover analysis was used. In order to study the effect of different shear connectors push over analysis is performed, load is given in terms of displacement so as to obtain an ultimate condition and also the deflection at which failure occurs [7].

3.2 Numerical Modeling

SCS sandwich shear walls with different shear connector configurations are considered [8] for the investigation of behavior of SCS wall under lateral loading. The cross section details of different connectors are given in Table 5.

Parametric study was conducted by making the total weight of connectors constant. Seismic capacity of SCS wall is evaluated using Push-over analysis on different

Table 4 Displacement protocol

Load step	Peak displacement (mm)	Drift ratio	Number of cycles
LS1	0.3556	0.02	2
LS2	1.778	0.12	2
LS3	2.667	0.18	2
LS4	3.6068	0.23	2
LS5	7.1882	0.47	2
LS6	10.795	0.70	2
LS7	14.401	0.93	2
LS8	18	1.17	2
LS9	21.59	1.40	2
LS10	25.197	1.63	2
LS11	28.80	1.87	2
LS12	32.41	2.10	2
LS13	35.99	2.33	2
LS14	43.205	2.80	2
LS15	50.393	3.27	2

Table 5 Cross section details of different connectors

Type of connector		L1	B1	W1	W2	t1	t2
Plate	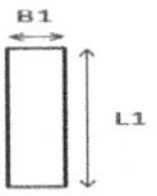	20	3.5	–	–	–	–
T-shaped	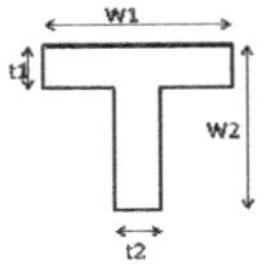	–	–	20	21.75	1.75	1.75
L-shaped	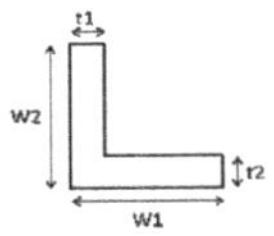	–	–	20	21.75	1.75	1.75

numerical models and their geometry are illustrated in Fig. 4. SCS wall with double truss (DT) and single truss (ST) reinforcement are provided with 10 and 12 mm diameter Fe500 bars respectively.

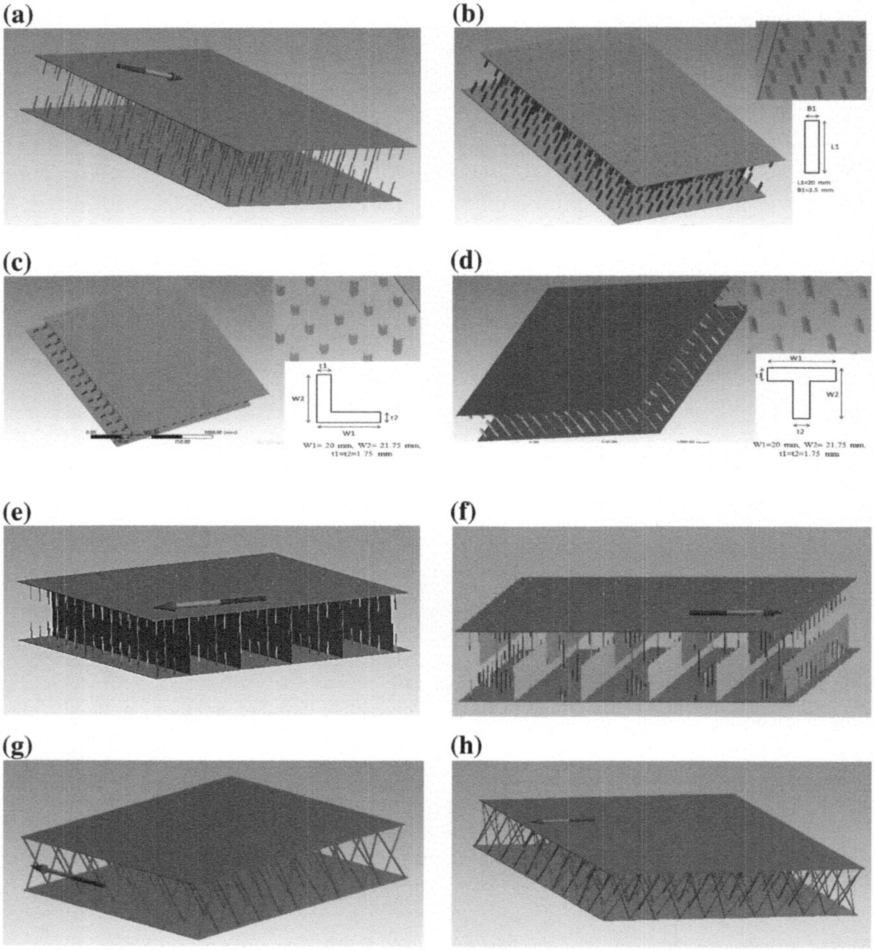

Fig. 4 SCS wall with **a** studs and tie rods, **b** rectangular plate connectors, **c** L-shaped connector, **d** T-shaped connector, **e** PST, **f** SPST, **g** ST, **h** DT

3.3 Meshing

It is the most important part of an analysis to discrete the structure. Coarse mesh is provided for all the models. Edge element size was fixed as 25 mm. Figure 5 shows the meshing.

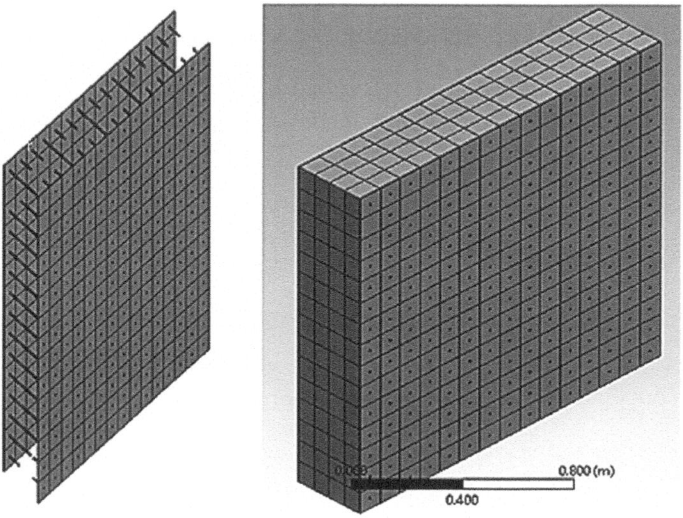

Fig. 5 Meshing of SCS wall

4 Results and Discussion

The results obtained by the comparison of a full model and half model SCS wall was found to be very less that is about 8.8% and is negligible. So it is advantageous to take the symmetry of the model to reduce the computational effort. Half model SCS wall is used for the study here.

During a severe earthquake the structure is likely to undergo inelastic deformations and has to rely on its hysteretic energy absorption capacity and ductility to avoid collapse. The results of non-linear static analysis of SCS sandwich shear wall under cyclic loading condition and results of push over analysis SCS wall with different connector configuration are presented here in Figs. 6 and 7.

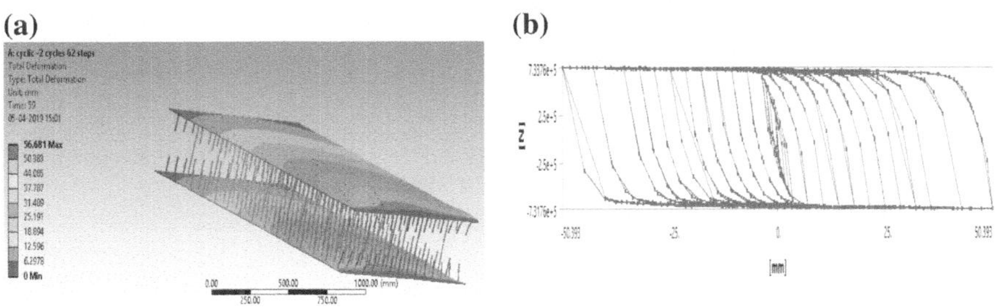

Fig. 6 **a** Total deformation of SCS wall under cyclic loading. **b** Hysteresis curve of model

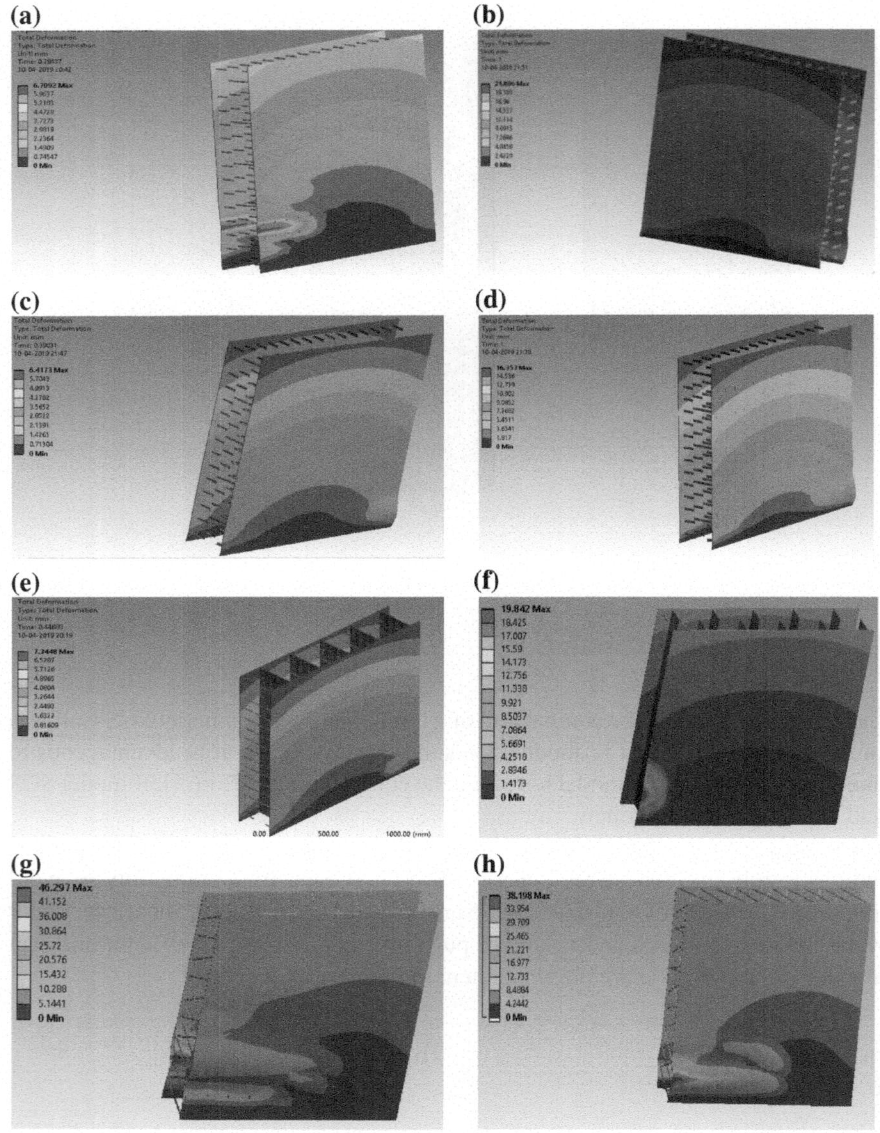

Fig. 7 Total deformation of SCS wall with **a** studs and tie rods, **b** rectangular plate connectors, **c** L-shaped connector, **d** T-shaped connector, **e** PST, **f** SPST, **g** ST, **h** DT

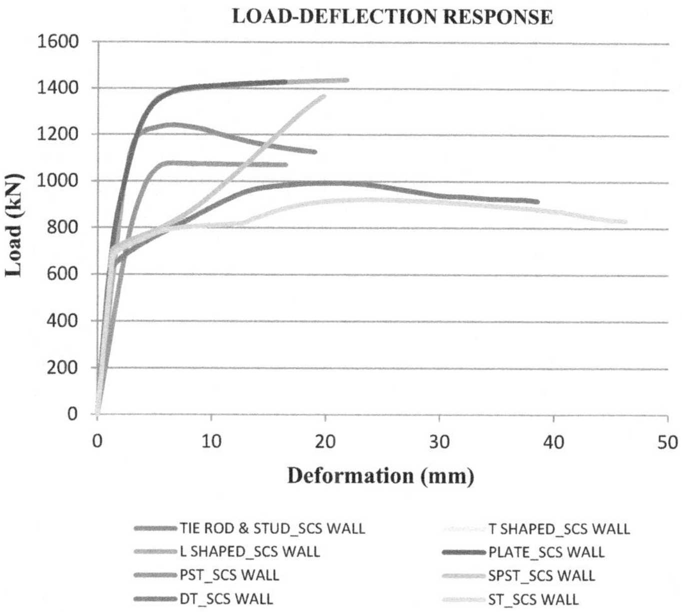

Fig. 8 Load-deformation plot obtained from push over analysis

From the deformation contour it was found that the steel face plate buckles at the bottom portion of the SCS wall. Staggered Plate inserted with studs and tie rods (SPST) shows better performance than that of PST.

The load-deformation response is plotted based on the results and are plotted as shown in Fig. 8.

4.1 Comparison of Results

Table 6 gives values of Maximum load, Total deformation, Energy Absorption Obtained by SCS wall with different shear connector configuration. Comparison of the energy absorption SCS with different shapes of shear connectors is shown in Fig. 9. The area under the load deflection graph gives the energy absorption. From the results compared it is found that the energy absorption is maximum for SCS with L-shaped plate connector and shows better performance.

Table 6 Maximum load, total deformation, energy absorption obtained by SCS wall with different shear connector configurations

Shape of shear connector	Maximum load (kN)	Total deformation (mm)	Energy absorption (kNm)
L-shaped	1437	21.806	12.484
Plate	1428	16.353	8.81
T-shaped	1380	6.4173	2.436
SPST	1368.3	19.842	5.375
Tie rod and stud	1240	6.7092	10.39
PST	1075.28	7.3448	6.66
DT	992.69	38.524	16.82
ST	922.52	46.297	19.93

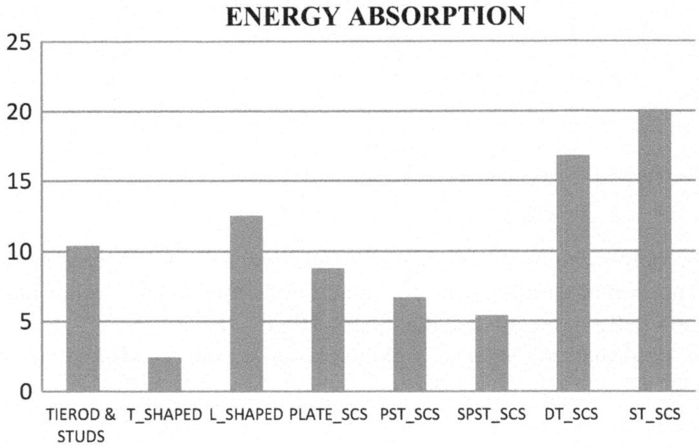

Fig. 9 Comparison of energy absorption of different shear connector configuration

5 Conclusions

As a new structural system, SCS shear wall features good integrity, ductility, and the ability to prevent leakage, impact and explosion. The study on behavior of different type of shear connectors suggested that L-shaped shear connector has more load carrying capacity. SCS wall with ST reinforcement shows good performance in terms of its higher energy absorption. Staggered Plate inserted with studs and tie rods (SPST) show good performance than SPT. On comparison, tie rod and studs absorb 15.20% more energy than plate connectors.

References

1. Yan J-B, Wang X-T, Wang Tao (2018) Compressive behaviour of normal weight concrete confined by the steel face plates in SCS sandwich wall. Constr Build Mater 171:437–454
2. Foundoukos N, Chapman JC (2008) Finite element analysis of steel–concrete–steel sandwich beams. J Constr Steel Res 64:947–961
3. Sohel KMA, Liewb JYR, Kohb CG (2015) Numerical modelling of lightweight steel-concrete-steel sandwich composite beams subjected to impact. Thin-Walled Struct 94:135–146
4. Liew JYR, Yan J-B, Huang Z-Y (2017) Steel-concrete-steel sandwich composite structures-recent innovations. J Constr Steel Res 130:202–221
5. Nguyen NH, Whittaker AS (2017) Numerical modelling of steel-plate concrete composite shear walls. Eng Struct 150:1–11
6. Epackachi S, Whittaker AS, Varma AH, Kurt EG (2015) Finite element modeling of steel-plate concrete composite wall piers. Eng Struct 100:369–384
7. Epackachi S, Nguyen NH, Kurt EG, Whittaker AS, Varma AH (2014) In-plane seismic behavior of rectangular steel-plate composite wall piers. ASCE
8. Bowerman H, Chapman JC (2015) Bi-steel steel-concrete-steel sandwich construction. In: Composite construction in steel and concrete IV, ASCE

Analytical Study On Modified Reduced Beam Section Connections Under Cyclic Behaviour

G. S. Greeshma and S. Usha

Abstract Steel frames with column and beam connections welded together show brittle nature during earthquakes. According to previous researches, the major damages were the cracks developed in the welded portion between the beam flange and the face of the column. Such cracking results in accumulation of unwanted stresses in the welded region. Reduced Beam Section (RBS) method is an effective way to limit the accumulation of stress in the vicinity of beam column joint. Out of different advanced beam to column design approaches developed after the earthquakes in Northridge in 1994 and Kobe in 1995, the reduced beam section connection has proven satisfactory ductility levels in numerous tests. The use of RBS in beams to column connections results in the formation of plastic deformations at a certain distance from the beam column joint. In the present study new types of reduced beam sections have been modeled using ANSYS 16 software by trimming the beam flange in different pattern and is compared each other under cyclic behavior. The results obtained from this study have shown that the reduced beam section with same holes dissipates more energy than other type of connections.

Keywords Steel moment connections · Reduced beam section · Plastic deformation

1 Introduction

Earthquakes generate one of most destructive loading scenarios to which a structure can be subjected, and if under these loads a structure fails then human life is undoubtedly harmed. The connection between beam and column is the key component by which a structure fails under seismic loading. This situation can lead to

G. S. Greeshma (✉) · S. Usha
Department of Civil Engineering, Sree Narayana Gurukulam College of Engineering, Ernakulum, Kerala, India
e-mail: greesh.gs@gmail.com

S. Usha
e-mail: ushaushus11@rediffmail.com

K. Dasgupta et al. (eds.), *Proceedings of SECON'19*,
Lecture Notes in Civil Engineering 46,
https://doi.org/10.1007/978-3-030-26365-2_36

a loss of structural stability and floor collapse, if not entire buildings. During the 1994 Northridge earthquake, significant impair to the beam-column joint welds was observed in various steel moment resistant frames. Ultimately, new strategies are employed for increasing the seismic ability of the steel connections used during Northridge earthquake. Altering the geometry of the access hole of weld, strengthening beam on the face of column by introducing a haunch, utilizing side plates, using external stiffeners and diminishing the beam strength at a certain distance from beam column junction are some of such strategies. Cropping the beam flanges, reducing the web of beam are some of the various methods used to weaken the beam section. The repairs after earthquake are limited to replacing the beams with reduced beam sections in such frames.

Between all the choices, the radius cut to reduce the beam flanges showed improved seismic reliability and which was regarded as reduced beam sections. Pachoumis et al. [1], presents the results of an experimental study of RBS connection with radius cut profile under cyclic behavior. From the experimental and numerical results it is found that the cyclic performance of the RBS connection is satisfactory and the plastic deformation was found in the RBS portion. Also there is no fracture of weld was noticed. Pachoumis et al. [2], was performed an experimental study in order to check the suggested values of the geometrical parameters of the RBS. From the study it can be concluded that RBS acts as an area capable of sustaining large inelastic strains while reducing stress in the less ductile portion vicinity to the column. Sofias et al. [3] studied performance of the RBS connections with the extended end plate under cyclic loading and the use of European profiles. During the study, RBS protected the connection and its' components from large deformations which leads to its failure. Swati et al. [4], conducted a study to discover about the benefits and importance of RBS connection for Indian profiles against connection without RBS. It can be observed that the specimen without RBS was poorly performed due to cracks on the bottom flange weld and the specimen with RBS was able to create a rotation of 0.02 radians without any distress in the welds.

In this study new types of reduced beam sections have been modeled using ANSYS 16 software by trimming the beam flange in different pattern and is compared each other under cyclic behavior. The purpose of this paper is to obtain numerical modeling results on eight RBS models. The objectives of this study include: (1) To shift plastic deformations at a particular distance from the beam column junction; (2) To examine the effect of equivalent plastic strain on all types of reduced beam section connections; (3) To acquire the effect of dissipated energy and stiffness on different types of reduced flange sections.

2 Numerical Study on RBS Connection

The analytical study includes the development of 8 finite element models to evaluate the influence of different parameters on the performance of connection. The RBS connection cut parameters follows the AISC-358 specifications. The straight distance from the beam column joint to the beginning of the radius cut (a) and the quantity of reduction in the beam flanges (c) must be proportional to the width of the beam flange. Radial cut and circular cut was adopted for the reduction of beam flanges. Figure 1 displays four types of the finite element models used in this study. Moreover the 8 models used for the analysis are summarized in Table 1.

WPB 290 was used for modeling of column and NPB 330 was used for the modeling of beam. The height of the column and the length of beam is 2500 mm. The column was assumed to be fixed at both ends and the beam is a cantilever beam. The material properties were assigned to the models and shown in Table 2. In order to create the plastic deformation in the connection components, a combined hardening rule (Isotropic-Bilinear) with Von Mises criterion was applied. Displacement is applied at the tip of the beam according to SAC loading protocol (Fig. 2).

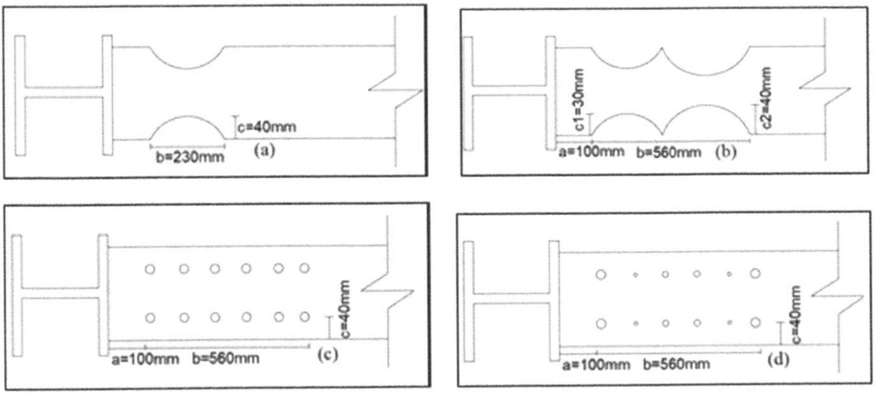

Fig. 1 Connection details **a** RBS, **b** DRBS, **c** RBS-SH3, **d** RBS-VH2

Table 1 Cut profiles used for modeling

Models	Reduce parameters (mm)
OBC	
RBS	a = 100, b = 280, c = 40
DRBS	a = 100, b = 280, C1 = 30, C2 = 40
RBS-SH1	8 circle with d = 60@71
RBS-SH2	7 circle with d = 50@85
RBS-SH3	6 circle with d = 40@104
RBS-VH1	7circle with d = 40, 50, 60, 60, 60, 50 & 40
RBS-VH2	6 circle with d = 60, 40, 50, 50, 40 & 60

Table 2 Material properties of model

Material	Fe 415 steel
Modulus of elasticity	2×10^5 N/mm^2
Poisson's ratio	0.3
Density	7850 kg/m^3

Fig. 2 SAC loading protocol

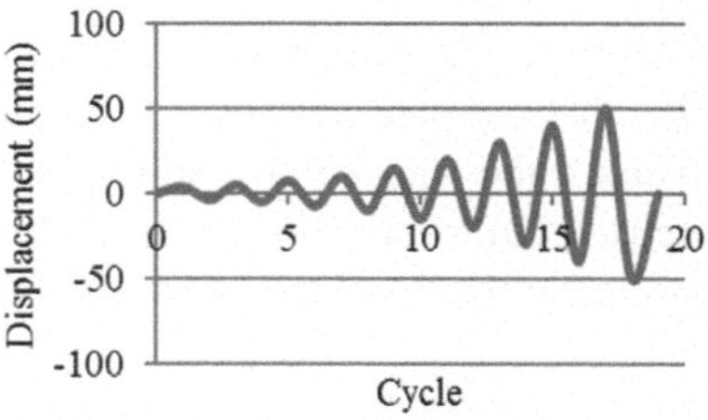

3 Result and Discussions

3.1 Equivalent Plastic Strain of Models

The equivalent plastic strain gives a measure of the amount of permanent strain in an engineering body and it is calculated from the component plastic strain. The equivalent plastic strain distribution in 3 of the models is shown in Figs. 3 and 4.

Fig. 3 Equivalent plastic strain of ordinary beam column connection

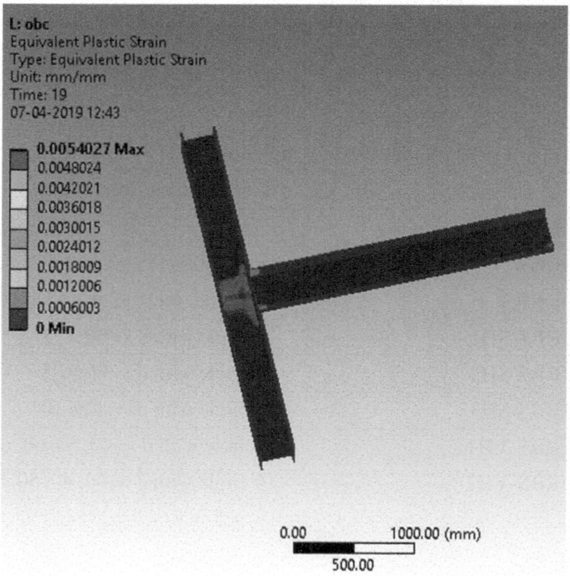

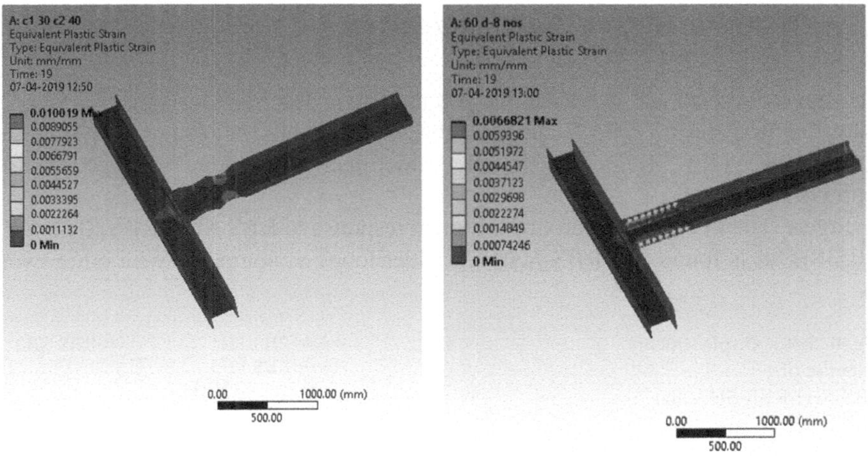

Fig. 4 Equivalent plastic strain of DRBS and RBS-SH1

Figure 3 shows the equivalent plastic strain distribution of OBC connection. In OBC connection, it can be seen that the strain is distributed in the region vicinity to the connection and faces of the column. Figure 4 shows the equivalent plastic strain distribution in DRBS and RBS-SH1 model. In DRBS model most of equivalent plastic strain is concentrated in the second radius cut and only a small amount is distributed in connection region. In RBS models with circle it is found that the strain concentration in all models is at the circles that are located at the centre. In all models the equivalent plastic strain is distributed in the beam faces except for OBC. It can be observed that all the reduced beam section models shifts the location of plastic hinge from connection region to the face of the beam. The equivalent plastic strain distribution is also balanced thereby reducing the beam local failure at RBS region. Figure 5 shows the maximum value of plastic strain for all models. It can be seen that the plastic strain value is slightly increasing from OBC to DRBS.

Fig. 5 Equivalent plastic strain

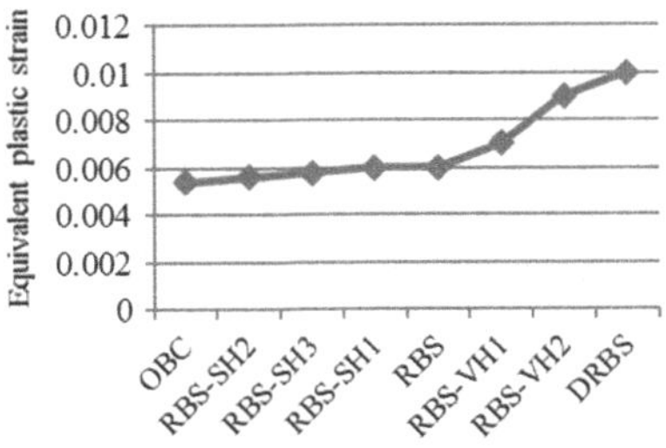

3.2 Hysteretic Response of the Models

The force-displacement response of the 8 models is described in Figs. 6, 7 and 8. With the increase in cyclic loading, all the models exhibit almost similar smooth curves except OBC. The area of hysteretic loop increases gradually in accordance with OBC.

Figure 6 describes the force-displacement response of RBS-SH1, RBS-SH2, and RBS-SH3.It is found that RBS-SH3 has wider loops as compared with other two.

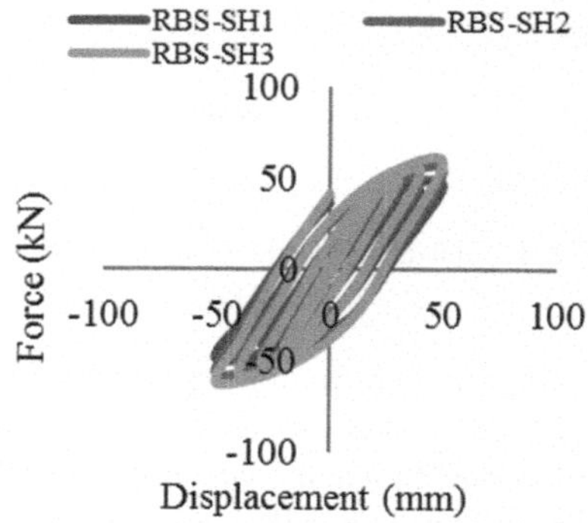

Fig. 6 Force-displacement response of RBS-SH1,RBS-SH2, and RBS-SH3

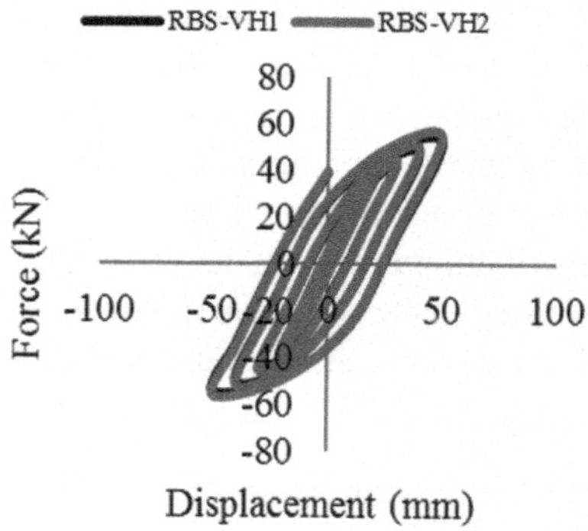

Fig. 7 Force-displacement response of RBS-VH1 and RBS-VH2

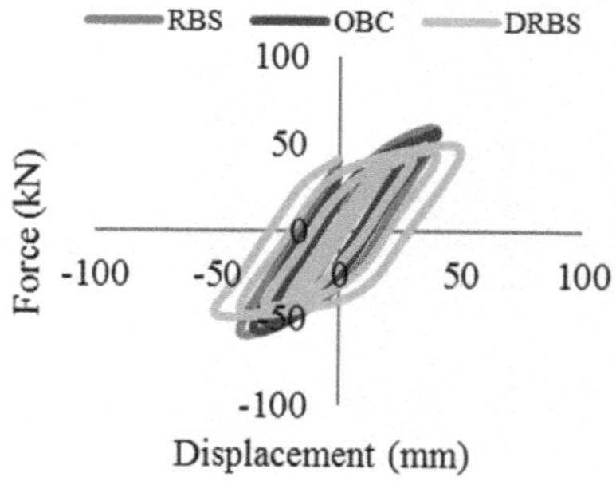

Fig. 8 Force-displacement response of OBC, RBS, DRBS

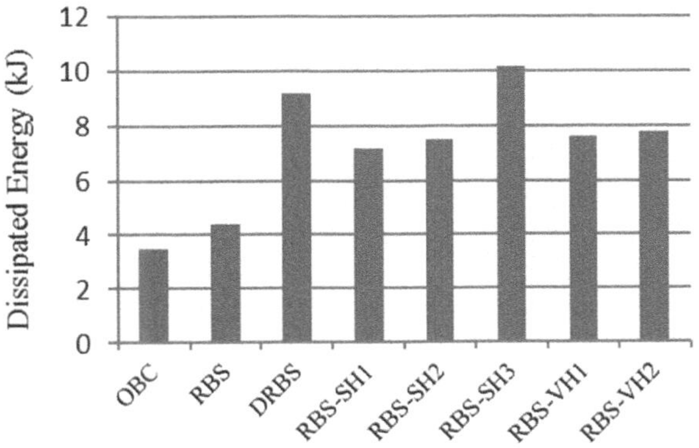

Fig. 9 Energy dissipated by the models

Figure 7 exhibits the hysteretic behavior of RBS-VH1 and RBS-VH2.Both the models shows relatively same hysteretic response. Figure 8 comprises of the hysteretic response of OBC, RBS and DRBS, DRBS has much wider loop as compared remaining models.

3.3 Energy Dissipation

The total energy dissipated by each model is described in Fig. 9. The models RBS-SH3 and DRBS have higher energy dissipation in accordance with others. RBS-SH1, RBS-SH2, RBS-VH1 and RBS-VH2 have nearly same level of energy dissipation. DRBS has 62% higher energy dissipation as compared to OBC whereas RBS-SH3 has 65% higher energy dissipation while comparing with OBC.

3.4 Stiffness

Stiffness is advantageous with respect to seismic damage because it can limit the deformation demands on a structure. Figure 10 exhibits the stiffness of the models. All the models have sufficiently good value of stiffness.

4 Conclusion

Reduced beam section exhibits excellent transformation of stress concentration from face of the column to the section region. It has significant increase in energy dissipation capacity as compared with normal end plate connection. From the result obtained from this paper it can be concluded as follows;

Fig. 10 Stiffness of models

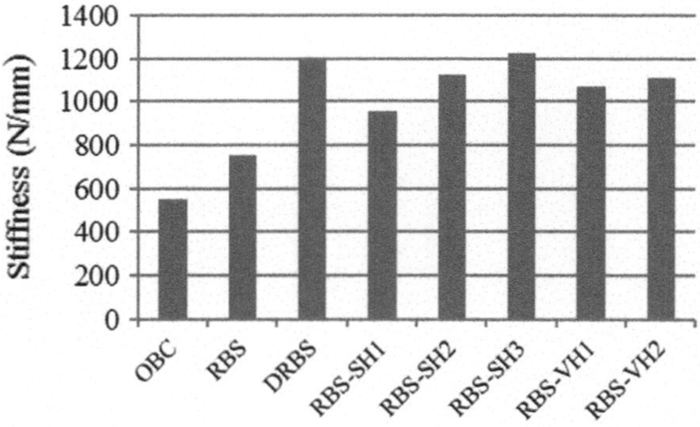

- RBS shifts the location of plastic hinge at a certain distance from the beam column junction.
- In DRBS, RBS with same holes and RBS with varying holes, plastic deformation took place in the beam.
- By the use of RBS, the connection region in all the models remains relatively safe.
- The model, RBS-SH3 (6 circle with d = 40@104) has higher rate of energy dissipation while comparing with others.

References

1. Pachoumis DT, Galoussis EG, Kalfas CN, Christitsas AD (2009) Reduced beam section moment connections subjected to cyclic loading: experimental analysis and FEM simulation. Eng Struct 31:216–223
2. Pachoumis DT, Galoussis EG, Kalfas CN, Efthimiou IZ (2010) Cyclic performance of steel moment-resisting connections with reduced beam sections experimental analysis and finite element model simulation. Eng Struct 32:2683–2692
3. Sofias CE, Kalfas CN, Pachoumis DT (2014) Experimental and FEM analysis of reduced beam section moment endplate connections under cyclic loading. Eng Struct 59:320–329
4. Swati AK, Gaurang V (2014) Study of steel moment connection with and without reduced beam section. Case Stud Struct Eng 1:26–31

Reliability Analysis of RC T-Beam Bridge Girder Subjected to Chloride-Induced Corrosion

A. Ranjith, K. Balaji Rao, Thripthi, A. Tanvi Rai and K. Manjunath

Abstract One of the main causes of reinforced concrete degradation is chloride-induced corrosion. Chloride-induced corrosion depends on many factors like properties of the material, loading on the structure and environmental condition. There is much research on the deterioration of concrete structures over the years; but there is relatively little research on changing climate affecting the deterioration, which is the topic of this paper. The present study is purely analytical wherein conceptual T-Beam bridge girder has been taken for the prediction of time to corrosion initiation and corrosion propagation, incorporating effect of climatic condition. For comparison, eight study locations are selected. Corrosion initiation time is predicted for each location considering the climatic data of the corresponding location. With the variation in the temperature and relative humidity data at different study location, large variation in corrosion initiation time is found.

Keywords Chloride-induced corrosion · Temperature · Relative humidity · Diffusion coefficient

---

A. Ranjith (✉) · Thripthi · A. Tanvi Rai
Department of Civil Engineering, N.M.A.M Institute of Technology, (Visvesvaraya Technological University, Belagavi), Nitte, Udupi District, Karnataka 574110, India
e-mail: ranjith.anand70@gmail.com

Thripthi
e-mail: karkerathripthi@gmail.com

A. Tanvi Rai
e-mail: tnvrai07194@gmail.com

K. Balaji Rao
Risk and Reliability Group, CSIR-Structural Engineering Research Centre, Chennai, Tamil Nadu, India

K. Manjunath
Department of Civil Engineering, Malnad College of Engineering, (Visvesvaraya Technological University, Belagavi), Hassan, Karnataka 577102, India
e-mail: kmnrpur@gmail.com

K. Dasgupta et al. (eds.), *Proceedings of SECON'19*,
Lecture Notes in Civil Engineering 46,
https://doi.org/10.1007/978-3-030-26365-2_37

1 Introduction

Reinforced concrete (RC) and concrete are widely used in construction of numerous structures all over the world. The safety of these structures is very important. If these structures are exposed to aggressive environment for a long time, deterioration occurs. Reinforcement corrosion is one of the major causes for deterioration. The durability of RC structure is threatened by chloride induced corrosion and carbonation-induced corrosion. Chloride ions may be present in the raw materials at the time of mixing or introduce from the outside aggressive environment (de-icing salts).

For any RC member, the service life is divided into two parts [1]: corrosion initiation period and corrosion propagation period. The time at which chloride concentration on the surface of the steel bar reaches the threshold chloride concentration value is called corrosion initiation time, *ti*. Corrosion reduces the load carrying capacity of RC member due to progressive loss of steel area and reduction in the bond between steel reinforcement and concrete, while the products of corrosion cause longitudinal cracking and spalling [2]. When the RC Structure is located in a coastal environment, it gets easily attacked by chloride ion penetration. Existing research has found that time, environment (such as temperature, relative humidity), and the performance of concrete influences chloride diffusion coefficient.

Parrott [3], Andrade and Castillo [4] proved experimentally the impact of environmental parameters on the corrosion process. Allampallewar and Srividya [5] studied the effects on t_i considering the local temperature and relative humidity of RC structures along the five Indian coastal cities, and the probability distribution of *ti* was analyzed. Hassan et al. [6] performed the reliability-based assessment of the RC structure subjected to chloride ingress considering the climatic conditions. Alhozaimy et al. [7] tested the corrosion of concrete at the bay area under chloride ion erosion with high temperature and high humidity. These study results indicated that it is necessary to consider the local microclimate to study the durability of the coastal RC structures. While there is much research on the deterioration of concrete structures, but there is relatively little focus on how deterioration is affected by climate change.

The present work emphasizes the impact of climatic conditions, namely temperature and relative humidity, on the two phases of corrosion. Finally, the model is applied to the different climatic conditions to better understand the impact of meteorological factors, especially temperature and relative humidity, on the corrosion process.

2 Methodology

2.1 Corrosion Prediction Model

The highly alkaline environment of the cement paste provides a passive condition around the reinforcing steel. However, when exposed to a severe environment where sea water and de-icing salts are present, corrosion takes place. Penetration of chlorides by diffusion is the dominant cause of corrosion initiation [2]. Chloride ions diffuse into the concrete and disrupt the passive layer after exceeding the threshold level. Rust which is the product of corrosion occupies a greater volume than the original volume of steel. This increase in volume exerts thrust on cover concrete resulting in cracks, spalling and delamination of concrete. As time progresses, the reinforcement cross-section progressively reduces affecting the strength and serviceability of the structure.

2.2 Corrosion Initiation

Corrosion mechanism in reinforced concrete members begin with corrosion initiation [8] and is modeled using Fick's second law of diffusion. The mechanism by which chloride induce corrosion is not entirely understood but the most popular theory is that chloride ions penetrate the protective oxide film easier than do other ions, leaving the steel vulnerable to corrosion. Hence, Chlorides are directly responsible for the initiation of corrosion.

From Fick's second law of diffusion [9], corrosion initiation time is given as:

$$ti = \frac{c^2}{4D_c}\left[erf^{-1}\left(\frac{c_s - c_{cr}}{c_s}\right)\right]^{-2} \tag{1}$$

Here c—concrete cover to reinforcement in mm; D_c—diffusion coefficient for chlorides in concrete in mm^2/year; c_s—surface chloride concentration in % weight of concrete; and C_{cr}—critical chloride concentration in % weight of concrete.

Various realistic diffusion models like the Mejlbro–Paulsen model [10], Duracrete model [11], and fib Model Code model [12] are available in the literature. In order to make use of these models a lot of laboratory and field data and/or information are necessary, this may prohibit the use of these models often. Hence, in the present study, Eq. 1 is used to estimate the time for corrosion initiation.

However, this model does not consider the effects of temperature, humidity, chloride binding and concrete ageing, the solution is only valid for saturated conditions. So in the present study, the diffusion model that can take care of all these parameters have been used (Eqs. 2–6).

2.3 Diffusion Co-Efficient

Diffusion coefficient depends not only on the material property such as concrete porosity, mix-proportions but also on the exposure time, temperature and relative humidity. Some of the researchers have taken concrete properties into account and expressed the diffusion coefficient as a function of w/c ratio [13]. Some of the researchers considered the effect of temperature [14], the effect of relative humidity [15]. Life-365 [16] suggests the relationship for diffusion coefficients for concrete exposed to chlorides attack. The reference diffusion coefficient for concrete cured about 28 days is given [13].

$$D_{ref} = 10^{\left[-12.06+2.4\left(\frac{w}{c}\right)\right]} \tag{2}$$

where D_{ref}—reference diffusion coefficient in m^2/s and *w/c*—water to cement ratio.

To obtain an effective diffusion coefficient this reference value is to be corrected as defined in the equation below [16]

$$D_{eff=D(t,T,h)} = D_{ref.f(t).f(T).f(h)} \tag{3}$$

Here, the first term is known as the reference diffusion coefficient taking into account the effect of w/c ratio and curing period of concrete.

The D_{ref} is evaluated for standard condition i.e., temperature 20 °C, relative humidity 100%, age of concrete 28 days. However, in actual condition, there will be variation in these conditions.

The second term is to taking into account the influence of concrete age on the chloride diffusion coefficient in concrete [17].

$$f(t) = \left[\frac{t_{ref}}{t}\right]^m \tag{4}$$

where t_{ref}—reference time in days; t—actual time of exposure in days and m—variable also called as age reduction factor (dimensionless).

The third term taking into account the effect of temperature on diffusion coefficient and is described using Arrhenius' law [13, 18];

$$f(T) = \exp\left[\frac{U}{R}\left(\frac{1}{T_{ref}} - \frac{1}{T}\right)\right] \tag{5}$$

where, U—activation energy of diffusion process (35,000 J/mol); R—gas constant (8.314 J/mol K) and T_{ref}—reference temperature 293 K (20 °C) and T—current absolute temperature of concrete in Kelvin.

$$f(h) = \left[1 + \left(\frac{1-h}{1-h_c}\right)^4\right]^{-1} \tag{6}$$

The fourth term takes into account, the effect of relative humidity on the chloride diffusion coefficient [18].

Where, h—current actual humidity and h_c—critical humidity value taken as 0.75.

2.4 Corrosion Propagation

The availability of oxygen and water at the surface of steel governs the corrosion rate. Also, water/cement ratio, concrete cover and environmental conditions influences the oxygen availability.

The variations in local temperature and relative humidity will change the vulnerability of the RC structure with regard to corrosion propagation. This micro-environment effect can be significant for bridges located in coastal areas.

The corrosion current density is larger when RH increases, T increases and c decrease [6].

Empirical relationship proposed by Breysse et al. [16] is used for corrosion current density:

$$\ln(i_{corr}) = 0.0312RH - \frac{4736}{T} + 1.695\frac{w}{c} - 0.391c + 14.589 \quad (7)$$

where RH—relative humidity in %, T—temperature in K, w/c—water/cement ratio and c—concrete cover in cm; the corrosion current density is in $\mu A/cm^2$.

Model proposed by Rodriguez et al. is widely accepted. From this model, the rate of corrosion can be determined as

$$r_{corr} = 0.0115 i_{corr}\alpha \quad (8)$$

The maximum pit depth of corrosion is given by

$$p(t) = r_{corr}(t - t_i) \quad (9)$$

where, r_{corr}—rate of corrosion in mm/year; i_{corr}—corrosion current density, in $\mu A/cm^2$, α—pitting factor, p(t)—maximum pit depth at time t and t_i—corrosion initiation time in years.

For pitting corrosion reduction in cross sectional area of the reinforcement can be calculated [19–22]. The loss in cross sectional area of the reinforcement by assuming the hemispherical pit formed due to corrosion propagation. Using this model, the remaining area of the steel A_{sr} (Fig. 1)

$$A_{sr} = \left\{ \begin{array}{cc} A_1 + A_2, & p(t) \le \frac{D_0}{\sqrt{2}} \\ \frac{\pi D_0^2}{4} - A_1 + A_2, & \frac{D_0}{\sqrt{2}} < p(t) \le D_0 \\ \frac{\pi D_0^2}{4}, & p(t) \ge D_0 \end{array} \right\} \quad (10)$$

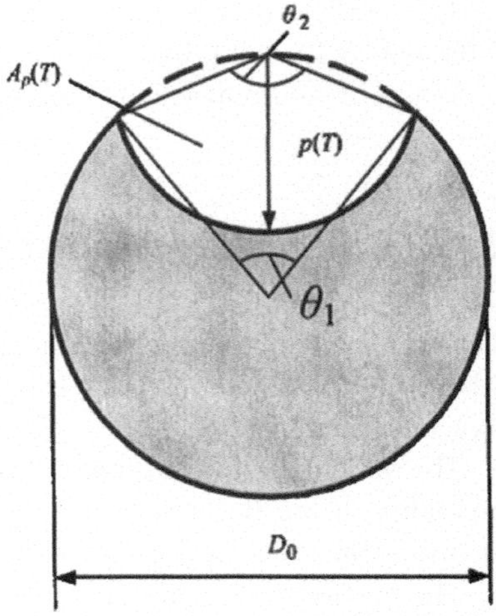

Fig. 1 Pit configuration [23]

where,

$$A_1 = \frac{1}{2}\left[\theta_1\left(\frac{D_0}{2}\right)^2 - \alpha\left(\frac{D_0}{2} - \frac{p(t)^2}{D_0}\right)\right] \tag{11}$$

$$A_2 = \frac{1}{2}\left[\theta_2(p(t))^2 - \alpha\left(\frac{p(t)^2}{D_0}\right)\right] \tag{12}$$

$$\alpha = 2p(t)\sqrt{1 - \left(\frac{p(t)}{D_0}\right)^2} \tag{13}$$

$$\theta_1 = 2\arcsin\left(\frac{\alpha}{D_0}\right) \tag{14}$$

$$\theta_1 = 2\arcsin\left(\frac{\alpha}{2p(t)}\right) \tag{15}$$

where D_o—initial diameter of reinforcing bar in mm and p(t)—maximum pit depth in mm.

2.5 Flexural Capacity for RC Section

The time-variant flexural resistance of an RC beam is given by

$$M_u = 0.542 f_{ck} b_f x_u(t)(d - x_u(t)) + (f_{sc}(t) - f_{cc})A_{sc}(d - d^1) \tag{16}$$

where, f_{cc}—stress in concrete in N/mm^2, $f_{sc}(t)$—stress in steel in N/mm^2, f_{ck}—concrete compressive strength in N/mm^2, b_f—width of flange in mm, d—effective depth in mm, $x_u(t)$—depth of neutral axis in mm.

The yield stress reduces linearly [20, 24] and is given by:

$$f_y(t) = \left(1 - \alpha_y \frac{A_{pit}(t)}{A_{stnom}} \times 100\right) f_{yo} \tag{17}$$

where, f_{y0}—yield stress of reinforcing bar (uncorreded), α_y—an empirical coefficient, and

$$A_{stnom} = n_b \times \pi \frac{D_0^2}{4} \tag{18}$$

Experimental results reported, α value was reported from 0.017 to 0.06 for pitting corrosion.

3 Climatic Profile of India

India Meteorological Department (IMD) [25] classifies four climatic seasons namely: S1—Winter Season / Cold Weather Season (January and February), S2—Pre-monsoon season/ summer season/ Hot weather season/Thunderstorm season (March, April and May), S3—South-west Monsoon/ Summer Monsoon (June, July, August and September), S4—Post-monsoon or Northeast monsoon or Retreating SW Monsoon season.

4 Illustrative Example

In the present study flexural strength and service life of the T beam bridge girder subjected to chloride ingress, placed at different location along the India is studied. The following location in India namely Bangalore, Bhubaneswar, Chennai, Goa, Mangalore, Rameshwaram, Rathnagiri and Visakhapatnam are considered in the present study.

RC T beam bridge girder built according to standard MORTH design, of span 10 m. The cross section of bridge is as shown in Fig. 2. As per IRC 6, carriageway width of 10.9 m is considered to have three lanes of traffic. This bridge is assumed to be located in the different region mentioned above. The bridge girder is assumed to be constructed using M30 and Fe415 grade concrete and steel respectively.

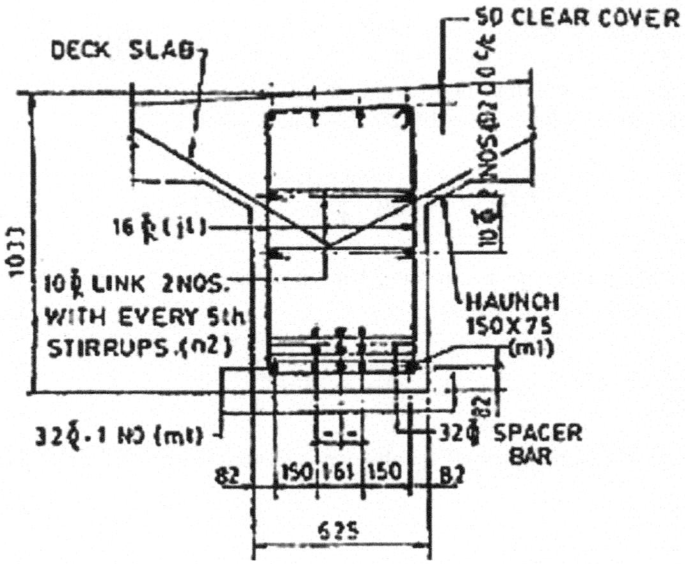

Fig. 2 Cross section of T-beam bridge girder for 10 m Span(MORTH)

4.1 Loading

For a carriage way width of 10.9 m, live load combination as per IRC 6-2017 is considered. The load intensities on the girder are computed. Worst loading arrangement for bending moment variation is considered. The maximum bending moment occurs at the center of the span in case of dead load and under the maximum load intensity in case of live load combination.

4.2 Calculation of Corrosion Initiation Time

The data and the input parameter used in the study are presented in Table 1.

The analysis for corrosion initiation time is done within a framework of Monte Carlo Simulation using relative frequency approach. The corrosion initiation time predicted is 19.8, 10.6, 10.1, 9.0, 5.6, 5.4, 11.1 and 8.8 years respectively at Bangalore, Bhubaneswar, Chennai, Goa, Mangalore, Rameshwaram, Rathnagiri and Visakhapatnam (Tables 2 and 3).

Table 1 Parameters used in the study

Surface chloride concentration	C_s	0.25	% chloride of concrete mass	Assumed COV = 0.2
Critical chloride Concentration	C_{cr}	0.125	% chloride of concrete mass	Assumed COV = 0.2
Water to cement ratio	w/c	0.45	–	
Concrete cover	c	50	mm	Assumed COV = 0.1
Diffusion decay index	m	0.525	–	
Pitting factor	α	6	–	
Reference temperature	Tref	293	K	
Reference time	tref	28	days	
Activation energy	Uc	35000	kg/mole	
Gas constant	R	8.314	J/mole k	
Critical humidity	hc	75	%	
Age	t	50	years	

Table 2 Average seasonal and annual temperature (in celsius)

City	Years of record	S1	S2	S3	S4	Annual
Bangalore	100	22.3	26.9	23.8	22.1	23.7
Bhubaneswar	49	23.6	30.6	29.0	24.5	26.9
Chennai	100	25.2	30.2	30.7	26.3	28.1
Goa	9	29.1	31.3	28.2	29.2	29.5
Mangalore	45	27.1	29.1	26.1	27.1	27.3
Rameshwaram	9	28.2	30.9	30.4	29.8	29.8
Rathnagiri	9	28.1	30.7	28.4	28.7	29.0
Visakhapatnam	29	24.9	29.0	29.0	26.3	27.3

Table 3 Average seasonal and annual relative humidity (in %)

City	Years of Record	S1	S2	S3	S4	Annual
Bangalore	20	56	52	76	70	64
Bhubaneswar	20	61	65	81	67	69
Chennai	20	73	67	65	77	70
Goa	9	57	68	85	67	69
Mangalore	10	67	74	89	76	76
Rameshwaram	9	77	78	78	77	78
Rathnagiri	9	56	67	85	65	68
Visakhapatnam	20	71	70	76	70	71

5 Result and Discussion

The corrosion initiation time calculated on seasonal basis is as presented in Table 4.

The corrosion initiation time is calculated considering the monthly mean temperature and monthly mean relative humidity.

As the corrosion progresses there will be reduction in the bar diameter of the reinforcing bar (Fig. 3).

The loss in diameter reduces the flexural capacity over the year. The reduction in flexural capacity is as shown in Fig. 4.

Figure 5 shows the variation of corrosion initiation time and corrosion current density with relative humidity and temperature respectively (Figs. 6 and 7).

The time to corrosion propagation is found to be around 7.5 years. Therefore, the total service life of the T-beam bridge girder is sum of the corrosion initiation time and corrosion propagation time. The total service life of the beam girder if it is assumed to be located at Bangalore, Bhubaneswar, Chennai, Goa, Mangalore,

Table 4 Calculated corrosion initiation time (in years)

City	S1	S2	S3	S4	Annual
Bangalore	46.78	51.80	7.60	13.71	19.8
Bhubaneswar	28.76	14.46	4.29	16.00	10.6
Chennai	9.07	12.28	14.43	6.27	10.1
Goa	31.22	10.67	3.77	12.90	9.0
Mangalore	14.18	6.95	3.83	6.49	5.6
Rameshwaram	5.74	4.73	4.82	5.31	5.4
Rathnagiri	35.52	12.01	3.74	15.79	11.1
Visakhapatnam	10.93	9.89	5.95	11.23	8.8

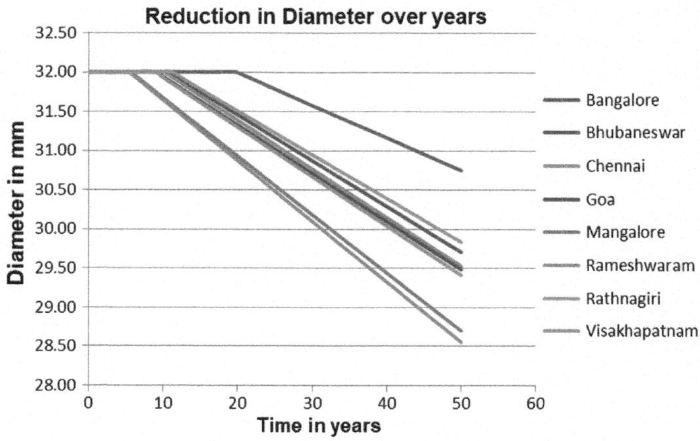

Fig. 3 Diameter loss in beam reinforcement due to corrosion

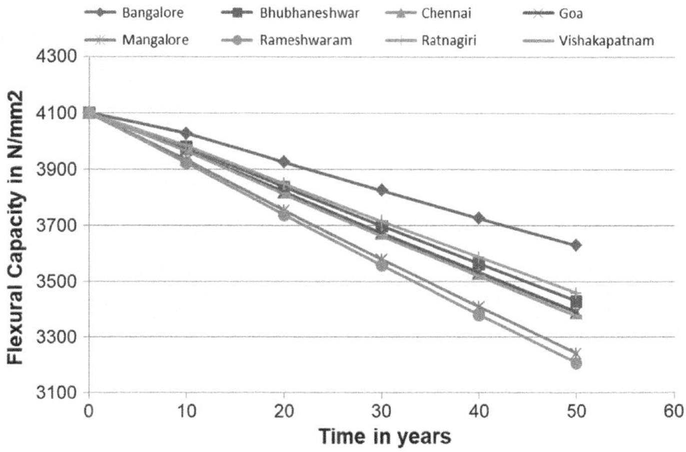

Fig. 4 Reduction in Flexural capacity of RC member over the years

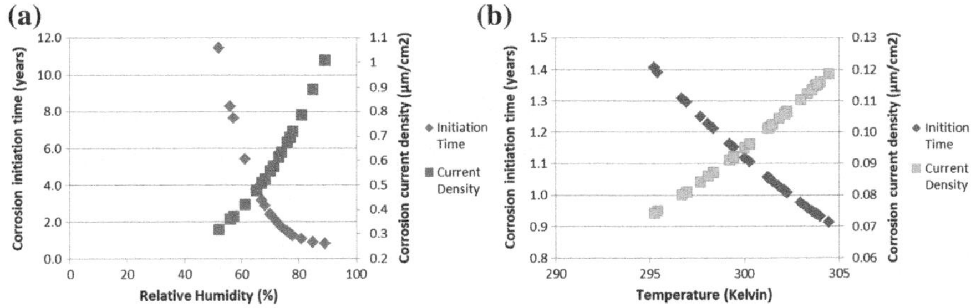

Fig. 5 **a** Variation of corrosion initiation time and corrosion current density with relative humidity, **b** variation of corrosion initiation time and corrosion current density with temperature

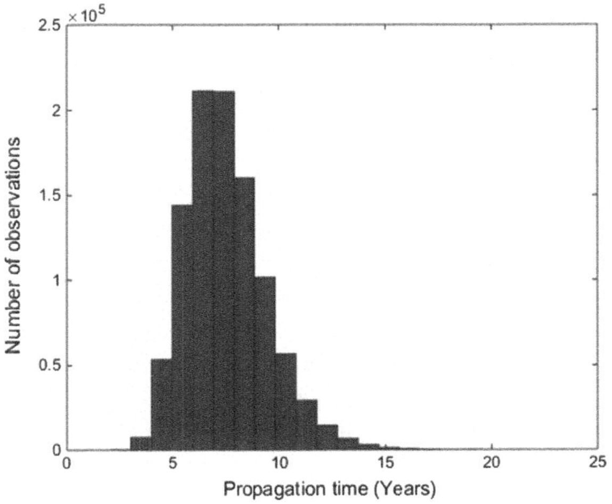

Fig. 6 Histogram of time to corrosion propagation

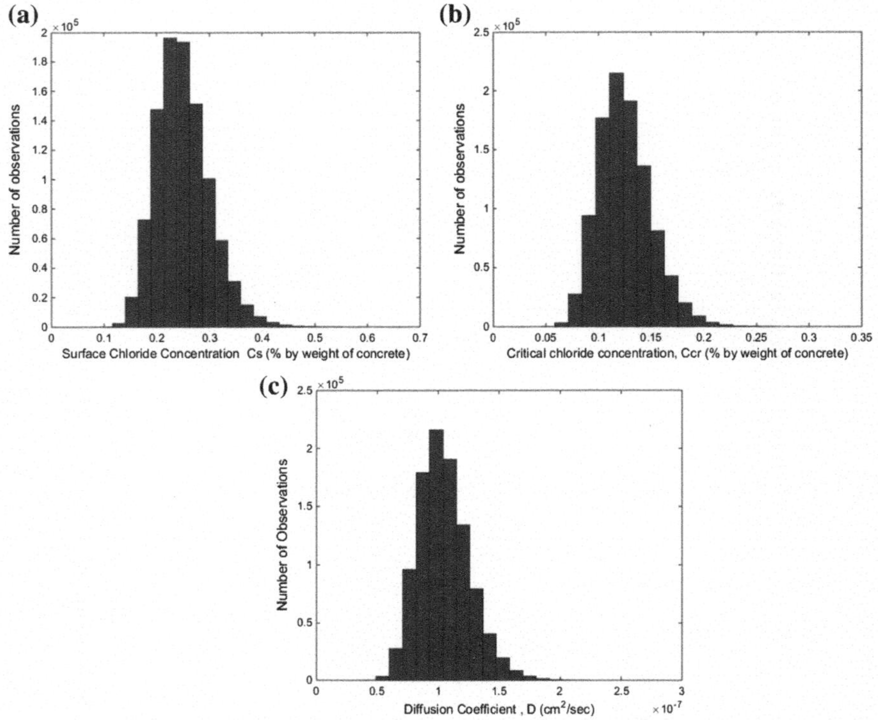

Fig. 7 **a** Histogram of surface chloride concentration, **b** Histogram of critical chloride concentration, **c** Histogram of Diffusion coefficient

Rameshwaram, Rathnagiri and Visakhapatnam, is 27.3, 18.1, 17.6, 16.5, 13.1, 12.9, 18.6 and 16.3 years respectively. The beam girder located in coastal region other than non-coastal region (Bangalore) is found to undergo corrosion at very faster rate, this is because of the ingress of chloride ions from surrounding environment over a period of time.

6 Conclusion

- The corrosion initiation time predicted is 19.8, 10.6, 10.1, 9.0, 5.6, 5.4, 11.1 and 8.8 years respectively at Bangalore, Bhubaneswar, Chennai, Goa, Mangalore, Rameshwaram, Rathnagiri and Visakhapatnam and the time to corrosion propagation is found to be around 7.5 years.
- The total service life of the beam girder if it is assumed to be located at Bangalore, Bhubaneswar, Chennai, Goa, Mangalore, Rameshwaram, Rathnagiri and Visakhapatnam, is 27.3, 18.1, 17.6, 16.5, 13.1, 12.9, 18.6 and 16.3 years respectively.

- It is evident from the investigation that the increase of either temperature or relative humidity leads to significant decrease in the time to corrosion initiation.
- With the increase in temperature and relative humidity, corrosion current density increases significantly.
- Reduction in bar diameter over a period of time due to chloride induced corrosion is significantly faster in coastal regions than in non-coastal regions (Fig. 3), this may be due to the fact that, if there were no chloride ions present at the time of mixing of raw materials also, the chloride ions may introduce into concrete from the outside aggressive environment.
- The amount of corrosion is determined based on the corrosion rate, propagation time period, and bar diameter. The reduction in flexural capacity after deterioration has been captured over the years.

References

1. Tuutti K (1982) Corrosion of reinforcement in concrete. CBI report, Swedish Cement and Concrete Research Institute, Stockholm, vol 4, p 82
2. Stewart MG, Rosowsky DV (1998) Structural safety and serviceability of concrete bridges subject to corrosion. J Infrastruct Syst 4(4):146–155
3. Parrott LJ (1991) Factors influencing relative humidity in concrete. Mag Concr Res 43(154):45–52
4. Andrade C, Castillo A (2003) Evolution of reinforcement corrosion due to climatic variations. Mater Corros 54(6):379–386
5. Allampllewar SB, Srivdya A (2008) Corrosion initiation time for reinforced concrete members along Indian coast: effects of temperature and relative humidity—A probabilistic approach. Constr Build Mater 15:539–553
6. El Hassan J, Bressolette P, Chateauneuf A, EL Tawil K (2010) Reliability-based assessment of the effect of climatic conditions on the corrosion of RC structures subjected to chloride ingress. Eng Struct 32:3279–3287
7. Alhozaimy A, Hussain RR, Al-Zaid R, Al-Negheimish A (2012) Coupled effect of ambient high relative humidity and varying temperature marine environment on corrosion of reinforced concrete. Constr Build Mater 28:670–679
8. Collepardi M, Marcialis A, Turriziani R (1972) Penetration of chloride ions into cement pastes and concretes. J Am Ceram Soc 55(10):534–535
9. Crank J (1975) Mathematics of diffusion. Oxford University Press, Oxford
10. Lindvall A (1999) Environmental actions and response—survey, inspection and measurement. Working Report, The European Community Brite EuRam Project BE95-1347, Task 7 report
11. Lindvall A (2001) Environmental actions and response: reinforced concrete structures exposed in wind and marine environments. Thesis for Degree of Licentiate of Engineering, Department of Building Materials, Chalmers University of Technology, Goteborg
12. Fib (2006) Model code for service life design. fib Bulletin 34, International Federation for Structural Concrete, Lausanne
13. Bentz E, Thomas M (2000) Manual of computer program for predicting the service life and life cycle costs of reinforced concrete exposed to chlorides, pp 1–55
14. Samson E, Marchand J (2007) Modeling the effect of temperature on ionic transporting cementitious materials. Cement Concr Res 37(3):455–468
15. Saetta AV, Scotta RV, Vitaliani RV (1993) Analysis of chloride diffusion into partially saturated concrete. ACI Mater J 90(5):441–451

16. Breysse D, Yotte S, Salta M, Schoefs F, Ricardo J, Pereira E (2008) Uncertainties in NDT condition assessment of corroding structures in marine environment. MEDACHS08, Construction heritage in coastal and marine environments. Damage, diagnostics, maintenance and rehabilitation (Lisbon, Portugal), LNEC
17. Bamforth PB (1999) The derivation of input data for modeling chloride ingress from eight-year UK coastal exposure trials. Mag Concr Res 51(2):87–96
18. Ababneh A, Benboudjema F, Xi Y (2003) Chloride penetration in non-saturated concrete. J Mater Civ Eng
19. Val DV, Melchers RE (1997) Reliability of deteriorating RC slab bridges. J Struct Eng 123(12):1638–1644
20. Stewart MG (2009) Mechanical behaviour of pitting corrosion of flexural and shear reinforcement and its effect on structural reliability of corroding RC beams. Struct Saf 31(1):19–30
21. Stewart MG, Al-Harthy A (2008) Pitting corrosion and structural reliability of corroding RC structures, experimental data and probabilistic analysis. Reliab Eng Syst Saf 93(3):273–382
22. Darmawan MS (2010) Pitting corrosion model for reinforced concrete structures in a chloride environment. Mag Concr Res 62(2):91–101
23. Val DV (2007) Deterioration of strength of RC beams due to corrosion and its influence on beam reliability. J Struct Eng 133(9):1297–1306
24. Du YG, Clark LA, Chan AH (2005) Residual capacity of corroded reinforcing bars. Mag Concr Res 57(3):135–148
25. Attri SD, Tyagi A (2010) Climate profile of India. India Meteorological Department, New Delhi

Probabilistic Analysis of Cracking Moment of RC T-Beam Bridge Girder

A. Ranjith, K. Balaji Rao, A. Tanvi Rai, Thripthi and K. Manjunath

Abstract Cracking of reinforced concrete flexural members is a highly random phenomenon. The evaluation of cracking resistance of a reinforced concrete structural element is important from a view point of its performance in limit state of serviceability i.e., in cracking. Deterministic analysis of cracking moment for all the beams is carried out. Probabilistic analysis of cracking from different cross sections of RC T-beam bridge girders is performed. For this purpose, different cross sections of RC T-beam bridge girders designed according to MORTH for seven spans namely 10, 12, 14, 16, 18, 21 and 24 m are considered. Probability of cracking of each cross sections of beam is determined using relative frequency approach within the framework of Monte Carlo Simulation. With respect to the distribution of basic variables considered in this study, cracking moment of RC T-beam is found to follow lognormal distribution. The probability of failure against cracking is found to be very high. Cracking moment is found to be independent of load combinations adopted in the study. The present article can be used as a reference for examining the beams for limit state of cracking.

A. Ranjith (✉) · A. Tanvi Rai · Thripthi
Department of Civil Engineering, N.M.A.M Institute of Technology, (Visvesvaraya Technological University, Belagavi), Nitte, Udupi District, Karnataka 574110, India
e-mail: ranjith.anand70@gmail.com

A. Tanvi Rai
e-mail: tnvrai07194@gmail.com

Thripthi
e-mail: karkerathripthi@gmail.com

K. Balaji Rao
Risk and Reliability Group, CSIR-Structural Engineering Research Centre, Chennai, Tamil Nadu, India

K. Manjunath
Department of Civil Engineering, Malnad College of Engineering, (Visvesvaraya Technological University, Belagavi), Hassan, Karnataka 577102, India
e-mail: kmnrpur@gmail.com

K. Dasgupta et al. (eds.), *Proceedings of SECON'19*,
Lecture Notes in Civil Engineering 46,
https://doi.org/10.1007/978-3-030-26365-2_38

Keywords Reinforced concrete · Cracking · Cracking moment · T beam girder · Basic variables

1 Introduction

A Bridge is a structure providing path over an obstacle without closing the way underneath. It must be strong enough to support its own weight as well as the weight of the people and vehicles that use it.

The T-beam Bridge is the most commonly adopted type in the span range of 10–25 m. The T-beam action is due to its monolithic construction with deck slab. To impart transverse stiffness to the deck, cross girders or diaphragms are provided at the regular intervals. The number of longitudinal girders depends on the width of the road. Three girders are normally provided for a two-lane road bridge.

A typical T-beam deck slab usually consists of longitudinal girder integrated with slab between the T-beams and cross girders to provide lateral firmness to the deck. The transfer of loads takes place from deck slab to the girders then to the substructure. Depending on the position of the live load on the deck slab the distribution of live load on each girder takes place.

Generally, the design moment is acquired by superimposing the maximum moments due to dead load and live load effects [1]. In the reinforced concrete bridges, a major portion of loads is due to dead loads which include self-weight of bridge, railing load, footpath, wearing course, etc. and it is considered uniformly distributed throughout the span [2, 3].

Serviceability refers to the behaviour of structures at working loads, with particular reference to cracking and deflection [1].

Cracking of reinforced concrete flexural members is a highly random phenomenon. The evaluation of cracking resistance of a reinforced concrete structural element Is Important from a view point of its performance in limit state of serviceability in cracking. The performance of structures working in heavy corrosion environment like elements of bridges and parking structures is affected by the cracking resistance of the structure. According to conventional elastic theory, cracking resistance of flexural members is evaluated as Cracking Moment 'M_{cr}' given by the expression [4];

$$M_{cr} = f_r I_g / Y_t \tag{1}$$

where f_r = Modulus of rupture, I_g = Gross moment of inertia, Y_t = distance to the extreme tension fibre from neutral axis.

Macregor et al. [5] conducted statistical analysis of resistance of reinforced and pre stressed concrete members. The variability in strengths of beams and columns are studied. Ahmed [6] carried out probabilistic analysis of the ultimate moment and cracking moment of doubly reinforced concrete beams using the Monte-Carlo simulation technique considering concrete strength and steel strength and geometric dimensions as variables. Based on the distribution of variables used in this study, ultimate moment and cracking moment is found to follow normal distribution. Kareem and Hsieh [7] carried out statistical analysis of tubular RC structures. Chandrashekar and Dayarathnam [8] conducted failure probability analysis of pre stressed concrete beams.

It can be noted from the expression of cracking moment that, M_{cr}, is purely deterministic in nature. But there will be large uncertainties due to the randomness of the terms involved in the cracking moment equation, hence probabilistic analysis of cracking moment for the statistical evaluation is important, the same has been taken as the study in the present article.

There was lot of attention given by the researchers to check the safety of the section against specified strength but little focus has been given to the statistical analysis of sections against serviceability, especially cracking of bridge girders. In the present study, an attempt has been made in carrying out the statistical analysis of RC T-beam bridge girders of different cross sections for different spans as suggested by MORTH against limit state of cracking, which is also compared with the external moment generated due to the possible load combinations the bridge can be subjected to.

2 Bridge Loading

Dead loads

Dead loads are loads enforced on a member by its own weight and the weight of other structural elements that it supports including rails, sidewalks, slabs and beams.

Live loads—IRC Loadings

Live loads are induced by the movement of the vehicles over the bridge and they are transitory in nature.

There are currently 4 types of IRC loading as per IRC: 6-2017 [9] which are considered as Live Load for Bridge design namely IRC class AA, IRC Class 70R, IRC Class A and IRC Class B loadings. IRC Class Special Vehicle (SV) Loading is included in IRC:6-2017 [9].

Class AA Loading [9]: In this type of loading two types of vehicles are considered one is tracked and another is wheeled vehicles. The load for tracked vehicle (simulating an army tank) is of 700 kN and for a wheeled vehicle (Heavy duty army truck) of 400 kN as shown in Fig. 1. Major bridges are situated on National Highways and State Highways and they are in turn need to be designed such heavy loading. The bridges which are located within the certain specified municipal limits are loaded with this type of loads. It is required that the bridges designed for IRC Class AA are checked for Class A loading as well. The IRC Class AA loadings have two patterns: (a) tracked type, and (b) wheeled type.

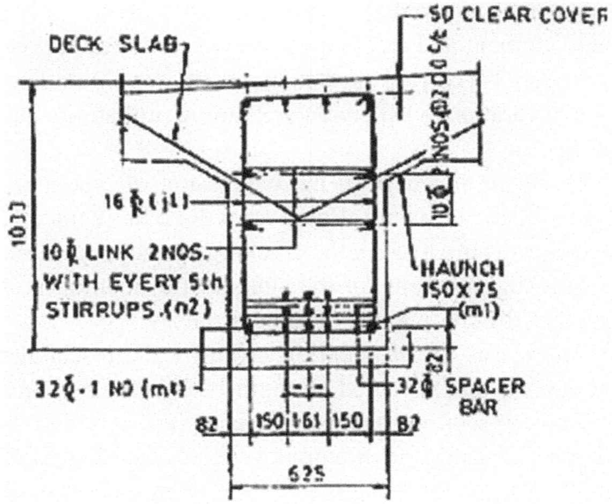

Fig. 1 Cross-section of T-beam bridge girder for 10 m span

IRC Class A and Class B Loading: IRC Class A loading is to be normally adopted on all roads on which permanent bridges and culverts are constructed [9, 3]. This is treated as standard loading. This load has eight axles with a total length of about 25 m. The heavy duty truck with two trailers transmits loads from 8 axles varying from a minimum of 27 kN to a maximum of 114 kN. The Class A loading is a 554 kN train of wheeled vehicles on eight axles.

IRC Class Special Vehicle (SV) Loading [9]: This loading is to be adopted for design of new bridges in select corridors as may be decided by the concerned authorities where passage of trailer vehicles carrying stator units, turbines, heavy equipment and machinery may occur occasionally.

In the present study two live load combinations are considered for number of lanes equal to 3, as per Table 6 of IRC:6-2017 [9] and also IRC class special vehicle loading is considered.

3 Deterministic Analysis of Cracking Moment

Deterministic analysis is carried out for a typical simply supported reinforced concrete T-beam bridge girder with cross-section. by considering the two live load combinations for a carriage way width of 10.9 m and also IRC [9, 10] class special vehicle loading: special multi-axle hydraulic trailer vehicle (Prime mover with 20 Axle Trailer—GVW = 385 Tonnes, as specified in IRC:6-2017 of different spans namely 10, 12, 14, 16, 18, 21 and 24 m).

Standard cross-section of T-beam bridge girder recommended by IRC [11] for different spans is shown in Figs. 1, 2, 3, 4, 5, 6 and 7 (Table 1).

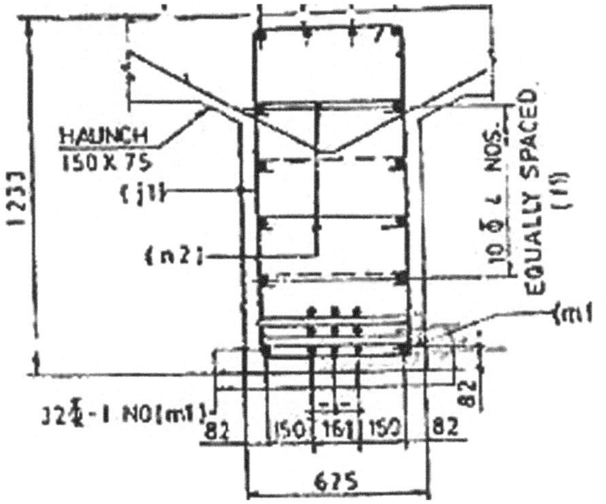

Fig. 2 Cross-section of T-beam bridge girder for 12 m span

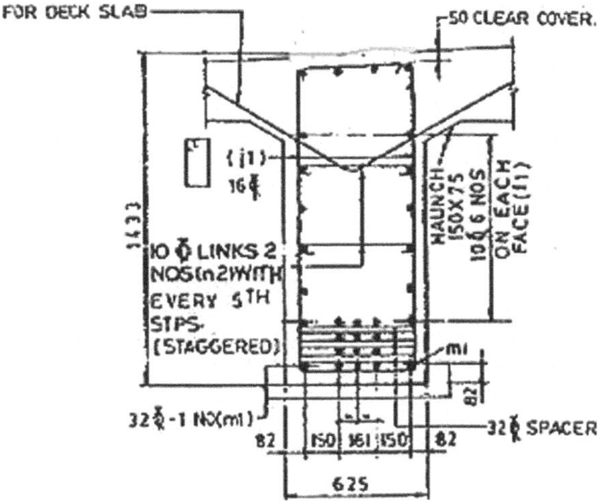

Fig. 3 Cross-section of T-beam bridge girder for 14 m span

Relative frequency approach

In this method the number of beams for which the external bending moment exceeds the cracking resistance of the beam is counted during the simulation process.

$$P_f = n_f/N \tag{2}$$

where, P_f is the probability of failure, n_f = number of beams which have cracked, N = total number of beams used in simulation = 10,000.

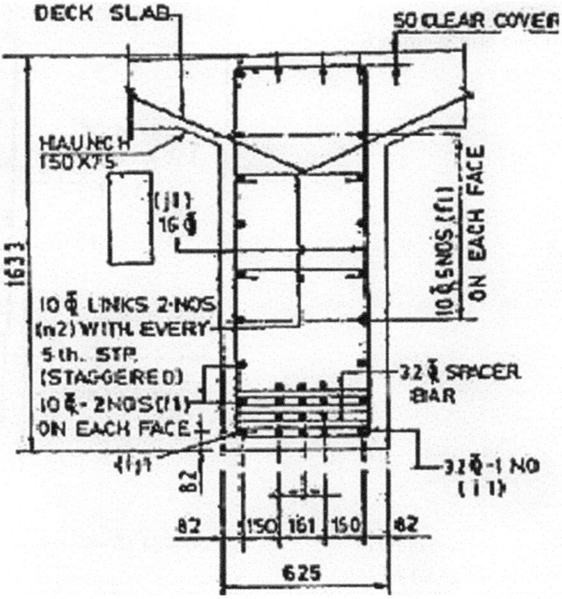

Fig. 4 Cross-section of T-beam bridge girder for 16 m span

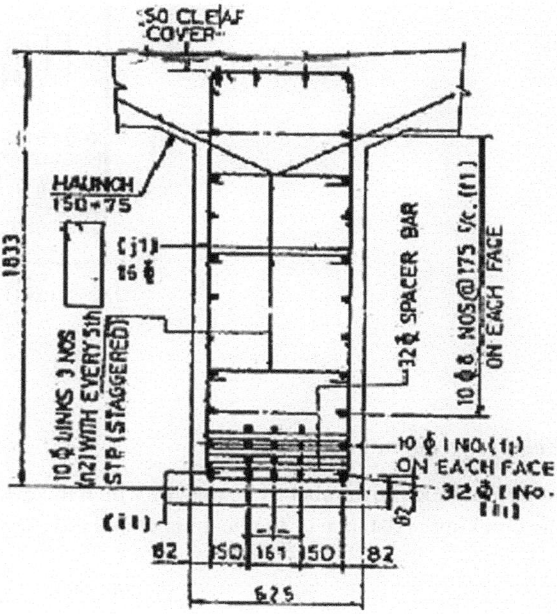

Fig. 5 Cross-section of T-beam bridge girder for 18 m span

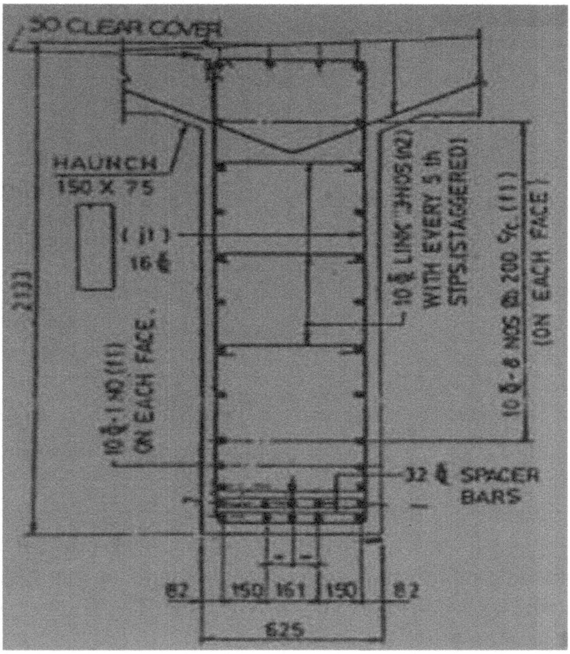

Fig. 6 Cross-section of T-beam bridge girder for 21 m span

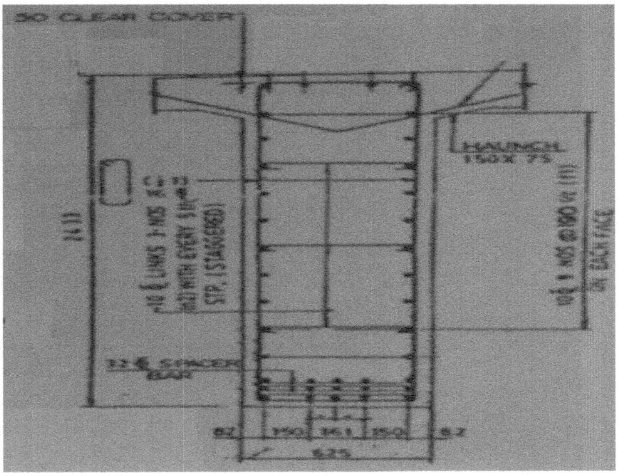

Fig. 7 Cross-section of T-beam bridge girder for 24 m span

Computational procedure

Monte Carlo simulation is carried out for T beam girder with number of simulation cycles equal to 10,000 (i.e., 10,000 random numbers are generated for each of the random variable, number of bins = 15).

Random numbers are generated for dead load moment. Ten thousand random external moments are obtained by adding the dead load and live load moments [12, 13]. Due to variations in cross-sectional dimensions of beam and the modulus of

Table 1 Deterministic analysis considering dead load moment alone as external moment acting on the beam

Span of the bridge (m)	Maximum moment M (kN-m)	Cracking moment M_{cr} (kN-m)	Is section cracked?
10	608.33	668.80	No
12	940.54	949.72	No
14	1364.80	1274.90	Yes
16	1890.48	1642.80	Yes
18	2526.98	2052.30	Yes
21	3710.12	2742.80	Yes
24	5195.16	3522.90	Yes

rupture of concrete, total moment acting on the beam considered to be a random variable [14, 15].

Ten thousand random cracking moments are calculated using the equation.

In each one of the ten thousand cases, the external moment is compared with the cracking moment, and if the external moment is greater than the cracking moment, the beam is considered to have cracked. The number of beams cracked at any stage of loading is determined (Table 2).

Probability of cracking is determined from Monte Carlo simulation using a relative frequency approach with 10,000 simulation cycles.

Statistical properties such as mean, standard deviation, skewness coefficient and kurtosis are determined for cracking moment. The analysis is repeated for all the spans considered (Tables 3, 4, 5, 6, 7 and 8).

Table 2 Basic random variables involved in the study

S. No.	Variable	Unit	Mean	COV	Type of distribution
1	Overall depth 'D'	mm	Specified	0.02	Normal
2	Effective depth 'd'	mm	Specified	0.02	Normal
3	Area of tension steel A_{st}	mm^2	Specified	0.05	Normal
4	Area of compression steel A_{sc}	mm^2	Specified	0.05	Normal
5	Area of stirrups A_{sti}	mm^2	Specified	0.05	Normal
6	Compressive strength of concrete f_{ck}	MPa	39.9	0.2	Lognormal
7	Yield strength of steel f_y	MPa	460	0.1	Lognormal
8	Dead load moment	kN-m	Deterministic	0.1	Normal
9	Live load moment	kN-m	Deterministic	0.25	Extreme type I largest

Table 3 Statistical properties of cracking moment for 3 lanes of class A loading, 'M_{cr}'

Variable	Mean (N-mm)	SD (N-mm)	COV	Skewness	Kurtosis
10	6.5078×10^8	1.4802×10^8	0.2274	0.7264	3.9378
12	9.2976×10^8	2.1247×10^8	0.2285	0.6916	3.8098
14	12.466×10^8	2.8752×10^8	0.2307	0.7157	4.0406
16	16.051×10^8	3.6460×10^8	0.2271	0.6898	3.7855
18	20.083×10^8	4.5959×10^8	0.2288	0.6545	3.7193
21	26.976×10^8	6.2865×10^8	0.233	0.7401	3.9286
24	34.634×10^8	7.9637×10^8	0.2299	0.7386	4.0559

Table 4 Statistical properties of external moment for 3 lanes of class A loading, 'S'

Variable	Mean (N-mm)	SD (N-mm)	COV	Skewness	Kurtosis
10	2.3069×10^9	4.2953×10^8	0.1862	1.1236	5.4057
12	3.1633×10^9	5.5802×10^8	0.1764	1.0525	4.9214
14	4.1294×10^9	7.0749×10^8	0.1713	1.1544	6.0728
16	5.3051×10^9	8.6488×10^8	0.1630	1.0324	5.3948
18	6.5986×10^9	1.0528×10^9	0.1596	1.1267	5.5175
21	8.5343×10^9	1.2556×10^9	0.1471	0.9503	4.6895
24	11.185×10^9	1.5618×10^9	0.1396	0.9796	4.9611

Table 5 Statistical properties of cracking moment for one lane of class AA (Tracked) vehicle for every two lanes with one lane of Class A loading for the remaining lane, 'M_{cr}'

Variable	Mean (N-mm)	SD (N-mm)	COV	Skewness	Kurtosis
10	6.5449×10^8	1.4995×10^8	0.2291	0.6979	3.8514
12	9.2836×10^8	2.1249×10^8	0.2289	0.6839	3.7404
14	12.473×10^8	2.8342×10^8	0.2272	0.6574	3.8167
16	16.123×10^8	3.7623×10^8	0.2333	0.7561	3.9636
18	20.043×10^8	4.5394×10^8	0.2265	0.7405	4.0675
21	26.774×10^8	6.1140×10^8	0.2284	0.7504	4.0126
24	34.685×10^8	7.9623×10^8	0.2296	0.7048	3.9077

4 Results and Discussion

Probability of failure

Probability of failure is calculated using Monte Carlo simulation (relative frequency approach) is as presented in Table 9. In the present study, for the sample size equal to 10,000 simulations, the number of intervals (bins) is works out to be 15 and same is used in all the cases while presenting histograms.

Table 6 Statistical properties of external moment for one lane of Class AA (Tracked) vehicle for every two lanes with one lane of Class A loading for the remaining lane, 'S'

Variable	Mean (N-mm)	SD (N-mm)	COV	Skewness	Kurtosis
10	1.7141×10^9	2.8239×10^8	0.1647	1.0472	5.146
12	2.3511×10^9	3.7438×10^8	0.1592	1.0547	5.2329
14	3.0848×10^9	4.5776×10^8	0.1484	0.9152	4.4859
16	3.9560×10^9	5.6973×10^8	0.144	0.9445	4.6092
18	4.9274×10^9	6.6286×10^8	0.1345	0.9230	5.0171
21	6.5244×10^9	8.1018×10^8	0.1242	0.7814	4.3763
24	8.6189×10^9	1.0237×10^8	0.1188	0.7521	4.4917

Table 7 Statistical properties of cracking moment for special vehicle, M_{cr}

Variable	Mean (N-mm)	SD (N-mm)	COV	Skewness	Kurtosis
10	6.545×10^8	1.500×10^8	0.2292	0.7021	4.0141
12	9.2795×10^8	2.1238×10^8	0.2289	0.7444	4.0975
14	12.481×10^8	2.8304×10^8	0.2268	0.6372	3.6462
16	16.154×10^8	3.6866×10^8	0.2282	0.7062	3.8933
18	20.133×10^8	4.6247×10^8	0.2297	0.7031	3.8552
21	26.822×10^8	6.184×10^8	0.2306	0.7436	4.1260
24	34.391×10^8	7.8669×10^8	0.2288	0.7586	4.1627

Table 8 Statistical properties of external moment for special vehicle, 'S'

Variable	Mean (N-mm)	SD (N-mm)	COV	Skewness	Kurtosis
10	2.6725×10^9	5.2317×10^8	0.1958	1.0706	5.2201
12	4.3398×10^9	8.7076×10^8	0.2006	1.1349	5.3502
14	6.2962×10^9	1.2468×10^9	0.1980	1.0837	4.9129
16	8.4741×10^9	1.6279×10^9	0.1921	1.0234	4.9399
18	1.2522×10^{10}	2.5293×10^9	0.2020	1.0813	5.0734
21	2.0037×10^{10}	4.1099×10^9	0.2051	1.1118	5.3208
24	2.9734×10^{10}	6.1234×10^9	0.2059	1.0705	5.0562

Histograms of Cracking Moment M_{cr}

Histogram particularly gives the mean value of the parameter under study. From histogram one can know the type of distribution that the study parameter may follow because all the random variables considered in the study to follow different types of distribution. In the present study the cracking moment is independent of loading

Table 9 Probability of failure for different spans of Rc T beam bridge girder for different combinations of live load using Monte Carlo Simulation

Span(m)	Probability of failure**—P_f		
	Combination of live load moment		
	Load combination 1–3 lanes of class A	Load combination 2—Class A + Class AA (Tracked)	Load combination 3—Special vehicle
10	0.9968	0.8495	0.996
12	0.9983	0.8170	1
14	0.9728	0.7849	1
16	0.9998	0.6625	1
18	0.9997	0.7204	1
21	0.9990	0.8540	1
24	0.9982	0.7788	1

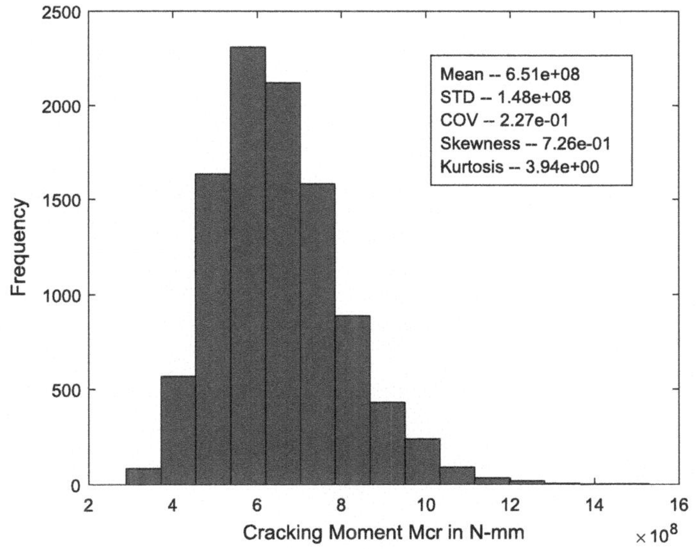

Fig. 8 Histogram of cracking moment for 10 m span

configuration as usual and it is following log normal distribution and is confirmed from chi-square test (chi square results are not presented in this paper) (Figs. 8, 9, 10, 11, 12, 13 and 14).

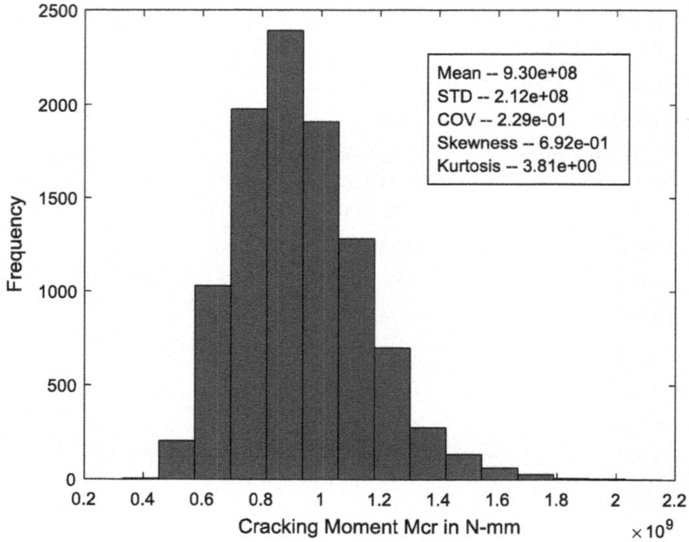

Fig. 9 Histogram of cracking moment for 12 m span

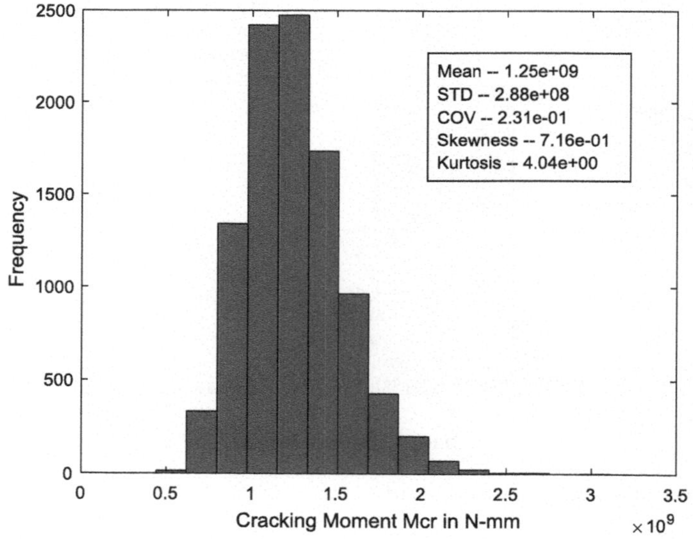

Fig. 10 Histogram of cracking moment for 14 m span

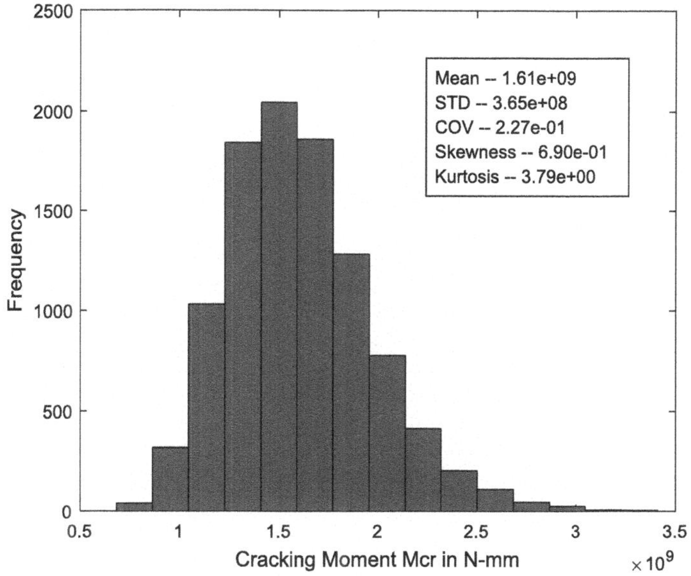

Fig. 11 Histogram of cracking moment for 16 m span

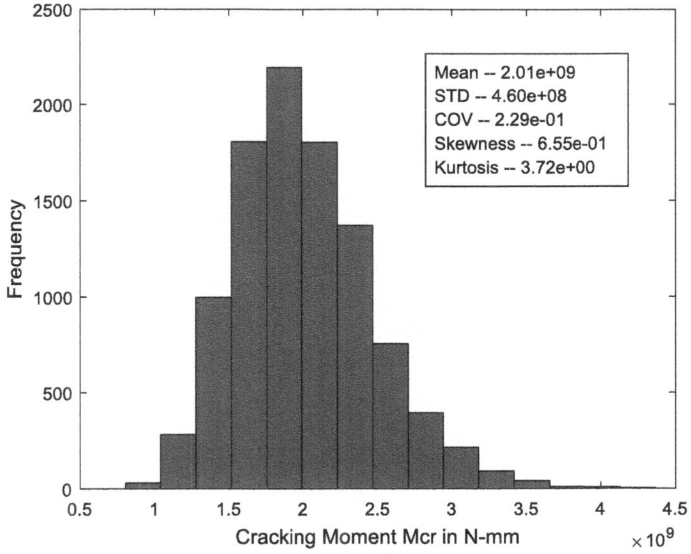

Fig. 12 Histogram of cracking moment for 18 m span

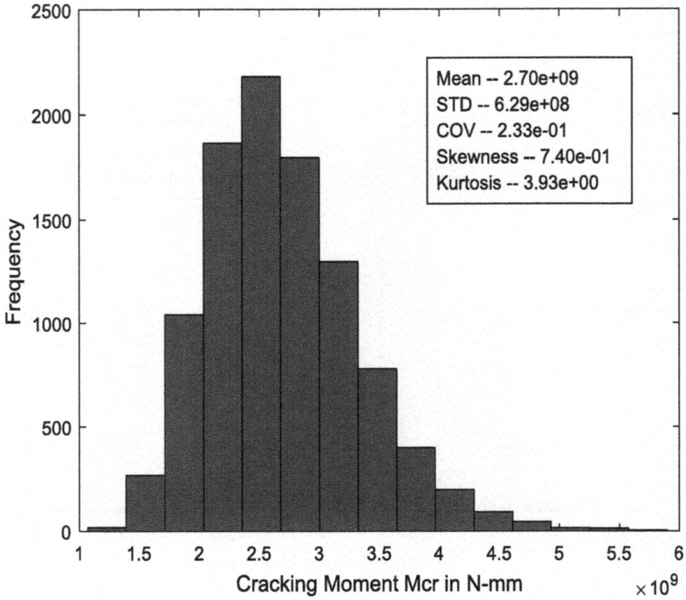

Fig. 13 Histogram of cracking moment for 21 m span

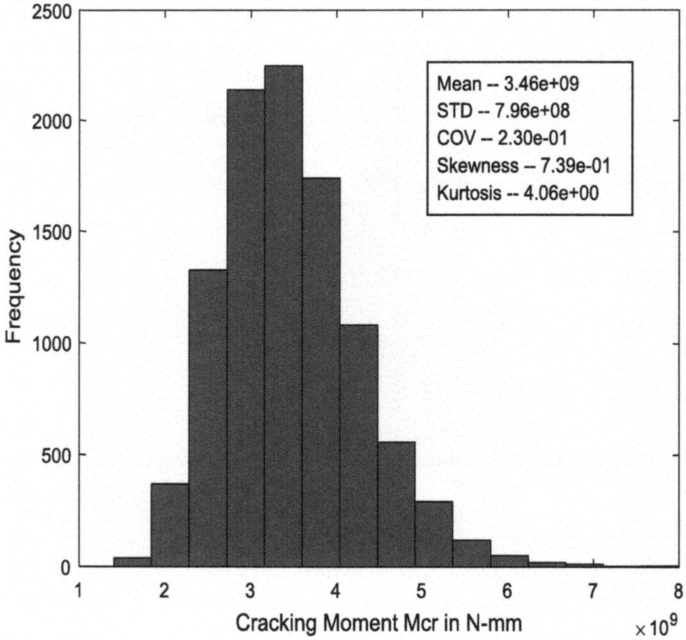

Fig. 14 Histogram of cracking moment for 24 m span

5 Conclusions

- The statics of M_{cr} and the probability distribution of cracking moment can be used to describe its randomness.
- The statistical evaluation of cracking moment can attribute in developing design philosophy for load resistant design format for the limit state of cracking.
- As in general, cracking moment is independent of the load applied to the section, the same can be found out from the statistical evaluation also. From Tables 3, 5 and 7 it can be found that, for all the combinations adopted in the present study, cracking moment is found to be independent of the external load applied.
- The cracking moment is found to be positively skewed and found to follow log-normal distribution.
- From the deterministic analysis itself (Table 1), one can find whether the section is cracked or not by comparing the external moment value with the cracking moment. This can give the reference for the designer to take care of the section at the design stage itself.
- The probability of failure of the sections (Table 9) considered against cracking is found to be very high for all the load combinations adopted in the present study and it is found that against special vehicle loading it is found that almost complete 100% failure of all the sections.
- In the recent code, IRC: 6-2017, Special vehicle loading has been suggested to check the safety of new bridges. So, present study reveals that existing RC bridges, against special vehicle loading are not safe against limit state of cracking. Emphasis can be given in future stages for the checking the safety of existing and new bridges, regarding this, the present study article can serve as a reference material.

References

1. The Egyptian Code for Design and Construction of Reinforced Concrete Structures, ECP 203-2007, Ministry of Housing, Egypt
2. Xanthakose PP (1994) Theory and design of bridges. Wiley
3. Johnson Victor D (2011) Essentials of bridge engineering, 6th edn. Oxford and IBH publications, New Delhi
4. Indian Standard 456 (2000) Plain and reinforced concrete—Code of practice, BIS
5. MacGregor JG, Mirza SA, Ellingwood B (1983) Statistical analysis of reinforced and pre-stressed concrete members. ACI J 80–16, 167–176
6. Ahmed MA (2003) Probabilistic analysis of ultimate moment of doubly reinforced concrete beam sections. J Eng Appl Sci 50(2):279–295
7. Kareem A, Hsieh J (1987) Statistical analysis of tubular RC structures. J Struct Eng 114:900–916
8. Chandrasekar P, Dayaratnam P Analysis of probability of failure of prestressed concrete beams. Build Sci 10, 161–167
9. IRC:6-2017 Standard Specification and Code of Practice for Road Bridges (2017) Indian Roads Congress

10. IRC:SP:37-2010 Guidelines for Evaluation of Load Carrying Capacity of Bridges (2010) Indian Roads Congress
11. Indian Roads Congress, Ministry of Surface Transport, Roads Wing (1993) Standard plans for highway bridges: RCC Beam and Slab Superstructure
12. Chandra Putcha S (1986) Closed form solution for reliability design of beams. Struct Saf 3:101–108
13. Ranganathan R (1985) Statistical analysis of floor loads in office buildings. DS and T Report No. A/1/83/STP-III/3, Civil Engineering Dept., IIT Bombay
14. All-Harthy AS, Frangopol DM (1993) Reliability analysis of prestressed concrete beams. In: Proceedings of ICOSSAR'93—The 6th international conference on structural safety and reliability, INNSBRUCK, Austria, vol II, pp 1367–1370
15. Selller A, Pinglot M, et al (1993) A comparative study of probabilistic methods in structural safety analysis. In: Proceedings of ICOSSAR*93—The 6 international conference on structural safety and reliability, INNSBRUCK, Austria, vol II, pp 1329–1336

Parameters to be Considered for Wind Analysis of Roofs with Openings—A Review

Abraham Grace Mary and Nizar Ruksa

Abstract Recent roof construction includes introduction of openings which is the intentional acceptance of ambient air into a space and it can also be used for purposes of thermal comfort or dehumidification. Choice of roof depends upon the architectural criteria as well as purpose of structure. Nowadays structure having large spans requires openings for various purposes. Providing openings can create both internal and external pressures which will require detailed analysis. Wind-induced internal and external pressures can be investigated by using a full-scale wind testing facility generically known as Wall of Wind (WOW), Wind tunnel experiments, rain flow count method and tornado vortex simulator. This paper reviews the test to be conducted, parameters to be considered for wind analysis of roofs with openings.

Keywords Wind load · External pressure · Internal pressure

1 Introduction

Roof is a part of building envelope. It is one of the major covering on the uppermost part of a building or shelter. It provides protection from animals and weather, rain or snow, heat, wind and sunlight. The word also denotes the framing or structure which supports the covering. Some roofs are also provided with openings. Opening is defined as the introduction of ambient air into any space. The major categories of openings are Mechanical ventilation which refers to any system that uses mechanical means, such as a fan, to introduce sub aerial air to a space. This includes positive pressure ventilation, exhausts ventilation, and balanced systems that use both supply and exhaust ventilation, Natural ventilation which refers to intentionally designed passive methods of introducing sub aerial to a space without the use of mechanical systems, Mixed mode ventilation (or hybrid ventilation) systems use both natural and mechanical processes and Infiltration is the uncontrolled flow of air from outdoors

A. Grace Mary · N. Ruksa (✉)
Amal Jyothi College of Engineering, Koovapally P.O., Kanjirapaly, Kottayam 686518, India
e-mail: ruksanizar@gmail.com

K. Dasgupta et al. (eds.), *Proceedings of SECON'19*,
Lecture Notes in Civil Engineering 46,
https://doi.org/10.1007/978-3-030-26365-2_39

to indoors through fissures and leaks (unplanned openings) in a building envelope. Another type of flow has been referred to as adventitious ventilation.

Wind loads are the major factors for the formation of both internal and external pressures. The effect of design wind loads in a building's envelope is due to the net combination of both external and internal pressures. The possibility of a building with or without openings to behave as a Helmholtz resonator under favorable external pressure forcing, results in enhanced internal and net pressure which finally results the moderating effects of envelope flexibility and leakage. Therefore, it is important to recognize various changes in the internal pressures at each stage of the opening failures to predict how the interactively updated loads are then transferred to the next point of a building. Different test methods are available for the prediction of the parameters like internal and external pressures. They are wind tunnel tests, wall of wind test and tornado vortex simulator. Out of these test most commonly used test is the wind tunnel test. This paper reviews the various parameters generated due to wind loads in various types of roofs with openings.

2 Experiments

2.1 Wind Tunnel Tests

In this experiment, before entering the test section with uniform wind speed and minimal turbulence across the test section, wind flows through a settling chamber, contraction cone, several screens and honeycombs. Test models are mounted on the 2.7 m diameter turn table, which allows full 360° rotation to test model for any desired wind azimuth, approximately 15 m from the test section entrance. Trip plates and spires are set up at the entrance to the test section, along with slant boards and roughness elements, arranged along the test section to initiate the growth of a thick simulated atmospheric boundary layer at model scale [1]. This test is used for many roofs with different shapes with and without openings (Fig. 1).

2.2 Wow Test Building

Wall of Wind (WOW) is a large scale wind engineering facility. It is usually carried out at Florida International University. The facility will allow for wind-driven rain testing building envelope as well as destructive testing of low-rise test buildings constructed with actual materials [2]. WOW testing will advance knowledge in the field of wind engineering for coastal hazard mitigation and enable the development of more effective design criteria, engineering solutions, and products helping to reduce risks associated with the use of conventional design provisions (Fig. 2).

Fig. 1 Wind tunnel test model [1]

Fig. 2 WOW in a testing position [2]

2.3 *Tornado Vortex Simulator*

This experiment is mainly used for testing buildings under various conditions of tornadoes. The roofs may or may not have openings. There is a suspended circular duct 1.5 m in diameter and 1.009 m in height, with an updraft diameter hole of 0.5 m in the Tongji tornado vortex simulator which houses a controlling fan to generate updraft. The simulator floor can be adjusted in the vertical direction and enables a range of heights for the inflow layer. Guide vanes are placed on the top of the simulator. It is used to generate a rotating forced downdraft to model a swirling flow [3].

3 Review on Various Studies for Wind Induced Parameters

3.1 *Based on Types of Roof*

3.1.1 Gable and Hip Roofs

A gable roof is the classic and commonly occurring roof shape in those parts of the world with cold or temperate climates. Hip-roof is a type of roof where all sides slope downwards to the walls and is usually with a fairly gentle slope there is no gables or other vertical sides to the roof. The design of hip roof is achieved using rafters, roof trusses or purlins. There will be great variation in the pitch of the roof and the height of the gutters.

Generically wind-induced internal and external pressures were investigated by using a full-scale wind testing facility known as Wall of Wind (WOW). The intensity of internal pressure was highly dependent on compartmentalization and the presence of openings in the partitioning wall. The opening of the window, together with a ceiling hatch, led to a 45% increase in the net wind load on the windward side of the gable roof model and a 20% increase for the windward side of the hip roof model. This observation reinforced the need to keep not only doors and windows covered with shutters during strong storms but also to close ceiling hatches. The peak internal pressure for the gable attic was also higher than the hip attic by more than 190% for the study cases. Furthermore, the worst net pressure coefficient at the eaves of the gable roof was found to be significantly higher than the eaves of the hip roof. For a building with significant compartmentalization, for example, due to partition walls in large residential buildings, there could be a significant reduction in internal pressure that warrants further future investigation. Relative increases of negative and positive pressures were observed due to the presence of vents (gable end, ridge, turbine, goose neck, and soffits). The peak internal pressure of the hip roof was lower than the gable roof by more than 190%. The worst net pressure coefficient at the eaves of the hip roof was found to be significantly lower than the gable roof (Figs. 3 and 4).

Fig. 3 Low-rise building with a gable roof [2]

Fig. 4 Low-rise building with a hip roof [2]

3.1.2 Sawtooth and Monosloped Roofs

A mono-sloped roof is a single-sloped roof surface, often not attached to another roof surface. This is in contrast to a dual-pitched roof, which is pitched in two different directions.

As the pressure coefficients are normalized to the mean wind velocity at mean roof height the results are directly comparable to previous results. There was a similarity and was observed that peak negative pressure coefficients in the high corner regions of the windward spans are nearly equal to previous. Holmes results obtained a lower value that may be explained because the pressure tappings were located relatively far away (2 m at full scale) from roof boundaries and be less likely to experience the vortex-induced peak suctions. Overall trends in the results are also consistent with the highest wind pressure coefficients occurring in all roof zones on the windward spans and reduced wind pressure coefficients on both middle and leeward span. It do not indicate any significant differences between the high corner peak wind loads on monosloped roofs and the windward span of sawtooth roofs of two through five spans. With increasing tributary area on the windward spans and monosloped roof the results revealed a faster fall off of peak pressure coefficients. It also found the area-averaged pressure coefficients along with the results. This study has also confirmed that there is no significant difference in extreme negative pressure coefficients on monosloped roofs and the windward spans of sawtooth roofs. The effect of downstream sections of the sawtooth roof and overall building aspect ratios may be a factor in the range of peak pressures. The results showed that the peak negative pressure coefficients in the higher corner regions of the monosloped roofs are nearly equal to the high corner peak negative pressure coefficients on the windward span of the sawtooth roof and low corner zone peak negative pressure coefficients on sawtooth are approximately 150% than on monosloped roofs (Fig. 5).

3.1.3 Hemi Ellipsoidal Roof

For of a hemi ellipsoidal roof, wind tunnel experiment is conducted in which the building prototype has an elliptical plan with semi major and semi minor axes. External pressures were measured at pressure taps which was set on the sealed roof whereas the internal pressures were measured inside the building.

Holmes (1979) described wind-induced internal pressure fluctuations from Helmholtz resonator theory when there is a dominant opening in the wall, which is as

$$\mathrm{Cpe} = \mathrm{Cpi} + \frac{\rho_{\mathrm{a}} l_{\mathrm{e}} V_{\mathrm{o}}}{\gamma A_{\mathrm{o}} P_{\mathrm{a}}} \ddot{C} pi + \frac{\rho_{\mathrm{a}} V_{\mathrm{o}}^{2} q}{2k^{2} \gamma^{2} A_{\mathrm{o}}^{2} P_{\mathrm{a}}^{2}} C\dot{p}i |C\dot{p}i| \tag{1}$$

The undamped Helmholtz frequency f of an oscillating air is shown as

Fig. 5 Four-span sawtooth roof building installed in vertical wall [4]

$$f = \frac{\sqrt{\gamma P_a A_o}}{\sqrt{\rho_a l_e V_o}} \tag{2}$$

In Eq. (1), internal and external pressure coefficients are obtained from

Cpi = Pi/q and Cpe = Pe/q, respectively, q = reference dynamic pressure, ρ_a and P_a = density and pressure of the air, respectively; γ = specific heat ratio; A_o = opening size; V_o = internal volume of the model; and k = discharge coefficient (Fig. 6).

Internal volume has insignificant effects on mean internal pressure but is negatively correlated with the standard deviation thereof. By considering variable inertial coefficient, a volume distortion method for testing internal pressure in the presence of a dominant opening on the roof was established. Compared to roof opening cases, a building with the same size of opening in its wall may exacerbate the net suction on the roof in windward directions but have a smaller Helmholtz resonance frequency. However, roof failure may generate a larger maximum internal suction than wall failure and is thereby more dangerous to the windward wall bearing positive external pressures. When in the presence of an interference model, the disturbed approaching flow results in large positive internal pressure acting through the opening in the roof. This produces a comparable positive internal pressure but larger internal suctions compared with those acting in the case of a wall opening. The internal pressure response controlled by Helmholtz resonance was an approximately Gaussian process. Interference from a neighboring building had a significant influence on Gaussian characteristics of internal pressure.

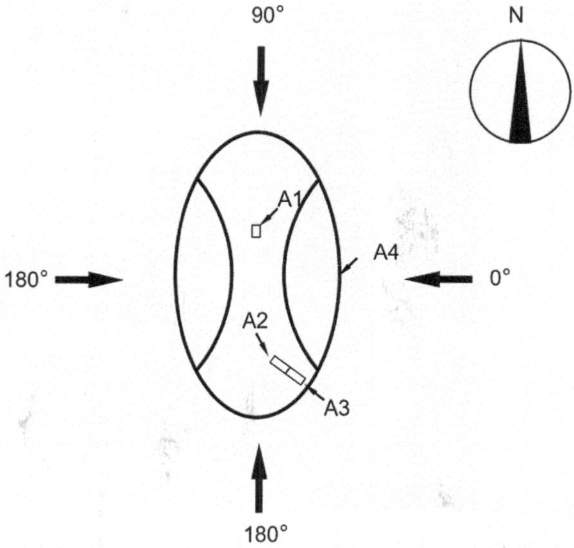

Fig. 6 Opening location and experimental wind azimuth [5]

3.2 *Based on Climate*

3.2.1 Tornado Induced Parameters

Tornadoes are one of the most violent storms, which may cause severe damage to structures. The most significant feature of a tornado is its three-dimensional funnel-shaped vortex structure that creates different vertical and horizontal profiles of wind velocity and pressure. Many buildings at the fringe of the tornado path may experience significant damage which means a tornado not only affects structures in its direct path [3]. For the cubic building model used in the tornado simulator experiments different opening configurations with three opening ratios and two azimuths of large opening were considered. Analysis of the effects of opening ratio, single central opening azimuth, and radial distance between the building model and tornado like vortex on external and internal wind pressure distributions were carried out.

The combined effects of pressure drop accompanying a tornado and aerodynamic interactions determine the final pressure distribution on the building envelope. Significant negative pressure drop dominates both the external and internal pressures on the building model when the building is located at the tornado vortex center. The relative contributions of the pressure drop of a tornado and the aerodynamic interactions to the overall effective pressures on the building envelope vary on the basis of the location of the building with respect to the tornado path. The findings of this study show that the current wind design provisions for boundary-layer winds may be modified to consider the tornado loading for structures. Both the opening: leakage ratio and the orientation of a single large opening on a wall (i.e., windward or side-wall) significantly influence the internal pressures. As expected, the building model experiences maximum wind forces when it is located at the tornado core radius.

Fig. 7 Hurricane-clip testing setup [6]

3.3 *Based on Connections*

3.3.1 Roof-to-Wall Connectors

The allowable capacities of metal connectors used in roof-to-wall connections are based on unidirectional component testing. Unidirectional testing protocol could potentially overestimate the connector's load capacity, as it fails to simulate the effects of triaxial aerodynamic loading such as simultaneous uplift, in-plane, and out-of-plane lateral loading that may occur during high wind events (Fig. 7).

Connection location, wind angle of attack, and building enclosure condition are based on the magnitude of the force components ratios. Therefore during wind events roof-to-wall connections could experience multiaxial force components. The presence of dominant openings in the windward direction caused high increase of load on roof-to-wall connections and also by increasing the internal pressure. This is due to breach in the building-envelope which is due to dominant windward opening. Opening location and orientation depends upon the uplift- and lateral-forces distribution among roof-to-wall connections. The asymmetric opening size on the windward face is occurred due to the asymmetry of external wind flow.

3.3.2 Roof Cladding

The rain-flow count method is used for certain basic load-cycle distributions of roof pressures, by which total load-cycle distributions are then computed on the basis of the long-term effects of wind climate. The total load-cycle distribution of roof pressure in temperate regions is significantly different from those in cyclonic regions. The distribution also depends on the location of the roof pressure. It is suggested that the statistical average of the cycle mean levels for a cycle range be used as an equivalent mean level. A conical "delta-wing" vortex from the upwind corner is developed when the wind direction is skewed from normal. From the results it was obtained that the roof pressures under these vortices are strongly negative and therefore fluctuating, dominating the design of roof cladding.

The results show that the long-term effects of wind climate on fatigue loading are significant. The number of cycles and the distribution of cycle ranges in temperate regions are quite different from those in cyclonic regions. The wind return period and the design life of roof cladding is proportional to number of cycles and cycle ranges that is when wind return period and design life increases, the number of cycles and cycle ranges also increases rapidly. When comparing the present results with the German recommendation the situation is reversed. Permissible fatigue design criterion is considered and the long-term effects of wind climate are included for cyclonic regions.

3.4 *Based on Roof Span*

3.4.1 Long-Span Roof Structure

Long-span roof structures are widely used in modern buildings to provide large, valuable, open spaces, usually required for gymnasiums, exhibition centers, airports, and railway stations. These types of roof structures are generally made of steel in space frame forms, resulting in small damping ratios, low natural frequencies, and sensitivity to dynamic wind actions. The dominant loading for the design of such flexible structures are wind forces. Large local fluctuations in wind forces must be considered to ensure adequate cladding design. However, the current design codes and standards may not provide relevant information for permitting accurate determination of the wind loads on complex long-span roof structures.

A numerical integration approach is proposed to calculate non-Gaussian peak factors for hardening pressure processes. Also an analytical formula for the non-Gaussian peak factor is obtained. Without loss of accuracy, it is used for evaluating the mean extreme of a hardening wind pressure process conveniently and efficiently. By comparing the results obtained using the numerical integration approach the validity of the proposed analytical non-Gaussian peak factor is investigated (Fig. 8).

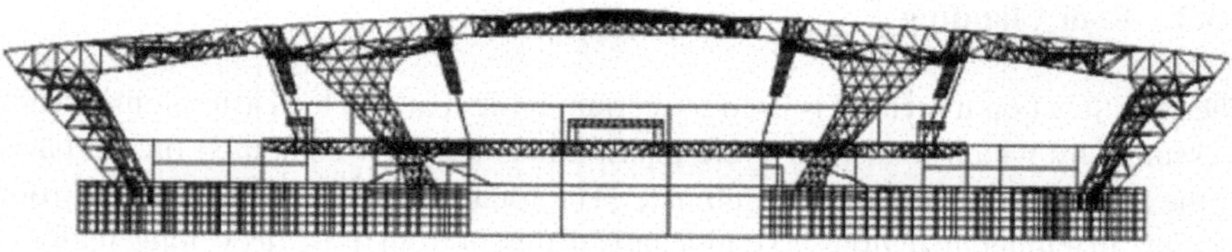

Fig. 8 Main cross section of Hangzhou East Railway Station (HERS) [1]

Hardening load processes were observed in samples of wind pressure acting on the roof of the HERS. An analytical formula based on the recently reported Hermite moment model was proposed for non-Gaussian peak factors of hardening load processes. By numerical integration and direct statistical analysis the accuracy and applicability of the newly proposed non-Gaussian peak factor formula were verified by numerical integration and direct statistical analysis. By employing various state-of-the-art methods and comparing with the direct statistical analysis of the simulated wind pressure data the peak factors for non-Gaussian wind load effects were estimated. The overestimating peak factors of hardening load processes are Davenport's peak factor and the TPP method.

3.5 Based on Opening and Opening Sizes

3.5.1 Dominant Windward Case with Various Size and Volumes

The Helmholtz resonance is generated due to the air flow in and out of the opening in the envelope, which can amplify the internal pressure fluctuations compared to the external pressures at the opening [7]. This showed that

- Both the mean internal pressure and mean external pressure at the dominant opening are equal.
- The characteristics of the internal pressure fluctuations depend upon the size of the dominant opening and the size of the volume.
- The relationship of internal pressure fluctuations resulting from a dominant windward wall opening to external pressure fluctuations at the windward wall opening has similar trends to that derived by Ginger et al. [8].
- Vickery and Bloxham (1992) standards are used for representing the relationship between the internal to windward wall external pressure fluctuations.

4 Conclusion

Wind tunnel studies, rain flow count method, tornado vortex simulator test and wall of wind were carried out for different types of roofs according to their shapes, openings ad their sizes and finally on the basis of wind and climatic conditions. On the basis of types of roofs, positive internal and external pressures developed in gable roofs are 150% more than that of hip roofs. Also the characteristics of the internal pressure fluctuations are significantly influenced by the size of the dominant opening and the size of the volume. On the basis of openings in the roof, the air flow in and out of the opening in the envelope generates the Helmholtz resonance, which can amplify the internal pressure fluctuations compared to the external pressures at the opening. On the basis of climatic conditions, the relative contributions of the pressure drop of a tornado and the aerodynamic interactions to the overall effective pressures on the building envelope vary on the basis of the location of the building with respect to the tornado path. The findings show that the current wind design provisions for boundary-layer winds may be modified to consider the tornado loading for structures. Both the opening: leakage ratio and the orientation of a single large opening on a wall significantly influence the internal pressures. Also the building model experiences maximum wind forces when it is located at the tornado core radius.

References

1. Huang MF, Lou W, Pan X, Chan CM, Li QS (2014) Hermite extreme value estimation of non-gaussian wind load process on a long-span roof structure. J Struct Eng 140(9):04014061. https://doi.org/10.1061/(asce)st.1943-541x.0000962
2. Tecle AS, Bitsuamlak GT, Chowdhury AG (2010) Opening and compartmentalization effects of internal pressure in low-rise buildings with gable and hip roofs. J Struct Eng 21(1):04014002. https://doi.org/10.1061/(ASCE)AE.1943-5568.0000101
3. Wang J, Cao S, Pang W, Cao J (2018) Experimental study on tornado-induced wind pressures on a cubic building with openings. J Struct Eng 144(2):04017206. https://doi.org/10.1061/(ASCE)ST.1943-541X.0001952
4. Prevatt DO, Cui B (2010) Wind tunnel studies on sawtooth and monosloped roofs. J Struct Eng 136(9):1161–1171. https://doi.org/10.1061/(ASCES)T.1943-541X.0000200
5. Xu H, Lou W (2016) Wind-induced internal pressures in building with dominant opening on hemi-ellipsoidal roof. J Struct Eng 142(7):1–10. https://doi.org/10.1061/(ASCE)EM.1943-7889.0001468
6. Chowdhury AG, Canino I, Mirmiran A, Suksawang N, Baheru T (2013) Wind-loading effects on roof-to-wall connections of timber residential buildings. J Eng Mech 139(3):386–395. https://doi.org/10.1061/(ASCE)EM.1943-7889.0000512
7. Wang YJ, Li QS (2015) Wind pressure characteristics of a low-rise building with various openings on a roof corner. J Wind Struct 21(1):1–23. https://doi.org/10.12989/was.2015.21.1.001
8. Ginger JD, Holmes JD, Kim PY (2010) Variation of internal pressure with varying sizes of dominant openings and volumes. J Struct Eng 136(10):1319–1326. https://doi.org/10.1061/ASCEST.1943-541X.0000225
9. Guha TK, Sharma RN, Richards PJ (2014) Dynamic wind load on an internal partition wall inside a compartmentalized building with an external dominant opening. J Struct Eng 19(2):89–100. https://doi.org/10.1061/(ASCE)AE.1943-5568.0000113
10. Karava P, Stathopoulos T (2009) Wind-induced internal pressures in buildings with large facade openings. In: Proceeding of 11th Americas Conference on wind engineering, American Association for Wind Engineering, Fort Collins, CO
11. Pan F, Cai CS, Zhang W (2013) Wind-induced internal pressures of buildings with multiple openings. J Eng Mech 139(3):376–385. https://doi.org/10.1061/(ASCE)EM.1943-7889.0000464
12. Xu YL (1995) Determination of wind-induced fatigue loading on roof cladding. J Eng Mech 121(9):956–963
13. Ginger JD (2000) Internal pressures and cladding net wind loads on full-scale low-rise building. J Struct Eng 126(4):538–543
14. Rizzo F, Ricciardelli F (2017) Design pressure coefficients for circular and elliptical plan structures with hyperbolic paraboloid roof. Eng Struct 139(2017):153–169
15. Holmes JD, Ginger JD (2012) Internal pressures–The dominant windward opening case–A review. J Wind Eng Ind Aerodyn 100(2012):70–76

16. Shice Yu (2016) Wind tunnel study on vortex-induced Helmholtz resonance excited by oblique flow. Exp Thermal Fluid Sci 74(2016):207–219
17. Rizzo F, D'Asdia P, Lazzari M (2017) Aerodynamic behavior of hyperbolic paraboloid shaped roofs: wind tunnel tests
18. Amin JA, Ahuja AK (2013) Effects of side ratio on wind-induced pressure distribution on rectangular buildings. J Struct 2013(Article ID 176739):12 p. http://dx.doi.org/10.1155/2013/176739

Numerical Studies on Strengthening of Continuous Steel-Concrete Composite Girders Using CFRP

Soorya M. Nair and Nithin Mohan

Abstract Steel-concrete composite girders are widely used in the construction of bridges and buildings. However when they are provided in continuous spans, a loss of strength and composite action will be occurring at the hogging moment region due to the development of tension in concrete and compression in steel. Limited works have been done to study this shortcoming in these widely used girders. Carbon fiber reinforced polymer (CFRP), is found to have lot of untapped potential for improving the strength of girders. This paper presents an analytical investigation into the behavior of composite girders strengthened in hogging moment region using CFRP. Effect of different arrangements and thickness of CFRP laminates are also being compared for different geometries. CFRP was found effectively improve the strength of composite girders in the negative moment regions especially when used in the form of strips.

Keywords Steel concrete composite · Hogging moment region · CFRP strengthening

1 Introduction

Steel–concrete composite girders have been widely used in construction due to their small size, light weight, high bearing capacity, rapid construction and better seismic, fire and corrosion resistance. For large spans in bridges and building, where continuous steel concrete composite girders are used, a negative moment region is developed over the intermediate supports. Due to the inactivity of concrete slab in this region, steel girder is supposed to act alone or compositely with the longitudinal slab reinforcement. The very purpose of composite construction is not fulfilled here and it leads to loss of strength and stiffness in this region.

S. M. Nair (✉) · N. Mohan
Vidya Academy of Science and Technology, Thrissur, India
e-mail: msooryanair@gmail.com

N. Mohan
e-mail: nithin.m@vidyaacademy.ac.in

K. Dasgupta et al. (eds.), *Proceedings of SECON'19*,
Lecture Notes in Civil Engineering 46,
https://doi.org/10.1007/978-3-030-26365-2_40

CFRP is one of the relatively new materials that have been extensively used in civil engineering applications. It is found to have lot of untapped potential for improving the strength of girders. Most of the research works in this area concentrates on strengthening girders by attaching CFRP to the steel girders at the sagging moment region [1]. Only few attempts have been done with limited parameters to explore the performance of girders at the hogging moment regions. Majid Mohammed [2] studied the behavior of continuous steel beam strengthened using CFRP in sagging and hogging moment regions which revealed that the strength of the girder improved by 70% compared to the plain beams. Sharif et al. [3] experimentally tested six 2 span continuous girders to study the effect of bonding CFRP laminates to the concrete slab at sagging and hogging moment region of steel concrete composite girders. This study confirmed that CFRP was also effective in improving the efficiency of continuous composite girders.

In view of the limited published work on the use of CFRP to strengthen composite girders at the negative moment zone, this paper aims to present an analytical study of continuous composite girders with CFRP bonded to the top of the concrete slab in the negative moment region. Effect of different arrangements and thickness of CFRP laminates are also being compared for different geometries.

2 Finite Element Analysis and Modeling

Experimental investigations are tedious, time consuming and sometimes practically impossible.FEM is a useful tool in predicting the failure load stress, strain distribution etc. in complex structures. ANSYS 16 Software was used to develop a non linear finite element model and study the performance of plain and strengthened steel-concrete composite girders. In ANSYS 16, Solid 65 was used to model concrete. Steel girder was defined using SOLID 186 element. LINK 180 was used for rebars and SHELL 181 defined CFRP laminates. A mesh size of 50 mm was adopted.

Sharif et al. [3] was chosen for validation. This study presents an investigation into the use of CFRP to maintain the composite action at the negative moment region of continuous composite girders by bonding CFRP sheets to the top of a concrete slab at the negative moment region. Six 2 span continuous girder specimens with each span of length 2500 were used. The girder Modeled and corresponding deformation are as shown in Fig. 1. Comparison of results obtained from the model and paper considered is represented using Fig. 2. 5% error was obtained during the validation.

3 Numerical Studies Conducted

In the present study five random specimens SB-1, SB-2, SB-3, SB-4 and SB-5 were included in the study. Each specimen has a span of 1250 mm. Schematic representations of the girders are as shown in Fig. 3. Load bearing stiffeners of 8 mm thickness

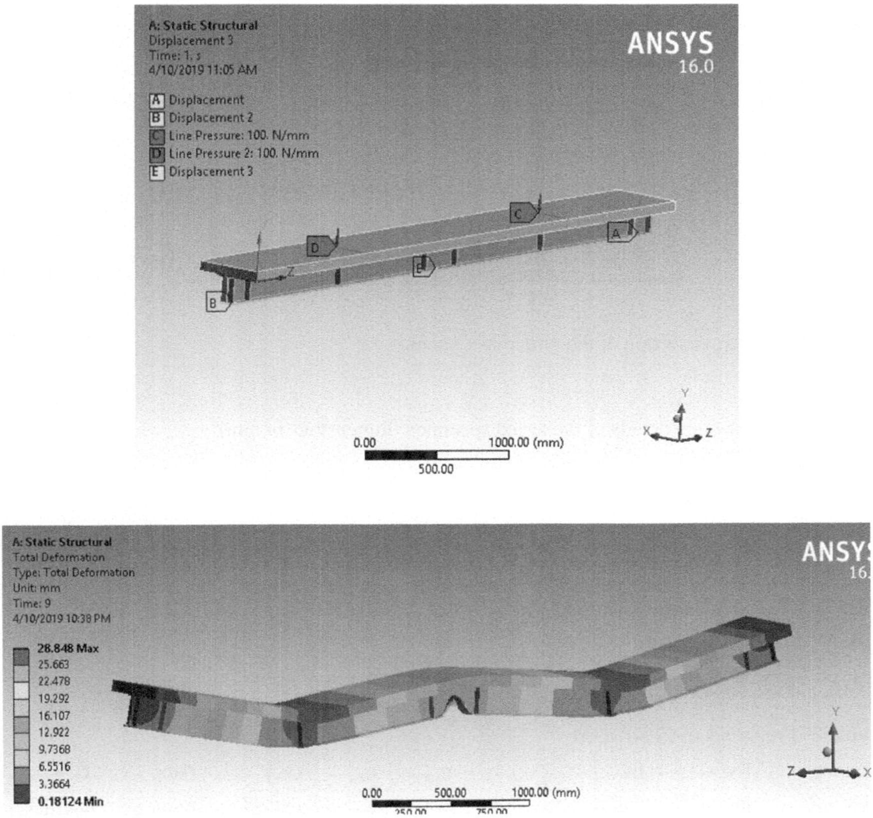

Fig. 1 Loading condition and deformed shape

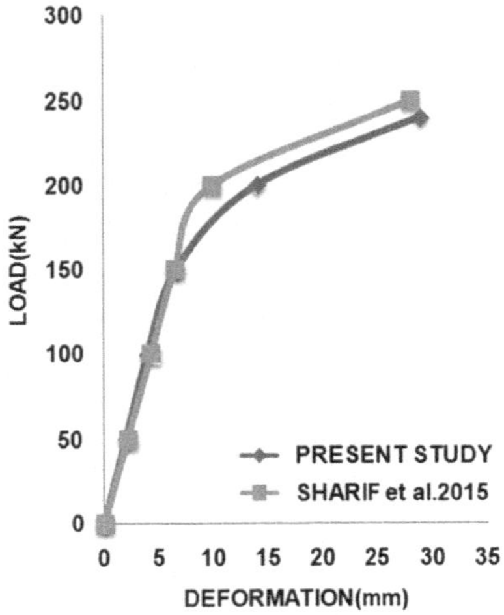

Fig. 2 Comparison of results for girder validated

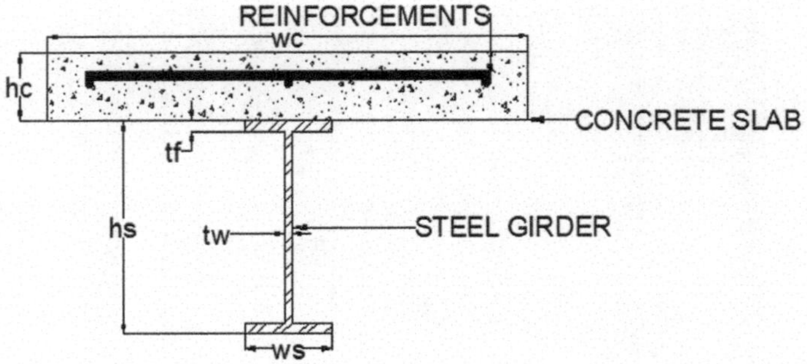

Fig. 3 Typical cross section of the girder specimens

Table 1 Cross section details of the tested specimen (dimensions in mm)

Specimen	w_c	h_c	w_s	h_s	t_f	t_w
SB-1	500	70	125	175	7.4	5.8
SB-2	500	70	110	196	8	8
SB-3	500	70	225	225	9.1	6.5
SB-4	400	65	90	200	11.3	7.5
SB-5	650	75	140	258	8	5.7

Table 2 Properties of CFRP used

Ex	Ey	Ez	υ_{xy}	υ_{yz}	υ_{zx}	Gxy	Gyz	Gzx
kN/m^2	kN/m^2	kN/m^2	–	–	–	kN/m^2	kN/m^2	kN/m^2
230,000	17,900	17,900	0.22	0.3	0.2	11,790	6880	11,790

were provided in order to avoid failure at loading points. Table 1 gives the cross section details of the specimens used.

25 Mpa concrete having modulus of elasticity 22,300 kN/m^2 and poison's ratio 0.3 is being used. Steel beams with yield strength. 278 Mpa and rebars with 360 Mpa were defined using bilinear relationships. The material property of CFRP given as input for the ANSYS model is given in Table 2. It was treated as an orthotropic material.

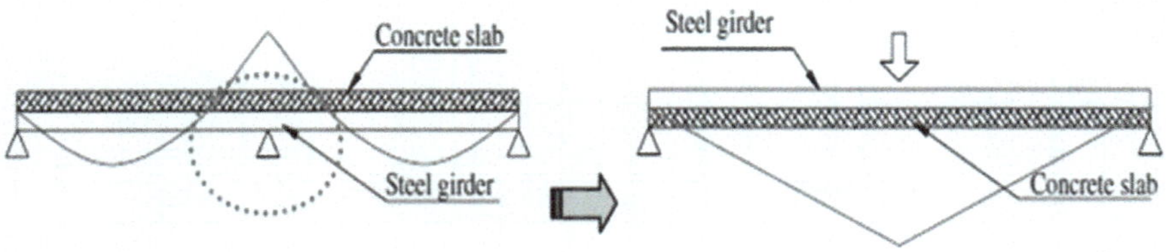

Fig. 4 Loading arrangement used in tested specimen

In analysis using the ANSYS workbench in order to accurately determine the effects on the hogging moment region, the negative moment region was isolated. The specimens were studied in inverted condition as shown in Fig. 4 in the rest of the study in order to simulate the behavior [4]. All specimens were subjected 4-point loading.

4 Results and Discussion

4.1 Effect of CFRP Arrangement

Specimen SB-1, SB-2 and SB-3 were used in this study. 0.5 mm thick CFRP sheets were attached to the concrete slab in 2 different ways: (i) attach it along the full width of the slab and (ii) as strips of width 100 mm at a spacing of 50 mm [5]. These arrangements have been schematically represented using Fig. 5.

Figure 6 shows the comparison of results obtained in the study. Strip arrangement of CFRP showed better performance while compared to full length cover. For 0.5 mm thick CFRP laminate, a 3% increase in the strength was obtained when provided in strip form while compared to the full length arrangement.

4.2 Effect of CFRP Thickness

CFRP was provided in strip arrangements in specimen SB-1 and SB-2. Different thickness ranging from 0.5 to 3 mm was provided.

The ultimate strength was found to increase with increase in the thickness of CFRP strips. 2 mm thick strip showed nearly 20% increase in the strength of the girder while compared to the control girder. Further increase in the thickness of the strip did not yield much significant results. The results obtained are represented using Fig. 7.

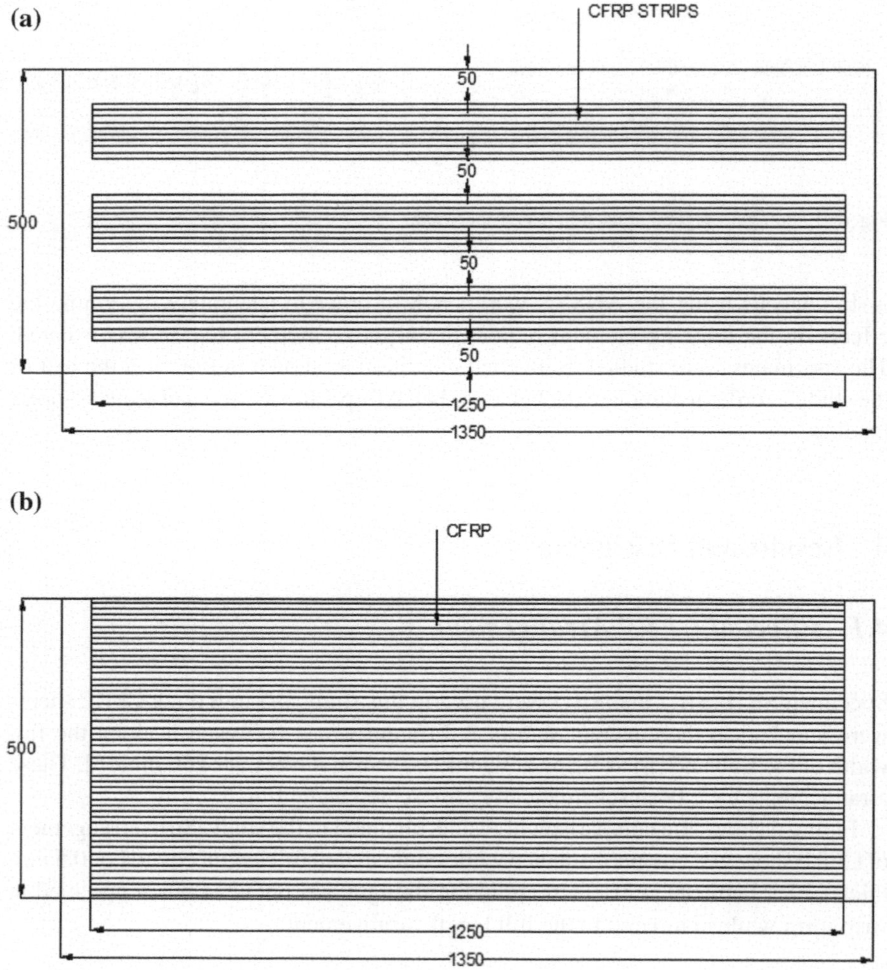

Fig. 5 Arrangement of CFRP: **a** strip form and **b** full length

4.3 *Effect of Strip Width*

SB-2, SB-4 and SB-5 were used in this study. Keeping the spacing between the strips constant as 50 mm, the width of the strips (b_s) was varied with respect to the width of the slab (w_c). Thickness of strip has been kept as 0.5 mm.

From the results obtained it could be inferred that ultimate strength increases when the wc/bs ratio was between 2 and 15. Further increase in the ratio leads to dip in the ultimate strength as shown in Fig. 8.

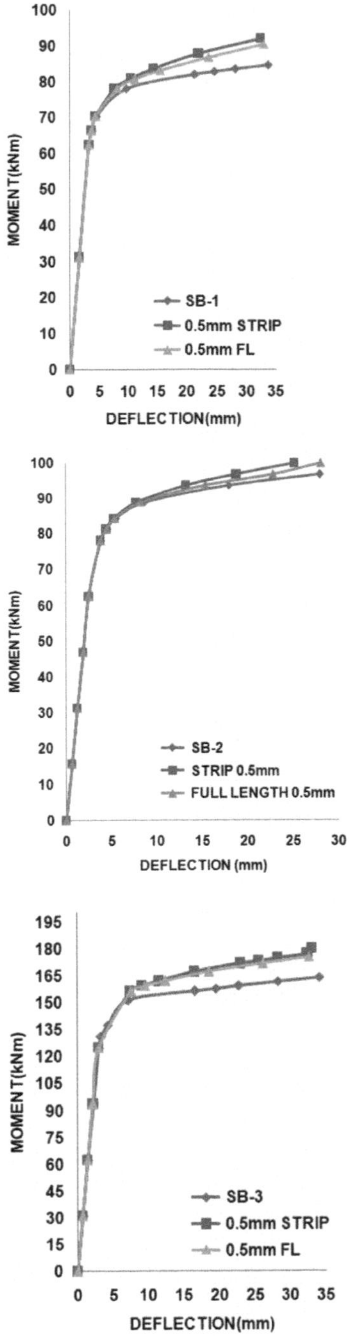

Fig. 6 Comparison of results for strip and full length for SB-1, SB-2 and SB-3

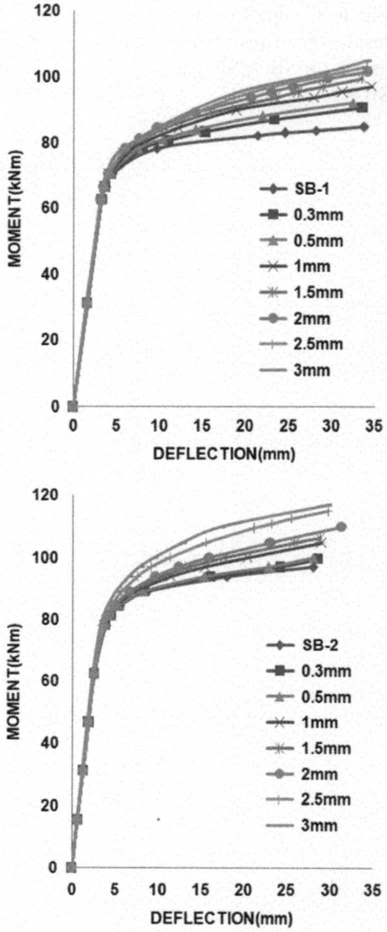

Fig. 7 Comparison of results for different thickness of CFRP for SB-1 and SB-2

5 Conclusion

The following conclusions can be drawn from this investigation:

- The ultimate strength of composite girder at the hogging moment region was found to improve with the usage of CFRP.
- Strip arrangement was concluded to be better than full width covering of the concrete slab. This may be probably due to the delamination effect occurring while providing CFRP along the full width of the slab. Further experimental studies needs to be conducted to confirm this effect.
- The strength was found to increase with increase in the thickness of the CFRP sheets used. 20% increase in strength at the hogging moment region was found to obtain when strip of 2 mm was used. Increase in ultimate strength obtained was not so significant for further increase in the thickness.

Fig. 8 Comparison of results for strip width ratio of CFRP for SB-2, SB-4 and SB-5

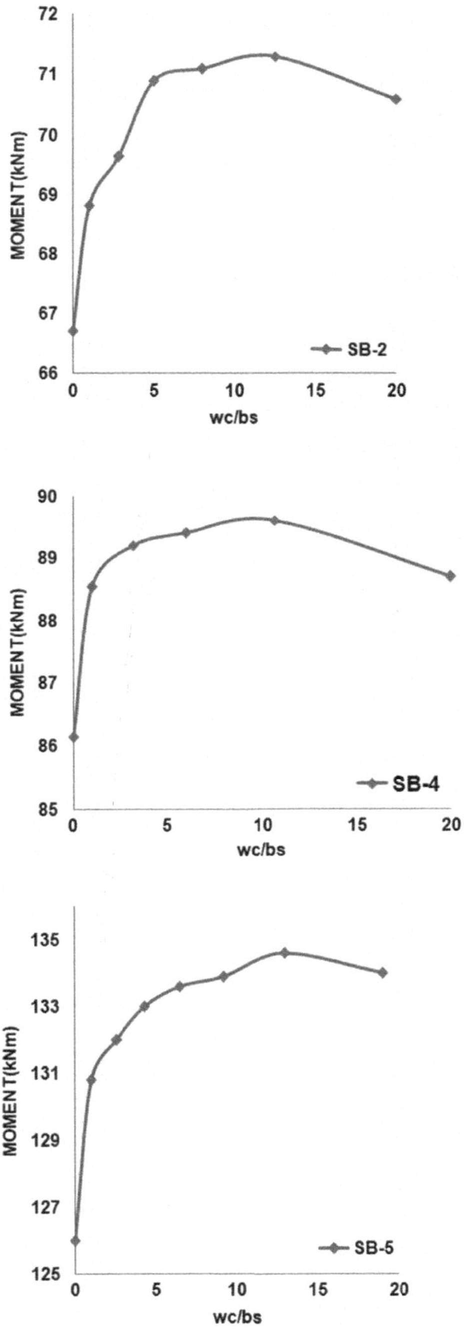

- The ultimate strength obtained was found to be more when the ratio of strip width (bs) to width of slab (wc) was kept between 2 and 15.

From the study it can be concluded that CFRP provided as strips attached to concrete at the hogging moment region of the continuous girders can be effectively proposed as an alternate to the conventional pre stressing techniques.

References

1. Tavakkolizadeh M, Saadatmanesh H (2003) Strengthening of steel concrete composite girders using carbon fiber reinforced polymers sheets. J Struct Eng 129(1):30–40
2. Khadim MMA (2012) Effect of CFRP plate length strengthening continuous steel beam. J Constr Build Mater 28(1):648–652
3. Sharif MA, Samaaneh MA, Azad AK, Baluch MH (2015) Use of CFRP to maintain composite action for continuous steel–concrete composite girders. J Constr Build Mater 28:648–652
4. El-Zohairy A, Salim HM, Fawzy SAA, Mustafa S, El-Shihy A (2015) Finite-element modeling of externally post tensioned composite beams. J Bridge Eng 20(12):04015018
5. Schnerch D, Rizkalla S (2008) Flexural strengthening of steel bridges with high modulus CFRP strips. J Bridge Eng 13(2):192–201

Effect of Shear Lag on Buckling Behavior of Hat Shaped Laminated Composite Box Sections

K. C. Praseeja and Nithin Mohan

Abstract A new class of materials, laminated composites is increasingly being used for a wide range of civil infrastructure applications and aerospace structures due to their high strength, stiffness, lightweight and durability. It is generally assumed in bending theory that plane sections remain plane after loading, this assumption does not hold for box beams with wide flanges. Shear lag effect can bring non uniform normal stress distribution on flanges; it would affect the strength design of thin-walled beams. The strength of thin walled members is governed by the buckling criterion. In this paper effect of shear lag on buckling behavior of laminated composites is examined. The present study investigates about analysis of hat shaped box beam model for buckling behavior and approach for finding out the shear lag effects on symmetrically laminated graphite epoxy thin walled composite box beams under flexural loading. A parametric study has been carried out using the homogeneous and orthotropic finite element models created by ANSYS16. Influences of orthotropic parameter and cross sectional parameter are studied.

Keywords Box beams · Buckling · Finite element models · Laminated composites · Shear lag · ANSYS16

1 Introduction

Fiber reinforced polymer composite (FRP) is a new construction material slowly gaining recognition from civil engineers. Composite materials are generally used in many fields of engineering like buildings, bridges, tunnel structures, spacecrafts including the applications of FRP composite beam, deck, and column. Box girders are widely used in the bridge engineering due to strong bending resistance and torsional

K. C. Praseeja (✉) · N. Mohan
Department of Civil Engineering, Vidya Academy of Science and Technology, Thrissur, India
e-mail: praseejachandran95@gmail.com

N. Mohan
e-mail: nithin.m@vidyaacademy.ac.in

K. Dasgupta et al. (eds.), *Proceedings of SECON'19*,
Lecture Notes in Civil Engineering 46,
https://doi.org/10.1007/978-3-030-26365-2_41

rigidity. Thin-walled composite beam structures are widespread in lot of engineering areas spatially because of their high strength-to-weight ratio. Thin structures are usually weak in buckling due to their slenderness and specific mechanical behavior. Therefore it is required to check the buckling strength of a structure before finalizing the designs. Buckling usually occurs by twisting or by a combination of bending and twisting, and the buckling failure will be responsive to the magnitude of the deflections. It is generally assumed in bending theory that plane sections remain plane after loading, this assumption does not hold for short span beams with wide flanges. Beams with wide flanges exhibit shear lag; i.e.—diminishing flange in-plane stresses away from the web [1]. Relatively few studies have been made on the effect of shearing strains along the middle surface of walls on the lateral buckling of thin walled open members. The reason is that when the shearing strains are taken into account the mathematical aspects of problems become considerably complicated. If the effect of the shearing strain on buckling is significant, a quick evaluation of possible shear strain effect is of importance to a practicing engineer at the early stage of design of thin walled structures.

Only a limited number of studies are conducted on the analysis of FRP box-girders. Upadhyay and Kalyanaraman [2] presented a simplified and computationally efficient procedure for the analysis of single cell FRP box-girder bridges made of blade angle or T stiffened panels, for efficient use at the design stages [3]. The paper studied the effect of shear lag of composite laminated plates with buckling form an applied analysis model for thin-walled composite box beams under bending loads. In this model, shear lag effect; shear deformation effect, and ply stresses; strains of the flanges in thin-walled composite box beams can be investigated and expressed explicitly.

The present study investigates about analysis of hat shaped box beam model for buckling behavior and approach for finding out the shear lag effects on symmetrically laminated thin walled composite box beams under flexural loading. A parametric study has been carried out using the homogeneous and orthotropic finite element models, the results obtained from the analyses of this paper can provide reference for the design of related engineering structures.

2 Finite Element Analysis

The problem is modeled in commercial software ANSYS16. It is an analysis software which makes use of FEA to study and analyze the behavior of structures under various boundary conditions. ANSYS FEA tools also offer unparallel ease to product developer's focus on the most important part of simulation process, understanding the results and the impact of design variations on the model. ANSYS incorporates parallel algorithms for faster computation time for these large models and used in a number of different engineering fields such as power generation, transportation and electronic devices for its easy applicability. Based on the capability of ANSYS the present work has been modeled in ANSYS by choosing shell element namely

Table 1 Validation result

Applied load (N)	Displacement (mm)		% Difference
	FE analysis		
	Reference study	Present study	
100	0.075	0.08	6.25

SHELL281 from ANSYS library under static loading conditions. Box beam model is discretized with 8 node isoparametric laminated shell element with six degrees of freedom at each nodes.

3 Validation Study

It is essential to ascertain the accuracy of modeling before it is used for the analysis. Therefore, results obtained from reference studies by Yaping et al. [3] were used to judge the accuracy of the planned finite element modeling. Considered a symmetric carbon epoxy-box beam (0°/+θ°/−θ°/0°) with geometrical parameters length = 210 mm, height = 17.5 mm, top flange width = 17 mm, thickness of flange = 1.5 mm, thickness of web = 1 mm. Material properties such as E_L = 114,576 N/mm^2, E_T = 9981.7 N/mm^2, G_{LT} = 4664.8 N/mm^2, μ_{LT} = 0.325. Vertical displacement of simply supported Beam subjected to concentrated load for the condition ply angle θ = 45° was evaluated in the analysis. Table 1 represents the displacement of composite beam. The difference between the results reported by Yaping et al. [3] and present study ranges as 6.25% and it shows good agreement.

4 Numerical Study

Beams were modeled and analyzed using ANSYS software. All elements are symmetric laminates each with 8 symmetric layers. Linear orthotropic properties for graphite epoxy material are taken as per Table 2. Element used is shell 281 which is an 8 nodded linear shell element with six degrees of freedom at each node. Shear lag and buckling load factor of models are calculated. Figure 1 represents buckled hat shaped box beam having size 500 × 250 × 3000 mm with ±45° fiber orientation. The normal stress distribution is not uniform in the wide flange, but the stress is maximum in general at the flange-web intersection, decreasing toward the middle

Table 2 Linear orthotropic properties

E_1 (GPa)	E_2 (GPa)	G_1 (GPa)	μ_1	μ_2
145	16.5	4.48	0.314	0.037

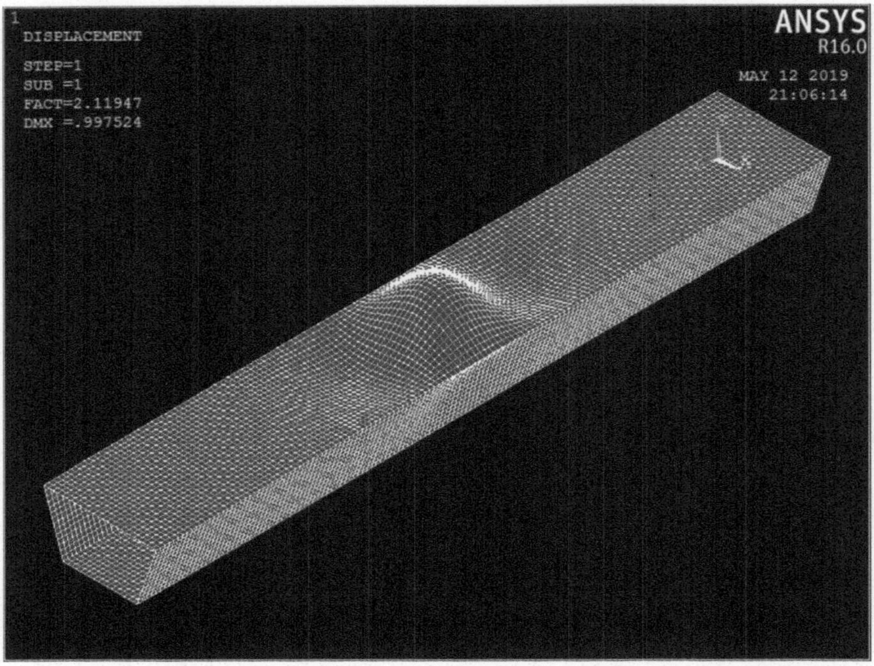

Fig. 1 Buckled hat shaped box beam having size 500 × 250 × 3000, tf = tw = 2 mm and fiber orientation for the flange as ±45°

of the flange. This phenomenon, known as the shear lag, has been studied for many years. In the present work a simplified approach, a three bar method for evaluating the shear-lag effects in graphite epoxy box sections adopted as defined by Upadhyay and Kalyanaraman [2]. The following expressions are used to obtain the shear lag.

$$E_{Z(i)} = ETS(i)/TE(i)$$

$$b_s = width\,of\ flange/10$$

$$k = \sqrt{A_{66,F}/b_s\left\{\frac{2}{(AFE \times E_{Z,F})} + \frac{1}{(AWE \times E_{Z,W})}\right\}}$$

$$N_X = M/(D(2AWE + AFE))$$

$$N_{X,MAX} = N_X(1 + (AFE/(AWE \times k \times LENGTH)))$$

TE(i) and $E_z(i)$ are the equivalent thickness and equivalent modulus of elasticity of the i'th panel respectively. AFE and AWE are the equivalent areas of flange and web.

D is the depth of section. $A_{66,F}$ is the in plane shear stiffness of the flange. Based on the elementary theory of bending, the in-plane stress, N_X, in the longitudinal direction of the panel is due to longitudinal bending moment M. $N_{X,MAX}$ is the maximum edge stresses in flanges of a simply supported girder subjected to central point load. 'k' is the shear-lag parameter. $N_X/N_{X,MAX}$ is signifying the shear lag effect. If the ratio is equal to one, it indicates, there will not be any shear lag effect. ω and K are the two parameters which governs the shear lag effect. The approach for finding out this parameters are adopted as defined by Chandak et al. [4].

$$\omega = \left\{\left(\frac{1}{2}\right) \times (A_{11}/A_{66})tf\right\} + \left\{\left(\frac{1}{6}\right) \times \left(\frac{A_{11}}{A_{66}}\right)W\right\}$$

$$K = [EA_{TF}/EA_G] + \left[(D_{11_{TF}})/(EI_G)\right]$$

$$EA_G = EA_{TF} + EA_{BF} + EA_W$$

$$EA_{TF} = ET_{TF} \times B$$

$$EA_{BF} = ET_{BF} \times b$$

$$EA_W = ET_W \times \left(h - t_{tf} - t_{bf}\right) \times 2$$

$$ET_{TF} = \left[A_{11_{TF}} - \left\{A_{12_{TF}} \times (A_{12}/A_{22})tf\right\}\right]$$

$$ET_{BF} = \left[A_{11_{BF}} - \left\{A_{12_{BF}} \times (A_{12}/A_{22})bf\right\}\right]$$

$$ET_w = \left[A_{11_w} - \left\{A_{12_w} \times (A_{12}/A_{22})w\right\}\right]$$

$$EI_G = EI_{TF} + EI_{BF} + EI_W$$

$$EI_{TF} = \left[\left\{D_{11_{TF}} \times B\right\} + \left\{EA_{TF} \times (n_c - t_{TF})^2\right\}\right]$$

$$EI_{BF} = \left[\left\{D_{11_{BF}} \times B\right\} + \left\{EA_{BF} \times (D - n_c - t_{BF})^2\right\}\right]$$

$$EI_W = \left[\left\{ET_W \times ((D - t_{TF} - t_{BF})^3/12)\right\} + \left\{EA_W \times \left\{\frac{D - t_{TF} - t_{BF}}{2} - n_c + t_{TF}\right\}^2\right\}\right]$$

EA_{TF}, EA_{BF} and EA_W are axial stiffness of top flange, bottom flange and web respectively. ET_{TF}, ET_{BF} and ET_W are flexural stiffness's of the elements. 'h' will be the length of web. ‘ω’ is the orthotropic parameter of top flange and web depends on extensional stiffness of elements. The parameter ‘K’ is the cross sectional shape parameter. Three fiber orientations 0°, ±45°, and 90° are considered to cover extreme cases of orthotropy. In all the cases, sections are preferred in such a manner that flange buckling will be the first mode of buckling.

4.1 Effect of Orthotropic Parameter

Many Beam models were analyzed using ANSYS software each with 8 symmetric layers. When flange has ± 45° fiber orientation, shear lag effect is lower than 90° because ±45° provides enough shear stiffness to laminate and the shear lag is almost low and constant. It can be observed that buckling coefficient decreases with increases of shear lag factor. In each combination, the distribution of stress differs considerably. Fiber orientation affects on the shear lag phenomenon noticeably shown in Tables 3, 4, 5, 6, 7 and 8.

The influence of ω, which is the orthotropic parameter, indicates that the effective width decreases with increase in value of ω. $(0/0/0/0)_s$ fiber orientation has lowest buckling factor and exhibit highest shear lag effect due to the significant influence

Table 3 Shear lag and buckling load factor for tf = tw = 2 mm, constant K = 0.37

B (mm)	D (mm)	L (mm)	Fiber orientation (°)	Buckling factor	Shear lag	ω
500	250	3000	0/0/0/0	1.84	0.756	21.822
			90/90/90/90	2.06	0.902	2.483
			45/−45/45/−45	2.12	0.964	0.836

Table 4 Shear lag and buckling load factor for tf = tw = 2 mm, constant K = 0.385

B (mm)	D (mm)	L (mm)	Fiber orientation (°)	Buckling factor	Shear lag	ω
450	200	2700	0/0/0/0	1.82	0.74	21.822
			90/90/90/90	1.94	0.894	2.483
			45/−45/45/−45	2.08	0.961	0.836

Table 5 Shear lag and buckling load factor for tf = tw = 2 mm, constant K = 0.407

B (mm)	D (mm)	L (mm)	Fiber orientation (°)	Buckling factor	Shear lag	ω
400	150	2800	0/0/0/0	1.36	0.752	21.822
			90/90/90/90	1.45	0.9	2.483
			45/−45/45/−45	1.67	0.963	0.836

Table 6 Shear lag and buckling load factor for tf = tw = 2 mm, constant K = 0.391

B (mm)	D (mm)	L (mm)	Fiber orientation (°)	Buckling factor	Shear lag	ω
350	150	2450	0/0/0/0	1.77	0.766	21.822
			90/90/90/90	1.93	0.906	2.483
			45/−45/45/−45	2.19	0.97	0.836

Table 7 Shear lag and buckling load factor for tf = tw = 2 mm, constant K = 0.323

B (mm)	D (mm)	L (mm)	Fiber orientation (°)	Buckling factor	Shear lag	ω
550	250	3300	0/0/0/0	1.62	0.801	21.822
			90/90/90/90	2.69	0.923	2.483
			45/−45/45/−45	2.04	0.972	0.836

Table 8 Shear lag and buckling load factor for tf = tw = 2 mm, constant ω = 21.822 and $(0/0/0/0)_s$

B (mm)	D (mm)	L (mm)	Buckling factor	Shear lag	K
350	100	1750	2	0.648	0.435
400	150	2000	2.31	0.684	0.407
450	200	2700	1.82	0.74	0.385
500	250	3000	1.84	0.756	0.37
550	300	3300	1.83	0.768	0.359

of orthotropy. For 0° fiber orientation has more Orthotropic parameter value. When ω increases, shear lag effect increases and buckling factor decreases.

4.2 *Effect of Cross Sectional Shape Parameter*

Beam models were analyzed using ANSYS software. Shear lag and buckling load factor calculated for constant ω.

Table 8 represents the shear lag and buckling factor of models with $(0/0/0/0)_s$ fiber orientation. Tables 9 and 10 shows models with $(90/90/90/90)_s$ and $(45/-45/45/-45)_s$ fiber orientations respectively. K is the cross sectional shape parameter and effective width decreases for larger values of K. Results prove that for constant ω value, shear lag parameter is constant for same value of K. When length and depth increases, shear lag effect decreases.

Table 9 Shear lag and buckling load factor for tf = tw = 2 mm, constant ω = 2.483 and $(90/90/90/90)_s$

B (mm)	D (mm)	L (mm)	Buckling factor	Shear lag	K
350	200	1750	4.491	0.894	0.354
400	200	2000	3.246	0.885	0.372
450	200	2250	2.45	0.875	0.385
500	300	2500	3.36	0.897	0.346
550	300	2750	2.67	0.891	0.359

Table 10 Shear lag and buckling load factor for tf = tw = 2 mm, constant ω = 0.836 and $(45/-45/45/-45)_s$

B (mm)	D(mm)	L (mm)	Buckling factor	Shear lag	K
350	100	2100	1.682	0.951	0.435
400	100	2000	1.548	0.938	0.446
450	300	2250	3.91	0.965	0.331
500	200	3000	1.67	0.958	0.398
550	150	2750	1.23	0.94	0.437

5 Conclusions

In the present work, effect of shear lag on flange buckling of box-beam has been discussed numerically with the help of commercially available software 'ANSYS'. Behavior of laminated box beams is affected by shear-lag, fiber orientation and cross sectional parameter. By an approximate analysis method, shear lag effect predicted. The shear-lag effect is higher in FRP panels, due to their orthotropy and lower shear Stiffness. The orthotropy of panels is due to fibre orientation. That is ±45° and 90° fiber orientation of flange gives low shear lag effect as a consequence of less uneven stress variation in flange width, and buckling coefficient increases. The two parameters which depend on shear lag are identified. When the value of orthotropic parameter increases, shear lag effect increases. Also It can be observed that for constant orthotropic parameter and cross sectional shape parameter, shear lag effect is constant. Effect of shear lag on buckling is significant in hat shaped box sections. Buckling factor value directly proportional to the shear lag parameter. Hence when shear lag effect increases buckling factor decreases. Precise prediction of structural response characteristics is a difficult problem in the analysis of laminated composites.

References

1. Sa-nguanmanasaka J et al (2006) Stress concentration due to shear lag in continuous box girders. Eng Struct 29(7):1414–1421 (Elsevier)
2. Upadhyay A, Kalyanaraman V (2003) Simplified analysis of FRP box gird ers. J Compos Struct 59:217–225 (Elsevier)
3. Yaping W et al (2003) Analysis of shear lag and shear deformation effects in laminated composite box beams under bending loads. J Eng Struct (Elsevier)
4. Chandak R et al (2007) Shear lag prediction in symmetrical laminated composite box beams using artificial neural network. Struct Eng Mech 29:77–89

Progressive Collapse Triggered by Column Loss: Detrimental Effect of Span Length and Beam Depth as Per GSA and DOD Guidelines

T. H. Divya and R. Nikhil

Abstract Failure of RC structure triggered by fracture is a much critical situation across various engineering disciplines. Progressive collapse, one of the most devastating type of building damage occurs when major structural load carrying member suddenly collapse due to accidental loads such as gas explosion, terrorist attack, fire, vehicle impact etc. Such accidents may occur at any floor and at any location. If neighboring members are not capable to resist and redistribute the additional load received, that part of the structure fails. This misbehavior of structure leads to costly damage, multiple injuries and possible loss of life. RC structure can resist progressive collapse through various mechanism such as Catenary action and Vierendeel action. The span length and beam depth required to satisfy the failure criteria for progressive collapse was obtained by using the virtual work method, as there is an equilibrium of the external work done by gravity load due to loss of column and the internal work done by plastic rotation of beam. The main focus of this paper is to assess the effect of location and span of RC building with an aspect ratio 1. In order to explain the behavior of structure elements, non-linear dynamic analysis was done using finite element software as per GSA and UFC-DoD guidelines. It was found that damage to corner column cause more torsional vibration compared to other cases.

Keywords Progressive collapse · Catenary action · Vierendeel frame action · Aspect ratio · L/D ratio · DCR value · Plastic hinge

1 Introduction

Progressive collapse is a chain reaction of failure, it happens when local failure of primary structural components leads to the failure of adjoining members and finally to the failure of partial or whole structural system. Progressive collapse is one of the most devasting type of structural failure, most often leading to costly damages, multiple injuries and moreover possible loss of human life. More factors contribute

---

T. H. Divya (✉) · R. Nikhil
Department of Civil Engineering, Universal Engineering College, Thrissur, Kerala, India
e-mail: divyasmithesh075@gmail.com

K. Dasgupta et al. (eds.), *Proceedings of SECON'19*,
Lecture Notes in Civil Engineering 46,
https://doi.org/10.1007/978-3-030-26365-2_42

to these progressive collapse such as construction errors, improper inspection, miscommunication, disqualified site engineers and poor design which have led to many changes in building codes throughout the world. According to ASCE standard 7-05 progressive collapse is defined as "the spread of initial local failure from element to element resulting, eventually, in the collapse of an entire structure or disproportionately of large part of it". Progressive collapse in structure mainly occur due to abnormal loads, gas explosion, vehicle impact, fire, earthquake or other manmade or natural hazards. When major structural load carrying members are suddenly collapsed by abnormal loads, remaining structural elements cannot support the weight of the building or gravity loads. As an ultimate result of this event, substantial part of the structure may collapse, causing greater damage to the structure than the initial impact. Therefore, this study focuses arresting or eliminating progressive collapse by varies guidelines. The U.S. General Service Administration (GSA) and Department of Defense (DoD) provide detailed stepwise procedure regarding methodology to resist progressive collapse of building structure. Produce castrophic effect on the structure and also inherently nonlinear event in which structural elements are stressed beyond the elastic limit to failure. Progressive collapse is a dynamic event, since it involves vibration of structural element and result in dynamic internal forces like an inertia force whose energy may or may not be absorbed by the structure.

Literature survey was done by referring and going through articles and journals published in the related area of the studies to get detailed knowledge about subject. Current case study developed by the following publications; Sasani et al. [3] presented the performance of the University of Arkansas Medical Center Dormitory an actual 10-storey reinforced concrete structure, conduct the removal of an exterior ground floor column removal scenario. In [5], Sasani and sagiroglu experimentally studied the resistance the progressive collapse of Hotel San Diego. Its frame structure, an actually 6-storey RC building, following instantaneous removal of two adjacent exterior column, one of which was a corner column. In Sasani [2] studied the same building analytically and compared the result with experimental data. Sasani et al. [4] studied the performance of crown plaza hotel against progressive collapse. Sagiroglu [1] presented the behavior of 7-storey RC building, first designed based on current standards and then analyzed under 15 different initial damage scenario that are select based on the suggestions in the USGSA and USDOD guidelines.

In this paper, buildings are designed in accordance with current Indian code, GSA and DoD guidelines. Current study resistance of progressive collapse of three different building with same aspect ratio whose span 3, 4, and 6 m and different L/D ratio and span such as 10, 12 and 15. Each building have total 9 column removal scenario. Column removal locations are corner, intermediate, perimeter of the building each location have three column removal in one by one respectively. 3C-G, 3C-M and 3C-R indicating that corner column removal locations at ground, medium (4th floor) and roof of 3m span. 3I-G, 3I-M and 3I-R indicating that intermediate column removal location at ground, medium and roof of 3 m span. 3P-G, 3P-M and 3P-R indicating that perimeter column removal locations at ground, medium and roof 3 m span.4C-G,4C-M and 4C-R indicating that corner column removal locations at ground, medium and roof of 4 m span. 4I-G, 4I-M and 4I-R indicating that intermediate column removal location at ground, medium and roof of 4 m span. 4P-G, 4P-M and 4P-R indicating that perimeter column removal locations at ground medium and roof 4 m span. 6C-G, 6C-M and 6C-R indicating that corner column removal locations at ground, medium and roof of 6 m span. 6I-G, 6I-M and 6I-R indicating that intermediate column removal location at ground, medium and roof of 6 m span. 6P-G, 6P-M and 6P-R indicating that perimeter column removal locations at ground, medium and roof 6 m span.

2 Glance of Model

The investigated structure is an 8-storey RC frame structure. All the floors are typical. Height of each floor 3 m and total building height 24 m and base of the building located −2.4 m (total height/10) below the round level. According to IS 875 part 2, live loads were provided as 3 kN/m^2 up to roof and 1.5 kN/m^2 at roof floor. A 1 kN/m^2 was applied as floor finish for all floors. Parapet load is 2 kN/m. The nominal compressive strength of concrete was assumed to be 34.5 MPa and unit weight of 23.5 kN/m^2 and modulus of elasticity of the concrete is estimated to be 29,368.35 MPa. The chosen rebar yield stress, fy 413 MPa and modulus of elasticity 200,000 MPa. Each building beam depth changed according to L/D = 10, 12, and 15. The size of beams sections were kept the same over the height in a building. Building aspect ratio, L/B = 1, L/H = 1 and B/H = 1 were maintained in all the buildings. Load combinations as per GSA codes depends on the analysis type, 2DL + 0.5LL for static analysis, DL + 0.25LL for dynamic analysis, where DL is dead load and LL is the live load. Seismic loading provided as per IS: 1893-2016. Zone: III, Soil type: II, Response Reduction Factor (R) = 5, Importance Factor (I) = 1.5. Load combinations for design of structural elements were considered as per IS 1893-2016. A mass proportional damping ratio $\xi = 0.05$ in the first vertical vibration model was used in ETABs 2017 assuming a complete removal of the specified columns as required by guidelines. Figure 1 shown the details of beam reinforcement as per GSA.

2.1 *Models*

Wall loads are changed according to beam depth. The aim of this work is to study the effect of progressive collapse in different span and depth of the beams. Arrest or reduce the progressive collapse of building and find out which location is more

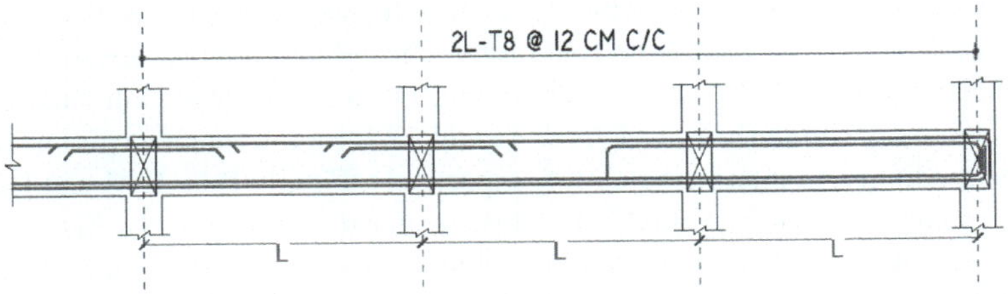

Fig. 1 Providing continuous bottom reinforcing steel across the connection is essential to accommodating the double-span condition

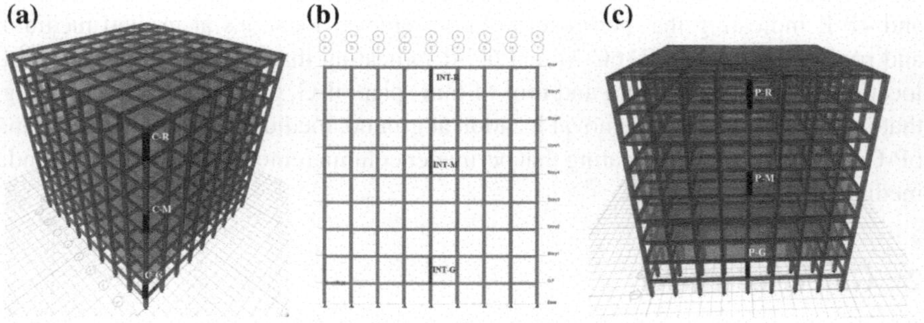

Fig. 2 **a**, **b**, **c** are column removal locations of 3 m span @ 8bay

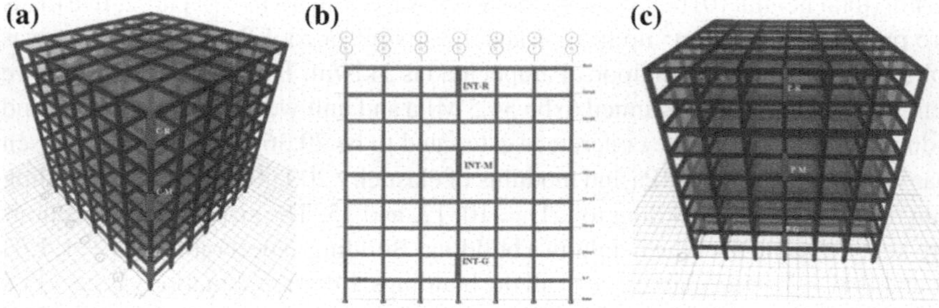

Fig. 3 **a**, **b**, **c** are column removal locations of 4 m span @ 6bay

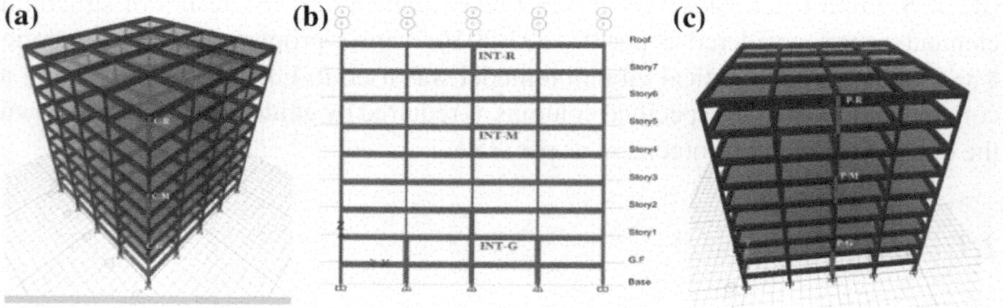

Fig. 4 **a**, **b**, **c** are column removal locations of 6 m span @ 4bay

prone to severe collapse in building. This study focuses 9 models were each span have 3 models with three different L/D ratios. The column removal location at corner indicated as C-G, C-M, and C-R are corner ground, corner medium and corner roof respectively. Similarly INT-G, INT-M and INT-R for intermediate ground, intermediate medium and intermediate roof respectively. Finally P-G, P-M and P-R for perimeter ground, perimeter medium and perimeter roof respectively. Figures 2, 3, and 4 shown the model and column removal locations of case study buildings.

3 Progressive Collapse Analysis Methodology [6]

As per unified facility criteria (UFC), there are three procedures used for analyzing structures subjected to progressive collapse: first one, Linear static procedure also called LSP which is simplest method in this method material is assumed to be linearly elastic; no geometric nonlinearity is considered; and structure is supposed to experience small deformation. Second one nonlinear static analysis also called NSP, in which both geometric and material nonlinearities are considered; third case is nonlinear dynamic procedure also called NDP which involves inertia and damping effects and it gives most accurate result. In the current study the method used is nonlinear dynamic procedure. Following procedure were carried out for this study:

1. Modelling in software
2. Apply load as per progressive collapse guidelines
3. Run the model with all column present
4. Find the gravity load carrying capacity of vertical members (i.e. column) to be removed
5. Remove the column and corresponding axial force of removed column s to be provided as point load
6. Remove the point load using time history ramp function
7. Evaluate the result based on demand capacity ratio (DCR) for beams.

4 Result and Discussion

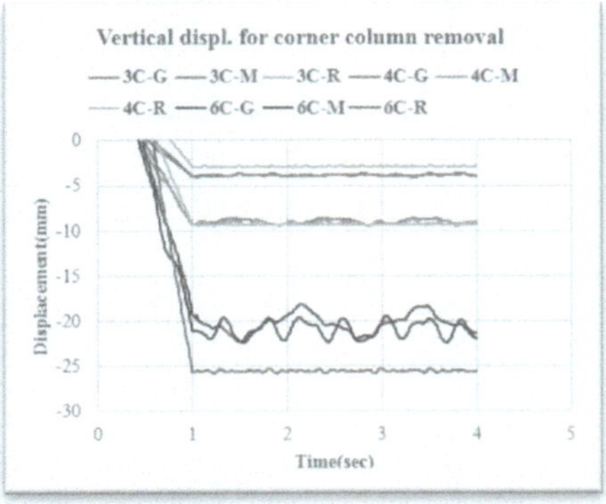

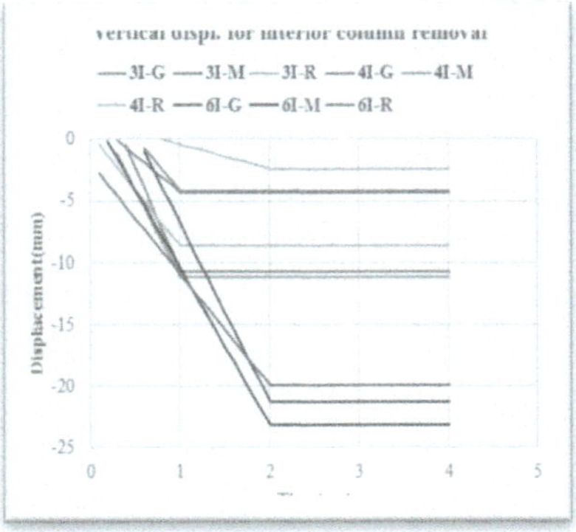

CORNERL/D=10 INTERMEDIATEL/D=10

As per the graph it was understood that maximum vibration occur in corner column removal case due to the torsional effect. Due to uniaxial moment small vibration is experienced in the perimeter column removal scenario. Maximum displacement mainly depends on the column removal locations, axial force and moment. However mostly observed maximum displacement is in the intermediate column removal scenario. Figure 5 showed the maximum displacement each column removal locations.

Other case is L/D = 12, the maximum displacement occurs in 6 m span. In the case of corner column removal, the maximum displacement is 30.393 mm at roof. For intermediate column removal, the maximum displacement at middle is 30.723 mm and perimeter column removal, the maximum displacement at ground is 25.983 mm. Figure 5 shown graphical representation of each column removal locations.

Due to large axial force at this location there is no vibration or minute vibration as the load is transferred to four adjacent columns. For L/D = 10, in the case of corner column removal scenario, the maximum displacement is 25.578 mm, at roof of 6 m span. In the case of intermediate column removal, the maximum displacement is at

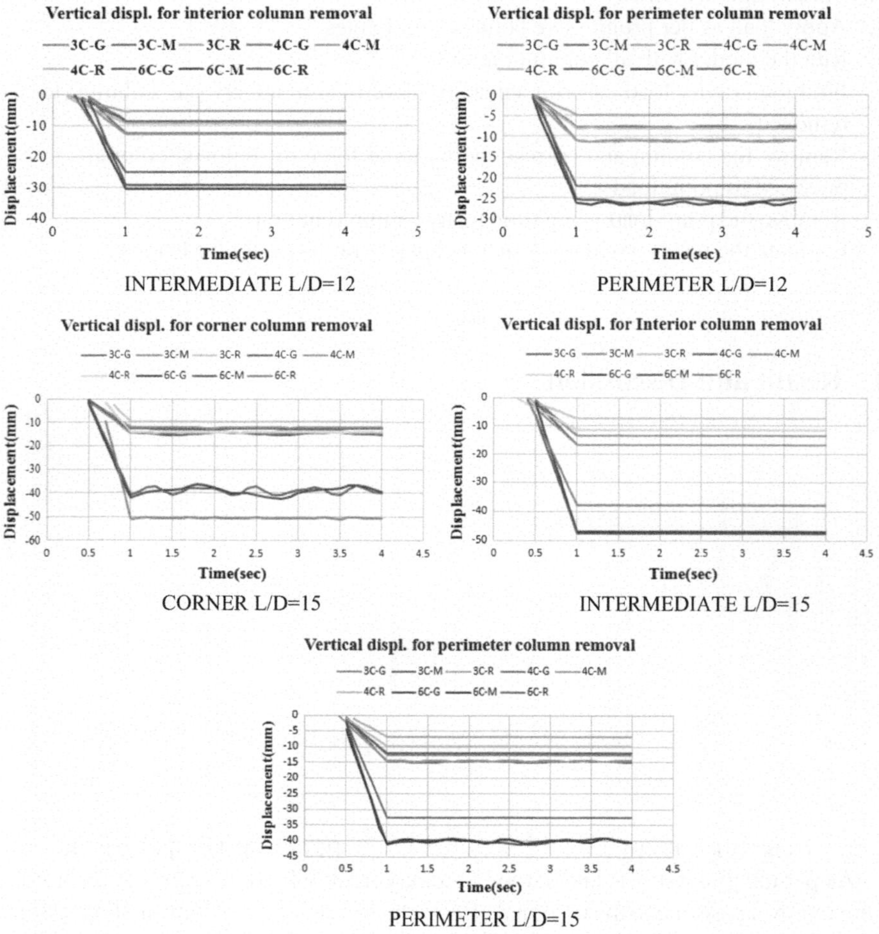

Fig. 5 Vertical displacement of different span in different location by column removal scenario

middle of 6 m span and is 23.179 mm and finally for the perimeter column removal, the maximum displacement is 20.3 mm at middle of 6 m span.

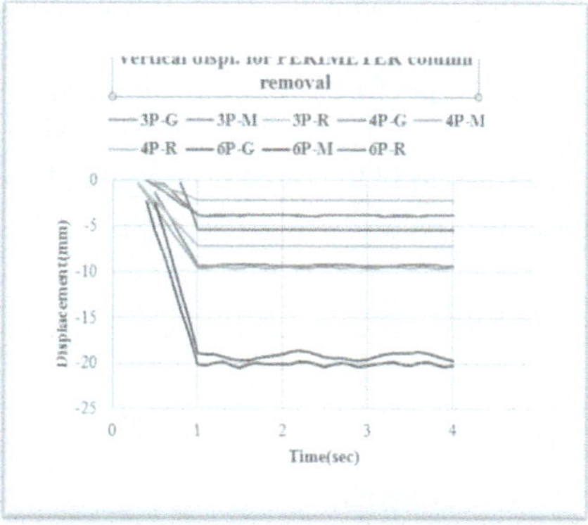

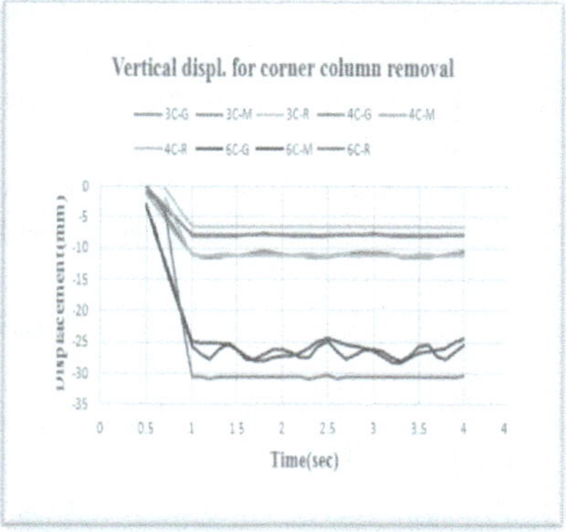

PERIMETER L/D=10 CORNER L/D=12

The same procedure was continued in the case of L/D = 15, the maximum displacement observed in 6 m span roof is 50.516 mm in the case of corner column removal. In the intermediate column removal scenario, the maximum displacement at middle is 47.554 mm. And maximum displacement at middle is 40.475 mm in the case of perimeter column removal scenario.

5 Demand Capacity Ratio [7]

According to GSA (20013) guidelines the DCR, of the member force and the member strength, is a measure to determine the failure of main structural member by linear dynamic procedure. Figure 6 shown the maximum bending moment diagram

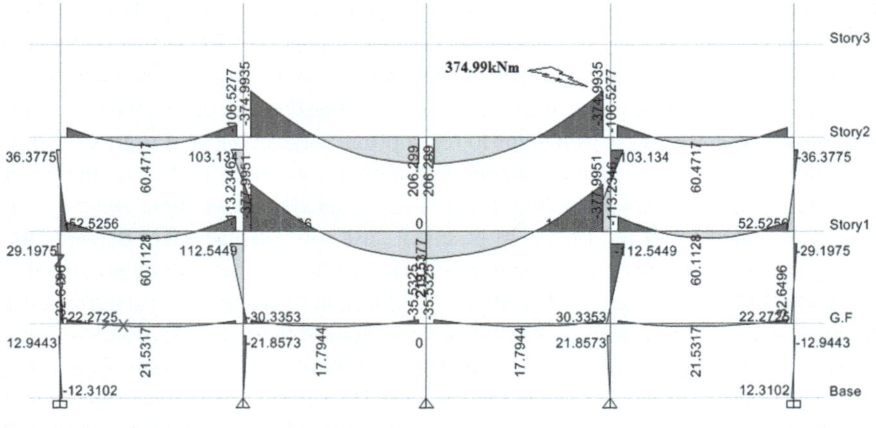

Fig. 6 Maximum bending moment under column removal scenario

$$DCR = QUD/QCE$$

where, QUD = Acting force or demand determine (shear, axial force, bending moment, and possible combined forces.

QCE = Apparent ultimate capacity (shear, axial force, bending moment, and possible combined forces.

DCR < 2.0 for typical structural configurations and DCR < 1.5 for atypical structural configurations. The current study model is typical

$$DCR = QUD/QCE$$

where, QUD = 377 kNm from the software.

QCE $\Rightarrow Mu_{\text{limit}} = 0.138 \times f_{ck} \times \text{b} \times \text{d}^2$.

$Mu_{\text{limit}} = 0.138 \times 34.5 \times 300 \times 365^2 = 190.288\,\text{kNm}$.

$$DCR \approx 2$$

The maximum DCR value obtained is for 6 m span L/D = 15 case, for column removal intermediate case at ground floor. The value of demand moment is 374.99 kNm obtained since 5 s. For the case study a maximum time of 5 s is considered, the number of output time steps were taken as 50 and output time step size 0.1 s. The versatile case is the above mentioned of the 81 column removal scenarios. In this case almost elements reached the collapse point. The maximum displacement is recorded since 5 s but first 1 s is not consider because the column is gradually removed after 1 s.

6 Conclusions

This paper conduct analytical modeling techniques for progressive collapse evaluation of RC structure. Progressive collapse resistance of 8-storey RC structure. Span to depth ratio have a significant effect on the performance of RC frame to resist progressive collapse, among the cases of column removal, most damage occurs when interior column is lost. Catenary action helps to resist progressive collapse. Progressive collapse of the RC concrete frame structure was ultimately controlled by the rupture of the reinforcing steel bars in the floor beam. Roof displacement is higher, when same load is provided in all the floors. In practical case roof displacement depends the depth of beam, span length and column removal location. Corner column removal is severe due to cantilever action. Variation of displacement is non uniform through the height of the structure, span, depth of beam, moment of inertia, deflection. After the study, 3 and 4 m span are found ideal and L/D ratio may vary from 10 to 12, remaining range is not safe. For 6 m span, only L/D = 10 is found safe. In seismic zone it is better not to consider L/D = 10 below. Progressive collapse reduce the rigidity of the structure. Provide adequate joint and adequate reinforcement detailing. Span of the structural element should be adequate as per the guidelines given by the standard codes. For future studies all spans may be given same depth i.e. here the versatile condition in L/D = 15 in 6 m span.

Acknowledgements The author acknowledges the co-authors for their extremely useful guidance and valuable presence in my work. The special acknowledgement is due to my guide, **Mr. Nikhil R.**, Assistant Professor, Universal Engineering College, Vallivattom, Thrissur.

References

1. Sagiroglu S, Sasani M (2013) Progressive collapse-resisting mechanisms of reinforced concrete structures and effects of initial damage locations. J Struct Eng 140:04013073. https://doi.org/10.1061/(ASCE)ST.1943-541X.0000854
2. Sasani M (2008) Response of a reinforced concrete in filled-frame structure to removal of two adjacent columns. J Struct Eng 30(10):248–2491
3. Sasani M, Bazan M, Sagiroglu S (2007) Experimental and analytical Progressive collapse evaluation of an actual reinforced concrete structure. J Struct Eng 104(6):731–739
4. Sasani M, Kazemi A, Sagiroglu S, Forest S (2011) Progressive Collapse resistance of an actual 11-story structure subjected to severe Initial damage. J Struct Eng. https://doi.org/10.1061/(ASCE)ST.1943-541X.0000418
5. Sasani M, Sagiroglu S (2008) Progressive collapse resistance of Hotel San Diego. J Struct Eng 134(3):478–488
6. Unified Facilities Criteria (UFC), Design of buildings to progressive collapse UFC 4-23-03 June 2013
7. U. S General Service Administration, Progressive collapse analysis and design guidelines for new federal office buildings and major modernization projects

Experimental and Analytical Study on Strengthening of Reinforced Concrete T-Beams Using External Prestressing

Krishnendhu Ajith and Ashok Mathew

Abstract A parametric study was carried out to investigate the structural behavior of reinforced concrete T beams under different external prestressing conditions. This paper adopted two external prestressing methods; to study and to compare the behaviour of externally prestressed tendons that is, 12 mm diameter steel bars and 12.7 mm steel strands by strengthening T-shaped reinforced concrete (RC) beams which is subjected to static loads. The influences of concrete strength, reinforcement ratio of the non-prestressed tensile steel bars and prestressing force on the these tendons in terms of ultimate load and load deformation were examined. The test results revealed that, compared with unstrengthened RC beams, the specimens externally strengthened with prestressed steel bars or steel strands showed a better performance in load carrying capacity. After that the better prestressing method was modelled and validated by finite element software ANSYS. It can provide a reference for the further promotion of the structure.

Keywords Experiment · RC beams · Flexure · Strengthening · External prestressing

1 Introduction

External prestressing technology is broadly used for the construction of highway, railway and urban bridges in prestressed concrete structures. In construction and designing, the theory formulas are mainly used for determining the prestressing force. Durability factors like creep, shrinkage and temperature causes many changes in prestress force. In the past 2 decades, there has been an increasing need around the world to strengthen reinforced concrete bridges due to heavier traffic loads, progressive structural aging and corrosion of steel reinforcement. One of the preferred

K. Ajith · A. Mathew (✉)
Sree Buddha College of Engineering, Pattoor, Padanilam, India
e-mail: fromashokmathew@gmail.com

K. Ajith
e-mail: krishnaindhu2017@gmail.com

K. Dasgupta et al. (eds.), *Proceedings of SECON'19*,
Lecture Notes in Civil Engineering 46,
https://doi.org/10.1007/978-3-030-26365-2_43

strengthening methods is external prestressing because of its speed and its possibility of monitoring, future retensioning and replacing the tendons. External prestressing was firstly used for strengthening of bridges, but nowadays it is used for built new structures and strengthening purposes. External prestressing becomes very important and popular technique for prestressing of concrete structures, as it improves the load carrying capacity of structural members. External prestressing is used in concrete, steel and timber structures also. Steel and fibre reinforced polymer (FRP) can be used as materials for external prestressing. The external pre-stressing actively inducts load by tendons to impose stress to the concrete, and the behavior remains predominantly flexural. External prestressing provides one of the most efficient solutions to increase the rating capacity of existing bridges when the infrastructures are in need of renewal. It is often useful for rehabilitation, for strengthening or stabilizing a structure by providing confining forces. The applications of external prestressing rehabilitation techniques have shown to not only increase the life expectancy of the member or system, but to increase the flexural strength considerably, resulting in reduced deflection and cracks widths. External prestressing techniques has been widely used in construction of new bridges as well as strengthening of existing bridges. Use of external prestressing has resulted in construction of several innovative bridges with large eccentricities and light weight concrete.

2 Research Significance

- To study and compare the load deflection characteristics.
- To compare the ultimate load carrying capacity.
- To investigate and compare the propagation of cracks experimentaly.
- To compare the effects of different tendons on external prestressing.

3 Methodology

- A detailed literature survey was carried out on studies on strengthening of reinforced T beams.
- Materials like cement, fine aggregates, coarse aggregates and tendons were collected.
- Laboratory tests were carried out on the raw materials used for this study. Physical properties of materials were tested and found that all properties are conforming to IS standards.
- Casting of cubes, cylinders and beams of M40 grade concrete to determine the hardened properties of concrete.
- Casting of reinforced concrete beams to study the flexural behavior.

- Testing the control beam specimen at 28 day to determine the ultimate load carrying capacity.
- Partially loading the reinforced concrete beams to be strengthened and retrofitted.
- Strengthening of the partially loaded beam by using external prestressing methods
- Testing of the retrofitted specimen.
- Conclusions and recommendations are finally made based on the findings and observations.

4 Materials Used

4.1 Cement

Ordinary Portland cement of 53 grades conforming to IS 12269-1987 was used in the study.

4.2 Fine Aggregate (FA)

Fine aggregate are soil particles passing through 4.75 mm IS sieve. Generally river sand, crushed stone, crushed gravel, M sand etc. are used as fine aggregate. In this study, M sand conforming to Zone II is used.

4.3 Coarse Aggregate (CA)

Coarse aggregates passing through 16 mm IS sieve and retained on 4.75 mm IS sieve is used for the study.

4.4 Water

Presence of organic or inorganic impurities in water will affect the strength of concrete. Potable water is generally considered as being acceptable.

Table 1 1 Mix proportion of control mix

Material	Cement	Fine aggregate	Coarse aggregate	Water
Ratio	1	0.76	2.13	0.45

5 Mix Proportioning

In this study M40 concrete was used. As per IS: 10262–2009, a mix proportion was suitably designed for M40 grade concrete after several trial mixes. The mix proportioning was done based on the properties of various materials like cement, fine aggregate and coarse aggregate. The mix proportion of control is given in Table 1.

6 Casting of Cubes, Cylinders and Beams

Concrete mix was prepared using cement, fine aggregate, coarse aggregate and water. Concrete mix for strengthening was prepared according to Indian standard. Concrete cubes of 15 cm × 15 cm × 15 cm, concrete cylinders of 15 cm × 30 cm and concrete beam of 50 cm × 10 cm × 10 cm are prepared and tested for compressive strength, split tensile strength and flexural strength respectively. M40 is used.

7 Testing of RCC T Beam

The beams designated as CB (Control Beam) was tested under two point loading system up to ultimate load and other three beams designated as VIII (For External Prestressed Beam) was tested under two point loading system up to 70% of the ultimate load of the Control Beam. The effective span of the beam was 100 cm and the distance between two point loads was 33.3 cm. The dial gauges was used to observe deflection at the center and under the load position. Load was applied through a hydraulically operated jack. The three beams designated as S (For External Prestressed Beam) after loading 70% of ultimate loading are used for rehabilitation (Fig. 1).

8 Rehabilitation of Beams

Out of six beams casted, three beams were loaded up to 70% of the ultimate load of control beam. Hence repairing and cleaning process of these beams was necessary before strengthening them. Then with the same procedure as of Control Beam (CB) testing of beam is done in order to determine ultimate load, corresponding deflections

Fig. 1 Experimental setup

are noted. After finding the ultimate load, the surface of the beam is cleaned to apply any of the bonding agents in order to get a perfect bond.

After this, the end plates are fixed using HUS3 screw anchor then, deviators are fixed using HITHY200 motar with HIT-V Rod and final post tensioning process are done. After the post tensioning process the retrofitted beams are tested and the results are observed and evaluated (Figs. 2, 3 and 4).

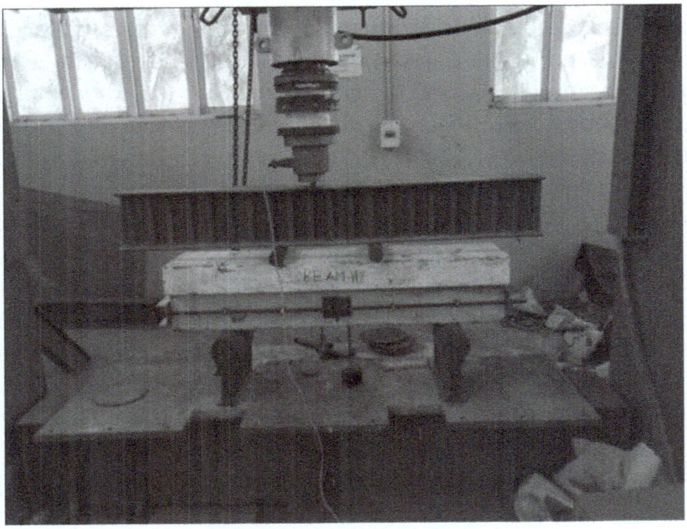

Fig. 2 Experimental setup of prestressed beams

Fig. 3 Crack pattern on externally prestressed T beam using 12 mm diameter steel bars

Fig. 4 Crack pattern on externally prestressed T beam using 12 mm steel strands

9 Results and Discussions

- Load Test on Reinforced Concrete T Beam

The T beam of size 250 mm × 250 mm × 1200 mm with breadth and depth of flange is 250 mm and 100 mm respectively breadth and depth of the web is 150 mm and 150 mm respectively was used in the experiment. A total of nine beam specimens, three of which were control beam, the load deflection curve is given in Fig. 5. And another six beams were given 70% ultimate load of control beams and then retrofitted with using external prestressing. The beam specimens are designed and cast as according to IS 456: 2000. The beams were tested in a 100T capacity Universal Testing Loading Frame. All the beams were tested to failure under two–point loading system. The load was applied using a hydraulic jack and controlled using a calibrated load cell. The load distribution was achieved by symmetrical steel rollers. The behavior of beam was observed form beginning to the failure. The first crack formation and its development and propagation are observed carefully. Each of the beam specimens were mounted on roller supports. In all the beam specimens, initiation of flexure cracks was from the bottom of beams. It is found that the flexural cracks were vertical and are towards the closest loading point (Table 2).

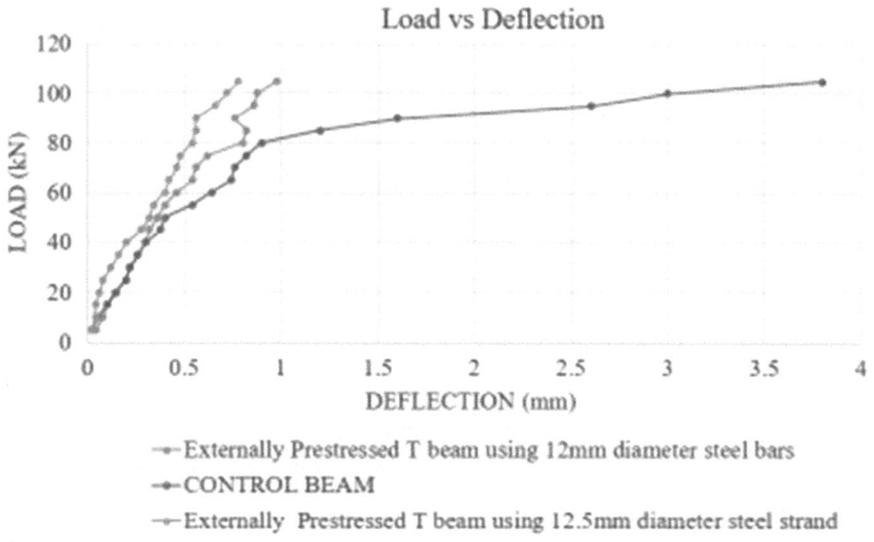

Fig. 5 Load deflection curve

Table 2 Results

Type of beam	Ultimate load (kN)	% Increase in load
Control beam (CB)	110	
Externally prestressed T beam using 12 mm diameter steel bars	129	19
Externally prestressed T beam using 12 mm steel strands	135	25

10 Analytical Validation

Results from experimental validation can be used to judge the quality, reliability and consistency of analytical results, it is an integral part of any good analytical practice.

Analytical methods need to be validated or revalidated

- Before their introduction into routine use
- Whenever the conditions change for which the method has been validated (e.g. an instrument with different characteristics or samples with a different matrix) and
- Whenever the method is changed and the change is outside the original scope of the method (Figs. 6, 7, 8 and 9).

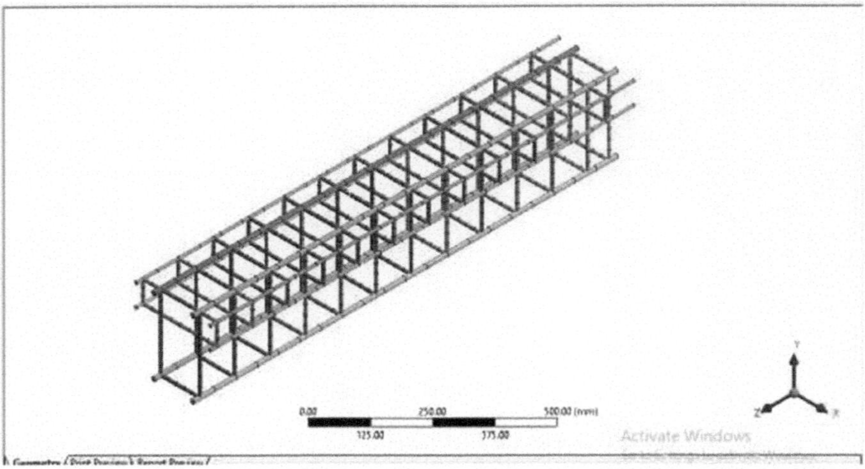

Fig. 6 Reinforcement detailing

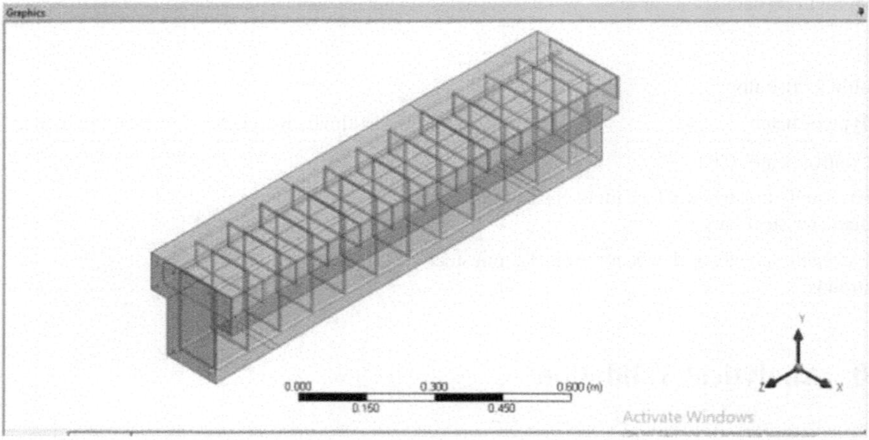

Fig. 7 Reinforcement detailing

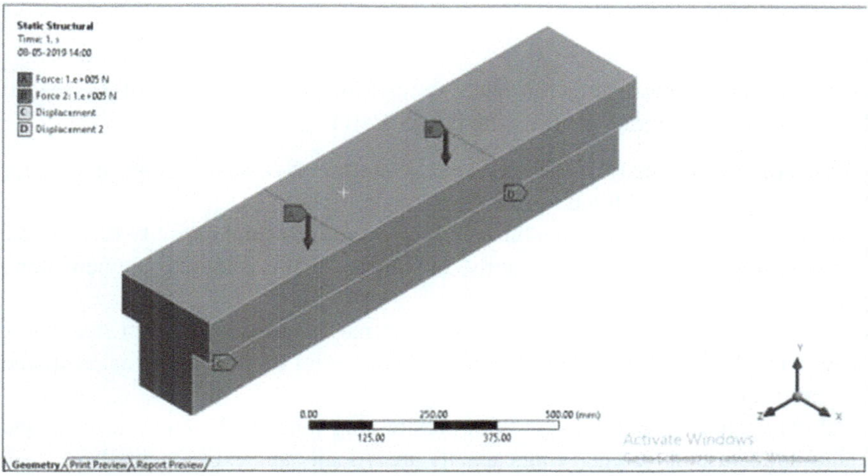

Fig. 8 Load application

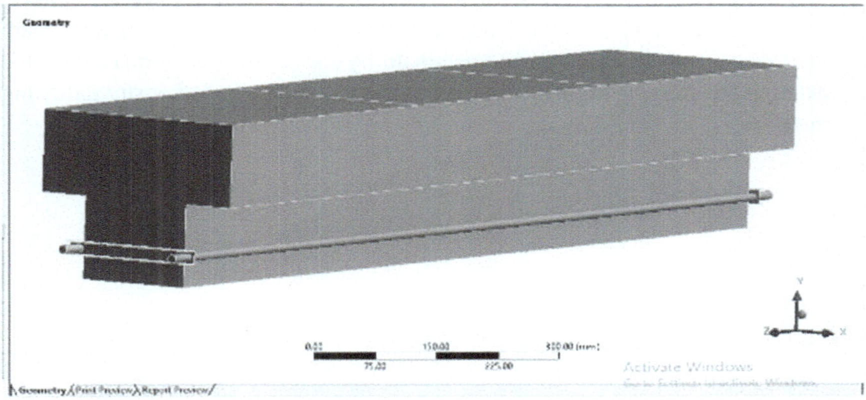

Fig. 9 External prestressed T beam on ANSYS

11 Conclusions

The experimental investigation is to check the possibility of Retrofitting of overloaded RC T beam using external prestressing is successful.

- The experimental analysis showed increase in ultimate load carrying capacity for both the types of retrofitted beams.
- External prestressed steel tendons can improve the flexural capacity of concrete simple supported beams significantly, and improvements is more prominent along with the initial effective stress.
- The ultimate load increased by 19% for external prestresssed 12 mm steel bar T beam and 25% for external prestressed 12.7 mm steel strand T beam as compared to control beam.
- Externally prestressed T beam with prestressing tendons 12.7 mm steel strand shows increase in load carrying capacity analytically and experimentally.
  - External prestressing on the T beam improves the stiffness of beams as the deflection decreased.
  - The deflection of the strengthened beam as compared to that of the control beam was less and also the load carrying capacity of the prestressed beam increased. From the economical, sustainability and application point of view, the external prestressing highly recommendable as compared to other strengthening methods.

Study of Viscoelastic Problems Utilizing Commercial Software's

Leema Mariam C. Mathew, R. Marimuthu and Afia S. Hameed

Abstract Objective of the present paper is to present the results obtained utilizing visco-elastic analysis with commercially available software such as FEAST^{SMT} and MARC. FEAST^{SMT} is a structural analysis software being developed at VSSC and marketed. Viscoelastic analysis capability is being regularly used in rocket industry for structural integrity analysis of solid propellant rocket motors. The finite elements required for such analysis are plane-strain, axi-symmetric and three dimensional solid elements. In this paper plane strain configuration is considered for the study with pressure, gravity and thermal loading. Results obtained with the two software are compared in terms of displacement and number of time steps required for the convergence of the exact solution where ever possible. From the study it is found that there are some mismatches in different softwares, in final answers and also the way in which it converges to the solution. The variation of results are presented in the paper.

Keywords Viscoelasticity · MARC · FEAST · Displacement · Time interval

1 Introduction

Viscoelastic material has both viscous and elastic behavior together, due to which for static loading structural configurations embedded or with such material will exhibit time dependent static response. Other than this time dependant behavior of material, these materials are nearly incompressible. Hence instead of displacement formulation mixed UP formulation is considered for the appropriate behavior of such structure. This paper contains comparison of results obtained utilizing visco-elastic analysis with commercially available software such as FEAST^{SMT} and MARC. The finite elements required for viscoelastic analysis are plane-strain, axi-symmetric and

---

L. M. C. Mathew (✉) · A. S. Hameed
Saintgits College of Engineering, Kottayam 686532, India
e-mail: leemacmariam@gmail.com

R. Marimuthu
VSSC, Thiruvananthapuram 695022, India

K. Dasgupta et al. (eds.), *Proceedings of SECON'19*,
Lecture Notes in Civil Engineering 46,
https://doi.org/10.1007/978-3-030-26365-2_44

three dimensional solid elements. Here plane strain configuration is considered for the study with pressure, gravity and thermal loading. Three numerical problems of journal paper by S. Yadagiri and C. Pappi Reddy (1985) "Viscoelastic Analysis of Nearly Incompressible Solids", *Computers and Structures* [1] are considered for the study.

2 MSC MARC

MSC Marc is a nonlinear finite elements analysis software used to simulate behavior of complex materials and interaction under large deformations and *strains*. MARC uses Herrmann element's for viscoelastic analysis.

3 FEAST$^{[Smt]}$

FEAST^{SMT} (Finite Element Analysis of Structures) is a structural analysis software based on Finite Element Method (FEM) realized by Structural Engineering Entity of Vikram Sarabhai Space Centre (VSSC), ISRO.

4 Numerical Discussion

Three plane strain numerical problems subjected to pressure, gravity and thermal loading are considered. All parameters are taken in SI units. For all problems Bulk modulus is taken as 10791×10^5 N/m^2.

4.1 Problem 1—Pressure Loading

First problem is a 8 noded plane strain problem subjected to pressure loading in mutually perpendicular directions as shown in Fig. 1. Because of symmetry, only quarter of the structure is considered and is divided into four plane strain 8 noded finite elements. Two edges are restrained. All dimensions are given in SI units. Applied loads are 98,100 N/m^2 in x direction and 49,050 N/m^2 in y direction.

Modulus of rigidity in N/m^2 is given as $G(t) = 19{,}620 + 96{,}1380\ e^{-1.5t}$.

Poisson's ratio will change with respect to time. This problem is a constant stress problem. Strains vary with time but are constant with geometry. The exact solution for the problem is

$$u(x,t) = 1.25341 \times 10^{(-2)}[49.866960 - 48.866959 \times e^{(-0.3t)} - 10^{(-6)}\ e^{(-1.49955t)}]x.$$

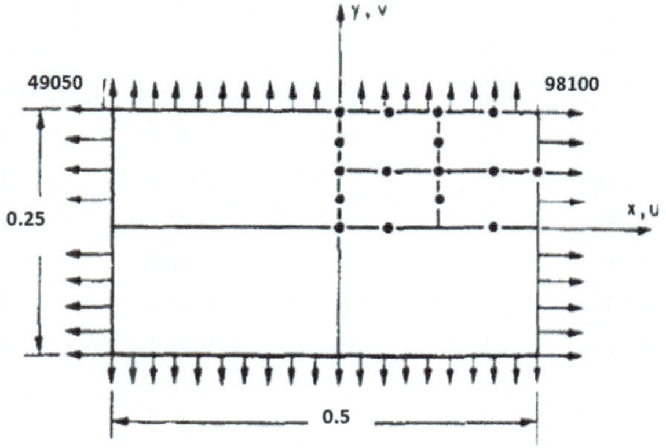

Fig. 1 Plane strain problem

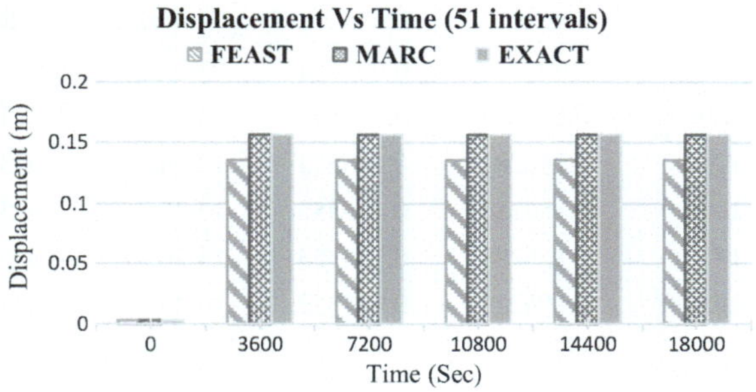

Fig. 2 Displacement versus time plot for 18,000 s @ 51 intervals

The solution at x = 0.25 m is considered for the study. Time considered for the study is 0–18,000 s (0–5 h) divided into 51 and 501 intervals.

Also 35 and 400 s divided into 100, 200, 400 and 800 intervals. The above problem is solved in MARC and FEAST[SMT] software. 8 noded plain strain elements is modeled in MARC and FEAST[SMT] as shown in Fig. 2.

4.1.1 Results

Case 1: 0–18,000 s divided into 51 intervals

In first case 18,000 s (5 h) is divided into 51 steps and results obtained in MARC matches with exact solution provided in Table 1. Table 2 provides the results obtained from MARC and FEAST assuming same data as in the previous case. Table 2 shows the difference in results, hence percentage error of values are calculated with respect to MARC. The variation of values in MARC and FEAST for displacement and strain is plotted in Figs. 2 and 3. In MARC convergence of values to the exact solution will occur at the second time step itself, while in FEAST convergence of value is not seen when less time intervals are considered. The variation of displacement contour at zero and 18,000 s are shown in Figs. 4a, b and 5a, b respectively.

Table 1 Comparison of values in MARC and Exact solution (problem 1)

Time	Displacement (m)			Major principal value of total strain		
	MARC	Exact	% error	MARC	Exact	% error
0	0.00313	0.00313	0.0002	0.01253	0.01253	0
3600	0.15628	0.15626	0.0138	0.62512	0.62504	0.0138
7200	0.15628	0.15626	0.0138	0.62512	0.62504	0.0138
10,800	0.15628	0.15626	0.0138	0.62512	0.62504	0.0138
14,400	0.15628	0.15626	0.0138	0.62512	0.62504	0.0138
18,000	0.15628	0.15626	0.0138	0.62512	0.62504	0.0138

Table 2 Comparison of values in MARC and FEAST for 51 intervals of time (problem 1)

Time	Displacement (m)			Major principal value of total strain		
	MARC	FEAST	% error	MARC	FEAST	% error
0	0.003134	0.003134	0.00	0.01253	0.01253	0
3600	0.156281	0.135553	13.26	0.62512	0.54221	13.26
7200	0.156281	0.135553	13.26	0.62512	0.54221	13.26
10,800	0.156281	0.135553	13.26	0.62512	0.54221	13.26
14,400	0.156281	0.135553	13.26	0.62512	0.54221	13.26
18,000	0.156281	0.135553	13.26	0.62512	0.54221	13.26

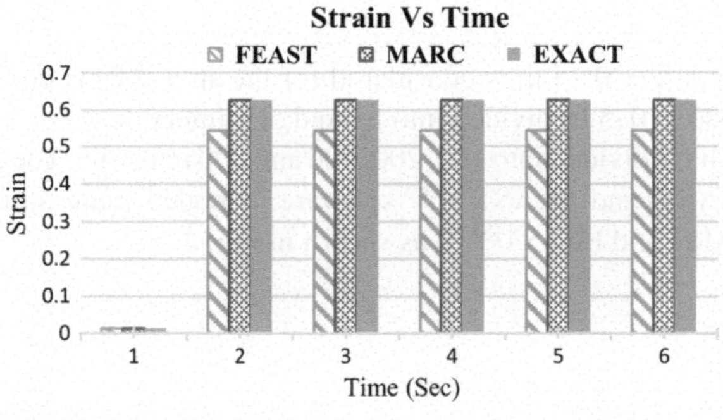

Fig. 3 Strain versus time plot for 18,000 s @ 51 intervals

Case 2: 0–18,000 s divided into 501 intervals

In second case number of time intervals increased for the given time interval as 501. Values are given in Table 3. In this case FEAST[SMT] converged to the exact value. The variation of values in MARC and FEAST[SMT] for displacement is plotted in Figs. 6 and 7. As the time intervals increased in FEAST[SMT] the values are matching with that of MARC, also values are converging. Displacement contour at

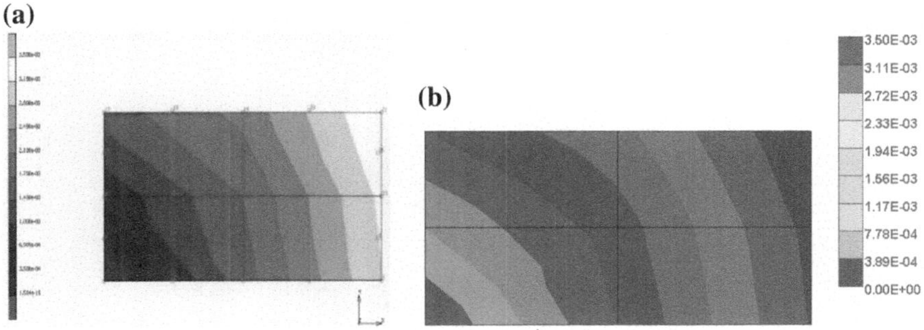

Fig. 4 Displacement contour at t = 0 in **a** MARC and **b** FEAST[SMT] @ 51 intervals

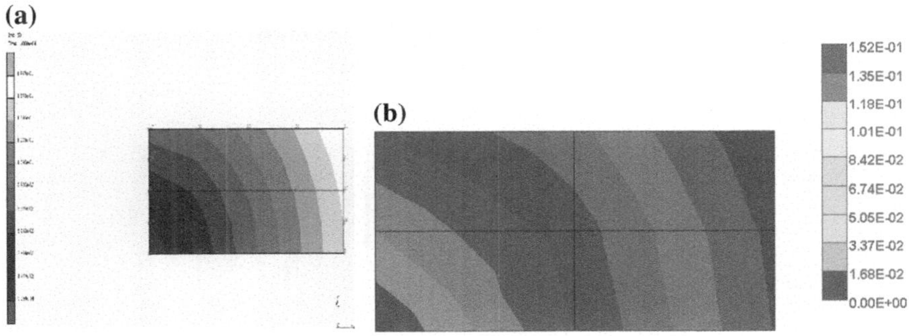

Fig. 5 Displacement contour at t = 18,000 in **a** MARC and **b** FEAST[SMT] @ 51 intervals

0 and 18,000 s is shown in Figs. 9a, b and 10 respectively shows a good match between them (Fig. 8).

Case 3: 0–35 s divided into different intervals

In this case 35 s of time is divided into 100, 200, 400 and 800 intervals of time and analyzed in both MARC and FEAST[SMT]. It is seen that the percentage of error is decreasing as the number of intervals are increasing which is shown in Fig. 10.

Table 3 Comparison of values in MARC and FEAST for 501 intervals of time (problem 1)

Time	Displacement (m)			Major principal value of total strain		
	MARC	FEAST	% error	MARC	FEAST	% error
0	0.003134	0.003134	0.000	0.01253	0.01253	0.0008
3600	0.156281	0.153473	1.797	0.62512	0.61389	1.7968
7200	0.156281	0.156227	0.035	0.62512	0.62491	0.0349
10,800	0.156281	0.156277	0.003	0.62512	0.62511	0.0026
14,400	0.156281	0.156278	0.002	0.62512	0.62511	0.0019
18,000	0.156281	0.156278	0.002	0.62512	0.62511	0.0019

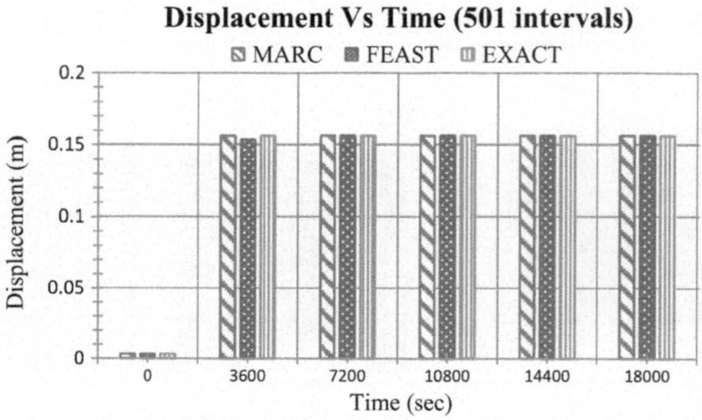

Fig. 6 Displacement versus time plot for 18,000 s @ 501 intervals

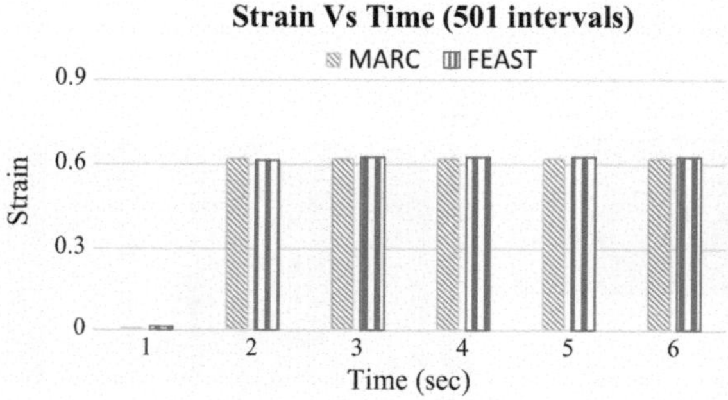

Fig. 7 Strain versus time plot for 18,000 s @ 501 intervals

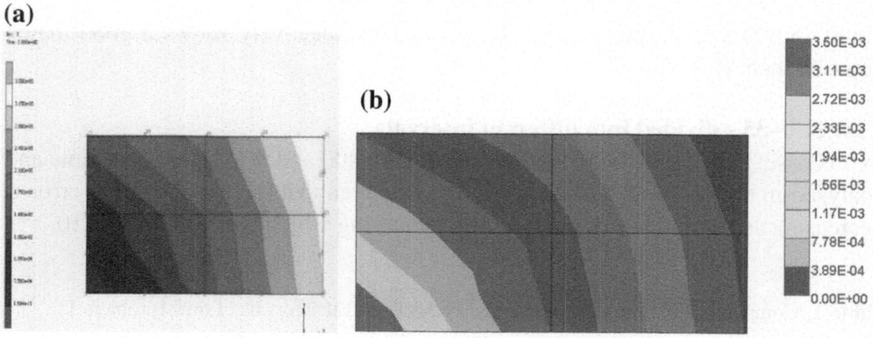

Fig. 8 Displacement contour at t = 0 in **a** MARC and **b** FEAST[SMT] @ 501 intervals

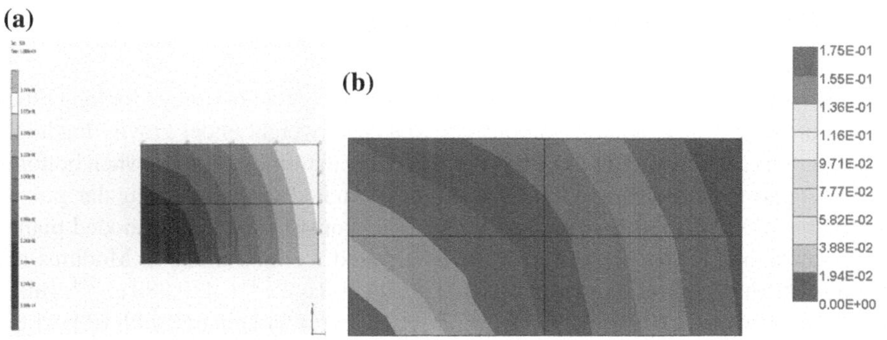

Fig. 9 Displacement contour at t = 18,000(501 int) in **a** MARC and **b** FEAST[SMT] @ 501 intervals

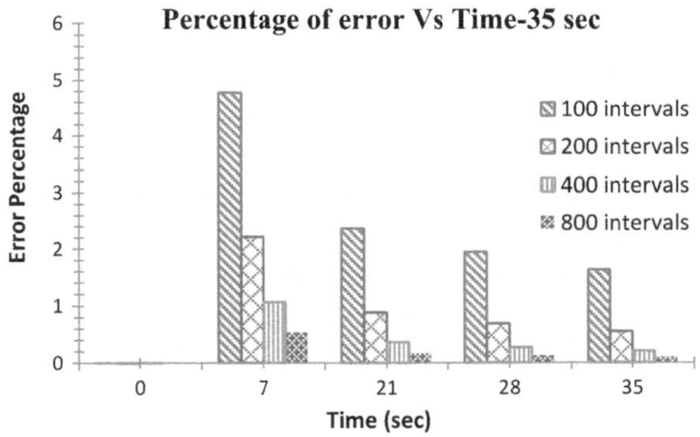

Fig. 10 Percentage of error in FEAST with MARC for 35 s

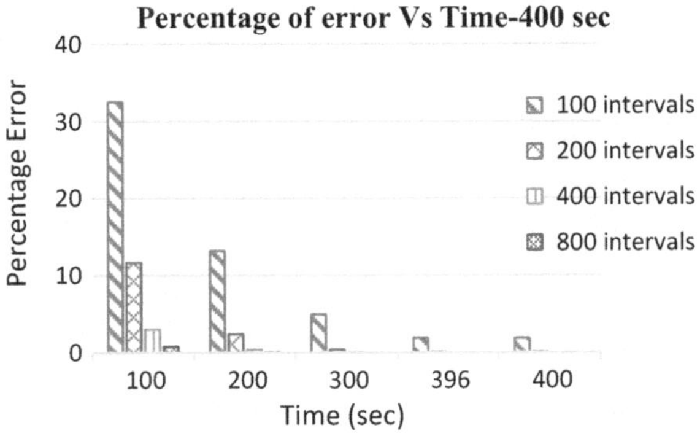

Fig. 11 Percentage of error in FEAST with MARC for 400 s

Case 4: 0–400 s divided into different intervals

When time interval considered increased the error in values also increased but it decreases as the number of intervals increased which is represented in Fig. 11.

4.2 *Problem 2—Gravitational Loading*

Due to viscoelastic nature of propellant grains, when it is kept in storage for long time it will undergo dimensional deviations due to its own weight under gravity loading. These grains are supported by casing (Fig. 12). Here extreme condition when bottom is not supported is considered. This problem is studied using a rectangular prism structure. It is analyzed for gravity loading. This consist of six, eight noded plane strain elements. Nodes along x = 0 line is arrested for both u and v. Modulus of rigidity in N/m^2 is given as $G(t) = 215{,}820 + 294{,}300\,e^{-0.25t} + 470{,}880\,e^{-0.5t}$. Bulk modulus is taken as $10{,}791 \times 10^5$ N/m^2. Poissons ratio will change with respect to time. Weight Density = 17,658 N/m^3. 8 nodded plain strain elements modelled in MARC and FEAST[SMT]. Two nodes, Node no. 27 (0.15, 0.9) and Node no. 29 (0.3, 0.9) are considered.

4.2.1 Results

Case 1: 0–18,000 s divided into 50 intervals
Here the time considered is 0–18,000 s divided into 51 intervals. Two nodes 27 and 29 were considered for the study. Vertical displacement which is more critical in case of slump problem is considered for the study. The variation in MARC and FEAST[SMT] is given in Table 4 and plotted in Figs. 13 and 14 for different intervals.

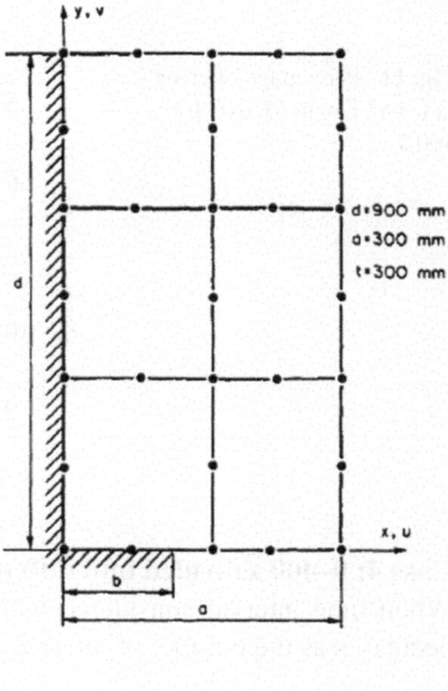

Fig. 12 Solid mass slump problem

Table 4 Comparison of values in MARC and FEAST for 51 intervals of time (problem 2)

Time (s)	Node 27 (0.15, 0.9)			Node 29 (0.3, 0.9)		
	FEAST Vert. Disp V (m)	MARC Vert. Disp V (m)	% error	FEAST Vert. Disp V (m)	MARC Vert. Disp V (m)	% error
0	0.000684	0.00069	1.3	0.00086	0.00096	10.3
3600	0.003092	0.00314	1.56	0.0039	0.00435	10.3
7200	0.003092	0.00314	1.56	0.0039	0.00435	10.3
10,800	0.003092	0.00314	1.56	0.0039	0.00435	10.3
14,400	0.003092	0.00314	1.56	0.0039	0.00435	10.3
18,000	0.003092	0.00314	1.56	0.0039	0.00435	10.3

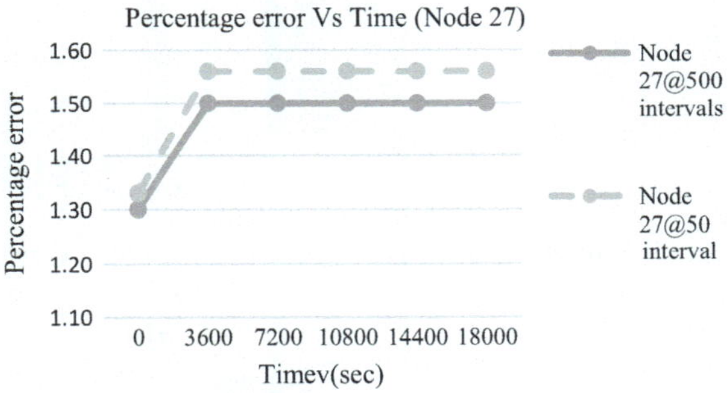

Fig. 13 Percentage of error in FEAST with MARC for slump problem @ Node 27

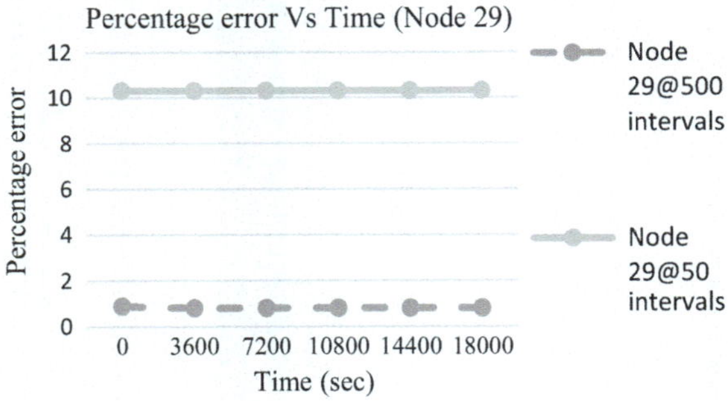

Fig. 14 Percentage of error in FEAST with MARC for slump problem @ Node 29

Case 2: 0–18,000 s divided into 500 intervals

When number of intervals increased percentage of decreased and reduced to admissible values. Also values converges fast. The results obtained is given in Table 5 and plotted in Figs. 13 and 14 for different intervals

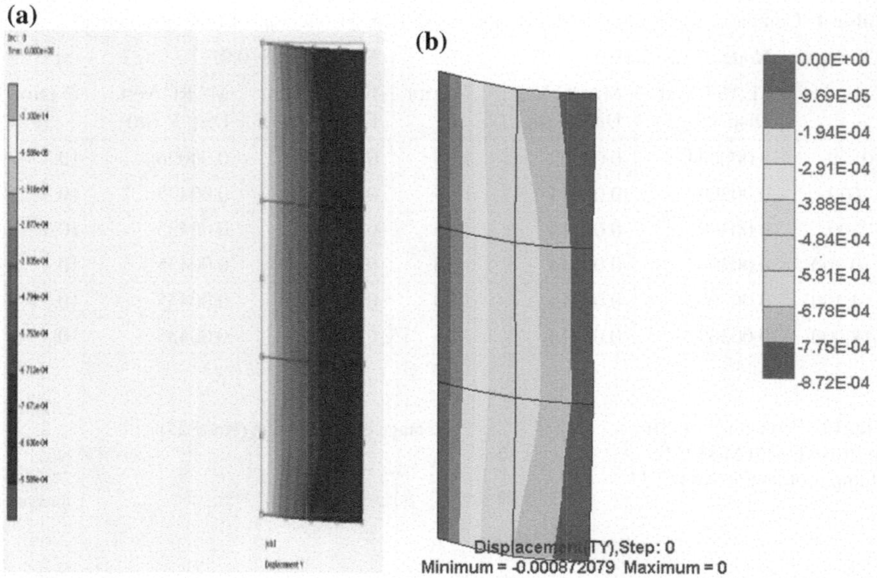

Fig. 15 Vertical displacement contour at t = 0 in **a** MARC and **b** FEAST[SMT]

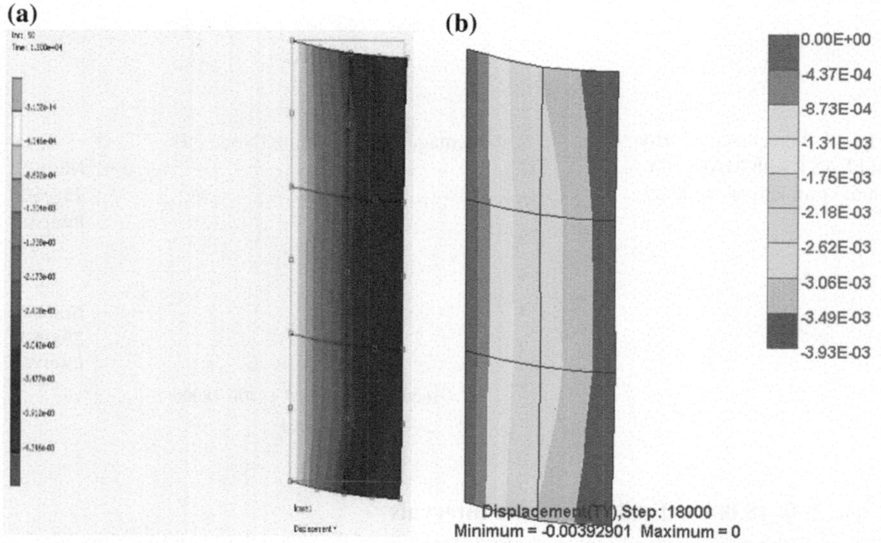

Fig. 16 Vertical displacement contour at t = 18,000 in **a** MARC and **b** FEAST[SMT]

Table 5 Comparison of values in MARC and FEAST for 501 intervals of time (problem 2)

Time (s)	Node 27 (0.15, 0.9)			Node 29 (0.3, 0.9)		
	FEAST Vert. Disp V (m)	MARC Vert. Disp V (m)	% error	FEAST Vert. Disp V (m)	MARC Vert. Disp V (m)	% error
0	0.00068	0.00069	1.33	0.00087	0.00096	9.27
3600	0.00309	0.00314	1.50	0.00393	0.004343	9.52
7200	0.00309	0.00314	1.50	0.00393	0.004343	9.52
10,800	0.00309	0.00314	1.50	0.00393	0.004343	9.52
14,400	0.00309	0.00314	1.50	0.00393	0.004343	9.52
18,000	0.00309	0.00314	1.50	0.00393	0.004343	9.52

4.3 *Problem 3—Thermal Loading*

Third problem is thermal analysis of a typical rocket grain given in Fig. 17. The long grain considered is encased in a rigid sheath. Analysis is done for a thermal shrinkage (from 60 to 30 °C). Because of symmetry only quarter of section is considered. Plane strain idealization is done and divided into 24 elements. The active degrees of freedom are 144 (Figs. 15 and 16).

Modulus of rigidity in N/m^2 is given as $G(t) = 215{,}820 + 294{,}300\,e^{-0.5t} + 470{,}880\,e^{-t}$. Bulk modulus is taken as 10791×10^5 N/m^2. Poissons ratio will change with respect to time. The coefficient of thermal expansion is $\alpha = 0.00011$ m/m °C. WLF constants $C_1 = 5.5$ and $C_2 = 140$. Reference temperature is taken as 298 K.

Grain configuration is modeled in MARC and FEAST[SMT]. 0–360 s (6 intervals) considered. Node 1(0.130, 0) and Node 11 (0, 0.236) are the locations considered.

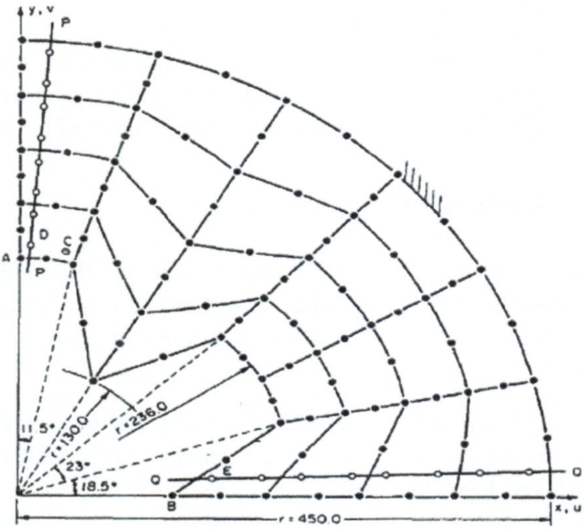

Fig. 17 Grain configurations (quarter section)

Table 6 Comparison of values in MARC and FEAST for 6 intervals of time (problem 3)

Time (s)	Node 1 (0.130, 0)			Node11 (0, 0.236)		
	FEAST Horiz. Dis U (m)	MARC Horiz. Dis U (m)	% error	FEAST Vert. Disp V (m)	MARC Vert. Disp V (m)	% error
0	0	0	0.00	0	0	0.00
60	0.000682	0.000639	6.66	0.000493	0.000503	1.99
120	0.001405	0.001278	9.92	0.001019	0.001007	1.27
180	0.002135	0.001918	11.31	0.00155	0.00151	2.65
240	0.002866	0.002557	12.08	0.002082	0.002013	3.40
300	0.003598	0.003196	12.57	0.002614	0.002516	3.87
360	0.00433	0.003835	12.90	0.003146	0.00302	4.19

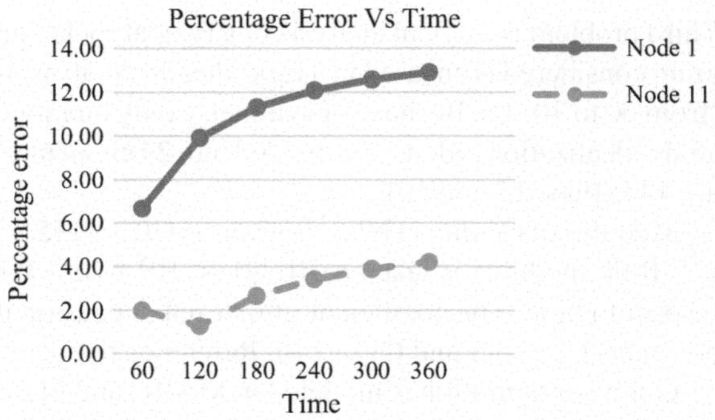

Fig. 18 Percentage of error in FEAST with MARC for thermal problem

4.3.1 Results

Results obtained are given in Table 6. The variation is plotted in Fig. 18. It is observed that percentage error is less in Node 11 compared to Node 1. The displacement contour obtained in both software are given in Figs. 19 and 20.

5 Conclusion

Behavior of viscoelastic materials are time dependent and their analysis require special analysis codes. Here in this paper viscoelastic analysis using two different commercially available software is done. Following conclusions are drawn from the study.

- For pressure and gravity loads FEAST converges to the solution when time interval considered sufficiently large, whereas for MARC convergence is achieved with lesser time intervals.
- In the case of thermal load with respect to MARC maximum error of 12% is observed for FEAST.

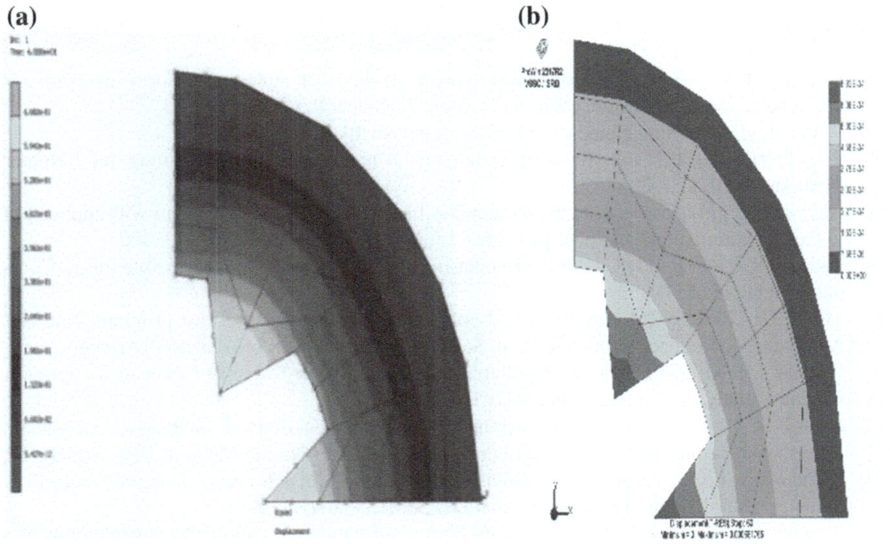

Fig. 19 Displacement contour at t = 60 in **a** MARC and **b** FEAST$^{[SMT]}$

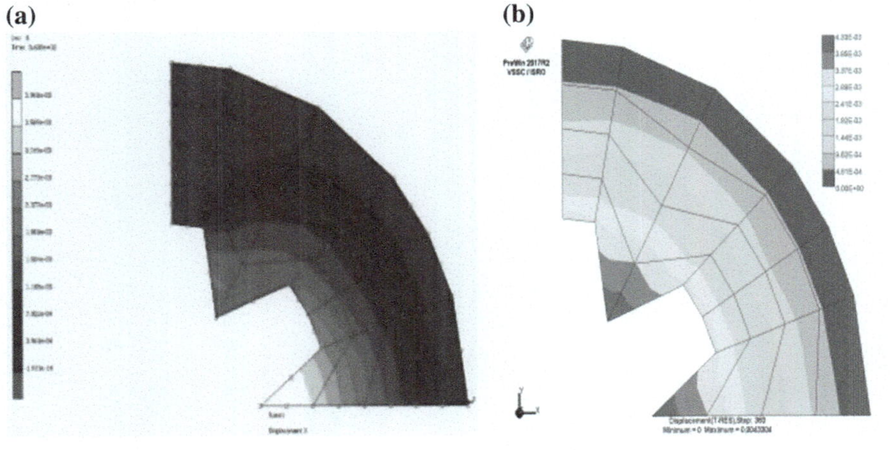

Fig. 20 Displacement contour at t = 360 in **a** MARC and **b** FEAST$^{[SMT]}$

References

1. Fried I (1974) Finite element analysis of incompressible material by residual energy balancing. Int J Sci Eng Res 10(9):993–1002
2. Hermann LR (1965) Elasticity equations for incompressible and nearly incompressible materials by a variational theorem. AIAA J 3:1896–1900
3. Lee EH (1955) Stress analysis in viscoelastic bodies. Department of Defense Contract, XIII(2)
4. Lee EH (1956) Stress analysis in viscoelastic materials. J Appl Phys 27

5. Morland LW, Lee EH (1960) Stress analysis for linear viscoelastic materials with temperature variation. J Rheol IV:233–260
6. Schapary RA (1961) Two simple approximate methods of laplace transform inversion for viscoelastic stress analysis. California Institute Technical Report, SM 61-23
7. White JL (1968) Finite elements in linear viscoelasticity. Int J Sci Eng Res 8
8. Pian THH, Tong P (1968) Basis of finite element methods for solid continua. Int J Numer Methods Eng 1(1):3–28
9. Naylor DJ (1974) stresses in nearly incompressible materials by finite elements with application to the calculation of excess pore pressures. Int J Numer Methods Eng 8:443–460
10. Yadagiri S, Pappi Reddy C (1985) Viscoelastic analysis of nearly incompressible solids. Comput Struct 20(5):817–882
11. Moore CJ (1988) SRM propellant and polymer materials structural test program. National Aeronautics and Space Administration, Scientific and Technical Information Division
12. Lajczok MR (1995) Effective propellant modulus approach for solid rocket motor ignition structural analysis. Comput Struct 56(I):101–105
13. Chen T-M (1995) The hybrid laplace transform/finite element dynamic analysis of viscoelastic Timoshenko beams method applied to the quasi-static. Int J Numer Methods Eng 38:509–522
14. Chyuan S-W (2002) Nonlinear thermoviscoelastic analysis of solid propellant grains subjected to temperature loading. Finite Elem Anal Des 38:613–680 (Elsevier)
15. Renganathan K, Rao BN Jana MK (2006) Slump estimation of cylindrical segment grains of a typical rocket motor under vertical storage condition. ISSN 97–104
16. Kanakaraju K, Rao BN, Marimuthu R (2007) Hybrid stress-displacement finite elements for viscoelastic analysis. World Scientific and Engineering Academy And Society

Numerical Study on Lateral Deflection and Flexural Capacity of RC Shear Wall With and Without Enlarged Boundary Element

Abhishek Kumar, Soumi Rajbanshi and Kaustubh Dasgupta

Abstract Reinforced Concrete (RC) special shear wall is widely used in the lateral load resisting system of multistoried buildings due to its large in-plane lateral strength and lateral stiffness. In the present study, the influence of enlarged boundary elements on the possible seismic behaviour of such shear wall is studied. Detailed finite element modelling of a slender RC shear wall is carried out with enlarged boundary elements. 8-noded solid elements and 2-noded truss elements are used for modelling the concrete and the reinforcement parts, respectively. Material nonlinearity is considered by assigning damaged plasticity model properties to concrete and bilinear strain-hardening constitutive model to reinforcement, respectively. For obtaining the effectiveness of the enlarged boundary elements, another shear wall with rectangular cross-section is modelled considering the total plan area to be the same for both the wall sections. Nonlinear static analysis is carried out for both the models under varying axial compression. Based on the lateral load-deformation response, strain distribution profiles and damage characteristics, the wall section with enlarged boundary elements is observed to have better peak flexural capacity and the lateral drift level at the instant of peak flexural capacity as compared to the rectangular wall.

Keywords Shear wall · Boundary element · Flexural capacity · Pushover analysis

1 Introduction

Shear walls are widely used in tall buildings to resist lateral loads imposed by wind or earthquakes by virtue of large in-plane lateral strength and lateral stiffness. Severe

---

A. Kumar (✉) · S. Rajbanshi · K. Dasgupta
Department of Civil Engineering, IIT Guwahati, Guwahati, India
e-mail: abhk311@gmail.com

S. Rajbanshi
e-mail: soumi7492@iitg.ac.in

K. Dasgupta
e-mail: kd@iitg.ac.in

K. Dasgupta et al. (eds.), *Proceedings of SECON'19*,
Lecture Notes in Civil Engineering 46,
https://doi.org/10.1007/978-3-030-26365-2_45

earthquake shaking leads to large bending moment and axial load demands at different sections along the height of a shear wall. Also, for rectangular shear wall sections, the vertical and horizontal steel reinforcement in the boundary part of the wall need to be designed for imparting the required ductility capacity. Conventionally, the walls are placed between building columns and the confinement reinforcement of the columns or enlarged boundary elements is used to provide the required ductility capacity [1, 2]. Although different studies have been carried out in the past investigating the behavior of shear walls, the effectiveness of enlarged boundary element for enhancing the lateral strength and ductility capacity of a rectangular shear wall section, has not been investigated. The objectives of this study are to observe the actual flexural capacity and lateral drift characteristics of the rectangular shaped wall with and without enlarged boundary element.

2 Modelling Methodology

In the present study, nonlinear static analysis has been carried out for shear wall with and without enlarged boundary element, using the ABAQUS program [3]. The concrete part is modelled using eight-noded linear hexahedral solid elements with reduced integration (C3D8R). Two noded linear truss element (T3D2) is used to model the steel reinforcement. Reinforcement is embedded in concrete and provided with embedded region constraint for proper interaction. Tie constraint is given to connect the different parts of the shear wall. Material nonlinearity is modelled using Concrete Damaged Plasticity (CDP) [4] model for concrete and strain hardening characteristic is used in the stress-strain curve for steel reinforcement. Force-controlled pushover analysis is carried out for obtaining the required response of the shear wall.

3 Model Validation

For validation of the analysis procedure, an RC wall from a past study is considered [5]. The cross sectional dimensions of the 2 m high rectangular wall are 1000 mm in length and 100 mm in thickness, respectively. The grades of concrete and rebar used are C80 and HRB400 respectively. The other parameters adopted for the model include an axial load of 1004 kN and design ductility demand of 3.

In the linear range of force-displacement curve, both numerical and experimental results are observed to be in close proximity (Fig. 1). Although the same trend is obtained for the numerical and experimental response in the nonlinear range, lateral shear capacity in numerical model is observed to be marginally higher as compared to the experimental model. That could be due to the unavailability of the actual material characteristics obtained during the experiment. Thus, the mentioned analysis

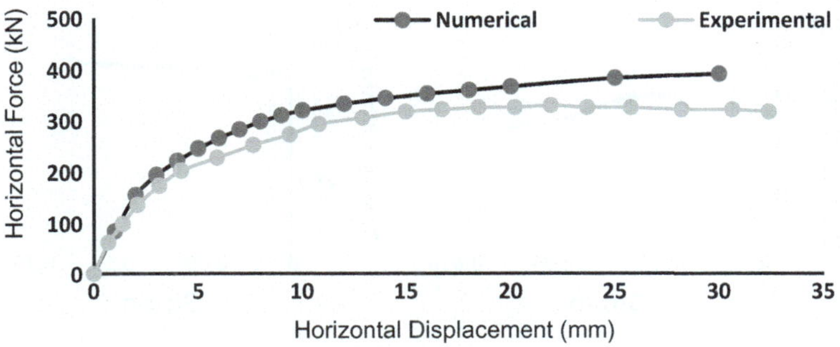

Fig. 1 Variation of lateral load with lateral deflection of shear wall for numerical and experimental investigations

Table 1 Lateral loads and displacements for validation

Method	Cracking		Yielding		Ultimate	
	Load (kN)	Δ (mm)	Load (kN)	Δ (mm)	Load (kN)	Δ (mm)
Experimental	160	3.1	264	8.5	326.3	20.4
Numerical	194	3.0	298	8.0	367.0	20.0

procedure can be adopted for further analysis of the wall models. The values of lateral load and lateral displacement (Δ) at a few salient points are shown in Table 1.

4 Comparative Analysis

The previously mentioned modelling and analysis are used for investigating the comparative behaviour of rectangular shear wall with and without enlarged boundary element. The program ABAQUS offers two techniques to solve the problems, namely (a) ABAQUS/Standard and (b) ABAQUS/Explicit. Both the techniques solve the equilibrium equations using numerical step-by-step procedure at discrete intervals of time Δt apart. Although convergence issues arise for ABAQUS/Standard version due to the selected small increments of time in analysis involving material nonlinearity, the same version is used in the present study. A realistic stress-strain curve for unconfined concrete of grade M30, is used for assigning material properties in the Concrete Damaged Plasticity (CDP) model (Fig. 2a). The stress-strain curve used for Fe500 grade of reinforcement is of the same nature as assigned in the validation study earlier (Fig. 2b). A mesh size of 50 mm is considered for both the shear wall models. All the translational and rotational Degrees of Freedom (DOFs) are restrained at the base of the wall. Under a constant level of external vertical load applied at the top of the wall, force-controlled lateral pushover analysis is carried out for both the models. For the adopted CDP model, the dilation angle, eccentricity, viscosity

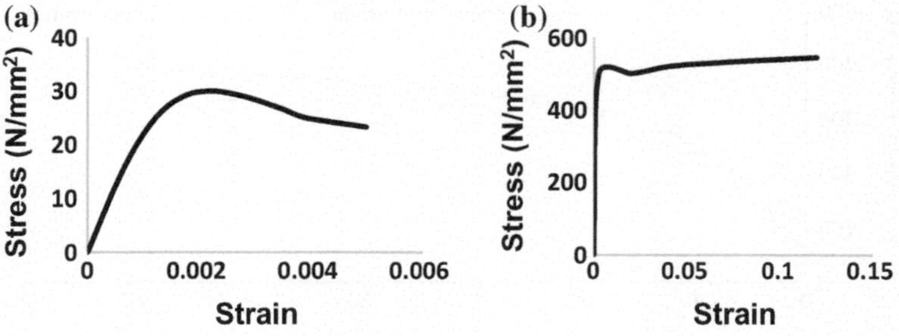

Fig. 2 Stress-strain curves for modelling of shear wall: **a** M30 concrete and **b** Fe 500 grade of reinforcement

parameter, shape factor (K_c) and the stress ratio (σ_{b0}/σ_{c0}) are assumed as 55°, 0.1, 0.01, 0.667 and 1.16 respectively [6].

The dimensions of shear wall with enlarged boundary element are decided in accordance with the minimum requirements as per the Indian Ductile Detailing Code IS 13920: 2016 [7]. The total length and the height of wall are 1.6 m and 3.1 m respectively for both the models. Vertical reinforcement ratios in boundary element and the web are considered as 0.8% and 0.25% respectively. Horizontal reinforcement ratio is taken as 0.25% and closed ties of 8 mm diameter with center to center spacing of 100 mm are provided in the boundary elements. To compare the effectiveness of enlarged boundary element, a rectangular shear wall is considered such that the overall cross-sectional area, cross-sectional area of boundary element and vertical reinforcement are the same as those for the wall model with enlarged boundary elements (Fig. 3).

5 Results and Discussion

Both the shear wall models, with and without enlarged boundary elements, are subjected to varying axial compression, namely 0.2, 10, 20, 30 and 50% of the design axial load carrying capacity (Fig. 4).

Ultimate lateral capacity is considered to be attained at the instant of yielding of longitudinal reinforcement. It is observed that the presence of enlarged boundary element increases the flexural capacity of the shear wall. The observed drift level of rectangular wall is higher as compared to the observed drift level for the wall with enlarged boundary elements, which implies that the presence of enlarged boundary element increases the lateral stiffness of the structural wall. Also for increasing axial load, the lateral shear capacity is observed to be increasing for both models. At low axial load, tensile strain is higher for walls without enlarged boundary element but at high axial load, both tensile and compressive strains are higher for wall without enlarged boundary element (Fig. 5).

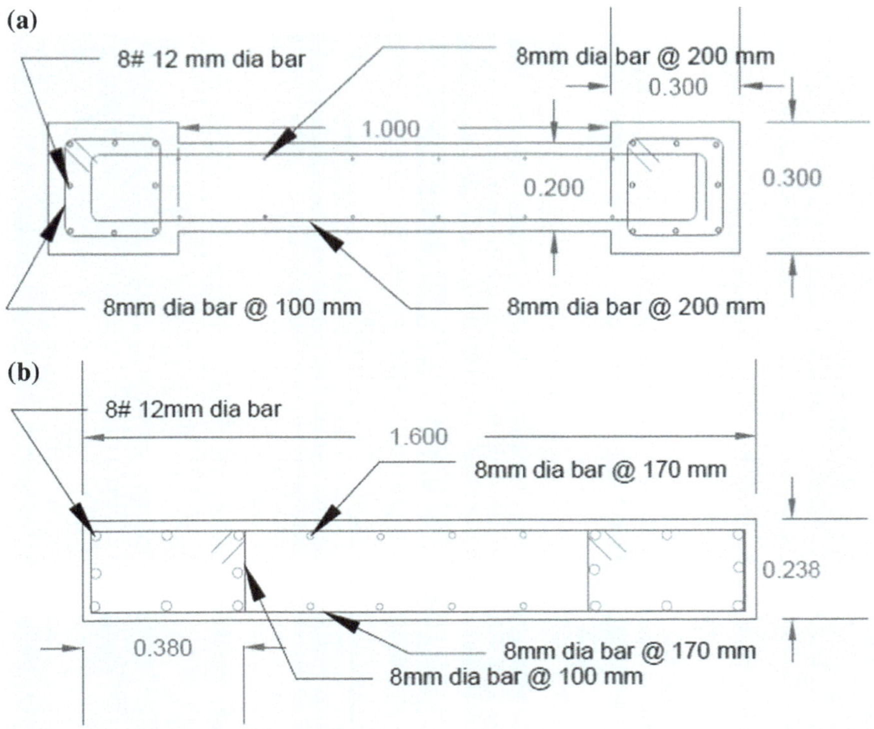

Fig. 3 Cross-section and reinforcement details of shear wall section **a** with enlarged boundary elements and **b** without enlarged boundary elements (all dimensions in m unless mentioned otherwise)

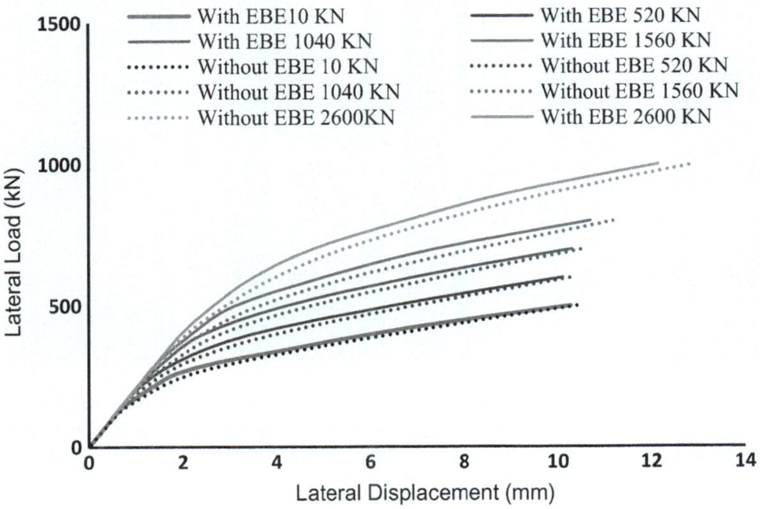

Fig. 4 Variation of lateral shear capacity with lateral deflection for the wall models at various vertical load levels

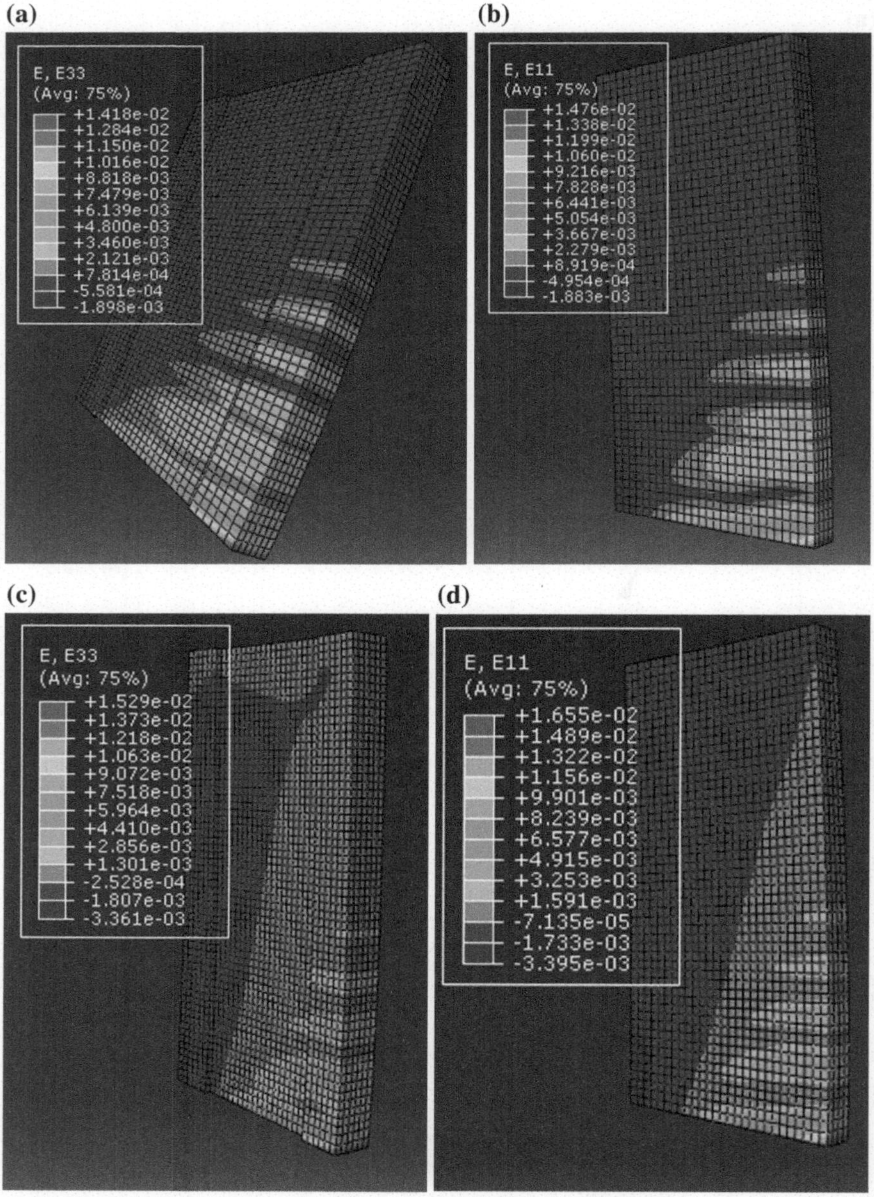

Fig. 5 Strain profile at 0.2% axial load ratio for wall **a** with enlarged boundary element and **b** without enlarged boundary element; strain profile at 50% axial load ratio for wall, **c** with enlarged boundary element and **d** without enlarged boundary element

6 Conclusions

From the present study, it is concluded that the presence of enlarged boundary element effectively enhances the wall performance by increasing its ultimate lateral load capacity and reducing damage level in the wall panel. The finite element model predicted damage pattern quite well, and hence predicted the ability of boundary elements in reducing damage level and crack distribution. This is attributed to the fact that boundary elements carry a large amount of axial force which reduces axial stress level in wall panels.

References

1. Taleb R, Kono S, Tani M, Sakashita M (2014) Effects of end regions confinement on seismic performance of RC cantilever walls. In: Proceedings of the 10th US National Conference on Earthquake Engineering, Anchorage, pp 21–25
2. Taleb R, Watanabe H, Kono S (2018) Numerical study on the ultimate deformation of RC structural walls with confined boundary regions. Periodica Polytech Civ Eng 62(1):191–199
3. Hibbit K (2010) Sorensen Inc.: ABAQUS/Standard user's manual (Version 6.11-3), Pawtucket, RI
4. Lubliner J, Oliver J, Oller S, Onate E (1989) A plastic-damage model for concrete. Int J Solids Struct 25(3):299–326
5. Deng MK, Liang XW, Yang K (2008) Experimental study on seismic behavior of high performance concrete shear wall with new strategy of transverse confining stirrups. In: Proceeding of the 14th World Conference on Earthquake Engineering, Xi'an University of Architecture & Technology, China, pp 1–8
6. Gulec CK, Whittaker AS (2009) Performance-based assessment and design of squat reinforced concrete shear walls. MCEER Technical Report-09-0010, MCEER, Buffalo
7. Bureau of Indian Standards (BIS) (2016) Indian standard code of practice for ductile detailing of reinforced concrete structures subjected to seismic forces, IS 13920:2016, Bureau of Indian Standards, New Delhi

A Comparative Study of Axial Force—Bending Moment Interaction Curve for Reinforced Concrete Slender Shear Wall With Enlarged Boundary Element

Soumi Rajbanshi, Abhishek Kumar and Kaustubh Dasgupta

Abstract Reinforced Concrete (RC) special shear walls are provided in the lateral load resisting system of multistoried buildings to improve the overall seismic performance of those buildings. For any RC wall section, the actual flexural capacity (M) is significantly influenced by the level of axial force (P), and the same is represented by the P-M interaction diagram. In the present study, finite element modelling is carried out for an isolated slender wall with enlarged boundary elements at its ends. 8-noded solid elements and 2-noded truss elements are used for modelling the concrete and the reinforcement parts, respectively. Material nonlinearity is considered by assigning damaged plasticity model properties to concrete and bilinear strain-hardening constitutive model to reinforcement, respectively. The effect of confinement in concrete is considered. The model is subjected to pushover analysis under varying levels of axial compression. The strain distribution profiles are noted to identify the four possible failure modes namely, (a) cracking of cover concrete, (b) spalling of cover concrete, (c) crushing of core concrete and (d) yielding of vertical reinforcement in boundary elements. The flexural capacities, observed for different failure levels, are compared with those obtained from the actual P-M interaction diagrams.

Keywords Shear wall capacity · Interaction diagram · Confinement effects

1 Introduction

In recent times, most of the multistoried buildings are provided with Reinforced Concrete (RC) shear wall as key elements in the lateral load resisting system. The effectiveness of RC shear walls in enhancing the stiffness and reducing the interstorey

S. Rajbanshi (✉) · A. Kumar · K. Dasgupta
Department of Civil Engineering, Indian Institute of Technology, Guwahati, India
e-mail: soumi7492@iitg.ac.in

A. Kumar
e-mail: kumar174104081@iitg.ac.in

K. Dasgupta
e-mail: kd@iitg.ac.in

K. Dasgupta et al. (eds.), *Proceedings of SECON'19*,
Lecture Notes in Civil Engineering 46,
https://doi.org/10.1007/978-3-030-26365-2_46

drift, i.e., improving the overall seismic performance of the building, is well studied. During earthquake shaking, the variation of axial force demand influences the actual flexural capacity of an RC shear wall section significantly. While the flexural design of a shear wall section is formally prescribed in the various codal provisions, the guidelines for estimation of actual flexural capacity are currently absent.

Among the past studies [1], axial force (*P*)—bending moment (*M*) diagram has been developed for rectangular RC shear walls, consistent with the earlier version of Indian Concrete Code IS 456:1978 [2]. For deriving the closed-form expressions, a uniform strip of steel was assumed as vertical reinforcement instead of the discrete bars. While another study [3] proposed modifications in strain limit states to make the design philosophy consistent with the current Indian Concrete Code IS 456:2000 [4], actual stress-strain characteristics of the concrete and steel reinforcement were considered later on for developing the actual *P-M* interaction curve [5]. It was observed that the curve gave a conservative estimate of the actual flexural capacity. The aim of the present study is to relate the actual *P-M* interaction curves to the corresponding strain profiles signifying the four possible principal failure modes namely, (a) cracking of cover concrete, (b) spalling of cover concrete, (c) crushing of core concrete and (d) yielding of vertical reinforcement in boundary elements.

2 Input Details and Modelling

In the present study, a slender RC shear wall with enlarged boundary elements (Fig. 1) is considered. The overall dimensions of the wall are assumed considering the guidelines provided in IS 13920:2016 [6]. To maintain the overall height to length ratio required for the wall to be classified as slender wall, the height of the wall is considered as 3.1 m. Since the horizontal reinforcement is considered in two layers, the thickness of the wall is kept as 200 mm. The vertical and horizontal steel reinforcements strictly conform to the IS 13920: 2016 [6], while the detailing for the enlarged boundary elements is in accordance with the special confining reinforcement clause as mentioned in the same document. All the degrees of freedom at the base of the

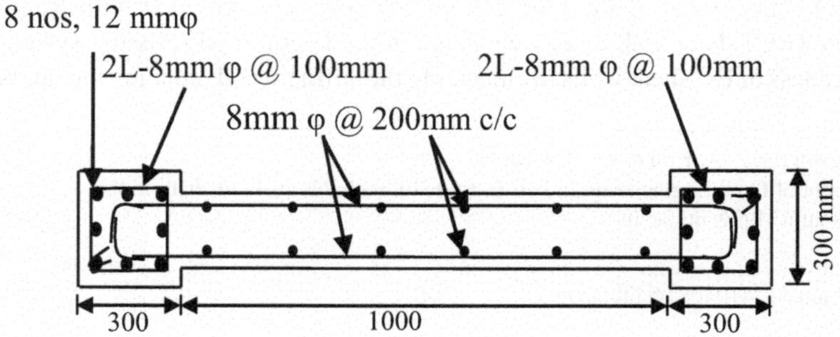

Fig. 1 Reinforcement details of the shear wall section

wall (including web and boundary element) are completely restrained. The grade of concrete and reinforcement are assumed to be M30 and Fe500 respectively.

For numerical simulation of nonlinear static behaviour of an RC shear wall with enlarged boundary element, the modelling is carried out using the program ABAQUS [7]. The concrete portion of the wall is modeled using 8-noded solid element with reduced integration (C3D8R), whereas, 2-noded truss element is used for modeling the steel reinforcement. Unconfined and confined concrete characteristics are considered for cover and core concrete respectively (Fig. 2a) along with the Concrete Damage Plasticity (CDP) model [8]. The constitutive characteristics of concrete are obtained considering Karthik-Mander model [9], while the salient parametric values for CDP model are adopted as per past study [10]. For reinforcement, bilinear stress-strain curve with strain hardening part is used with actual yield strength (Fig. 2b).

The model is subjected to force-controlled nonlinear static analysis under varying levels of axial loading. The control points assumed to identify the various points of

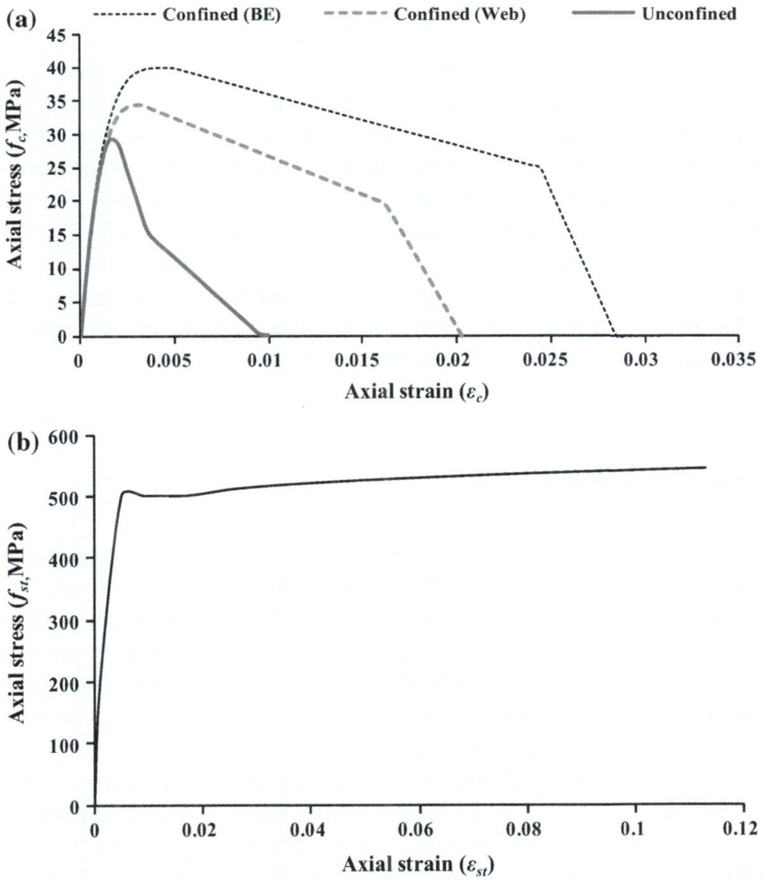

Fig. 2 Stress-strain curve for **a** unconfined and confined concrete, and for **b** steel reinforcement

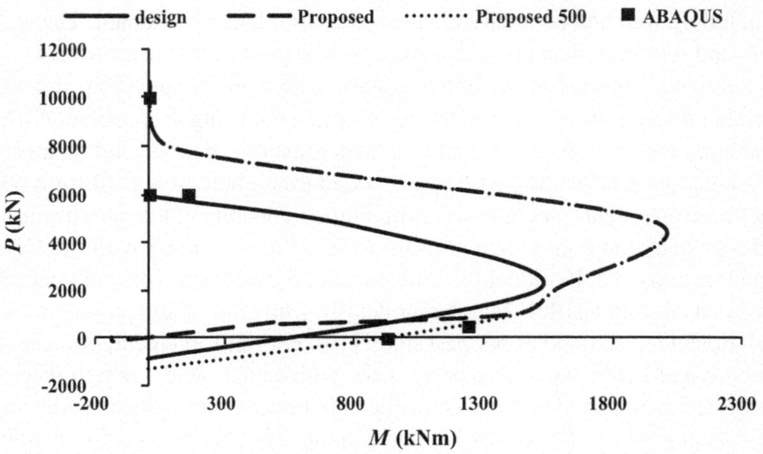

Fig. 3 Comparison of P-M interaction curve generated using various methods

P-M curve are obtained as: (a) the maximum strain at the extreme tension edge of cover concrete being numerically equal to the spalling strain (the ultimate strain of unconfined concrete or the strain at which the cover concrete starts to spall off); (b) the maximum tensile stress in the extreme layer of vertical reinforcement in boundary element being equal to the actual yield strength of steel, (c) maximum compressive strain in confined concrete being equal to the failure strain, and (d) the maximum tensile stress in the extreme layer of vertical reinforcement in boundary element being equal to the ultimate tensile stress. The obtained points are compared with the design as well as the actual *P-M* characteristics of the section (Fig. 3) generated by adopting the methodology proposed in literature [5].

3 Results and Discussion

From Fig. 3, it is observed that for regions with higher axial compression on the section, the obtained points are located closer to the design curve as compared to their distances from the actual curve. Similarly, for lower axial compression, the obtained points are closer to the actual *P-M* curve. These observations can be attributed to the fact that the lower portion of the *P-M* curve is characterised by under-reinforced behavior, hence the actual material properties particularly those of reinforcement, tend to influence the failure mode of the wall.

Since, the applied axial load on the wall section significantly influences its flexural capacity, the axial load is varied to observe its influence on the strain profile of the section. Three levels of axial loads are considered, namely (i) 10 kN (low axial loading), (b) 1000 kN (medium axial loading) and 4000 kN (high axial loading).

The variation of longitudinal strain in concrete with the relative distance from the extreme tension edge (where the relative distance is the ratio of distance of the strain location from the tension edge to the total length of the cross section, is shown in Fig. 4. The strain profiles are obtained by keeping axial load constant and by varying the lateral load levels (F450, F600 etc.). The strain profile across the depth of the section tends to become steeper with increase in the lateral load (Fig. 4). However, it is noticed that as soon as the strain value on the tension side approaches spalling strain value, the location of the neutral axis almost remains the same with further increase in the strain value at the outermost edges. That violates the conventional assumption of keeping the maximum strain as constant and increasing the depth of neutral axis while generating *P-M* interaction curve. However, the location of NA is observed to shift towards the outermost fiber on compression side with increase in axial load. It can also be noted that when a section reaches its spalling strain the confined concrete of web as well as BE, both have already reached the peak strains, i.e., the section tends to mobilize its maximum capacity.

Further for a constant lateral load (450 kN), the section is subjected to varying axial load. It is observed from Fig. 5 that with increase in the axial load, the neutral axis progresses towards the outermost tension fiber, i.e., the section gradually behaves as an over-reinforced section with possibility of crushing failure of concrete. The observation is consistent with the fact that with increasing the axial load on a section, its behaviour changes to an over-reinforced one (also obtained by moving upwards along the *P-M* curve). However, for a constant level of lateral load, the obtained points lie towards the inner region bounded by the *P-M* curve.

4 Summary and Conclusions

In the present study, the influence of combined axial and lateral loadings on the non-linear static behavior of an RC shear wall section with enlarged boundary elements is investigated. The effects have been highlighted by the means of strain profiles at different loadings and correlating them to different points on the *P-M* interaction curve. A brief comparison with the actual and design *P-M* curve is also presented. It can be concluded that as the section is subjected to a constant axial load along with an increasing lateral load, the neutral axis progresses towards outermost compressive fiber, while the reverse trend is observed in case of a section subjected to an increasing axial load in addition to a constant lateral load. Also, when the actual and the design *P-M* characteristics are correlated, the points show proximity to both the curves individually for the different load combinations.

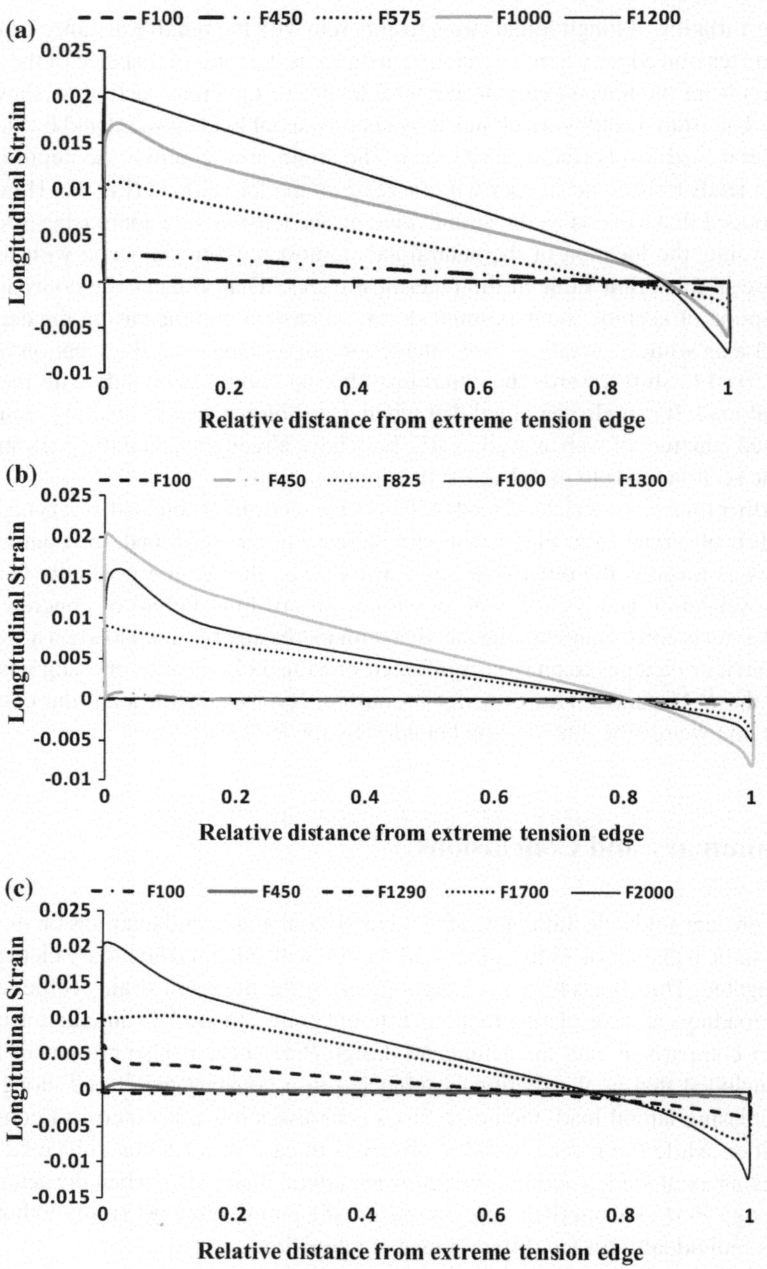

Fig. 4 Strain profiles at the bottom section of shear wall under varying lateral load and constant axial loads of **a** 10 kN **b** 1000 kN **c** 4000 kN

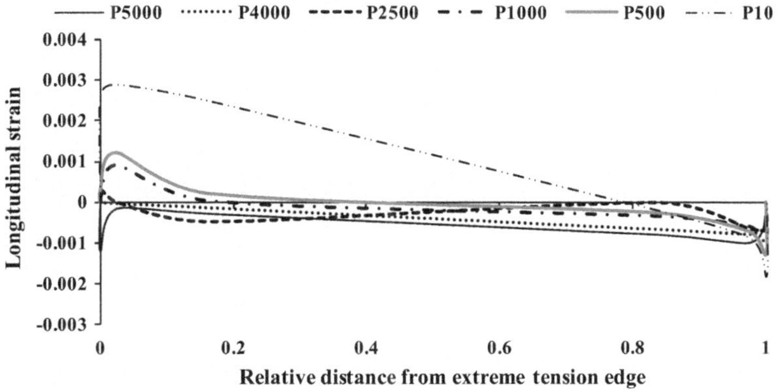

Fig. 5 Strain profile of bottom section under varying axial load and constant lateral load (450 kN)

References

1. Medhekar MS, Jain SK (1993) Seismic behaviour, design and detailing of RC shear walls, Part I: Behaviour and strength. Indian Concr J 67:311–318
2. BIS (1978) Plain and reinforced concrete—Code of Practice IS 456:1978. Bureau of Indian Standards, New Delhi
3. Dasgupta K, Murty CVR (2005) Seismic design of RC columns and wall sections: Part I-Consistent limit state design philosophy. Indian Concr J 79(3):33–42
4. BIS (2000) Plain and reinforced concrete—Code of practice IS 456:2000, Fourth Revision. Bureau of Indian Standards, New Delhi
5. Rajbanshi S, Dasgupta K (2018) Axial force—Bending moment interaction curve for reinforced concrete shear wall sections using actual material characteristics. In: SEC18: Proceedings of the 11th Structural Engineering Convention 2018, pp 132–137
6. BIS (2016) Ductile design and detailing of reinforced concrete structures subjected to seismic forces—Code of practice IS 13920:2016. Bureau of Indian Standards, New Delhi
7. Hibbit K (2010) Sorensen Inc.: ABAQUS/Standard user's manual (Version 6.11-3), Pawtucket, RI
8. Lubliner J, Oliver J, Oller S, Onate E (1989) A plastic-damage model for concrete. Int J Solids Struct 25(3):299–326
9. Karthik MM, Mander JB (2011) Stress-block parameters for unconfined and confined concrete based on a unified stress-strain model. J Struct Eng ASCE 137(2):270–273
10. Kaushik S, Dasgupta K (2016) Seismic damage in shear wall-slab junction in RC buildings. Procedia Eng 144:1332–1339. https://doi.org/10.1016/j.proeng.2016.05.162

Structural Optimisation of Helideck Structure Using FEM

Anitta Jose, Rajesh P. Nair, B. Sanoob and Jose Paul

Abstract Offshore structures and vessels have the helideck platforms for facilitating helicopter landing as well as take-off phases. The weight reduction for a specific well-known design without altering its main dimensions offers best possible alternative for the sake of economy. Reducing the structural weight of helideck is the best way possible alternative for the sake of economy. Optimising the structural weight is of paramount importance in marine applications. Hence, it becomes necessary to focus on the response of the structure using finite element analysis. From the FEA results the structure can be optimized so as to reduce the weight of the structure. The structure is analysed for normal and emergency landing of helicopter at and for two materials, Steel AH36 and Aluminium alloy of grade 5083-H116. Also, the structure is analysed for different scantling of secondary stiffener. The results are documented and compared so as to obtain an optimised design for the structure.

Keywords Helideck · Sikorsky S-92 · ABAQUS 6.16 · Steel AH36 · Aluminium alloy 5083-H116

1 Introduction

Helicopters occupy important niches in both military and civil aviation. These are platforms on which helicopters may take off and land and are essential structures installed in offshore plants and abroad ships. Helicopters have been the main method of transporting personnel to and from offshore installations for more than 60 years. Since there's always the probability of a helicopters emergency landing in a lifetime of such structures, evaluation of structural performance to achieve structural capacity becomes necessary.

A. Jose (✉) · R. P. Nair
Department of Ship Technology, CUSAT, Cochin, Kerala 682022, India
e-mail: anittatreesajose@gmail.com

B. Sanoob · J. Paul
Cochin Shipyard Ltd., Perumanoor P.O., Kochi, Kerala 682015, India

K. Dasgupta et al. (eds.), *Proceedings of SECON'19*,
Lecture Notes in Civil Engineering 46,
https://doi.org/10.1007/978-3-030-26365-2_47

Helidecks are of different variety based on various factors such as the material used for construction. The structure is usually made of steel although there is a very small number of wooden helidecks on old marine structures in the Arctic Ocean. Steel AH36 belongs to high strength category and is used primarily in shipbuilding or repair and other marine applications. The high strength characteristics service the structural requirements of ships, barges, and other types of marine equipment. The properties of grade AH36 high strength structural shipbuilding steel is given in ASTM A131: Part 4 specification. Helideck structures are generally constructed by using steel material. However, aluminium is widely used for some reasons such as weight saving, no-maintenance, anti-corrosion, convenience of assembly, longest life time and non-sparking. Aluminium 5083 is known for exceptional performance in extreme environments. It has the highest strength of the non-heat treatable alloys.

Helidecks can found in two ways in offshore structure as either mounted on deck or as cantilevered from deck, depending on the purpose and availability of space. Cantilever helidecks are usually supported by means of trusses whereas in the case of mounted helidecks pillars form the main substructure. The helideck consist of deck plate, stiffeners and girders mounted over substructure, the substructure consist of bulkheads and pillars. Doo-Hwan et al. [1] conducted finite element analysis of a steel helideck mounted on the upper deck of a ship (shuttle tanker) considering the emergency landing of the Sikorsky S-92 helicopter. The superstructure and substructure were designed, and the influence of various design parameters was analyzed on the basis of the FEA results. Mohammad et al. [2] investigated the structural response of an offshore helideck using the nonlinear static pushover analysis under vertical and lateral loads caused by helicopters emergency landing.

The entire structure is mounted over an oil recovery vessel. The dimension of deck plate and type of stiffeners are determined based on design code DNV-GL. This paper focuses on the response of this structure using the dynamic analysis under normal and impact loads and the structure is optimized by varying material properties.

2 Methodology

The thickness of the deck plate is determined using Class rules DNVGL-OS-E401. In the present study, finite element analysis (FEA) is conducted with regard to a helideck mounted on the upper deck of an oil-recovery vessel considering the normal and emergency landing of the Sikorsky S-92 helicopter. The helideck structure is optimised by comparing FEA results for different models.

3 Scantling Calculation

The helideck is to be positioned on an oil recovery vessel having an overall length of 108 m, rule length of 104.76 m, greatest moulded breadth of 16 m and moulded depth

Table 1 Material properties

	Steel A36	Aluminium alloy 5083-H116
Young's modulus (GPa)	210	71
Poisson's ratio	0.3	0.33
Density (kg/m^3)	7810	2660

of 7.5 m. The structure is designed for the safe landing of Sikorsky S-92 helicopter with a maximum take off mass of 11.86 t. The design is based on DNVGL-OS-E401, Part 6, Chap. 5, Sect. 5. The thicknesses obtained from scantling calculation are 12 mm and 17 mm for Steel AH36 and Aluminium alloy of grade 5083-H116 respectively. The scantling of primary T-Stiffener is chosen to be $440 \times 8 + 120 \times 10$ which is spaced at 2400 mm. The section modulus for chosen T-stiffener section is 824 cm^3 which more than the required amount. The section modulus of L-stiffener is 871 cm^3. The additional scantling is provided to account for routing of pipes, cables, etc., through the deep girders. Providing deep section is advantageous in reducing structural vibration.

4 Material Properties

The properties of the material used are listed in Table 1.

5 Finite Element Modelling

Analysis of helideck structure involves tedious calculations and analysis with classical methods is difficult. The finite element method is well suited for the analysis. The plate along with supporting structure is modeled for Steel AH-36 and Aluminium alloy of grade 5083-H116 using ABAQUS 6.16.

The model consists of:

- Deck plate of dimension 22.2 m $\times$ 16 m
- Primary stiffeners: T $440 \times 8 + 120 \times 10$, spaced at 2400 mm
- Secondary stiffeners: L-Stiffener equivalent to HP 200×9, spaced at 600 mm
- Circular hollow pillars of 0.219 m diameter, 2.8 m height
- Bulk heads, 2.8 m height.

The assembly of structure showing internal arrangements is shown in Fig. 1. In the present study two cases of helicopter landings are considered, with the normal landing in the center and emergency landing at edge sections of the helideck as shown in Fig. 2. The helideck is generally installed by fixing the pillars on the upper

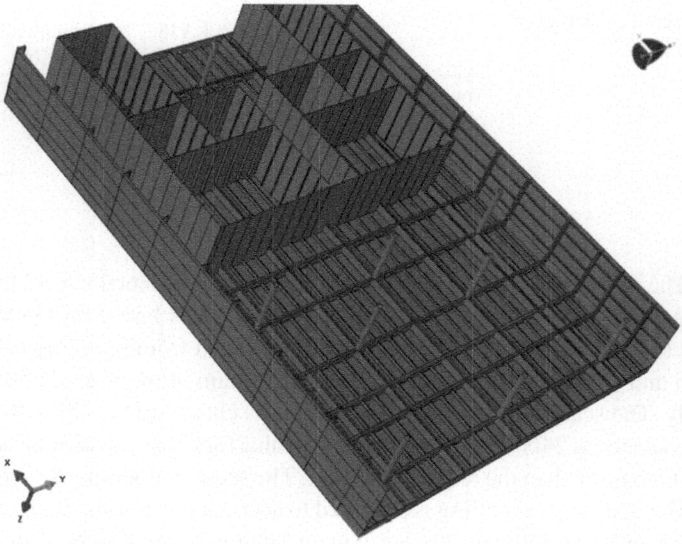

Fig. 1 Assembly of the Helideck structure (inverted view)

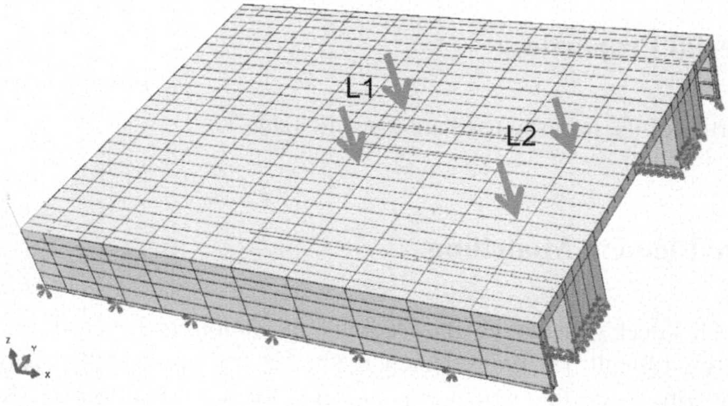

Fig. 2 Landing positions of helicopter

deck of a vessel. Therefore, the fixed boundary condition is applied to the bottom of the pillars and bulkheads under the deck structure to reflect the actual installation conditions. The envelope acceleration of the structure is assigned. The landing load of the helicopter is assigned as two point load. The structure is hexagonally meshed into 118,272 non-linear elements. The Von-Mises stress of the structure is determined. The structural responses of helideck for different materials are compared to obtain an optimised design.

6 Analysis of Helideck Structure

The analysis with varying parameters, materials and landing position are done to optimise the structural weight of the helideck. The model 1 with HP Steel AH-36 and equivalent L-stiffener for 200 × 9 is analysed for two landing cases. The model is analysed for one case of emergency landing, which is selected at the center of the deck plate. The emergency landing of helicopter is taken at one of the edge assuming the worst landing. For normal landing of helicopter gradually applied load is considered, whereas for emergency landing sudden applied is taken. The Steel AH-36 helideck with L-stiffeners equivalent to HP 200 × 9 and lower scantling HP 180 × 8 are analysed. Similarly Aluminium alloy 5083-H116 helideck with the optimized scantling of equivalent L-stiffeners is analysed. The Von-Mises stresses are tabulated and compared to optimise the structural weight.

7 Acceptance Criteria

The maximum Von-Mises stress obtained for each structural member is compared with the allowable stress under each set of landing conditions. The allowable stresses are determined in accordance with the design criteria of DNV-GL OS E401. The obtained stresses for the static plus dynamic (S + D) design load scenarios shall comply with the following criteria:

$$\sigma \leq C_s R_{eH}$$

where, σ is the Von-Mises stress obtained by FE analysis.

C_s is the coefficient for load, 0.9 for normal landing and 1 for emergency landing

R_{eH} is the yield stress of the material.

Yield stress for Steel A36 and Aluminium alloy 5083-H116 are 355 MPa and 228 MPa respectively. The maximum stress allowed for each material is listed in Table 2.

Table 2 Allowable stresses for structural member

Material used	Allowable stress	
	Normal landing (MPa)	Emergency landing (MPa)
Steel AH-36	319.5	355
Aluminium 5083-H116	205.2	228

8 Analysis Result of Model with Steel AH-36

The preliminary analysis was carried out on helideck plate with Steel AH-36. The model was analysed for two landing cases. The Von-Mises stresses in normal and emergency landing cases are noted and compared with allowable stress. The allowable stress for normal and emergency landing cases are computed by using acceptable criteria. The stress plot for normal and emergency landing of helicopter on Steel AH-36 with L-stiffener equivalent to HP 200 × 9 and HP 180 × 8 are shown Figs. 3 and 4 respectively . The maximum stresses are within acceptable limit. The structure with

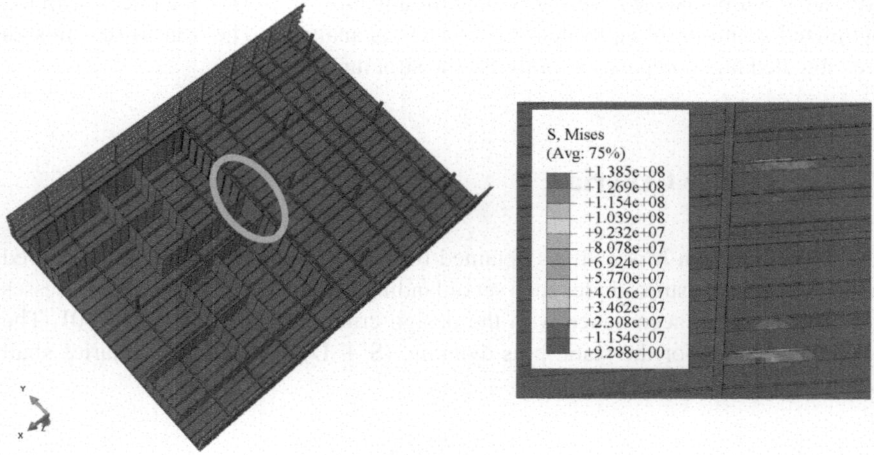

Fig. 3 Stress plot for steel AH-36 Helideck (normal landing)

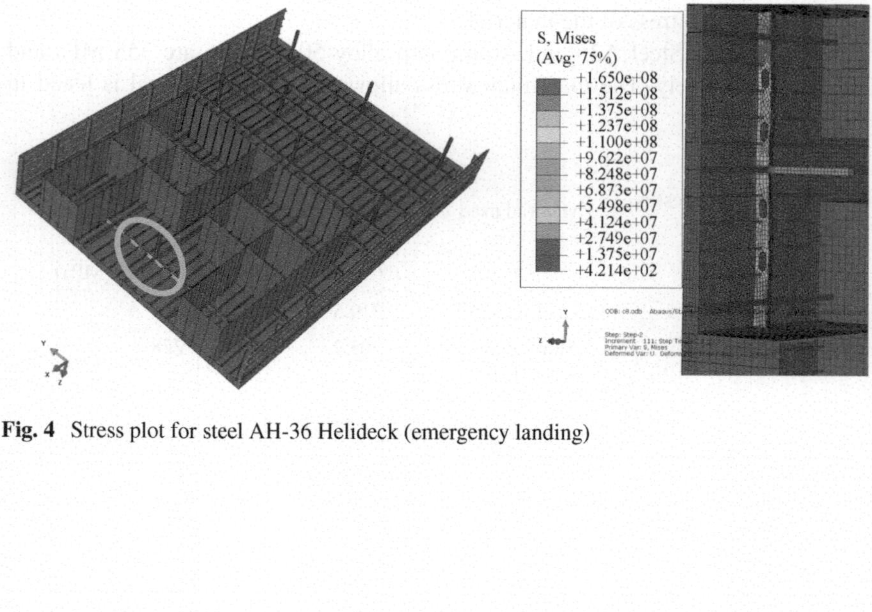

Fig. 4 Stress plot for steel AH-36 Helideck (emergency landing)

Table 3 Stress results of steel AH-36 Helideck

	Normal landing			Emergency landing		
		HP 200 × 9	HP 180 × 8	$\sigma_{allowable}$ (MPa)	HP 200 × 9	HP 180 × 8
	$\sigma_{allowable}$ (MPa)	σ_{max} (MPa)	σ_{max} (MPa)		σ_{max} (MPa)	σ_{max} (MPa)
L-stiffeners	319.5	138.5	152.1	355	41.24	28.55
T-stiffeners		34.62	59.69		165.0	171.3
Plate		57.70	63.36		41.24	28.55
Pillars		11.54	12.67		54.98	71.38
Bulkheads		34.62	38.02		27.49	14.28

L-stiffener equivalent to HP 180 × 8 is analysed with properties of Aluminium alloy 5083-H116. The maximum stress in this optimized model is within the acceptable limit (Table 3).

9 Analysis Result of Model with Aluminium Alloy 5083-H116

The optimised design obtained after preliminary analysis with Steel AH-36 is analysed with Aluminium alloy 508-H116 for emergency landing. The stresses obtained in this analysis are within acceptable limit of allowable stresses. Hence the helideck is safe for emergency landing of helicopter. The stress plot from FEA is shown in Fig. 5. Table 4 shows the stress result of Aluminium helideck.

10 Conclusions

The parameters that affect the design performance of the helideck based on FEA results are determined. The maximum stress of the helideck is dependent on the landing position, with the maximum stress occurring when the landing position was aligned at the edge of the deck plate.

The weight of Steel AH-36 helideck is 78 tonnes, The weight of Aluminium 5083-H116 helideck is 30 tonnes, Even though stress concentration is high for aluminium model, the weight reduction achieved is 60% compared with that of steel. The maximum structural optimisation is achieved while the structure is designed for Aluminium alloy 5083-H116. Steel AH-36 helideck structure can structurally optimised by substituting with Aluminium 5083-H116.

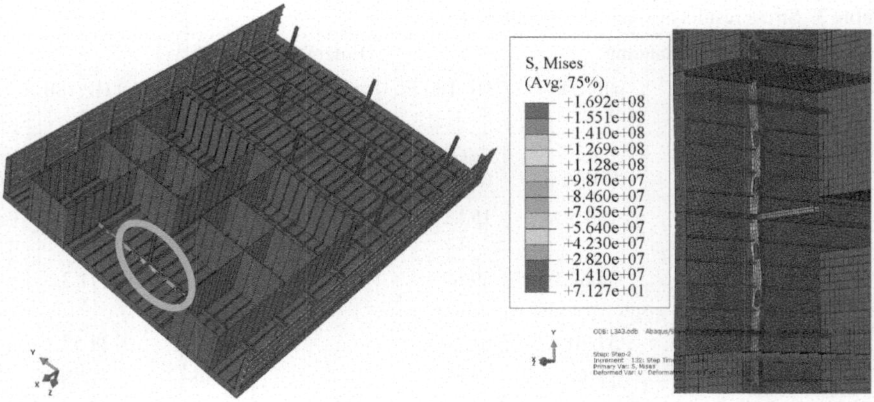

Fig. 5 Stress plot for aluminium Helideck (emergency landing)

Table 4 Stress results of aluminium alloy 5083-H116 Helideck

	Emergency landing	
	$\sigma_{allowable}$ (MPa)	σ_{max} (MPa)
L-stiffeners	228	28.20
T-stiffeners		169.20
Plate		28.20
Pillars		56.40
Bulkheads		14.10

References

1. Park DH, Kim JH (2016) Parametric study for suggestion of the design procedure for offshore plant helideck subjected to impact load. Struct Eng Mech 60(5)
2. Vaghefi M, Bagheri H (2013) Nonlinear analysis of offshore helidecks due to the helicopter emergency landing loads. Middle-East J Sci Res 13(10)

Finite Element Analysis of Hybrid Bonded FRP Strengthen System

T. A. Sreelakshmi and B. R. Beena

Abstract Importance of fiber reinforced polymer (FRP) strengthened technique is significant in the case of reinforced concrete slabs. Even though a wide variety of FRP strengthening technique is available, but premature debonding is the main failure model in ordinary bond technique, and the strengthening effect is limited. The present study concentrated on hybrid bonded FRP strengthen technique and it is a combination of externally adhered and mechanically fastened fiber reinforced polymer system. This study focused on finite element analysis by using ANSYS and compare the structural behavior of externally bonded (EB) and hybrid bonded (HB) FRP strengthening systems.

Keywords Fiber-reinforced polymer · De-bonding · Hybrid · Strengthening · FE model

1 Introduction

Application of fiber reinforced polymer (FRP) sheets to strengthen the existing reinforced concrete members have emerged as a better solution in past few years. Change in earthquake zones, corrosion in reinforcements, change in building type, loading conditions, and use of older versions of standard codes etc. may result to deficient design which leads to the deterioration of members. Strengthening of concrete with CFRP results in an increase in load capacity as well as an increase in stiffness. The material properties of the CFRP which shows good tensile strength, modulus of elasticity, and elastic strain are capable enough to enhance the capacity of RC slabs.

The external bonding of fiber-reinforced polymers (FRP) has become a current method for strengthening reinforced concrete (RC) structures. The performance of

T. A. Sreelakshmi (✉) · B. R. Beena
Department of Civil Engineering, Federal Institute of Science and Technology, Angamaly, India
e-mail: tasreelakshmi@gmail.com

B. R. Beena
e-mail: beenabr@gmail.com

K. Dasgupta et al. (eds.), *Proceedings of SECON'19*,
Lecture Notes in Civil Engineering 46,
https://doi.org/10.1007/978-3-030-26365-2_48

FRP-strengthened RC structures is often controlled by the bond strength of the interface between the FRP and the concrete. However, due to the early debonding failure, the strength utilization ratio of FRP is often only 15–35% of the material strength of the FRP sheet, depending on the cause of the delamination [1].

Another recently developed method of interfacial bonding is the mechanically fastened FRP (MF-FRP) technique [2]. The mechanical fastening of steel plates with anchor bolts is a traditional and effective method of attaching steel plates to concrete structures in which the tension resistance of the steel plate is transmitted into the concrete through the bolts and the bearing resistance of the holes in the steel plate. However, conventional commercially available FRP laminate—plates, strips, or sheets-does not have sufficient bearing strength for such fastening, and thus the bolts or mechanical fasteners easily cut through the laminate and cause it to split longitudinally. Mechanical fastening is, therefore, not applicable to conventional FRP laminate.

Hybrid-bonded FRP (HB-FRP) strengthening system has recently been developed at the City University of Hong Kong [3, 4]. It is a combination of both externally bonded and mechanically fastened FRP system. Experimental testing has demonstrated that this new HB-FRP technique can increase the interfacial bond strength several folds. In this study a finite element (FE) model for hybrid bonded FRP strengthened slab is developed and study the performance of structural behavior of the system.

2 Structural Model and Material Properties

Thirteen flexural members model were simulating in finite element analysis, including one reference specimen without any FRP strengthening and three others strengthened with the conventional EB-FRP technique, and remain specimens that were strengthened with the new HB-FRP technique. Details of the RC specimen are shown in Figs. 1 and 2 and the material properties of the specimens are given in Table 1. The mechanical fastener was made of two nails (4 mm in diameter and 33 mm in length) and a piece of mild steel cover plate (3 mm × 30 mm × 70 mm). The nominal thickness of one ply (layer) of FRP was 0.165 mm, and the width of the FRP strip was 50 mm. Hence, the area of FRP under one mechanical fastener is 30 mm × 50 mm. A relatively higher strength of concrete was used for the specimen to avoid

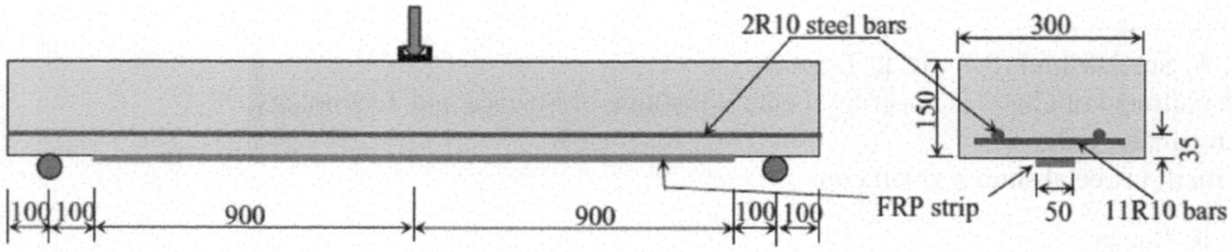

Fig. 1 Specimen details of EB FRP

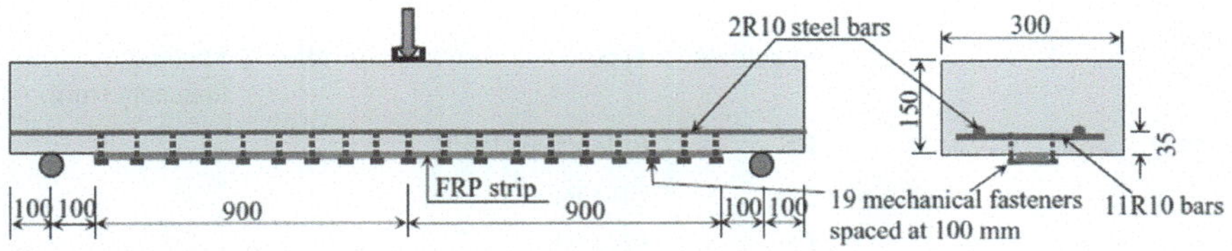

Fig. 2 Specimen details of HB FRP

Table 1 Material properties

Concrete	Compressive strength f_{ck} (MPa)	73.8
	Elastic modulus E_c (GPa)	36
	Poisson's ratio	0.2
	Tensile strength f_t (MPa)	4.2
Mild steel and plate	Yield strength f_y (MPa)	300
	Elastic modulus E_s (GPa)	200
	Poisson's ratio	0.3
Carbon fibre reinforced polymer	Elastic modulus E_{frp} (GPa)	227

concrete crushing when the flexural strength of the beam was significantly increased by the application of several plies of FRP in the tension zone [3, 4]. Parametric study has to be conducted, changing in number of layers of FRP plies and also changing the spacing of mechanical fasteners. Details of parametric studies are tabulated in Table 2.

3 Geometric Modelling

The two-dimensional FE analyses are conducted using the program ANSYS Workbench. ANSYS offers different types of elements as per the requirements. In this paper concrete, steel bars, and CFRP are modelled by SOLID 65, LINK 180, and SHELL 181 elements respectively. SOLID 65 is an eight nodded solid element with three translation degrees of freedom at each node i.e. x, y, and z directions. It is capable of cracking in three orthogonal directions, plastic deformations, crushing, and creep. LINK 180 is a two nodded line element with three translation degree of freedom at each node i.e., x, y, and z directions. This element is capable of plastic deformations. SHELL 181 is a four–node element with six degrees of freedom at each node. i.e., translations in the x, y and z directions and rotations about the x, y, and z axes. The finite element model of reference slab, EB FRP strengthened slab and HB FRP strengthened slabs are shown in Figs. 3, 4 and 5 respectively.

Table 2 Test specimen details

Specimen	Type	No. of plies	Spacing of fasteners (mm)
RS	Reference slab	–	–
EB_1	EB FRP	2	–
EB_2	EB FRP	4	–
EB_3	EB FRP	6	–
HB_1	HB FRP	2	100
HB_2	HB FRP	2	200
HB_3	HB FRP	2	300
HB_4	HB FRP	4	100
HB_5	HB FRP	4	200
HB_6	HB FRP	4	300
HB_7	HB FRP	6	100
HB_8	HB FRP	6	200
HB_9	HB FRP	6	300

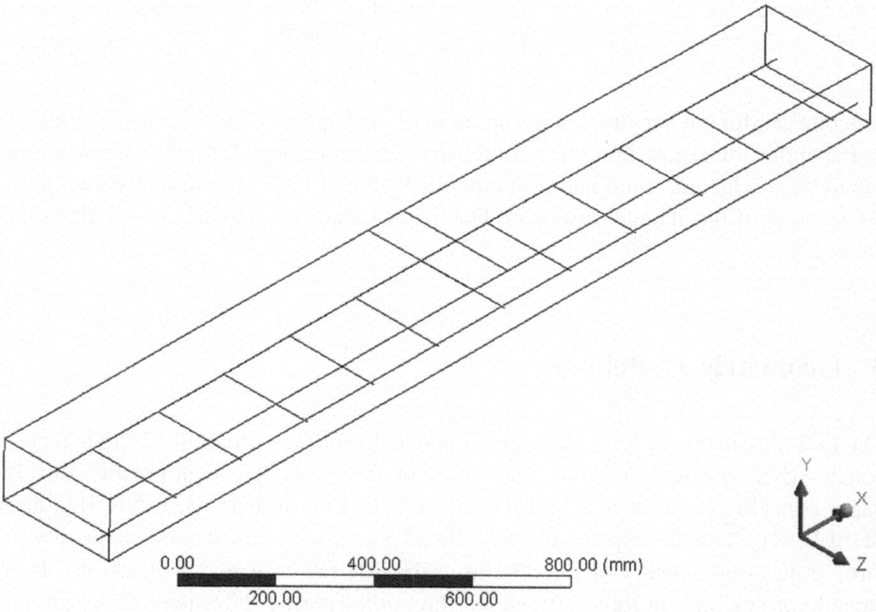

Fig. 3 Finite element model of reference slab

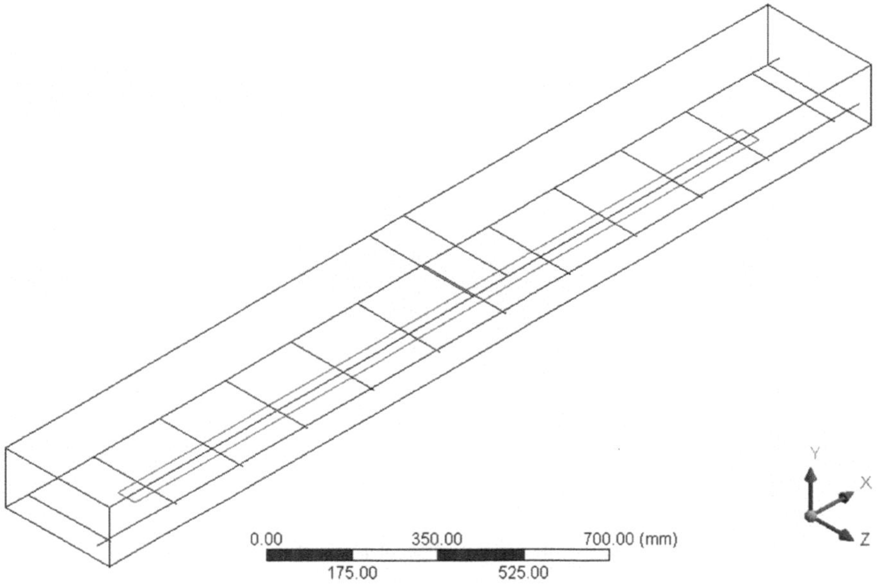

Fig. 4 Finite element model of EB FRP strengthened slab

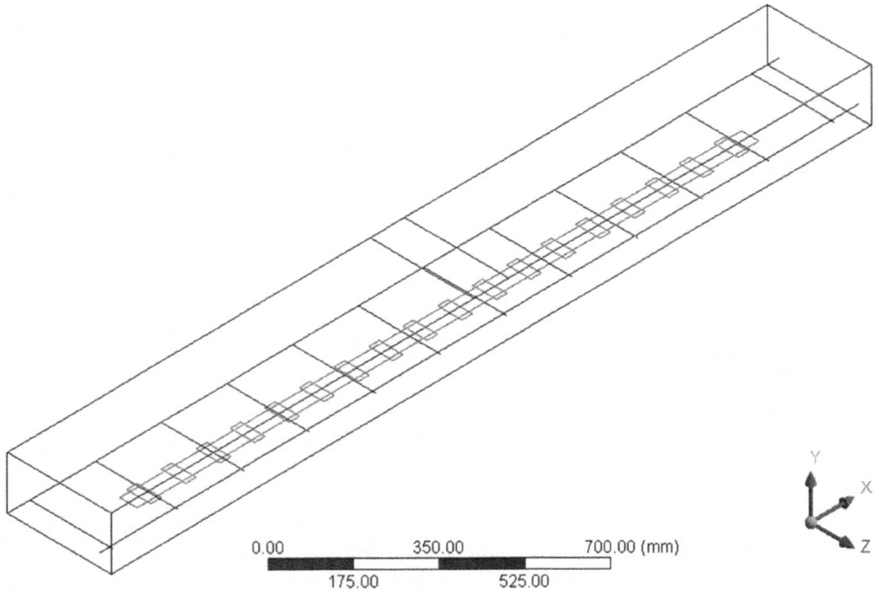

Fig. 5 Finite element model of HB FRP strengthened slab

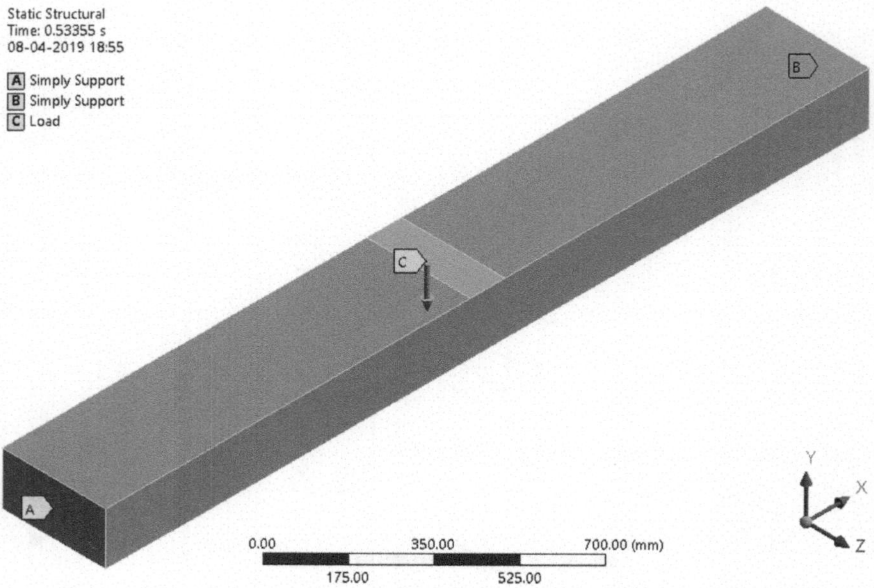

Fig. 6 Boundary conditions of specimen

4 Boundary Conditions

Boundary conditions provided for the specimen for this study as simply supported with roller support at one side and hinged support at the other side as shown in Fig. 6. Load was provided as displacement-controlled mode. A plate is provided at the loading location to distribute the load evenly. The element size has been taken as 50 mm. This size is achieved by checking the convergence criteria.

5 Results and Discussions

5.1 Load Versus Mid-span Deflection of the Specimen

The ultimate load and the corresponding deformations are tabulated in Table 3. Figure 7 shows the results of the FE analysis of the load versus the deflection at the mid-span of externally bonded FRP strengthened slab with varying number of layers of FRP. The graph gives the ultimate load carrying capacity of the EB FRP system increases up to four plies. But the load carrying capacity decrease in the case of RC slab with six layers of FRP adhered externally. The failure of the system indicates the insufficient bond strength between CFRP and the concrete.

Table 3 Test results

Specimen	Type	No. of plies	Spacing of fasteners (mm)	Ultimate load (kN)	Deformation (mm)
RS	Reference slab	–	–	10.538	0.5709
EB_1	EB FRP	2	–	18.208	0.3385
EB_2	EB FRP	4	–	28.785	3.4788
EB_3	EB FRP	6	–	28.139	8.4907
HB_1	HB FRP	2	100	31.45	1.9716
HB_2	HB FRP	2	200	28.139	1.6187
HB_3	HB FRP	2	300	28.139	1.4621
HB_4	HB FRP	4	100	41.382	7.5711
HB_5	HB FRP	4	200	36.416	3.9147
HB_6	HB FRP	4	300	35.864	3.2125
HB_7	HB FRP	6	100	67.866	11.760
HB_8	HB FRP	6	200	54.624	11.238
HB_9	HB FRP	6	300	56.624	11.090

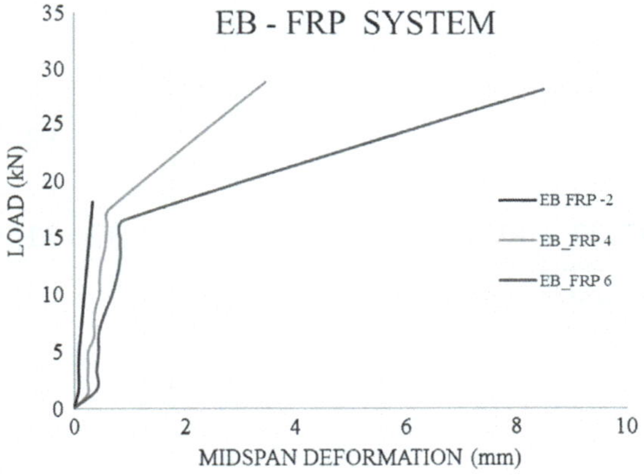

Fig. 7 Load–deflection response curve of EB FRP strengthened specimen

Figure 8 shows the load–deflection response spectrum of HB FRP strengthened RC specimen with 2 layers of plies. The spacing of mechanical fasteners such as 100, 200 and 300 mm are compared. The ultimate load carrying capacity of HB FRP strengthened with 2-layer plies and 100 mm spacing of mechanical fasteners shows the highest value compared to other two.

Figure 9 given that the ultimate load carrying capacity of HB FRP strengthened specimen with 4 layers of plies and spacing of mechanical fasteners changes similar to 2 layered system. The hybrid bonded with 100 mm spacing of mechanical fasteners gives the highest value of ultimate load capacity of 41.38 kN.

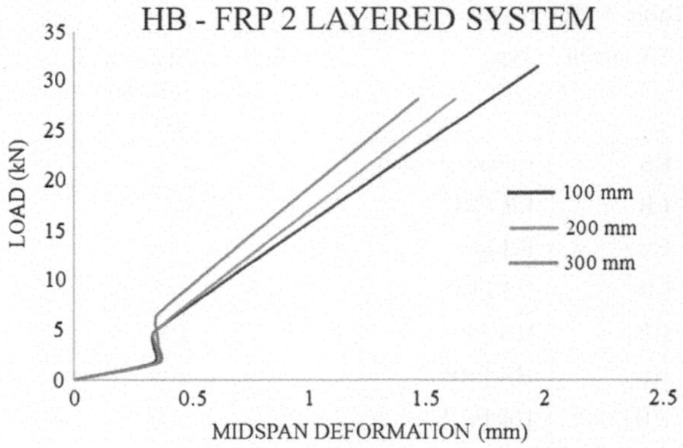

Fig. 8 Load—deflection response curve of HB FRP strengthened specimen with 2 layer of plies

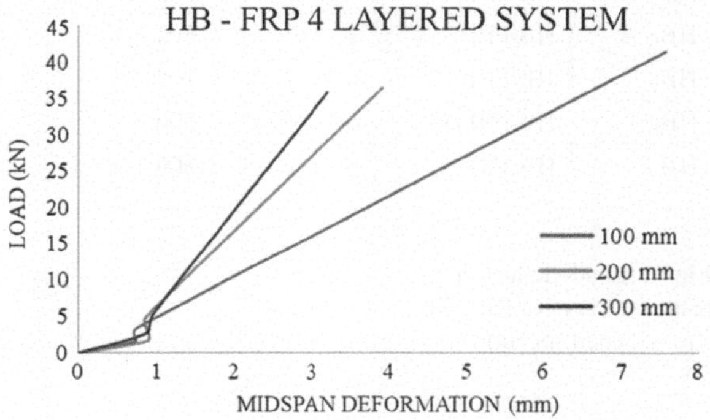

Fig. 9 Load—deflection response curve of HB FRP strengthened specimen with 4 layer of plies

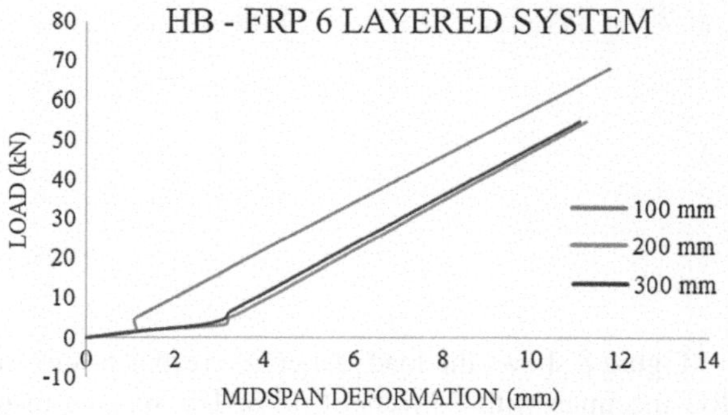

Fig. 10 Load—deflection response curve of HB FRP strengthened specimen with 6 layer of plies

Figure 10 shows that the ultimate loading capacity is increased in the case of hybrid bonded FRP strengthened specimen with 6 layers of plies and spacing of mechanical fasteners as 100 mm compare to other two. The ultimate strength of HB_7 reaches to 67.86 kN.

Fig. 11 Load—deflection response curve

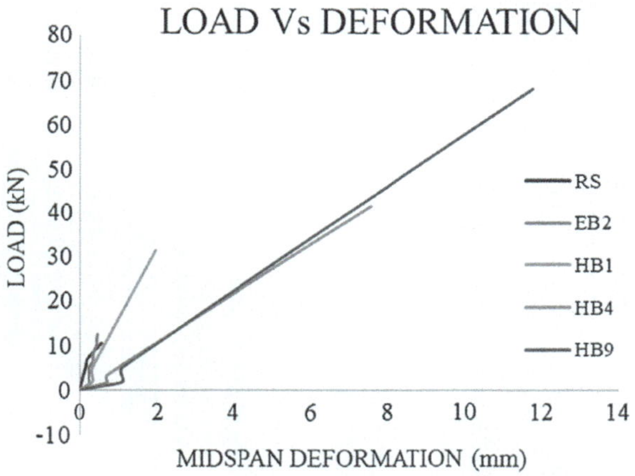

Figure 11 summarize the load deformation response spectrum of peak strength values from externally bonded with varying number of layers and hybrid bonded with varying number of layers and also varying spacing between mechanical fasteners. From which it can be seen that the HB-FRP system significantly increased the interfacial bonding strength of the members compared to the EB-FRP system. The specimen HB_7 had a flexural strength that was 273% greater than that of EB_1. This is clearly demonstrates the effectiveness of the hybrid FRP system.

The ultimate load of the RC specimen without FRP strengthening was calculated to be 10.5 kN. The EB-FRP system increased the strength from 10.5 to 28.78 kN, which effectively means that the EB-FRP system contributed a strength of 18.28 kN. The ultimate strength of the HB-FRP system was 67.86 kN. Taking away the 10.5 kN that was contributed by the original RC specimen without FRP strengthening, the additional strength that was due to the HB-FRP system was 57.36. This is 3.13 times (57.36/18.28) the bond strength that was contributed by the conventional EB-FRP system. Figures 12, 13, 14, 15 and 16 shows the deformation diagram of RC specimen without FRP strengthening, EB_2, HB_1, HB_4 and HB_7. Figure 11 shows the deformation diagram of reference slab, from which it can clearly said that the specimen failed with lower load carrying capacity and corresponding lower deformation but the specimen with CFRP strengthened specimen shows higher strength and deformation. The deformation diagram shows that the load is taken completely by CFRP layers and the RC specimen is completely safe.

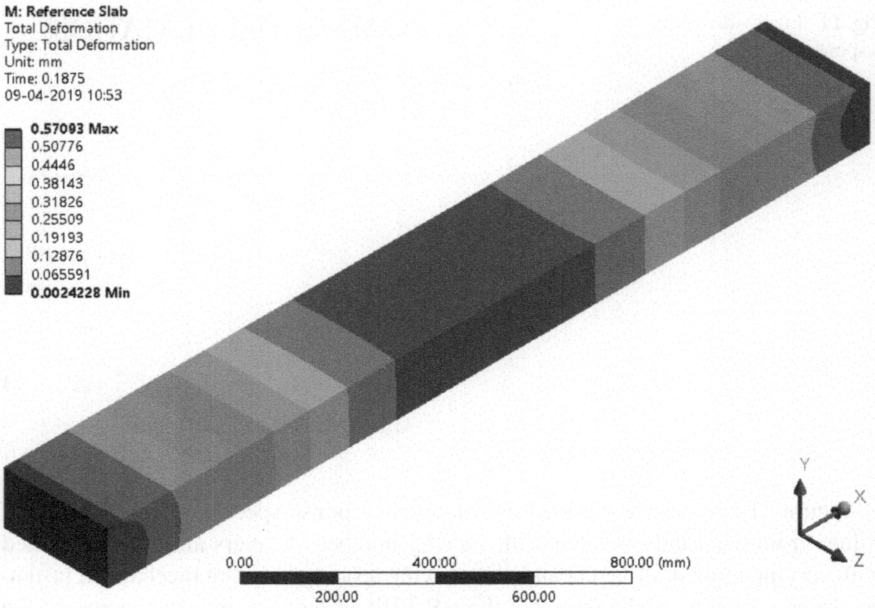

Fig. 12 Deformation diagram of reference slab

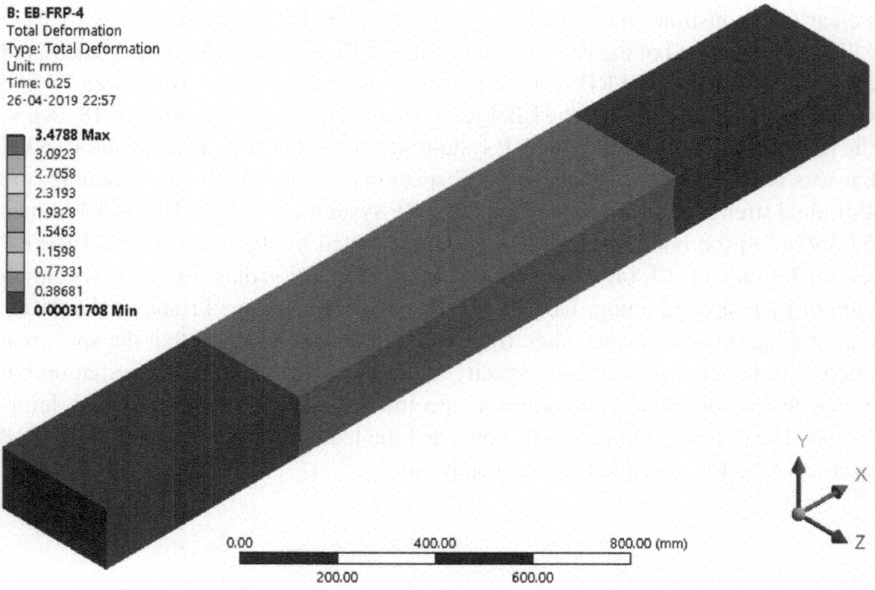

Fig. 13 Deformation diagram of EB_2

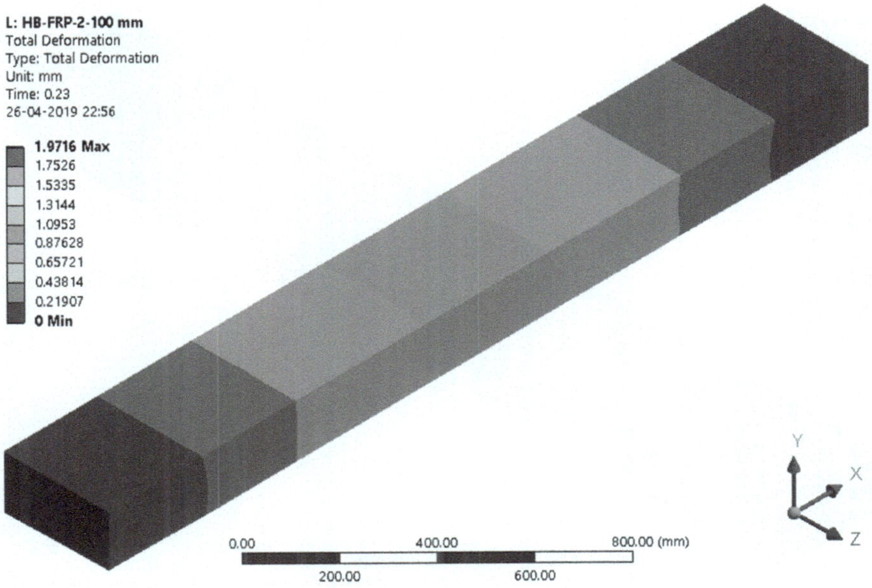

Fig. 14 Deformation diagram of HB_1

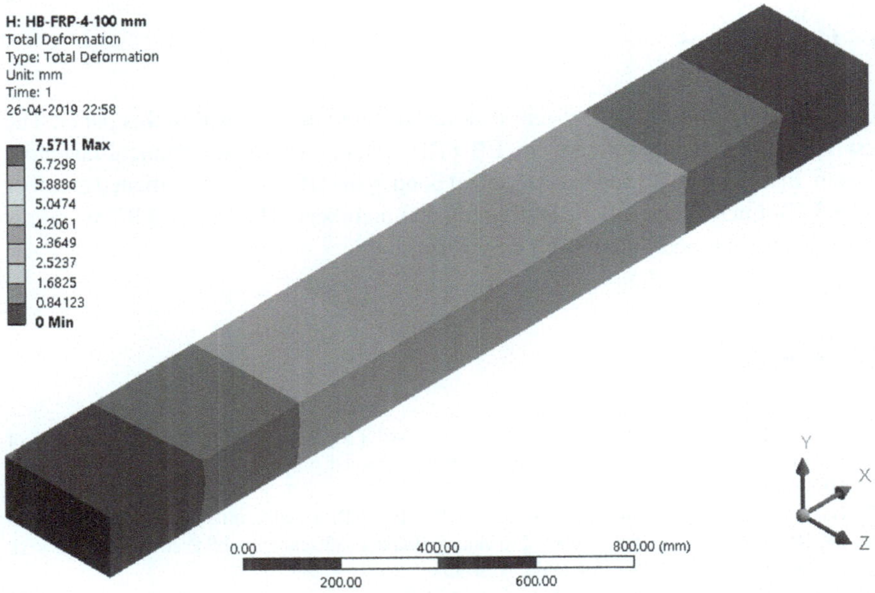

Fig. 15 Deformation diagram of HB_4

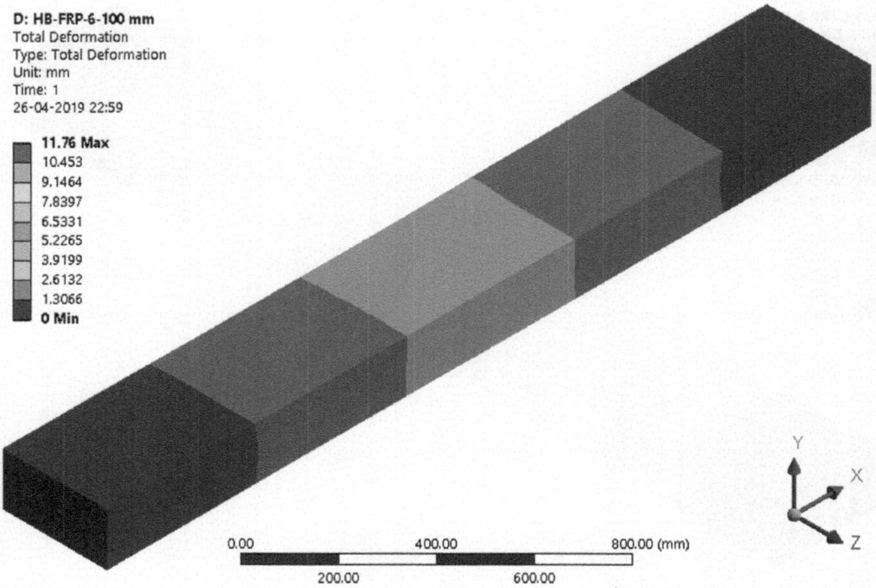

Fig. 16 Deformation diagram of HB_7

6 Conclusions

A new hybrid bonding FRP system is developed and investigated in this paper. The new HB-FRP system combines the EB-FRP system with the mechanical fastening system. In this work showed the interfacial bond in the HB-FRP strengthened member to be 3.13 times that in EB-FRP strengthened members. The hybrid FRP system is more reliable method than EB-FRP System.

References

1. Brena SF, Bramblett RM, Wood SL, Kreger ME (2003) Increasing flexural capacity of reinforced concrete beams using carbon fiber-reinforced polymer composites. ACI Struct J 100(1):36–46
2. Bank LC (2004) Mechanically fastened FRP (MF-FRP) a viable alternative for strengthening RC members. In: Proceeding 2nd international conference on FRP composites in civil engineering (CICE 2004), Adelaide, Australia
3. Wu YF, Huang Y (2008) Hybrid bonding of FRP to reinforced concrete structures. J Compos for Construct 12(3):266–273
4. Yun Y, Wu YF, Tang WC (2008) Performance of FRP bonding systems under fatigue loading. Eng Struct 30(11):3129–3140
5. Wu YF, Wang Z, Liu K, He W (2009) Numerical analysis of hybrid-bonded FRP strengthened concrete beams. Comput Aided Civ Infrastruct Eng 24(5):371–384
6. Wu ZM, Hu CH, Wu YF, Zheng JJ (2011) Application of improved hybrid FRP technique to FRP de-bonding prevention. Construct Build Mat 25(6):2898–2905

7. Guan Y, Jiang B, Song X (2011) Experimental study and numerical simulation on bonding behavior of the FRP-bolt strengthening technology. Geotechnical special publication, No. 219, ASCE
8. Guan Y, Jiang B, Song X (2011) Experimental study on RC beams strengthened in shear with the FRP-bolt strengthening technology. Geotechnical special publication, No. 219, ASCE
9. Ebead U (2011) Hybrid externally bonded/mechanically fastened fiber reinforced polymer for RC beam strengthening. ACI Struct J 669–678
10. Wu ZM, Wu YF, Hu CH, Zheng JJ (2011) Improved hybrid bonding technique for attaching FRP to reinforced concrete beams. Mag Concr Res 63(11):861–869
11. Niu P, Li S, Li S (2011) Researches on the flexing resistance of RC defective beams of hybrid-bonding of FRP. Geotechnical special publication, No. 219, ASCE
12. Guan Y, Jiang B, Song X (2012) Experimental study and numerical simulation on bonding behavior of the new HB-FRP strengthening technology. J Perform Construct Facil 26(2):220–227
13. Ebead UA et al (2013) Finite element modelling of hybrid MF/EB FRP strengthened RC beams. FRPRCS1
14. Ebead U, Saeed H (2013) Numerical modelling of shear strengthened reinforced concrete beams using different systems. J Compos Constr 18(1)
15. Ebead U, Saeed H (2013) Hybrid shear strengthening system for reinforced concrete beams. Eng Struct 49:421–433
16. Zhou Y, Gou M, Zhang F, Wang D (2013) Reinforced concrete beams strengthened with carbon fibre reinforced polymer by friction hybrid bond technique: experimental investigation. Mat Des 50:130–139
17. Baghi H, Barros JA, Rezazadeh M (2017) Shear strengthening of damaged reinforced concrete beams with hybrid composite plates. Compos Struct 178:353–371
18. Chen C, Sui L, Xing F, Li D, Zhou Y, Li P (2017) Predicting bond behaviour of HB FRP strengthened concrete structures subjected to different confining effects. Compos Struct 187:212–225

Evaluation of Structural Response of Ground Supported Cylindrical Water Tanks to Static and Harmonic Loading

Shilja Sureshkumar and Asha Joseph

Abstract Earthquake forces and hydrodynamic pressure are important parameters to be considered in the design of water tanks. In this study, response of ground supported cylindrical water tanks were studied under static and harmonic loading at different water heights and different aspect ratios using finite element software ANSYS. The main objective is to determine the influence of soil structure interaction on dynamic response of water tanks. Soil is modelled as three layers consist of clayey gravel, silty sand and hard rock. It is concluded that, soil structure interaction amplify response of water tanks under static and harmonic loading.

Keywords Fluid structure interaction (FSI) · Soil structure interaction (SSI) · Harmonic loading · Hydrostatic pressure · Aspect ratio

1 Introduction

Based on different material, location, shape, size, support type etc., liquid storage tanks may be classified into various categories. Among different types of liquid storage tanks, steel tanks and concrete tanks are very common in lifeline applications. Response of elevated tanks and ground supported tanks are different, based on their support conditions provided. Seismic safety of liquid-filled container is of great concern because of the potential adverse of economic and environmental impacts associated with failure of the containers and liquid spillage on the surrounding area.

S. Sureshkumar
Department of Civil Engineering, Albertian Institute of Science and Technology, Kalamassery 682022, India
e-mail: shilja1193@gmail.com

A. Joseph (✉)
Department of Civil Engineering, Federal Institute of Science and Technology, Angamaly, Kochi 683577, India
e-mail: ashameledath@gmail.com

K. Dasgupta et al. (eds.), *Proceedings of SECON'19*,
Lecture Notes in Civil Engineering 46,
https://doi.org/10.1007/978-3-030-26365-2_49

As a result, a considerable amount of research effort has been devoted for a better determination of the seismic behaviour of liquid tanks and the improvement of associated design codes [1].

A wide variety of failure mechanisms are possible for liquid storage tanks. It usually depends on tank geometry, fluid structure interaction, soil structure interaction and a lot of other factors such as the tank material, type of support structure, sloshing of liquid etc. Failure patterns of rectangular tank are significantly different from those of cylindrical, spherical, and conical tanks for that matter. Similarly, flexible tanks and rigid tanks have different mode of failure patterns [2]. Different combinations of above possible parameters make the failure mechanism more complex. Many tank failures were occurred during past earthquakes. Which revealed that dynamic study of water tanks are essential for providing safety of tanks, hence it will reduce the impact of damages. Some of the failure modes of tanks are shell buckling, inlet-outlet pipe breaks, support failure, sloshing damages, base sliding, damages to floating roof etc. [3].

In 1963, George W. Housner developed spring mass model for the analysis of water tanks. In a liquid storage tanks, liquid in the lower region of tank is impulsive liquid mass which is rigidly connected to tank wall and induces impulsive hydrodynamic pressure on tank wall and tank base [4]. Fluid which undergoes sloshing motion and located on the upper region of tank is convective liquid mass and it exerts convective hydrodynamic pressure on tank wall and base. Impulsive and convective liquid masses are suitably represented in the spring mass model of tank-liquid system. Fluid structure interaction (FSI) is an important parameters considered in the analysis of the tanks [5–7]. Sloshing is one of the phenomenon which makes the behaviour of the tanks much complicated. Sloshing induced hydrodynamic load may adversely affect the dynamic behaviour and structural safety of liquid storage tank. Dynamic characteristic of these systems is greatly affected by the dynamic behaviour of the liquid free surface motion and it is very important regarding to the safety of transportation systems, human life and environmental issues [8].

In seismic analysis, soil structure interaction (SSI) should be considered. Dynamic response of soil-structure system is a function of two factors, dynamic parameters of site and forces. Dynamic parameters of the site include soil modulus of elasticity, soil Poisson's coefficient, and damping in soil. Veletsos and Meek (1974) investigates the difference in the seismic behaviour between the same structure placed on firm soil and on soft soil [9]. They identified two main factors responsible for the difference in tank response. The first of these aspects was that structures on a flexible base have more degrees of freedom and therefore, different dynamic characteristics other than the structures on a rigid base. The second aspect indicated that a part of the vibrational energy of a structure placed on a flexible base will be dissipated by the radiation of waves into the supporting medium and by damping in the foundation material [10]. Type of soil has considerable effect on behaviour of soil and structure interaction. Based on research studies it is clear that, fine-grained or softer soils have higher interaction with structure. When shear modulus of the surrounding soil decreases or the surrounding soil is softened, displacements and the term of damping motion of tank is elongated. Therefore, it is recommended not to build heavy structures in soft soils and if it is necessary to do so, dynamic soil and structure interaction should be considered based on non-linear behaviour of soil [11–13].

According through the research study conducted by Hirde and Hedaoo [14] it was concluded that, earthquake forces for soft soil is about 18–19% greater than that of medium soil, earthquake forces for medium soil is about 26–27% greater than that of hard soil, earthquake forces for soft soil is about 40–41% greater than that of hard soil at all earthquake zones with tank full and tank empty condition. The responses for soft soil are more because it have higher structural response factor (Sa/g) [14]. Mohr-Coulomb model and Drucker Prager model are two types of model used for analysis of soil behaviour.

2 Finite Element Simulation with Ansys

Approximate solutions to boundary value problems for partial differential equations are determined by a numerical technique called finite element method (FEM). It subdivides a large problem into smaller, simpler parts that are called finite elements. FEM is best understood from its practical application, known as finite element analysis (FEA). Main steps in FEA are structural idealization, discretization, finite element formulation, assemble the elemental equilibrium equation to get global equilibrium equation, application of boundary conditions, selection of equilibrium equation and computation of stress, strain etc.

ANSYS is a well-developed software for the finite element analysis. The programs have the ability to include fluid elements, fluid-structure interactions, soil-structure interaction and other options such as geometric and material nonlinearity such as large deformation assumptions provides many different analysis options. The element SOLID 187 is used for modelling tank and soil and FLUID 220 element used for modelling water inside the tank. Contact between tank and fluid, tank and soil and between soil layers are modelled by using the elements CONTA 174 and TARGE 170.

3 Modelling in Ansys Workbench 16

In this present study, ground supported cylindrical water tanks are analyzed as two steps. In the first case FSI effect is considered and SSI effect is ignored. In the second case both FSI and SSI effects are considered. Hence two different types of models are required for the analysis.

In the case of analysis considering FSI effect, three concrete ground supported circular water tanks with four different water fill condition are modelled in ANSYS workbench 16. Tank aspect ratios are 0.5, 1, and 1.5 and different water heights are 3, 6, 9 and 11 m. Tank bottom thickness is same as wall thickness and it is taken as 0.5 m. Height of the tanks are kept constant as 12 m and diameter of tanks changes with respect to aspect ratios. Density, poison's ratio and young's modulus of concrete taken as 2500 kg/m^3, 0.16, and 27.386 GPa respectively. Density and bulk modulus of water are 1000 kg/m^3 and 2.2 GPa. Fixed support is provided at tank bottom and soil-structure interaction is ignored. Water body is modelled as an acoustic element with frictionless contact between tank and fluid. Tank surface which is in contact with water is defined as Acoustic FSI interface and Acoustic free surface is defined to determine sloshing characteristics. Finite element model of tanks having aspect ratio 0.5 at different water heights are shows in Fig. 1.

In the case of analysis considering both FSI and SSI effects, in addition to above modelling procedure soil is modelled as three layers and tank is placed on a slab having 0.5 m thickness which is placed on top clayey gravel layer. Drucker-Prager model is adopted for soil modelling. Viscous boundary is provided at four sides of the soil layers and fixed support is provided at the bottom of the tank. Properties of soil layers are given in Table 1 and Finite element model of tank with soil layer profile is shown in Fig. 2. Soil block is rectangular in geometry having width equal to 5 times tank diameter and total depth equal to 3 times tank diameter.

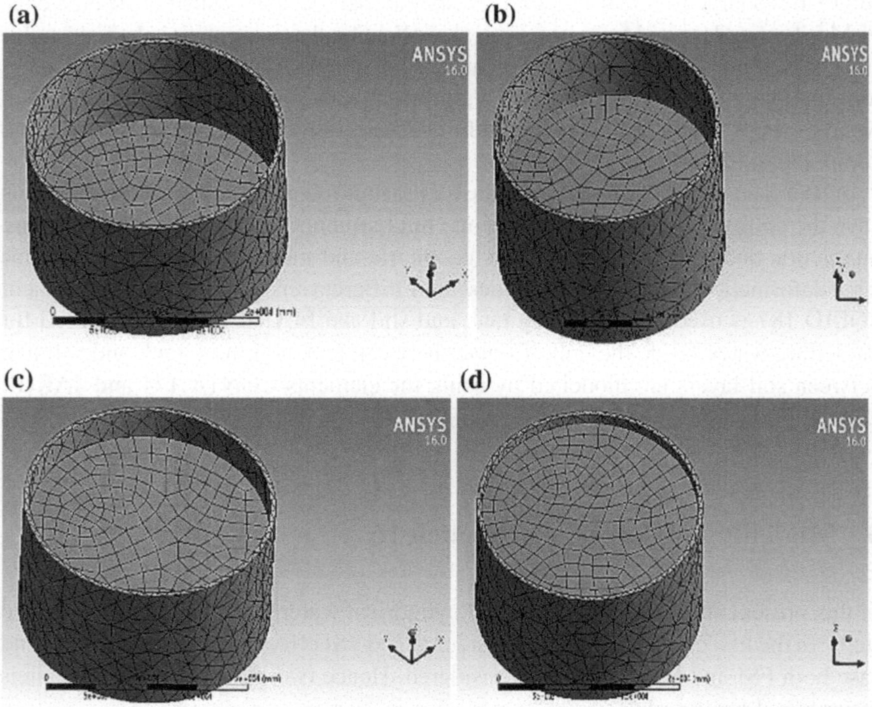

Fig. 1 FE model of tank (aspect ratio: 0.5) for different water heights **a** 3 m, **b** 6 m, **c** 9 m, **d** 11 m

Table 1 Properties of soil layer

Soil properties	Clayey gravel	Silty sand	Hard rock
Density (kg/m^3)	2140.672	2344.546	2792.29
Young's modulus (MPa)	90	30	85,000
Poisson's ratio	0.35	0.4	0.38
Cohesion (MPa)	0.020	0.020	4
Angle of friction (°)	35	40	30

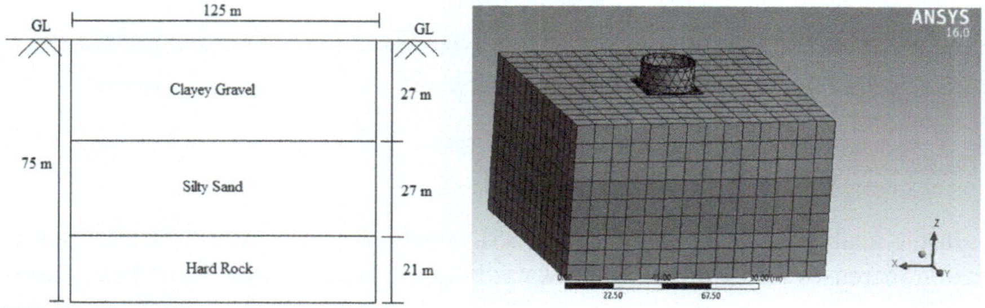

Fig. 2 FE model of soil structure with water tank having aspect ratio 0.5 and water height 11 m

4 Static Structural Analysis

Analyses are performed for all tank models having different aspect ratios and different water heights. Hydrostatic pressure is applied on tank wall and bottom. Variation of total deformation, equivalent stress and hoop stress are determined in both cases of analysis. In both cases of analysis it is observed that, for a particular aspect ratio deformation and stress values are increases with increase in water heights. But when aspect ratio increases from 0.5 to 1.5 deformation and stress values are decreases for all water heights.

4.1 Variations of Total Deformation

Total deformation of tank varies with respect to water heights and aspect ratios. Variation of total deformation of tank is shown in Fig. 3. In the analysis considering FSI effect, at aspect ratio 0.5, total deformation increases 17.06 times, when water height increases from 3 to 11 m. Corresponding increase in total deformation for aspect ratio 1 and 1.5 are 10.642 times and 8.388 times respectively. By comparing results, total deformation decreases into 69.959 and 85.339%, when aspect ratio increases from 0.5 to 1 and 1.5 respectively.

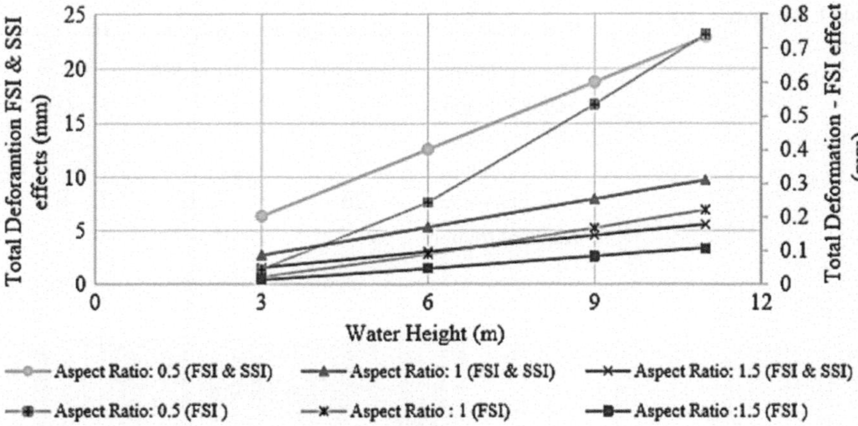

Fig. 3 Variation of total deformation with water heights

In the analysis considering FSI and SSI effects, at aspect ratio 0.5, total deformation increases 3.619 times, when water height increases from 3 to 11 m. Corresponding increase in total deformation for aspect ratio 1 and 1.5 are 3.623 times and 3.633 times respectively. It is shows in Fig. 3. By comparing results, total deformation decreases into 57.565 and 75.598%, when aspect ratio increases from 0.5 to 1 and 1.5 respectively.

4.2 Variation of Equivalent Stress

Variation of equivalent stress with respect to water heights and aspect ratios are tabulated in Table 2. In the analysis considering FSI effect, at aspect ratio 0.5, Equivalent stress increases 6.71 times, when water height increases from 3 to 11 m. Similarly Equivalent stress increases 7.48 times and 6.376 times at aspect ratio 1 and 1.5 respectively. By comparing results, Equivalent stress decreases by 45.158 and 60.545%, when aspect ratio increases from 0.5 to 1 and 1.5 respectively.

In the analysis considering FSI and SSI effects, at aspect ratio 0.5, Equivalent stress increases 4.152 times, when water height increases from 3 to 11 m. Similarly Equivalent stress increases 3.719 times and 3.776 times at aspect ratio 1 and 1.5 respectively. By comparing results, Equivalent stress decreases to 41.015 and 54.552%, when aspect ratio increases from 0.5 to 1 and 1.5 respectively.

Table 2 Variation of equivalent stress

Aspect ratio	Water height (m)	Equivalent stress (MPa)	
		FSI effect	FSI & SSI effects
0.5	3	0.321	1.058
	6	0.946	2.249
	9	1.679	3.538
	11	2.155	4.393
1	3	0.158	0.696
	6	0.531	1.394
	9	0.919	2.107
	11	1.182	2.591
1.5	3	0.133	0.528
	6	0.395	1.080
	9	0.662	1.624
	11	0.850	1.996

4.3 *Variation of Hoop Stress*

Variation of hoop stress with respect to water heights and aspect ratios is tabulated in Table 3. In the analysis considering FSI effect, at aspect ratio 0.5, Hoop stress increases 8.9 times, when water height increases from 3 to 11 m. Corresponding increase in Hoop stress for aspect ratio 1 and 1.5 are 9.344 times and 7.8 times

Table 3 Variation of hoop stress

Aspect ratio	Water height (m)	Hoop stress (MPa)	
		FSI effect	FSI & SSI effects
0.5	3	0.193	0.795
	6	0.604	1.823
	9	1.268	2.259
	11	1.722	2.547
1	3	0.104	0.545
	6	0.407	1.098
	9	0.745	1.647
	11	0.972	2.014
1.5	3	0.088	0.452
	6	0.298	0.896
	9	0.533	1.340
	11	0.699	1.636

respectively. By comparing results, Hoop stress decreases to 43.53 and 59.373%, when aspect ratio increases from 0.5 to 1 and 1.5 respectively.

In the analysis considering both FSI and SSI effects, at aspect ratio 0.5, Hoop stress increases 3.202 times, when water height increases from 3 to 11 m. Corresponding increase in hoop stress for aspect ratio 1 and 1.5 are 3.691 times and 3.617 times respectively. By comparing results, Hoop stress decreases to 20.919 and 35.786%, when aspect ratio increases from 0.5 to 1 and 1.5 respectively.

It is observed that, the higher values of total deformation, Equivalent stress and Hoop stress are obtained in the analysis considering FSI and SSI together as compared with the results obtained from the analysis considering only FSI effect. Hence it is concluded that soil structure interaction have a significant effect on the response of tank under static loadings.

5 Modal Analysis

The modal analysis determines the vibration characteristics (natural frequencies and corresponding mode shapes) of the structure. The natural frequencies and mode shapes are important parameters in the design of a structure for dynamic loading conditions. When the excitation frequency is equal to the natural frequency of the system leads to resonance and causes various damages to the structure.

In modal analysis, natural frequencies of tanks having different aspect ratios with full water fill condition are determined. Impulsive frequency and sloshing frequency of tanks are obtained from analysis considering FSI effect and from analysis considering combined FSI and SSI effects. Results are tabulated in Table 4.

In the analysis considering FSI effect, by observing the variation of fundamental modes of vibrations it is clear that, when aspect ratio increases from 0.5 to 1.5 impulsive frequency increases 36.47% and sloshing frequency increases 79.25%. In the analysis considering FSI and SSI effects, When aspect ratio increases from 0.5 to 1 and 1.5 impulsive frequency increases 6.85% and 44.68% respectively and sloshing increases 88.6% and 79.54% respectively.

Soil structure interaction effect has significant influence on impulsive frequency. As compared to FSI results, impulsive frequency decreases 25.69 times at aspect ratio 0.5 by considering both FSI and SSI effects. Corresponding decrease in impulsive frequency at aspect ratio 1 and 1.5 are 26.62 times and 24.23 times respectively. Also

Table 4 Modal analysis results

Aspect ratio	Impulsive frequency		Sloshing frequency	
	FSI effect	FSI & SSI effects	FSI effect	FSI & SSI effects
0.5	26.876	1.046	0.18869	0.1887
1	29.761	1.1177	0.35567	0.3559
1.5	36.679	1.5134	0.33823	0.3388

it is concluded that soil structure interaction effect did not have any significant effect on sloshing characteristics. Because there is no variations in sloshing frequency by considering SSI effect as compared to FSI analysis.

6 Harmonic Analysis

Analysis are performed for tanks having different aspect ratios with full water fill conditions. Harmonic analysis determines the response of the structure under a steady-state sinusoidal loading at a given frequency. In harmonic analysis, the loading and response of the structure is cyclic. Tanks are subjected to 50 kN load at different frequency range, i.e. 5–50 Hz. Directional deformation, Normal stress, hoop stress are determined and variations of results with respect to aspect ratios are studied for both cases of analysis. Results are tabulated in Table 5.

In the analysis considering FSI effect, at aspect ratio 0.5 the peak values are obtained at natural frequency 26.3 Hz. Variation of directional deformation, hoop stress and normal stress of tank having aspect ratio 0.5 at full water height is plotted in Fig. 4. When aspect ratio increases from 0.5 to 1 directional deformation, Normal stress and hoop stress increases 2.17 times, 14.14 and 40.246% respectively. Similarly when aspect ratio increases from 0.5 to 1 directional deformation, Normal stress and hoop stress increases 3.136 times, 3.949 times and 2.176 times respectively.

In the analysis considering both FSI and SSI effects, Stress and deformation of the tanks are increases up to natural frequency and then shows decreasing. Variation of directional deformation, hoop stress and normal stress of tank having aspect ratio 0.5 at full water height is plotted in Fig. 5. Here also stress and deformation values of the tanks are increases with increase in aspect ratios. When aspect ratio increases from 0.5 to 1 directional deformation, Normal stress and hoop stress increases 77.132%, 79.263% and 17.322% respectively. Similarly when aspect ratio increases from 0.5 to 1.5 directional deformation, Normal stress and hoop stress increases 2.30 times, 4.01 times and 1.579 times respectively. At all aspect ratios stress and deformation are obtained higher in combined FSI & SSI analysis as compared to FSI analysis. Hence it is concluded that soil structure interaction effect has significant effect on harmonic response of the water tanks.

Table 5 Harmonic analysis results (FSI effect)

Aspect ratio	Directional deformation (mm)		Normal stress (Z direction) (Pa)		Hoop stress (Pa)	
	FSI effect	FSI & SSI effects	FSI effect	FSI & SSI effects	FSI effect	FSI & SSI effects
0.5	0.00954	0.04819	122.38	138.02	26.805	93.17
1	0.020715	0.08537	139.69	247.42	37.593	109.309
1.5	0.029919	0.1110	483.39	554.76	58.3327	147.154

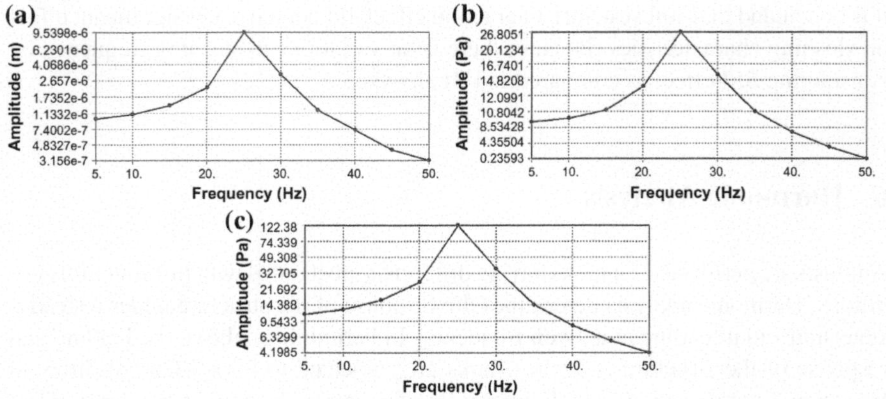

Fig. 4 Variation of **a** directional deformation, **b** hoop stress, **c** normal stress with respect to frequency (FSI effect)

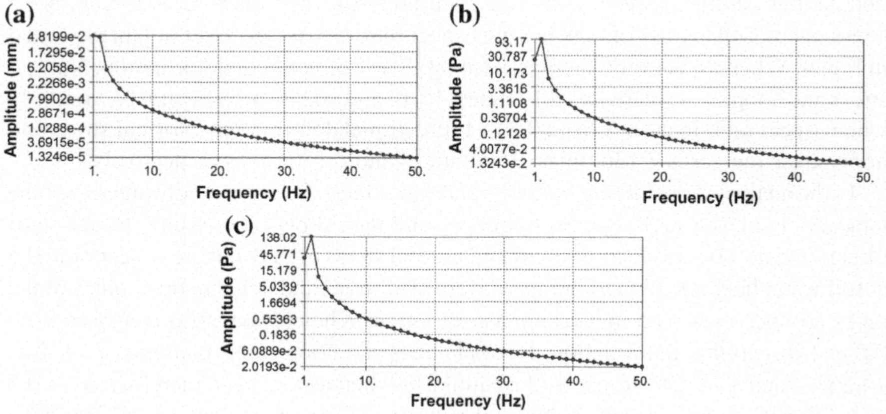

Fig. 5 Variation of **a** directional deformation, **b** hoop stress, **c** normal stress with respect to frequency (FSI and SSI effects)

7 Conclusions

Analysis of ground supported cylindrical water tanks are carried out. Analysis are carried out as two sections. In the first case FSI effect is considered and in the second case FSI and SSI are considered together. Static analysis and harmonic analysis are performed for all tanks using ANSYS software. Influence of FSI and SSI on static and harmonic response of the tank are studied. Based on the results the following conclusions are obtained.

1. In static analysis, stress and deformation values increase as water height increases for all aspect ratios. This is same for analysis considering FSI effect and analysis considering FSI and SSI effects.
2. In static analysis, as aspect ratio increases stress and deformations values are decreases. This is same for analysis considering FSI effect and analysis considering FSI and SSI effects.
3. In static analysis, higher values of stress and deformations are obtained in the analysis considering FSI and SSI together. Hence it is concluded that, SSI has significant effect on response of tanks under static loadings.
4. In harmonic analysis, stress and deformation values increase as aspect ratio increases. This is same for analysis considering FSI effect and analysis considering FSI and SSI effects.
5. In harmonic analysis, higher values of stress and deformations are obtained in the analysis considering FSI and SSI together. Hence it is concluded that, SSI has significant effect on harmonic response of tanks.
6. Hence soil structure interaction should be considered in the dynamic analysis of water tanks for providing much safety and reduce the damages of tanks.

References

1. Chaduvulaa U, Patela D et al (2013) Fluid-structure-soil interaction effect on seismic behaviour of elevated water tanks. Procedia Eng 51:84–91
2. Joseph A, Joseph G (2016) Response of ground supported cylindrical water tanks to harmonic loading. Civ Eng Urb Plan Int J (CiVEJ) 3(2):183–190
3. Jain Sudhir K, Sameer US (1990) Seismic design of frame staging for elevated water tank. In: Ninth symposium on earthquake engineering, vol 1, pp 14–16
4. George W (1963) Housner, the dynamic behaviour of water tank. Bull Seismol Soc Am 53:381–387
5. Sezen H, Livaoglu et al (2008) Dynamic analysis and seismic performance evaluation of above ground liquid-containing tanks. Eng Struct 30(3):794–803
6. Waghmare MV, Madhekar SN (2013) Behaviour of elevated water tank under sloshing effect. Int J Adv Technol Civ Eng 2:51–54
7. Dona Rose KJ, Sreekumar M et al (2015) A study of overhead water tanks subjected to dynamic loads. Int J Eng Trends Technol 344–348
8. Anumod AS et al (2014) Finite element analysis of steel storage tank under seismic load. Int J Eng Res Appl (IJERA) 47–54
9. Veletsos AS, Meek JW, (1974) Dynamic behaviour of building-foundation systems. Earthq Eng Struct Dyn 3(2):121-138
10. Anestis S, Veletsos et al (1992) Dynamic response of flexibility supported liquid-storage tanks. J Struct Eng-ASCE 118:264–283
11. Medhat A, Haroun et al (1992) Parametric study of seismic soil-tank interaction In: Horizontal excitation. J Struct Eng-ASCE 118:783–797
12. Livaoglu R (2008) Investigation of seismic behaviour of fluid-rectangular tank-soil/foundation systems in frequency domain. Soil Dyn Earthq Eng 28(2):132–146

13. Livaoglu R, Dogangun A (2007) Effect of foundation embedment on seismic behaviour of elevated tanks considering fluid–structure-soil interaction. Soil Dyn Earthq Eng 27(9):855–863
14. Hirde S, Hedaoo M (2011) Seismic performance of elevated water tanks. Int J Adv Eng Res Stud 1:78–87

Response of Self Centering Steel Moment Resisting Frames Against Cyclic Loading

Meeval Maria Bibin and Asha Joseph

Abstract A Self Centering Moment Resisting Frame (SCMRF), an alternative to a conventional Moment Resisting frame, is characterized by gap opening and closing at the beam-column interface under earthquake loading. The beams are post tensioned to columns by high strength post tensioning (PT) strands oriented horizontally to provide self-centering forces when gap opening occurs. A Self Centering Moment Resisting Frame for earthquake resistant structures has the attributes of ductility, requires no field welding and returns the structure to its pre earthquake position which prevents permanent deformations when subjected to seismic loading. Energy dissipation is provided by supplemental elements that deform under the gap opening behavior. Unlike special steel moment resisting frame with welded connections a SCMRF can be designed to survive earthquake without much structural damage which provide immediate occupancy performance. Nonlinear static analysis is performed in ANSYS to investigate the effect of posttension load and bracing configuration on seismic performance of frame. The results indicated that an increase in PT force increases the residual drift and maximum load carrying capacity. Bracing in SCMRF is effective in increasing the load bearing capacity and energy dissipation. Bracing shifts the maximum stress position from beam column connection to gusset plate so that it act as sacrificial element which can easily be replaced without damaging the beams or columns.

Keywords Self centered moment resisting frame · Post tensioned strand · Energy dissipation · Nonlinear static analysis

---

M. M. Bibin (✉) · A. Joseph
Department of Civil Engineering, Federal Institute of Science and Technology, Angamaly, India
e-mail: meevalitty@gmail.com

A. Joseph
e-mail: ashameledath@gmail.com

K. Dasgupta et al. (eds.), *Proceedings of SECON'19*, Lecture Notes in Civil Engineering 46,
https://doi.org/10.1007/978-3-030-26365-2_50

1 Introduction

Conventional moment resisting frames use fully restrained welded connections between the beams and columns. This design method used for conventional moment resisting frames leads to significant inelastic deformations and the formation of plastic hinges in the beams under the earthquake. Plastic hinges may cause significant damage which may result in residual drift. The amplitude of residual story drift is the most important contributor to economic losses of buildings following an earthquake and leads to the demolition of the structure after an earthquake which is very expensive [1].

A Self-Centering Moment Resisting Frame (SCMRF) is a new seismic load resisting system that offers the potential for greatly reducing time and cost associated with post-earthquake repair of the structure by reducing residual drifts [2]. These frames are capable of retaining its structural integrity after a seismic load. The beam column joints are not fixed as in a welded or bolted joints, hence relative movement occurs without much hindrance. SCMRF mainly consists of Post tensioned (PT) strands and long slotted shear tabs as in Fig. 1. The PT strands run parallel on either side of beam and are anchored to outer flange of column to precompress the beams to the columns. This PT strand offers recentering capability to frame. The recentering behavior of an SCMRF is characterized by connection gap opening and closing at the beam column interface. The gap opening allows the beam to rotate relative to the column without damaging the beams or columns. The Shear tabs are designed in such a manner that only the shear forces are transferred and the rotation of members at the joint is not restricted [3].

Gap opening and the force in the PT strand provide the bilinear elastic self-centering behavior but do not provide energy dissipation to the structure. To produce sufficient energy dissipation, an energy dissipating element such as brace is incorporated into self-centering systems. These energy dissipating devices should be replaceable after large earthquake inorder to minimize repair cost and time.

This study aims to investigate the effect of posttension load and bracing configuration on seismic performance of SCMRF by conducting nonlinear static analysis. Seismic performance of SCMRF is compared to conventional Welded moment resisting frame (WMRF).

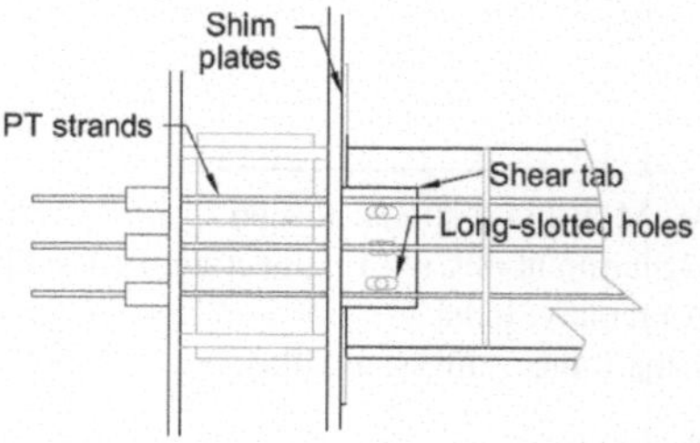

Fig. 1 Details of SCMRF

2 Finite Element Formulation

In order to investigate the effect of post tensioned load and bracing configuration on the seismic performance of SCMRF, finite element analysis was done on SCMRF using ANSYS 16.0. The model and its configuration are chosen as per the details provided in the prior research paper [4].

2.1 Geometric Description

The center to center distance between columns of SCMRF measures 3235 mm wide with story height of 1724 mm. The beam and column members were designed to remain elastic. The beams in these tests were W18 × 106 and the columns were W14 × 132. The sectional details are given in Table 1 where d is the overall depth, t_w is the width of web, b_f is the breadth of flange and t_f is the thickness of flange in the beam and column.

The PT Strands spans throughout the length of the beam. It is anchored to the outer flanges of the column. These high strength steel Post tensioned strands help to resist the moment at the beam-column interface. The PT Strands is the major component that provide for the re-centering property of the frame. Each post tensioned strand was a seven-wire strand having a diameter of 12.7 mm. The area of the strand was 132.7 mm^2. PT forces are varied between 120 and 246 kN [5]. The Shear tab which is 100 × 250 mm with 15 mm thickness helps to connect the beam and column at the joint. Its sole function is to transfer the shear forces acting at the joint. The shear tab is bolted on to the beam. The long slotted holes in the shear tab helps to ensure that the rotation is not arrested [3]. Slotted holes are elliptical in shape with 25 and 50 mm dimension in minor and major axis. Three number of high strength bolts of yield stress 660 MPa are used to connect shear tab to beam on each side [6]. Bolt head is having diameter of 40 mm and shank is of 25 mm diameter. Shim plates which are 12.7 mm thick were tack welded to the columns to ensure that beams bear only at their flanges to prevent beam web yielding because of bearing stresses [6]. Stiffeners are provided in columns to prevent local buckling of column.

Table 1 Section details

Dimensions (mm)	Beam W18 × 106	Column W14 × 132
d	475.74	372.364
t_w	14.986	16.383
b_f	284.48	374.015
t_f	23.876	26.162

Table 2 Material properties and elements used in the modeling

Properties	Density (kg/m^3)	Young's Modulus (GPa)	Poisson's ratio	Yield stress (MPa)	Ultimate stress (MPa)
Beam and column	7850	200	0.3	345	450
PT strand	7850	196.5	0.3	1305	1861
Bolts	7850	200	0.3	660	–
Shear tab	7850	200	0.3	345	450
Shim and stiffeners	7850	200	0.3	895	–
Braces	7850	200	0.3	345	450

2.2 Material Properties

A bilinear stress strain curve was assumed for modeling the materials. Table 2 gives the material property used in the present study. The modulus of elasticity, E for steel material used in all components except for PT strand was 200 GPa while the modulus of elasticity of PT strand was 196.5 GPa. Material yielding was set to be predicted using Von Mises yield criterion [4].

2.3 Finite Element Modeling

Modeling in ANSYS includes providing appropriate elements, defining geometry and assigning suitable material models. The SCMRF model consists of the column, beams, reinforcing plates, shim plates, bolts, steel PT strands as shown in Fig. 2. Beams and columns are modeled using SOLID186 element. SOLID186 is a 3-D 20-node solid element that exhibits quadratic displacement behavior. The element is

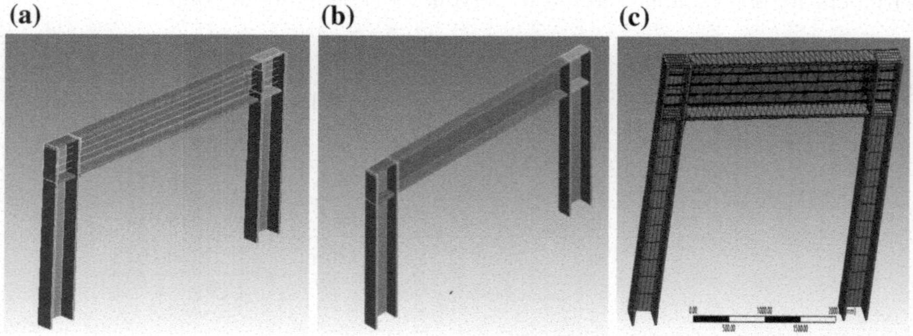

Fig. 2 Geometrical model of **a** SCMRF, **b** WMRF, **c** finite element model of SCMRF

defined by 20 nodes having three degrees of freedom per node: translations in the nodal x, y, and z directions. It is well suited to model irregular meshes. PT strand is modeled using BEAM188 element. The element is a linear two-node beam element in 3-D. It is having six degrees of freedom, translations in the x, y, and z directions and rotations about the x, y, and z directions. This element is well suited for large strain nonlinear problems. Bracings are modeled using SHELL181 element, which is suitable for analyzing thin to moderately thick shell structures. It is a four-node element with six degrees of freedom at each node: translations in the x, y, and z directions, and rotations about the x, y, and z-axes. It accounts for load stiffness effects of distributed pressures. In this study, symmetry conditions were employed for vertical plane and, thereby, half of the connection assembly was modeled.

2.3.1 Connection Details

In SCMRF, PT strand running parallel to beam are anchored at the outer flange of column. Shim plates are welded to column. Frictionless connection is provided between shim plate and beam. Shear tab is welded to column and bolted to beam. The connections are shown in Fig. 3.

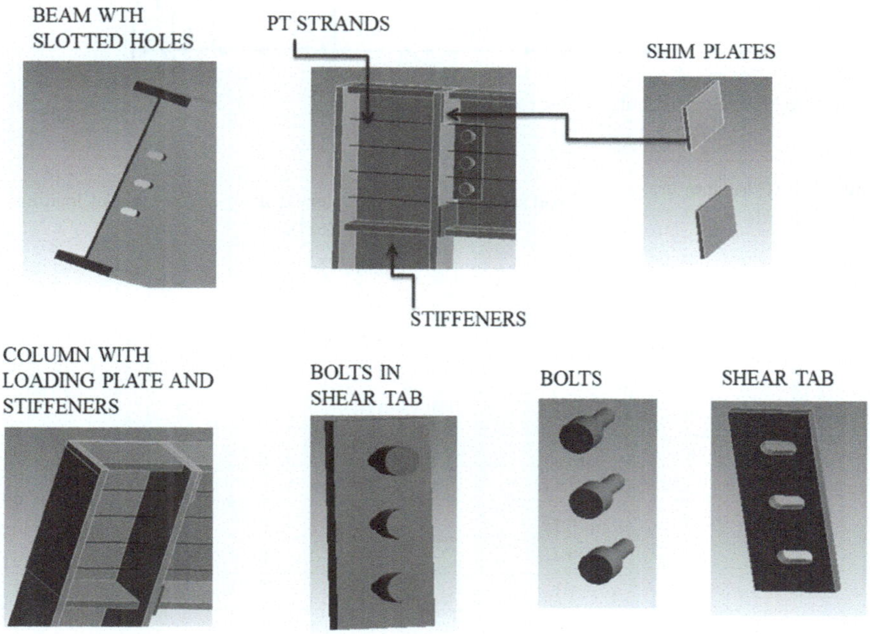

Fig. 3 Connection details of SCMRF

2.3.2 Support and Loading

Pin support conditions were created for the column base by restraining the vertical and horizontal displacement of the nodes on the centreline. For determining seismic performance of SCMRF, cyclic loading was applied to the column flange. The loading is applied in accordance with SAC loading protocol [7] as shown in Fig. 4. This loading protocol was developed to obtain the behaviour of beam column moment connections. The details of loading sequence for SAC loading protocol are presented in Table 3.

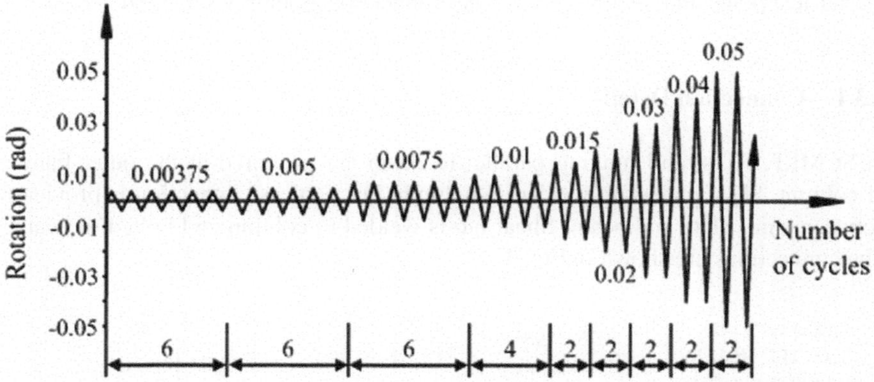

Fig. 4 SAC loading protocol

Table 3 SAC loading protocol details

Load step number	Inter story drift angle (rad)	Number of loading cycles
1	0.00375	6
2	0.005	6
3	0.0075	6
4	0.01	4
5	0.015	2
6	0.02	2
7	0.03	2
8	0.04	2
9	0.05	2
10	0.06	2

3 Response of WMRF and SCMRF to Cyclic Loading

Cyclic analyses were carried out in SCMRF and WMRF with comparable load capacity to show recentering capability of self-centering frame. After cyclic analysis the residual drift of WMRF was observed to be 30.21 mm while that for SCMRF is only 11.68 mm. The self-centering behavior of SCMRF can be seen in Fig. 5.

The major stress in WMRF is concentrated at the beam-column junction as shown in Fig. 6 indicating that the plastic hinge is formed at the column which is not good for a seismic connection.

For a good seismic connection, the formation of plastic hinge should be away from the beam-column joint. In SCMRF, plastic hinge is formed at shim plate as shown in Fig. 7.

3.1 Parametric Analysis

The Post tensioned load is varied between 120 and 246 kN [2] and cyclic analysis is carried out to determine the best PT force for considered model and this load is used for further studies. Effect of post tensioning force on self-centering capacity of frame can be studied by comparing energy dissipation capacity, residual drift and maximum load that frame can withstand. SAC loading is provided in frame as displacement at top of column. Area within the hysteresis loop was integrated up to story drift of 3% to determine E_d, which is the accumulated energy dissipated as shown in Table 4. Comparing the energy dissipation and residual drifts given in Table 4, it can be seen that PT force of 156 kN gives better seismic performance than others. Therefore PT force of 156 kN was used for further studies.

Fig. 5 Hysteresis loop of WMRF and SCMRF

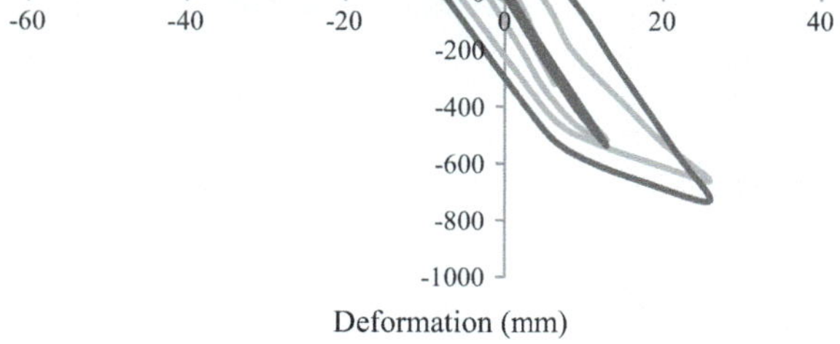

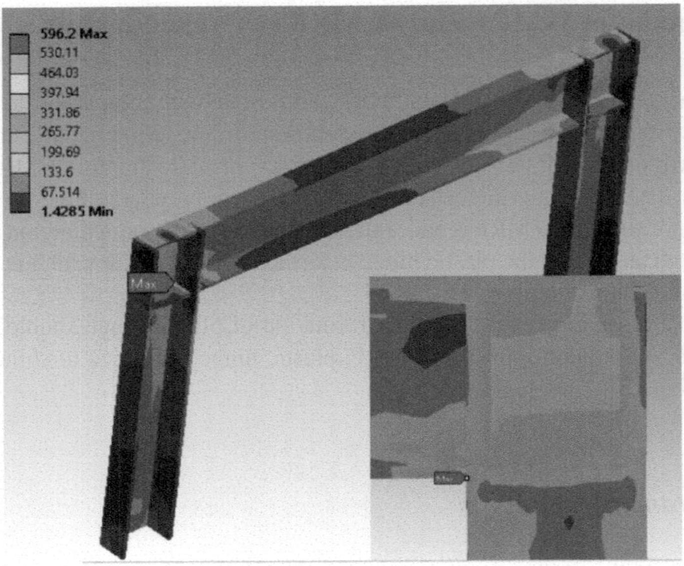

Fig. 6 Von Mises stress distribution in WMRF

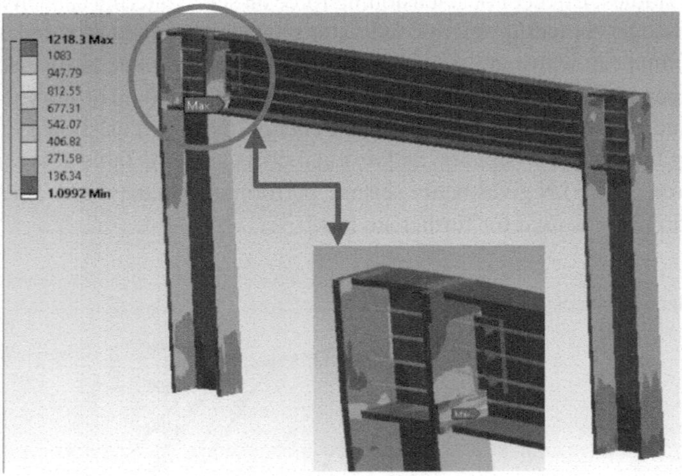

Fig. 7 Von Mises stress distribution in SCMRF

Table 4 Response values after 3% drift

PT force (kN)	Residual displacement (mm)	Max load (kN)	Max moment (kNm)	Energy dissipated, E_d (kNm)
Welded	30.215	885.34	1526.3	80.07
120	4.845	934.88	1611.7	39.45
138	6.424	953.38	1643.6	40.82
156	8.43	971.44	1674.7	111.4
174	10.562	988.06	1703.4	44.61
192	11.686	1001.06	1725.8	46.1
210	12.229	1012.84	1746.1	46.84
228	12.722	1020.24	1758.8	48.98
246	13.534	1030.02	1775.7	49.95

Effects of bracing on Dynamic response of SCMRF were investigated with diagonal and V bracing. Channel sections are used for bracing. Diagonal bracing (100 × 50 × 6) and V bracing (100 × 50 × 4) are used in study. Bracings are welded to gusset plate of 100 × 200 mm dimension at both ends. The gusset plate is welded to beam alone at the top so that it does not interfere with gap formation. The results of study conducted on the bracing configuration are shown in Table 5. Area within the hysteresis loop was integrated up to story drift of 1.5% to determine E_d, which is the accumulated energy dissipated as shown in Table 5.

It can be observed from Table 5 that the energy dissipated and load capacity is more for SCMRF with diagonal bracing. The maximum stress is shifted form shim plate as shown in Fig. 7 to gusset plate as shown in Fig. 8, making gusset plate a sacrificial element which can easily be replaced without damaging the beam or column.

Table 5 Response values after 1.5% drift

	Max load (kN)	Max stress on beam (MPa)	Max stress on column (MPa)	Max residual drift (mm)	Energy dissipated, E_d (kNm)
SCMRF	638.6	439.58	535.29	4.46	5.92
SCMRF with V bracing	698.8	486.58	545.36	1.37	5.63
SCMRF with diagonal bracing	1923	463.8	533.9	4.22	8.41

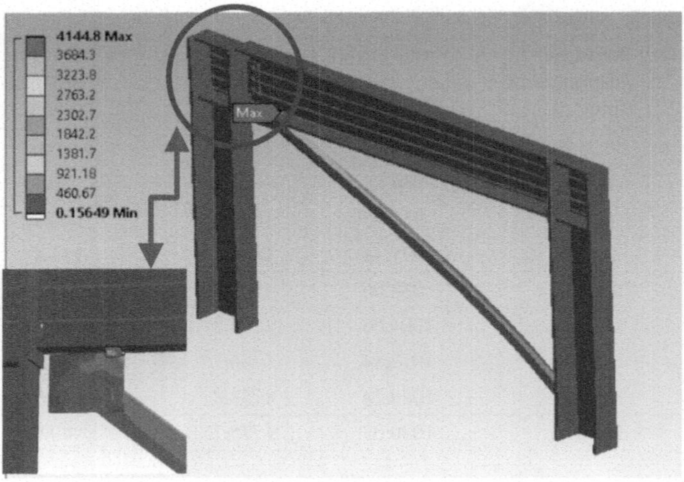

Fig. 8 Von Mises stress distribution in SCMRF with diagonal bracing

4 Conclusions

Nonlinear static analysis were carried out in single bay single floor SCMRF to investigate the effect of post tensioned load and the bracing configuration on recentering capability, energy dissipation of frame. The results of analysis shows that Self Centering Moment Resisting Frame (SCMRF) with bracing shows better seismic performance by reducing the residual drift and improving energy dissipation capacity.

The test results on the study of PT force shows that the stiffness, load bearing capacity increases with increase in PT force. The examined study of bracing include different configuration of bracing introduced in SCMRF. In the case of Diagonal bracing, the energy dissipation was better with good recentering capability than V bracing. Furthermore results shows that damage was concentrated on replaceable gusset plate that can easily be replaced after earthquake. Therefore SCMRF with Diagonal bracing can be considered as structure with better seismic performance.

References

1. Ahmadi O, Ricles J, Sause R (2016) Seismic collapse assessment of self—centering steel moment resisting frame system with web friction devices. Ph.D. thesis, Lehigh University, Pennsylvania
2. Winkley T (2011) Self-centering steel plate shear walls: large scale experimental investigation. Ph.D. thesis, University of Washington
3. Sreekumar A, Asraff A, Ramanujan J (2011) Finite element analysis of self centering moment resistant frames with and without steel plate shear wall. Procedia Tech 24:161–168
4. Clayton P, Berman J, Lowes L (2012) Experimental investigation of self-centering steel plate shear walls. J Struct Eng ASCE 138:952–960

5. Research Council of Structural Connection (2009) Specification of structural joints using high strength bolts. Chicago 16(2):52
6. Ricles J, Sause R, Peng W (2002) Experimental evaluation of earthquake resistent post tensioned connections. J Struct Eng ASCE 128:850–859
7. Federal Emergency Management Agency (FEMA) (2000) FEMA 350: seismic design criteria for new steel moment frame buildings. California Universities for research in Earthquake Engineering

Comparison of Seismic Performance of Knee Braces in Steel Frames with Y Shaped Eccentric Braces

Fathima Farheen and S. P. Akshara

Abstract Today steel is the most useful building material in the construction field. Steel structures tend to be more economical than concrete structures for tall and large span buildings and bridges due to its large strength to weight ratio. The steel structures in the areas of high seismic activity should be stiff enough to prevent structural damage and have sufficient ductility to prevent collapse. Incidentally, steel bracing provides an effective and economical solution for resisting lateral forces in a framed structure. Among the bracings, Knee braced steel frame has got excellent ductility and lateral stiffness. Since the knee element properly fuses, yielding occurs only on knee element leaving the structural elements safe. In this paper, a study on the seismic effect of knee braced steel frames of different configurations were performed and compared with Y shaped Eccentric bracings, having a vertical shear link. Performance of the frames were studied using non-linear static analysis and non-linear time history analysis for various influencing factors.

Keywords Non-linear static analysis · Non-linear time history analysis · Steel bracing · Stiffness

1 Introduction

Buildings undergoing seismic activity are designed based on two criteria: the structure should have sufficient strength and stiffness to prevent damage and control deflection during low to moderate earthquakes whereas during severe earthquakes, the structure must have adequate ductility and energy absorption capacity. Addition of lateral load steel bracing system provides adequate strength for high rise buildings.

F. Farheen (✉) · S. P. Akshara
Department of Civil Engineering, Federal Institute of Science and Technology, Angamaly, Kochi 683577, India
e-mail: fathimafarheen95@gmail.com

S. P. Akshara
e-mail: akshara027@gmail.com

K. Dasgupta et al. (eds.), *Proceedings of SECON'19*,
Lecture Notes in Civil Engineering 46,
https://doi.org/10.1007/978-3-030-26365-2_51

Bracing is a secondary member that helps to transfer the lateral loads to its support and makes the structure stable.

Presently, in designing seismic resistant steel buildings, systems such as moment frames, concentrically braced frames and eccentrically braced frames are widely used. The excessive drift due to higher structural flexibility and inevitable stress concentration at the welds of connecting columns and beams are essential parameters that limit the applicability of these structures. Hence a bracing system known as knee braced frame (KBF) was proposed by Aristizabal Ochoa and then revised by Balendra to overcome the limitations [1]. In a knee braced frame, a knee element is provided diagonally across the beam elements, thus main cross braces are connected to these elements. Knee braced frame has got good ductile strength and lateral stiffness. When the connections are properly fused, knee will yield first and the damages on the main members can be prevented. The knee elements are designed to remain elastic and yield before the structural members during small earthquakes. During a moderate excitation, the energy dissipation within the knee elements should be sufficient to prevent damages in the main frame. It is likely that some damage to the main structure occur, but can be reduced by the presence of the knee elements, thus prevents the structure from collapse.

In earlier days, knee braces were used for wind resistant design [2]. However, with extensive studies knee braces exhibited good seismic properties. Simple connections are generally incorporated for ease of construction and reparability after a seismic activity. The knee braces compared to other bracing systems are less obstructive, and thus are architecturally attractive [3]. Seismic strengthening of the existing buildings can be done using knee braced frames (KBF).

2 Objectives

- To study the seismic performances of knee braces and Y shaped eccentric braces in steel frames.
- To conduct non-linear static analysis and non-linear time history analysis in knee braces and Y shaped eccentric braces and identify the better configuration.

3 Numerical Analysis

In order to investigate the seismic performance of knee braces and Y shaped eccentric braces, finite element analysis was done using ANSYS 16.0. Non-linear static and time history analysis of various types of knee brace configurations at five storey and number of bays were performed to obtain the above mentioned objective. The configuration portraying the better performance was then compared with Y shaped Eccentric Bracing. The knee brace configurations analysed were (a) diagonal knee

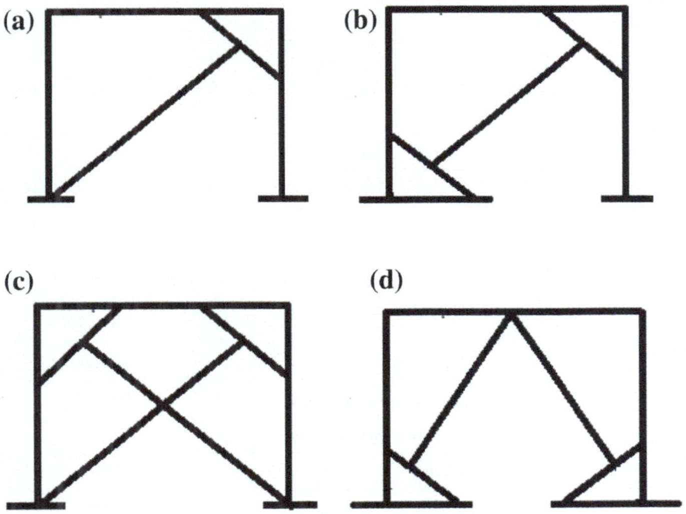

Fig. 1 Different knee brace configurations **a** diagonal knee brace with 1 knee, **b** diagonal knee brace with 2 knee, **c** X KBF, **d** chevron KBF

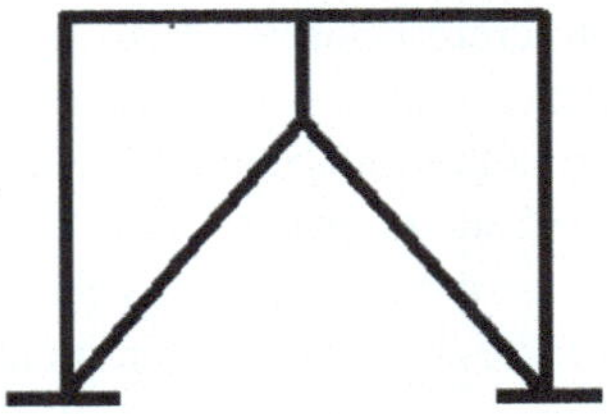

Fig. 2 Y shaped eccentric bracing

brace with 1 knee, (b) diagonal knee brace with 2 knee, (c) X KBF, (d) chevron KBF. Figures 1 and 2 shows the Knee Braces and Y shaped eccentric braces respectively.

A storey height of 2.8 m with bay width 3.2 m was chosen for the study. The base of the frame was considered fixed. Welded connections was used for connecting the members. The meshing was provided as default during the analysis. The beams, columns, braces and knee chosen were I-sections of sizes ISHB 150 × 150 × 27.1 kg/m, ISHB 20 × 200 × 37.3 kg/m, ISMB 100 × 75 × 11.5 kg/m, and ISLB 125 × 75 × 11.9 kg/m respectively. The geometric description of the members are given in Table 1.

The material properties given as input for the analysis are given in Table 2.

3.1 Non Linear Static Analysis

Non-linear static Analysis can be used to evaluate the seismic performance of new and existing structures. As the use of linear static analysis is restricted with respect

Table 1 Geometric description

Element	Section	Depth of the section, d (mm)	Width of flange, b_f (mm)	Thickness of flange, t_f (mm)	Thickness of web, t_w (mm)
Column	ISHB 200	200	200	9.0	6.1
Beam	ISHB 150	150	150	9.0	5.4
Brace	ISMB 100	100	75	7.2	4.0
Knee	ISLB 125	125	75	6.5	4.4

Table 2 Material properties

Properties	Values
Modulus of elasticity	200 GPa
Poisson's ratio of steel	0.3
Density	7850 kg/m^3
Yield stress	250 MPa

to high seismic zones and height and also the modes of vibration may be more than the fundamental mode, non-linear static analysis was used. In this analysis, the structure is subjected to a monotonic displacement controlled lateral load pattern, that continuously increases through elastic and inelastic behaviour until an ultimate condition can be attained. A non-linear relationship can be thus be obtained between load and displacement [4].

Five storey one bay steel frame and five storey five bay steel frame were considered for the analysis. An incremental displacement of 5 mm was given to the right top-most corner of the frame. The displacement was applied until an ultimate load leading to failure of the frame was attained. Henceforth, a graph with the load and displacement was formulated.

3.1.1 Five Storey One Bay Steel Frame

Five storey one bay steel frame with different bracing configurations were analysed using non-linear static analysis. The displacement was provided at the right top-most corner of the frames. Figure 3 shows maximum displacement obtained for the configurations considered.

Figure 4 shows the load carrying capacity curve of the different bracing configurations considered and their ultimate load and maximum displacement values are tabulated in Table 3.

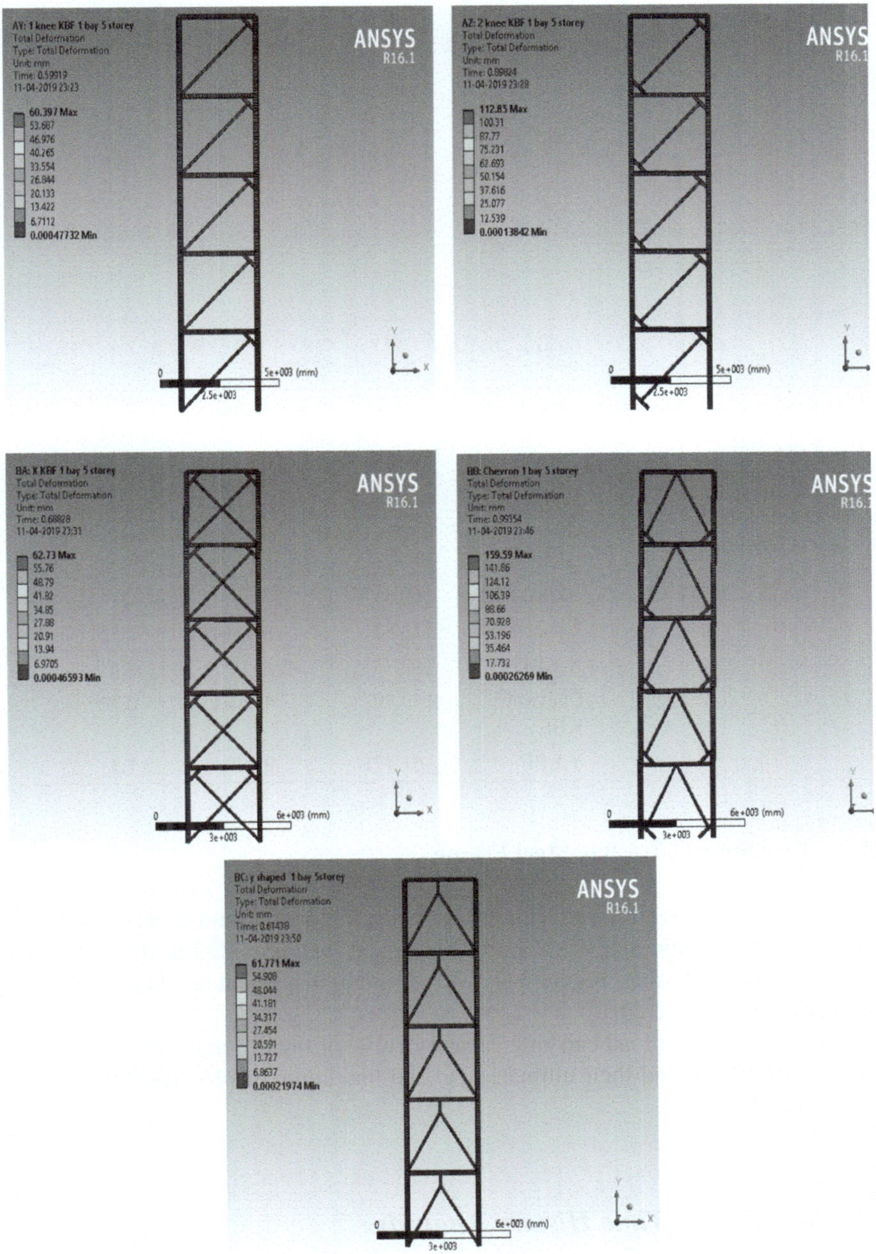

Fig. 3 Maximum displacement of different bracing configurations

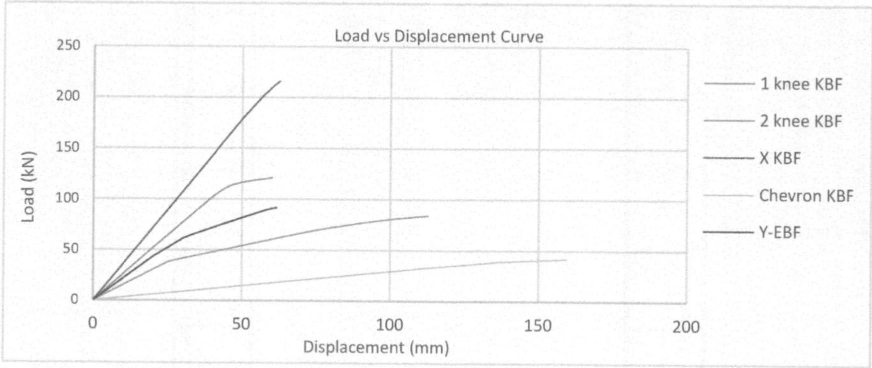

Fig. 4 Load versus displacement curve for five storey one bay steel frame

Table 3 Load and displacement values of different configurations

Configuration type	Maximum displacement (mm)	Ultimate load (kN)	Stiffness (kN/mm)
1 knee KBF	60.397	121.1	2
2 knee KBF	112.85	84.244	0.75
X KBF	62.73	216.09	3.5
Chevron KBF	159.59	41.763	0.26
Y EBF	61.771	91.793	1.5

3.1.2 Five Storey Five Bay Steel Frame

Five storey five bay steel frame with different bracing configurations were analysed using non-linear static analysis. The displacement was provided at the right top-most corner of the frames. Figure 5 shows maximum displacement obtained for the configurations considered.

Figure 6 shows the load carrying capacity curve of the different bracing configurations considered and their ultimate load and maximum displacement values are tabulated in Table 4.

3.2 *Non Linear Time History Analysis*

Time history analysis can be performed to obtain displacement under transient loading. An earthquake data is given as input loading. In this analysis, El Centro Earthquake was chosen as the input loading.

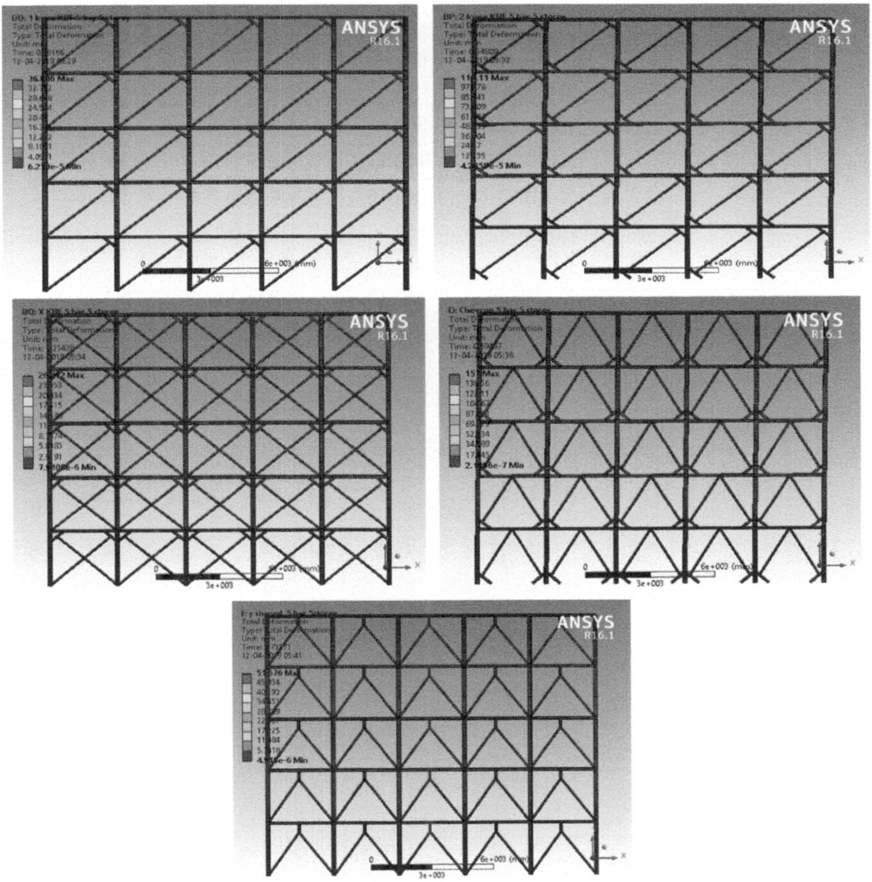

Fig. 5 Maximum displacement of different bracing configurations

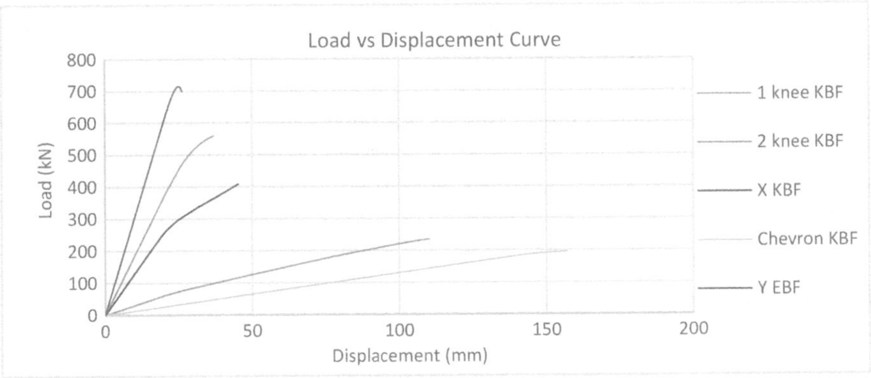

Fig. 6 Load versus displacement graph of different bracing configurations

Table 4 Load and displacement values of different configurations

Configuration type	Maximum displacement (mm)	Ultimate load (kN)	Stiffness (kN/mm)
1 knee KBF	36.846	559.7	15
2 knee KBF	110.11	235	2.1
X KBF	24.616	715.4	29
Chevron KBF	157	195.02	1.24
Y EBF	51.616	434.56	8.4

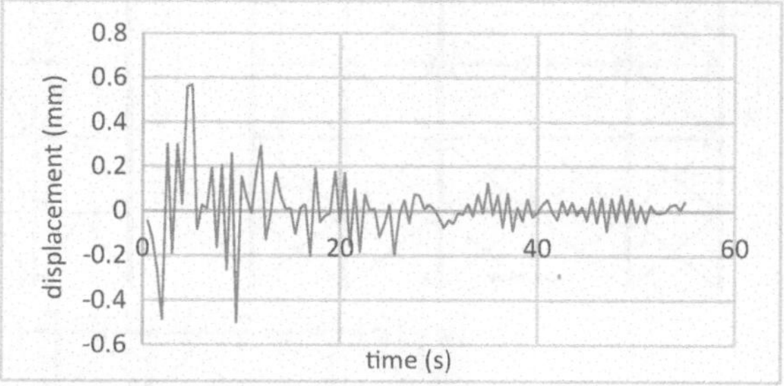

Fig. 7 Displacement graph for diagonal knee bracing with 1 knee

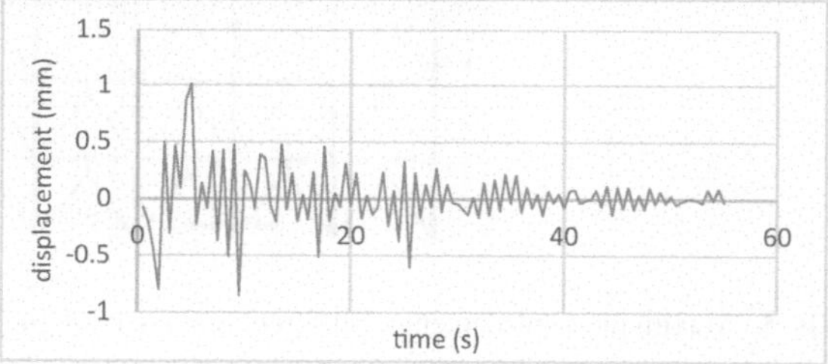

Fig. 8 Displacement graph for diagonal knee bracing with 2 knee

3.2.1 Five Storey One Bay Steel Frame

A five storey one bay steel frame of various bracing configurations were subjected to the values of El Centro Earthquake and displacement graph were obtained as shown in Figs. 7, 8, 9, 10 and 11 respectively (Table 5).

3.3 Five Storey Five Bay Steel Frame

A five storey one bay steel frame of various bracing configurations were subjected to the values of El Centro Earthquake and displacement graph were obtained as shown in Figs. 12, 13, 14, 15 and 16 respectively (Table 6).

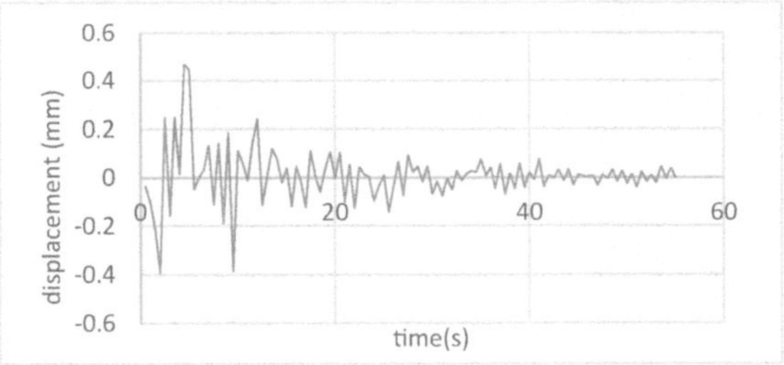

Fig. 9 Displacement graph for X knee bracing

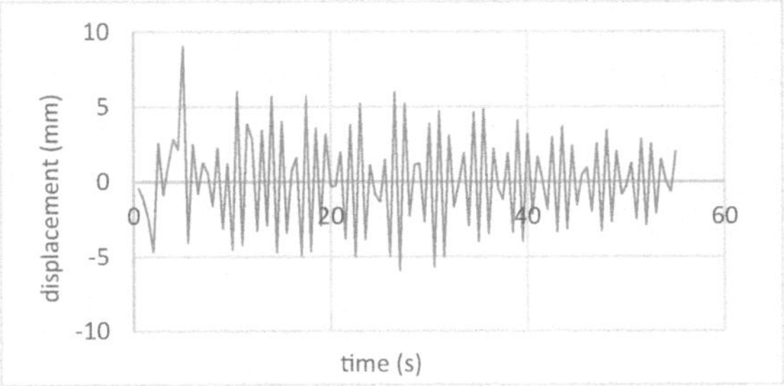

Fig. 10 Displacement graph for chevron knee bracing

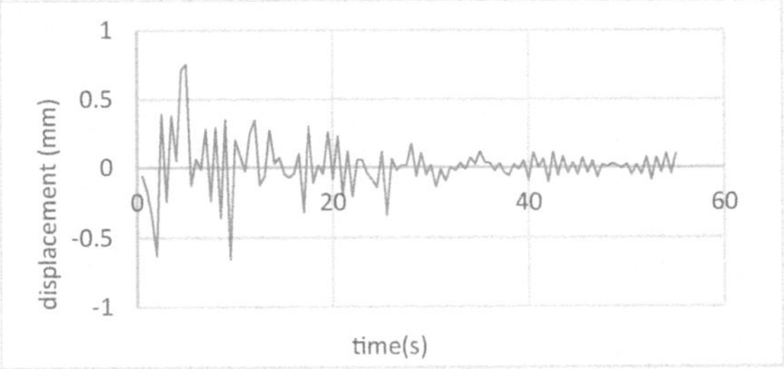

Fig. 11 Displacement graph for Y shaped eccentric bracing

Table 5 Comparison of displacement of frames [5]

Configuration type	Displacement (mm)
1 knee KBF	0.32446
2 knee KBF	2.1162
X KBF	0.20387
Chevron KBF	7.8204
Y EBF	0.45946

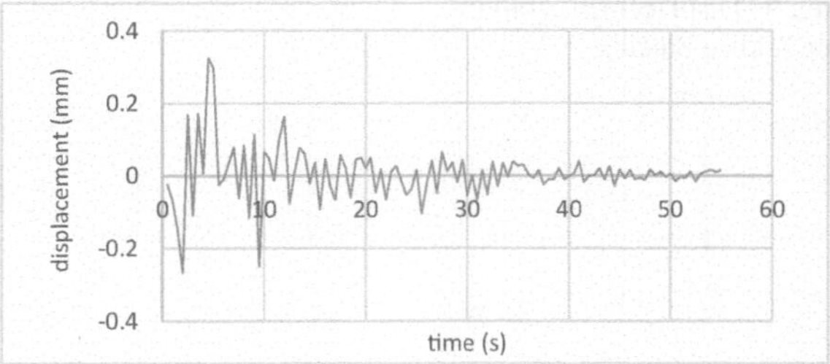

Fig. 12 Displacement graph for diagonal knee bracing with 1 knee

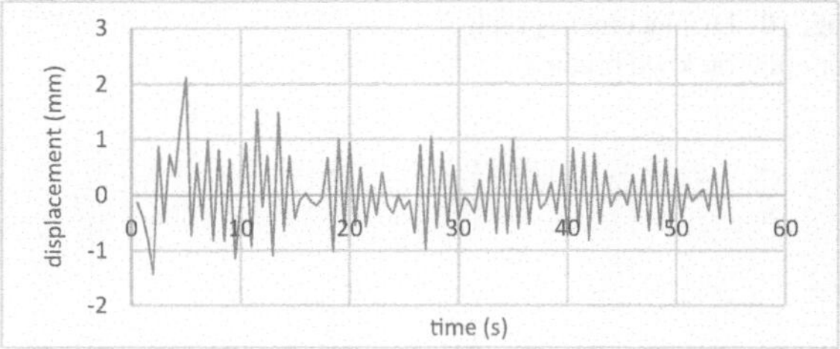

Fig. 13 Displacement graph for diagonal knee bracing with 2 knee

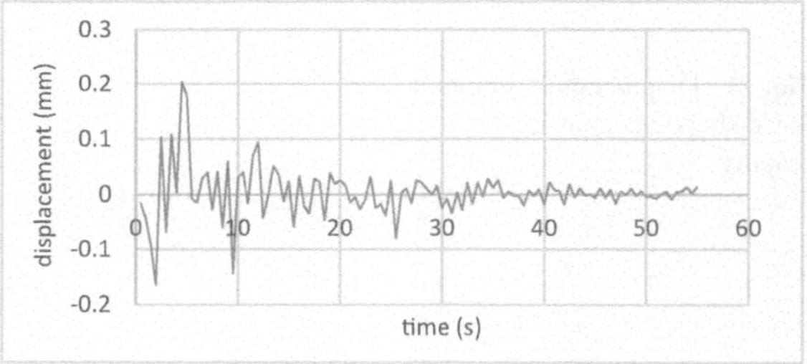

Fig. 14 Displacement graph for X knee bracing

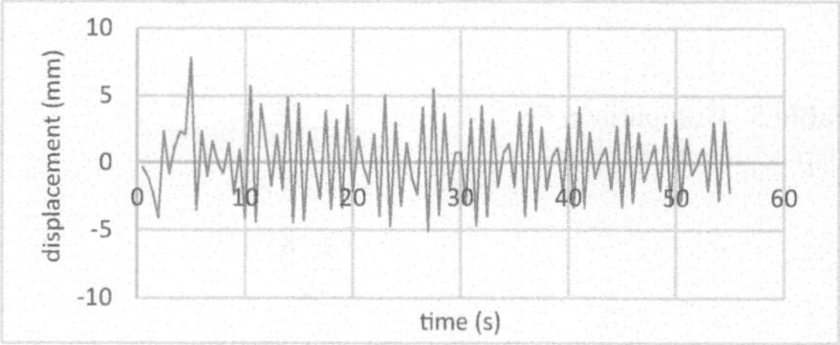

Fig. 15 Displacement graph for chevron knee bracing

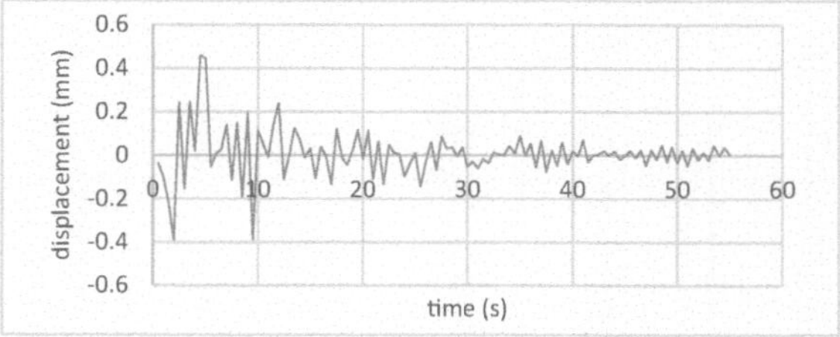

Fig. 16 Displacement graph for Y shaped eccentric bracing

Table 6 Comparison of displacement of frames [6]

Configuration type	Displacement (mm)
1 knee KBF	0.56842
2 knee KBF	1.0156
X KBF	0.46809
Chevron KBF	9.052
Y EBF	0.75311

4 Conclusions

From the non-linear studies that were conducted on different configurations of bracing systems, the following conclusions were made:

- Seismic performance of knee brace with 1 knee, knee brace with 2 knee, X KBF, Chevron KBF and Y Shaped Eccentric braces were studied.
- From the non-linear static analysis and non-linear time history analysis conducted, it was found that the X KBF showed very good behavior during a seismic activity. The ultimate load for X KBF is very much higher compared to other configurations. X KBF showed more lateral stiffness compared to other configurations. Since the braces are arranged in X shape, it will provide structural stiffness and reduces the maximum inter-storey drift.
- Diagonal braced frame with 1 knee have the less resistance to seismic forces than X KBF because provision of one diagonal element reduces the lateral stiffness during earthquake.
- The two knee diagonal knee bracing and Chevron Bracing exhibited the lowest ultimate load values, hence reducing its value for stiffness.
- From the time history analysis, it was noted that the maximum stiffness showcased by X KBF was more than the other three knee bracing configurations and Y shaped Eccentric bracing. In both cases, diagonal Braced frame with 1 knee exhibited a close behaviour to X KBF whereas Chevron Knee bracing showed least stiffness.
- Y shaped eccentric bracing shows less ductility and load carrying behaviour. Henceforth, knee bracing is recommended over Y shaped eccentric braced frame.
- Main beams are restrained from rotation by the provision of knee elements. The rotation may lead to interruption in the performance of structures like power plants where heavy equipment are braced by floor beams.

References

1. Balendra T, Sam MT, Liaw CY (1990) Diagonal brace with ductile knee anchor for a seismic steel frame. Earthq Eng Struct Dyn J 19(6):847–858
2. Arisizabal-Ochoa JD (1986) Disposable knee bracing: improvement in seismic design of steel frames. J Struct Eng ASCE 112(7):1544–1552

3. Sam MT, Liaw CY, Balendra T (1995) Earthquake-resistant steel frames with energy dissipating knee elements. Eng Struct 17(5):334–343
4. Balendra T, Lim EL, Liaw CY (1997) Large-scale seismic testing of knee-brace frame. J Struct Eng ASCE 123(1):11–19
5. Williams MS, Blakeborough A, Cle´ment D, Bourahla N (2002) Seismic behaviour of knee braced frames. Proc Inst Civ Eng Struct Build 152(2):147–155
6. Clement DE, William M (2004) Seismic design and analysis of a knee braced frame building. J Earthq Eng 8(4):323–543
7. Balendra T, Sam MT, Liaw CY (1991) Design of earthquake resistance steel frame with knee bracing. Earthq Eng Struct Dyn J 19(6):847–858
8. Balendra T, Sam MT, Liaw CY, Lee SL (1991) Preliminary studies into the behavior of knee braced frames subject to seismic loading. Eng Struct 13(1):67–74
9. Balendra T, Lim EL, Lee SL (1994) Ductile knee braced frame with shear yielding knee for seismic resistance structures. Earthq Eng Struct Dyn J 19(6):847–858
10. Khosravi P, Mofid M (2000) Non-linear analysis of disposable knee bracing. Comput Struct J 75:65–72
11. Huang Z, Qing-Song L, Long-Zhu C (2005) Elastoplastic analysis of knee bracing frame. J Zhejiang Univ Sci 6.4(8):784–789
12. Mofid M, lotfollahi M (2006) On the characteristics of new ductile knee bracing system. J Constr Steel Res 62:271–281
13. Khosravi P, Mofid M (2006) On the design of new ductile knee bracing. J Constr Steel Res 62:282–294
14. Miri M, Zare A, Hossein Abbas zadeh (2009) Seismic behaviour of steel frames investigation with knee brace based on pushover analysis. World Acad of Sci Eng Technol Int J Civil Environ Eng 3(2):122–128
15. Hsu HL, Lee CY (2011) Experimental evaluation on the seismic performance of steel knee braced frame structures with energy dissipation mechanism. Steel Compos Struct 11(1):77–91
16. Leelataviwat S, Suksan B, Srechai J, Warnitchai P (2011) Seismic design and behaviour of ductile knee-braced moment frames. J Struct Eng ASCE 137(5):579–588

Seismic Performance Evaluation of Hybrid Coupled Walls

M. R. Krishnapriya and Gayathri Krishna Kumar

Abstract Hybrid coupled walls (HCWs) are comprised of two or more reinforced concrete wall piers connected with steel coupling beams distributed over the height of the structure. Extensive research over the past several decades suggests that such systems are well suited in regions of high seismic risk. Steel coupling beams in a hybrid coupled shear wall provide a viable alternative for concrete coupling beams coupling individual reinforced concrete wall piers [1]. In such a system, coupling beams are designed to undergo inelastic deformation and dissipate seismic energy by means of capacity design principles. The coupling beam is installed with a steel plate with slits which acts as the sacrificing element. The study takes into account different slit configurations. Based on non-linear static and cyclic analysis the behaviour of coupling beams under different configuration is assessed and the best slit configuration is optimised. Time history analysis is carried out on the HCW model with the best slit configuration in ANSYS and compared with the conventional one. Thus the study proposes a new configuration of coupling beams with slits, which endows the coupling beam with damage controllability, a large capacity to consume energy, and a design flexibility regarding stiffness and strength.

Keywords Hybrid coupled wall (HCW) · Steel coupling beam · Slit configurations · Nonlinear dynamic analysis

1 Introduction

Recent major earthquakes, have demonstrated that our built environment and infrastructure, particularly in the urban context, need to be more resilient to earthquakes. Novel type of hybrid coupled wall (HCW), which consists of reinforced concrete

M. R. Krishnapriya (✉) · G. Krishna Kumar
Department of Civil Engineering, Federal Institute of Science and Technology, Angamaly, India
e-mail: krishnapriyamr94@gmail.com

G. Krishna Kumar
e-mail: gaya3krishnakumar92@gmail.com

K. Dasgupta et al. (eds.), *Proceedings of SECON'19*,
Lecture Notes in Civil Engineering 46,
https://doi.org/10.1007/978-3-030-26365-2_52

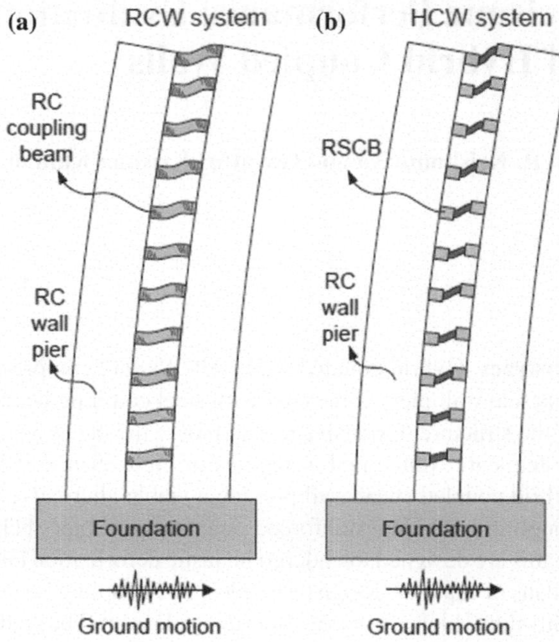

Fig. 1 Sketch of coupled walls. **a** RC coupled wall. **b** Hybrid coupled wall [2]

(RC) wall piers and replaceable steel coupling beams (RSCBs), has been proposed for enhancing the seismic resilience of high-rise buildings [2]. In such a system, coupling beams are designed to undergo inelastic deformation and dissipate seismic energy, as shown in Fig. 1. By means of capacity design principles, the inelastic deformation can concentrate in the fuse shear links. Coupling beams are employed to better distribute load and deformation demands throughout the wall system rather than concentrate it at the plastic hinge region [3, 4]. This solution also evolved as a practical means to provide openings otherwise shear walls remains solid [5].

Xiaodong et al. presented a paper on the assessment of the seismic performance of the novel hybrid coupled walls. The performance of a central "fuse" shear link connected to steel beam segments at its two ends is illustrated. By means of capacity design principles, the inelastic deformation can concentrate in the "fuse" shear links, while the steel beam segments remain elastic. In addition, the greater energy dissipation capacity of the RSCBs than the RC coupling beams can further decrease the lateral drifts of the HCW building. The maximum interstory drift ratio of the HCW building is up to 24.5% smaller than the RCW building. When subjected to severe earthquakes, the HCW have stable hysteretic responses and develop large over strength in the cyclic reversal after yielding of the shear links, which ensures an adequate coupling ratio. On the contrary, the conventional RC coupling beams have considerable degradation of stiffness and strength after yielding, which in turn leads to a decrease in the coupling ratio of the coupled wall system. The results indicate that most of the damage is concentrated in the coupling beams and non-structural components [2].

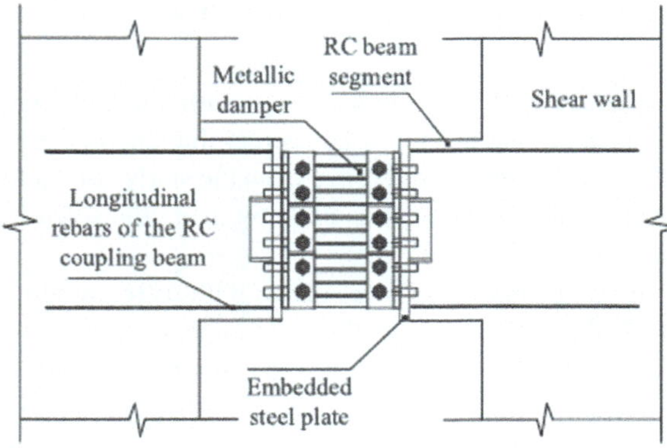

Fig. 2 Configuration of hybrid coupling beam [6]

Tao et al., proposed hybrid coupling beams installed with metallic dampers with slits to significantly improve the seismic performance of reinforced concrete (RC) shear wall structures. This study began by proposing a particular configuration installed with a metallic damper with slits. The proposed hybrid coupling beam is configured in typical manner with a metallic damper at the mid-span and an RC beam segment at each end of the metallic damper, as shown in Fig. 2 Therefore, the damper relies upon the bending plasticity to dissipate energy. The advantage is that the slits decouple the strength and stiffness designs of the shear-type damper, endowing it with a large design flexibility. Therefore, the strength of the hybrid coupling beam is determined by the component with the smallest strength, which is the damper. The damper yields first during an earthquake, thus absorbing a large amount of energy, and protecting the RC part of the coupling beam [6].

The hysteretic curve of the metallic damper indicates a very large capacity to dissipate energy. The damper dissipated more than 90% of the total energy consumption of the hybrid coupling beam. It is therefore concluded that the metallic damper with slits can effectively protect the RC components by consuming a large portion of the seismic energy [6].

2 Finite Element Analysis

The principal objective is to investigate the seismic behavior of hybrid coupled wall systems with slits using detailed and comprehensive analytical tools that account for all the major aspects of structural behavior.

2.1 Description of the Specimen

A hybrid coupling beam specimen with a metallic damper is designed with dimensions, as shown in Fig. 3. The actual thickness of the steel plate for the damper was taken as 10.1 mm as reference from the study conducted by Tao et al., Finite Element Analysis of the model was done by using the software ANSYS 16. Figure 4 shows the typical FE model.

Five FE models of steel plates with different slit configuration was considered. Steel plates with slits at different angles are modelled. A steel plate without slits was also modelled as reference to compare the performance of steel plates with slits.

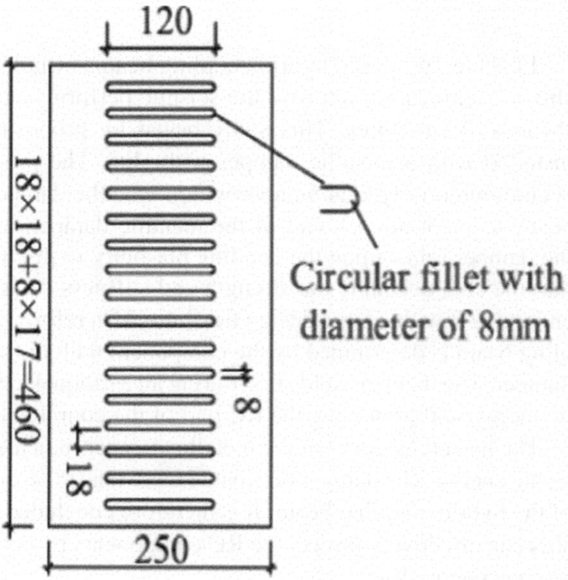

Fig. 3 Design of steel damper with slits

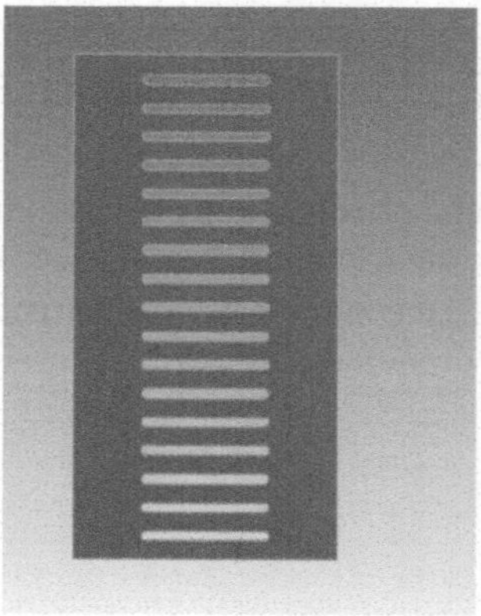

Fig. 4 Model of steel plate with slits [6]

Model 1 (N0_SL) consists of no slits. Model 2 (SL_15) has its slits inclined at an angle 15°. Model 3 (SL_30) has its slits inclined at an angle 30°. Model 4 (SL_45) has its slits inclined at an angle 45°. Model 5 (SL_0) has its slits inclined at an angle 0°. The specimens were loaded horizontally. One end of the steel plate was fixed and other end was displacement-controlled.

2.2 *Analysis Results*

The effect of variation of slit configuration in steel plates was assessed by conducting non-linear static pushover analysis and cyclic analysis. The total deformation and force is given in Table 1 which concludes that the model SL_45 undergoes more deformation. This deformation indicates the ductile performance of the steel plate. Figure 5 shows the pushover curve. The ductility ratio and over strength factor for

Table 1 Results of pushover analysis

Model	Load (kN)	Total deformation (mm)
N0_SL	818.88	1.549
SL_15	151.21	20.324
SL_30	173.31	23.245
SL_45	262.67	31.676
SL_0	135.47	20.111

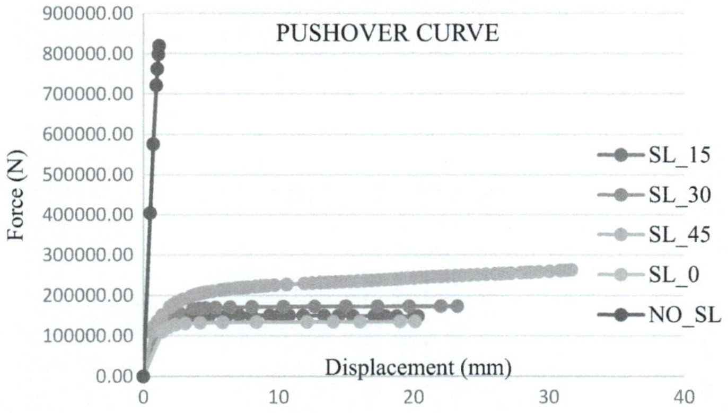

Fig. 5 Pushover curve

each model is found out. The ratio of ultimate displacement to the yield displacement is taken as the ductility ratio and the ratio of ultimate force to the yield force is taken as the over strength factor. Table 2 shows the results summary of ductility ratio and over strength factor. Taking into consideration all these factors we can conclude that SL_45 is the better model compared to others.

2.3 *Hysteretic Behavior*

The seismic performance of all the configurations are studied by applying a cyclic loading. Figure 6 shows the combination of hysteretic curves of all the models considered. The force-displacement hysteretic responses of the different configuration shows that SL_45 shows better seismic performance compared to others. Table 3 shows the results summary of cyclic analysis of all the models. The results shows that the models SL_30 and SL_45 takes more cycles compared to other models.

Table 2 Results summary of ductility ratio and over strength factor

Model	Ductility ratio	Over strength factor
N0_SL	1	1
SL_15	14.085	1.244
SL_30	22.944	1.611
SL_45	52.22	2.439
SL_0	14.328	1.233

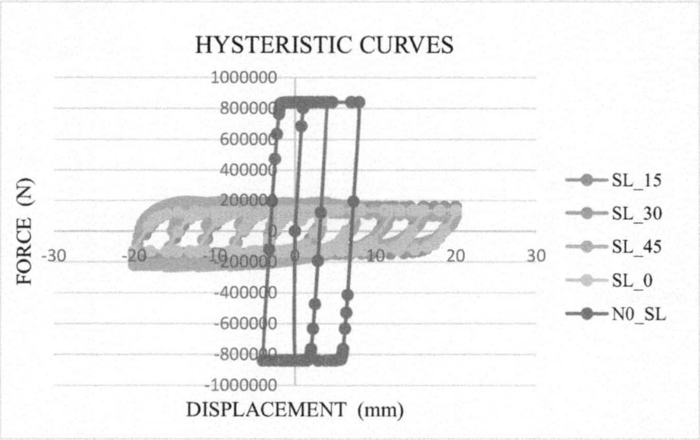

Fig. 6 Combination of hysteretic curves of all models

Table 3 Results summary of cyclic analysis

Model	Force (kN)	Deformation (mm)	Time (s)	Cycles	Area (m^2)
NO_SL	839.87	8.9072	5	1.25	21.51
NO_SL	−839.77	4.4601	3	0.75	
SL_15	159.15	21.342	17	4.25	26.73
SL_15	−150.88	2.4036	10.156	2.539	
SL_30	168.44	11.031	15.52	3.88	27.77
SL_30	−188.96	24.938	19	4.75	
SL_45	190.54	20.845	19.334	4.83	32.97
SL_45	−228.01	23.318	15	3.75	
SL_0	133.90	12.036	9	2.25	26.62
SL_0	−133.89	12.029	11	2.75	

3 Hybrid Coupled Wall

The seismic performance of reinforced concrete (RC) shear wall structures can be significantly improved by providing coupling beams installed with metallic dampers. The damper yields first during an earthquake, thus absorbing a large amount of energy, and protecting the RC part of the coupling beam. The study proposes the model SL_45 as the best configuration. Three models were considered, one with steel plates with slits (Fig. 7), other without slits (Fig. 8) and a conventional coupling beam. Pushover analysis of the three models was carried out. Push over analysis helps in understanding the deformation and cracking of a structure in case of earthquake and gives a fair understanding of the deformation of building and formation of plastic hinges in the structure. Table 4 shows the results of pushover analysis. In hybrid coupled walls the steel plates with slits dissipated more energy. This can be clearly observed from the stress distribution as shown in Fig. 9. In conventional system more stress was transferred to the columns.

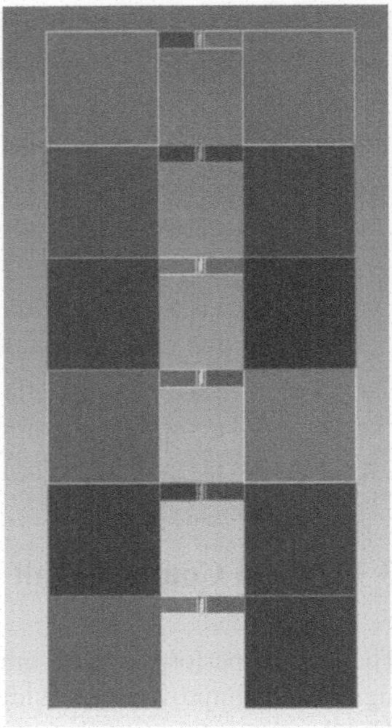

Fig. 7 Coupling beam with slits

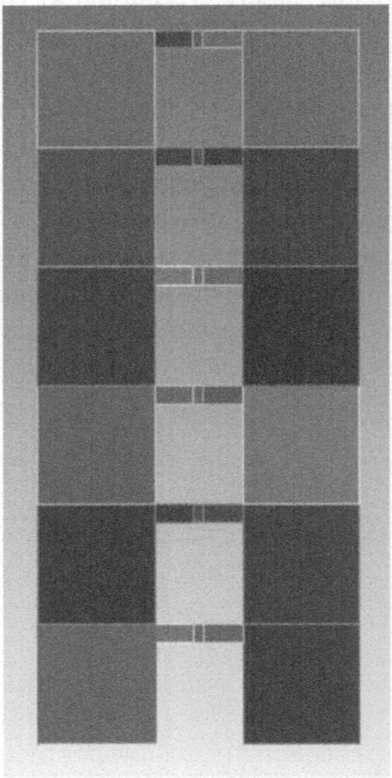

Fig. 8 Coupling beam with no slits

Table 4 Results of pushover analysis

Model	Max. load (kN)	Total deformation (mm)
Conventional	1486.6	142.23
With no slits	1544.7	156.59
With slits	1562.7	161.43

(**a**) With slits (**b**) without slits (**c**) conventional

Fig. 9 Stress distribution of HCW

4 Conclusions and Recommendations

The objective of this work has been to investigate the seismic performance of hybrid coupled wall systems. The study proposes a new configuration of coupling beams using steel plates with slits, which endows the coupling beam with damage controllability, a large capacity to consume energy, and a design flexibility regarding stiffness and strength. The seismic performance of the different configurations of steel plates with slits were examined under cyclic loading. The different configurations used for the study are coupling beams with no slits, slits at an angle 15°, 30°, 45° and 0°. On examining these configurations, considering the energy dissipation and seismic performance coupling beams with slits at an angle 45° found to be the better model.

The performance of the HCW with slits is assessed against an equivalent HCW without slits and also a conventional model. Compared with the HCW building with slits the load carrying capacity was increased by 5.11% than conventional model. Most importantly, damage to the RC components was well controlled in coupling beam with slits, significantly improving the seismic performance. In light of observations and findings obtained from this research, HCW with slits helps to mitigate the damage occurring in wall piers and other structural components (Fig. 10).

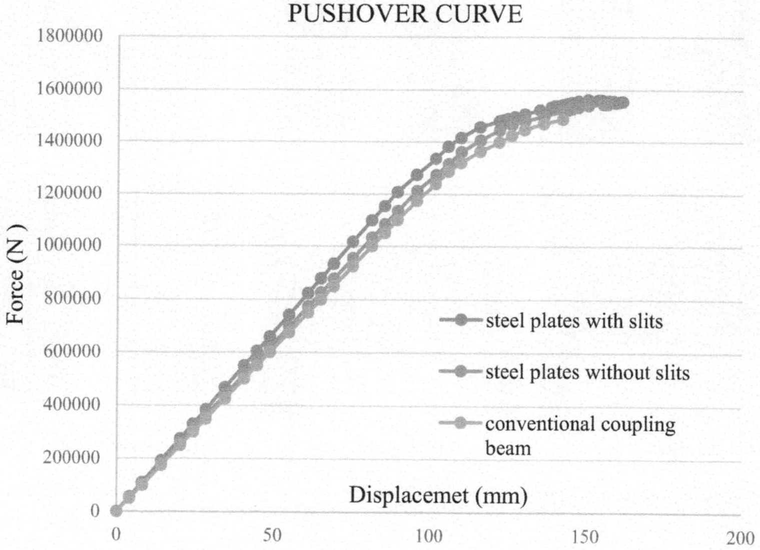

Fig. 10 Pushover curves of hybrid coupled walls

References

1. Chairunnisa N, Satyarno I, Muslikh, Aminullah A (2014) Various existing methods of coupling beams and a new alternative hybrid method (Sci Direct)
2. Ji X, Liu D, Hutt CM (2018) Seismic performance evaluation of a high-rise building with novel hybrid coupled walls. J Eng Struct (Sci Direct)
3. El-Tawil S, Harries KA, Fortney PJ, Shahrooz BM, Kurama Y (2010) Seismic design of hybrid coupled wall systems: state of the art. J Struct Eng ASCE
4. Eljadei AA, Harries KA (2014). Design of coupled wall structures as evolving structural systems. J Eng Struct (Sci Direct)
5. Hitaka T, Matsui C (2003) Experimental study on steel shear wall with slits. J Struct Eng ASCE
6. Wanga T, Shang Q, Wang X, Li J, Kong Z (2018) Experimental validation of RC shear wall structures with hybrid coupling beams. J Soil Dyn Earthq Eng (Sci Direct)
7. Rezapour M, Ghassemieh M (2018) Macroscopic modelling of coupled concrete shear wall. J Eng Struct (Sci Direct)
8. Farsi A, Keshavarzi F, Pouladi P, Mirghaderi R (2016) Experimental study of a replaceable steel coupling beam with an end-plate connection. J Constr Steel Res (Sci Direct)

Effect of Bracings and Shearwalls on Seismic Performance of Buildings Situated on Sloping Region

Sooraj Babu and Reshma Prasad

Abstract Hill towns of India are vulnerable to different types of natural hazards, which may cause enormous destruction during their occurrence. It leads to substantial loss of precious human life and resources. In this study, two different configurations of buildings have been modeled and analyzed using ETABS 2016 software. The building can be made strong to resist lateral loads by adding bracings or shear wall to the structure. A parametric study has been carried out, in which the buildings are geometrically varied with different bracing and shear wall configuration. All analytical models have been subjected to seismic forces along and across hill slope direction and analyzed by using Time History Analysis. Various building configurations including combination of shear wall and bracings were considered and their behavior in terms of maximum storey displacements, storey drifts and base shear in buildings were discussed. After studying the behavior of buildings on slopes, notable increase in seismic performance is obtained on combined configurations using both bracings and shear walls.

Keywords Step back buildings · Step back set back buildings · Bracings · Shear walls · Time history analysis · Storey displacement · Storey drift · Base shear

1 Introduction

Structure subjected to seismic/earthquake forces are always vulnerable to damage and if it occurs on hills the chances of damage increases much more due to increased lateral forces on short columns on uphill side and thus leads to the formation of plastic hinges. Structures on slopes differ from those on plains because they are irregular horizontally as well as vertically. It has been observed from the past earthquakes, that

S. Babu (✉) · R. Prasad
Department of Civil Engineering, Federal Institute of Science and Technology, Angamaly, Kochi 683577, India
e-mail: soorajbabu1995@gmail.com

R. Prasad
e-mail: reshma.prasad1@gmail.com

K. Dasgupta et al. (eds.), *Proceedings of SECON'19*,
Lecture Notes in Civil Engineering 46,
https://doi.org/10.1007/978-3-030-26365-2_53

buildings in hilly regions have experienced high degree of demand leading to collapse though they have been designed for safety of the occupants against natural hazards [1]. Hence, while adopting practice of multi-storey buildings in these hilly and seismically active areas, utmost care should be taken, for making these buildings earthquake resistant. Therefore, this paper aims to study the different building configurations and construction methods regarding the seismic behavior enforced in hill towns for safety against disasters thus providing different techniques to reduce and regulate the seismic activities produced by nature [2].

2 Modelling

The base models considered for the seismic analysis in this study are Step back type building and step back set back type buildings [3]. The software used for modeling and analysis is ETABS 2016. Slab thickness defined in the software is about 125 mm. The foundation depth taken on bottom bay is 1.5 m for all the models. The slope inclination assumed is of 26°. Base models created are shown in Fig. 1.

The loads taken into consideration are dead load of the structure and imposed load of 3 kN/m^2. The seismic parameters considered in dynamic analysis of all the models are assumed as per IS 1893 (Part 1): 2002. The hill buildings are assumed to be in Zone V with the peak ground acceleration value of 0.36 g. The importance factor, I is taken as 1.5 (for important building). Also, the response reduction factor R taken as 5 for SMRF system of the buildings. The soil strata beneath the foundation are assumed as medium soil [4, 5]. In this study different geometrical variation of bracings and shear wall configurations are introduced along and across the sloping region in the respective building configurations and their seismic performance is been evaluated [6]. The time history function data entered for analysis process was that of El Centro Earthquake (Imperial Valley earthquake) which occurred in 1940 at southern California [7].

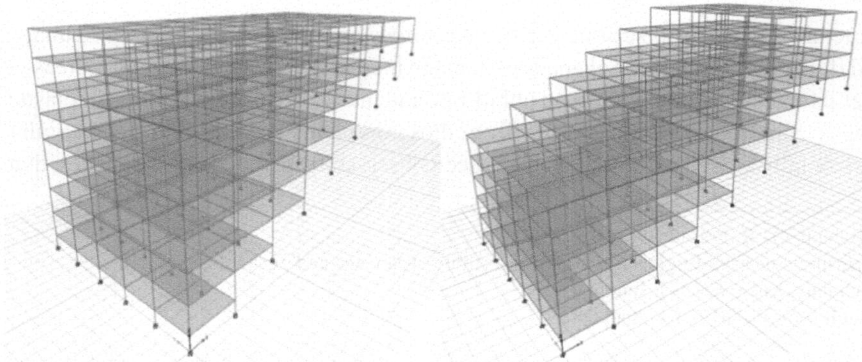

Fig. 1 Step back and step back set back building model created

2.1 Shearwall Configurations Created

On Step back buildings: Shear walls of Concrete M30 grade with thickness of 250 mm were adopted. The nomenclatures given for the configurations are,

(a) SB_S_Y1 (d) SB_S_Y4 (g) SB_S_C3
(b) SB_S_Y2 (e) SB_S_C1 (h) SB_S_C4
(c) SB_S_Y3 (f) SB_S_C2 (i) SB_S_C5

Different shear wall configurations are provided across the slope direction (Y direction) from (a) to (d) and from (e) to (i) shear walls are provided along and across the slopes (both X & Y dir) [8] as shown in Fig. 2.

On Step back set back buildings:

(a) SS_S_Y1 (d) SS_S_C1 (g) SS_S_C4
(b) SS_S_Y2 (e) SS_S_C2
(c) SS_S_Y3 (f) SS_S_C3

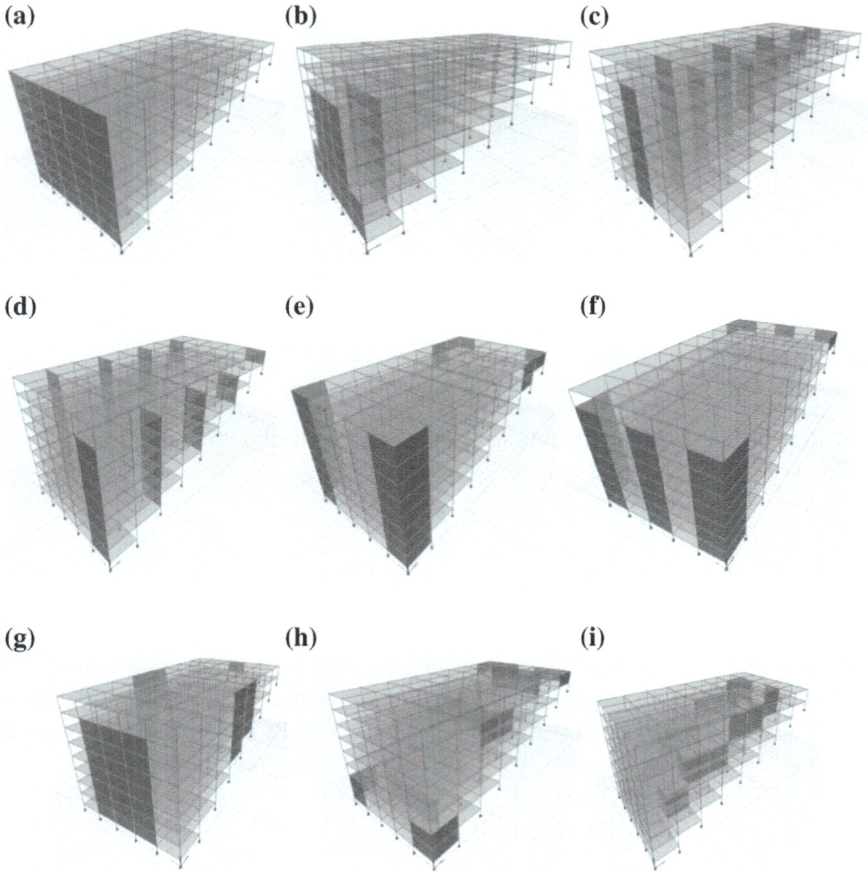

Fig. 2 Shear wall configurations created on step back buildings

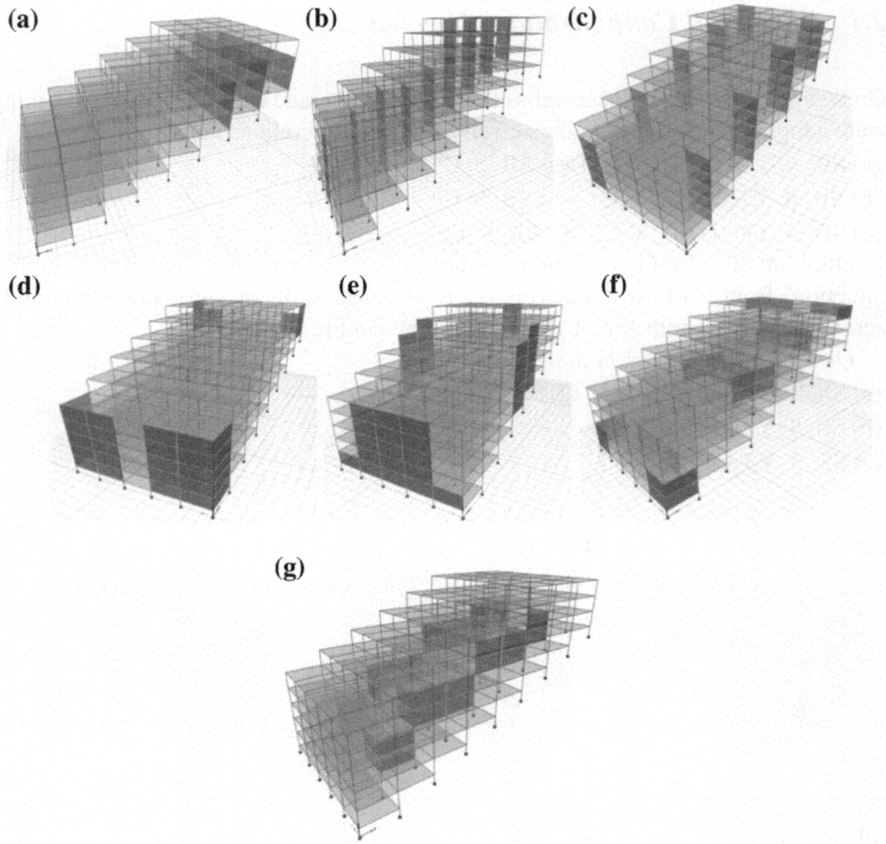

Fig. 3 Shear wall configurations created on step back set back buildings

Different shear wall configurations adopted on step back set buildings are shown in Fig. 3.

2.2 Bracing Configurations Created

On Step back buildings: Steel bracing of ISA 150 mm × 150 mm × 18 mm Fe250 is selected. 'X' Type bracings were adopted on the frames [9]. Same configuration types that were created on above models with shear walls are replaced by bracing units here. Nomenclatures given for the different bracing configurations are

(a) SB_B_Y1 (d) SB_B_Y4 (g) SB_B_C3
(b) SB_B_Y2 (e) SB_B_C1 (h) SB_B_C4
(c) SB_B_Y3 (f) SB_B_C2 (i) SB_B_C5

On Step back set back buildings:
(a) SS_B_Y1 (d) SS_B_C1 (g) SB_B_C4
(b) SS_B_Y2 (e) SS_B_C2
(c) SS_B_Y3 (f) SS_B_C3

3 Analysis and Results

As per IS 1893:2002 cl 7.11.1, Max storey displacement must be less than H * 0.004, where H is the height of the building. In the case of step back buildings H = 25.5 m therefore the max storey displacement must be less than 102 mm. For Step back set back H = 31.5 m so the max storey displacement must be less than 126 mm for it to be seismically stable. In this study the seismic parameters along x direction are not considered as it is already seismically stable for both buildings (52.06 and 111.72 mm for step back and step back set back respectively). Therefore shear walls and bracings are only provided across the slope direction.

3.1 Shear Wall Configurations

Time history analysis results of Step back buildings with shear walls provided across and along slope are tabulated in Tables 1 and 2.

SB_S_Y1 is the preferred configuration as the max storey displacement shows 75.13% reduction while providing shear walls across the slope with base shear of 26,070.32 kN whereas SB_S_C3 provides 45.68% displacement reduction along slope and 82.04% across slope. Overall best seismic performance is for SB_S_C3 but providing with higher reinforcements for the larger base shear.

Time history analysis results of step back set back buildings with shear walls provided along and across slope are tabulated in Tables 3 and 4.

SS_S_Y3 is the preferred configuration as the max storey displacement shows 64.78% reduction while providing shear walls across the slope with base shear of

Table 1 Time history analysis of step back buildings with shear walls across slope

Model	Across slope (Y DIR)			
	Max storey displacement	% difference	Max storey drift	Base shear (kN)
SB	180.441	1	0.0112	14,550.56
SB_S_Y1	44.87	75.13	0.0082	26,070.32
SB_S_Y2	60.656	66.38	0.0087	20,241.03
SB_S_Y3	84.192	53.34	0.0045	29,996.56
SB_S_Y4	85.383	52.68	0.0047	30,859.98

Table 2 Analysis of step back buildings with shear walls along and across slope

Model	Time history analysis							
	X				Y			
	Displacement (mm)	% difference	Max storey drift	Base shear (kN)	Displacement	% difference	Max storey drift	Base shear (kN)
SB	52.062	1	0.0061	20,956.76	180.441	1	0.0112	14,550.56
SB_S_C1	22.41	56.95	0.0011	32,317.98	74.08	58.94	0.0057	30,472.11
SB_S_C2	29.092	44.12	0.0018	28,464.55	82.86	54.07	0.0062	30,275.43
SB_S_C3	28.286	45.68	0.0024	34,805.1	32.4	82.04	0.0043	31,474.67
SB_S_C4	27	48.13	0.0037	26,746.55	90.58	49.8	0.01	21,990.73
SB_S_C5	14.356	72.42	0.0021	26,941.6	85.492	52.62	0.0056	26,559.32

Table 3 Time history analysis of step back set back buildings with shear walls across slope

Model	Across slope (Y DIR)			
	Max storey displacement (mm)	% difference	Max storey drift	Base shear (kN)
SS	156.505	1	0.0139	16,914.24
SS_S_Y1	75.577	51.7	0.0072	25,713.97
SS_S_Y2	60.092	61.6	0.0054	29,213.96
SS_S_Y3	55.112	64.78	0.0043	32,523.26

Table 4 Analysis of step back set back buildings with shear walls along and across slope

Model	Time history analysis							
	X				Y			
	Displacement (mm)	% difference	Max storey drift	Base shear (kN)	Displacement (mm)	% difference	Max storey drift	Base shear (kN)
SS	111.723	1	0.0106	14,689.59	156.505	1	0.0139	16,914.24
SS_S_C1	43.56	61.01	0.0047	22,312.192	77.589	50.42	0.0123	31,749.67
SS_S_C2	40.84	63.44	0.0081	29,175.53	74.748	52.23	0.0082	20,770.61
SS_S_C3	73.93	33.82	0.0094	21,889.82	77.26	50.63	0.0105	15,711.12
SS_S_C4	65.9	41.01	0.0095	30,289.39	78.64	49.75	0.01	27,240.1

32,523.26 kN whereas SS_S_C2 provides 63.44% displacement reduction along slope and 52.23% across slope. Overall best seismic performance is for SB_S_C2 as it is seismically stable on both directions effectively with moderate base shear.

3.2 *Bracing Configurations*

Time history analysis results of step back buildings with bracings provided along and across slope are tabulated in Tables 5 and 6.

Table 5 Time history analysis of step back buildings with bracings across slope

Model	Across slope (Y DIR)			
	Max storey displacement (mm)	% difference	Max storey drift	Base shear (kN)
SB	180.441	1	0.0112	14,550.56
SB_B_Y1	77.77	56.9	0.0051	27,287.69
SB_B_Y2	100.77	44.15	0.0112	18,084.08
SB_B_Y3	101.53	43.73	0.0052	20,513.9
SB_B_Y4	102.08	43.42	0.0054	19,749.87

Table 6 Time history analysis of step back buildings with bracings along and across slope

Model	Time history analysis							
	X				Y			
	Displacement (mm)	% difference	Max storey drift	Base shear (kN)	Displacement (mm)	% difference	Max storey drift	Base shear (kN)
SB	52.062	1	0.0061	20,956.76	180.441	1	0.0112	14,550.56
SB_B_C1	33.124	36.37	0.0022	27,067.95	107.4	40.47	0.0056	22,234.54
SB_B_C2	48.716	6.42	0.0039	28,168.05	103.28	42.76	0.0062	19,093.43
SB_B_C3	26.64	48.83	0.0022	25,944.63	94.68	47.52	0.0057	21,263.45
SB__B_C4	50.51	2.98	0.0059	29,693.41	138.74	23.11	0.0096	18,083.46
SB_B_C5	25.73	50.57	0.0027	22,672.16	119.76	33.62	0.0082	17,836.26

SB_B_Y1 is the preferred configuration as the max storey displacement shows 56.9% reduction while providing bracings across the slope with base shear of 27,287.69 kN whereas SB_B_C3 provides 48.83% displacement reduction along slope and 47.52% across slope. Overall best seismic performance is for SB_B_C3 as it has lower base shear along Y dir compared to that of SB_B_Y1. Time history analysis results of step back set back buildings with bracings provided along and across slope are tabulated in Tables 7 and 8.

SS_B_Y3 is the preferred configuration as the max storey displacement shows 51.45% reduction while providing bracings across the slope with base shear of 13,833.84 kN whereas SS_S_C3 provides 29.62% displacement reduction along

Table 7 Time history analysis of step back set back buildings with bracings across slope

Model	Across slope (Y DIR)			
	Max storey displacement (mm)	% difference	Max storey drift	Base shear (kN)
SS	156.505	1	0.0139	16,914.24
SS_B_Y1	77.29	50.61	0.007	20,551.92
SS_B_Y2	79.501	49.2	0.0079	12,763.33
SS_B_Y3	75.98	51.45	0.0062	13,833.84

Table 8 Analysis of step back set back buildings with bracings along and across slope

Model	Time history analysis							
	X				Y			
	Displacement (mm)	% difference	Max storey drift	Base shear (kN)	Displacement	% difference	Max storey drift	Base shear (kN)
SS	111.723	1	0.0106	14,689.59	156.505	1	0.0139	16,914.24
SS_B_C1	70.553	36.85	0.0061	29,017.35	91.24	41.7	0.0095	23420.84
SS_B_C2	71.229	36.24	0.0064	19,966.88	92.267	41.04	0.0086	20,520.97
SS_B_C3	78.63	29.62	0.0089	17,721.58	86.77	44.55	0.0118	13,619.67
SS_B_C4	80.27	28.15	0.0093	18,625.62	93.4	40.32	0.0085	15,890.31

slope and 44.55% across slope. Overall best seismic performance is for SB_S_C3 as it is seismically stable on both directions effectively with lowest base shear.

4 Combined Bracings and Shearwall Configurations Created

Same type and material as of the above bracings and shear walls are used in the combined configurations. Nomenclatures given for the different combined configurations on step back buildings are,
(a) SB_BS1 (d) SB_BS4 (g) SB_BS7
(b) SB_BS2 (e) SB_BS5 (h) SB_BS8
(c) SB_BS3 (f) SB_BS6 (i) SB_BS9

Different combined bracings and shear wall configurations created on step back buildings are shown in Fig. 4.

On Step back set back buildings:
(a) SS_BS1 (d) SS_BS4 (g) SS_BS7
(b) SS_BS2 (e) SS_BS5
(c) SS_BS3 (f) SS_BS6

Different combined bracings and shear wall configurations created on step back set back buildings are shown in Fig. 5.

5 Analysis of Combined Bracings and Shear Wall Configurations

Time history analysis results of step back buildings with combined configurations using bracings and shear walls along and across slope are tabulated in Table 9.

In the case of combined shear wall and bracing configurations in step back buildings, SB_BS5 is the most preferred configuration as the max storey displacement shows 49.15% reduction along slope and 74.71% across slope which has the highest reduction value on either directions than other configurations. The base shear is found to be higher so necessary structural reinforcements must be provided.

Time history analysis results of combined configurations created on step back set back buildings are tabulated in Table 10. Here SS_BS7 is the most preferred combined configuration as the max storey displacement shows 29.81% reduction along slope and 49.62% across slope which has the lowest value for base shear than other configurations. Shear walls and bracings are provided at inner section of the building in a combined C section manner.

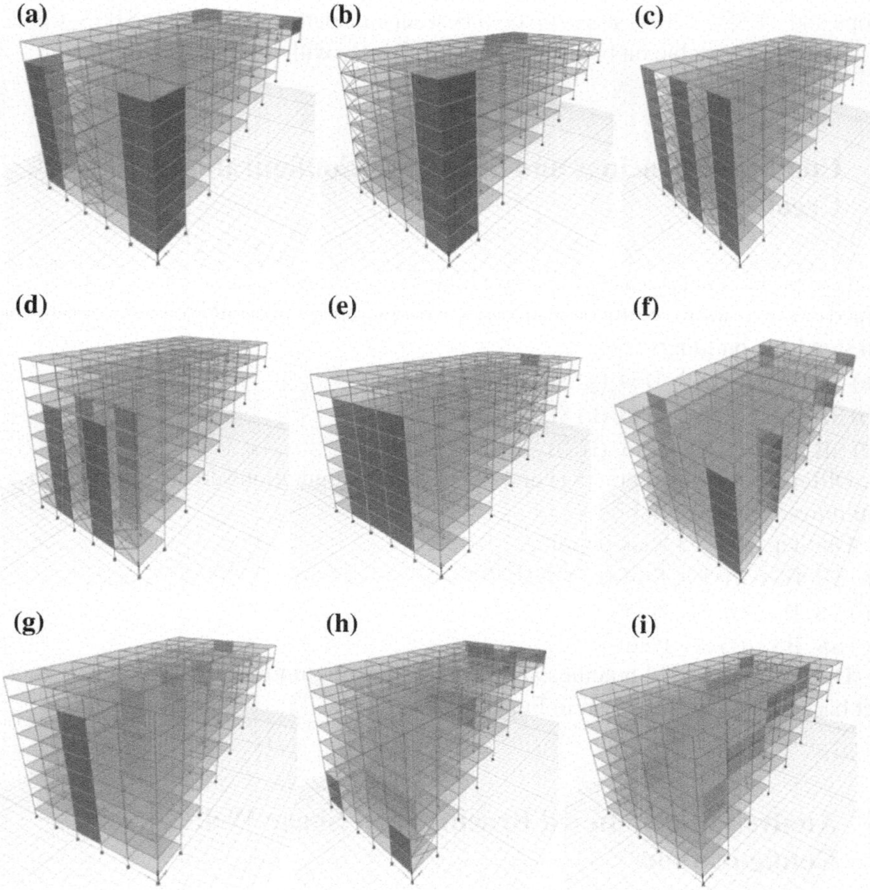

Fig. 4 Combined bracings and shear wall configurations created on step back buildings

6 Conclusions

From the present study the following conclusions are drawn.

1. Including shear walls and bracings with different configurations independently and in combined manner showed improvement in the seismic performance on both step back and step back set back buildings since lateral displacements reduces considerably than other typical conventional buildings that are situated on sloping regions.
2. For better seismic stability construction, the mentioned configurations in the respective building types can be adopted accordingly
3. Combined configurations using shear walls and bracings provides greater overall performances than the individual configurations. Combined configuration

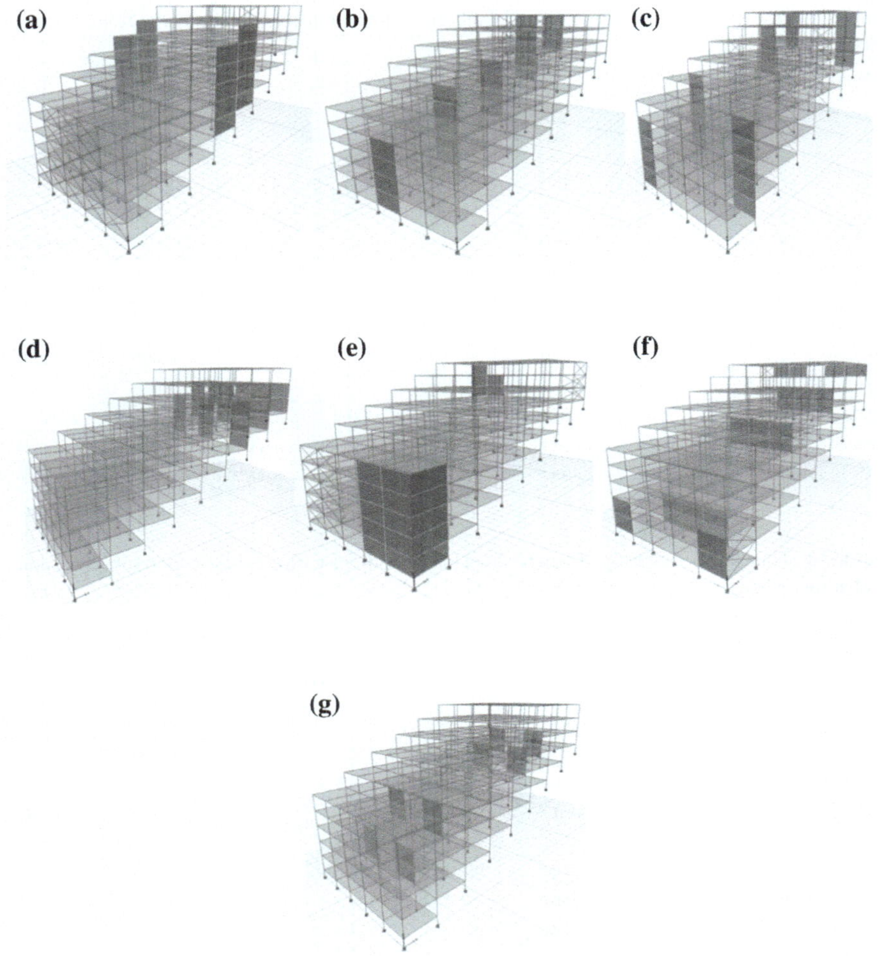

Fig. 5 Combined configurations on step back set back buildings

achieves 20% extra stability in the case of step back buildings and 5% more performance in step back set back buildings.

4. Base shear of the buildings increases with increase in seismic weight and reduction in storey displacements. Due to the provision of shear walls and bracings base shear get increased in both X and Y direction.
5. Higher storey drift is found at the lower sections of the building, this is generally caused due to the open storey effect which occurs at lower area. Providing efficient reinforcements for the columns in those regions can reduce this defect.
6. Fundamental time period decreases with decrease in overall storey displacement as increase in lateral stiffness occurs due to provision of bracings and shear walls.

Table 9 Time history analysis of step back buildings with combined configurations along & across slope

Model	Time history analysis							
	X				Y			
	Displacement (mm)	% difference	Max storey drift	Base shear (kN)	Displacement (mm)	% difference	Max storey drift	Base shear (kN)
SB	52.062	1	0.0061	20,956.76	180.441	1	0.0112	14,550.56
SB_BS1	28.065	46.09	0.0017	27,260.17	87.08	51.74	0.0062	27,068.26
SB_BS2	27.37	47.42	0.0021	25,197.17	94.35	47.71	0.0047	20,196.97
SB_BS3	54.308	4.31	0.0063	22,101.46	59.6	66.96	0.0099	35,059.42
SB_BS4	52	0.11	0.0062	21,796.27	96.28	46.64	0.0124	21,553.27
SB_BS5	26.473	49.15	0.0022	26,739.42	45.621	74.71	0.0055	29,408.97
SB_BS6	48.274	7.27	0.0059	21,070.55	89.258	50.53	0.0055	20,449.55
SB_BS7	49.385	5.14	0.0057	21,724.97	87.854	51.31	0.0049	24,390.74
SB_BS8	48.89	6.09	0.0057	33,501.06	97.47	45.98	0.0093	16,598.31
SB_BS9	14.545	72.06	0.0019	27,113.65	100.6	44.24	0.0056	22,465.06

Table 10 Time history analysis of step back set back buildings with combined configurations along and across slope

Model	Time history analysis							
	X				Y			
	Displacement (mm)	% difference	Max storey drift	Base shear (kN)	Displacement	% difference	Max storey drift	Base shear (kN)
SS	111.723	1	0.0106	14,689.59	156.505	1	0.0139	16,914.24
SS_BS1	41.196	63.12	0.0083	28,421.56	92.432	40.93	0.0075	17,271.57
SS_BS2	111.985	0.23	0.0106	15,058.26	58.905	62.36	0.0059	24,496.19
SS_BS3	109.667	1.84	0.0103	16,376.23	59.208	62.16	0.0056	26,896.01
SS_BS4	93.513	16.3	0.0085	18,011.21	74.27	52.54	0.0069	22,334.46
SS_BS5	67.834	39.28	0.0064	24,262.84	84.147	46.23	0.0116	24,171.47
SS_BS6	78.366	29.85	0.0086	18,532.75	78.875	49.6	0.01	16,058.09
SS_BS7	78.412	29.81	0.0092	19,144.85	78.84	49.62	0.0101	11,631.32

References

1. Mevawala NB, Desai DAK (2016) Seismic non-linear time history analysis of building resting on sloping ground with special study on Nepal Earthquake. IOSR J Mech Civil Eng (IOSR-JMCE)
2. Kumar A (2018) Review of building regulations for safety against hazards in Indian hill towns. J Urban Manage
3. Nagargoje SM, Sable KS (2012) Seismic performance of multi-storeyed building on sloping ground. Elixir Int J
4. Kiran T, Jayaramappa N (2017) Seismic performance of RC frame buildings resting on sloping ground. IOSR J Mech Civil Eng (IOSR-JMCE) 14(2)
5. Patel M, Umar Farooque Patel M (2014) A performance study and seismic evaluation of RC frame buildings on sloping ground. IOSR J Mech Civil Eng (IOSR-JMCE)
6. Birajdar BG, Nalawade SS (2004) Seismic analysis of buildings resting on sloping ground. In: 13th world conference on earthquake engineering, Vancouver, B.C., Canada, paper no. 1472

7. Pardhasaradhi A, Raj Kumar V (2017) Dynamic analysis of sloped buildings: experimental and numerical studies. Int J Innov Res Technol 3(10)
8. Pawar SP (2016) Effect of positioning of RC shear walls of different shapes on seismic performance of building resting on sloping ground. Int J Civil Eng Technol (IJCIET) 7(3)
9. Prasath R, Joshua Daniel A (2016) Study for improving the performance of the step-back building with bracings. Int Res J Eng Technol

Seismic Analysis of Transmission Tower Line Systems

P. V. Vaisakh and Neeraja Nair

Abstract Electricity is one of the most important necessity of any growing economy. The transmission tower line system consisting of transmission towers and conductors are the main element of an electrical system. Failure of this system leads to greater economic loss. However most of the researches are concentrated on static load, impulsive load and wind loads but not on seismic effects of the system. This study mainly focuses on the seismic behaviour of the combined system of transmission towers and conductors. The model is created based on Indian standard codes. A three tower model connected by cables are used for analysis as they give more realistic results than a single tower model. Four basic towers 66, 132, 220 and 400 kV are created in SAP2000 and time history analysis is conducted on these models to study their seismic behaviour. Analysis is also conducted to find out their behaviour under critical conditions of seismic loading.

Keywords Transmission tower line system · Seismic analysis · Time history analysis · Critical condition

1 Introduction

An electricity transmission system usually consists of a group of steel lattice towers which supports transmission lines including conductors and ground wires that distribute electricity from a power plant to electrical substations. As the modern lifestyle is mostly dependent on electricity, electricity transmission systems are required to cover almost all kinds of terrains which also include areas of seismic activities. Overhead high-voltage electric transmission lines play an important role in the operation of are liable electrical power system, whose damage can cause great economic

P. V. Vaisakh (✉) · N. Nair
Department of Civil Engineering, Federal Institute of Science and Technology, Angamaly, Kochi 683577, India
e-mail: vaisakhvijayan.p@gmail.com

N. Nair
e-mail: neeraja15@gmail.com

K. Dasgupta et al. (eds.), *Proceedings of SECON'19*, Lecture Notes in Civil Engineering 46,
https://doi.org/10.1007/978-3-030-26365-2_54

loss and bring inconvenience to human life. Due to the importance to have electricity after or during the occurrence of a disaster, it is necessary that the transmission tower line system withstand the catastrophic effects of the disaster. In recent years large numbers of transmission lines were damaged by catastrophic earthquakes around the world. The transmission tower tilt or collapse, conductor breakage, foundation subsidence and insulator destruction were the major types of failure.

Although there are some codal provisions to improve the seismic performance of transmission tower- line system, they are not enough to make the system seismic resistant. So in recent years many researches were conducted to study the seismic properties of transmission tower systems. These researches were conducted in two separate ways. In some of the researches, the researchers were only interested in seismic performance of the tower structure that supports the transmission lines. So the researches were mainly done on single tower model subjected to seismic simulations. This idea was objected by some researchers as they said that the transmission lines and towers act as a coupled system during earthquake so the effect of the cables and conductors attached to the towers should also be used during analysis. For this purpose the researchers will make use of a model consisting of three towers and two spans of cables and conductors attached between them. During analysis the researchers will take into account the results given by the central tower which would simulate the field conditions more accurately.

2 Modelling

Four basic transmission tower configurations were selected for the study which are classified according to the electrical current capacity carried by them as 66, 132, 220 and 400 kV. The three dimensional model of the towers were created using AUTOCAD software and these models were used as reference to create the analysis models in SAP2000. The ground wire and conductors used are different for each category of towers. The towers are designed using the guidelines described in IS 802-1995 and Indian Electrical Rules 1956. The specifications of the towers, ground wires and conductors are given in Tables 1, 2 and 3. Mild steel angle sections are used to model the towers and the conductors. The conductors are called aluminium conductor steel reinforced cable (ACSR) a combination of aluminium and steel of which the outer strands are high purity aluminium and the inner core is steel.

The models were created in SAP2000 using the three dimensional models created in AUTOCAD as reference. The model consists of three towers connected by conductors and ground wire. Conductors were modelled as tendons by inputting their features such as modulus of elasticity, coefficient of linear expansion, weight and diameter. This type of modelling were chosen because they were proven to give more realistic results than the analysis of a single tower under loading. The models created are shown in Fig. 1.

Table 1 Specifications of the towers

Tower (kV)	Height (m)	Sections	Maximum span (m)
66	19	110 × 110 × 12 mm 100 × 100 × 10 mm 100 × 100 × 8 mm	250
132	30	130 × 130 × 10 mm 110 × 110 × 10 mm 100 × 100 × 10 mm	320
220	35	130 × 130 × 12 mm 110 × 110 × 10 mm 100 × 100 × 10 mm 100 × 100 × 8 mm	350
400	30	150 × 150 × 12 mm 150 × 150 × 10 mm 110 × 110 × 10 mm 100 × 100 × 8 mm	400

Table 2 Specifications of ground wire

Properties	66 kV	132 kV	220 kV	400 kV
Strand (mm)	7/2.5	7/3.15	7/3.15	7/4
Diameter (mm)	6	9.45	9.45	12
Weight (kg/m)	0.18	0.428	0.428	0.72
Ultimate tensile strength (kg)	1610	5710	5710	6450

Table 3 Specifications of conductors

Properties	66 kV	132 kV	220 kV	400 kV
Name	ACSR dog	ACSR panther	ACSR zebra	ACSR moose
Strand (mm)	7/1.57	7/3	7/3.18	7/3.53
Diameter (mm)	14.15	21	28.62	31.77
Sectional area (mm^2)	118.5	261.5	484.5	597
Weight (kg/mm)	0.394	0.976	1.621	1.998
Ultimate breaking load (kN)	32.41	89.67	130.32	159.60

3 Loads

The loads applied on the towers are mainly the tension due to the conductors. The weight of the conductors as well as the ground wires will act as a tension on the cross arm ends of the towers. Apart from the weight of the cables the sag of the conductors also causes the tension in the tower. The basic combinations described in the Indian codes are used for the analysis.

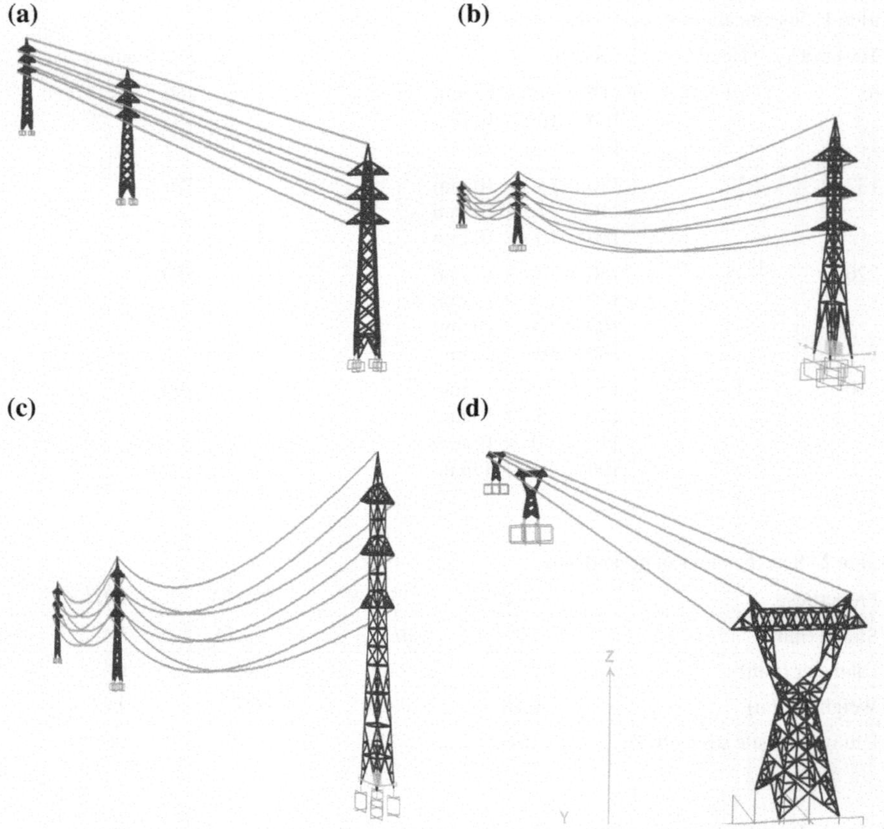

Fig. 1 Models created in SAP2000. **a** 66 kV, **b** 132 kV, **c** 220 kV **d** 400 kV

4 Factors of Safety

The factor of safety of a conductor is defined as the ratio of the ultimate strength of the conductor to the load imposed under assumed loading condition. According to Rule 76 (1)(c) of the Indian Electricity Rules, 1956, the minimum factor of safety for conductors is taken as two. Also the conductor tension at 32 °C without external load should not exceed 35% of ultimate strength of conductor during initial unloaded tension and 25% of ultimate strength of conductor during final unloaded tension. The rule does not specify any loading conditions to which the minimum factor of safety should correspond. Generally, these loading conditions are taken as the minimum temperature and the maximum wind in the area considered. Factor of safety adopted for the towers during the design were aimed at making the structure economical, safe and reliable. As per Rule 76 (1)(a) of the Indian Electrical Rules, 1956, the factors of safety, to be adopted in the design of steel transmission line towers is 2.0 under normal conditions and 1.5 under broken-wire conditions.

5 Analysis

Time history analysis is used to analyse the models. The data used for the seismic analysis is the time history data of the Elcentro earthquake. The time history graph of the Elcentro earthquake for a period of 54 s is shown in Fig. 2.

5.1 *Permissible Deflections*

Sufficient data are not available with regard to the permissible limits of deflection of towers, as specified by the various authorities. However, one practice given below is followed in the USSR:

Assuming that there is no shifting of the foundation, the deflection of the top of the support in the longitudinal direction from the vertical should not exceed the following limits:

For heavy-angle structure (1/120) H
For small angle and straight line structures with strain insulators (1/100) H
For supports with heights exceeding 160 m and intended to be used at crossing locations (1/140) H

The models analysed in this paper are heavy angle structures so permissible top deflections of the tower model are shown in Table 4.

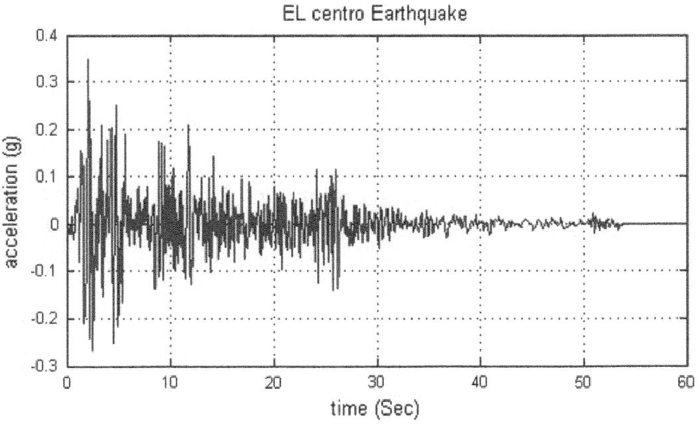

Fig. 2 Time history function of Elcentro earthquake

Table 4 Permissible deflections of the towers

Tower (kV)	Permissible deflection (cm)
66	15.83
132	25
220	29.16
400	25

5.2 *Analysis Results*

The analysis of the models have been done using time history analysis and it was found that the top displacement of the transmission towers were as follows (Table 5).

It was found from Table 5 that the deflections of the considered models are well within the permissible limits specified in Table 4 and hence there is no failure of the tower system under normal conditions.

6 Critical Conditions

As the towers are found to be safe under normal seismic loading conditions, they also need to be checked for seismic safety under certain critical conditions. These critical conditions arise in situations where the tower experiences unequal forces due to the conductors on two different sides of the tower. In real life these conditions arise in two different situations which are the terminal towers at the end of the transmission system and the towers which experiences conductor breakage. Terminal towers refers to the towers which are located at the end of the transmission tower system. Usually towers in which are located within the system that is not at the ends have almost equal forces acting on both sides due to the conductor. This helps them to be in a state of equilibrium and hence the effects of dynamic forces are reduced. In the case of terminal towers the tower has normally laid conductors on one side and a set of conductors that run into a substation or transformer. Here the equilibrium is not maintained and the tower experiences unequal forces which makes them more susceptible to action of dynamic forces.

Second type of condition arises when a conductor breaks in the system due to some reasons. The sudden breakage of wires in a transmission line system leads to

Table 5 Top displacements obtained for the towers

Tower (kV)	Displacement(cm)
66	8.49
132	13.57
220	21.42
400	17.19

Table 6 Top displacement of the towers in the case of terminal towers

Tower (kV)	Displacement	Permissible displacement
66	20.5	15.83
132	28.18	25
220	31.23	29.16
400	43.54	25

Table 7 Top displacement of the towers under broken wire condition

Tower (kV)	Displacement	Permissible displacement
66	16.80	15.83
132	27.53	25
220	22.19	29.16
400	37.46	25

development of sudden unbalanced forces in the system which results in dynamic movements. The subsequent dynamic loads acting on the system can be very large in magnitude, and can result in the failure of one or more support structures. These dynamic loads may cause a cascading type of failure. The collapse of the first tower results in suddenly applied high unbalanced forces in the longitudinal line direction on the adjacent towers, due to the tension in the wires in the adjacent spans.

6.1 Analysis Results

Time history analysis was conducted on the models to get the effect of unbalanced forces on two critical conditions that is the terminal tower condition as well as the broken wire condition. The displacements obtained in terminal towers and broken wire conditions are shown in Tables 6 and 7 respectively.

It is observed that under critical conditions the top displacements of the towers exceeds the permissible limits.

7 Scope for Future Study

The above researches indicates that the transmission tower line systems are vulnerable to the effects of seismic forces under certain conditions. Since these structures are considered as important components supporting daily life the study opens up for further studies into integration of seismic mitigation methods into the transmission towers and its components. Also the analysis was only done using Elcentro earthquake. The study can be extended to find out the effects of various other earthquakes using different modes and components of excitation.

8 Conclusions

The following conclusions can be made from the studies conducted on transmission tower systems.

1. The analysis was done using Elcentro earthquake which is one of the highest earthquake ever recorded and the transmission tower line systems in India were found to be safe. The deflections at the top are found to be well within the permissible limits.
2. Two critical conditions arise while analysing the seismic vulnerability of transmission tower systems which are the terminal towers and broken wire condition of towers.
3. The top displacement of the towers under critical conditions were found to exceed the permissible limits. So the transmission tower systems are found to be not safe under the critical conditions considered.

Seismic Performance Analysis of Vertically Irregular Structures with Diagrid

Merry James and Neeraja Nair

Abstract A large number of vertically irregular buildings exist in modern urban infrastructures, and so the area of vertically irregular type of building is now having a lot of interest. When such buildings are located in a high seismic zone, the structural engineer's role becomes more challenging and consideration of lateral load is very much important. Recently the diagrid structural system is widely used as lateral load resisting system due to its structural efficiency and aesthetic potential provided by the unique geometric configuration of the system. This paper presents a brief study about diagrids on a geometrically irregular structure. For this particular study a building with base dimension 36 m × 36 m and 129.6 m height is taken and vertical geometric irregularity is given to the base structure. Each storey height is 3.6 m. Diagrid with uniform angle throughout the height is provided as lateral load resisting system. Time history analysis is done using ETABS 2016. Seismic performance of geometrically irregular building provided with diagrids is studied by varying diagrid angles. The results in terms of maximum storey displacement, maximum storey drift, time period, structural weights and base shear are compared.

Keywords Vertical irregularity · Diagrid structures · Time history analysis

1 Introduction

In multi-storeyed framed buildings, damage from earthquake ground motion generally initiates at locations of structural weaknesses. Vertical irregularities in building are one of the major reasons of failures during earthquakes. A common type of vertical geometrical irregularity in building structures, is the presence of setbacks i.e. the presence of abrupt reduction of the lateral dimension of the building at specific levels

M. James (✉) · N. Nair
Department of Civil Engineering, Federal Institute of Science and Technology, Angamaly, India
e-mail: merrykudiyirickal@gmail.com

N. Nair
e-mail: neeraja15@gmail.com

K. Dasgupta et al. (eds.), *Proceedings of SECON'19*,
Lecture Notes in Civil Engineering 46,
https://doi.org/10.1007/978-3-030-26365-2_55

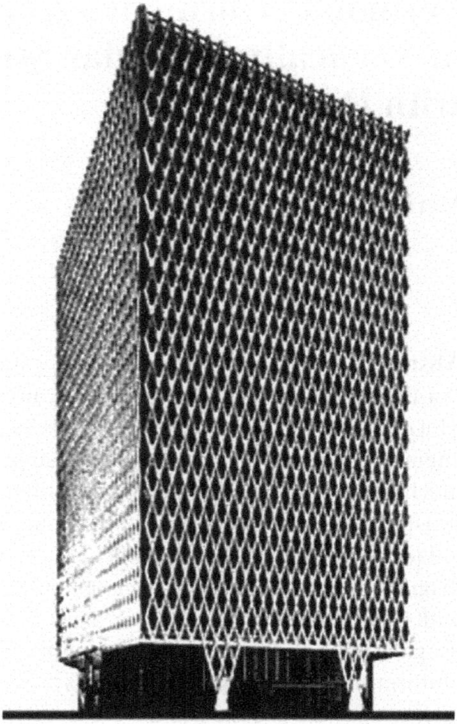

Fig. 1 IBM building Pittsburg, USA

of the elevation. In such types of building the lateral load resisting system becomes more important than the structural system that resists the gravitational loads. The lateral load resisting systems that are widely used are mainly rigid frame, shear wall, wall-frame, braced tube system, outrigger system, diagrid system and tubular system. Recent trend shows that the diagrid structural system is becoming popular in the design of buildings due to its inherent structural and architectural advantages. Figure 1 shows the first modern application of diagrid was IBM building in Pittsburgh (now known as the United Steel Workers Building). In this paper, the behaviour of geometrically irregular building with diagrid is studied and the optimum angle of diagrid for a regular geometry diagrid is found.

2 Literature Review

When performing seismic analysis and design on a regular and vertically irregular building the absolute displacements of irregular structures at respective nodes were found to be greater than that in case of regular structure [1]. Vertically irregular building tend to attract more seismic forces, therefore a study about the behaviour of lateral load resisting systems on vertically irregular structure is having good scope.

Diagrid has good appearance and it is easily recognized. The configuration and efficiency of a diagrid system reduce the number of structural element required on the façade of the buildings, therefore less obstruction to the outside view. The structural efficiency of diagrid system also helps in avoiding interior and corner columns, therefore allowing significant flexibility with the floor plan. Perimeter "diagrid" system saves approximately 20% of the structural steel weight when compared to a conventional moment-frame structure From the analysis and design of diagrid structural system for high rise steel buildings it is observed that most of the lateral load is resisted by diagrid columns in the periphery, while gravity load is resisted by both the internal columns and the peripheral diagonal columns [2]. Finding the optimal diagonal angle for a diagrid structure is a nontrivial and essential step in the design process. Diagonal angle can be efficiently adjusted to increase the lateral stiffness and improve the linear and nonlinear response of the structure under extreme and service loads and also different geometrical patterns can be provided for the diagrid structures [3]. A stiffness-based methodology for determining preliminary design sizes for the diagonals was introduced [4]. Various researchers had carried out study for geometrically regular shaped building by incorporating diagrid structural system. There is an absence of scientific research dealing with performance of diagrid structural system in high rise building especially with geometrically irregular and unsymmetrical shape in plan.

3 Modeling and Analysis

3.1 Building Configuration

The 36 storey tall building is having 36 m × 36 m plan dimension. The storey height is 3.6 m. The design dead load and live load on floor slab are 3.75 kN/m^2 and 2.5 kN/m^2 respectively. Member sizes were taken as given in the journal. All structural members are designed using IS 800:2007 [5]. A symmetrical setback is given from 18th storey to top by removing the corners. The support conditions are assumed as hinged. Figure 2 shows structural plans of building upto 18th storey and above 18thstorey, Fig. 3 shows the rendered 3D view of building and Fig. 4 shows dimensions of interior column; in Table 1 all member sizes are mentioned.

The angle of diagrid is calculated for a 2 storey, 4 storey, 6 storey and 18 storey module based on height and spacing of diagrid. Odd number modules are eliminated because it would not complete the pattern at the top and having an incomplete uppermost module make diagrids vulnerable to soft-story failure in those uppermost stories. Special considerations may be required in these modules [6]. The angle for 2 storey diagrid is 50.2° (D2), 4 storey diagrid 67.5° (D4), 6 storey diagrid 74.5° (D6), 18 storey diagrid 84.4° (D18). Figure 5 shows elevation of different models.

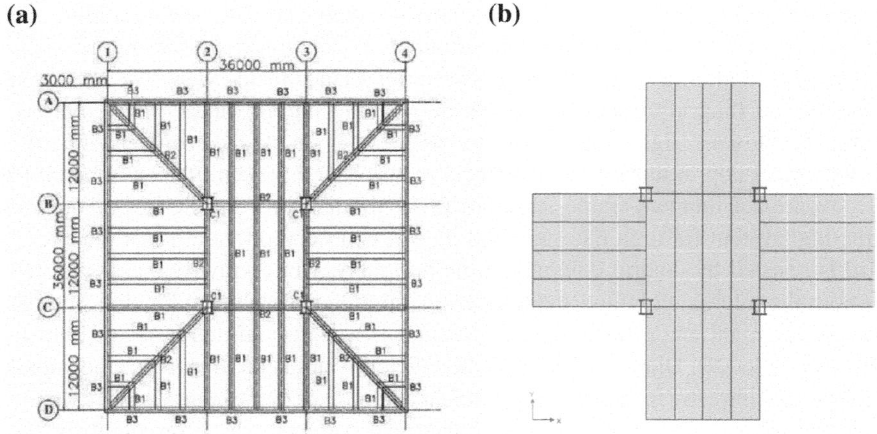

Fig. 2 Structural plans (**a**) up to 18th storey, (**b**) for 19th–36th storey

Fig. 3 3D view of building

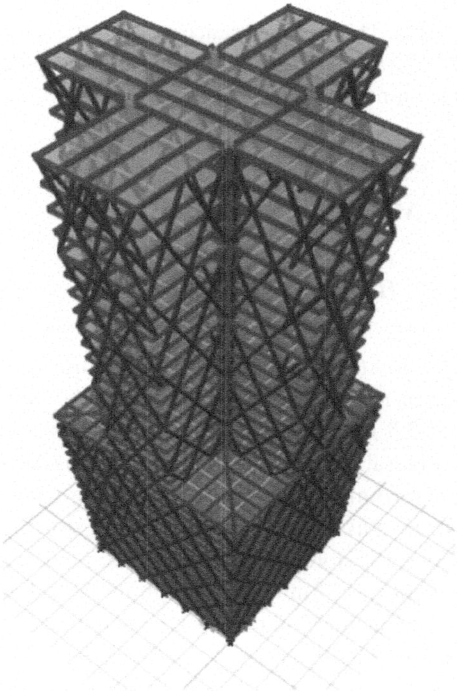

Fig. 4 Interior column

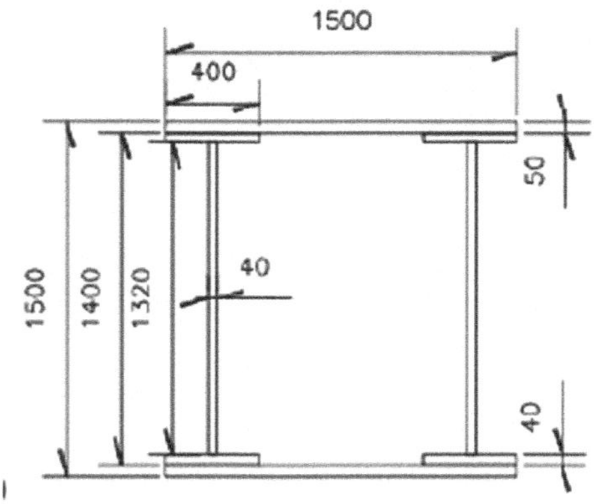

Table 1 Typical member sizes

Diagonal column	Interior column	Beams
375 mm pipe sections with 12 mm thickness (from 19th to 36th storey)	1500 mm × 1500 mm	B1 and B3 = ISMB 550
450 mm Pipe sections with 25 mm thickness (from 1st to 18th storey)		B2 = ISWB 600 with top and bottom plate of 220 mm × 50 mm

3.2 Seismic Analysis

In this paper focus is given to seismic analysis. It is recommended to use nonlinear dynamic analysis, since nonlinear dynamic analysis is the only method which describes the actual behaviour of structure during earthquake. Also in order to study the nonlinear behaviour of diagrids it is very much useful. In the case of diagrids with incomplete module at the top it is recommended to use nonlinear dynamic analysis [7].

Seismic analysis of all models is conducted in ETABS 2016 software using time history analysis. Time history analysis is a step-by step analysis of the dynamic response of a structure to a specified loading that may vary with time. The time history function data entered for analysis process was that of El Centro Earthquake (Imperial Valley earthquake) which occurred in 1940 at southern California. It had a moment magnitude of 6.9 and a maximum perceived intensity of X (Extreme) on the Mercalli intensity scale. The acceleration—time history graph is shown in Fig. 6 Comparison of results for diagrids with different angle is done in terms of maximum top storey displacement, storey drift and time period. Optimum angle for diagrid is found by comparing the results.

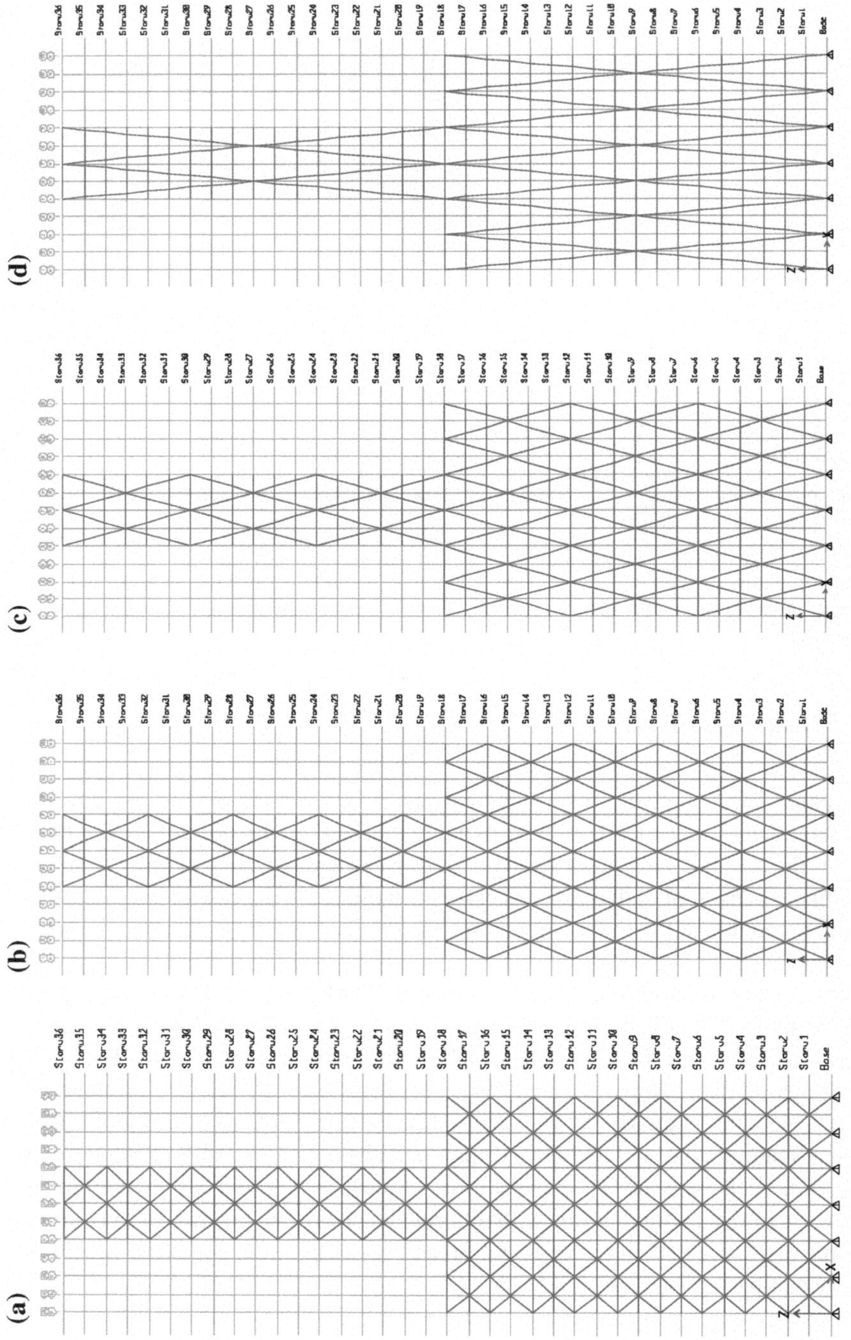

Fig. 5 Elevational view of diagrid with different module sizes. **a** D2, **b** D4, **c** D6, **d** D18

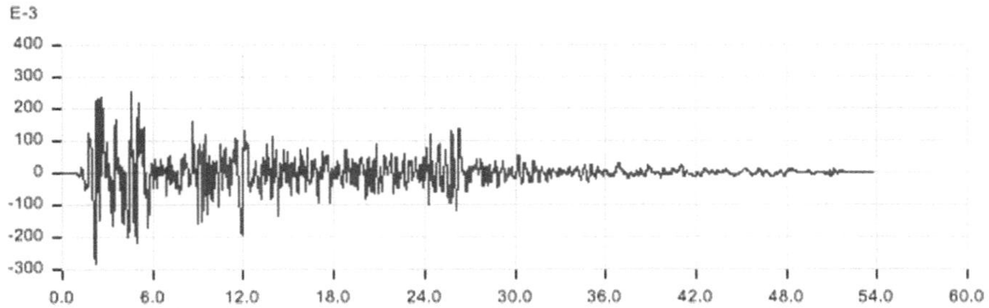

Fig. 6 Acceleration time history graph

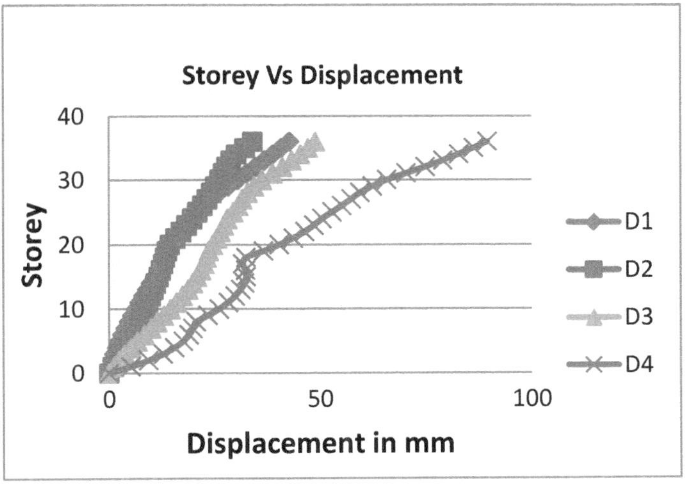

Fig. 7 Comparing max storey displacement for different diagrid configuration

3.2.1 Storey Displacement

As per IS1893:part1 (2002) Maximum storey displacement shall not exceed 0.004 times the height. Max displacement for this building is 518.4 m. Figure 7 shows a graph between maximum storey displacements versus storey for different diagrid models.

3.2.2 Storey Drift

See Fig. 8.

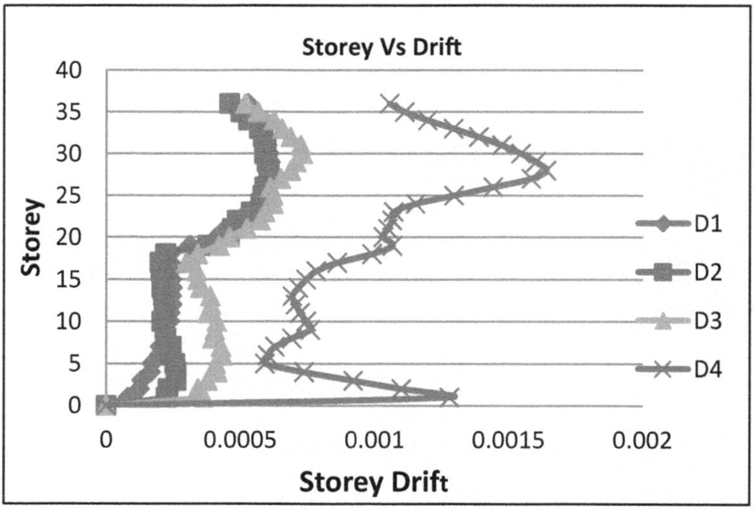

Fig. 8 Comparing maximum storey drift for different diagrid configuration

Table 2 Analysis results

Type	Modal analysis	Time history analysis			
	Time period (s)	Max displacement (mm)	Max drift	Base shear (kN)	Structural weight (kN)
D2	2.45	42.6	0.000613	5454.04	18.78×10^4
D4	2.48	33.97	0.000594	5343.64	18.57×10^4
D6	2.78	8.86	0.000732	4750.08	18.48×10^4
D18	4.46	89.51	0.00164	3304.0	18.48×10^4
Conventional model	6.374	72.68	0.002261	3206.42	17.84×10^4

3.2.3 Comparison of Results

See Table 2.

4 Conclusions

From the time history analysis of symmetric model, it is observed that D2 diagrid structure with angle 67.22° shows better performance than other angles with respect to maximum displacement and drift. Max displacement and storey drift is less for D2 diagrid. Base shear and structural weight slightly reduces as diagrid angle increases. There is only small reduction for structural weight, since we kept same member sizes for all models.

References

1. Himanshu Bansal G (2014) Seismic analysis and design of vertically irregular RC building frames. Int J Sci Res 3(8)
2. Jani K, Patel PV (2013) Analysis and design of diagrid structural system for high rise steel buildings. Procedia Eng 51:92–100
3. Montuori GM, Mele E, Brandonisio G, De Luca A (2014) Geometrical patterns for diagrid buildings: exploring alternative design strategies from the structural point of view. Eng Struct 71:112–127
4. Moon KS, Connor JJ, Fernandez JE (2007) Diagrid structural system for tall buildings: characteristics and methodology for preliminary design. Struct Des Tall Spec Build 16(2):205–230
5. Moon KS (2011) Diagrid structures for complex-shaped tall buildings. In: The 5th international conference of Euro Asia civil engineering Forum (EACEF-5). Procedia Eng 14:1343–1350
6. Asadi E, Li Y, Heo YA (2018) Seismic performance assessment and loss estimation of steel diagrid structures. J Struct Eng 144(10):04018179
7. Mohod MV, Karwa NA (2014) Seismic behaviour of setback buildings. Int J Innov Res Sci Eng Technol 3(9)

Effect of Mid-Storey Isolation in Regular and Stiffness Irregular Buildings

Anisha George and Mariamol Kuriakose

Abstract Seismic isolation is a well-known seismic strengthening method and mid-storey isolation, which is providing isolation at middle stories, is one among those techniques. Mid-storey isolation is a less explored strengthening measure compared to the popular base isolation method. It has many advantages over the base isolation in terms of construction efficiency, cost, space use, maintenance, etc. Since buildings with discontinuities are more prone to serious damage or failure during earthquakes due to their weak points, study on the effect of these isolation techniques on irregular buildings is relevant. This paper presents the time-history analysis of a regular and a stiffness irregular G+20 storey building and their seismic responses when incorporated with isolators at intermediate stories rather than at the base. The structural analysis software SAP2000 has been used for the analysis process. Storey shear, storey drift and acceleration of these high rise structures are plotted in comparison with that of fixed base.

Keywords SAP2000 · High rise structures · Mid-storey isolation · Vertical irregular buildings · Seismic isolation

1 Introduction

Complexities in plan dimension or lateral force resisting system of multi-storey buildings are becoming common now a days. Usually structures are built with some irregularities like change in stiffness, mass, geometry etc. Introduction of these vertical irregularities reduces their seismic performance compared to regular ones due to the formation of weak points along their height. According to IS 1893 (Part 1): 2016, a soft storey is a storey whose lateral stiffness is less than that of the storey above. Mass irregularity shall be considered to exist, when the seismic weight of any floor is more than 150% of that of the floors below. Vertical geometrical irregularity exist,

A. George (✉) · M. Kuriakose
Department of Civil Engineering, SJCET, Palai, India
e-mail: anishageorge7295@gmail.com

K. Dasgupta et al. (eds.), *Proceedings of SECON'19*,
Lecture Notes in Civil Engineering 46,
https://doi.org/10.1007/978-3-030-26365-2_56

when the horizontal dimension of the lateral force resisting system in any storey is more than 125% of the storey below.

Different types of earthquake resisting construction techniques are being practiced today. Seismic isolation is a field of research for over 25 years. It is a process of decoupling the structure from ground motions using devices called isolators. Most common form of seismic isolation is base isolation, which consists of isolating the structure at the foundation level. But the effectiveness of base isolation decreases with increase in the height of the building. Its application may sometimes get limited due to technical and economical problems. It paved way for mid-storey isolation, where isolation is installed in an intermediate storey of the building. It is more effective in medium to high rise buildings, due to their increase in flexibility. Its main application is retrofitting, as it eliminates the efforts of excavations and supports at foundation level, thus an advantage for existing buildings. This consists of cutting the column mid section of the storey to be isolated, and installing the isolators. It allows buildings to have separate structural design and functions for the substructure and superstructure, divided by the isolation interface. Due to the mass damper effect of superstructure, forces acting on the substructure reduces. The main design target of mid-storey isolation is to reduce the storey shear of the building, thereby reducing other responses and protect the structural and non-structural components.

This study aims to find out the effect of mid-storey isolation when incorporated in high rise buildings, both regular and with vertical stiffness irregularity; thereby, compare the seismic parameters on providing isolators at different locations along the height. Structural analysis software SAP2000 is used for the study.

2 Literature Review

A number of literatures are available presenting their works in the area of mid-storey isolation. A review of literatures is presented in brief summarizing works done by various scholars in this field.

Irregular buildings are a major section to be considered in the study of seismic performance. On analyzing different building types of reinforced concrete, it was found that the buildings with discontinuous columns in the ground storey showed the worst seismic performance with undesirable failure patterns [5]. Isolation has proved to be a faithful earthquake resistant technique in structures located in high seismic zones. Comparing the isolation systems, the dynamic behaviour of a mid-storey isolated structural system is more complex than a base isolated system [2]. The design of a good retrofitting solution will need the choice of the suitable number of isolation levels to introduce in a building, the elevations to place these isolation levels and also the properties of each of the isolators to be provided. An optimization procedure and a genetic algorithm was specially developed, which can automatically and effectively explore huge set of potential retrofit solutions, formed by all possible combinations of isolator numbers, locations and properties [1]. Comparing the effects of different inter-story isolation schemes as a function of their location, it has been found that relative to base isolation, isolation at mid-height leads to comparable reduction in superstructure responses above the isolation level but an increase below the isolation level. Moving the isolation system one or more levels above would eliminate the need for a seismic gap at the base [3]. A middle-story isolated structure, when considering the building as a whole, is affected by higher mode vibrations; so the building characteristics are governed not only by the stiffness of the isolation layer and the number of isolators, but also by the stiffness and mass ratio of the superstructure and substructure. Degree of freedom of architectural planning can be expanded and the seismic performance increased by the adoption of a middle-story isolated structure [4]. Comparison with other strengthening methods, like framed K-shape steel bracing, the externally cladding SRC frame bracing, the energy absorbing vibration damping, and the base isolation; it was found that the mid-story seismic isolation can fully meet the seismic requirements and is able to ensure the concurrent occupation during retrofitting work at a lower cost [6].

It is found that the method of mid-storey isolation is effective in reducing the seismic effects on buildings, but the effect of the same in the case of irregular buildings is not seen. Thus, a comparison between the effect of this isolation technique in regular and in a case of vertical irregular building is presented in this study.

3 Aim

To find the effect of mid storey isolation in the seismic responses of regular and vertical stiffness irregular high rise buildings.

4 Model Description

Regular building used is of G+20 stories with plan dimension 30 × 16 m. Each story is of 3.5 m height. Beams used are 0.35 × 0.4 m and columns are 0.4 × 0.6 m. Slabs are 150 mm thick. Live load of 4 kN/m^2 on floors and 2 kN/m^2 on roof is given. Stiffness irregularity is created in the second model by increasing the bottom storey height to 4.5 m.

Isolator used in common is high damping rubber bearing with an effective horizontal stiffness of 3.27 kN/mm and vertical stiffness 2796 kN/mm. There are a total of 35 bearings installed, each with a height of 136 mm and 10% damping. These are installed in the middle of columns in different storey locations to study the effect of varying the isolation interface along the height of the building. The 1940 El Centro earthquake is used for the time history analysis.

5 Results and Discussions

The basic aim of seismic isolation is to increase the time period of the structure such that the resonance condition is prevented. The time periods on isolating the regular building at two intermediate locations; one third (storey 7) and two third (storey 14) of height are shown in Table 1.

The results shows an increase in time periods of X, Y and rotational mode of building. On isolating storey 7, maximum storey shear which was at storey 1 of fixed building, reduced 30.6% from 207.7 kN to 144 kN. Peak base shear value reduced 27.9% from 6257.2 kN to 4505.8 kN. The roof level which experienced maximum acceleration of 4.25 m/s^2 in fixed condition was reduced by 31.7% to 2.9 m/s^2. The time history comparison of these responses are shown in Fig. 1.

Maximum values of storey shear got reduced by 12.9% and 24% on isolating 14th and 7th storey respectively in the regular building. Also, maximum values of acceleration got reduced by 15.7% and 18.1% respectively. Table 2 shows the maximum values of storey shear and acceleration occurred in the regular building.

Thus it is found that the mass ratio (mass of superstructure to substructure) of a mid-storey isolated building affects the efficiency of isolation layer. Buildings with higher mass ratio, i.e., isolation at lower stories, behaves more efficient than those with lower mass ratio or higher isolation storey. Buildings with higher mass ratio will have peak storey shear and acceleration in substructure while buildings with lower

Table 1 Comparison of time period of regular building in fixed and isolated condition

Mode	Fixed	MSI—14	MSI—7
X	3.17	3.26	3.41
Y	3.44	3.53	3.67
Rotational	2.96	3.03	3.18

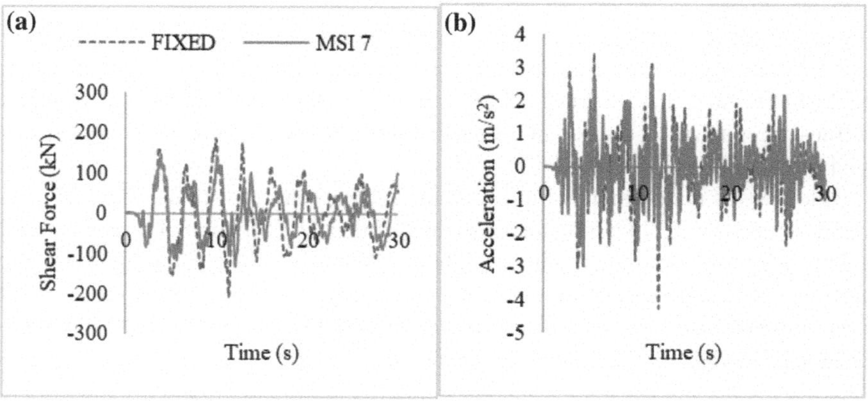

Fig. 1 Comparison of responses of regular building on isolating 7th storey **a** storey shear, **b** roof acceleration

Table 2 Maximum values of storey shear and acceleration in regular building

Isolation storey	Storey shear (kN)			Acceleration (m/s^2)		
	Max	Time (s)	Storey	Max	Time (s)	Storey
Fixed	207.69	11.39	1	4.25	12.2	Roof
14	180.9	10.06	1	3.58	11.5	19
7	157.7	3.76	6	3.48	11.77	6

mass ratio will have peak acceleration in superstructure which is above the isolation layer. Table 3 shows the mass ratio values of isolation at one third and two third of the building.

For the stiffness irregular building, the increase in time periods X, Y and rotational modes are shown in Table 4. Time periods of stiffness irregular building are found to be more than regular one.

Figure 2 shows the comparison of time history responses of stiffness irregular building on isolating 7th storey. Maximum base shear reduced 28% from 5811.7 kN to 4181.2 kN. Maximum storey shear which occurred at storey 1 reduced 26.7% from 193.4 kN to 141.6 kN. Roof which experienced peak acceleration effect was reduced 27.8% from 3.77 m/s^2 to 2.72 m/s^2.

Peak values of storey shear and acceleration of stiffness irregular building are shown in Table 5. Maximum values of storey shear got reduced by 12.9% and 19.5% on isolating 14th and 7th storey respectively. Also, maximum values of acceleration got reduced by 5.3% and 10.8% respectively.

Storey drift variation on change of isolation height at a time of 11.5 s is shown in Fig. 3. Storey drift of building model with isolation level at storey 14 (lower mass ratio) tend to overlap with that of the fixed case. Storey drift is found to reduce as the isolation level decreases.

Table 3 Mass ratio values of regular building

Isolation storey	Mass ratio (μ)
7	2.38
14	0.57

Table 4 Comparison of time period of stiffness irregular building in fixed and isolated condition

Mode	Fixed	MSI—14	MSI—7
X	3.23	3.32	3.47
Y	3.53	3.61	3.75
Rotational	3.02	3.1	3.24

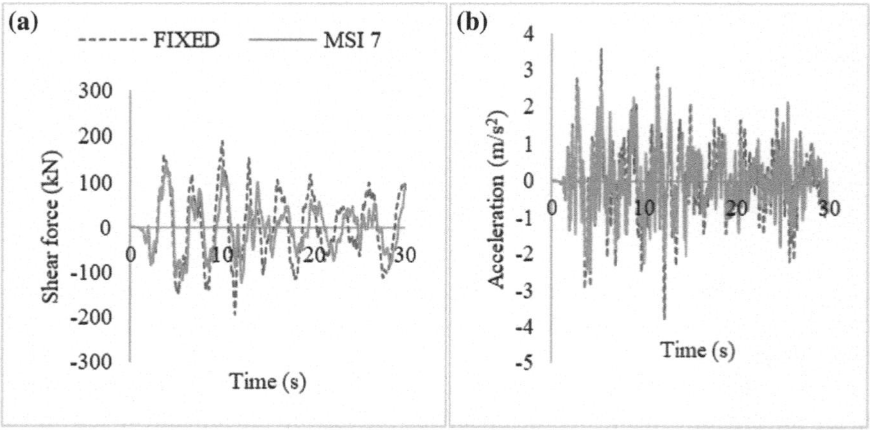

Fig. 2 Comparison of responses of stiffness irregular building on isolating 7th storey **a** storey shear, **b** roof acceleration

Table 5 Maximum values of storey shear and acceleration in stiffness irregular building

Isolation storey	Storey shear (kN)			Acceleration (m/s^2)		
	Max	Time (s)	Storey	Max	Time (s)	Storey
Fixed	193.4	11.4	1	3.77	11.5	18
14	168.3	10.07	1	3.57	11.51	19
7	155.7	3.78	6	3.36	11.77	5

6 Conclusions

Mid-storey isolation can be adopted as an economical method to protect high rise buildings from the earthquake ground motions. Efficiency of this type of isolation depends on mass of structures above and below isolation, type and properties of isolators, etc.

- Mid-storey isolation design achieved a reduction of storey shear and acceleration.
- Reduction of shear and acceleration in stiffness irregular building was found to be slightly less than the regular building on mid-storey isolation.
- The superstructure acceleration is reduced by about 31% on isolating middle storey, but reduction of acceleration at the base is limited to about 4% for regular building, which is slightly more than that of the building with the soft storey.
- Storey drift reduced as the isolation height is decreased. Also, base shear reduced by about 28% for both regular and irregular model.
- Max storey shear and acceleration occurs below the isolation layer for lower level isolation (higher mass ratio) and max acceleration occurs above the isolation layer for higher level isolation (lower mass ratio).

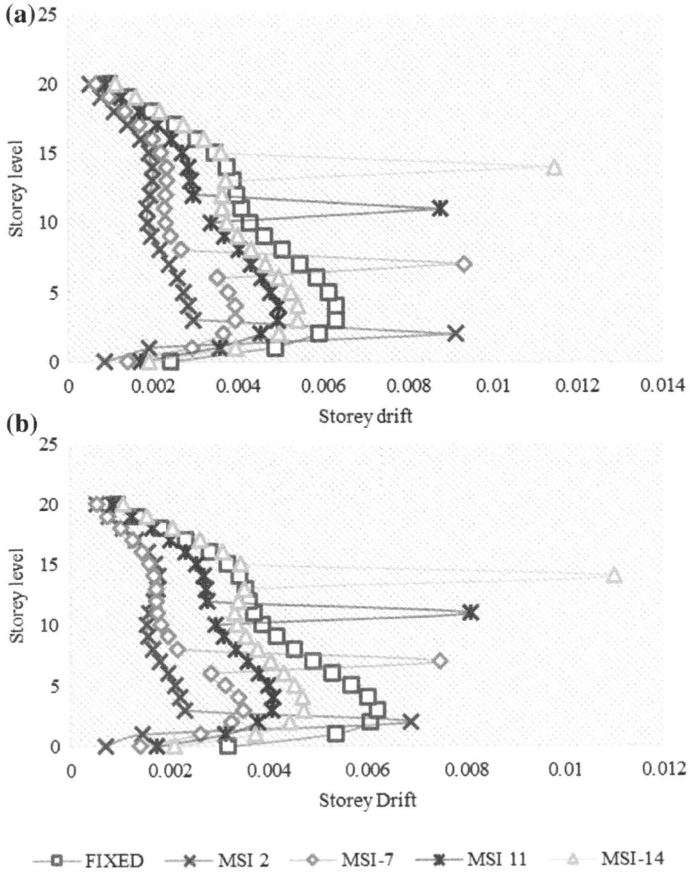

Fig. 3 Storey drift variation for various isolation levels of **a** regular, **b** stiffness irregular model

- Effectiveness of storey isolation decreases with increase in isolation height in both buildings. Isolating at higher stories decreases the mass of superstructure and thus design targets are not fulfilled. Isolating lower stories results in reduction of shear forces, acceleration, displacement, axial loads and bending moments.

References

1. Charmpis DC, Phocas MC, Komodromos P (2015) Optimized retrofit of multi-storey buildings using seismic isolation at various elevations: assessment for several earthquake excitations. Bull Earthq Eng 13:2745–2768
2. Wang SJ, Chang KC, Hwang JS, Hsiao JY, Lee BH, Hung YC (2012) Dynamic behavior of a building structure tested with base and mid-story isolation systems. Eng Struct 42:420–433
3. Ryan KL, Earl CL (2010) Analysis and design of inter-story isolation systems with nonlinear devices. J Earthq Eng 14:1044–1062
4. Tsuneki Y, Torii S, Murakami K, Sueoka T (2008) Middle-story isolated structural system of high-rise building. In: Proceedings of 14th world conference on earthquake engineering, Beijing, China, 12–17 Oct
5. Repapis C, Zeris C, Vintzileou E (2006) Evaluation of the seismic performance of existing rc buildings: ii. A case study for regular and irregular buildings. J Earthq Eng 10(3):429–452
6. Sueoka T, Torii S, Tsuneki Y (2004) The application of response control design using middle-story isolation system to high-rise building. In: Proceedings of 13th world conference on earthquake engineering, paper no. 3457, Vancouver, B.C., Canada, 1–6 Aug

Study of the Effect of Seismic Pounding on Tall Buildings

Swathy S. G. Nathan and S. Bincy

Abstract The building structures are vulnerable to severe damage and collapse due to strong ground motion. Among the various structural damages, pounding damage is the most serious one. This is the most serious seismic hazard usually occur in highly populated areas during earthquake. Pounding is mainly observed when there is insufficient gap between the adjacent buildings. This scenario mainly is in urban areas where multi-storey buildings are closely occupied due to non-availability of sufficient land or may due to high land value. The adjacent buildings with insufficient gap may collide each other under strong ground motion which results in the damage of the structure. Hence, it is necessary to study of the effect of pounding on buildings. Here, the dynamic analysis of the effect of pounding on building is studied using the software SAP 2000. Two adjacent RC frame structures are considered for the analysis. The building structure is modelled and analyzed for pounding and without pounding case. Non-linear time history method of dynamic analysis has been used in this study. Later the effect of shear wall and cross braces on seismic pounding is studied. The parameters considered for the study are storey displacement and drift. This study mainly focuses on the effect of mitigation techniques on pounding to determine the efficient method for minimizing the damages under seismic effect.

Keywords Pounding · Adjacent building · Mitigation · Seismic gap · SAP2000

1 Introduction

Earthquakes have always been a source of great devastation for mankind. It is evident from the past and recent earthquake damages records, that the building structures are subjected to severe damages and collapse during earthquakes. Nowadays with the fast growth of metropolitan cities, land limitation has become a critical issue, which results in the construction of high rise buildings very close to each other. Such buildings are prone to seismic pounding. Pounding is a phenomenon, in which

S. S. G. Nathan (✉) · S. Bincy
Department of Civil Engineering, SJCET, Palai, India
e-mail: swathysgn1995@gmail.com

K. Dasgupta et al. (eds.), *Proceedings of SECON'19*,
Lecture Notes in Civil Engineering 46,
https://doi.org/10.1007/978-3-030-26365-2_57

two buildings strike due to their lateral movements induced by lateral forces, where earthquake is one of the major causes for lateral forces on the buildings [1]. An efficient and durable structural design is always required to prevent pounding effect [2]. The simplest method to avoid pounding damage is to provide enough separation gaps. Spectral difference method based on DDC (Double Difference Combination) rule was used to determine the separation distance [3].

In order to study the behavior and response of pounding, various studies were conducted. These helps in getting a good perspective about the nature of pounding. Usually pounding is considered to be a non-linear problem. The pounding effects are simulated using spring dash pot model [4], depending on SDOF (single degree of freedom system) or MDOF (multi-degree of freedom system) [5]. Various survey had also been conducted to determine the intensity of damages incurred during pounding [6]. The dynamic analyses were similar to lumped mass system [7], where torsional effects are neglected to avoid the complication in analysis. Also assumptions were made that the floor levels are same so that the effect of pounding will concentrate on the lumped mass level of the structure. It was found that adjacent buildings with different materials exhibit different hysteretic behavior [8]. The effect of pounding can be reduced by adopting various techniques like use of dampers, shear wall, braces and interconnections between the buildings like cables [9].

The present study mainly concentrates on the effect of cross braces and RC wall on the pounding between adjacent building. The nonlinear time history analysis was carried out in SAP2000 by using El-Centro earthquake excitation.

2 Model Description

The main objective of this study is to analyze the effect of pounding on adjacent building under earthquake. Two RC frame structures, (Fig. 1) one of 16 storey and other of 20 storey have been considered for the study. The gap provided between two building is 85 mm. The base of the structure is assumed to be fixed in order to neglect the soil structure interaction. Gap elements are used to connect the two structures. The stiffness of the link element is taken to be greater than the stiffness of the buildings. The stiffness here taken is 64,000,000 kN/m. The stiffness is calculated using the following equation.

$$K = \frac{AE}{L}$$

A and *L* is the area and length of the diaphragm of building in contact and E is the modulus of elasticity of concrete. The geometric properties of the building models are given as follows (Table 1).

The model of the building drawn in SAP2000 is shown below.

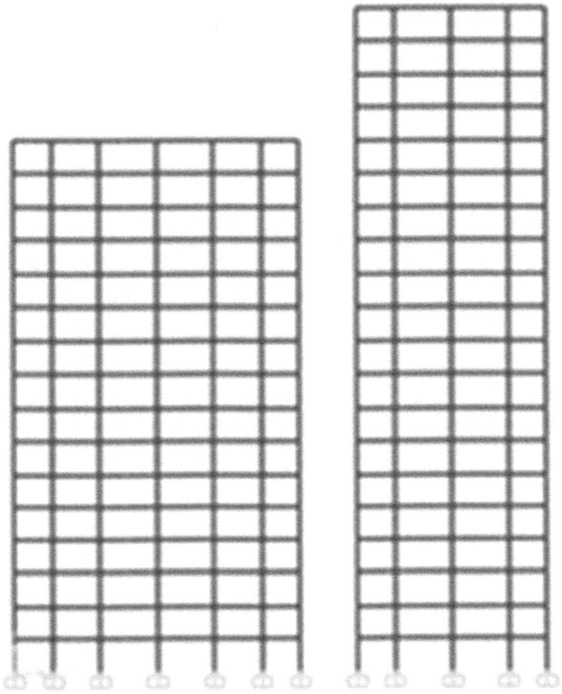

Fig. 1 Pounding model

Table 1 Geometric property of the building model

Model	Outer column (mm)	Inner column (mm)	Beam (mm)	Slab (mm)
G+16	300 × 500	300 × 650	300 × 450	150
G+19	300 × 600	300 × 700	300 × 450	160

3 Results and Discussions

The displacement of both buildings at pounding level (Fig. 2) is plotted against time scale. The graph is unstable which shows that pounding will cause drastic changes in

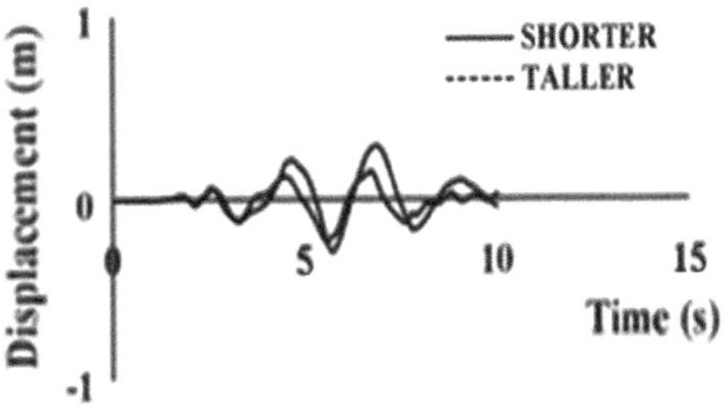

Fig. 2 Displacement—time graph

the behavior of the buildings. The maximum positive displacement of left building is 157 mm and the negative displacement of other is 295 mm. Hence, the out of phase movement obtained is 452 mm which is greater than the gap provided. Thus, there is chance of pounding between the two buildings under the earthquake motion.

The pounding force will be the compressive force acting on the gap element, which determines the collision force between the buildings. Here, a force of 21,430 kN (Fig. 3) is obtained at time of 5.7 s where the negative displacement is maximum. This has been validated by analysing an adjacent seven and four storey building [10].

The building model have been incorporated with braces and shear wall to study their effect on pounding. Inclusion of braces and shear wall increases the stiffness of the whole system, thereby reduces the displacement of the system on pounding. This will help in reducing the damages under pounding, thereby keeps the building safe. ISMB 175 is used as cross braces (Fig. 4a) which is placed at all the edges of the both buildings. Shear wall of 160 mm thickness has been provided to both the buildings in such a way that the overall displacement at the pounding level is reduced. The

Fig. 3 Pounding force

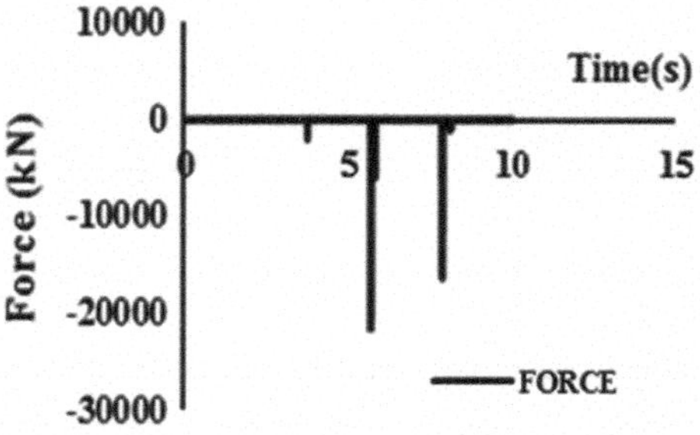

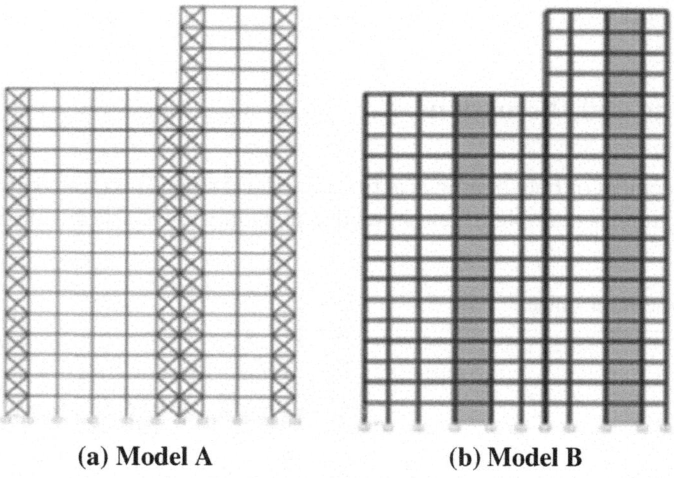

Fig. 4 Model of building with **a** bracing system and **b** shear wall

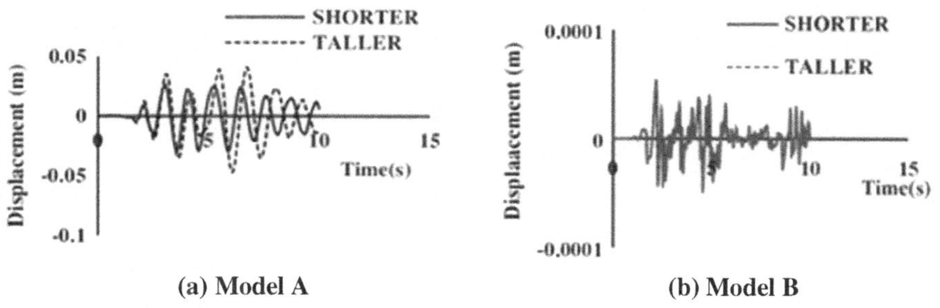

Fig. 5 Displacement graph at the pounding level

position of shear wall in the corner position has increased the displacement compared to the inner core position (Fig. 4b).

The out of phase movement of buildings with cross braces is 74 mm which is less than the expansion joint provided between the buildings. Hence, there is no chance of pounding between them. When shear wall of 160 mm thick is incorporated in the buildings, the displacement has reduced to 95% in both the shorter and taller building due to additional stiffness provided to them. And the out of phase movement of the two building is 0.11 mm which is obtained from the displacement-time graph (Fig. 5).

3.1 *Comparison of Displacement in Shorter and Taller Buildings*

(a) **When cross braces are provided**

The displacement of shorter building (Fig. 6a) at pounding level is found to be 26.5 mm which is 82% less compared with building without bracing system. Similarly, displacement of taller building (Fig. 6b) is 30 mm which is 85% less than building without bracings. This reduction in displacement is due to the fact that the shorter building will prevent the taller building from further movement.

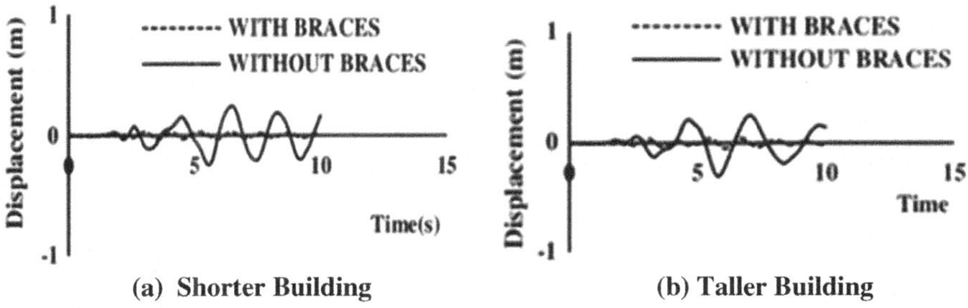

Fig. 6 Comparison of displacement at pounding level with and without braces

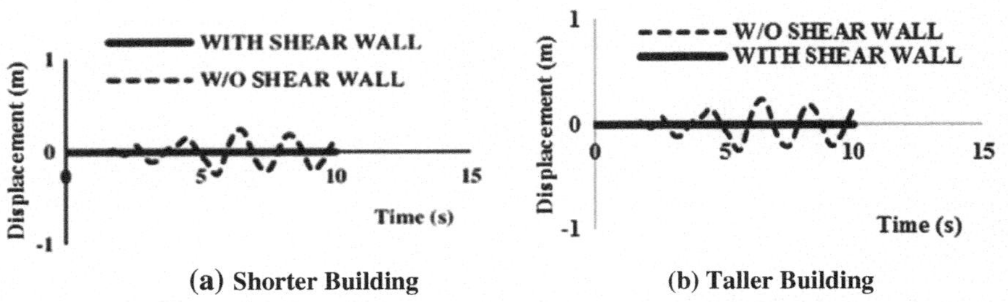

Fig. 7 Comparison of displacement at pounding level with and without shear wall

Table 2 Storey drift in shorter and taller building

Conditions	Shorter	Taller
Braces	0.00347	0.00423
Without braces	0.00628	0.00554
Shear wall	0.0000257	0.0000186
Without shear wall	0.00628	0.00554

(b) **When shear wall is provided**

Displacement of shorter building (Fig. 7a) at pounding level is 0.052 mm which is 98% reduction in displacement compared to building without shear wall. This decrease in displacement is due to increase in stiffness of the building. Similarly, for taller building the displacement is 0.049 mm (Fig. 7b).

3.2 Comparison of Drift in Shorter and Taller Building

The storey drift is reduced to 45% when cross braces are provided to shorter building while it is reducing to 23% in case of taller building. While providing shear wall to both shorter and taller building, the drift is reduced to more than 95%. This implies that providing the shear wall will help in increasing the stiffness of building and thereby reduces the damages. The storey drift obtained for both shorter and taller building are given below (Table 2).

4 Conclusions

This paper mainly focused on studying the effect of pounding on adjacent buildings and to determine the effect of braces and shear wall on pounding. It was found that adjacent building will collide with each other when not provided sufficient gap between them. The simplest way to reduce the pounding is to provide sufficient gap between the buildings. In case of unavailability of sufficient land, braces and shear wall can be used. It shows a massive reduction in displacement under pounding when these braces and shear wall are provided. This reduction in displacement is mainly due to the fact that there will be increase in stiffness of the buildings in addition to the stiffness of the system when it comes in contact under earthquake excitation. Use of shear wall helps in reducing the displacement to 90% compared to those with cross braces. Hence, it is clear that shear wall is more effective in reducing the pounding damages when it is located at the inner core of the buildings compared to braces. If shear wall is located at all the edges of the building, it will not be effective. In that case, cross braces is more suitable to reduce the damages.

References

1. Rojas F, Anderson J (2015) Pounding of an 18-story building during recorded earthquakes. J Struct Eng ASCE
2. Maison, B, Kasai K (1990) Analysis for type of structural pounding. J Struct Eng ASCE
3. Jeng V, Kasai K, Jagiasi A (1992) The separation to avoid seismic pounding. Earthq Eng
4. Hao H, Liu X, Shen J (2000) Pounding response of adjacent buildings subjected to spatial earthquake ground excitation. Adv Struct Eng
5. Huo Y, Zhang J (2013) Effects of pounding and skewness on seismic responses of typical multispan highway bridges using the fragility function method. J Bridge Eng
6. Abdel M, Ghouneim M (2009) Modelling and analysis of factors affecting seismic pounding of adjacent multi storey building. Earthq Resistant Eng Struct
7. Licari M, Sorace S, Terenzi G (2015) Nonlinear modeling and mitigation of seismic pounding between RC frame buildings. J Earthq Eng
8. Jankowski R (2009) Non-linear FEM analysis of earthquake induced pounding between the main building and the stairway tower of Olive view hospital. Eng Struct
9. Abdel M, Ahmed N (2017) Pounding mitigation in buildings using localized interconnections. Adv Struct Eng Mech
10. Jameel M, Islam ABMS, Hussain RR, Hasan SD, Khaleel M (2013) Non-linear FEM analysis of seismic induced pounding between neighbouring multi-storey structures. Lat Am J Solids Struct 10(5):921–939

Analysis and Comparative Study of Seismically Detailed Column Under Blast Loading

P. K. Hridhya and Ance Mathew

Abstract In India, the structures are not usually designed for extreme loading conditions and when they subjected to such load conditions, it can be led to catastrophic failure. Recently, terrorist attacks target our important building such as school buildings, shopping complex, government offices etc. led to their structural failure. Structure all over the world is exposed to terrorist attacks. According to the National Bomb data center, India had the highest bomb blast in the world in the past two years. Blast resistant design of the structure is uneconomical and it involves tedious calculations. But the seismic design of the building is a part of the conventional design. If the dynamic response of the building meets the requirements then we can avoid the complex calculations associated with the blast-resistant design and we can obtain an economical section. In this study, the column was analysed using the high fidelity physics based finite element code, LS DYNA.

Keywords Seismic detailing · Conventional detailing · Blast loading · LS DYNA

1 Introduction

Columns are the major element in frame structures. They are designed to support all loads coming from the upper floor and the weight of the roof. Column failure is the primary cause that leads to the collapse of the structure. The exterior columns are more vulnerable to terrorist attack. The better understanding of the capacity of a column would aid in the prediction of the building performance. Building all over the worlds has the probability to the treats of a terrorist attack. Public buildings, government offices, monuments, educational buildings, and the buildings, which are near to the exposure manufacturing or storage, are susceptible to terrorist attack.

P. K. Hridhya (✉) · A. Mathew
Department of Civil Engineering, SJCET, Palai, India
e-mail: hridhyapk42@gmail.com

K. Dasgupta et al. (eds.), *Proceedings of SECON'19*,
Lecture Notes in Civil Engineering 46,
https://doi.org/10.1007/978-3-030-26365-2_58

According to the national bomb data center (NBDC), India had the highest bomb blast in the world (National Consortium for the Study of Terrorism and Response to Terrorism [9]. In 2016, 337 Improvised Explosive Devices (IED) blast occurred in India. In 2015, there was 268 IED blast. In 2014,190 blast and 283 blasts in 2013 and 365 in 2012. The latest IED blast reported as Pulwama attack, one of the deadliest attack in 2019 in which 660 lb of explosive. Which includes 180 lb of RDX, high explosive, and ammonium nitrites. Due to the uncertainty in predicting accidental attack and explosion, a few information is available on the behavior of the concrete structure under blast loading. Research work on the blast phenomenon is carried out by military authorities and the data is not available for the public. Structural engineers are still seeking the possible best solution to prevent the total or partial collapse of the structure. In this study the effect of charge weight, scale distance and the spacing of transverse reinforcements were studied.

2 Literature Review

Blast loading on structures can result from accidental or planned activities. The accidental explosion may occur during the manufacturing or during the transportation of the explosion. During the detonation time, high pressure is released due to rapid chemical reactions. This will lead to total or partial progressive collapse. Progressive collapse referred to as the failure of column and tributary beam and floor [3, 8, 11]. In an RC frame structure, columns are the one which dissipates ductile energy [5]. RC column transfer floor load and roof load to the foundation. It is an important component of a load resisting system [4]. Columns are also undergo lateral deformation when subjected to cyclic loading such as earth quake. Many researchers [2, 5, 6] have done many works on seismically detailed structural element subjected to blast loading. Conventional design procedure would not consider the load from explosives. To prevent the progressive failure of the column, the component should design to resist blast loading by providing adequate strength and ductility. The conventional structural design does not consider the loading from the detonation of explosives. Thus, few buildings are designed to resist the blast loading. The structures designed to withstand one load type can often have the capacity to resist another type of load [6]. The behavior of the column under blast loading often depends upon the axial compressive force [1]. By applying the axial load, the column resistance can be increased significantly.

2.1 Explosion and Blast Phenomenon

Explosions are one of the most frequent sources of accidental and man-made catastrophes which is responsible for relevant economic losses. Explosions are a very fast chemical reaction which is associated with the release of large scale hot gases with high energy. When the explosives get detonated, the expanded gas compresses the surrounding air. Due to this disequilibrium between gases, the blast wave travels. The shock wave propagates with a given speed and magnitude [10].

2.2 *Blast Scaling Laws*

Most of the data available for blast analysis is from experimental investigation and which involves various parameters like charge weight, type of explosives and scale distance. Blast scaling law enables the researcher to compare their test results. Hopkinson- Cranz scaling law, also known as cube-root law is the most commonly using scaling law. According to Hopkinson-Cranz scaling law the detonation of a charge weight W_1, at a set scale distance R_1 will produce blast parameters like Incident pressure, impulse, blast duration which is similar to the parameters produced by another charge weight W_2, at a corresponding scale distant R_2, detonated in the same atmospheric conditions. Hopkinson-Cranz scaling law is given below.

$$\frac{R_1}{W_1^{1/3}} = \frac{R_2}{W_2^{1/3}} = Z \tag{1}$$

3 Validation

Validation of numerical modeling was done with the results reported by Kyei and Braimah [7].

The explosive charge of about 100 kg TNT (trinitrotoluene), detonated at a scale distance of 1 m/kg$^{1/3}$. The RC column of the cross-sectional dimension of 300 mm × 300 mm and the height of the column is about 3000 mm. The column consists of 4 numbers of 25 mm diameter longitudinal bars and 10 mm diameter transverse bars. Special confining bars are provided at 75 mm center-to-center spacing in the plastic hinge region and 150 mm center-to-center spacing is provided at the region where special confining bars are not provided. The parameters considered are given in Table 1. The numerical model was developed in LS DYNA software. The result obtained is shown in Fig. 1. From the graph, we can see that the displacement obtained from the numerical model is 61.9 mm and the displacement reported by Kyei and Braimah [7] from the experimental program is 66.2. The variation is about 6.6%. Which is acceptable.

Table 1 Summary of displacements

Scale distance (m/kg$^{1/3}$)	Charge weight (kg)	Displacement (mm)	
		Conventional column	Ductile column
0.62	100	72.8	69
	250	186	169
	500	332	304
	1000	730	550
0.80	100	39.5	25
	250	96	84.1
	500	182	158
	1000	421	157
1.0	100	25	21.9
	250	39.3	36.7
	500	77.1	70
	1000	157	144

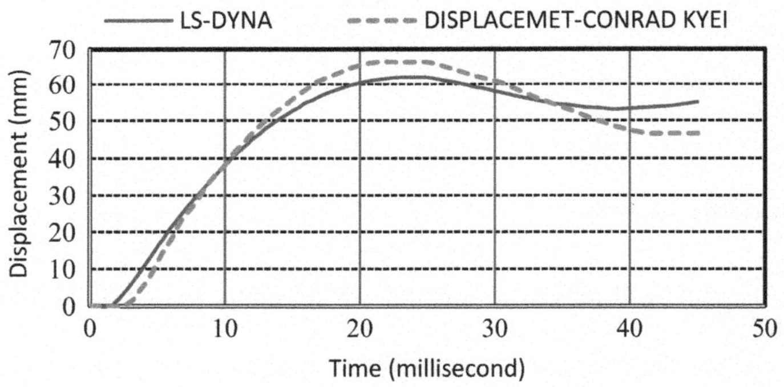

Fig. 1 Displacement-time history curve

4 Column Detailing

As per IS 13920-2016, the minimum dimension of the column shall not be less than 20 d_b, where d_b is the diameter of the largest longitudinal reinforcement bar in the beam passing through or anchoring into the column at the joint or 300 mm. Here cross-sectional aspect ratio of the column also considered. As per clause 7.1.2 the cross-sectional aspect ratio ie. The ratio of smaller dimension to larger dimension shall not be less than 0.45. The reinforcement bars in a column provide tensile strength for the column. In this paper, the reinforcement bars selected according to IS 13920.

As per the code, the rectangular column should have a minimum of 4 longitudinal bars. Here 2.2% (A_S/A_C) of the area of steel is adopted i.e. 4 no. of 25 mm diameter bars. The minimum cover provided for column 40 mm.

5 Finite Elemental Modeling

LS DYNA is a general-purpose finite element code for analyzing large deformation static and dynamic response of structures. For modeling concrete element 8 nodded under-integrated solid hexahedron element was selected. In addition, Hughes-Liu element is used to model both transverse and longitudinal reinforcement. Piecewise-Linear-plasticity material is used to model both longitudinal and transverse reinforcement. The Material_Piecewise_Linear_Plasticity (MAT 24) is an elasto-plastic material for which a subjective pressure strain bend can be characterized (LSTC 2013). There are four methods to simulate blast loading in LS DYNA, ARBIRTARY_LAGRANIAN_ULERIAN (ALE) method, PARTICLE method, LOAD BLAST ENHANCE (LBE) and the combination of ALE and LBE. In this paper LOAD_BLAST_ENHANCE (LBE) keycard was used to simulate blast loading. By using this keycard, the empirical pressure load directly applied to the structure.

6 Parametric Studies

Parametric studies are conducted after validating the numerical model with the results obtained reported by Kyei and Braimah [7]. The analysis studied the effect of charge weight where scale distance is constant, the effect of scale distance and the effect of ductile detailing of the column.

6.1 Effect of Charge Weight

To study the effect of charge weight, the explosion carried out at same scale distance with different charge weight. Hopkinsons-Cranz scaling law is used to find the range of the explosion. The scale distance 0.8, 1 and 1.5 $m/kg^{1/3}$ are used to study the effects. The charge weight of 100, 250, 500 and 1000 kg used to simulate blast loading on the structure.

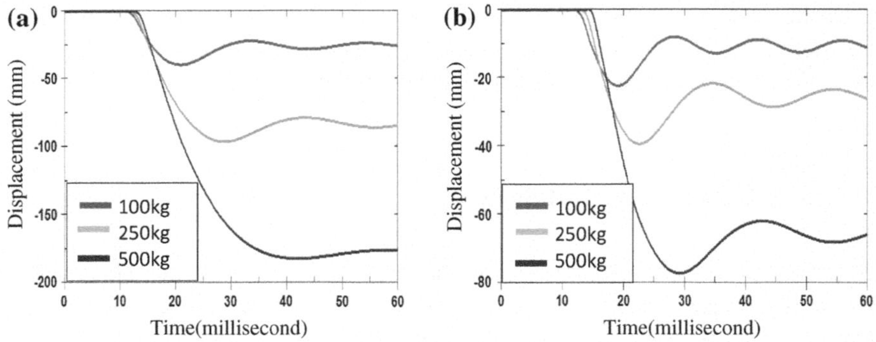

Fig. 2 **a** Scale distance-0.8/kg$^{1/3}$, **b** scale distance-1 m/kg$^{1/3}$

According to Hopkinson-Crans scaling law, the blast pressure produced by different charge weight will be the same if the scale distance kept constant. The impulse and duration of blast loading will vary with respect to charge weight. This implies that the higher magnitude charger produces higher impulse. Figure 2 shows the effect of charge weight on the displacement of the column element.

From Fig. 2a and b we can see that the displacement increases for increasing charge weight. This is due to the increase in impulse with the increase in charge weight. For example, explosion of 100 kg explosives, at a scale distance of 1 m/kg$^{1/3}$, will results in a displacement of 25 mm and for the same scale distance, 250 kg will results in a displacement of 40 mm. If we compare Fig. 2a and b, the displacement increases with decrees in scale distance. For example, the displacement produced by explosive of weight 500 kg at a scale distance of 1 m/kg$^{1/3}$ was obtained as 77.1 mm and for the same charge weight, at a scale distance of 0.8 m/kg$^{1/3}$, resulted in a displacement of 182 mm.

6.2 Effect of Ductile Detailing

Parametric studies conducted to find out the effect of ductile detailing. The columns detailed according to IS 13920. The detonation takes place at a scale distance of 0.62, 0.8 and 1 m/kg$^{1/3}$.

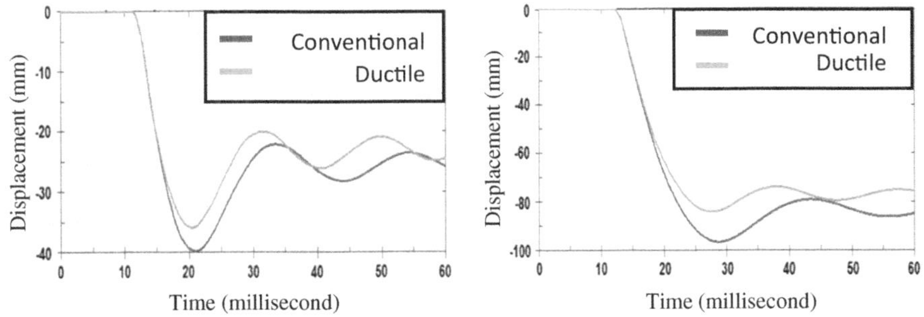

Fig. 3 **a** 100 kg of TNT at a scale distance of 0.8 m/kg$^{1/3}$, **b** 250 kg of TNT at a scale distance of 0.8

From Fig. 3a we can see that the displacement obtained when 100 kg weight of charge detonated at a scale distance of 0.8 m/kg$^{1/3}$ was 40 mm and for the same charge weight and scale distance ductile detailed column rested in a displacement of 35 mm. The percentage increase in lateral displacement observed as 14.28% with respect to the ductile detailed column. Figure 3b shows the displacement time history of charge weight 250 kg at a scale distance of 0.8 m/kg$^{1/3}$. For conventionally detailed column, the displacement resulted from the explosion of 250 kg weight of charge at a scale distance of 0.8 m/kg$^{1/3}$ was 98.2 mm and for the same charge weight and scale distance, ductile detailed column resulted in a displacement of 84.1 mm. which was 14.36% lower than the conventional column. Table 1 summarises the lateral displacement resulted from a no. of analysis.

Table 1 shows that the resulted displacement increases with the increasing charge weight and decreases with the increase in scale distance.

Table 1 also gives the effect of transverse reinforcement bar. Conventionally detailed column undergoes large displacement than the ductile detailed column. This is due to the increase in stiffness. As the ductility increases, the displacement also decreases. The column with ductile reinforcement undergoes large strain before the collapse. The damage profile of the conventional column under 1000 kg of TNT explosion at a scale distance of 0.8 m/kg$^{1/3}$ is shown in Table 2. From Table 2 The damage profile can be compared. The conventional column undergoes more damage than the ductile detailed column.

Table 2 Damage profile

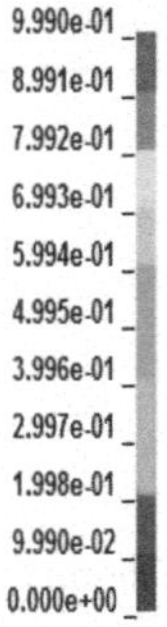	Conventional column Scale distance 0.8 m/kg$^{1/3}$ Charge weight 1000 kg			
Fringe level	Ductile detailed column Scale distance 0.8 m/kg$^{1/3}$ Charge weight 1000 kg			
		20 s	40 s	60 s

7 Conclusion

From this study, we can conclude that the damage of the column can be minimized by providing seismic detailing. By reducing the spacing of transverse reinforcement, we can reduce the lateral displacement of the column. The effect of special confining bars is more effective at low scale distance. There is only a minimum effect for higher scale distance. For the same scale distance, the charge weight plays a key role in the whole damage of the RC column. There will be increasing lateral displacement for increasing charge weight. For the same charge weight, lateral displacement can be reduced by increasing the scale distance.

References

1. Astarlioglu S, Krautammer T, Morency D, Tran TP (2013) Behavior of reinforced concrete column under combined effects of axial and blast-induced transverse loads. Eng Struct
2. Bao X, Li B (2010) Residual strength of blast damaged reinforces concrete column. Int J Impact Eng 37:295–308
3. Byfield M (2006) Behaviour and design of commercial multistory buildings subjected to blast. J Perform Constructed Facil, 324–329
4. Fujikura S, Bruneau M, Lopez-Garcia D (2008) Experimental investigation of multihazard resistant bridge piers having concrete-filled steel tube under blast loading. J Bridge Eng 13(6):586–594

5. Fujikura S, Bruneau M (2011) Experimental investigation of seismically resistant bridge piers under blast loading. J Bridge Eng16(1):63–71
6. Hayes JR, Woodson S, Pekelnicky RG, Poland CD, Corley G, Sozen M (2005) Can strengthening for earthquake improve blast and progressive collapse resistant? J Struct Eng 131:1157–1177
7. Kyei C, Braimah A (2017) Effects of transverse reinforcement spacing on the response of reinforced concrete columns subjected to blast loading. Eng Struct 142:148-164
8. Li J, Hao H (2013) Numerical study of structural progressive collapse using substructure technique. Eng Struct 32:1691–1703
9. National Consortium for the Study of Terrorism and Response to Terrorism (2018) Global Terrorism Database. Retrieved from https://www.start.umd.edu/gtd university of Maryland
10. Paris F, Augenti N (2012) Influence of seismic design criteria on blast resistance of RC framed building: a case study. Eng Struct 44:78–93
11. Shi Y, Li ZX, Hao H (2010) A new method for progressive collapse analysis of RC frame under blast loading. Eng Struct 32:1691–1703

Seismic Analysis of Steel Diagrid Structures Using Triple Friction Pendulum Isolator (TFP)

P. R. Althaf Hyder and E. K. Amritha

Abstract Today new technique of construction as well as aspect of design is coming to forefront as method of conventional design has failed to give the proper result. Base isolation is one of the most widely accepted seismic protection system used in earthquake prone areas. The base isolation system separates the structure from its foundation and primary moves it relate to that of the super structure. Diagrid structure is new trending concept in the field of structural engineering taking into account the factors of structural stability, aesthetic appearance and economic consideration. The scope of the paper is related to the seismic analysis of steel diagrid structure in combination with base isolation. Here Triple Friction Pendulum Isolator is used for base isolation. Dynamic linear response spectrum analysis and dynamic linear time history analysis is performed in the isolated diagrid building. Further a comparative study of performance of base isolated diagrid building has been carried out by response spectrum and time history analysis by changing the bracings and bracing arrangement pattern. i.e., X, V, inverted V, eccen forward and eccen backward bracings in whole, alternate, horizontal and vertical pattern arrangements. For this a 22 storey and 11 storey steel diagrid building is designed and the above mentioned analysis is carried out. Base shear and top storey displacement are used as parameters for this study. From the results it concluded that for both 22 storey and 11 storey diagrid building, building having V bracing with alternate arrangement pattern showing better results.

Keywords Diagrid structure · Base isolation · Triple friction pendulum isolator · Bracing arrangement pattern · Response spectrum analysis · Time history analysis

---

P. R. Althaf Hyder (✉) · E. K. Amritha
Department of Civil Engineering, Universal Engineering College,
A. P. J. Abdul Kalam University, Thiruvananthapuram, India
e-mail: althafhyder918446914@gmail.com

E. K. Amritha
e-mail: ekamritha@gmail.com

K. Dasgupta et al. (eds.), *Proceedings of SECON'19*,
Lecture Notes in Civil Engineering 46,
https://doi.org/10.1007/978-3-030-26365-2_59

1 Introduction

Earthquake can be defined as a process where there is sudden release of stress waves and large amount of energy due to violent tremor caused to earth's crust. Severity of earthquake depends upon the amount of energy released. Size and type of rupture area also influence the magnitude of an earthquake. Effects of an earthquake depend on factors like rock and soil media through which the wave propagates. Damages caused by an earthquake is further dependent on site conditions such as characteristics of soil, ground conditions, water table and topography etc. Sudden release of stored energy in earth's crust leads to destruction of natural environment as well as manmade infrastructure; it also causes loss of life and money. Hence to avoid these problems proper mitigation techniques should be adopted. Earthquake effect mitigation can be achieved by proper seismic studies and analysis. Adopting suitable retrofitting techniques after these studies help to reduce damage caused during an earthquake.

Two different load types considered during a seismic analysis of a structure are static and dynamic loads, static loads does not vary with respect to time whereas dynamic loads are time varying. Due to rapid industrialization and urbanization tall lighter structures are constructed around the world. As the height of the structure increases they become more flexible and slender. Low inherent damping of building makes the structure more prone to vibrations under dynamic loads. Dynamic loading are sometimes neglected during design process due to its complexity, this in turn leads to sever damage to the structure during an earthquake. Structural stiffness and durability is affected by Periodic dynamic loading. Resonance occurs if frequency of vibration coincides with structure's natural frequency, leading to total collapse of structure.

A Diagonal Grid or Diagrid Structure is a framework of diagonally intersecting section (steel section used in this paper) that is used in the construction of skyscrapers buildings and rooftops. These structures offer unique supremacy to high risers because of structural efficiency and pleasant aesthetics. It conveniently eliminates the dependence on column of a structure and also requires comparatively lesser structural steel and thus optimize the cost. The Steel Diagrid Structure are more popular than other conventional materials such as wooden beams and concrete as they are quickly erected. These buildings are energy efficient, environmentally sensitive and a clear winner in the run for what future sustainable buildings may look like.

Base isolation is one of the most widely accepted seismic protection system used in building in Earthquake prone areas. The base isolation system separates the structures from its foundation and primarily moves it relative to that of the super structure. The scope of the paper is related to the seismic analysis of steel diagrid structure in combination with base isolation. Here Triple Friction Pendulum Isolator is used for base isolation. Dynamic linear response spectrum analysis and dynamic linear time history analysis is performed in the isolated diagrid building. Further a comparative study of performance of base isolated diagrid building has been carried out by response spectrum and time history analysis by changing the bracings and bracing arrangement pattern. Base shear and top storey displacement are used as parameters for this study (Figs. 1 and 2).

Fig. 1 A typical diagrid structure [4]

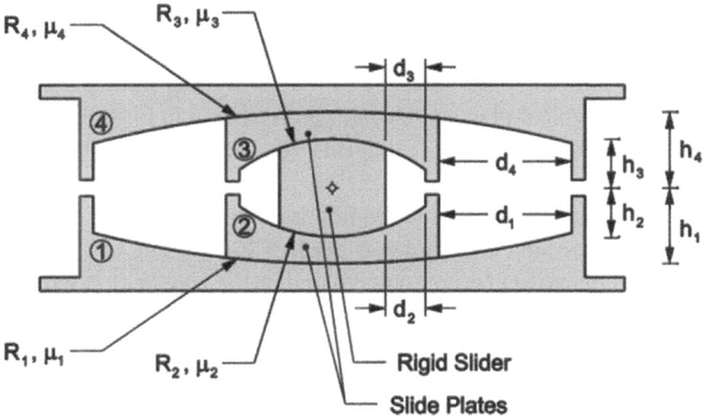

Fig. 2 Cross section of TFP [5]

2 Objectives and Scope

- Diagrid structures are efficient in lateral load resistance and esthetic effect and TFP is cost efficient isolator and good energy dissipater, so combining these two will expected give good seismic resistance
- For this, selected two types of steel buildings, one short building (11 storey) and one tall building (22 storey) based on the specified height to width ratio of base isolated building [1]
- Provided different types of bracings, namely X, V, Inverted V, Eccen forward and Eccen backward etc. in different arrangement pattern namely Whole, Alternate, Horizontal and Vertical etc. in these building and made the building diagrid structure
- Provided TFP isolator in these diagrid structures based on total load coming on the base
- Done seismic analysis in these structures
- Compared the results and find out which type of diagrid structure providing maximum earthquake resistance in the application of TFP.

3 Analysis and Design of 22 Storey Diagrid System and Properties of TFP Isolator

Building Configuration In structures different types of bracings namely X, V, inverted V, eccen forward and eccen backward are given and made the structure diagrid. The angle of inclination is kept uniform throughout the height. The 22 storey tall diagrid buildings are having 24 m × 24 m plan dimension. The storey height is 3.5 m. The typical plan and elevations are shown in Figs. 3, 4, 5, 6, 7 and 8 respectively. The interior frame of the diagrid structures is designed only for gravity load. The live loads on floor slab are 3 kN/m^2 and dead loads are auto generated by software. The dynamic along wind loading is computed based on the basic wind speed of 55 m/s and terrain category IV as per IS:875 (III)-1987.

The design earthquake load is computed based on the zone factor of 0.36, medium soil, importance factor of 1 and response reduction factor of 4 (IS: 1893 (Part-I), 2002). Modeling, analysis and design of diagrid structure are carried out using ETABS software. For linear static and dynamic analysis by time history and response spectrum method the beams and columns are modeled by flexural elements and braces are modeled by truss elements. The support conditions are assumed as hinged. All structural members are designed using IS 800:2007. Secondary effect like temperature variation is not considered in the design, assuming small variation in inside and outside temperature.

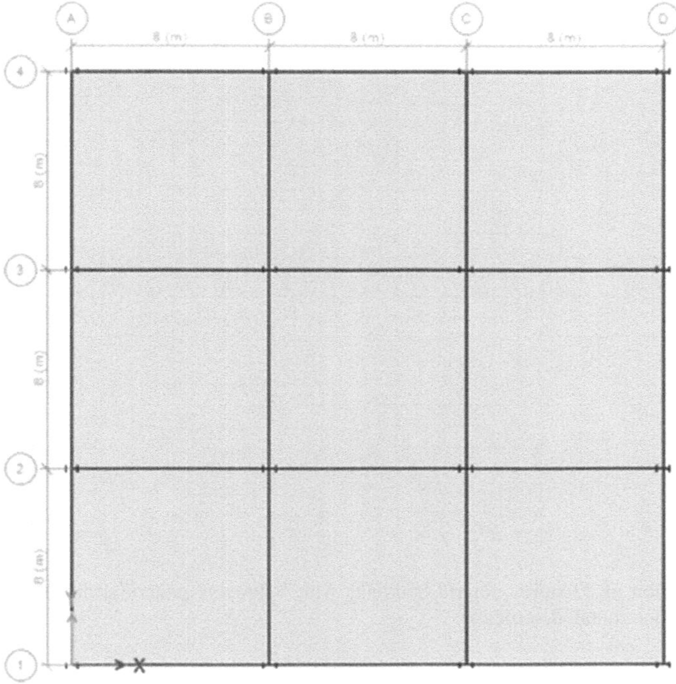

Fig. 3 Typical plan common to all 22 storey diagrid building

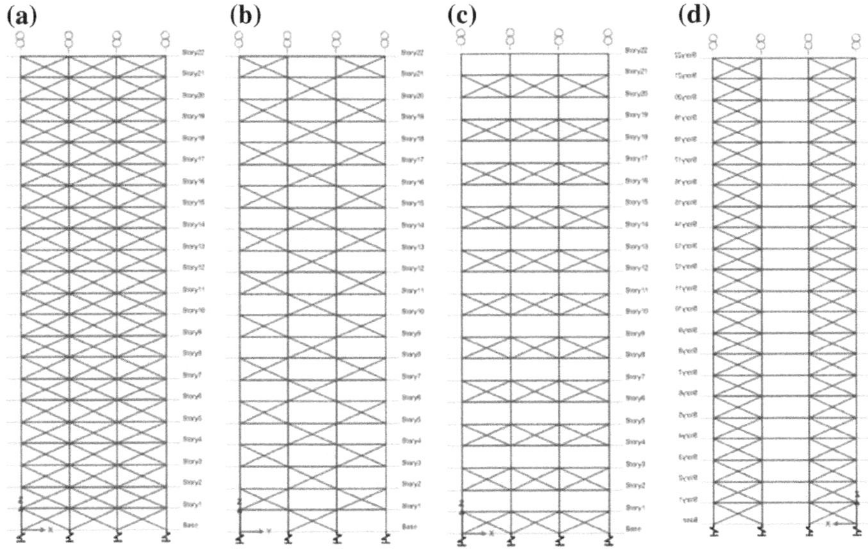

Fig. 4 Elevation of 22 storey diagrid buildings with X bracing pattern arrangement (**a** whole, **b** alternate, **c** horizontal, **d** vertical)

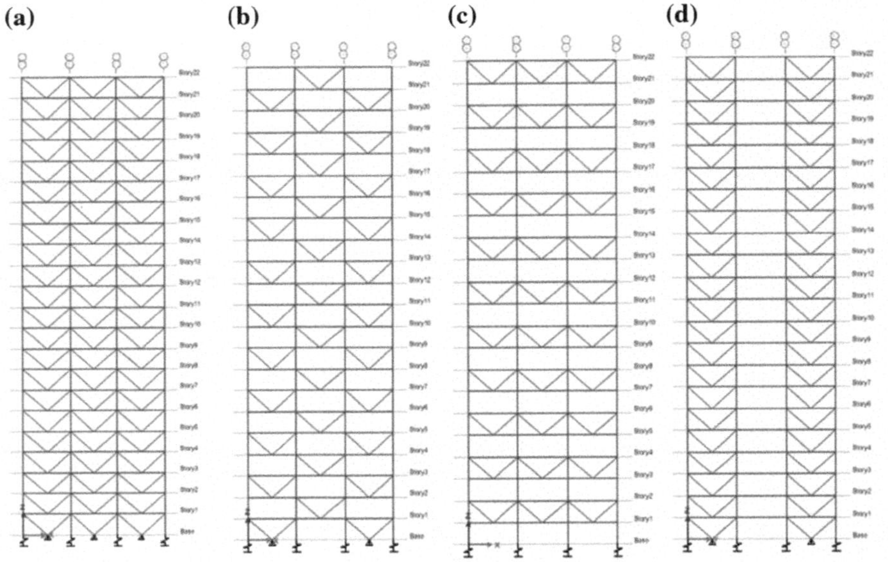

Fig. 5 Elevation of 22 storey diagrid buildings with V bracing pattern arrangement (**a** whole, **b** alternate, **c** horizontal, **d** vertical)

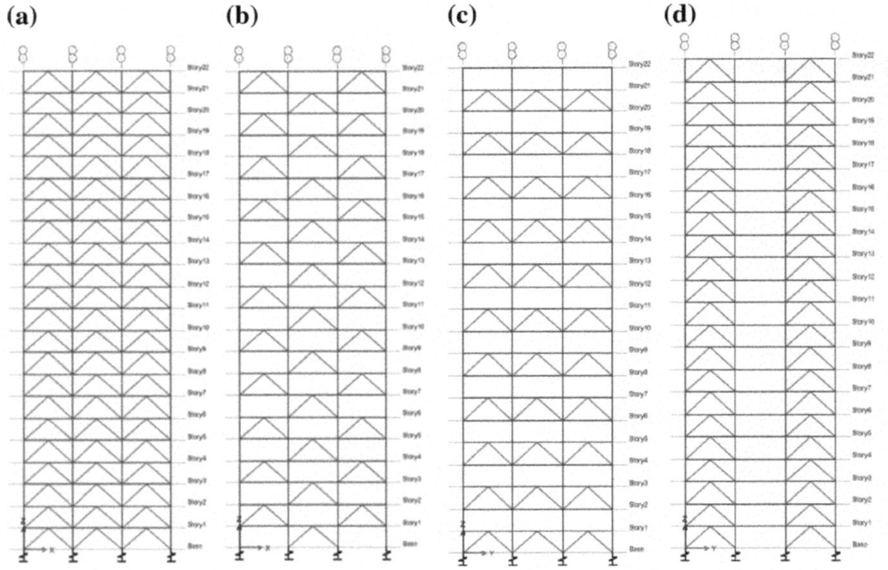

Fig. 6 Elevation of 22 storey diagrid buildings with inverted V bracing pattern arrangement (**a** whole, **b** alternate, **c** horizontal, **d** vertical)

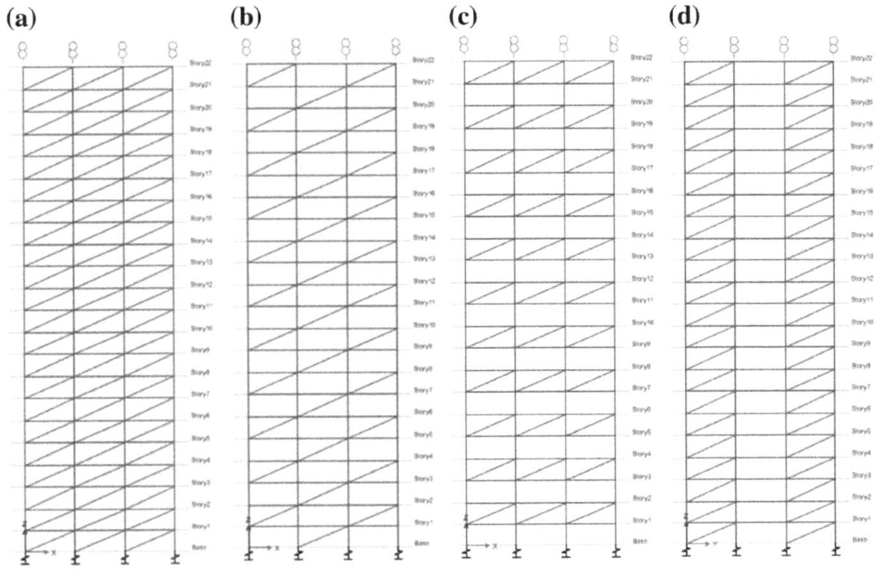

Fig. 7 Elevation of 22 storey diagrid buildings with eccen forward bracing pattern arrangement (**a** whole, **b** alternate, **c** horizontal, **d** vertical)

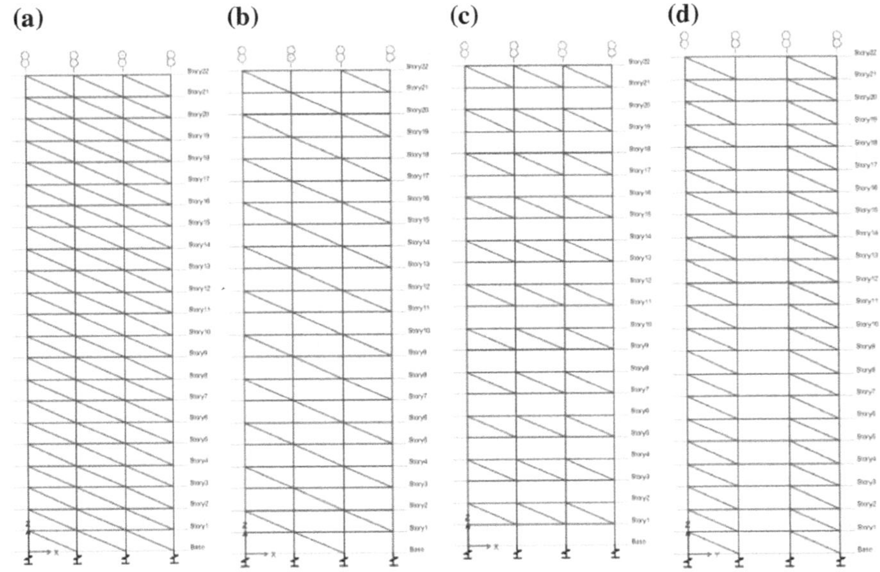

Fig. 8 Elevation of 22 storey diagrid buildings with eccen backward bracing pattern arrangement (**a** whole, **b** alternate, **c** horizontal, **d** vertical)

Datas used for time history analysis and response spectrum analysis of 22 storey diagrid structure

Dead load	auto generated by software
Live load	3 kN/m^2

Seismic coefficients (as per IS 1893)

R (response reduction factor)	4
Z (zone factor)	0.36
I (importance factor)	1

Wind coefficients [as per IS 875 (part 3)]

Cp (wind ward coefficient)	0.8
Cp (leeward coefficient)	0.5
K1 (risk coefficient)	1
K3 (topography factor)	1

Properties of TFP isolator

The above mentioned TFP isolator was kept on the base of the 22 storey diagrid building and analysis was done by dynamic linear time history and dynamic linear response spectrum analysis (Table 1).

Analysis results of 22 storey buildings with TFP isolator The analysis results in terms of top storey displacement and base shear are presented in this section. From the analysis diagrid structure with alternate bracing pattern showing better results. Figure 9 shows the top storey displacement and base shear comparison graph of 22 storey diagrid structures with alternate bracing pattern respectively.

Table 1 Isolation bearing properties

Geometric properties		Frictional properties	
Property	Value (mm)	Property	Value
$R_{1eff} = R_{4eff}$	2133	$\mu_1 = \mu_4$ lower bound	0.067
$R_{2eff} = R_{3eff}$	330	$\mu_2 = \mu_3$ lower bound	0.212
$d_{1*} = d_{4*}$	339.8	μ lower bound	0.045
$d_{2*} = d_{3*}$	41.5	$\mu_1 = \mu_4$ upper bound	0.080
		$\mu_2 = \mu_3$ upper bound	0.254
		μ upper bound	0.053

where

R_{1eff}, R_{4eff}—effective radii of outer regimes

$R_{2eff} = R_{3eff}$—effective radii of inner regimes

μ_1, μ_4—coefficient of dynamic frictions of outer regimes

μ_2, μ_3—coefficient of dynamic frictions of outer regimes

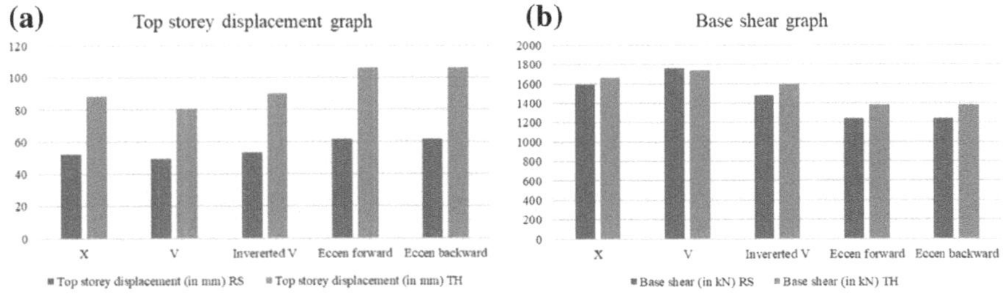

Fig. 9 Top storey displacement and base shear comparison graph of 22 storey diagrid structures with alternate bracing pattern (RS—by response spectrum analysis TH—by time history)

4 Analysis and Design of 11 Storey Diagrid System and Properties of TFP Isolator

Building Configuration Except plan, elevation and building components rest of the properties are same to 11 storey diagrid buildings as compared to 22 storey diagrid buildings. The 11 storey tall diagrid buildings are having 12 m × 12 m plan dimension. The storey height is 3.5 m. The typical plan of the building is shown in Fig. 10. The bracings used and bracing pattern arrangements are as same as done in 22 storey diagrid building.

Datas used for time history analysis and response spectrum analysis of 11 storey diagrid structure

Dead load	auto generated by software
Live load	3 kN/m^2

Seismic coefficients (as per IS 1893)

R (response reduction factor)	4
Z (zone factor)	0.36
I (importance factor)	1

Wind coefficients [as per IS 875 (part 3)]

Cp (wind ward coefficient)	0.8
Cp (leeward coefficient)	0.5
K1 (risk coefficient)	1
K3 (topography factor)	1

Properties of TFP isolator

The above mentioned TFP isolator was kept on the base of the 11 storey diagrid building and analysis was done by dynamic linear time history and dynamic linear response spectrum analysis (Table 2).

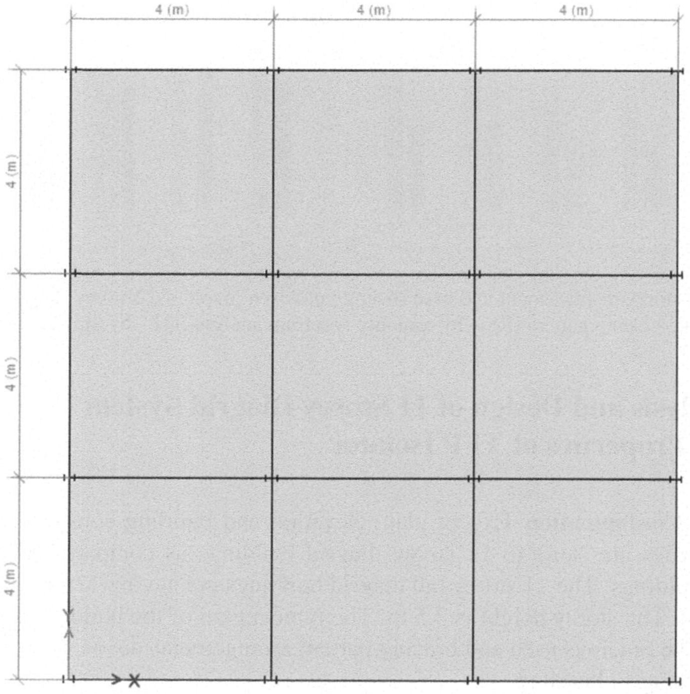

Fig. 10 Typical plan common to all 11 storey diagrid building

Table 2 Isolation bearing properties

Geometric properties		Frictional properties	
Property	Value (mm)	Property	Value
$R_{1eff} = R_{4eff}$	3395	$\mu_1 = \mu_4$ lower bound	0.078
$R_{2eff} = R_{3eff}$	526	$\mu_2 = \mu_3$ lower bound	0.066
$d_{1*} = d_{4*}$	540.40	μ lower bound	0.076
$d_{2*} = d_{3*}$	65.90	$\mu_1 = \mu_4$ upper bound	0.093
		$\mu_2 = \mu_3$ upper bound	0.080
		μ upper bound	0.091

where
R_{1eff}, R_{4eff}—effective radii of outer regimes
$R_{2eff} = R_{3eff}$—effective radii of inner regimes
μ_1, μ_4—coefficient of dynamic frictions of outer regimes
μ_2, μ_3—coefficient of dynamic frictions of outer regimes

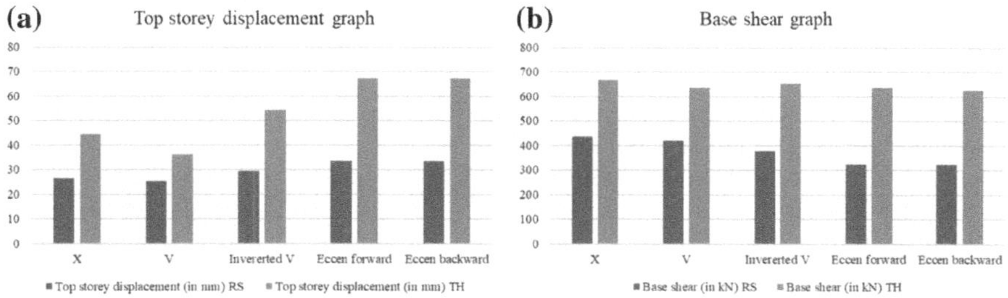

Fig. 11 Top storey displacement and base shear comparison graph of 11 storey diagrid structures with alternate bracing pattern (RS—by response spectrum analysis TH—by time history)

Table 3 Size of typical members of 22 and 11 storey diagrid structures

Storey	Diagonal bracing sections	Interior column sections	Beam sections
22 storey	375 mm pipe sections with 12 mm thick	ISMB 500	ISMB 600
11 storey	375 mm pipe sections with 12 mm thick	ISMB 225	ISMB 200

Analysis results of 11 storey buildings with TFP isolator The analysis results in terms of top storey displacement and base shear are presented in this section. From the analysis diagrid structure with alternate bracing pattern showing better results. Figure 11 shows the top storey displacement and base shear comparison graph of 11 storey diagrid structures with alternate bracing pattern respectively (Table 3).

Here the material property of all sections are in Fe 250 grade and the plan dimensions for 22 and 11 storey structures are different, because the diagrid buildings are designed as per specified height to width ratio of base isolated building. That is the height to width ratio of base isolated building should be within 2–4.3.

5 Summary and Conclusions

Based on study carried out in this paper following conclusions are derived for diagrid structural system with TFP:

- From the analysis, for both 22 and 11 storey buildings V bracing with alternate arrangement pattern showing better results
- Increase of slope of braces increases the shear lag effect and lateral strength in diagrid structures [2]
- The TFP found to be excellent seismic control device for the diagrid structures having incomplete module in frames in controlling forced responses [3].

References

1. Li H-n, Wu X-X (2006) Limitations of height-to-width ratio for base-isolated buildings under earthquake. Struct Des Tall Spec Build 15:277–287
2. Kim J, Lee Y-H (2010) Seismic performance evaluation of diagrid system buildings. J Struct Eng © ASCE 143(12):1–7
3. Bhuva LJ, Vachani MV (2018) Seismic performance of irregular building in plan using triple friction pendulum. Int J Adv Eng Res Dev V:5(04):2296–2303
4. https://www.google.com/imgres?imgurl=https%3A%2F%2Ftheconstructor.org%2Fwp-content%2Fuploads%2F2016%2F10%2Fdiagrid-structural-system.jpg&imgrefurl=https%3A%2F%2Ftheconstructor.org%2Fstructural-engg%2Fdiagrid-structural-system%2F13731%2F&docid=PFtCmAMwunbGhM&tbnid=g0BwOqeNUc8EOM%3A&vet=10ahUKEwi2kZflievkAhXJ-GEKHcdCDi8QMwhxKAAwAA..i&w=552&h=371&bih=587&biw=1350&q=diagrid%20structures&ved=0ahUKEwi2kZflievkAhXJ-GEKHcdCDi8QMwhxKAAwAA&iact=mrc&uact=8
5. https://www.google.com/imgres?imgurl=https%3A%2F%2Fwww.researchgate.net%2Fprofile%2FMichael_Constantinou%2Fpublication%2F238439732%2Ffigure%2Ffig7%2FAS%3A668623861456907%401536423759632%2FCross-section-of-the-triple-FP-bearing-labeled-with-parameters-that-dictate-behavior-R-i.png&imgrefurl=https%3A%2F%2Fwww.researchgate.net%2Ffigure%2FCross-section-of-the-triple-FP-bearing-labeled-with-parameters-that-dictate-behavior-R-i_fig7_238439732&docid=5YJbcnHMNxT5KM&tbnid=LRRqWlaeOwAr0M%3A&vet=10ahUKEwjt-M7Rj-vkAhV_8HMBHdttC0sQMwhCKAIwAg..i&w=610&h=353&bih=538&biw=1350&q=cross%20section%20of%20triple%20friction%20pendulum%20isolators&ved=0ahUKEwjt-M7Rj-vkAhV_8HMBHdttC0sQMwhCKAIwAg&iact=mrc&uact=8

Review of IS 1893-1(2002): Effect of Unreinforced Masonry Infill Walls on Seismic Response of Framed Structures

Gayathri Krishna Kumar and M. G. Airin

Abstract Earthquake, its occurrence and effects, its impact and structural response have been studied for many years in earthquake history and is well documented. The structural engineers have tried to examine the various method, with an aim to determine the complex dynamic effect of seismically induced forces in structures, for designing of earthquake resistant structures in a advanced and easy manner. From the study conducted it was found that more precise results are found from nonlinear static analysis method. An overview of the past researches conducted on the modelling of masonry infilled frame issues, it was found that macro model which consider the effect of masonry weak links is used for modelling the infill panels. Different factors governing the period of vibration was checked, and the result shows the effect of stiffness of the building is the most important factor influencing the period of vibration. Parametric study was conducted to determine the most influential factor that affects the period of vibration of a structure. From the observations it was clear that the effect of stiffness is the most important factor influencing the period of vibration. Therefore a curve with dimensionless height and lateral displacement were plotted using nonlinear static analysis obtained from SAP2000. From the above mentioned curve, the effective stiffness of the building under consideration is calculated, which route to find the period of vibration of the structure that is considered for the seismic analysis [1].

Keywords Dynamic · Infill · Vibration · Parametric

G. Krishna Kumar (✉)
Federal Institute of Science and Technology, Angamaly, India
e-mail: gaya3krishnakumar92@gmail.com

M. G. Airin
SCMS School of Engineering and Technology, Karukutty, India
e-mail: airinmg@scmsgroup.org

K. Dasgupta et al. (eds.), *Proceedings of SECON'19*,
Lecture Notes in Civil Engineering 46,
https://doi.org/10.1007/978-3-030-26365-2_60

1 Introduction

When a structure is imperiled to ground motions in an earthquake, it responds by vibrating. The random motion of the ground caused by an earthquake can be resolved in any three mutually perpendicular directions. Generally, however, the inertia forces generated by the horizontal components of ground motion require greater consideration in seismic design. Structural analysis is mainly used for finding out the behavior of a structure when subjected to some act. This action can be in the form of load due to the weight of things such as people, furniture, wind, snow, etc. or some other kind of excitation such as an earthquake, quivering of the ground due to a blast nearby, etc. since all these loads are dynamic including the self-weight of the structure because at some point in time these loads were not there [1].

2 Analysis Methods

Static analysis describes a series of forces acting on a building to represent the effect of earthquake ground motion, typically defined by a seismic design response spectrum It claim that the building responds in its fundamental mode. The response is recite from a design response spectrum, given the natural frequency of the building (either calculated or defined by the building code.

Static procedures are appropriate when higher mode effects are not significant. This is generally true for short, regular buildings. Therefore, for tall buildings, buildings with torsional irregularities, or non-orthogonal structures, a dynamic procedure is required. In the linear dynamic procedure, the building is modelled as a multi-degree-of-freedom (MDOF) system with a linear elastic stiffness matrix and an equivalent viscous damping matrix.

In general, linear procedures are applicable when the structure is expected to remain nearly elastic for the level of ground motion or when the design results in almost uniform distribution of nonlinear response throughout the structure. As the performance objective of the structure implies larger inelastic demands, the uncertainty with linear procedures increases to a point that requires a high level of conservatism in demand assumptions and acceptability criteria to elude unintended performance. Therefore, procedures incorporating inelastic analysis can reduce the uncertainty and conservatism.

2.1 Effect of Infill Wall

The infill wall is known as the supported wall that closes the perimeter of a building. It is generally made of steel or reinforced concrete or masonry. The infill wall has the unique static function to bear its own weight. Though it has been understood that the

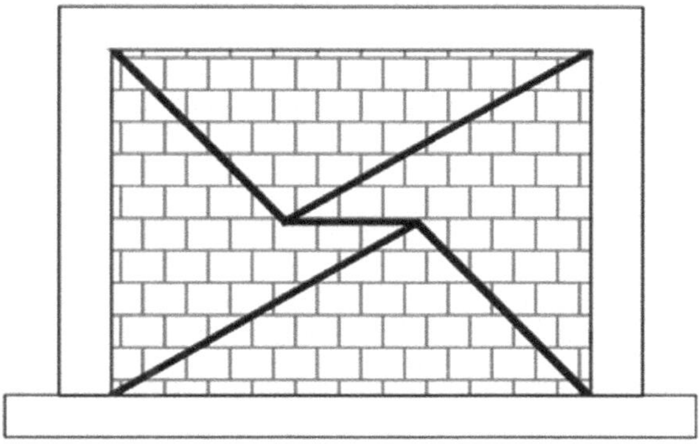

Fig. 1 Recent modelling Technique (Furtado 2015)

infills play significant role in enhancing the lateral stiffness of complete structure, serves to the past knowledge in various earthquakes have proved that the partially infilled framed structures somehow are affected adversely. This paper intends to highlight the need of knowledge on infilled frames.

In the literature, different modelling proposal techniques that simulate the behaviour of the infills' panels can be found and are divided in two different groups, namely micro-models and simplified macro-models. The first of them includes models in which the panel is divided into numerous elements taking into account the local effects in detail, and the second includes simplified models based on the physical understanding of the behaviour of the infill panels submitted to earthquakes loadings and past experimental tests. In the case of the last group, a few of struts are used to represent the effect of this non-structural element on the structural response. The same technique is used in SAP2000 for modelling (Fig. 1).

3 Nonlinear Static Analysis Using SAP2000

Nonlinear static analysis were carried out for five storey and ten storey building, their period of vibration was used for comparison it was clear from the comparison the equation formulated in IS 1893-1 (2002) needs modification (Fig. 2).

It is clear from the consolidated results Table 1, the effect of infill is not fully considered in IS 1893 part 1 2000 therefore modification is required for the time period equation given in IS 1893 part 1 2000 [2, 3].

The value of time period decreases for with infill structures it is because time period is inversely proportional to the stiffness of the building by the relation $T = 2\pi\sqrt{\frac{M}{K}}$, as we know the infill impart stiffness to the building.

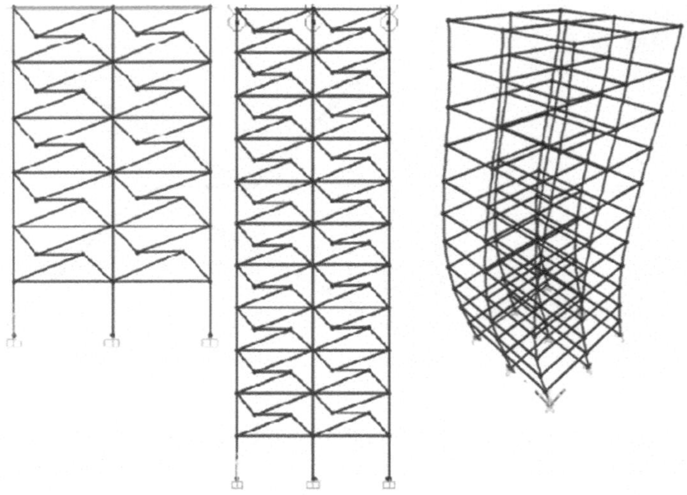

Fig. 2 Ten and five storey model (SAP2000)

Table 1 Comparison of results for ten and five storey

Category		Without infill		With infill	
No. of storey		5	10	5	10
Time period (s)	IS 1893	0.655	1.032	0.567	0.953
	SAP2000	1.2423	2.2526	0.82508	1.2454
Base Shear (kN)	IS 1893	5148.56	6710	7290	9564
	SAP2000	4608	6912	4032	6336

4 Nonlinear Static Analysis Using SAP2000

4.1 Period of Vibration

The fundamental period is a vital design parameter that plays a significant role in the computation of design base shear. The inclusion of the masonry panels rigidity in structural modelling cause variation in the fundamental period by stiffening the structure, and in turn affects considerably the overall response of the building to earthquake ground motion. Factors influencing period of vibration. Height of the building, Base dimension of the building, No: of panels in both directions, Amount of infill and properties of infill, Effect of stiffness of the structure. The work carried out takes into account the stiffness, height and dimension of the building. The other factors are neglected because the variation of time period is less (Tables 2 and 3) when compared to the other factors [4].

Table 2 Time period for different Infill materials

Model No	Material	Time period (s)
Model 1	Brick	0.79461
Model 2	Solid block	0.78774
Model 3	Hollow block	0.79259

Table 3 Time period for Infill thickness

Model No	Thickness (mm)	Time period (s)
Model 4	230	0.79461
Model 5	250	0.80503
Model 6	300	0.81676

4.2 Stiffness

Infilling panels are found to increase stiffness of the structure, increase in initial stiffness, obtained for small strains, can reach 7 times that of bare frame. The stiffness ratio is calculated in each direction at individual story level. First, obtain the inverse of the angle of the inter-story deflection, r_s at each story level, and then compute the numerical average of r_s of all the stories above grade, r_S. The r_s value at individual story level is divided by the r_s value, and the result is defined as the stiffness ratio. The computation of stiffness on each floor and the stiffness ratio of the building was by using the displacement at each level and the corresponding height at that level [5].

Effect of beam-to-column stiffness ratio (ρ) on lateral displacement patterns [2, 3]

The beam-to-column stiffness ratio, ρ, as a parameter that oversees the behavior of frame type building systems [6]. It is the ratio of sum of the beam stiffness to column stiffness at the story that is nearest to the mid-height of the building, and it is constant for structures that have uniform lateral stiffness along their height. With the same modulus of elasticity for girders and columns, the general form of ρ is given by

$$\frac{\sum \frac{I}{L}\text{beam}}{\sum \frac{I}{L}\text{column} + \sum \frac{I}{L}\text{infill}} \tag{1}$$

I—Moment of inertia
L—Length

Relation between dimensionless height, ρ, displacements

- Height of the building = 10.5 m, ρ = 1.155, Time period = 0.28611 s
- Height of the building = 16.5 m
- ρ = 1.040
- Time period = 0.40383 s (Tables 4 and 5).

Table 4 Model and deflection for each level

Model	Dimensionless height	Deflection
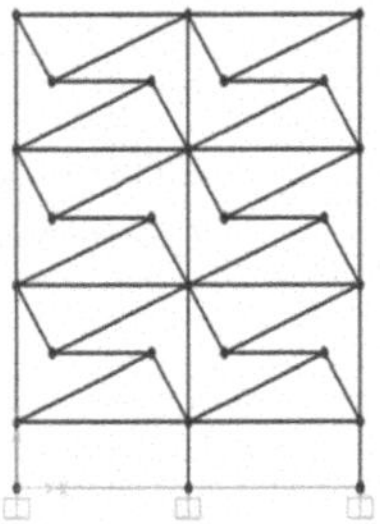	1.0	0.0085
	0.714	0.0069
	0.428	0.0049
	0.142	0.0025

Table 5 Model and deflection for each level

Model	Dimensionless height	Deflection
	1.0	0.0501
	0.86	0.0447
	0.73	0.0383
	0.6	0.0314
	0.46	0.0242
	0.33	0.0173
	0.2	0.0111
	0.05	0.0054

Dependence of Stiffness ratio and period of vibration
The inclusion of masonry panels rigidity in the structural modelling changes the fundamental periods of the model by stiffening the structure, and in turn affects considerably the overall response of the building to earthquake ground motion. The stiffness ratio is calculated in each direction at each story level. First, obtain the inverse of the angle of the inter-story deflection, r_s at each story level, and then calculate the numerical average of r_s of all the stories above grade, r_S. The r_s value at each story level is divided by the r_S value, and the result is defined as the stiffness ratio. For obtaining the curve between the time period of vibration and stiffness ratio different infill wall configurations are modelled using SAP2000, therefore every possibility is considered which give accurate time period for any infill configuration [7].

Dependence of lateral deformation on ρ for first mode
See Fig. 3

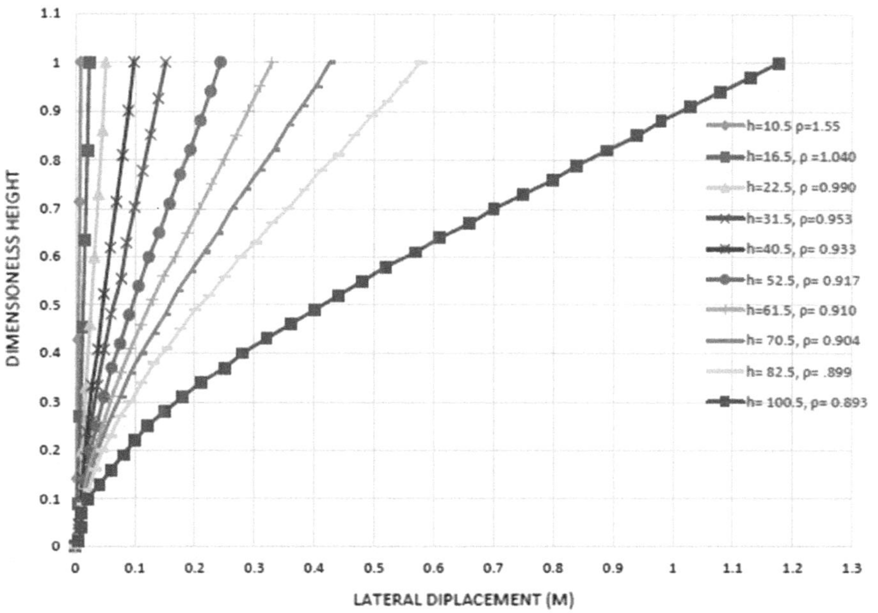

Fig. 3 Dependence of lateral deformation on ρ for first mode

4.3 Computation of Stiffness Ratio

The stiffness ratio is calculated in each direction at each story level. First, obtain the inverse of the angle of the inter-story deflection, r_s at each story level, and then compute the numerical average of r_s of all the stories above grade r_{s1}. The r_s value at each story level is divided by the value r_{s1}, and the result is defined as the stiffness ratio. The computation of stiffness on each floor and the stiffness ratio of the building was by using the displacement at each level and the corresponding height at that level. [Storey stiffness ratio table (http://maual.midasuser.com)] (Table 6).

5 Testament of the Results

Different models where used for validating the obtained results, 10, 19.5, 52.5, 80 m high and different plan dimensions where used for validation. Each storey level of 10 m high building was converted into its dimensionless height and deflection for each level is found out from the dimensionless height—deflection curve.

The results obtained from the stiffness ratio method is almost equal to the value obtained from the analysis therefore the stiffness ratio method is validated. Column (4 and 5) shows the percentage variation of time period obtained from analysis with stiffness ratio method and IS 1893-1 (2002). The percentage variation of time period

Table 6 Testament of the results for different heights and plan dimension

Stiffness ratio	Time period (s)	Stiffness ratio	Time period (s)
0.45	5.5503	1.14	0.619
0.55	3.883	1.18	0.284
0.602	3.1815	1.19	0.40383
0.709	1.37316	1.2483	0.28251
0.74	4.1653	1.25	1.361
0.82	3.262	1.2547	0.28251
0.89	2.69	1.2865	0.286
0.91	2.2663	1.3	0.296
0.9405	1.05	1.39	1.0694
0.99	2.206	1.4	0.87
1.087	0.214	1.55	0.718

obtained from stiffness method is from 0.6 to 7.0% and the variation of time period from IS 1893 is from 6 to 44.7%. Since the percentage variation for stiffness method is less when compared to IS code method (IS 1893-1 2002) we can use stiffness method to find time period accurately for seismic analysis of framed structures.

6 Conclusions

Analytical study was carried out to investigate the effect of changes in latest revision of IS: 1893 on Time Period calculations for multistorey buildings. Though the number of buildings analyzed is a few to make generalization about the effect of revisions yet some important conclusions have been arrived. The infill wall impart lateral stiffness to the structure, lateral stiffness is an important factor that influence the period of vibration. Therefore if the stiffness of the building is known we can directly the find the time period of vibration.

It is clear from the consolidated results the effect of infill is not fully considered in IS 1893 part 1 2000 therefore modification is required for the time period equation given in IS 1893 part 1 2000.

The value of time period decreases for with-infill structures. It is because time period is inversely proportional to the stiffness of the building by the relation $T = 2\pi\sqrt{M/}$, as we know the infill impart stiffness to the building.

From Table 7, column (4 and 5) shows the percentage variation of time period obtained from analysis with stiffness ratio method and IS 1893-1 (2002). The percentage variation of time period obtained from stiffness method is from 0.6 to 7.0% and the variation of time period from IS 1893 is from 6 to 44.7%. Since the percentage variation for stiffness method is less when compared to IS code method (IS

Table 7 Testament of the results for different heights and plan dimension

Height (m) (1)	Time period (s) (SAP2000) (2)	Time period (s) (Present study) (3)	Time period (s) IS 1893-1 (2002) (4)	Percentage variation (2 & 3) (5)	Percentage variation (2 & 4) (6)
10	0.286	0.284	0.25	**0.6**	12
19.5	0.4697	0.4620	0.5	**1.49**	6
52.5 (unsymmetrical)	1.80	1.53	1.11	**3.8**	24
52.5 (symmetrical)	2.009	2.03	1.11	**1.03**	44.7
80	3.594	3.32	2.07	**7**	42

Bold indicates that percentage variation is less for proposed study

1893-1 2002) we can use stiffness method to find time period accurately for seismic analysis of framed structures.

References

1. Shinde DN (2015) Pushover analysis of multi story building. 32(3):417–442
2. IS 1893 (Part 1): 2002 Indian standard criteria for earthquake resistant design of structures, part 1 general provisions and buildings (Fifth revision)
3. IS 875 (Part 2): 1987 Indian standard code of practice for design loads (other than earthquake) for buildings and structures, part 2 imposed loads (Second revision)
4. Rai DC (2001) Review of documents on seismic evaluation of existing buildings. Interim report 1: a—earthquake codes, IITK –GSDMA project on building codes, IITK-GSDMAEQ03-V1.0, pp 1–32
5. Das A, Guha P (2016) Comparative study of the static and dynamic seismic analysis of RC regular and irregular frame structures. J Sci Eng Res 7(4)
6. Chatpan C, Chopra AK (2013) Evaluation of modal pushover analysis using generic frames. Earthq Eng Struct Dyn 32(3):417–442
7. Meshram VR (2015) Study of base shear and storey drift by dynamic analysis. Int J Eng Innov Technol (IJEIT) 4(8)
8. Akkar S, Gülkan P, Yazgan U (2015) A simple procedure to compute the interstorey drift for frame type structures. 3(3):446–453

Psychosocial and Occupational Hazards in Kerala Construction Industry

R. Abhijith, C. P. Deepika, P. N. Mirfath and Saraswathi Menon

Abstract Construction industry has accomplished extensive growth worldwide particularly in past few years. For a project to be successful, safety of the structure as well as that of the workers is of great importance. Safety issues are to be considered from the planning stage to the completion of the structure. Construction industry employs both skilled and unskilled labour subjected to accidents and risks. The number of fatal accidents taking place at the construction sites is quite increasing. According to a recent study by the International Labour Organization (ILO) 165 out of every 1000 workers have faced injuries from the job. From the results of Occupational Safety and Health Administration examination, the causes of construction fatalities, it was shown that 39% of fatalities in construction were caused by falls, 8.4% were struck by objects, 1.4% were caught in between incidents, and 8.5% were caused by electrocution. Very little research has been done on the occupational health, hazards and psycho-social problems of workers especially in different countries like India. Safety in construction is important as any fatal accident will cause irreparable loss of human life, money and progress of work leading to reduction in morale and demotivation of work force. A psychosocial hazard is an occupational hazard that affect the psychological as well as the social well-being of workers, including their ability to participate in a work environment among other people. Exposure to psychosocial hazards in the workplace produces both psychological and physiological damage to the employees. It also produces repercussions within the society, thereby reducing the productivity in local/state economies, corroding familial/interpersonal relationships, and producing negative behavioural outcomes. In this study, we would

---

R. Abhijith · C. P. Deepika · P. N. Mirfath (✉) · S. Menon
Department of Civil Engineering, Viswajyothi College of Engineering and Technology, Vazhakulam, India
e-mail: mirfathp@gmail.com

R. Abhijith
e-mail: abhijithr@gmail.com

C. P. Deepika
e-mail: deepikacp28@gmail.com

S. Menon
e-mail: Saraswathimenon17@gmai.com

K. Dasgupta et al. (eds.), *Proceedings of SECON'19*,
Lecture Notes in Civil Engineering 46,
https://doi.org/10.1007/978-3-030-26365-2_61

like to conduct a questionnaire survey across Kerala construction industry and analyse the result using AMOS 22.0 to evaluate the reasons behind accidents in Kerala construction industry.

Keywords Occupational hazard · Safety · Psychosocial hazards

1 Introduction

A possible source of danger is known as hazard. Hazards can be classified into Occupational hazards and Psychosocial hazards. Occupational hazard is any hazard that is experienced in the workplace. Occupational hazards can be classified as physical, chemical and biological hazards. It represent any type of condition, situation, object, procedure or requirements which can injure you at work. It includes both long term and short term risk associated with the workplace environment and is a field of study within occupational safety and health and public health. Short term risks may include physical injury, while long-term risks may be increased risk of developing cancer or heart disease.

Psychosocial hazard is any type of hazard that affects the psychological well being of workers including their ability to participate in a work environment. Factors affecting psychosocial factors are stress, workload, worktime, home work interface, interpersonal relationship at work.

2 Significance

2.1 Peruman Rail Accident

In the Peruman railway accident, the (Train No:26) Bangalore–Thiruvananthapuram Central Island Express train derailed on the Peruman bridge over Ashtamudi Lake, near Perinadu in the Kollam district of Kerala. 105 people were died in the accident. Even though the main reason of the accident is still unknown, over speed, track alignments and faulty wheels, track maintenance work.

2.2 Collpase of Building at Kozhikode

Collapse of building in Kozhikode took place in 03 May 2018. Two people were died and six were injured. The main *reason* behind the accident was improper study on the geological conditions of the area.

Accidents due to psychological and occupational safety is increasing day by day, thus the significance of the study also increases.

Table 1 IS codes and articles

Item	Description	Remarks
IS 14489:1998 Code of practice on occupational health and safety audit	Provision of safety equipments, safety training, etc.	Major reason that contributes to accidents
Article 24-Child Labour	Prohibition of children under the age of 14	Occurrence of accidents due to lack of knowledge
Article 39-Equality in Wages	Equal wages for men and women	Inequality arises
Article 42-Maternity Leave	Provision of maternity leave and humane working conditions	Leads to lack of immunity
Article 43 A-Living Wage	Living wage, ensuring a decent standard of life	Poverty
Article 21-Protection to Life	No person shall be deprived of his personal life	Injuries and fatalities
Article 47-Public Safety	Improvement of public health and safety	Destruction of public properties

3 Hypothesis

See Table 1.

From the prevailing studies, accidents in construction industry increases due to increase in stress.
From the current studies, it can be founded that hazards occurs due to unsafe working condition and overburden.
From the present conditions, it can be said that hazards increases due to home work conflicts.
From the current condition, it can be founded that increased accidents are due to lack of safety and awareness.
From the prevailing studies, it can be founded that increased working time reduces social commitments.
From the present condition, it can be founded that increase in accidents are due to increase in carelessness.

4 Questionnaire Preparation

Based on the journal studied questionnaire were prepared covering all aspects like demographic details, behavioral problems, safety and health conditions. Separate

Table 2 List of companies

Company name	Date of response	Designation of responder
Ambuja Cements	30/01/2019	Safety Officer
Kochi Metro Rail Ltd.	05/02/2019	HR
Mary Matha Constructions	05/02/2019	Site Engineer
G&G Constructions	12/02/2019	MD
SOLBA Construction	12/02/2019	Site Engineer
Kukoos Nest Constructions	19/02/2019	Site Engineer
Nate Constructions	26/02/2019	Safety Officer
Kumar & Kumar	26/02/2019	MD
Autokast Ltd.	02/03/2019	HR
GJ Builders	03/03/2019	Site Engineer
MENONS Constructions	03/03/2019	Site Engineer

questionnaire was prepared for HR level and for labour level. The questionnaire consists of descriptive questions, yes or no questions and five-point scale questions.

5 Data Collection

Data were collected from thirty-five companies across Thodupuzha, Alappuzha, Ernakulam and Muvattupuzha. The companies include major construction companies to small scale construction units. It also includes manufacturing units of civil engineering field (Table 2).

6 Analysis

The data collected is analysed and bar graphs and piecharts are drawn to depict the data.

From the pie chart 1 we can confirm that majority of the companies doesn't conduct safety committee. About 75% of the companies doesn't conduct any type of safety audit or meetings.

From the pie chart 2 we can infer that 55% of the companies doesn't have a professional code.

From the pie chart 3 it can be analysed that 50% of the companies have instructions to the workers and safety boards displayed in the site premises.

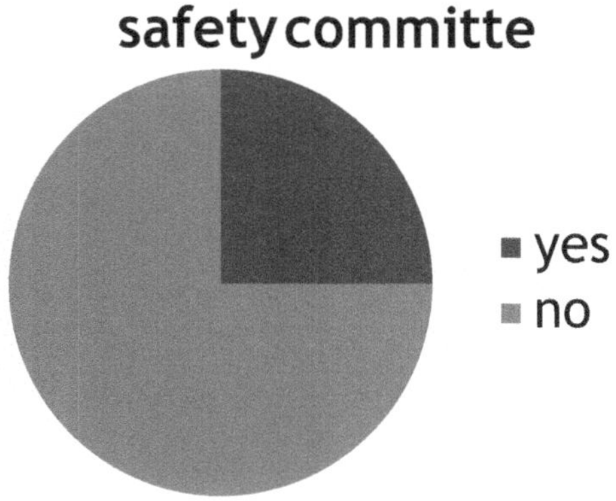

Fig. 1 Pie chart for question 7

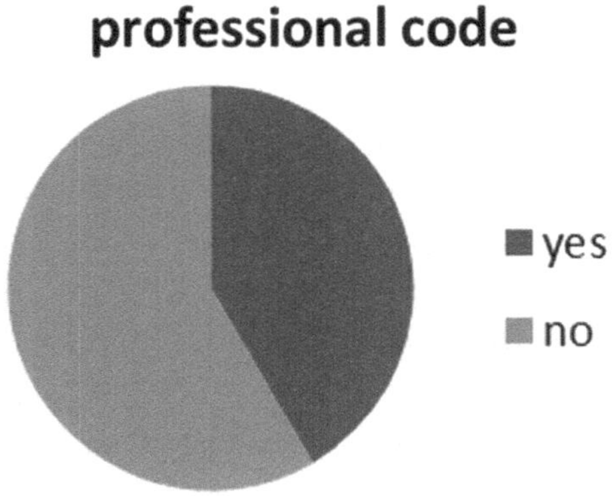

Fig. 2 Pie chart for question number 1

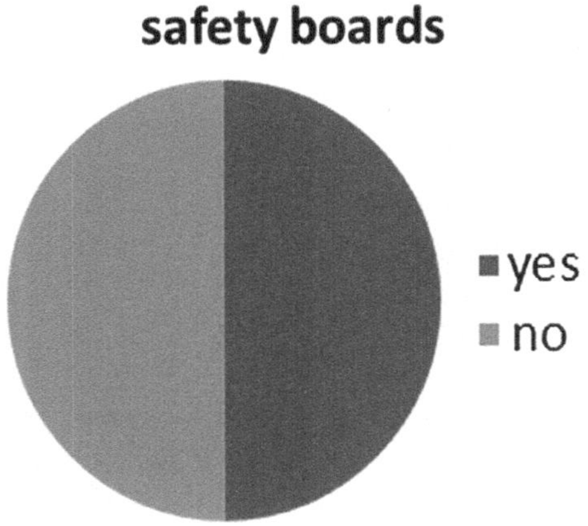

Fig. 3 Pie chart for question 6

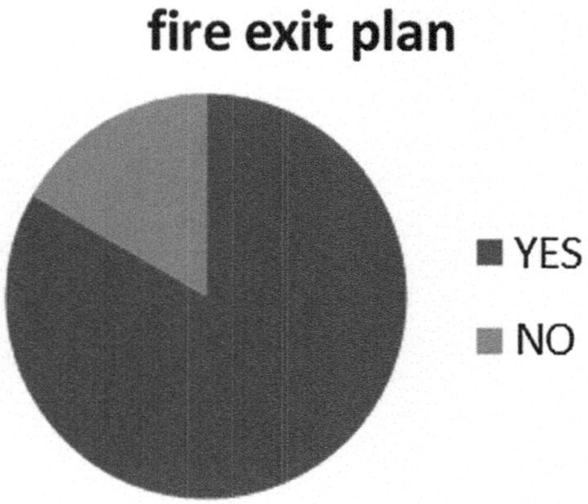

Fig. 4 Pie chart for question number 20

From the pie chart 4 it can be concluded that 15% of the companies doesn't have a fire exit plan.

From the bargraph 5 it can be seen that majority of the companies doesn't follow any kind of safety measures and doesn't adopt any kind of safety policies.

7 Inferences

Majority of the companies doesn't have:

- Professional code
- Safety committee
- Fire exit plan and route
- Doesn't conduct mock drills

Due to overtime (12–14 h):

- Stress increase
- Homework conflicts
- Divorce rates increases

Some other observations:

- It is observed that workers or employees are suffering from cancer due to the consumption of tea made from over brewed tea leaves
- Studies showed that increased working time reduces the social mingling and commitment of workers
- Increased home-work conflict reduces the familial commitment.

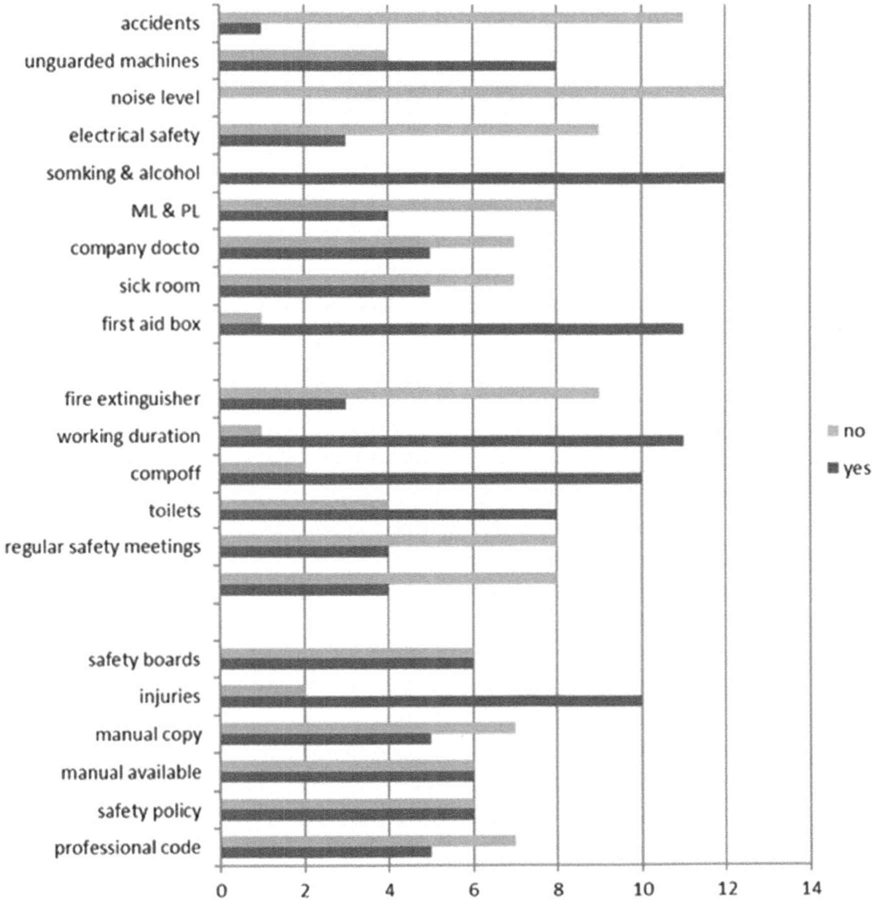

Fig. 5 Bargraph for yes/no questions

Annexure 1: Questionnaire—MD/HR Manager

1. Mention the IS code practiced in the company…………………
2. Does the company have a professional code? (YES/NO)
3. Does the company have a written health and safety policy? (YES/NO)
4. Is the manual readily available to your employees? (YES/NO)
5. Whether a copy of the manual is kept at the worksite? (YES/NO)
6. Whether the company is insured? (YES/NO)
7. Whether the company provide corporate insurance coverage to employees? (YES/NO) [1]
8. Does the company have a formal system for reporting, recording and investigation of injuries and illness? (YES/NO) [2]
9. Does the company have safety boards or displays? (YES/NO)

10. Does the company conduct mock drills? (YES/NO)
11. If yes, how often..................
12. Does the company have a safety committee? (YES/NO)
13. Does the company hold regular safety meetings? (YES/NO)
14. If yes, frequency of meetings ....
15. Does the company gives adequate safety training to the workers? (YES/NO)
16. Does the company have LOTO system? (YES/NO) [3]
17. Does the site have emergency contact number boards? (YES/NO)
18. Does the site have enough toilets separate for men and women? (YES/NO)
19. Kindly mention the regular working hours of the company........
20. Kindly mention the actual working hours of the company........
21. Kindly specify the type of appointment (Regular/Contract)
22. If contract based appointment mention the time period.......
23. Whether the workers of the company have ESI benefits? (YES/NO)
24. Whether the workers of the company have PF benefits? (YES/NO)
25. Whether any compensation given to the workers? (YES/NO)
26. Is the site working during national holidays? (YES/NO)
27. Is there any fire extinguisher? (YES/NO)
28. Is there any fire exit plan? (YES/NO)
29. Does the company have a first aid box at the work site? (YES/NO)
30. Does the company have a sick room at work site? (YES/NO)
31. If yes, does the company have separate sick rooms for men and women? (YES/NO)
32. Does the company have a company doctor? (YES/NO)
33. Is there any provision for ML? (YES/NO)
34. Is there any provision for PL? (YES/NO)
35. Whether the workers working above a height of 1.8 m is provided with safety equipments? (YES/NO)
36. Whether there is any penalties or punishments for delay of the employee? (YES/NO)

Nature	Machine failure	Human failure	Fatality
Frequency			

Type of accidents	Procedural	Engineered	Systematic
Frequency			

Annexure 2: Questionnaire—Labour

1. Gender F/M/T
2. Kindly mention the designation of the responder…………………
3. Mention the salary per year…………………………..
4. Height …….
5. Weight……
6. Nature of job………………..
7. Risk level

1	2	3	4	5
Very low	Low	Medium	High	Very high

8. What are the risks involved in your work?
   ………………………………………………………………………………
9. Educational qualifications:………………………………..
10. Stress level

1	2	3	4	5
Very low	Low	Medium	High	Very high

11. Have you ever faced any kind of physical harassment? (YES/NO) [4]
12. Have you ever faced any kind of mental harassment? (YES/NO) [4]
13. Disease background

DIRECT ILLNESS	INDIRECT ILLNESS

14. Have you ever suffered from any kind of injuries from the work place? (YES/NO)
15. Are you provided with any kind of protection equipments? (YES/NO)
16. Mention the number of leave per year

References

1. Salim SM, Romli FI, Besar J, Aminian NO, A study on potential physical hazards at construction sites. J Mech Eng

2. Govindankutty, Priyadarshan S, A study on occupational hazards and the safety measures taken by the tea plantation workers of Kerala. Int J Sci Eng Technol Res (IJSETR)
3. Hasan MM, Khanam R, Mostafa Zaman AKM, Ibrahim M, Occupational health and safety status of ongoing construction work, Science and Technology University, Dumki, Patuakhali. J Health Environ Res
4. Mohammed Sirajuddeen I, Risk management in labour psychosocial issues in construction projects. Int Res J Eng Technol (IRJET)

Evaluation of 5S Conformity in Residential Building Sites Using RADAR Charts

Mancy Sunny and V. V. Anu

Abstract 5S is an essential tool in lean assembling frameworks. It is a tool for cleaning, arranging, sorting out and giving the vital basis to work place quality enhancement. This methodology can manage the space, human effort, time, quality and capital to make the end product with minimum faults and make the site as well ordered, disciplined and clean place to work. 5S application improves personal standards and motivation of workers in their workplace and has a high impact on work area, safety, quality, and efficiency. Efficient material management is essential in managing a productive site. The purpose of this paper is to study the principle of 5S methodology for identification of waste in construction organization to improve the labor productivity, safety and quality of works on residential building sites. This project intends to create a framework for 5S audits for construction sites to assess the current condition and also the conformity towards 5S.

Keywords 5S tools · Lean construction · Check list · Quality improvement · Productivity

1 Introduction

Lean manufacturing includes a set of tools and practices, which when implemented properly and fully to improve system performance. The 5S lean tool is one of them. It is acronyms for sort, set in order, shines, standardize and sustain. 5S helps to reduce the non-value adding time also increase productivity and improve quality. It has been used in the design of efficient facilities. 5S techniques have been integrated with other lean tools we can reduce the changeover time. 5S concept is intended

M. Sunny (✉)
Toc H Institute of Science and Technology, Arakkunnam, India
e-mail: mancysunny@gmail.com

V. V. Anu
CE Department, Toc H Institute of Science and Technology, Arakkunnam, India
e-mail: anuillus@gmail.com

K. Dasgupta et al. (eds.), *Proceedings of SECON'19*,
Lecture Notes in Civil Engineering 46,
https://doi.org/10.1007/978-3-030-26365-2_62

to organize, clean, standardized and maintain the discipline at workplace in pursuit of sustainable improvements in the productivity, efficiency cost optimization and reduction of waste in an organization [1]

Sort

The first element of 5S is Sort is about sorting or keeping the necessary things or items at their appropriate places. To introduce the principle of organization called Sort, there is need of combined effort of organization and self-discipline.

- Sort calls for effective utilization of workplace space and also promotes that the goods or items should be segregated strictly in accordance to the relevance and frequency of utilization to create efficient workplace.
- Sort is helpful element for estimating the material or goods requirements at present and future, which is needed and stored at their respective workplace.
- Sort facilitates avoiding the stack of irrelevant things at workplace thereby eliminating hindrance in the flow of work.

The benefits of 5S first element sort include: saving of space, searching time is short, safe and clean workplace, detection of damage is easy.

Set in order

Items should be very much clear so that anyone can find the useful items at any time. The objective of the second element of 5S is to develop the economical use of workspace with neat and orderly storage of goods.

- Set in Order requires prioritization of the necessity and importance of goods/equipment to maximize ease of location. The key questions who, what, why, where, when and how should be asked of oneself in respect of each item.
- Set in Order involves ensuring designated locations for all items in the workplace, thereby facilitating employees to have efficient control over all the operations.

The benefits of Set in order are rapid processing, reduction of error, discipline and creative ideas generated along with high moral [2]

Shine

The third S, Shine means Cleaning, which emphasizes self-inspection, cleanliness and creating a faultless workplace.

- Shining at the workplace eliminates dirt, dust, fluids, and other debris. An effective maintenance schedule should be developed to sweep or clean the dirt from the equipment, machines and free the workspace area from contaminated particles.
- A checklist can facilitate the employee about the work in future in that activity. It focuses on cleanliness with a formation of teams targeting specific area for cleaning which include machinery, surrounding and storage area.
- It also helps in identifying unplanned breakdown of the equipment.

The benefits of Cleaning include: reduced equipment failure, improved product quality, improved safety at work and efficient work environment.

Standardize
The fourth S standardize means 'Standardization' that is, maintaining one S workplace so that it is productive and comfortable by repeating Sort-Set in order-Shine. During this phase of implementation, the team develops the standard operating procedures for establishing the improvement of workplace practices [3].

- There is need of uniformity in the fourth element of 5S for establishing and ensuring of standards, to guarantee the tidiness and cleanliness of workplace.
- Visual management is an effective mechanism for the continuous improvement because it can play important role in production, quality, safety and all customer services by putting up appropriate labels.

The benefits of Standardization include: low maintenance and overhead cost, loyalty to the organization and increase of process efficiency.

Sustain The fifth S sustains stands for sustaining the entire previous S. Th e fifth S is critical to understand and implement because it requires overall changes in the behavior of employees at various levels within an organization.

- Sustain means ingraining the ability of doing things the way they are supposed to be done. It helps and encourages the employees of creating good habits.
- This element also play important role to make the continuity of daily routine. It is a process of repetition and act as integral part of the safety.
- In this step there is always a need of practice to maintain the self-discipline in their daily life for 5S system and make habit of it within the organization.

Benefits of sustain involves increase of labour productivity, quality products with no accident at workplace [4].

1.1 Aim

To evaluate the 5S conformity in existing construction site.

1.2 Objective

The main objective of this study is:

- To propose a methodology and evaluate construction sites with existing audit framework using checklist.
- To prepare radar chart for 5S.

1.3 Scope of Project

Implement 5S in construction projects.
Increase the productivity of firm.
Improve the safety on working environment.
Manage space, human effort, time, quality and capital to make the end product with less faults.
Provide awareness about 5S implementation among staff and workers.

1.4 Scope of the Study

- The 5S implementation is limited to a part of works at the time of evaluation of construction projects, not to the whole organization.
- Only building works (residential) will be taken into consideration with budget not exceeding 70 lakhs rupees.
- Dealing with all building works involving plastering, flooring, and brickwork, shuttering and concreting.
- 10 sites will be evaluated to find the conformity of the site towards 5S.

1.5 Methodology

The methodology adopted for the study is illustrated in Fig. 1

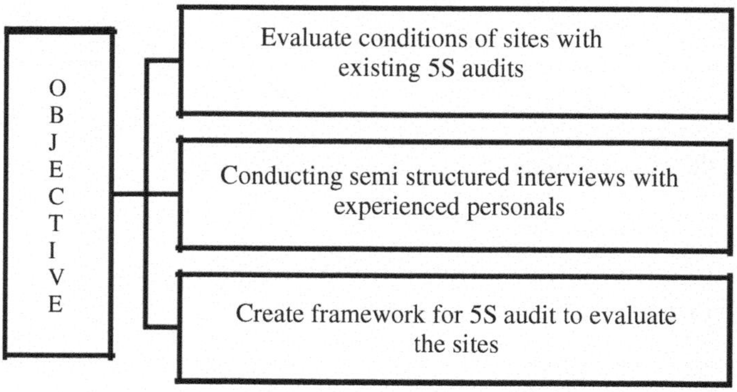

Fig. 1 Methodology of work

2 Critical Review

The literature review reveals that 5S implementation leads to realization of remarkable results for the organizations. 5S approach can easily be applied in various organizations due to its simplicity and easy recognition. 5S is a universal tool which provides huge benefits with no losses by holding improvement initiatives. 5S has proven to be a tool contributing to the sustained growth of organizations and maintain its development in the long term. It is evident that it is not just a pure clean up or housekeeping process. In the past very less studies have been done on evaluation of 5S conformity on construction sites. Being a relatively new and tough to implement, this method is not widely used for construction projects. The detailed theoretical background of 5S method is still under study. This project aims to evaluate 5S conformity in existing sites and also proposes a methodology for 5S implementation.

3 Interview with Experienced Personnels

The sites when audited for 5S conformity give indications on the areas to be improved so as to bring in quality at workplace. Sorting of workplace helps to distinguish between what is needed and not needed. Set in order creates a place for everything and in its adequate position. Therefore by implementing 5S in a site creates a visual control. Anything deviating from 5S plan can easily be pointed out once everyone is aware of 5S. Every activity must be 5S oriented so that it turn out as a culture [5]. This requires training of workers in the site to adapt 5S method. The key performance indicators are derived based on the nature of work at the site and emphasis is given to these key works while implementing 5S. Involvement of supervisors and workers are essential in identifying the current problems and their cooperation is essential in framing a action plan. The 5S has to be defined with regards to construction industry for clear identification of problems and better framework of 5S implementation. Semi structured interviews, are the best used, when we won't get more than one chance to interview someone and when we will be sending several interviewers out into the field to collect data. The semi-structured interview provides a clear set of instructions for interviewers and can provide reliable, comparable qualitative data. The interviews were done after keen observations and thorough study about the 5S principles. The questions were asked to get more insight into the site conditions and to relate it with 5S's [6].

4 5S Audits

The level of 5S adaptability is ascertained by performing 5S audits in the sites. This helps to understand the level of conformity of the workplace towards the already defined basic 5S activities. Checklists help in evaluating a particular job area where the change is to be made. Moreover to sustain the changes regular audits have to be performed to ensure the change is maintained and makes room for further improvement. Audits were done on construction sites before adopting 5S methodology [4]. The checklist mainly used to evaluate the first three Ss. Since the 5S training is not given to workers, so the standardization and sustain category is not measured at the initial stage. 10 sites are selected for the audit. The scores are plotted on a radar chart. The score for each item were given based on the number of problems in the site. For each category scores were given for different items. The sort, set in order, shine categories were first taken into account. The values obtained from different sites are tabulated as follows. The scores were given out of 20.

The data obtained was plotted in the radar chart. The axis shows the scores for each S plotted. The radar chart is obtained as a pentagon when the scores are made for all 5S's. The axes show the scores of each S. The use of radar chart is extended to set targets to achieve at different level of 5S implementation. This chart (Figs. 2, 3, 4, 5, 6, 7, 8, 9, 10, 11 and 12) shows the general trend of construction sites towards sort, set in order and shine aspects.

A	B	C	D	E	F	G	H	I
SL.NO	SITE	SORT	SET IN ORDER	SHINE	STANDARDIZE	SUSTAIN	total	
1	A residential building, Vaikkom	10	13	2	18	8	51	
2	A residential building, Vallakam	12	15	14	16	8	65	
3	A commercial building, Vallakam	13	7	9	8	16	53	
4	A residential building, keezhoor	9	8	8	12	12	49	
5	A residential building, muttiyara	12	14	13	13	14	66	
6	A residential building, peruva	12	15	12	14	14	67	
7	A residential building, namakuzhy	14	11	8	12	9	54	
8	A residential building, palachuvadu	8	11	10	13	11	53	
9	A residential building, perumpadavom	7	8	5	6	10	36	
10	A residential building, andhyal	9	11	10	13	10	53	

Fig. 2 5S score for each site

Fig. 3 Radar chart for Site 1

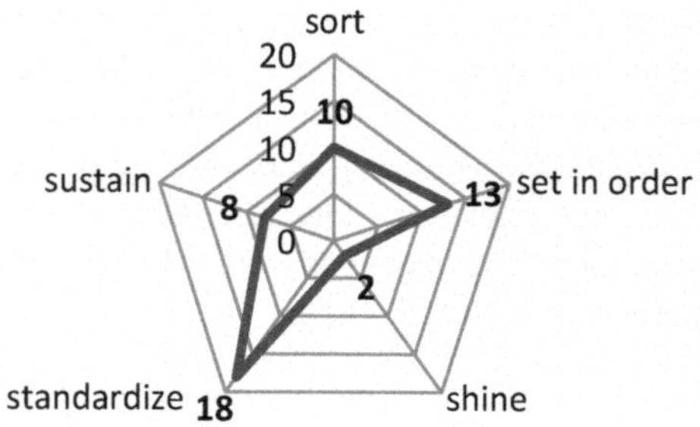

Fig. 4 Radar chart for Site 2

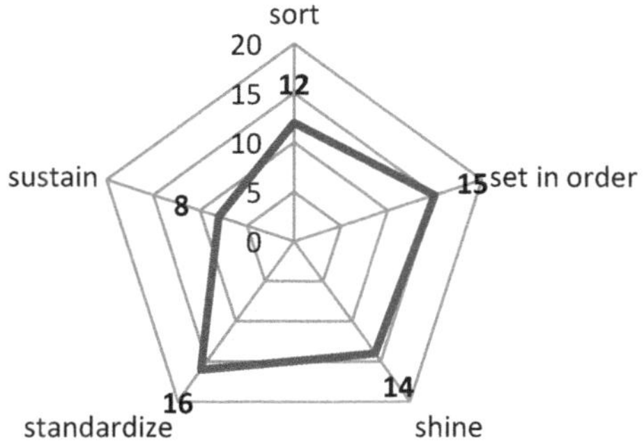

Fig. 5 Radar chart for Site 3

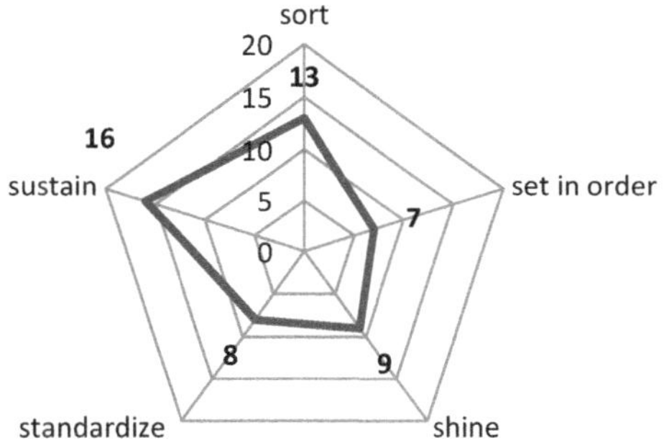

Fig. 6 Radar chart for Site 4

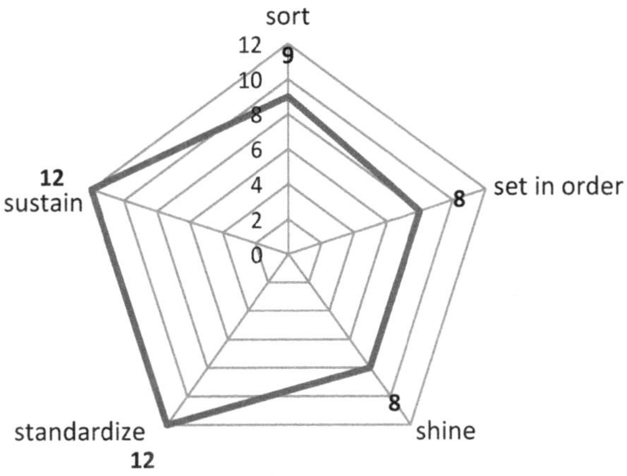

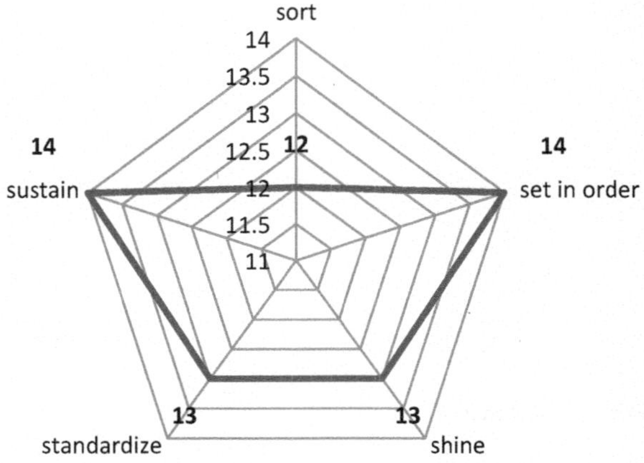

Fig. 7 Radar chart for Site 5

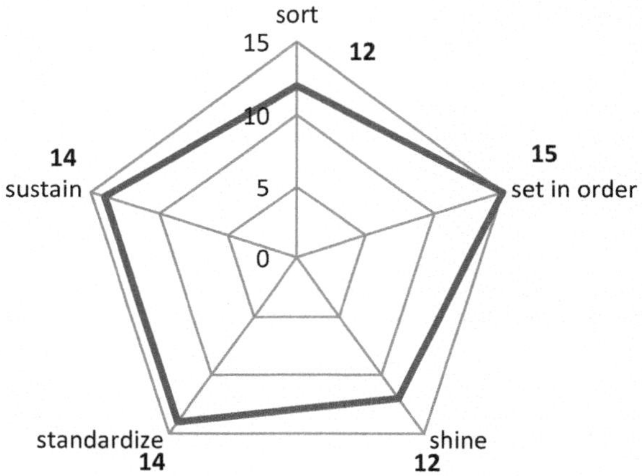

Fig. 8 Radar chart for Site 6

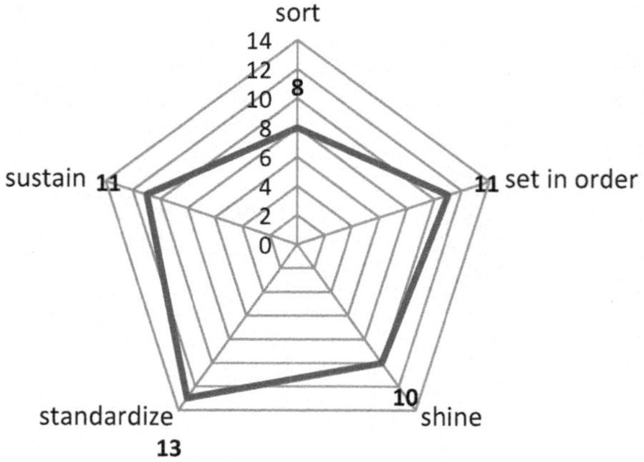

Fig. 9 Radar chart for Site 7

Fig. 10 Radar chart for Site 8

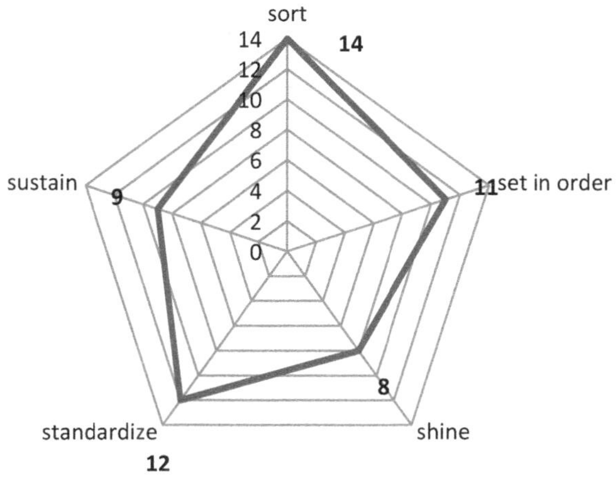

Fig. 11 Radar chart for Site 9

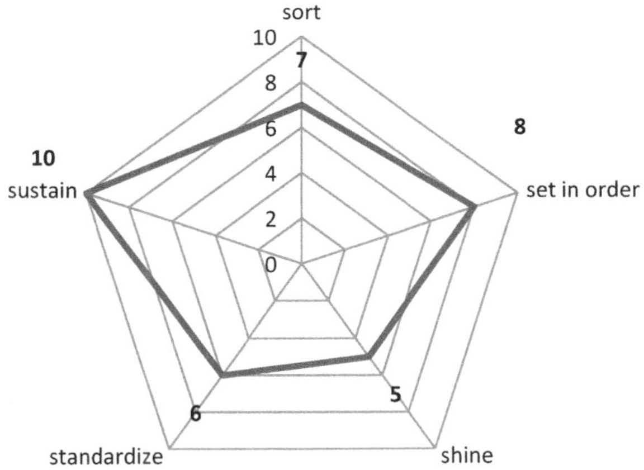

Fig. 12 Radar chart for Site 10

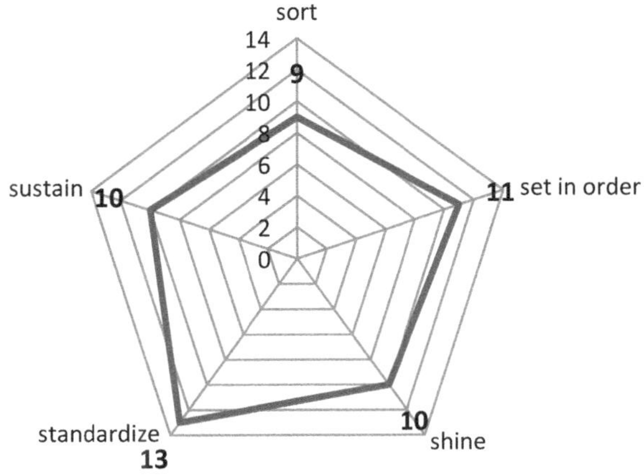

5 Conclusion

The Radar chart pattern shows that the sites lack conformity towards set in order and shine aspects of 5S. Most of the sites were unclean and the debris were scattered. This in turn made difficulties in easy moving around the sites and are a possible threat to slips trips and falls. The safety measures were not met and the equipment were dealt without caution. The amount of waste generated in construction sites are high but cannot be cleaned as efficiently as in a manufacturing unit. Materials were stacked around the sites without any proper arrangement making it difficult to find and assemble parts. No standard procedure of operations was maintained and the site supervisors had little control over the workers.

The effective implementation requires training of workers and convincing top management about the 5S. This in effect consumes huge amount of time. This may lead to disruption and even the workers may feel a sense of being measured which in turn affects the productivity. From literature review on 5S implementation, the strong support and commitment from the top management is the most essential success factor for the implementation of 5S in any organization. There should be commitment of providing resources for intense training and improvement by the top management.

So as to overcome the current difficulties it was clear that 5S audit framework for construction sites is essential in carrying out effective assessment for sites. Therefore framework for evaluating sites is essential. The framework is intended to assess the conditions of construction sites and its conformity towards 5S practices

References

1. Anderson S, Mitchell P (2005) What is 5S really all about? A white paper, Kaizen Solutions Inc., Canada. www.kaizenimprovement.ca
2. Ahuja IPS, Khamba JS (2008) Total productive maintenance—literature review and directions. Int J Qual Reliab Manag 25(7):709–756
3. Albert M (2003) This shop really shines…and sorts, simplifies, standardizes and sustains. Modern Machine Shop. Website: http://www.mmsonline.com/articles/this-shop-reallyshinesand-sorts-simplifies-standardizes-and-sustains
4. Ablanedo-Rosas JH, Alidaee B, Moreno JC, Urbina J (2010) Quality improvement supported by the 5S, an empirical case study of Mexican organizations. Int J Prod Res 48(23):7063–7087
5. Aziz ARA, Nishazini MB, Fareza, Azizan NA (2014) Survey to see the impact of 5S implementation among staff of KPJ Seremban specialist hospital Malaysia. IOSR J Bus Manag (IOSR–JBM) 16(3), 82–96
6. Veres C, Marian L, Monica S (2018) Case study concerning 5S method impact in an automotive company. In: 11th international conference inter disciplinarity in engineering, INTER-ENG 2017, Tirgu–Mures, Romania, vol. 22, pp. 900–905. (Elsevier)
7. Brayer P, Walsh M (2002) Facilitating change–implement 5–S: an Australian case study. Manag Audit J 17(6):329–332
8. Ghodrati, Zulkifli N (2003) A review on 5S implementation in industrial and business organizations. IOSR J Bus Manag (IOSR-JBM)

9. Gupta S, Jain SK (2014) The 5S and Kaizen concept for overall improvement of the organization: a case study. Int J Lean Enterp Res 1(1)
10. Randhawa JS, Ahuja IS (2017) 5S—a quality improvement tool for sustainable performance: literature review and directions. Int J Qual Reliab Manag 34(3):334–361

Forecasting of Material Cost in Road Construction Through Material Control Techniques

V. Deepa, Eldhose Sahimol and V. V. Anu

Abstract In a construction project, materials account for more than 40% of the total project cost. A small saving in material cost through efficient control of materials can result to a large saving in the total project cost. Most organizations face challenges of overstocking of materials, duplication, and resource wastage. The difficulty is to acquire approaches that would enable minimization of such wastages and losses to improve on materials control performance. The objective of this study is to find the benefits of the material control techniques by improving quality and minimization of cost on effective material variety. The material data's are collected from various sites which is then quantified, measured and analyze through material control techniques thereby constraints of material productivity can be found out. For this purpose material requirement planning techniques such as DOQ, EOQ and PPB methods are used. Then it is compared with conventional method. The study helps in reducing total material cost and there by reduction of total project cost.

Keywords Material control techniques · Material requirement planning · Effective material control · Cost comparison

1 Introduction

The construction of road gives the better transportation facility around the country. For better development of the country in the construction assertive, the project management is necessary. Time and cost are the two basic parameters to control work

V. Deepa (✉) · E. Sahimol
Department of Civil Engineering, Toc H Institute of Science & Technology, Arakunnam, India
e-mail: deepamaniyur@gmail.com

E. Sahimol
e-mail: sahimol@tistcochin.edu.in

V. V. Anu
CE Department, Toc H Institute of Science and Technology, Arakunnam, India
e-mail: anuillus@gmail.com

K. Dasgupta et al. (eds.), *Proceedings of SECON'19*,
Lecture Notes in Civil Engineering 46,
https://doi.org/10.1007/978-3-030-26365-2_63

in the execution of the road construction. Optimization is a systematic effort made to improve profit margins and obtain the best results under given circumstances. Flow of cost and its usage is very important aspect for beneficial point of view. It is necessary to develop the planning in terms of material to easy the works and risks that arise in the projects. For the material cost analysis of the road works, Material control technique is used to overcome the problems raised during execution.

Material control is a system which ensures that right quality of material is available in the right quantity at the right time, right place with the right amount of investment. It calculates material requirements and schedules supply, to meet demand across all products and parts in one or more plants. It focuses on optimizing inventory. This project demonstrates the functioning of Material control techniques using Discrete Order Quantity (DOQ) and Economic Order Quantity (EOQ) lot sizing techniques. DOQ orders just as much material as required, EOQ determines the ideal order quantity, so as to minimize the total cost of inventory management and only ordering material in that quantity, whenever required. Part Period Balancing (PPB) method is similar to the DOQ method, but it carries some amount of material as safety stock. In case of any delay in purchasing, safety stock can be used instead of that order and it avoid delay in the construction work. Each of these methods is suitable for specific demand scenarios.

In conventional methods, the project budgeted total cost is determined by the difference, between the actual cost, and planned cost. That means the project manager's focus is on only for planned cost and expenditure cost as actual cost. In modern days many schedule properties and cost parameters are considered. Because it is very important in every construction project, losses are due to inadequate construction management and cost performances done by the contracts in the road construction projects. So it is necessary to develop the planning technique to easy the works and risks arises in the projects.

1.1 Objective

- To analyze material control techniques during the construction of flexible pavement.
- To optimize cost by effective material management through material control techniques in road construction.

1.2 Scope

- Reduction of cost through minimizing duplication, variety reduction, wastage and overstocking of materials during construction site.
- It can improve cost effectiveness in road construction site.

2 Methodology

The methodology adopted for the present study is illustrated in Fig. 1.

Literature survey includes citing to previous works that dealt with road construction and material control techniques were also acquired. Field data collection consists of material management at a road construction site, identification of major materials used in road construction and types of pavements. Details on cost of materials, lead time and transportation cost between source and site, quantity of materials and number of labours would thus lie as secondary concerns. Analysis of factors through material control techniques includes calculation of material cost by EOQ, PPB and DOQ methods. Conclusion consists of major findings that have been obtained through this study.

2.1 Literature Survey

Literature survey includes citing to previous works that dealt with road construction, Material Requirement Planning and Material control techniques were also acquired.

2.1.1 Critical Review

The basic objective of construction project is to create a unique facility, product or service within the specified scope, quality, time and cost. Some road construction projects face losses due to inadequate construction management and cost performances done by the contracts. It is necessary to probe into previous studies that have focused upon material management and process modeling. Though material control studies in road construction projects were not purely found, some key points like major road construction materials and sections relevant to study are referred.

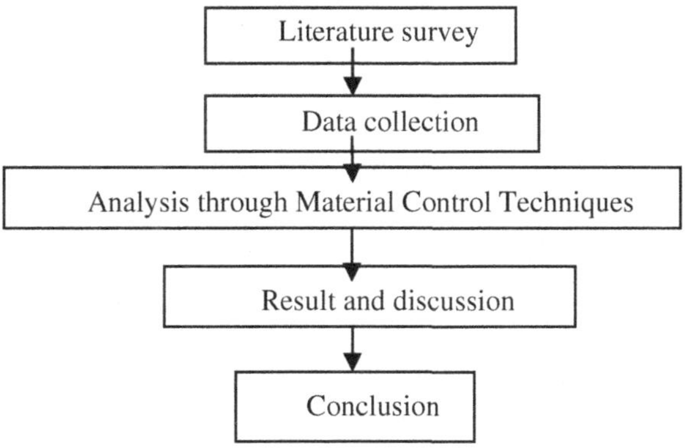

Fig. 1 Methodology for the work

Table 1 Material details

Name	Material used	Demand (cu ft)	Unit cost (Rs.)	Purchasing cost (Rs.)
Site 1	6 mm aggregate	1899.38	27.63	52,479.87
	10 mm aggregate	492.52	27.21	13,401.47
	12 mm aggregate	11,396	27.21	310,085
	20 mm aggregate	1278.455	27.6	35,285.36
	40 mm aggregate	3979.149	35.24	140,225.21
	WMM	7198	38.10	274,243.80
	GSB	5505.84	33.34	183,564.71
	Bitumen S-65	63.528 barrel	4630	294,134.64
	Emulsion SS1	4.876 barrel	7243	35,316.87
	Emulsion RS1	6.020 barrel	6807	42,203.4

Source: [1, 2, 3]

In this study, analysis of material control techniques and cost optimization of materials used in road construction are done. It is comparatively a new technique in road construction firms. It is more beneficial for road construction firms to improve cost effectiveness in their projects.

2.2 Data Collection

The data relevent to this study are identified through literature survey, field data collection and questionnaire survey. The questionnaire survey is carried out to gather information from technical professionals and contractors who were involved in the road construction projects.

Case Study

The case study is carried out in the construction of village roads in Calicut district. For the analysis of total material cost the value of identified factors affecting the material cost are necessary. For this purpose a field data collection study is conducted on the sites of village roads (Table 1).

2.3 Analysis Through Material Control Techniques

Material Control system ensures that the right quality of material with right quantity is available with right amount of money at right time. And it also calculates material requirements, schedule and cost optimization. For this study MRP techniques are used. Collected datas are evaluated and material cost is optimized using MRP techniques. EOQ, PPB and DOQ methods are used for this purpose. The calculated costs are compared with the conventional cost.

2.3.1 Conventional Method

In conventional methods, the project budgeted total cost is determined by the difference between the actual cost and planned cost. Total is calculated as the sum of purchasing, carrying, ordering and transportation cost. There is no consideration for ordering the material in an economical way. There by total cost of material during purchasing is too high.

Sample calculation for 6 mm aggregate for Site 1

Annual demand (cu ft)	= 1899.38
Quantity of material per order (cu ft)	= 450
Number of order	= 1899.38/450
	= 4
Unit cost of material (Rs.)	= 27.63
Cost per order (Rs.)	= 100
Purchasing cost (Rs.)	= 1899.38 × 27.63
	= 52,479.87
Carrying cost (Rs.) (Assume 10% of purchasing cost)	= 5247.98
Ordering cost (Rs.)	= 4 × 100
	= 400
Total cost of material (Rs.) (Purchasing + ordering + carrying cost)	= **58,127.85**
Transportation cost per order (Rs.)	= 115
Distance covered (km)	= 26
Total transportation cost (Rs.)	= 4 × 115 × 26
	= 11,960
Total cost of material including transportation cost (Rs.)	= **70,087.9**

The total material cost calculated by using conventional method is Rs. 70088. The same procedure is repeated for each material in each site for calculating total cost of materials by conventional method. Then total cost of materials used for the entire project is estimated by adding total cost of each material.

2.3.2 EOQ Method

Economic Order Quantity (EOQ) is used to calculate the optimal quantity of material that can be procured to minimize both the holding cost and ordering cost associated with the material.

EOQ is calculated using the equation,

$$\mathbf{EOQ = (2AO/C)^{(1/2)}}$$

where,

A = Annual Demand in units
O = Cost incurred for placing a single order
C = Holding Cost per unit per year
N, Number of orders per year = A/EOQ
AA, Average annual ordering cost = (Number of orders per year) × O
AI, Average Inventory = EOQ/2
AH, Annual holding cost = (Average Inventory) × C
TC, Total Annual Cost = Annual average ordering cost + Annual holding cost.

Sample calculation for 6 mm aggregate for Site 1

Annual demand, A (cu ft)	= 1899.4
Unit cost of material (Rs.)	= 27.63
Purchasing cost (Rs.)	= 1899.4 × 27.63 = 52,479.9
Carrying cost, C (Rs.) (10% of cost of material per unit)	= 2.8
Cost per order, O (Rs.)	= 100
Quantity of material per order (cu ft)	= 400
EOQ (cu ft)	= 370.8
Number of orders, N (Rs.)	= 5
Average annual ordering cost, AA (Rs.)	= 500
Average inventory, AI (Rs.)	= 185.4
Annual holding cost, AH (Rs.)	= 512.2
Total cost, TC (Rs.)	= Annual average ordering cost + Annual holding cost + purchasing cost = **53,492.1**
Transportation cost per order (Rs.)	= 105
Distance covered (km)	= 26
Total transportation cost (Rs.)	= 5 × 105 × 26 = 13,650
Total cost of material including transportation cost (Rs.)	= **67,142.1**

The total material cost obtained by using EOQ method is Rs. 67,142. The same procedure is repeated for each material in each site for calculating total cost of materials by EOQ method. Then total cost of materials used for the entire project is estimated by adding total cost of each material.

2.3.3 DOQ Method

It is one of the most common lot sizing technique. In this technique, ordering only the exact amount of required material. Discrete Order Quantity (DOQ) is a technique used, where the amount of order each time taken is equal to the net requirements for the product for that time period. This should reduce the holding cost, but it maximizes the ordering cost.

- Item: In MRP, it is the name or code number used for the material in the schedule.
- Lot Size: This is the quantity of units ordered for construction.
- Lead Time (LT): This is the time needed to assemble or manufacture an item from beginning to end.
- Gross Requirements (GR): This is the total demand for an item during a specific time period.
- Net Requirements (NR): This is the actual, estimated quantity to be furnished in a particular time period, schedule.

$$\text{NR} = (\text{GR} - \text{inventory on hand}) + \text{safety stock}$$

Sample calculation for 6 mm aggregate for Site 1

The total cost required for purchasing 6 mm aggregate by using DOQ method is Rs. 66,153. Total material cost is the sum of purchasing, ordering and total transportation cost. DOQ method requires no holding cost. The same procedure is repeated for each material in each site for calculating total cost of materials by DOQ method. Then total cost of materials used for the entire project is estimated by adding total cost of each material (Table 2).

2.3.4 PPB Method

Part Period Balancing method is similar to the DOQ method, but it carries some amount of material as safety stock. In case of any delay in purchasing, safety stock can be used instead of that order and it avoid delay in the construction work.

Sample calculation for 6 mm aggregate for Site 1

See Table 3.

Table 2 DOQ calculation of 6 mm aggregate

Site 1
Item: 6 mm aggregate
Transportation distance: 26 km

							PC (Rs.)	OC (Rs.)	TTC (Rs.)	Total material cost (Rs.)
Time in days	1	2	3	4	5	6				
GR		500	300	300	500	300				
Stock available	0	0	0	0	0	0				
NR		500	300	300	500	300				
Planned order release	500	300	300	500	300					
Transportation cost	133	80	80	133	80					
Cost (POR × unit cost)	13,815	8289	8289	13,815	8289					
							52,497	500	13,156	**66,153**

Purchasing cost, PC = Σ (POR x unit cost)
Ordering cost, OC = Number of orders x cost incurred for placing a single order
Total transportation cost, TTC = Σ Transportation cost x Transportation distance
Total material cost = PC+OC+TTC
PPB method requires holding cost for the safety stock, so extra cost required for holding the material (HC) is added with total material cost.
Holding cost, HC = 10% x Stock available x unit cost of material
Source: [1, 2, 3]

Table 3 PPB calculation of 6 mm aggregate

Site 1
Item: 6 mm aggregate
Transportation distance: 26 km

							PC (Rs.)	OC (Rs.)	HC (Rs.)	TTC (Rs.)	Total material cost (Rs.)
Time in days	1	2	3	4	5	6					
GR		500	300	300	500	300					
Stock available	0	300	300	300	300	0					
NR		500	300	300	500	300					
Planned order release	800	300	300	500							
Transportation cost	210	80	80	133							
Cost (POR x unit cost)	22,140	8289	8289	13,815							
							52,497	500	828.9	13,078	**66,903.9**

Purchasing cost, PC = Σ (POR x unit cost)
Ordering cost, OC = Number of orders x cost incurred for placing a single order
Total transportation cost, TTC = Σ Transportation cost x Transportation distance
Total material cost = PC+OC+TTC
PPB method requires holding cost for the safety stock, so extra cost required for holding the material (HC) is added with total material cost.
Holding cost, HC = 10% x Stock available x unit cost of material
Source: [1, 2, 3]

The total cost required for purchasing 6 mm aggregate by using PPB method is Rs. 66,904. Total material cost is the sum of purchasing, ordering, holding and total transportation cost. PPB method requires holding cost for the safety stock.

2.4 Result

The total material costs obtained through the analysis of different methods are tabulated in Table 4.

From this result it is observed that the material cost can be reduced by using EOQ, PPB and DOQ methods (Fig. 2). EOQ determines the ideal order quantity. In case of EOQ method about 1–6% reduction is observed as compared to the conventional method. DOQ is based on Just in time (JIT) method, i.e. it requires no holding and storage cost, it only requires purchasing, ordering and transportation cost. In case of DOQ about 6–14% reduction in total material cost is observed. PPB is similar to the DOQ method but it requires holding cost for the safety stock stored for urgent requirements and for avoiding delay during placing an order. About 6–12% cost can be saved through PPB method.

EOQ method is not suitable for less quantity of materials about lesser than 250 cu.ft, because it increases the transportation cost, there by a slight increase in the total material cost is also observed. In EOQ, sometimes total number of orders is greater than the conventional method orders. EOQ method is more suitable for higher demands.

Table 4 Cost differences for road construction

Description	Using conventional method	Using EOQ technique	Using DOQ technique	Using PPB technique
Total cost for materials	1,744,336.7	1,644,867.3	1,587,262.7	1,597,878.9

Source: [1, 2, 3]

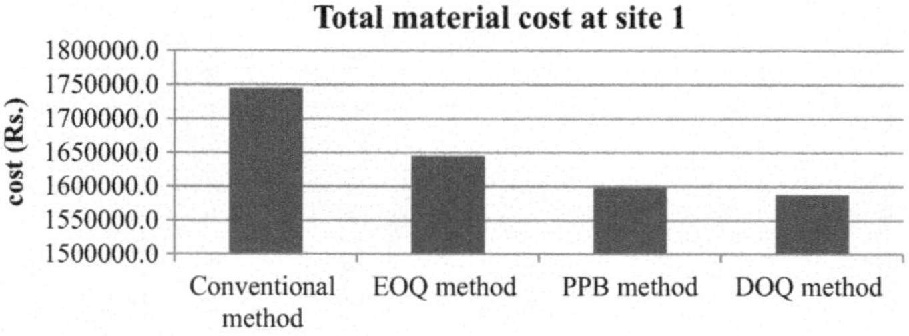

Fig. 2 Comparison of total material cost at site 1 Source: [1, 2, 3]

3 Conclusion

The planning of materials in road construction is considerably a complex process. In this paper, the use of EOQ, PPB and DOQ methods are discussed to minimize inventory, ordering cost and holding cost in road construction through a case study. EOQ decides the absolute order quantity, so as to minimize the total cost of inventory management and only ordering material in that quantity, whenever required and EOQ lot sizing technique was chosen for site, where the demand was considered to be uniform. The material was ordered in a specific lot size which made a understanding between the holding costs and ordering costs. EOQ method is chosen because it is the best way to simplify and explain the basic calculations involved in MRP. DOQ orders only the exact amount of materials for construction. Which minimizes wastage of materials, excess inventory etc. DOQ method is suggested for the sites where the exact demand for the material is known before making an order. PPB method very useful as compared with the other two methods, because while suggesting DOQ method any delay in placing the material at correct time cause delay in the construction work. Instead of this reason PPB method stores an amount of material as safety stock. In case of EOQ it maximizes the number of orders, PPB is based on the requirements of the material on specific day so it reduces the wastage of material. Compared with conventional method EOQ, PPB and DOQ method reduces the total material cost and there by a small savings can be achieved in total project cost.

3.1 Recommendations

- Material Control Techniques can be applied in all construction sites.
- Materials like bitumen and emulsions have some limitations during purchasing through material control techniques.
- EOQ method can be preferred for the purchase of large quantity of materials to achieve cost reduction.

3.2 Future Scope

Material control techniques help to achieve a significant level of cost effectiveness in road construction projects. For future research, the study recommends expanding the use of material control techniques in district and highway projects as well. A model can be used for developing an application for predicting total material cost.

References

1. Balaji C (2015) An analysis on inventory management at Whirlpool India limited, Pondicherry, Int J Commer Manage Res 1:58–60
2. Singh S (2017) Evaluation of different lot sizing techniques in a MRP system, Int J Adv Eng Res Dev (IJAERD) 1–7
3. Gosrani V, Kolekar A (2017) MRP using DOQ and EOQ lot Sizing Techniques, Int J Innovative Res Sci Eng Technol 6:4110–4117

Assessment of Factors Causing Delays in Construction for Indian Residential Building

Akshay Saji, Aldrin Peter, Anand Ajith, Fibin Mathew and K. K. Smitha

Abstract Construction Industry plays an important role in socio-economy development of any developing country. Construction time serves as a benchmark for assessing the performance of any project. Unexpected problems encountered during construction phase causes unwanted delays in project completion. Construction delays can be defined as the late completion of work compared to the planned schedule or contract schedule. It can be minimized only when their causes are identified and suitable preventive measures are adopted. The objective of this paper is to identify the factors affecting delays in a construction project and presents the initial stages of development of a software to predict the possible delays.

Keywords Construction · Delay · Planned schedule · Socio-economy development

1 Introduction

Every construction project has their own objectives to be achieved such as meet the client requirement at minimum cost, attain the required quality and complete the work as planned. Delay in the construction can be considered as one of the most recurring problems in construction industry. Construction Industry plays an important role in socio-economy development of any developing country. Construction time serves as a benchmark for assessing the performance of any project. Unexpected problems encountered during construction phase causes unwanted delays in project completion. Construction delays can be minimized only when their causes are identified and suitable preventive measures are adopted. Construction delays can be defined in many ways. In general, delay can be said as the situation whereby the actual progress of construction is slower than the planned schedule. Delay in construction project can give high impact to the stakeholders including owner, designer, contractors, users and others. Delay can give negative impact to achieve the objective,

A. Saji (✉) · A. Peter · A. Ajith · F. Mathew · K. K. Smitha
Department of Civil Engineering, Toc H Institute of Science & Technology, Arakkunam, Ernakulam, India
e-mail: akshay003saji@gmail.com

K. Dasgupta et al. (eds.), *Proceedings of SECON'19*,
Lecture Notes in Civil Engineering 46,
https://doi.org/10.1007/978-3-030-26365-2_64

whereby, if delay happen, the extension of project time are needed, that can lead to the increasing of cost. So, delay can be categorize as serious problem that should not be ignored althrough the planning phase of construction project.

Construction delays in residential and light construction are often the result of miscommunication between contractors, subcontractors, and property owners. These types of misunderstandings and unrealistic expectations are usually avoided through the use of detailed critical path schedules, which specify the work, and timetable to be used, but most importantly, the logical sequence of events which must occur for a project to be completed.

Before analysing construction delays, a clear understanding of the general types of delays is necessary. There are four basic ways to categorize delays:

- Critical or Non-Critical
- Excusable or Non-Excusable
- Concurrent or Non-Concurrent
- Compensable or Non-Compensable.

1.1 Critical Delays or Non-critical Delays

Whether the type of schedule delay is excusable or not, it's important to also assess whether the delay is Critical or not. A Critical schedule delay is one that impacts the project's delivery date. If a delay has no effect to activities on the project's Critical Path, then the delay may not warrant too much attention, unless of course there is some substantial money involved.

1.2 Excusable/Non-excusable Delays

Excusable delays are caused by conditions that are reasonably unforeseen and not within the contractor's/owner's control.

Examples of this kind of delay include:

- Labour strikes
- Fires, floods, earthquakes and most natural disasters
- Changes initiated by the owner
- Errors and omissions in the plans, design docs and specifications
- Differing site conditions or concealed conditions
- Lack of action by governmental or oversight bodies
- Intervention by outside agencies

Often the construction contract will outline valid excusable delay causes. Excusable delays are those that are beyond the control of the owner/contractor and leaves them without fault or negligence.

Non-excusable delays are a result of a delay that was within control of the Owner/Contractor. The Owner/Contractor is fully responsible for the activity delays. Examples of non-excusable delays include:

- Delayed mobilization
- Delayed submission of submittals
- Overall late performance and execution
- Late performance of subcontractors
- Late performance by suppliers
- Faulty workmanship by the contractor or subcontractor

A project specific labour strike caused by either the contractor's unwillingness to negotiate or by unfair job practices is an example of non-excusable delays. Non-excusable delays can be either compensable or non-compensable or can result in an extension of time.

1.3 Concurrent Delays

There are many moving parts between an owner and a contractor (and subs or suppliers) on a construction project and, at times, a concurrent delay can occur. This type of schedule delay happens when two or more parties are at fault. The complexity here lies in determining to what extent each party contributed to the delay. There are some technical methods of analysing schedules that can assist in this determination. The overlapping of delays makes it difficult to analyse and factors like the duration of the delay, time of occurrence of the delay and float ownership have to be carefully considered in the technical analysis.

1.4 Compensable Delays

This is where the rubber hits the road. A compensable delay is one where there's going to be some compensation involved for the delay to the project. That means that the owner or contractor is liable for an extension of time or cost compensation or both. The cost compensation is to account for damages or extra costs associated with the delay. A compensable delay will be a non-excusable delay, so the list of causes above applies here.

Ajibade and Henry [1] have done research on construction delays and their causative factors in Nigeria. 44 delay factors were identified and questionnaire was prepared on a 5 scale basis. The obtained data was then analyzed using Statistical Package for Social Sciences (SPSS).

Majid [2] outlined that the delays can be minimized when their causes are identified. Delay is a situation when the contractor, consultant, and client jointly or severally contributed to the non-completion of the project within the original or the stipulated or agreed contract period.

Muhweri and Acai [3] identified 81 project delay factors through literature study. The research classified the causes of delay under four main groups such as consultant related, contractor related, client related and externals related and then assessed their impact on delay using Relative Importance Index (RII).

Pablo et al. [4] analysed cause of delay and Time Performance in Construction Industry. The delays can be assessed both at activity level and project level. The activity level examines delay that affects completion of an activity that may or may not have any influence on succeeding activities. Project level analysis focused on impact of activity delays on completion of project. The analysis contributed a methodology to examine delay causes (qualitative) and time performance (quantitative) and to relate these two dimensions. The statistical analysis used two indicators to examine quantitatively and qualitatively the attributes of delay causes. The qualitative dimension was analysed using Reason for Non Compliances (RNCs) of delay at the activity level. The quantitative dimension used Delay Index (DI) as an indicator for time performance and its impact on critical and non-critical activities. The methodology differentiates impact of delay on critical and non-critical activities. The study provides information to take better decision during delay and managerial action for mitigating delay to project managers.

2 Preparation of Questionnaire

After finding out the major common delays occurring in a construction site a questionnaire was prepared containing 15 delays with marking scale of 0 to 10 according to the priority at the particular construction project. Table 1 shows the questionnaire developed in order to collect the details.

3 Software Development

Based on the data collected using the details in Table 1, an Android software is proposed to be developed which take input as the factors that is presented in Table 1 and the output as the construction delays caused by various factors provided in Table 1.

The software developed in android studio, is an official integrated development environment for Google's Android operating system designed specifically for Android development.

Table 1 Collection of details

Name and designation	
Contact No.	
Email Id	
Organization details	
Location	
Type of project	
No. of floors	
Total Sq. Feet area	
Estimated time of construction	
Elapsed time	
Signature/seal	

(continued)

Table 1 (continued)

Major delays	Delay occurred										
	0	1	2	3	4	5	6	7	8	9	10
Delay in revising and approving documents (design, drawings, submittals etc.)											
Changes to the project											
Delay of financing and payments by owner											
Difficulties in financing the project by contractor											
Poor site management and supervision											
Lack of experience of contractor/Poor qualification of contractor's staff											
Shortage of construction materials											
Delays due to material delivery											
Inflation and escalation of material prices											
Shortage of labours											
Unqualified workforce/Low productivity of labour											
Shortage of equipment and/or equipment failure											
Unexpected site conditions (Unexpected subsurface conditions e.g. soil, high water table, etc.)											
Restriction at job site (Poor site access, traffic congestion)											
Lack of site utilities or services such as (water, electricity, etc.)											
Others (specify if any)											

0 No delay
1–10 Less to extreme delay

3.1 *Android Project View*

All files related to the project is stored under this view. This view does not reflect the actual file hierarchy on disk, but is organized by modules and file types to simplify navigation between key source files of your project, hiding certain files or directories that are not commonly used. Some of the structural changes compared to the structure on disk include the following:

- Shows all the project's build-related configuration files in a top-level Gradle Script group.
- Shows all manifest files for each module in a module-level group (when you have different manifest files for different product flavours and build types).
- Shows all alternative resource files in a single group, instead of in separate folders per resource qualifier. For example, all density versions of your launcher icon are visible side-by-side.

Within each Android app module, files are shown in the following groups:

3.2 *Manifests*

This folder contains the AndroidManifest.xml file. The manifest presents essential information about the application to the Android system, information the system must have before it can run any of the application's code, like:

- Package Name—This represents Unique Identifiers of the application.
- Components—They are the essential building blocks of an Android app. Each component is an entry point through which the system or a user can enter your app.
- Permissions—The permissions that the app needs in order to access protected parts of the system or other apps. It also declares any permissions that other apps must have if they want to access content from this app.

Among other things, the manifest does the following:

- It names the Java package for the application. The package name serves as a unique identifier for the application.
- It describes the components of the application—the activities, services, broadcast receivers, and content providers that the application is composed of. It names the classes that implement each of the components and publishes their capabilities (for example, which Intent messages they can handle). These declarations let the Android system know what the components are and under what conditions they can be launched.
- It determines the processes which will host application components.
- It declares which of the permissions the application must have in order to access protected parts of the API and interact with other applications.

- It also declares the permissions that others are required to have in order to interact with the application's components.
- It lists the Instrumentation classes that provide profiling and other information as the application is running. These declarations are present in the manifest only while the application is being developed and tested; they're removed before the application is published.
- It declares the minimum level of the Android API that the application requires.
- It lists the libraries that the application must be linked against.

3.3 Java

This contains the source code files, separated by package names, including Unit test code. It contains a folder naming the package name of the app provided. Inside the folder containing codes related to each activity.

3.4 Launcher Page

The launcher page of the developed software is shown in Fig. 1. It includes the name of the software, logo and the loading bar.

Name of the developed software is D-ANALYZER which is briefed form of DELAY ANALYZER.

3.5 Map Page

The Launcher page is followed by the Map page which is shown in Fig. 2, that include the map of various locations. The clients can select their required location from this page. The map page facility is being powered by google map.

3.6 Input Page

After the selection of the location the software will move on to the input page which is shown in Fig. 3. The data in hand which is used for analysing the software is entered in this page. It includes location, time of the estimated project in days and the square feet of the project.

Fig. 1 Launcher page

3.7 *Output Page*

The output page is the final page of the software. It shows the final output of the delay analysis, it includes the percentage value of major delays that can occur in the selected location dand also shows the estimated time delay in days.

Analysis of Data

The collected data which is obtained from the questionnaire survey is analysed by SPSS Software. SPSS (The Statistical Package for the Social Sciences) is a software developed by IBM and it is widely used to analyse data and make predictions based on specific collections of data.

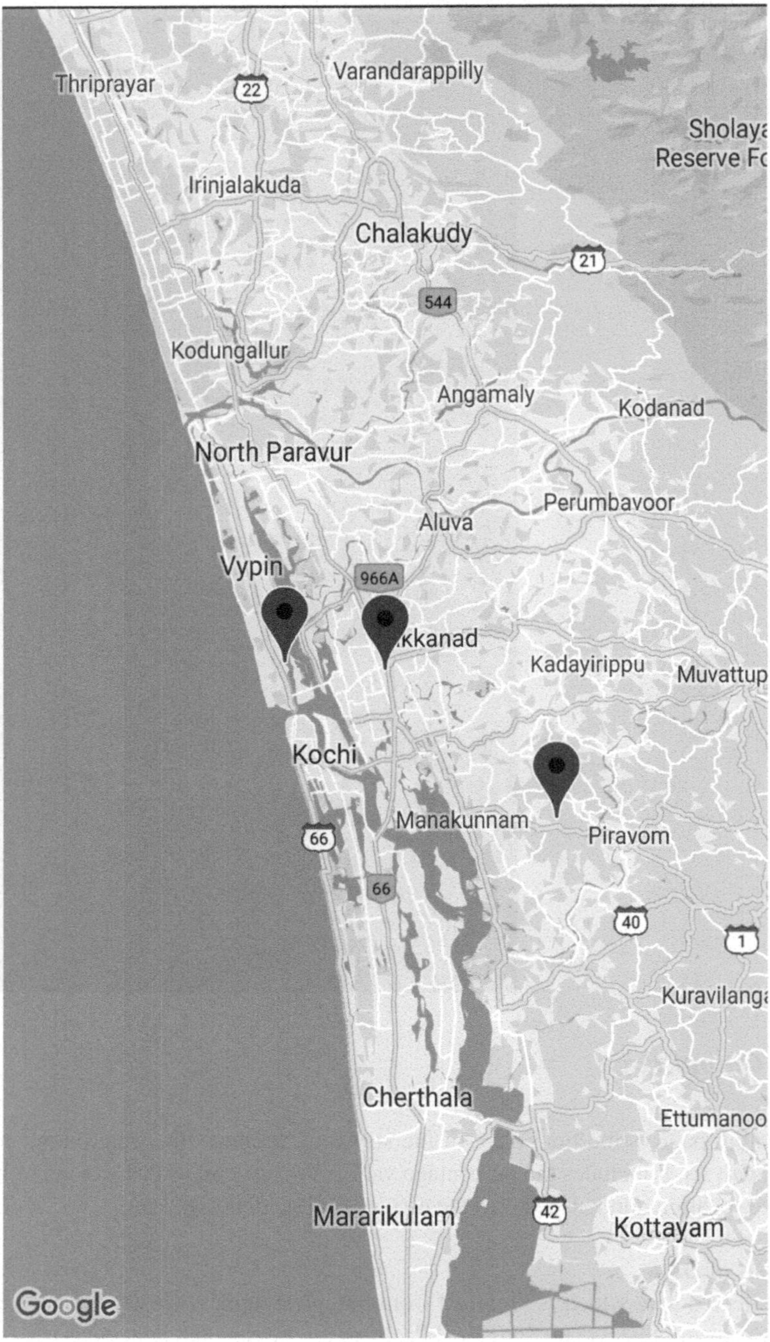

Fig. 2 Map page

Fig. 3 Input page

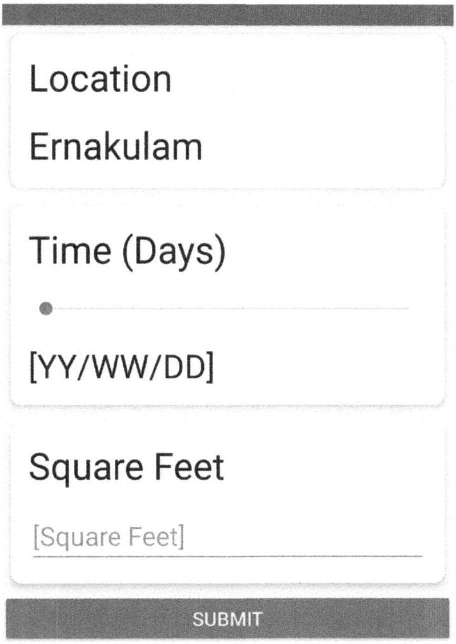

Data Flow Diagram

A data-flow diagram (DFD) is a way of representing a flow of a data of a process or a system (usually an information system). The DFD also provides information about the outputs and inputs of each entity and the process itself. Figure 4 shows the data flow diagram.

Activity Diagram

Activity diagram is basically a flowchart to represent the flow from one activity to another activity. The activity can be described as an operation of the system (Fig. 5).

Algorithm

1. Start
2. Read Username, Password
3. If Username, Password validated then step 5 else step 4
4. Go to registration page
5. Read location, square feet, Time
6. Delay = Location (square feet, Time)
7. Display Delay
8. Stop

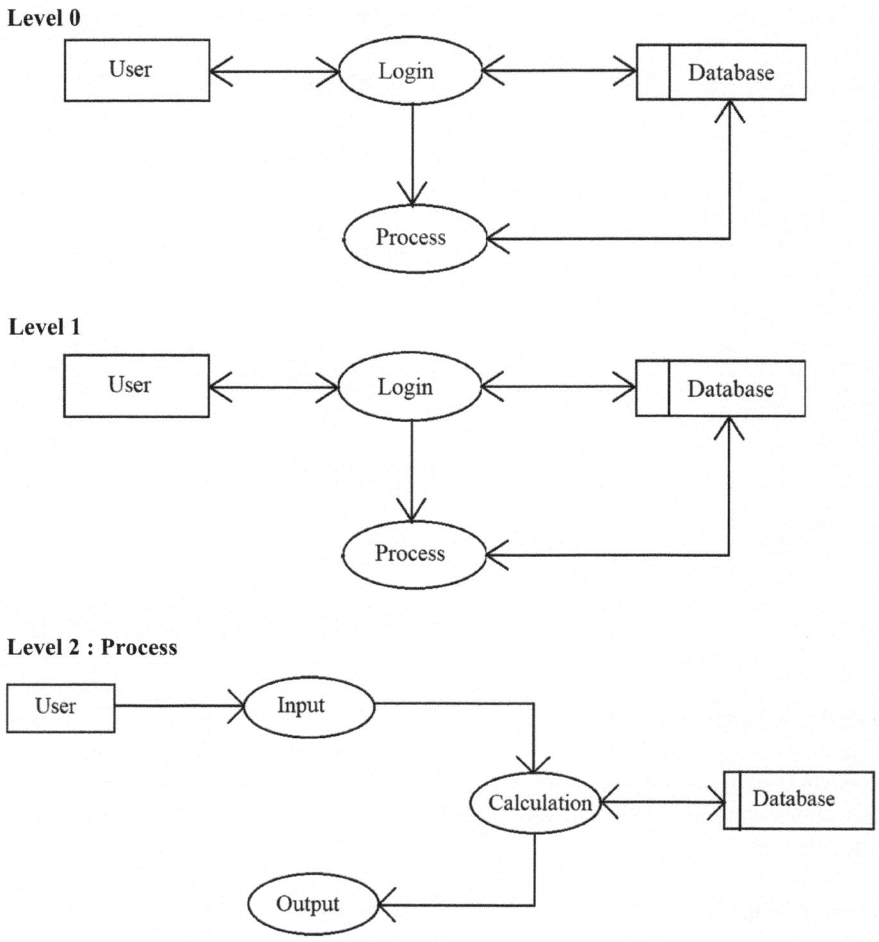

Fig. 4 Data flow diagram

4 Conclusions

Delays occur often on a construction site due to the result of mistakes and miscommunication of contractors, subcontractors, and property owners. Construction delays can be minimized only when their causes are identified and suitable preventive measures are adopted. A new method for the evaluation of delays caused in construction industry is developed in this study. Once the factors affecting the construction delays are identified, the method to curb those delays can be devised and can be implemented.

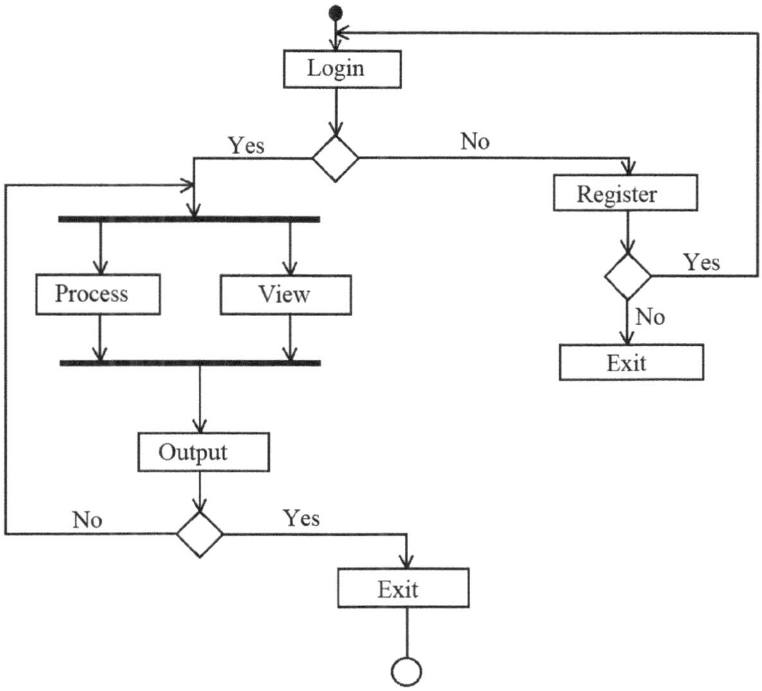

Fig. 5 Activity diagram

References

1. Ajibade AA, Henry AO (2001) Construction delay computation method. J Constr Eng Manag ASCE127:60–65
2. Majid IA (2006) Causes and effects of delays in Aceh construction industry. Master of Science in Construction Management. University Technology Malaysia
3. Muhweri L, Acai GO (2014) An assessment of the factory causing delays on building construction projects in Uganda. Int J Constr Eng Manag, 13–23
4. Pablo G, Vicente G, Keith M, Francisco O (2014) Analysis of causes of delay and time performance in construction projects. J Constr Eng Manag 140:1–9

Influence of Cycle Time on the Productivity of Transit Mixers

Abitha Varghese and Annie Sonia Xavier

Abstract Ready mix concrete is an advancement over the old site mix concrete and is preferred due to its flexibility, time, quality, cost effectiveness etc. Ready mix concrete is transported from plant to site using transit mixers and volumetric mixers. Among which transit mixers are widely used now-a-days. Productivity of mixers therefore contributes towards the success of ready-mix concrete plant and construction site. Productivity of transit mixers depends on the cycle time involved during the process of concrete delivery. So it is necessary to identify and quantify those factors to minimize the cycle time and to increase productivity. This paper deals with the identification of factors affecting cycle time of transit mixers and analyzing the impact of these factors on productivity. Hauling time and time for transit mixer weighment were found to be the major factors that affect the productivity.

Keywords Cycle time · Productivity · Ready-mix concrete · Transit mixer

1 Introduction

Ready Mix Concrete (RMC) batch plant is an industrialized plant, where concrete is made by mixing cement, coarse aggregates, fine aggregates, water and admixtures. Then, it is delivered to the construction site using transit mixers. Ready mix concrete preferred over other types of concrete due to its versatility in placing, precision of the mixture and reduced work site confusion. Increased usage to ready mix concrete has increased the importance of transit mixers. Productivity of transit mixers depends on combined operation of concrete mixing in the plant and the delivery to the site. An improvement in productivity of transit mixers can increase the speed of concreting

---

A. Varghese (✉)
Toc H Institute of Science and Technology, Arakkunnam, Kochi, India
e-mail: abithavarghese93@gmail.com

A. S. Xavier
CE Department, Toc H Institute of Science and Technology, Arakkunnam, Kochi, India
e-mail: anniesonia12@gmail.com

K. Dasgupta et al. (eds.), *Proceedings of SECON'19*,
Lecture Notes in Civil Engineering 46,
https://doi.org/10.1007/978-3-030-26365-2_65

process and it also increases the number of works per day an RMC batch unit can undertake.

Productivity of transit mixers can be checked by cycle time analysis. Cycle time is the time required to complete an operation from start to finish. Productivity is defined as the output per unit of time. Productivity of transit mixer can be defined as the amount of concrete batched and loaded in transit mixer during a cycle to the time required complete that cycle. During a cycle transit mixers have to undergo a number of operations or factors and the time taken to complete these operations together constitute the cycle time.

2 Operations During the Cycle Time of Transit Mixer

Transit mixers have to undergo many operations during a cycle. Time taken by the transit mixer to complete these operations gives the overall cycle time. The operations start in the batch plant and ends in the batch plant. Figure 1 shows the sequence of operations.

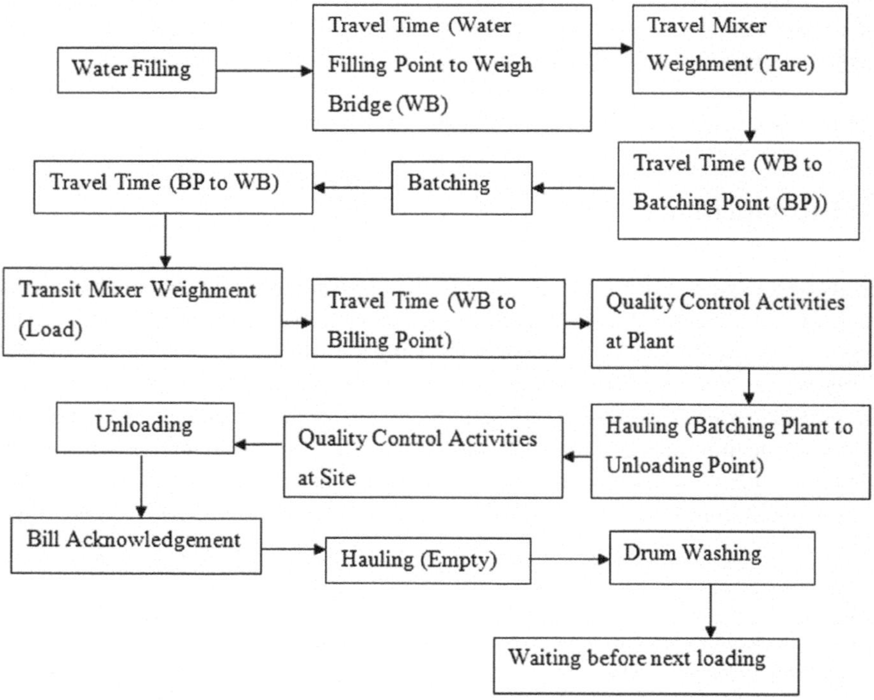

Fig. 1 Operations during the cycle of transit mixer

Water filling: Transit mixer is equipped with a tank of varying capacity. Water is required inside transit mixers for purpose of remixing the concrete once it reaches the site, for pump washing, drum washing etc. It is the first operation during the cycle. Time of water filling depends on the availability of water in plant, pipe diameter, pumping efficiency, tank capacity etc.
Travel time in plant: Travel time in plant depends on layout of plant, obstructions, ground condition etc. Point inside the batch plant may be at a distance. For example water filling point and weigh Bridge may be two extreme locations.
Weighment: Weighing of mixers is done to determine the amount of concrete inside. This is done to provide evidence to the client that the required quantity of concrete is supplied. Tare weight and load weight is taken to calculate the amount of concrete. Time for weighing depends on the availability of quality control person who is in charge of carrying out the operation.
Batching time: Batching starts with the process of measuring concrete mix ingredients either by volume or by mass and introducing them into the mixture, followed by mixing and finally discharging the mix into the mixer. Quality and appropriate maintenance of batching equipment and system error during operation can affect the batching time. Time for batching varies with the capacity of batching unit. If the batching unit is of higher capacity less time is needed for batching.
Billing and quality control activities in plant: Bill contain the details like quantity of material used, weight of transit mixer (tare and load), tax invoice, type of mix, grade of concrete, address of client etc. Bill is issued to the client as evidence. Quality control activities include slump checking, sampling for cube casting, slump correction if required etc. If slump correction is required it takes additional time.
Hauling: hauling of transit mixer from batch plant to construction site and back to plant. Hauling time directly affects cycle time of transit mixers. Hauling time varies due to speed variation, haul road condition, haul road material, driver's experience, traffic, driver's visibility, like rain climatic factors, mist, fog etc.
Bill acknowledgement and quality control activities in site: Bill acknowledgement is done by the client when mixer has unloaded the concrete in the site. Quality control activities in site include slump correction because due to the haul distance the mix can be become tight or loose which has to be corrected. If sampling for cube casting not done at the plant it will be done at the site.
Unloading: unloading can be done either manually or using pumps. Unloading time has direct impact on the cycle time thus it a dependent factor. It varies due to type of structure to be concreted, length of pipeline, labors experience in concreting, number of vibrators, mix design etc. Unloading time changes for the type of structure to be concreted, it takes little time for slab concreting compared to wall and beam.
Drum washing: When the mixer return to the plant after unloading drum washing is done prior to the next loading.
Waiting before next loading: when transit mixer after drum washing, there can be delay due to batching of previous truck or due to lack of order.

3 Factors Affecting the Cycle Time of Transit Mixers

There are direct and indirect factors affecting the cycle time of transit mixers in RMC batching plant. Direct factors directly contribute towards the cycle time whereas indirect factors indirectly influence the cycle time. Direct factors are measurable and they are included in the operations during the cycle of transit mixers. Indirect factors arise due to the external environment and they are beyond the control of management teams. Measurable cycle time influencing factors are discussed in the above section "operations during the cycle time of transit mixers" and indirect factors are discussed in Table 1.

Table 1 Factors affecting the cycle time of transit mixers

No.	Indirect factor	Direct factor
1	Haul road condition	Water filling
2	Haul road material	Travel time in plant
3	Plant and site ground condition	Transit mixer weighment (tare and load)
4	Climate	Batching time
5	Driver's experience	Time for billing and quality control activities
6	Labours effectiveness in concreting	Hauling time (to and fro)
7	Type of structure to be concreted	Waiting time before unloading
8	Pipe line length	Unloading time (pump)
9	Workers health condition	Time for bill acknowledgement
10	Plant efficiency	Drum washing
11	Obstacles in site and plant	Waiting time before next loading
12	Supervision and management	–
13	Traffic	–
14	Building height	–
15	Capacity and type of pump	–
16	Mode of unloading	–
17	Haul distance	–
18	Hauling speed	–
19	Fill degree	–
20	Mix design	–
21	Plant layout	–
22	Capacity of plant	–

4 Calculation of Cycle Time and Productivity

A checklist survey was conducted in a RMC batching plant of capacity of 0.5 m^3 at Muppathadam near Ernakulam. A construction site, at Chunaganveli near Rajagiri Hospital Aluva which is about 14 km away from the RMC is fixed as the delivery point. The type of structure concreted was raft foundation which was same throughout the study. Capacity of transit mixer is 6 m^3 i.e., volume capacity of the transit mixer (V_{TR}). Transit mixer is of the model 2518 Tata/Leyland. Time study was conducted using GPS and stopwatch to measure the time taken by transit mixer to complete each activity/factor and summation of those reading gives the cycle time. Productivity is calculated by dividing volume capacity of transit mixer with cycle time. 106 transit mixer cycle time readings in the month of November, December (2018) and January (2019) was recorded (Fig. 2).

Transit mixer productivity is defined as

$$\text{Productivity } (P_{TR}) = \frac{V_{TR}}{CT_{TR}} \tag{1}$$

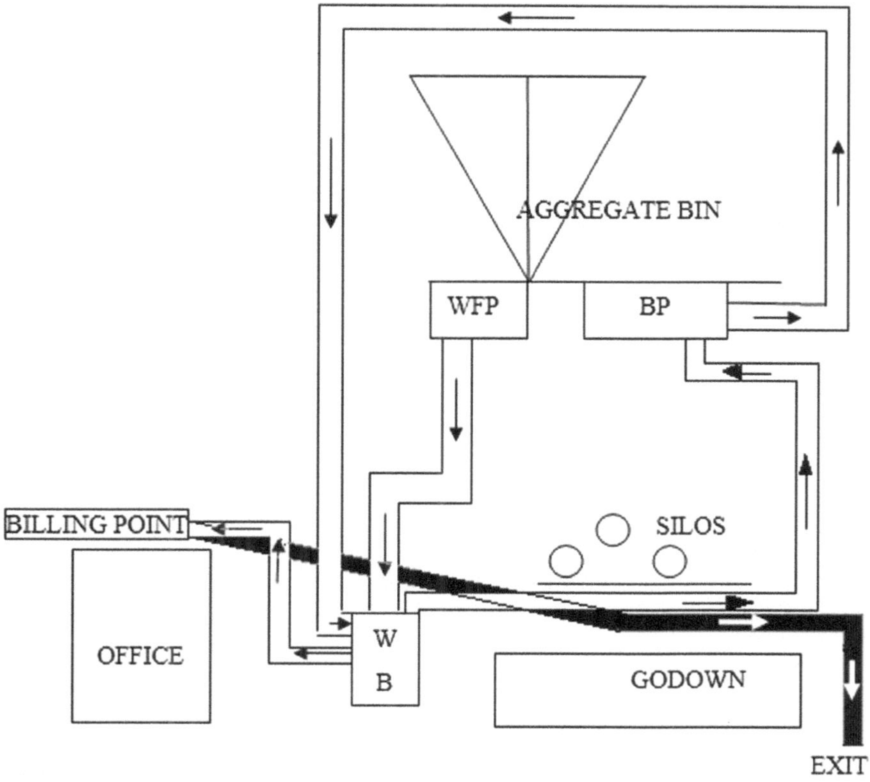

Fig. 2 Transit mixer movement inside the plant

where

P_{TR} is the actual transit mixer productivity (m^3/h)
V_{TR} is the volume capacity of the transit mixer (m^3)
CT_{TR} is the transit Mixer Cycle Time (h). It is the sum of time taken by each factors/activity during a cycle.

These abbreviations are used in analyses using SPSS software (Table 2). Sample data collection through checklist is shown in Table 3.

Productivity of transit mixer was calculated using this cycle time readings. 106 cycle time and productivity readings were calculated.

Table 2 Factors included in the calculation of cycle time

S.No.	Factors	Abbreviations
i	Time for water filling	wf
ii	Travel time (water filling point to weigh bridge)	tt
iii	Transit mixer weighment (tare)	tmwtare
iv	Travel time (weigh bridge to batching plant)	ttwb
v	Waiting time before batching (if any)	wtbb
vi	Batching time	bt
vii	Travel time (batching plant to weigh bridge)	ttbw
viii	Transit mixer weighment (load)	tmwload
ix	Travel time(weigh bridge to billing point)	ttwtb
x	Time for billing and quality control activities	
	(a) Slump check and billing	scb
	(b) Slump correction (if necessary)	sc
xi	Hauling time (batching plant to unloading point)	ht
xii	Waiting time before unloading	
	(a) Time for slump check and correction if necessary	tfscc
	(b) Sampling for cube casting	
xiii	Unloading time	ut
xiv	Time for bill acknowledgement	tba
xv	Hauling time (empty)	hte
xvi	Drum washing	dw
xvii	Waiting time before next loading (if any)	wtbnl

Table 3 Sample checklist

Checklist	
Plant location: Muppathadam	Date: 5-12-018
Site location: Chunaganveli	Truck capacity: 6 m^3
Distance: 14 km	Plant capacity: 0.5 m^3
Truck model: 2518 Tata/Leyland	Start time: 5.30 am
Type of structure concreted: raft foundation	Quantity of concrete: 6 m^3
	Truck No: 12

S. No.	Factors	Duration (min)
1	Water filling	5.16
2	Travel time (water filling point to weigh bridge)	0.22
3	Transit mixer weighment (tare)	0.18
4	Travel time(weigh bridge to batching point)	1.05
5	Waiting time before batching (if any)	
6	Batching time (6 m^3)	18.16
7	Travel time (batching point to weigh bridge)	0.54
8	Travel time (weigh bridge to billing point)	0.22
9	Transit mixer weighment (load)	0.27
10	Time for billing and quality control activities	
	(a) Slump check and billing	4.1
	(b) Slump correction (if necessary)	
11	Hauling time (batching plant to unloading point)	30.45
12	Waiting time before unloading	
	(a) Time for slump check and correction if necessary	10
	(b) Sampling for cube casting	
13	Unloading time (pump)	20.31
14	Time for bill acknowledgement	5.0
15	Hauling time (empty)	33.58
16	Drum washing	5.2
17	Waiting time before next loading (if any)	
	Average cycle time	134.44

5 Influence of Measured Cycle Time Factors on the Productivity

Factors affecting the cycle time of transit mixers can be categorized as direct and indirect factors. Direct factors are the measured factors which directly influence the cycle time whereas indirect factors directly indirectly affect the cycle time and these factors are difficult to control because they depend on the external environment.

Influence of those measured cycle time factors on productivity of transit mixer was analyzed using automatic linear modeling in IBM SPSS software.

Figure 3 shows the influence of measured factors on the productivity. More the line width, more the influence of factors on productivity. The orange colour of line indicates that the factors have negative influence on cycle time i.e., as cycle time increases, productivity decreases. Cycle tm means the cycle time of transit mixers. The abbreviations shown in figure is mentioned in Table 2. Hauling and travelling inside the batch plant takes more time and it influence the productivity the more. Hauling is a factor which is not uncontrollable by management team because the time for hauling can vary with distance from plant to site or due to heavy traffic. But factors like travel time inside the batch plant can be controlled by proper arrangement and management inside the plant or unnecessary movements can be avoided by doing more activities parallelly.

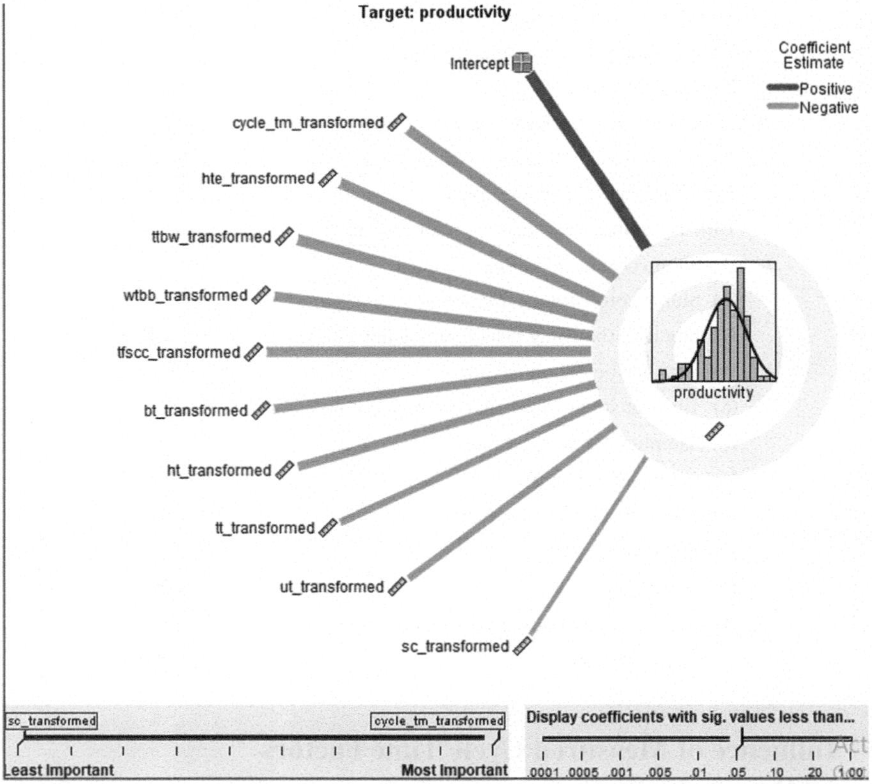

Fig. 3 Influence of measured factors on the productivity

Table 4 Transit mixer productivity results

Average cycle time (h)	Productivity (m^3/h)
2.30	2.608

6 Result and Discussion

Average cycle time and productivity were calculated using the above Eq. (1) for 106 transit mixers are shown in Table 4.

In order to find out the relationship between the measured cycle time factors and productivity automatic linear regression was done using SPSS software. The width of line is directly proportional to its relation with productivity. From the analysis it is clear that major factors that influence the productivity of transit mixer is the hauling time and travel time. Hauling depends on the speed, haul road condition, haul road material, driver's experience, traffic, driver's visibility, climatic factors like rain, mist, fog etc. travel time can be reduced by proper arrangement and management inside the batch plant.

Another factor that influence to the productivity is time for weighment. It depends on the availability of quality control person who is in charge of carrying out the operation and sometimes tare weight will not be taken before loading. This time can be reduced by proper management in the plant.

Time for slump check and batching time are other important factors which affects the productivity. Batching time varies according to the capacity and condition of batching plant. Slump checking is done both in batching plants before hauling and in site before unloading. Slump check and correction is done as a part of quality control activities.

7 Conclusion

This study was conducted to analyze the influence of cycle time factors on productivity of transit mixer. Transit mixer is a kind of frequently used construction equipment for transporting building concrete. It brings a lot of convenience to the construction industry by reducing the manpower, material resources etc. and has improved the work efficiency which results in project progress. Therefore it is an indispensable tool in the construction industry. Productivity of transit mixer contributes towards the success of ready mix concrete plant and construction site.

The major factors influencing the productivity were found to be hauling time and travel time inside the plant. Though hauling time is the major factor affecting the productivity we cannot reduce it. Because hauling time increase or decrease accordingly with distance between batching plant and construction site. About half of transit mixer cycle occurs inside the batching plant and the balance occurs outside the batching plant. The processes occurring inside the batching plant can be controlled to some extent while the outside factors are difficult to control due to external influence. For example the batching time can be reduced by increasing the capacity of batching unit which in turn increases the productivity. Proper management and co-ordination inside the batching plant can reduce the cycle time and thereby increasing the productivity.

Probabilistic Evaluation of the Stability in Earned Value Management Forecasting on Topologically Diverse Project Networks

V. H. Ladeeda and Jacob Jeevan

Abstract "When will the project performance indices in my project become stable?" is an ever interesting fundamental question in earned value management (EVM) but mostly unresolved as well, especially for the highly dynamic construction projects which call for frequent updating and rescheduling. The stability of the performance indices assures the credibility of EVM as an effective tool to analyse the project performances as well as to forecast the requirements. This paper presents an assessment of time and cost performance stability in EVM based on an analytical stability simulation model which evaluates the probability of the stable project performance indices during the project execution for topologically diverse construction project networks. The model is used to identify the effects of percentage completion, network topology as well as its compounding effects on the stability of EVM forecasting and topology–stability relations were developed for schedule performances. Finally, the performance indices are compared for the stability behaviour.

Keywords Earned value management · Cost performance index · Schedule performance indices · Network topology · Stability simulation model · Stability analysis

1 Introduction

Although a number of project management techniques were in practice until the Earned Value Management (EVM) was adopted to construction in 1990s as a project monitoring and control tool, the EVM gained popularity within a short period of time owing to its ability to integrate basic project constraints (known as the "project management triangle" which consists of time, cost and scope) as a single integrated

V. H. Ladeeda (✉) · J. Jeevan
Department of Civil Engineering, Mar Athanasius College of Engineering, Kothamangalam, Kerala, India
e-mail: vhladeeda@gmail.com

J. Jeevan
e-mail: jeevanjacob@mace.ac.in

K. Dasgupta et al. (eds.), *Proceedings of SECON'19*,
Lecture Notes in Civil Engineering 46,
https://doi.org/10.1007/978-3-030-26365-2_66

system. By comparing the time and cost performances with the actual work progress, the EVM provides the accurate assessment of scheduling and costing efficiency using the respective performance indices namely, the Schedule Performance Index or SPI (the ratio of cumulative values of earned value to the planned value at the status date) and the Cost Performance Index or CPI (the ratio of cumulative values of earned value to the actual cost at the status date) and are used to forecast the respective project performances to get the reliable early warning. But, the inconvenience of using the cost metrics to assess the schedule performance (using SPI) paved way to the introduction of a time based new metric called "Earned Schedule" (ES) as an extension to the traditional EVM. Hence, the assessment of both the time and cost performances can be performed using the respective metrics in terms of SPI(t) (the ratio of earned schedule to the actual time at the status date) and CPI respectively. For the reliable forecasting, it is necessary to have some degree of forecasting stability in EVM, which in turn is the derivative of the stability of its key performance indices. Hence, any stability related issue associated with the performance indices will influence the subsequent EVM forecasting and the unstable performance indices will lead to the unreliable as well as misleading warnings.

The EVM was proved to be an effective tool for the monitoring and control of construction projects even with its long term nature, high degree of uncertainties and complexities. But, the complexity associated with the project networks is much more significant as compared to that of other industries owing to the highly dynamic nature of construction projects which call for frequent updating and rescheduling. Hence, the objective of this paper is to study the stability of various performance indices in EVM for time and cost as well as to analyze the effect of project network topology on the EVM forecasting stability.

The stability of EVM was an unnoticed aspect until the researches revealed that the accuracy of EVM was a function of stability [1]. The later studies began to pay attention to define a stable CPI as that CPI not varied from the completion points of 50% [2] and 20% [3] by a threshold of 10%. The project cost assessment studies revealed the forecasting accuracy dependence on project progress [4]. The latter definition for CPI was modified by US Department of Defence (DoD) and proposed a thumb rule [5] which states that "the cumulative CPI will not change by more than 0.1 from its value at 20% completion point, and in most cases it only worsens". But, the DoD stability rule was proved to be unsatisfactory even for the DoD portfolio [6]. The recent EVM stability studies proposed a new metric called 'mean lags' to evaluate the stability of deterministic EVM forecasting methods on topologically diverse project networks and observed a stability–accuracy trade-off for time and cost forecasting methods in EVM [7]. A probabilistic evaluation of CPI stability using simulation based analytical model reported the importance of CPI stability profile as an indicator of project plan quality [8]. The evaluation of the stability moment of performance indices identified potential influencing factors [9] using statistical methods.

The remainder of this paper is structured as follows. The next section outlines an analytical stability simulation model for the probabilistic evaluation of stability of performance indices followed by the experimental framework. The results are observed for the computational experiment on the topologically diverse project networks. Finally, a summary and conclusions are presented.

2 Stability Simulation Model

For a construction project having N activities, three point estimation techniques can be applied for time and cost to introduce variability [10]. Hence, the budgeted cost of each activity is assumed to be the product of the means of the three point estimates of individual activity cost per unit time and activity duration. The earned value (EV) is determined by assuming that the actual work progresses at a constant rate with time (as monetary units per unit time). The individual activity cost is computed from the simulated values for activity cost per unit time and activity duration, assuming the time dependency of cost for each individual activity. The performance indices SPI and CPI are assessed at every review of the project upon the completion of each activity, and the stability rule for the performance indices are defined with a specified threshold for stability variation. Hence, the input parameters for the model are the actual and planned values for the activity cost per unit time and actual duration, and the output parameters are the budgeted cost of activity, actual cost of activity, EV, SPI, CPI and SPI(t). The planned activity duration, work progress, and the number of activities in the network are the control variables.

The stability rule applied in the formulated analytic stability model for both CPI and SPI from the previous work of author is extended to incorporate the stability of earned schedule (ES) based performance index SPI(t), in which a stable performance index is defined as that performance index with its value at the status date is not exceeded by a specified threshold of 10% from its value at project completion. The previous work of author defines the stability of performance indices (PIs) as given in Eq. (1).

$$\text{PI stability} = \frac{1}{n}\sum_{i=1}^{n} 1\left\{\left|\text{PI}_{\text{x}}^{\text{k}} - \text{PI}_{100}^{\text{k}}\right| < 0.1\right\} \tag{1}$$

where $\mathbf{1}\{A\} = \begin{cases} 1, & if\ A\ is\ true \\ 0, & if\ A\ is\ false \end{cases}$, is a binary function.

The earned schedule at the status date,

$$\text{ES(t)} = \text{t} + \frac{EV - PV_t}{PV_{t+1} - PV_t} \times 1(\text{calendar unit}), \tag{2}$$

where, PV is the planned value for the status date and t is such that EV $\geq PV_t$ and EV $\leq PV_{t+1(calendarunits)}$.

Hence, the schedule performance index with respect to time at the status date,

$$\text{SPI(t)} = \frac{ES(t)}{AD}, \tag{3}$$

where AD is the actual duration of the project until the status date.

By applying the stability definition for SPI(t) similar to Eq. (1),

$$\text{SPI(t) stability} = \frac{1}{n}\sum_{i=1}^{n} 1\left\{\left|\text{SPI(t)}_{\text{x}}^{\text{k}} - \text{SPI(t)}_{100}^{\text{k}}\right| < 0.1\right\}. \tag{4}$$

3 Experimental Framework

This section describes the framework for the computational experiment using the stability simulation model, which consists of network generation, Monte Carlo simulation and stability analysis.

3.1 Network Generation

Thirteen activity on node (AON) project networks each having six activities and with varying topology were generated by varying the value of a topological parameter called "Serial/parallel indicator" abbreviated as "SP indicator" in step size of 0.2. The SP indicator [11] is defined as in Eq. (5),

$$\text{SP indicator} = \frac{\text{m} - 1}{\text{n} - 1} \text{ for n} > 1, \quad \text{SP} \in [0, 1] \tag{5}$$

where m is the maximal progressive level and n is the number of non-dummy activities in the project network. The values of SP indicator 1 and 0 correspond to the completely serial and parallel network topology respectively and any intermediate value can be related to its closeness to the serial or parallel topology. Based on this fact, the project networks can be classified as 'serial', 'serial parallel' and 'parallel' for the ranges of 1–0.6, 0.6–0.4 and 0.4–0 respectively [11].

Even for the networks with same SP value, the activity distribution along the project levels can also be varied according to another topological parameter called "Activity distribution ratio (AD)" [11], which can be represented as in Eq. (6),

$$\text{Activity distribution ratio, AD} = \begin{cases} 0; & \text{m} \in \{1, n\} \\ \frac{\alpha_{\text{w}}}{\alpha_{\max}} = \frac{\sum_{\text{a}=1}^{\text{m}} \left|\text{w}_{\text{a}} - \text{w}'\right|}{2(\text{m}-1)(\text{w}-1)}; & \text{otherwise, AD} \in [0, 1] \end{cases} \tag{6}$$

where w is the weight of the progressive level and the average deviation, $w' = n/m$. α_w is the measure of total absolute deviation of activity distribution and α_{max} is the maximum value of α_w for a network with n activities and m progressive levels. The value of AD = 0 represent the networks having equal width progressive levels and those with AD = 1 have one progressive level with maximal width and the others with a single activity. The topological characteristics for the generated networks are given in Table 1, in which each network is named based on its SP indicator value and the number of the network within the same SP level. The *PL* in Table 1 indicates the number of progressive levels in the network.

3.2 Monte Carlo Simulation

The actual duration and the actual cost per unit time were simulated for thousand iterations for each activity using the generalized beta distribution using FlexSim 18.2.2 for the coefficient of variation (CV) of 0.25 for the duration. The variability of time and costs were introduced using the three point estimation technique, for which the set of optimistic, most likely and the pessimistic values are {8, 13, 29} and {35, 50, 70} respectively. The probability distribution function for the generalized beta distribution [7] can be given by Eq. (7), the mean μ and mode m of the distribution were used to define the shape parameters θ_1 and θ_2 as given in Eq. (8).

$$f(x) = \frac{\Gamma(\theta_1 + \theta_2)}{\Gamma(\theta_1)\Gamma(\theta_1)(b - a)^{\theta_1+\theta_2-1}}(x - a)^{\theta_1-1}(b - x)^{\theta_2-1};\ x \in [a, b] \quad (7)$$

$$\theta_1 = -\frac{(b + a - 2m)(a - \mu)}{(m - \mu)(a - b)} \text{ and } \theta_2 = \frac{(b + a - 2m)(b - \mu)}{(m - \mu)(a - b)} \quad (8)$$

3.3 Stability Analysis

The stability analysis was performed in Microsoft Excel for the performance indices in EVM [CPI, SPI and SPI(t)] using the formulated analytic stability model, assuming that both the costs and work progress are time dependent. The activity level deviations for time and cost introduced by Monte Carlo simulation were translated to project level by EVM measurement. The projects were reviewed after the completion of every activity. Table 1 represents the results of stability analysis along with the topological characteristics of each network and "R" represents the stability after each review.

Table 1 The results of stability analysis

No.	Network	Network characteristics			CPI stability (%)					SPI stability (%)				
		SP	AD	PL	R1	R2	R3	R4	R5	R1	R2	R3	R4	R5
1	SP 1	1	0	6	38	59	73	89	97	100	100	100	100	100
2	SP 0.8-1	0.8	1	5	54	36	57	79	95	38	100	100	100	100
3	SP 0.8-2	0.8	1	5	33	58	57	79	95	100	46	100	100	100
4	SP 0.8-3	0.8	1	5	33	51	64	79	95	100	100	59	100	100
5	SP 0.8-4	0.8	1	5	29	35	35	24	44	100	100	100	73	100
6	SP 0.8-5	0.8	1	5	33	51	62	68	72	100	100	100	100	81
7	SP 0.6-1	0.6	1	4	29	32	36	65	91	100	30	34	100	100
8	SP 0.6-2	0.6	1	4	29	35	31	28	91	100	100	37	42	100
9	SP 0.6-3	0.6	0.67	4	29	42	44	9	91	100	43	46	8	100
10	SP 0.4-1	0.4	1	3	22	11	18	23	86	100	19	22	22	100
11	SP 0.4-2	0.4	0	3	1	15	4	11	8	28	36	38	45	11
12	SP 0.2	0.2	0	2	27	36	40	5	40	19	25	38	7	4
13	SP 0	0	0	1	59	55	49	38	21	5	8	8	5	3

(continued)

Table 1 (continued)

SPI(t) stability (%)					Stability variations (%)			Stability rank			
R1	R2	R3	R4	R5	CPI	SPI	SPI(t)	CPI	SPI	SPI(t)	Overall
0	1	21	60	99	60	0	100	1	1	4	1
0	48	24	58	95	45–65	20–50	100	4	2	3	2
0	2	54	59	95				3	4	2	2
0	0	20	62	95				2	6	5	4
0	0	7	38	59				11	5	7	8
0	0	1	1	10				5	3	12	7
0	0	17	26	87	70–90	60–90	100	6	7	6	6
0	0	6	22	35				7	8	8	8
10	2	78	78	80				8	9	1	5
0	0	9	15	8	85–95	70–80	85–100	9	10	10	10
5	21	13	34	4				13	11	9	11
1	6	19	1	0	88	90	100	12	12	11	12
0	0	0	0	0	64	70	100	10	13	13	13

4 Observations and Results

Table 1 exhibits the results of the stability analysis which indicate that the stability of the performance indices CPI, SPI and SPI(t) are significantly influenced by the percentage completion as well as the project network topology in terms of project seriality, activity distribution and the progressive level. In general, there is a continuous improvement in the stability of performance indices with the project progress (percentage completion) unless dominated by the effects of project network topology. The effect of project network topology can be explained in terms of the topological parameters—SP indicator, activity distribution ratio and progressive level.

4.1 Effect of Network Topology

It can be observed that the networks with the three distinct SP ranges have similar stability performances, which is in line with the topological classification of project networks as mentioned in Sect. 3.1. Also, the stability of the performance indices is found to be deteriorating as the networks assume from more serial to more parallel (SP indicator varying from 1 to 0) topology, and is severe for the networks with SP indicator value less than or equal to 0.4 (i.e., parallel networks). The highest stability attainments for CPI, SPI and SPI(t) are nearly 60%, 50% and 40% respectively for the parallel networks (except for the CPI and SPI of SP 0.4-1 network).

Even the networks with same SP value can have variable stability owing to the distribution of activities within the network. For the serial as well as serial parallel networks, the effect of percentage completion can also be explained in topological terms such that the stability improves as the project progress approaches the higher progressive levels. But, the effect of percentage completion is not pronounced for the parallel networks ($0 \leq SP < 0.4$) if all the progressive levels are having equal widths (i.e., $AD = 0$). The higher CPI stability at the initial stages and its further deterioration is owing to the adverse effect due to the combination of all of the above mentioned topological features.

The nature and the degree of influence of the above mentioned factors are different for the performance indices. The stability deterioration of SPI is observed to be strictly following the serial/parallel characteristics, the effect is less pronounced for CPI and the least for SPI(t). For the same SP level, the better overall stability performance for CPI is observed when the wider progressive level is close to the middle of the network, and is at the finish of the network for SPI. But, the severe stability deterioration for SPI(t) is observed as the wider progressive level shifts from the middle to finish position of the network. The stability fall for SPI due to the progressive level weights is local in nature (Refer Table 1) while that for SPI(t) affect the overall stability performance with higher initial lag of stability with a shift of stability curve towards

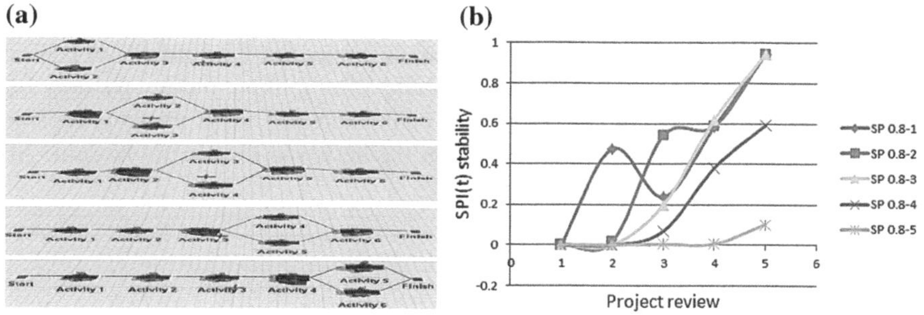

Fig. 1 **a** SP 0.8 networks; **b** SPI(t) stability curves for SP 0.8 networks

right (Fig. 1 represents the overall stability performance for SP 0.8 networks). The effect of percentage completion on the stability of CPI and SPI(t) was ruled out in parallel networks ($0 \leq \text{SP} < 0.4$) for AD = 0. But, the SPI stability improves unless AD $\neq$ 0.

4.2 Relative Stability Performance

By analyzing the stability variations from Table 1, it can be observed that the larger stability deviations occur when the projects reviewed at the wider progressive level, and the effect is much more pronounced when it is close to the project finish counteracting the effect of percentage completion. The computed stability variation of the network based on the maximum and minimum stability (Sect. 4.2) does not account the above mentioned facts and cannot be used as a basis to evaluate the overall stability of the project network as well as to compare the stability performances of different networks (because the stability variations for the networks are observed at the different reviews and the higher level reviews deserve higher weightage to indicate the higher stability contributed by the percentage completion). Hence, a common stability parameter is required to assess the overall stability of each of the networks for the stability of different indices and to have a comparison on the relative stability performance. So a stability parameter (P_S) is introduced which satisfy the effects of the above mentioned factors, and can be defined as in Eq. (9), where, R_i is the ith review of the project, St_i is the stability at ith review and r is the total number of reviews. The parameter takes the ideal maximum value of $\frac{r(r+1)}{2}$.

$$\text{Stability parameter, } \text{P}_\text{S} = \sum_{i=1}^{r} (R_i \times St_i) \tag{9}$$

Higher the value of the stability parameter indicates the better stability performance and the ideal maximum value for this analysis having five reviews is 15.

Based on the stability parameter, the networks can be ranked for the stability performances of each performance index and can be compared. The parameter is used to rank the networks based on the overall stability after ranking them separately for each of the performance index stability as given in Table 1. The parameter is capable of differentiating the stability performance of the networks even with the same SP indicator level.

4.3 Topology–Stability Relation

From the ranking of networks, it is obvious that the SPI stability strictly follows the influence of SP indicator in which the stability deteriorates with the decrease in SP indicator value. Similarly, the SPI(t) stability strictly follows the parameter $R_{SPI(t)}$, except for the serial networks, as given in Eq. (10),

$$R_{SPI(t)} = \frac{\sum_{i=1}^{pl} \frac{PL_i}{W_i}}{W_{max}} \tag{10}$$

where PL_i is the ith progressive level, W_i is the width of corresponding progressive level and W_{max} is the maximum progressive level width of the project network. Higher the value indicates the higher stability performance. This relation differentiate the stability performance among the parallel and serial parallel networks (the values for the networks SP 0 to SP 0.4-2 are 0.03, 0.33, 1.125 and 1.5; the values for SP 0.6-2, SP 0.6-1 and SP 0.6-3 are 2.67, 2.89 and 3.75 respectively) and follows the stability ranking for SPI(t) as given in Table 1.

4.4 Time and Cost Performance Stability

Upon comparing the stability of SPI and SPI(t) using Fig. 2a, SPI shows the overestimated stability, which is due to the inherent drawback of traditional EVM, in which SPI shows its value as unity for the finished activities, the disadvantage of using cost metric to assess the schedule performance. But, the SPI(t) which is a time based metric shows the gradual development with the progress, and is more close to the practical scenario. Hence, SPI(t) is more reliable and convenient as compared to SPI to assess scheduling efficiency.

The comparison of cost and time performance stabilities (CPI and SPI(t) performance) based on Table 1 and Fig. 2b shows that the cost performance is more stable and is having some degree of initial stability as compared to the latter. Although the schedule performance is more sensitive to the effects of project network topology

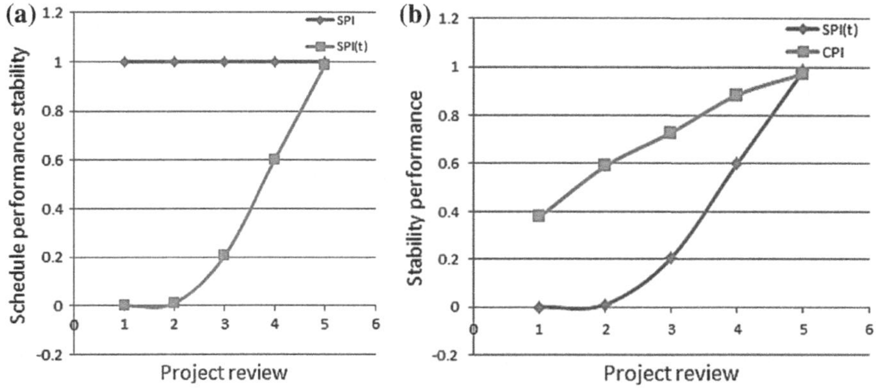

Fig. 2 **a** Schedule performance stability versus project review; **b** Stability performance versus project review

(in general), there exists more evident topology–stability relation. However, the stability performance of CPI and SPI(t) are more similar in nature as compared to that of SPI's.

5 Conclusions

For the reliable forecasting of project performances in EVM, the stability of project performance indices in EVM is an important concern to avoid any misleading early warnings. Since the EVM forecasting is based on its performance indices CPI, SPI and SPI(t), any stability related issue associated with the performance indices will directly in the subsequent forecasting. The project network topology too has a significant bearing on the stability as the construction industry is concerned, which is subjected to frequent updating and rescheduling owing to its highly dynamic nature. Hence, the objective of this paper is to evaluate the stability of performance indices in EVM by probabilistic approach as an extension of author's previous work of analytic stability model using simulation. It is observed that the percentage completion and the project network topology (in terms of project seriality, activity distribution and the progressive level) have significant influence on the EVM forecasting stability, and the effect of latter is capable of counteracting the effect of the former. The compounding effects of the above mentioned factors are studied by a newly defined 'stability parameter' in order to assess the overall stability performance of each performance index as well as to compare the performances of the topologically diverse construction projects by stability ranking approach, and stability–topology relations for the schedule performances are figured out. The comparison of performance indices based on the stability aspects highlights the sensitivity of schedule performances with a more evident topology–stability relation as compared to the cost performance as well as the superiority of SPI(t) over SPI. The influence of performance index stability on the stability of EVM forecasting methods and the evaluation of findings in real life projects can be regarded the future works.

References

1. Schnaars S (1984) Situational factors affecting forecast accuracy. J Market Res 21(3):290–297
2. Payne K (1990) An investigation of the stability of the cost performance index. Technical Rep., Air Force Institute of Technology, WrightPatterson AFB, OH
3. Christensen D, Heise S (1993) Cost performance index stability. Nat Contract Manage J 25:7–15
4. Zwikael O, Globerson S, Tzvi R (2000) Evaluation of models for forecasting the final cost of a project. Project Manage J 31:53–57
5. Christensen D, Templin C (2002) EAC evaluation methods: do they still work? Acquis Rev Q 9:105–116
6. Henderson K, Zwikael O (2008) Does project performance stability exist? A re-examination of CPI and evaluation of SPIðtÞ stability. CrossTalk 21:7–13
7. Wauters M, Vanhoucke M (2014) Study of the stability of earned value management forecasting. J Constr Eng Manag 141(1992):1–10. https://doi.org/10.1061/(ASCE)CO.1943-7862.0000947
8. Kim B (2015 May) Probabilistic evaluation of cost performance stability in earned value management. J Manag Eng 32:1–13. https://doi.org/10.1061/(ASCE)ME.1943-5479.0000383
9. Koning P, Vanhoucke M (2016) Stability of earned value do project characteristics influence the stability moment of the cost and schedule performance index. J Mod Proj Manag
10. PMI (Project Management Institute) (2013) A guide to the project management body of knowledge, 5th edn. Newton Square, PA
11. Vanhoucke M (2009) Measuring time—improving project performance using earned value management, vol 136. International series in operations research and management science. Springer, New York

Development of Frame Work for Residential Building Construction Using Agile Management

Anand Jose Paul and Sahimol Eldhose

Abstract Agile Project Management is an iterative approach to planning and guiding project processes. It is suited for large complex projects where it is difficult to specify the product in advance. This thesis has focused on what opportunities there might be in implementing the Agile project management approach on the basis of Scrumban method in the residential building construction. The major advantages found with implementing the Scrumban approach is an improve productivity and resource allocation. The Scrumban approach almost forces the client to increase their participation in the project compared to the situation today. By the use of time management and specific meetings it will also be beneficial for keeping track of the project's progression and status. This project intends to create a methodology for improvement of residential building construction by Agile management software.

Keywords Agile management · Scrum · Kanban · Scrumban · Lean principles · Minimizing the waste

1 Introduction

Project management is the science which applies skills, tools and techniques to fulfil project activities in a way that the expectations and requirements of stakeholders are fulfilled or exceeded. Traditional projects are performed in a single continuous flow, with sequential steps that include project initiation, project planning, project execution, and project closeout [1]. Typical construction projects have an initiation/planning phase, a design phase, a construction phase, a testing phase, and a turnover to the user phase, followed by project closeout. There are more elaborate models with multiple phase gates and bid phases, but the main point in common is the

A. J. Paul · S. Eldhose (✉)
Department of Civil Engineering, Toc H Institute of Science and Technology, Arakunnam, India
e-mail: sahimol27@gmail.com

A. J. Paul
e-mail: anandjosepaul@gmail.com

K. Dasgupta et al. (eds.), *Proceedings of SECON'19*,
Lecture Notes in Civil Engineering 46,
https://doi.org/10.1007/978-3-030-26365-2_67

sequential nature, with user input happening primarily in the planning and or design phase. During the construction phase, another sequential work plan is put together, typically in a project schedule, built up by trade [2]. The typical large construction project has multiple contractors working for a general contractor, with a separate designer and a separate owner, and in some cases separate end users (tenants), with complex contracting methodologies in place

In a perfect world, the construction scheduling, planning, and execution processes would be detailed and robust enough to avoid project issues and delays. However, a significant number of projects, especially large and complex ones, have delays and cost increases. Aside from material costs, the most significant cost element in construction project is labour [3]. Project labour costs increase from the baseline plan when delays occur. It may seem obvious, but to avoid delays the project needs the right people, at the right location, with the right material and tools, and with the right work instructions (Fig. 1).

In this study, data collected on-site were analysed to calculate the defined indicators. Delays and overruns can result from not having the right material on hand at the right time in site, which may occur due to supplier backlogs, shipping delays, funding restrictions, and so forth. Delays can occur from not having adequate work instructions on hand at the right time (work packages), which may result from incomplete or inaccurate design and documentation, delays in decision making or instructions, or changing scope. Finally, delays from labour shortages can occur due to an inability to find the right trades, scheduling issues with vendors or contractors, and so on [4]. The scrumban framework and agile management software must aim at providing a continuous workflow where the movement of processes, information, and resources are undisturbed and controlled along the end-end construction process.

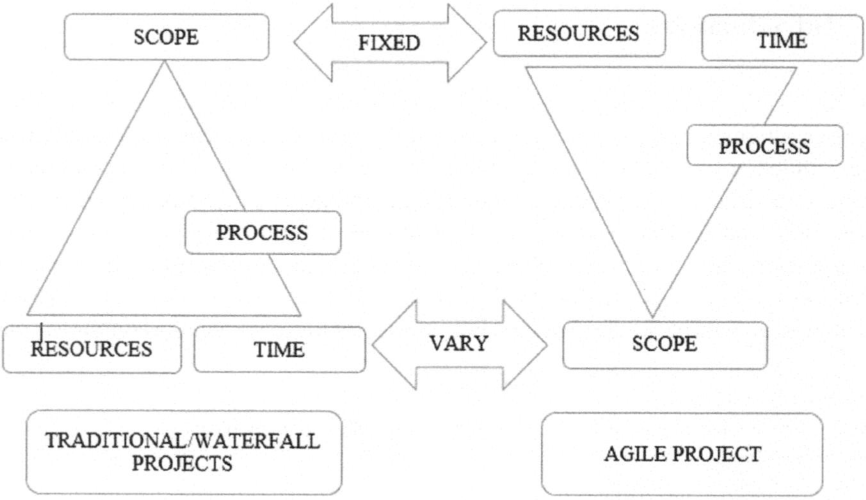

Fig. 1 Traditional versus Agile flow chart

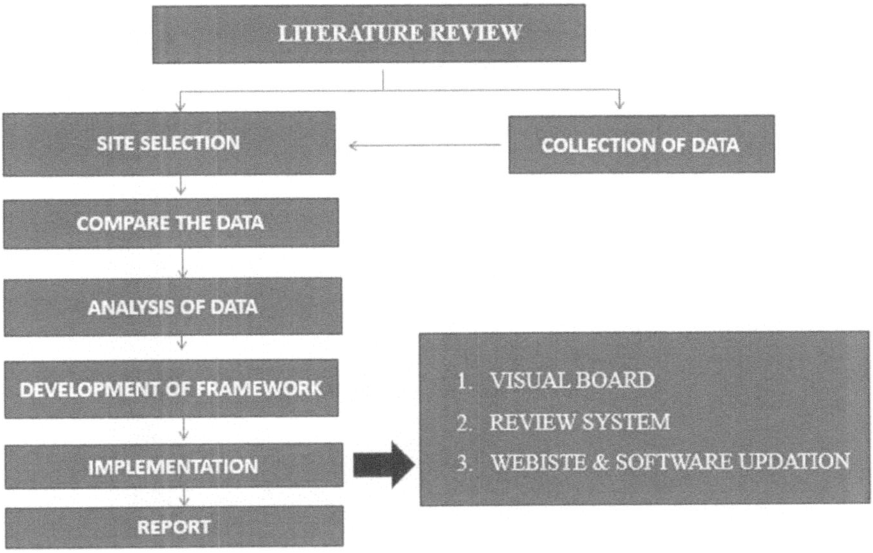

Fig. 2 Methodology

2 Methodology

A detailed methodology regarding the workflow of the project is drawn out prior to the commencement of the work. Tentative schedules were prepared for the same. The project, having fixed the goals, planned to start with a detailed literature review, referring to various International as well as National journals in the field agile management. The workflow is as shown in Fig. 2.

3 Site Selection and Investigation

The site is located in a residential area of Thodupuzha, in the Royal garden villas Mylacombu, forming a rectangular shape with overall area of approximately 10 cents. Data were collected directly from the involved people in this work via conducting semi-structured interviews. The data collection and analysis was based on the knowledge obtained from reviewing the literature, available samples and from expert advice. The interview was conducted to determine the opinion of client, contractors and consultant regarding the project progress (Figs. 3 and 4).

Actual duration of previous work completion of project—300 days (Fig. 5).

Goal: Construction of Royal Garden villas Plot no.10.
Scrumban Master: The Scrumban master is the team member who has the final say over what going into the project and what the result is supposed to look like.

Fig. 3 Front view of royal garden villas

Fig. 4 Front 3D elevation of construction building

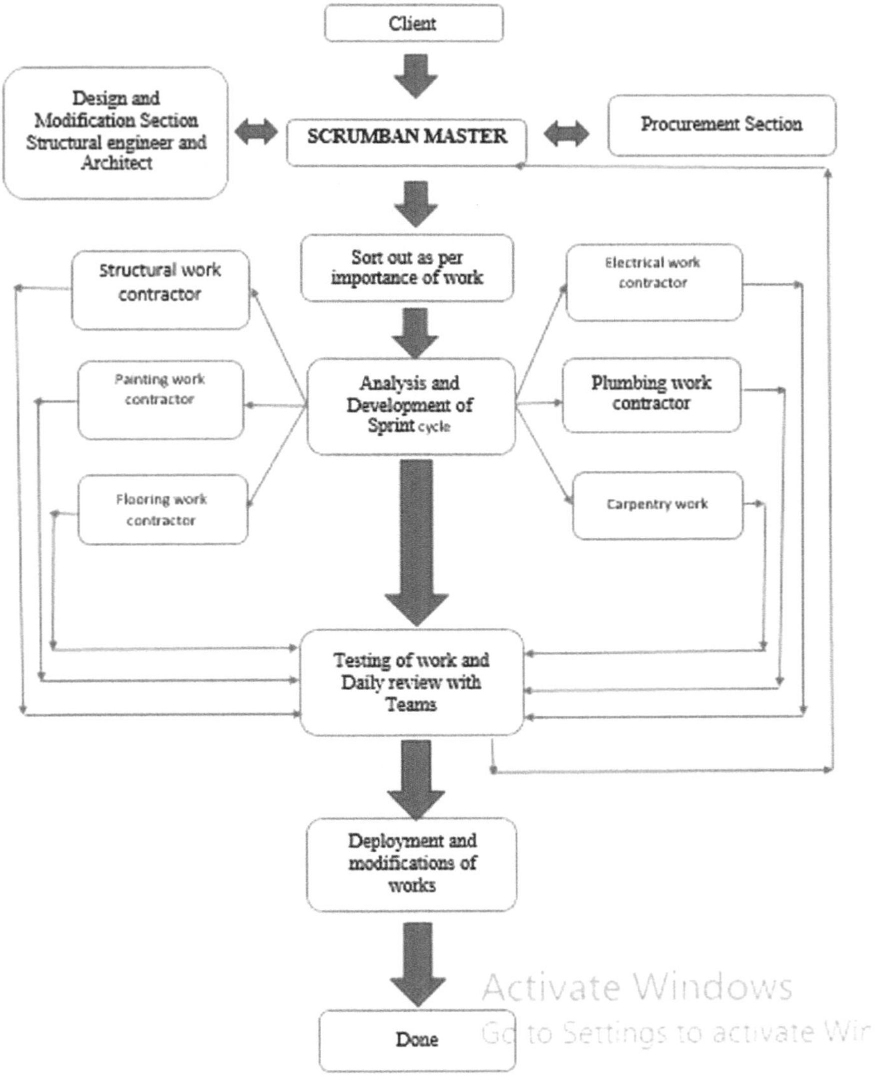

Fig. 5 Framework

Make a Backlog: The Backklog is the prioritized list of tasks that make up the complete project.

The Scrumban master is in charge of the backlog, but it can be developed with input from the whole team.

Sprint Planning: This is where figure out the goals and the workload capacity for the upcoming Sprint. The whole team needs to have a clear understanding of the goals for each Sprint.

Assign Tasks: Each task within the sprint should be accounted for by contractors who will be responsible for getting it done. A review sheet is arranged for assigning daily task and analysing daily progress.
Begin the Sprint: It is the time to start the work, with each team member tackling the tasks the were assigned, collaborating when possible.
Hold Daily Review: Meet every day to discuss yesterday's work, todays work, and any impediments in the team's way.
Finish the Sprint and Hold a Retrospective: When all the tasks are completed, its time to have a retrospective meeting to present and review the work that was done, refining the backlog to prepare for the next sprint.
Quality Check: Quality check is conducted cyclically preliminary and NDT are used assure the quality of construction.

4 Implementation of Scrumban Frame Work

Review System

Review focusing on what each person accomplished yesterday and will accomplish today, the team gains an excellent understanding of what work has been done and what work remains. The daily review meeting is also a status update meeting in which a scrumban master is collecting information about who is behind schedule. Scrumban master has assigned works in every morning, responsible contractors carried their task. After the completion of the daily task, contractors are entered actual daily works done in sheet. Sometimes delay obtained in progress of work, but review chart helps reallocate and improve project performance. Rather, it is a meeting in which team members make commitments to each other (Fig. 6).

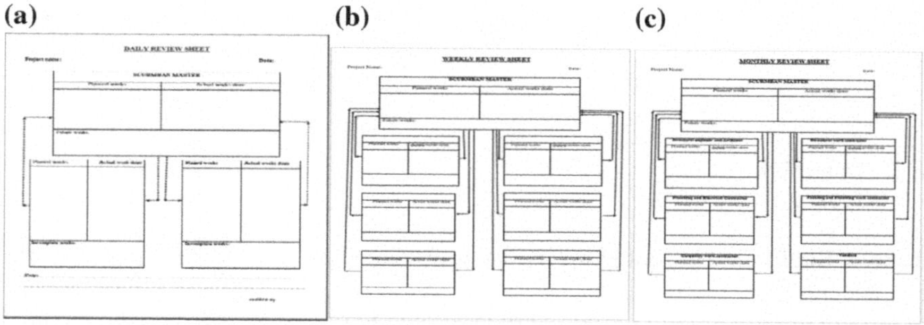

Fig. 6 Screen shot of the review sheet **a** Daily review, **b** Weekly review, **c** Monthly review

Visual Board

A Scrumban visual board is one of the tools that can be used to implement Scrumban to manage work at a personal or organizational level. Scrumban boards visually depict work at various stages of a process using cards to represent work items and columns to represent each stage of the process. Cards are moved from left to right to show progress and to help coordinate teams performing the work. In this project main works likes civil, electrical, plumbing, painting, flooring, carpentry, finishing and quality check are again divided into according to priority. Separate colour code are provide for easy identification of their daily task from scrumban work flow board (Tables 1 and 2).

5 Software Control

A management software is developed for project controlling. Software have admin and user interface platform. Client have a option to view the workflow (Fig. 7).

Intercommunication platform used to track the daily progress. Scrumban master and members have separate user name and password, to access their window. In the morning team head has assign the daily task and in the evening collect their feedback of the progress work. From the daily review sheet master has analyzed the workflow. If any delay obtained, master has full option to re-allocate the work chart. Admin panel is shown in Fig. 8.

Cost updating platform is arranged for analyzing the cost fluctuations as shown in Fig. 9.

Table 1 Work splitting chart

Work chart

Civil works	Electrical	Carpentry	Plumbing	Painting	Flooring	Finishing works & quality checking
1. Earth work	1. Conduit placing	1. Door and window frame	1. Pipe placing	1. First coat	1. Granite and tile	1. Materialn test
2. Foundation						
3. Basement						
4. Plinth beam						
5. Brick work						
6. Lintel beam						
7. Roof Slab	2. Wire pulling Checking			2. Putty		
8. Plastering	3. Fixing light and fan	2. Door and window	2. Checking	3. Second coat		2. NDT

Table 2 Scrumban work flow board

Scrumban visual flow board							
Back log	To do	Development		Testing		Deployment	Done
		Ongoing	Done	Ongoing	Done		

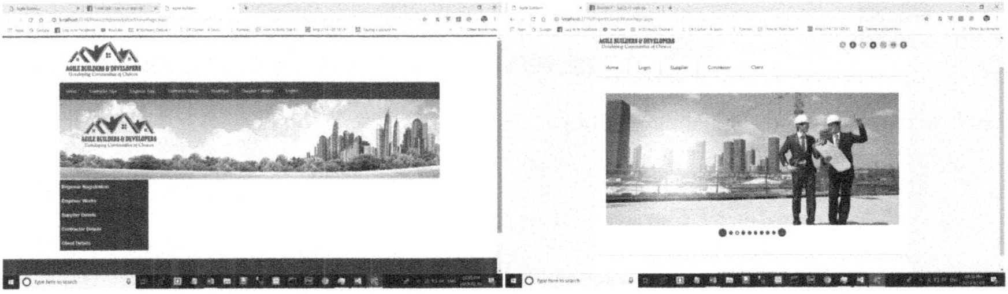

Fig. 7 Screen shot of user interface

5.1 Results and Discussions

- **Tests on Cement at Construction Site**

Color Test of Cement—The color test of cement is conducted. light greenish shade is identified.
Presence of Lumps—No lumps.
Cement Adulteration Test—Feel smooth.
Temperature Test of Cement—Feel cool.
Float Test—Flot some time in water.
Setting Test—Partially set.

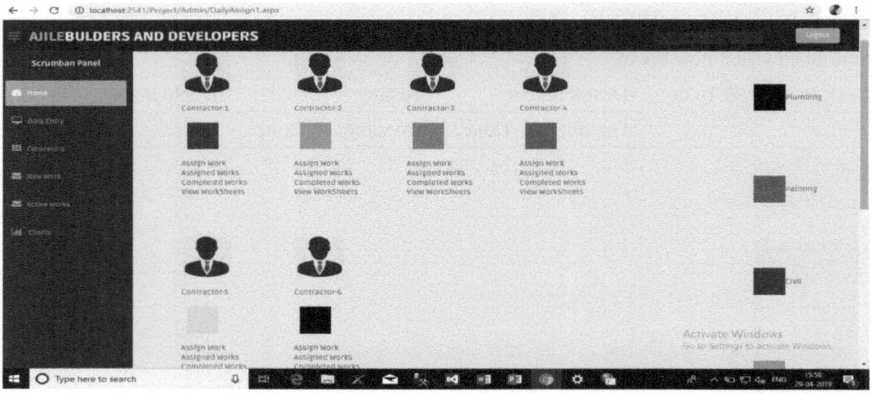

Fig. 8 Screenshot of admin interface Scrumban panel

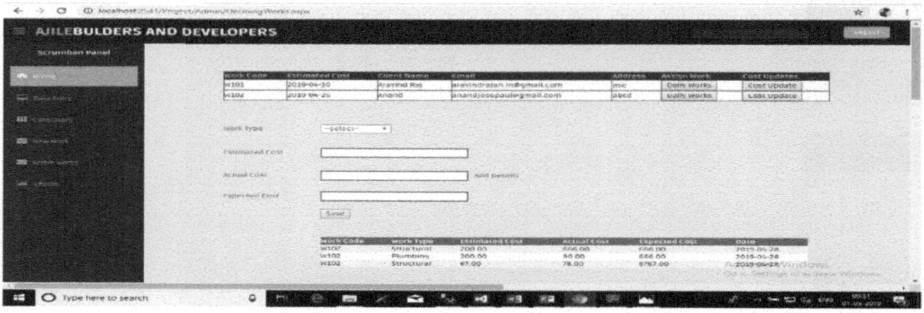

Fig. 9 Screenshot of cost updating platform

- **Tests on Aggregates at Construction Site**

Silt content test—There is no clay particles found in fine aggregates (Fig. 10).

Fig. 10 Silt content test

Fig. 11 Bending Test

- **Tests on Steel at Construction Site**

Bending test

Steel has high tensile strength. There is no crack obtained in bended region of steel Fig. 11.

- **Tests on Hardened Concrete**.

Compressive strength

Sl. No.	Specimen	Compressive strength N/mm^2	
		7th day	28th day
	Normal concrete	13.04	19.1

Tapping method

Strength of concrete (N/mm^2)	Test results		
	Blow of hammer (0.4 kg) upon concrete surface	Blow of hammer (0.4 kg) upon chisel placed at right angles to concrete surface	Scratching by Chisel
	Sound clear whitish mark remains	Thin scales split off round the mark	Visible scratches no deeper than 1 mm

Rebound hammer test

Sl. No.	Test surface	Angle	Rebound number	Average value of rebound number	Compressive strength (N/mm^2)	Standard value (N/mm^2)
1	Column	90	27,25,26	25	22	20
2	Beam	90	22,25,24	23	19	20
3	Slab	90	22,23,24	23	19	20

6 Conclusions

This study was carried out to development and implement agile management in construction industry using scrumban method. Moreover, agile management actions, scrumban framework development and their effectiveness and usage were settled on. The relevance of scrumban frame work and software in construction project management was found out by conducting literature review and personal interviews. Identification and analysis of the previous residential buildings data are done. A simple cost effective project control web application was developed.

- Agile values are beneficial to the project development process and promote communication both horizontally and vertically throughout the entire project life cycle.
- Review system helps to improve project performance and project progress status.
- The Scrumban board gives an excellent overview of project progress. Board helps to visualize the daily progress in site, contractors and labors are got awareness about past, present and future stages of work. Visualizing work in a team environment simplifies communication and leads to improved productivity in this project.
- Using of agile management software enhances communication, teamwork, collaboration, and organizational change improve the adoption of higher value products during the preliminary stages and throughout the project lifecycle.
- Now residential building construction going, almost 80% works are complete.
- Cost of the project hiked due to material rate fluctuation.
- Time optimization can done successfully by daily review system. Quality of the entire project maintained by conducting site test.

References

1. Aibinu AA, Jagboro GO (2002) The effects of construction delays on project delivery in Nigerian construction industry. Int J Proj Manag 20:593–599
2. Matthies C (2018) Scrum2Kanban: integrating Kanban and scrum in a university software engineering capstone course
3. Albarqi AA, Qureshi R (2018) The proposed L-Scrumban methodology to improve the efficiency of agile software development. Int J Inf Eng Electron Bus 3:23–35
4. Banijamali A, Dawadi R, Ahmad MO, Similä J, Oivo M, Liukkunen K (2016) An empirical study on the impact of Scrumban on geographically distributed software development. In: Proceedings of the 4th international conference on model-driven engineering and software development, pp. 567–577

Measurement of Aggregate Size and Shape Using Image Analysis

Parth Thaker and Narendra Arora

Abstract Aggregate quality influences concrete mix design and properties. It is necessary to measure the quality of aggregate. The most popular field test to measure the quality of aggregate is the sieve test. It is not possible to measure shape characteristic and other parameters by sieve analysis. Therefore, it is essential to develop rapid assessment techniques for quality control of aggregate. Selection of methodology for aggregate shape characteristic measurement depends on parameters such as aggregate size, accuracy, reliability of the method, time required to analyze the sample, human efforts required, measurement of other characteristics of the particle apart from the size, the robustness of testing equipment, the initial cost of the equipment, its maintenance and operational costs. Digital image analysis is a process to gather information regarding a characteristic of particle through computer programming. Digital image analysis method has a potential to estimate the characteristics namely, the size and shape of aggregates rapidly and accurately. Two-dimensional image analysis of aggregate gives a relative idea of aggregate properties in a more accurate manner as compared to the procedure suggested in Indian standards. This paper summarizes the developments and research in the area of image analysis.

Keywords Aggregates · Digital image analysis · Particle shape · Particle size

1 Introduction

Concrete mix design is significantly influenced by the shape and size of particle. Aggregate occupies approximately 60–80% of the volume of concrete; so it is important to identify the characteristics of aggregate. Aggregate characteristic does not only affect the strength of concrete but also influences the structural performance and

P. Thaker (✉)
Faculty of Technology, CEPT University, Ahmedabad, Gujarat, India
e-mail: parth.thaker@cept.ac.in

N. Arora
L.E. College, Morbi, Gujarat, India

K. Dasgupta et al. (eds.), *Proceedings of SECON'19*,
Lecture Notes in Civil Engineering 46,
https://doi.org/10.1007/978-3-030-26365-2_68

durability of concrete. Natural aggregate is either formed by weathering and abrasion process or by the artificial crushing of bigger parent mass [1]. The microstructure of parent rock, prior exposure condition, and processing factor decide the characteristic of aggregate that significantly affects concrete properties [2]. Petrographic classification, chemical and mineral composition, strength, hardness, specific gravity, color, pore structure, physical and chemical stability depend on parent rock properties. Aggregate properties such as particle shape, size, surface texture, and absorption are not associated with the parent rock. In general, mortar or concrete properties have been preferably found by characteristics of aggregate because they influence both fresh and hardened properties.

Galloway [3] has defined the aggregate shape by sphericity, form, and roundness. If three axes or dimensions of the particle are nearly the same, then the particle is classified as spherical. A form of particle which is also known as shape factor is used to differentiate between particles which have same sphericity [4]. If longest axis, intermediate axis, and shortest axis are represented by L, I, and S respectively, then;

$$Sphericity = \sqrt[3]{\frac{S.I}{L^2}} \tag{1}$$

$$Shape\ factor = \frac{S}{\sqrt{LI}}\ \ or\ \ Shape\ factor = \frac{LS}{I^2} \tag{2}$$

However, various researchers have given different definitions of shape parameters which are also not correlated with one another. Roundness and angularity are also two other characteristics to define particle shape [5]. Particles losing edges and corners by attrition process become well rounded such as wind-blown sands, as well as sand and gravel from seashore or riverbeds. Aggregates manufactured by crushing of rocks have sharp and well-defined edges which are called angular particles [2]. The shape of the particle can also be classified as angular, sub-angular, sub-rounded, rounded, and well-rounded [6]. Aggregate shape characteristics have been classified as rounded, irregular or partly rounded, angular, and flaky as per IS-383:1970 (reaffirmed-2002) [7]. Surface characteristics of aggregates are based on visual examination of hand specimens. As per IS-383:1970 (reaffirmed-2002) [7], Surface characteristics of aggregates has been classified into five groups such as glassy, smooth, granular, crystalline, honeycombed, and porous.

Concrete mix is prepared by particles of various sizes. Particle size distribution is represented by grading. Grading of coarse and fine aggregate is given in IS-383:1970 (reaffirmed-2002) [7]. It is important to select suitable grading and maximum size of aggregate. For example, use of coarser sand leads to the production of harsh and unworkable mix while the use of very fine sand increases water requirement for the desired workability (which also increases cement requirement for given water-cement ratio) and becomes uneconomical. Well-graded aggregate reduces the void ratio and produces the most workable and economical concrete.

2 Measurement Techniques for Particle Size and Shape

It is necessary to develop an understanding of aggregate characteristic. Many efforts have been made to discover rapid, automated, and accurate techniques for the measurement of particle shape characteristic. Concise review of various principles and measurement techniques has been given by McCave and Syvitski [8]. Figure 1 gives the schematic view of four techniques. Out of all the four techniques, only automated sieve analysis is developed for the construction industry.

Selection of a suitable method for measurement depends on various parameters such as estimated range of particle size, solubility, ease of handling, flowability, cost, specific requirements, and time. Figure 2 shows the graphical representation of the accuracy of various methods for a range of particle size [9]. Fluid sedimentation and sieve analysis require comparatively longer time than the other two methods. Laser diffraction technique is used to analyze particle size rapidly but it is applicable for certain particle size range (fine particles). Suspension of the coarse particle to capture diffraction images for a long time is difficult. An image analysis technique covers larger range (0.04–50 mm) as shown in Fig. 2.

3 Sieve Analysis

Sieve analysis is a traditional method to find the particle sizes and gradation of material. For obtaining particle gradation, it is required to arrange an appropriate set of sieves with mesh size increasing from bottom to top and shake the sieves by putting the sample on the top sieve. The passing of the particle through sieve depends on the orientation of particle, the ratio of particle size and sieve opening, various movements (horizontal and vertical) along with the duration of sieve shaker.

Figure 3 represents the various sieving methods versus particle size range. Sieve analysis is inexpensive, easy to perform in the field, and covers a wide range of particle sizes, but it is labor intensive and time-consuming. It also requires cleaning and maintenance of sieves. Automated sieve analyzer is a device which can perform sieve analysis accurately and give computer-generated reports of the sieve test. It also has a facility for emptying and cleaning of the sieves. RETSCH- solutions in milling and sieving is one of the companies which provides a range of automated sieving device for air jet, wet, and dry sieving.

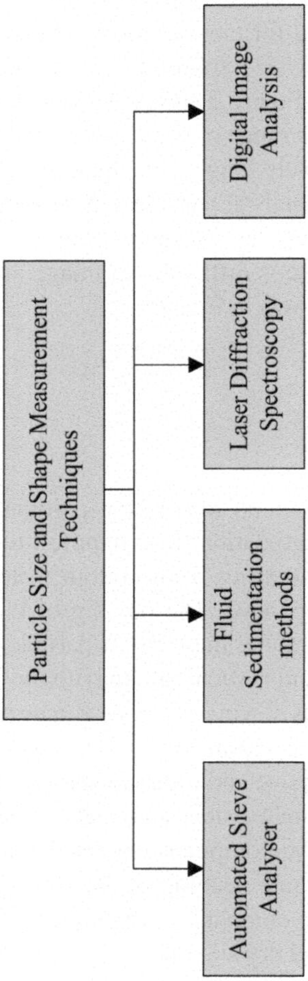

Fig. 1 Measurement techniques for particle size and shape

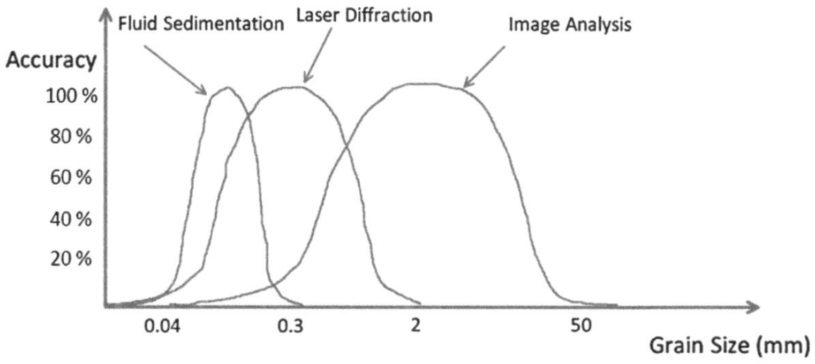

Fig. 2 Graphical representation of accuracy of various technologies for a wide range of particles by Rauch et al. [9]

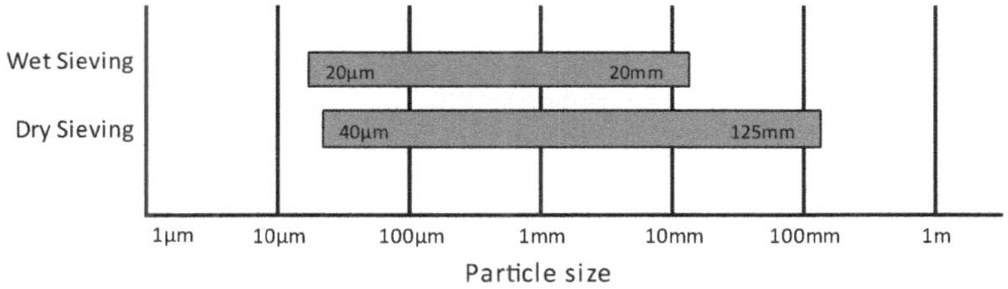

Fig. 3 Particle size versus the method of sieve analysis

4 Fluid Sedimentation Method

In fluid sedimentation method, particle settlement can be either gravitational or centrifugal. Gravitational sedimentation is applicable to relatively larger particles because the rate of settlement for small size particle is too low to give a practical analysis time under gravitational method. Sometimes very small sized particle never settles by gravitational sedimentation unless particle density is extremely high. Therefore centrifugal sedimentation analysis is used for the smaller particle.

IS-5282:1969 (reaffirmed-2000) [10] is an Indian standard which covers liquid sedimentation methods for determination of particle size of the powder. IS-5282:1969 [10] has given the following methods for particle size measurement.

1. Fixed position pipette incremental method—pipette at the top
2. Fixed position pipette incremental method—pipette at side
3. Centrifugal sedimentation method

Liquid fluid sedimentation method is applicable to particles of size from 0.01 μm to about 50 μm. Fluid sedimentation method requires simple and inexpensive equipment. It is also used to measure a wide range of particle with considerable accuracy and reproducibility. A major limitation of this approach is that it measures the particle size in terms of the equivalent diameter of a sphere falling through the fluid. Therefore particle shape characteristic is not obtained directly by this approach. This method requires temperature control to avoid suppressing convectional currents. Particles also have to be insoluble in suspending liquid.

"CPS Disc Centrifuge, Models DC12000, DC18000, DC20000, and DC24000" manufactured by Sinsil International and "SediGraph 5100" manufactured by Micromeritic Instrument corporation are the devices used for fluid sedimentation methods.

5 Laser Diffraction Spectroscopy

Particle interacts with light by reflection, diffraction, adsorption, and refraction. Laser diffraction spectroscopy is a well-known technique used for the assessment of particle size. The core idea behind laser diffraction spectroscopy is the measurement of scattering light angle to evaluate particle size. Large size particle scatters light at a small angle and small size particle scatters light at a wide angle. Loizeau et al. [11] and Buurman et al. [12] have studied the application and accuracy of laser diffraction techniques for earth materials.

ISO 13320:2009 [13] has covered particle size analysis by laser diffraction. This method covers submicron to millimeter size range (general range—0.02 μm to 2 mm) for particle sizing. Laser diffraction spectroscopy analyzes particles much faster than the conventional sedimentation method for fine particles. Large numbers of particles sampled in each measurement give freedom of repeatability. A major limitation of this approach is suspending coarse particle for the required time to capture diffraction images for analysis. It is also an expensive method of particle sizing. This method gives outputs such as volume and specific surface area of the particle, considering particle as a sphere.

Many companies manufactured laser diffraction equipment, including Malvern Instruments. Malvern Instrument's Mastersizer 3000 uses laser diffraction techniques to measure particle sizes ranging from 0.01 μm to 3.5 mm. Mastersizer 3000 particle size range also depends on sample and sample preparation.

6 Image Analysis

Image analysis is a process to gather information regarding a characteristic of a particle through computer programing. It is vital to capture an aggregate image with proper lighting and background for digital image analysis. Each particle boundary is obtained by digital image analyzer programs. These programs are designed to sense boundary of a particle by tracing the outline, pixel counting or line scanning. Digital image analysis gives useful data and information on the shape characteristics of particle apart from its size. Digital image analysis techniques such as photogrammetry, X-ray tomography, and laser profiling have been tried by various researchers.

Kennedy et al. [14], Ferniund [15], and Kuo et al. [16, 17], have analyzed earth materials through digital image analysis. Digital image analysis is a quick technique to obtain a characteristic of aggregate and real-time quality control is possible with it [18]. In the recent era, a large number of particle analysis is possible within a fraction of time with digital image analysis. It is also possible to develop an automated method for digital image analysis to reduce time and efforts of a human. Kumara et al. [19] have studied particle size distribution of granular materials by sieve analysis and image analysis. They also suggested the use of black color sheet to minimize the effect of particle shadow for digital image analysis. Rodriguez et al. [20] have carried out a study of image analysis for soil particles. They emphasized on the establishment of a methodology for digital image processing because results are affected by various parameters such as image acquisition procedure, image processing, and the choice of quantity.

Digital image analysis is also useful to investigate shape properties of natural and crushed aggregate. Correlation between shape characteristic of aggregate and concrete compressive strength has been carried out by Rýza Polat et al. [21]. Digital imaging systems are available as commercial products or tools for research laboratories to evaluate aggregate shape characteristics. Malvern Instrument's Morphologi G3 is one of the products which measures particle shape and size characteristics by static image analysis technique.

Kumara et al. [22] have proposed the method for the volume-based gradation curves for sand and sand crushed stone mixtures using two dimensional image analysis. The results of the image analysis show that the crush sand particles have more angularity as compared to natural sand particles. Vohra and Thaker [23] have summarized work related to the use of digital image analysis to optimize the cement mortar mix. Yang et al. [24] have developed a system to detect the size of the manufactured sand particles. This system can be applied for various materials and plays an important role to assure the quality of the aggregate in the field. The particle characterization theory and methods, morphological analysis, and application was discussed by Beddow [25].

7 Summary

Aggregate shape, size, surface texture, specific gravity, and grading affect fresh and hardened concrete properties. Aggregate size and shape measurement is comparatively difficult but has a larger impact on concrete properties. Four methods are available to estimate the size and shape characteristics of particles. The accuracy of each method depends on the particle sizes. Aggregate size varies from 0.15 to 25 mm for normal concrete. Image analysis covers the range 0.04–50 mm. Therefore, it is possible to analyze the aggregate size and shape characteristics using image analysis. Moreover, digital image analysis is an accurate, reliable, and robust method to determine the aggregate characteristics. It also has a potential to estimate size and shape characteristics of aggregate rapidly and accurately with lesser human efforts and cost. In general, a camera and a computer are required for two-dimensional image analysis techniques. Two-dimensional image analyses of aggregate give a relative idea of aggregate properties in a more accurate manner as compared to the procedure suggested in Indian standards.

References

1. Neville AM, Brooks JJ (1987) Concrete technology. Longman Scientific & Technical, England, pp 242–246
2. Mehta PK, Monteiro PJM (2008) Concrete: structures, properties and materials. IBRACON, Sao Paulo
3. Galloway JE (1994) Grading, shape, and surface properties. ASTM Spec Tech Publ 169:401–410
4. Hudson B (1999) Modification to the fine aggregate angularity test. In: Proceedings, seventh annual international center for aggregates research symposium, Austin, TX
5. Kwan AK, Mora CF, Chan HC (1999) Particle shape analysis of coarse aggregate using digital image processing. Cem Concr Res 29(9):1403–1410
6. Popovics S (1979) Concrete-making materials. Hemisphere Pub, Corp
7. IS: 383 (1970) Specification for coarse and fine aggregates from natural sources for concrete
8. McCave IN, Syvitski JP (1991) Principles and methods of geological particle size analysis. Principles, methods and application of particle size analysis, pp 3–21
9. Rauch A, Hass C, Kim H (1999 Apr) State of the art in aggregate classification: review of aggregate gradation technologies. In: Proceedings of 7th annual symposium, international center for aggregates research
10. IS 5282 (1969) Indian standard liquid sedimentation methods for determination of particle size of powders
11. Loizeau JL, Arbouille D, Santiago S, Vernet JP (1994) Evaluation of a wide range laser diffraction grain size analyser for use with sediments. Sedimentology 41(2):353–361
12. Buurman P, Pape T, Muggler CC (1997) Laser grain-size determination in soil genetic studies 1. Practical problems. Soil Sci 162(3):211–218
13. ISO 13320 (2009) Particle size analysis—laser diffraction methods. International Standards Organization
14. Kennedy SK, Mazzullo J (1991) Image analysis method of grain size measurement. In: Principles, methods, and application of particle size analysis, pp. 76–87
15. Fernlund JM (1998) The effect of particle form on sieve analysis: a test by image analysis. Eng Geol 50(1–2):111–124
16. Kuo CY, Frost JD, Lai JS, Wang LB (1996) Three-dimensional image analysis of aggregate particles from orthogonal projections. Transp Res Rec 1526(1):98–103
17. Kuo CY, Rollings RS, Lynch LN (1998) Morphological study of coarse aggregates using image analysis. J Mater Civ Eng 10(3):135–142
18. Maerz NH (2004) Technical and computational aspects of the measurement of aggregate shape by digital image analysis. J Comput Civ Eng 18(1):10–18
19. Kumara GHAJJ, Hayano K, Ogiwara K (2012) Image analysis techniques on evaluation of particle size distribution of gravel. Int. J. Geomate 3(1):290–297
20. Rodriguez J, Johansson J, Edeskär T (2012) Particle shape determination by two-dimensional image analysis in geotechnical engineering. In: Nordic geotechnical meeting: 09/05/2012-12/05/2012 (pp. 207-218). Danish Geotechnical Society
21. Polat R, Yadollahi MM, Sagsoz AE, Arasan S (2013) The correlation between aggregate shape and compressive strength of concrete: digital image processing approach. Int J Struct Civ Eng Res 2(3)

22. Kumara JJ, Hayano K, Kikuchi Y (2017) Evaluation of area-and volume-based gradations of sand-crushed stone mixture by 2d images. KSCE J Civ Eng 21(3):774–781
23. Vohra C, Thaker P (2019) Optimization of cement mortar mix using digital image analysis : state of art. In Emerging research and innovations in civil engineering, pp 198–203
24. Yang J, Yu W, Fang HY, Huang XY, Chen SJ (2018) Detection of size of manufactured sand particles based on digital image processing. PLoS ONE 13(12):e0206135
25. Beddow JK (2018) Particle characterization in technology: Volume II: Morphological analysis. CRC press

Strengthening of Concrete Block Masonry Walls Using Steel Wire Mesh

S. Suraj and S. Unnikrishnan

Abstract In most of developed and developing countries, masonry is still widely used for building construction. Unreinforced masonry walls (URM) are widely used all around India. Seismic forces have a large impact on buildings. Masonry walls are more vulnerable during earthquakes. Hence there is a necessity to find solutions to strengthen URM walls. Application of steel wire mesh and mortar is one of the commonly used techniques for seismic retrofitting of such buildings. URM walls include those made of clay bricks, concrete blocks, hollow bricks etc. The effect of strengthening on in-plane shear behaviour of URM walls, including failure modes, shear strength and ductility are investigated. The panels are subjected to in-plane diagonal tensile test. Test is done according to ASTM E519-02. The present study investigates the effect of welded wire mesh and mortar on shear strength parameters of masonry wall panels. Four set samples of masonry wall panels made of locally available concrete blocks in kerala are selected and strengthened using varying spacing of wire mesh and hexagonal wire mesh. Shear strength parameters are clearly studied using bilinear idealization and adaptability of different wire meshes in strengthening of existing masonry walls is clearly investigated.

Keywords In-plane behaviour · Shear strength · Pseudo ductility

1 Introduction

Unreinforced masonry (URM) structures are the most common and oldest form of building construction technique existing in the world. In most of the developed and developing countries masonry is still being widely used in practice due to its low cost and easy construction technique. URM is unquestionably recognized as the type of construction most vulnerable to earthquakes. Most of the existing URM buildings

S. Suraj (✉) · S. Unnikrishnan
Sree Buddha College of Engineering, Pattoor, India
e-mail: surajsurendran18@gmail.com

S. Unnikrishnan
e-mail: unnisubhash89@gmail.com

K. Dasgupta et al. (eds.), *Proceedings of SECON'19*,
Lecture Notes in Civil Engineering 46,
https://doi.org/10.1007/978-3-030-26365-2_69

seem to be the oldest buildings which tend to be at great risk during earthquake. In most cases masonry structures are constructed without any consideration for seismic loading resulting in huge loss of life as experienced in the past earthquakes (Bhuj 2001, Kashmir 2005, Uttarkashi 1991, Killari 1993). During earthquake, URM buildings experience seismic loading both in-plane and out-of-plane. However, their relative magnitude depends on the type of diaphragm i.e., how the wall is connected with the roof.

Masonry constructions are composed of brittle or quasi-brittle materials, and generally have low resistance to seismic events. The current performance requirements prescribed by the design codes are frequently not accomplished by existing constructions, either because these requirements became more demanding, or because the negative effects of aging in the long-term behaviour of the materials resulted in a substantial decrease of the load carrying capacity of the masonry elements. Therefore, the development of effective procedures to retrofit existing masonry constructions, in order to upgrade their load bearing capacity and increase their ductility response, is still of great importance.

The recent earthquakes have created a necessity to review the capability of existing structures during earthquake, and to find a suitable strengthening technique to strengthen a newly constructed masonry structure or to retrofit an existing old structure. Various rehabilitation and retrofitting techniques are available to enhance the seismic performance of URM buildings. These techniques include application of fibre reinforced polymers (FRP), Ferro-cement overlay, shotcrete overlay, grout injection, application of steel elements, post tensioning, etc. These well-established techniques need to be verified for local materials and building system commonly used in practice. Among all the options available, Ferro-cement overlay is a technique which can be applied easily, rapid in construction and very low in cost, especially in developing countries with no heavy machinery and high-level skilled workers. In this technique, steel welded wire mesh (WWM) is connected or anchored to the surface of masonry through bolts/screws/steel rods subsequently covered with plaster coating. Strengthening of masonry using FRP, steel cord, steel grid, polymer grid etc. has been widely used. In this study an attempt has been made to strengthen the URM using WWM and 1:3 mortar. Ferro cement is a commonly used strengthening system. This is a cementitious composite layer laminated with metallic mesh and has advantages such as a high tensile strength-to weight ratio and superior cracking behaviour. Very few studies are available in strengthening of masonry with Ferro cement, but a considerable number of researches have been carried out in reinforced concrete structures with Ferro cement. The present study aims to evaluate and analyze the strength characteristics of URM solid concrete block walls strengthened with steel wire mesh.

2 Literature Review

Ashraf et al. [1] studied Seismic Behavior of Unreinforced and Confined Brick Masonry Walls before and after Ferro cement Overlay Retrofitting. The effectiveness of the retrofitting method was assessed from the failure pattern, energy dissipation, and force-deformation behavior of the walls that are tested before and after retrofitting. The test results before the application of ferrocement shown that the capacity of confined masonry wall is almost double to that of unreinforced masonry wall. The test results after retrofitting shown that the applied retrofitting scheme significantly increased the lateral load capacity of the unreinforced masonry wall, however it was marginally beneficial in the confined masonry walls.

Kadam et al. [2] studied Strengthening of unreinforced masonry using welded wire mesh and micro-concrete—Behaviour under in-plane action. The study demonstrates that the use of welded wire mesh with micro-concrete is an effective method for strengthening URM walls. An increase up to seven times in shear strength, and up to twenty four times in ductility has been observed for reinforcement ratio of 0.29% in both the directions.

Gattesco et al. [3] carried out Experimental and analytical study to evaluate the effectiveness of an in-plane reinforcement for masonry walls using GFRP meshes. The principal tensile strengths were derived from the experimental results obtained and an analytical formulation was proposed for the reinforced mortar resistance prediction. The formulation shows that the contribution of the reinforced mortar coating was influenced by the characteristics of both the masonry and the reinforcement.

Shermi et al. [4] studied In-plane behaviour of unreinforced masonry panel strengthened with welded wire mesh and mortar. The behaviour of URM specimens in diagonal shear test has been observed as combination of diagonal failure and sliding shear failure modes. The WWM reinforced samples showed significant increase in shear strength and ductility compared to unreinforced masonry specimen.

Shabdin et al. [5] studied Experimental diagonal tension (shear) test of Unreinforced Masonry (URM) walls strengthened with textile reinforced mortar (TRM). One face strengthened walls were sensitive to the applied load and can buckle or experience out-of-plane deformations. Textile reinforced mortar improved diagonal load carrying capacity and deformation capacity of the masonry wall which also caused the strengthened walls to fail in a ductile manner.

3 Objectives

1. To study the effect of size and type of steel wire mesh on failure modes and ultimate capacity of URM walls.
2. To compare the experimental lateral load capacity of URM solid concrete block walls with different types of wire mesh.
3. To study the in-plane shear behavior of URM walls including failure modes, shear strength and ductility.

4 Methodology

1. A detailed literature survey was carried out on Studies on strengthening of unreinforced masonry walls.
2. Physical properties of materials were tested and found that all properties are conforming to IS standards.
3. Fixing the mortar mix ratio and calculation of mortar cube strength.
4. Choosing different types of wire meshes for strengthening of masonry panels.
5. Casting of masonry panels of 660 mm * 660 mm * 230 mm.
6. Application of wire meshes on concrete block masonry panels.
7. In plane diagonal loading of masonry panels according to ASTM E-519.
8. Conclusions and recommendations are finally made based on the findings and observations.

5 Materials

5.1 Cement

Ordinary Portland cement of 53 grades conforming to IS 12269-2013 was used in the study.

5.2 Fine Aggregate (FA)

Fine aggregate are soil particles passing through 4.75 mm IS sieve. Generally river sand, crushed stone, crushed gravel, M sand etc. are used as fine aggregate. In this study, M sand conforming to Zone II is used.

5.3 Concrete Blocks (CB)

Locally available concrete blocks of size 300 * 200 * 150 was chosen for the study. As per IS 2185-1 (2005) Table 2, the solid concrete block unit conforms to Grade C (5.0) with block density not less than 1800 kg/m^3.

5.4 Steel Wire Mesh

Reinforcing meshes for use in Ferro cement shall be evaluated for their susceptibility to take and hold shape as well as for their strength performance in the composite system. Generally, it consists of thin wires, either woven or welded into a mesh, but main requirement is that it must be easily handled and if necessary, flexible enough to be bent around sharp corners. The wire meshes are usually 0.5–1.0 mm in diameter and spaced at 5–25 mm apart and the volume of the mesh ranges from 1 to 8% of the total volume of the structural element. Types of wire used in Ferro cement include:

- Hexagonal wire mesh
- Square mesh of 12 mm spacing
- Square mesh of 25 mm spacing.

5.5 Mortar Mix

Mortar mix was prepared using cement, fine aggregate and water. Different ratios of mortar mix such as 1:3, 1:4 and 1:6 are prepared and tested. Mortar mix for plastering was prepared according to IS 2250 (1981). Mortar cubes of 10 cm * 10 cm * 10 cm are prepared and tested for compressive strength. Mortar mix of 1:4 was chosen for plastering in the present study.

6 Casting of Concrete Block Masonry Panels

To cast concrete block masonry panels, the surface of solid blocks are slightly wet to prevent the absorption of water from mortar mix. A smooth surface is prepared and mortar mix is applied in a layer of 10 mm thick. Solid concrete blocks are placed over the mortar layer and slightly pressed. Broken blocks are placed in alternate layers to eliminate vertical joints. Different layers are placed and joints are filled with mortar. Excess mortar is swiped off from the surface and wall is finished ensuring that the blocks are placed level. Various stages of laying of concrete block panels are shown in Figs. 1 and 2.

7 Application of Steel Wire Mesh

Steel wire mesh of varying spacing and hexagonal steel wire mesh are applied on clay brick and concrete block panels with the use of steel pins and steel washer. Another coat of plaster is applied over steel wire mesh.

Fig. 1 Casting of concrete block panels

Fig. 2 concrete block panels

1. Application of 12 mm spacing wire mesh (CB-I) is shown in Fig. 3.
2. Application of 25 mm spacing wire mesh (CB-II) is shown in Fig. 4.
3. Application of Hexagonal wire mesh (CB-III) is shown in Fig. 5.

Fig. 3 Application of 12 mm wire mesh

Fig. 4 Application of 25 mm wire mesh

8 Testing of Concrete Block Panels

8.1 In Plane Diagonal Tension in Masonry Assemblages

In-plane diagonal test for masonry panels are done according to ASTM E519-02. Masonry panel is scaled down and is loaded in compression along one diagonal in universal testing machine thus causing a diagonal tension failure with the specimen splitting apart parallel to the direction of load. This test method was developed to measure more accurately the diagonal tensile or shear strength of masonry than other available methods. The specimen size was selected as being the smallest that would

Fig. 5 Application of hexagonal wire mesh

be reasonably representative of a full-size masonry assemblage and that would permit the use of testing machines such as are used by many laboratories. Test setup is shown in Fig. 6. Loading shoes made of steel were used for in-plane loading as per ASTM E519-02. Loading shoes are shown in Fig. 7.

Fig. 6 Loading of concrete block panel

Fig. 7 Loading shoe

9 Bilinear Idealization

Behaviour of masonry is highly nonlinear, in order to find out the main aspects of the inelastic behaviour of masonry panels, the actual behaviour is idealized with the help of bilinear curve [6]. Bilinear idealization is recommended by all codal provisions in order to assess the existing masonry structures by non-linear staticprocedure. In the present study the ultimate strength and ductility is evaluated for the masonry wall panels applied with different types of steel wire mesh. The bilinear idealization is obtained by ensuring that the areas below actual and bilinear idealized curve were equal and in agreement with [6]. For all reinforced panels an equivalent bilinear curve was defined with reference to Fig. 8 [5].

10 Results and Discussions

Experimental investigation was conducted on 4 set of specimens. First set consists of unreinforced concrete block panel (CB-N), second set was concrete block panel strengthened with 12 mm spacing wire mesh (CB-I), third set was concrete block panel strengthened with 25 mm spacing wire mesh, fourth set was concrete block panel strengthened with hexagonal wire mesh. Calculations of shear stress and shear strain is done according to the formula

$$\text{Shear stress S} = \frac{0.707P}{A}$$

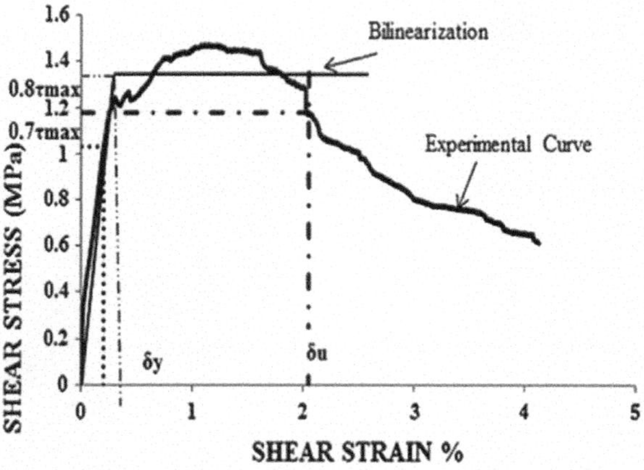

Fig. 8 Bilinear curve

where

S = shear stress on net area, MPa
P = applied load, N
A = net area of the specimen, mm^2, calculated as follows:

$$A = \frac{(W + h)t}{2}$$

where

w = width of specimen, mm
h = height of specimen, mm
t = total thickness of specimen, mm
n = percent of the gross area of the unit that is solid, expressed as a decimal

i. Shear Strain, $\gamma = \frac{\Delta V + \Delta H}{g}$

where

γ = shearing strain, or mm/mm
ΔV = vertical shortening, mm
ΔH = horizontal extension, mm
g = vertical gage length, mm

ii. Modulus of Rigidity $G = \frac{S}{\gamma}$

where

G = modulus of rigidity, MPa.

Figure 9 shows the failure mode of masonry panel. Tests conducted shows that the application of wire meshes holds the concrete block layers and thereby increases their resistance to failure.

It was observed that the masonry panels strengthened with wire mesh behaved as a single composite material having improved ductile behaviour and shows resistance to cracking along joints of masonry panel. Figures 10, 11, 12 and 13 shows the shear stress-shear strain curve obtained based on experimental test data. It can be seen that the reinforced masonry samples with steel wire mesh shows a linear elastic behaviour initially and then reduced as failure initiated.

Table 1 shows the calculations of ductility of specimens strengthened with steel wire meshes. Calculations are done according to ASTM E519-02.

Fig. 9 Failure of concrete block specimen

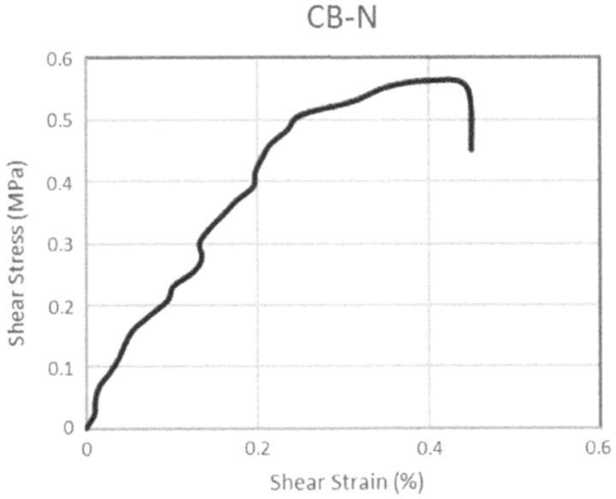

Fig. 10 Unreinforced concrete block specimen

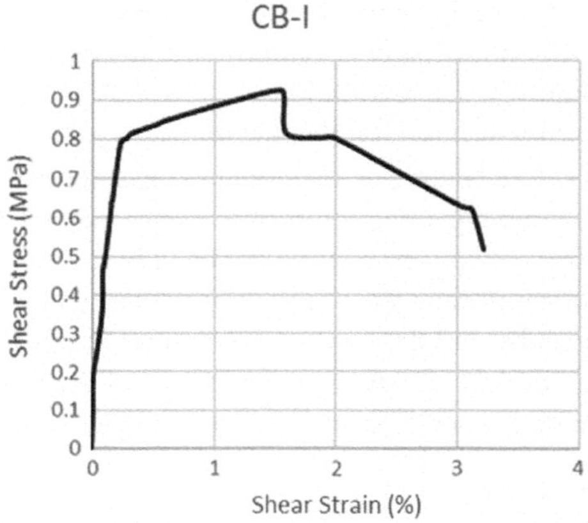

Fig. 11 Specimen with 12 mm spacing wire mesh

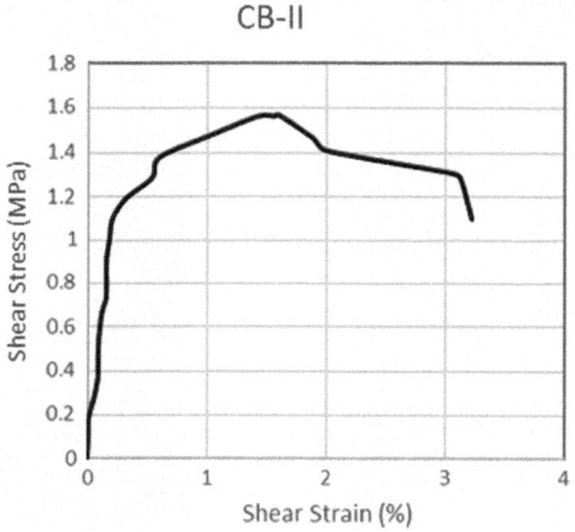

Fig. 12 Specimen with 25 mm spacing wire mesh

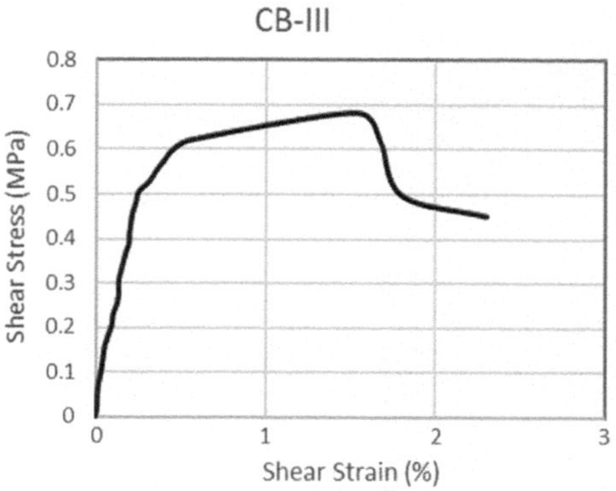

Fig. 13 Specimen with hexagonal wire mesh

Table 1 Experimental results of in-plane diagonal test

Panel	P_{max} (kN)	τ_{max} N/mm^2	τ/τ_o	d_y (%)	d_u (%)	μ
CB-N	122	0.56	–	–	–	–
CB-I	201	0.92	1.64	0.59	2.38	4.03
CB-II	342	1.57	2.81	0.57	3.06	5.37
CB-III	149	0.68	1.21	0.43	1.69	3.93

P_{max} maximum load applied; τ_{max} maximum shear stress; d_y yield drift; d_u ultimate drift corresponding to $0.8\tau_{max}$; μ ductility of specimen

11 Conclusions

This experimental study was conducted to find out a solution to strengthen existing load bearing masonry structures made of locally available concrete blocks in the locality. Use of steel wire mesh was chosen for strengthening as it is a cost effective technique and are easily available in the locality and can be applied without any practical difficulty on the face of walls. Three different locally available wire meshes of 12-gauge thickness were used. They include 12 mm spacing mesh, 25 mm spacing mesh and hexagonal wire mesh. Based on observations following conclusions were made:

- Failure modes were clearly observed, for CB-N specimen failure has been a combination of diagonal failure and sliding shear failure manner. For strengthened wire mesh failure occurred along diagonal and shows resistance against other failure modes as the wire mesh along with 1:4 mortar holds the concrete blocks thereby showing resistance from failure.
- The proposed method of strengthening of concrete block masonry panels confines the masonry well and it behaves like a composite material with improved properties.
- The shear strength of specimen improved up to 3 times by the application of steel wire mesh.
- Ductility ratio of URM specimen after strengthening increased from 3 to 5 times.
- The technique is more effective under the application of 25 mm spacing wire mesh and has shown more diagonal load carrying capacity almost 3 times that of URM specimen.
- The steel wire mesh reinforced samples showed significant increase in shear strength and ductility compared to unreinforced masonry specimen.

References

1. Ashraf M et al (2012) Seismic behavior of unreinforced and confined brick masonry walls before and after ferrocement overlay retrofitting. Int J Archit Heritage 6:665–688
2. Kadam SB et al (2014) Strengthening of unreinforced masonry using welded wire mesh and micro-concrete—behaviour under in-plane action. Constr Build Mater 54:247–257
3. Gattesco N et al (2015) Experimental and analytical study to evaluate the effectiveness of an in-plane reinforcement for masonry walls using GFRP meshes. Constr Build Mater 88:94–104
4. Shermi C et al (2018) In-plane behaviour of unreinforced masonry panel strengthened with welded wire mesh and mortar. Constr Build Mater 178:195–203
5. Shabdin M et al (2018) Experimental diagonal tension (shear) test of Un-Reinforced Masonry(URM) walls strengthened with textile reinforced mortar (TRM). Constr Build Mater 164:704–715
6. Marcari G et al (2007) In plane shear performance of masonry panels strengthened with FRP. Compos B Eng 38(7):887–901

Shear Strengthening of RC Deep Beam Using Steel Plates

K. Chandrakanth and P. E. Kavitha

Abstract Deep beams are structural elements that can carry a significant amount of load to the supports by compression force as a combination of load and reactions. They are loaded as simple beams. Deep beams have large depth to thickness. Because of the geometry of deep beams, their behaviour is different with slender beam or intermediate beam. Shear failure is the most common mode of failure in deep beams. The shear failure mode occurs in the absence of flexural cracks. The shear behaviour of deep beams are very complex and there is still no agreement on the role of size effect in shear due to lack of information. Sudden failure due to low shear strength is not desirable mode of failure. The reinforced concrete beams are primarily designed for flexural strength and shear strength. Among various methods of shear enhancement, replacement of stirrups using steel plates is a new technique. Providing holes in these steel plates brings better bonding of concrete between the steel plates and it is expected that the shear strengthening of deep beams can be achieved and hence the load carrying capacity can be increased. This paper aims at finding the best suited thickness for steel plate having suitable configuration of holes so that maximum load carrying capacity can be attained.

Keyword RCC

1 Introduction

Reinforced concrete, or RCC, is concrete that contains embedded steel bars, plates or fibres that strengthen the material. The capability to carry loads by these materials is magnified, and because of this RCC is used extensively in all construction. In fact, it has become the most commonly used construction material. Reinforced materials

K. Chandrakanth (✉) · P. E. Kavitha
Department of Civil Engineering, Federal Institute of Science and Technology, Angamaly, India
e-mail: chndrknthk@gmail.com; chandrakanth.k.8245@gmail.com

P. E. Kavitha
e-mail: pe_kavitha@yahoo.com

K. Dasgupta et al. (eds.), *Proceedings of SECON'19*,
Lecture Notes in Civil Engineering 46,
https://doi.org/10.1007/978-3-030-26365-2_70

are embedded in concrete in such a way that the two materials resist the applied forces together. These stresses are resisted by the combined action of compressive strength of concrete and tensile strength of steel. Plain concrete on the other hand, cannot withstand the stresses created by vibrations, wind, or other forces and hence it is not suitable for most construction projects [1].

Slabs, beams, columns and foundations are made up reinforced cement concrete. Beams can be defined as a structural member which carries all loads and resists it from bending. There are various types of materials used for beam such as steel, wood, aluminium, etc. But reinforced cement concrete is more commonly used.

2 Literature Review

Transverse external posttensioning is a shear strengthening method, achieved by vertically applying post tension along the shear span. In this method, two steel plates are placed at the top and bottom of the beam and the pretension is applied vertically through these plates along the beam. The member attains higher ductility and stiffness resulting in ductile flexural failure. The ultimate shear capacity of the beam is improved due to aggregate interlocking and it also prevents splitting cracks [2].

In Steel plated concrete composite method of shear strengthening, holes are bored at the bottom or the sides of the beam and rebars are embedded into the holes using modified epoxy resin adhesives. Then steel plates with studs are placed at bottom or sides of the beam and concrete is poured between the plate and the beam. The new added layer enhances the shear carrying capacity of the beam [3].

Ferrocement composite can also improve the shear strength of beams. Welded and expanded wire meshes are lighter, easier to handle, easier to cut and easier to bend than steel reinforcements. Increasing the number of layers of expanded and welded wire meshes leads to increase in load carrying capacity, shear stresses, stiffness and toughness of the beam. But cracks with greater number and narrower widths were observed in beams reinforced with steel meshes [4].

Pre-tensioned stainless steel ribbons is another method for improving the shear capacity of beams. Shear capacity can be improved by using prestressed stainless steel ribbons and steel angles. This method is easy to install and usually reversible. Here the beam is pre-stressed using steel ribbons, which run across the beam through full depth or mid-depth and is held in place using steel angles [5].

3 Deep Beams

Deep beams are structural elements loaded as simple beams in which a significant amount of load is carried to the supports by compression force combining the load and the reaction. As a result, the strain distribution is no longer considered linear, and the shear deformations become significant when compared to pure flexure.

Beams with large depths with respect to spans are called deep beams. As per IS-456 (2000) Clause 29, a simply supported beam can be classified as deep beam when effective span L to overall depth D ratio is less than 2. Continuous beams can be considered as deep beam if the effective span to overall depth ratio is less than 2.5. Some of the examples of deep beam are floor slabs under horizontal load, short span beams carrying heavy loads, and transfer girders.

3.1 Failure Modes in Deep Beams

The parameters which are more important in failure modes are the amount of top and bottom reinforcement, the concrete strength, the shear span to depth ratio, the amount of web reinforcement, the overall geometry and the position of loading. Thus, the failure modes in deep beams are described in the following [6, 7].

3.1.1 Flexure Plus Shear Failure Mode

This type of failure occurs in the beam with minimum amount of top reinforcement about 0.002 bd, where b and d are beam thickness and effective depth, respectively. The failure occurs when the major diagonal and flexural cracks are formed together with crashing concrete at the ends of major diagonal cracks [6].

3.1.2 Bearing Failure Mode

Reinforcement concrete deep beams with fixed ended supports vulnerable to bearing failure under the load batch, when the area of loading is not enough, and the stress will exceed the capacity of concrete. This failure is basically a local failure and it is preventable, with suitable detailing around the loading area, so the load can be distributed into the body of beam without overstressing locally [6].

3.1.3 Shear Failure Mode

The shear failure mode occurs in the absence of flexural cracks at the fixed ended supports. In this mechanism, after the formation of initial flexural cracks at the mid span of beam, the diagonal cracks appeared and caused the excessive spalling and crashing of the concrete at the edges of the loading patch and the lower corner of beam [7].

3.2 Shear Strength of Deep Beams

The shear behaviour of deep beams is very complex and there is still no agreement on the role of size effect in shear due to lack of information. Deep beams are classified as non-flexural members, in which plane sections do not remain plane in bending. Therefore, the principles of stress analysis developed for slender beams are neither applicable nor adequate to determine the strength of deep beams. An important characteristic of RC deep beams is their high shear strength. The greater shear strength of deep beams is due to internal arch action, which transfers the load directly to support through concrete struts. Since the reinforcement acts as a tie, RC beams are analogous to steel trusses. Deep beams are also classified as distributed regions, which are characterised by non-linear strain distribution. Elastic solutions of deep beams provide good descriptions of their behaviour before cracking. Yet, after cracking major redistribution of strains and stresses takes place and the beam strength must be predicted by non-linear analysis. For a simple deep beam with concentrated load on top, the top load and bottom reactions create large compressive stresses at a right angle to the beam axis. These stresses interact with the shear stresses to form complicated stress field in the web. Because of short horizontal distance between the top and bottom load points i.e. small a/d ratios, the effect of such stresses result in arch action which is unique in deep beams. Because of theses complexities, study of deep beams has become a special interest.

4 Numerical Analysis of RC Deep Beams Using Steel Plates

In order to investigate the different options to improve the shear strength of deep beams, concrete beams with steel plates as shear reinforcement is adopted in the present study and is compared with the conventional steel reinforced beams. Non-linear finite element analysis is performed in ANSYS 16.0 for the present study (Table 1).

4.1 Validation of Numerical Model

A deep beam of overall depth 0.5 m, 0.2 m wide and length being 0.9 m was considered for validation. Effective span of the deep beam is 0.6 m. Grade of concrete was taken as M20 and the grade of steel was taken as Fe415 with a minimum cover of 25 mm.

The reinforcement detailing is as shown in Fig. 1. 5 numbers of 6 mm diameter bars at 170 mm centre to centre spacing were used as stirrups. 6 numbers of 6 mm diameter bars at 120 mm centre to centre spacing were provided as longitudinal reinforcement. 6 numbers of 16 mm diameter longitudinal bars were provided at equal spacing from a distance of 0.095 m from soffit (Figs. 2, 3 and 4).

Table 1 Model configurations

S. No.	Figure	Remarks
1.	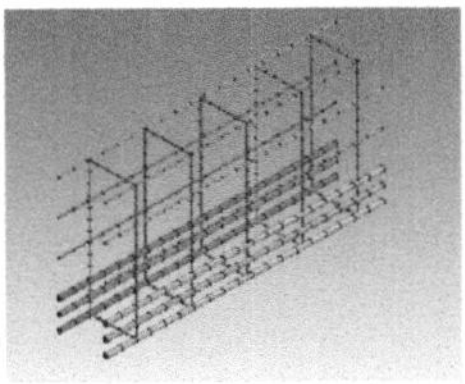	Conventional beam with stirrups as shear reinforcement
2.	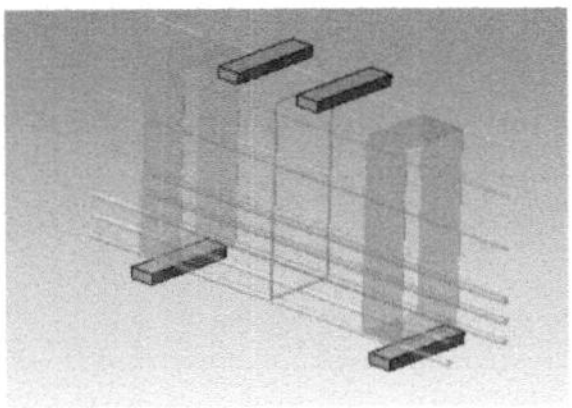	Replacement of stirrups with mild steel plates at both ends
3.	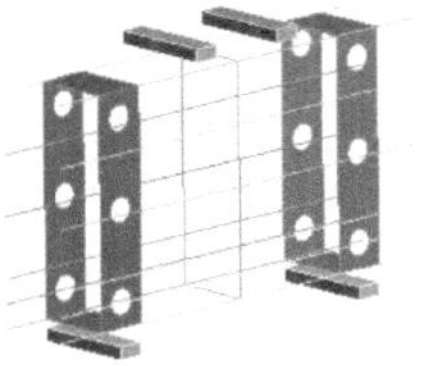	Steel plates having six circular holes at sides
3.	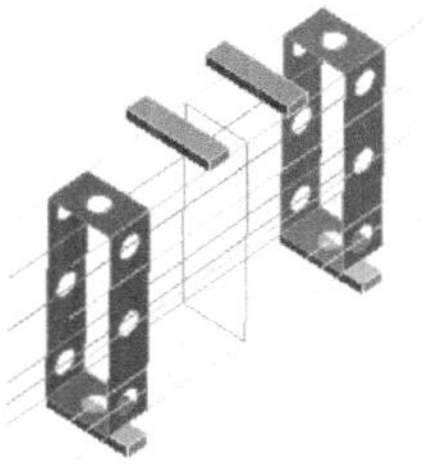	Steel plates having eight circular holes at top, bottom and sides combined
4.	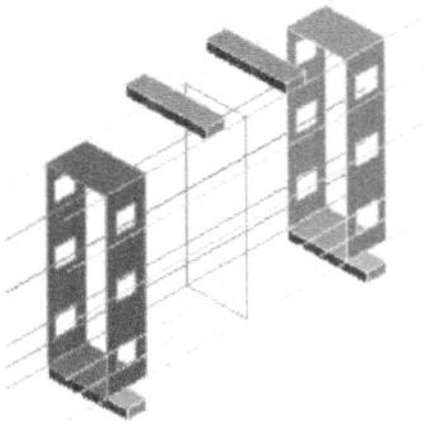	Steel plates having six rectangular holes at sides

(continued)

Table 1 (continued)

S. No.	Figure	Remarks
5.	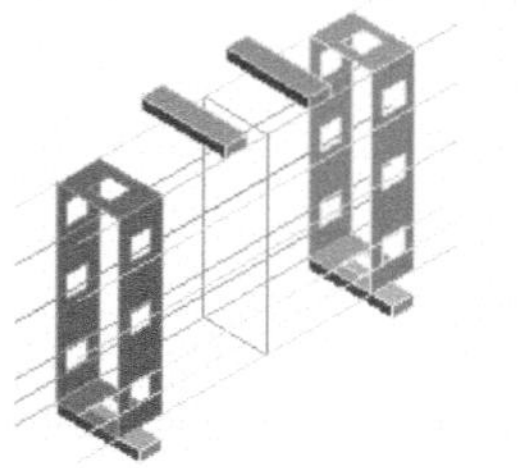	Steel plates having eight rectangular holes at top, bottom and sides combined
6.	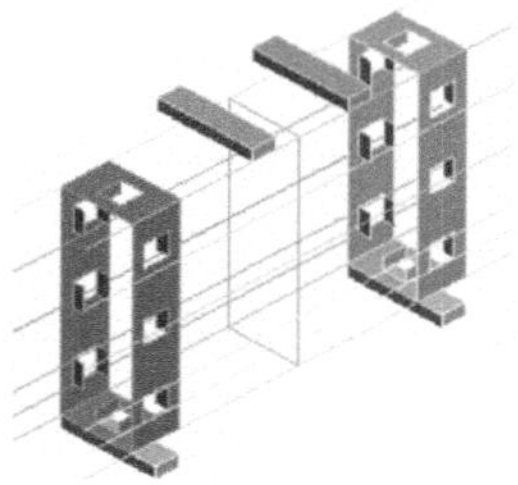	Steel plates having eight rectangular holes with windows

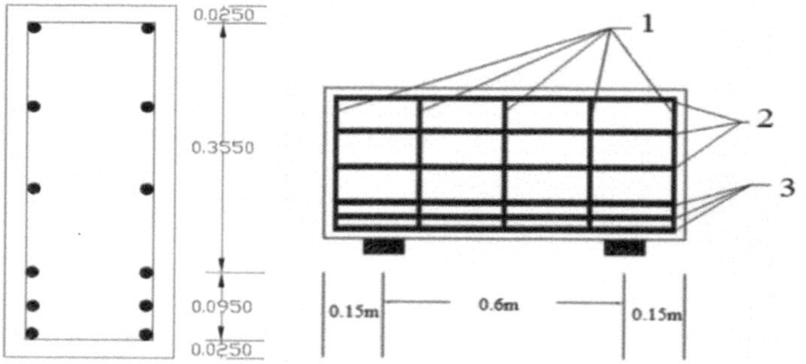

Fig. 1 Reinforcement details

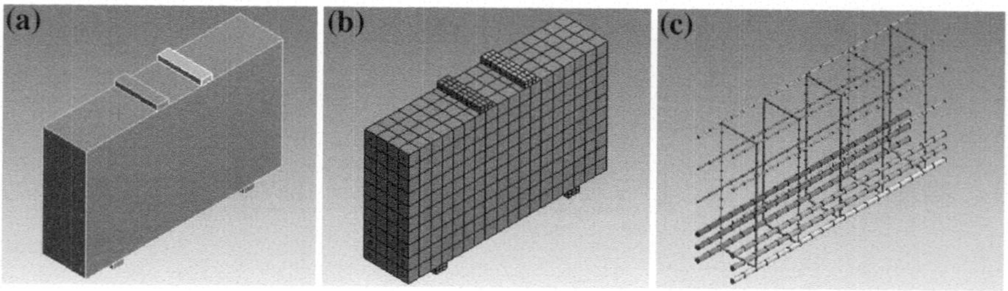

Fig. 2 Geometry of model

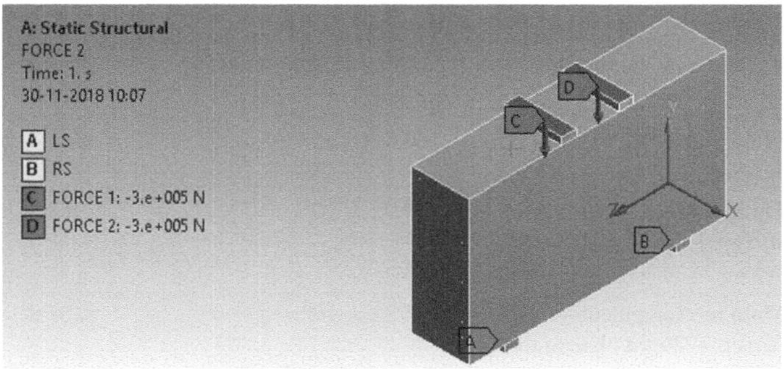

Fig. 3 Boundary conditions

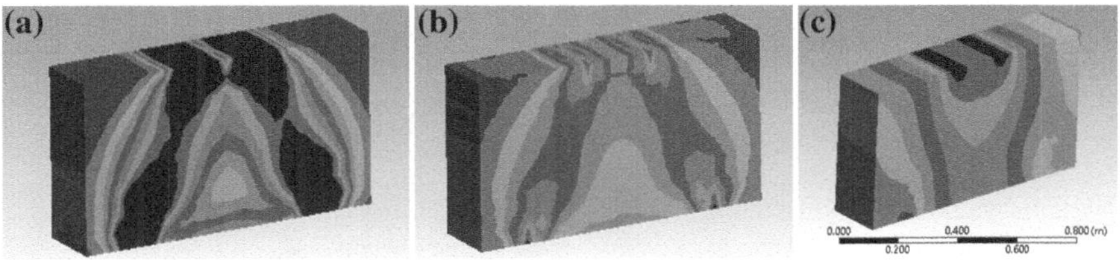

Fig. 4 Equivalent total **a** strain, **b** stress, **c** deformation

4.2 Replacement of Stirrups Using Steel Plates

The validated deep beam is having five stirrups for shear reinforcement. As a trial, two stirrups at each ends of the beam is replaced using steel plates. Steel plates of 1 and 2 mm are chosen. While analysing using ANSYS, it was found that the ultimate load carrying capacity of the steel plates are much lower than that of the original beam having stirrups. This may be due to the lack of bonding or contact of the concrete inside the steel plate and concrete outside the steel plate. Also, the steel plate having 1 mm thickness is having very low load carrying capacity due to its low stiffness. The values are tabulated in Table 2 (Fig. 5).

4.3 Replacement of Stirrups with Steel Plates Having 6 Circular Holes (50 mm Diameter)

Since there is no enough contact between the concrete in the inner face and outer face of the steel plates, holes are provided in the steel plates. Six holes having 50 mm diameter is provided in the steel plates at the side faces of 1 mm and 2 mm thickness respectively. It showed effectiveness in ultimate load carrying capacity. The ultimate load has increased to a small margin from the original beam having stirrups. It was

Table 2 Comparison of deep beam with stirrups and steel plates

Sl. No.	Reinforcement details	Volume of steel for shear (mm^3)	Ultimate load (kN)
1.	Main reinforcement + 5 stirrups	169.6×10^3	1380.1
2.	Main reinforcement + Mid stirrup + 1 mm plate at the sides	153.92×10^3	788.8
3.	Main reinforcement + Mid stirrup + 2 mm plate at the sides	273.92×10^3	1207.8

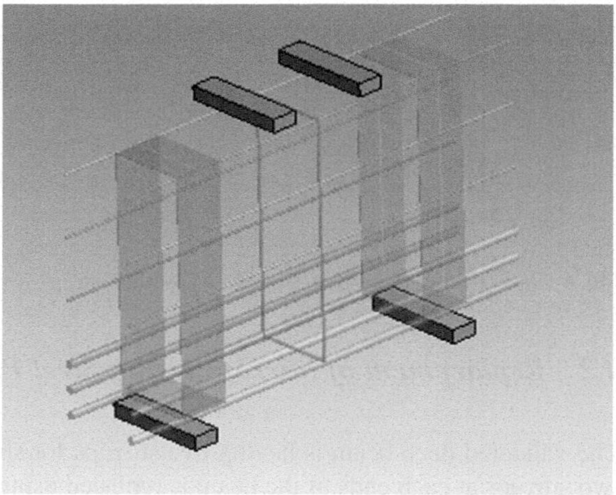

Fig. 5 Steel plates as shear reinforcement

also found that, the thickness of the stirrups is not having any significant effect at all. The 1 mm thickened steel plate and steel plate having 2 mm thickness is having almost same amount of load carrying capacity (Figs. 6, 7; Table 3).

4.4 Replacement of Stirrups with Steel Plates Having 8 Circular Holes

Since providing holes increased the load carrying capacity of the load carrying capacity of the beam, the steel plates with 8 holes are attempted. A 50 mm diameter hole is provided in the top and bottom face of the steel plate. It decreases the volume of steel also. It was found that the load carrying capacity is significantly improved when compared with the original deep beam with stirrups. The values are tabulated in Table 4.

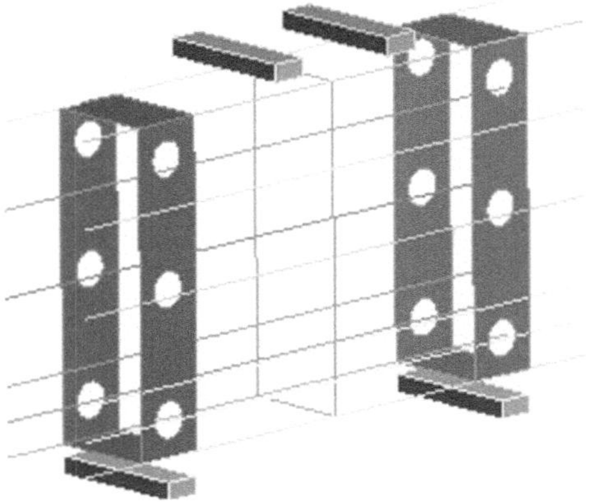

Fig. 6 Steel plates having circular holes

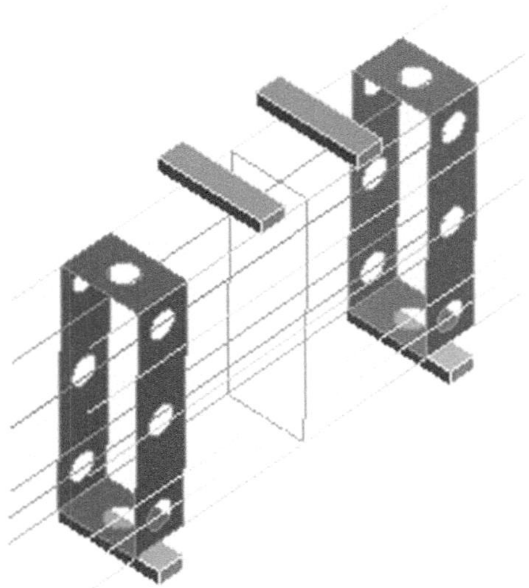

Fig. 7 Steel plates having circular holes at top, bottom and sides

Table 3 Comparison of deep beam with stirrups and steel plates with 50 mm diameter holes

Sl. No.	Reinforcement details	Volume of steel for shear (mm^3)	Load (kN)
1.	Main reinforcement + 5 stirrups	169.6×10^3	1380.1
2.	Main reinforcement + Mid stirrup + 1 mm plate with 6 holes at sides	130.38×10^3	1441.7
3.	Main reinforcement + Mid stirrup + 2 mm plate with 6 holes at sides	226.84×10^3	1445.8

Table 4 Comparison of deep beam with stirrups and steel plates with 50 mm diameter holes

Sl. No.	Reinforcement details	Volume of steel for shear (mm^3)	Load (kN)
1.	Main reinforcement + 5 stirrups	169.6×10^3	1380.1
2.	Main reinforcement + mid stirrup + 1 mm plate with 8 holes	122.56×10^3	1444.2
3.	Main reinforcement + mid stirrup + 2 mm plate with 8 holes	211.2×10^3	1455.9

Table 5 Comparison of deep beam with stirrups and steel plates with 60 and 40 mm diameter holes

Sl. No.	Reinforcement details	Volume of Steel for Shear (mm^3)	Load (kN)
1.	Main reinforcement + 5 stirrups	169.6×10^3	1380.1
2.	Main reinforcement + Mid stirrup + 1 mm plate with 8 holes	133.92×10^3 (40 mm) 131.30×10^3 (60 mm)	1354.3 1365.7
3.	Main reinforcement + Mid stirrup + 2 mm plate with 8 holes	233.71×10^3 (40 mm) 228.68×10^3 (60 mm)	1381.2 1379.2

4.5 Replacement of Stirrups with Steel Plates Having Circular Holes (40 and 60 mm Diameter)

By providing 50 mm diameter holes in steel plates, we got satisfying results of load carrying capacity. So, next we have to find the best hole diameter configuration for these steel plates so that maximum load carrying capacity is attained. 40 and 60 mm diameter holes were analysed. But the load carrying capacity is much lower than that of the steel plates with 50 mm diameter holes. In fact, the results are lower than that of the beam with stirrups. This may be due to lack of bonding of concrete between the steel plates in case of 40 mm diameter holes as the hole diameter is much lower, and due to insufficient amount of steel content in case of steel plate with 60 mm diameter holes (Table 5).

4.6 Replacement of Stirrups with Steel Plates Having Rectangular Openings

After providing circular holes, rectangular holes having size of 50 × 50 mm size was also analysed to check whether there is an increase in the ultimate load carrying capacity. Six holes (at sides only) and 8 holes (at sides including top and bottom) were provided in steel plates of sizes 1 and 2 mm. it was observed that the load carrying capacity is not up to the level of circular holes in any of the cases, but also slightly less than that of original beam.

Since rectangular hole does not make any impact in the load carrying capacity as compared to that of the circular holes, rectangular openings of 6 holes having windows were analysed. Windows were provided in the steel plates of thicknesses 1 and 2 mm. but windows also did not make any positive impact on the ultimate load carrying capacity in the deep beam when compared to circular holes (Fig. 8; Table 6).

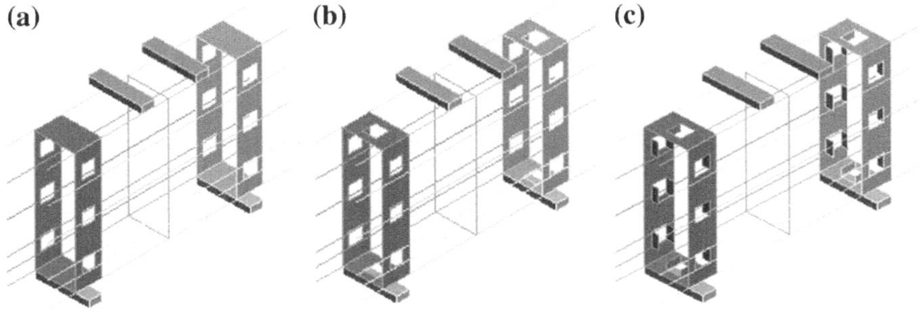

Fig. 8 Steel plates with rectangular openings and windows

Table 6 Comparison of deep beam with stirrups and steel plates

Sl. No.	Reinforcement details	Volume of steel for shear (mm^3)	Load (kN)
1.	Main reinforcement + 5 stirrups	169.6×10^3	1380.1
2.	Main reinforcement + mid stirrup + 1 mm plate with 6 rectangular openings	138.92×10^3	1255.5
3.	Main reinforcement + mid stirrup + 2 mm plate with 6 rectangular openings	243.92×10^3	1269.7
4.	Main reinforcement + mid stirrup + 1 mm plate with 8 rectangular openings	133.92×10^3	1296.6
5.	Main reinforcement + mid stirrup + 2 mm plate with 8 rectangular openings	233.92×10^3	1323.7
6.	Main reinforcement + mid stirrup + 1 mm plate with 8 rectangular window openings	153.92×10^3	1304.6
7.	Main reinforcement + mid stirrup + 2 mm plate with 8 rectangular window openings	273.92×10^3	1337.8

5 Conclusions

Deep beams are subjected to shear failure and hence they are to be strengthened in shear by providing better shear reinforcements. The present study focuses on providing steel plates as a replacement to conventional stirrup reinforcements. When steel plates were provided as shear reinforcements, it was observed that the load carrying capacity of the beam specimen under two point loading, reduced compared to that of beams using stirrups reinforcements. That may be due to the lack of contact between the concrete volume inside and outside the steel plates provided. Hence a modified model with steel plates having holes was adopted and tested. The model showed a better load carrying capacity, compared to conventional stirrupped beams. The study was also extended to optimize the shape and number of holes in the steel plates.

Bonding between steel plate and concrete was improved when 6 numbers of 50 mm diameter holes were provided in 1 mm thick steel plate. Bonding increased further when 8 number of 50 mm diameter holes were provided. Load carrying capacity of specimens having steel plates with 8 numbers of holes was obtained as 1444.2 kN and was higher than that of specimens having steel plates with 6 numbers of holes which was obtained as 1441.7 kN. Steel plate with 8 numbers of holes possess less volume of steel compared to steel plate having 6 numbers of holes. Specimens with rectangular holes were also tested for load carrying and is obtained as 1323.3 kN. Another set of specimen were analysed in which rectangular openings were provided in the form of windows. The results showed higher load carrying capacity of 1337.8 kN than those with rectangular holes. From this study, it can be concluded that circular openings were more efficient than rectangular and windowed rectangular openings. Moreover, from the test results it was found that eight circular holes of 50 mm diameter is the optimum steel plate configuration for the specimen tested.

References

1. Patil SS, Swami PS, Kore PN, Behaviour of concrete deep beams with high strength reinforcement. Int J Curr Eng Technol. E-ISSN 2277–4106, P-ISSN 2347–5161
2. Sirimontree S, Witchayangkoon B, Khaosri N, Teerawong J (2011) Shear strength of reinforced concrete beam strengthened by transverse external post-tension. Am J Eng Appl Sci 4(1):108–115. ISSN 1941-7020
3. Wang J-J, Tao M-X, Zhou M, Niea X (2018) Force transfer mechanism in RC beams strengthened in shear by means of steel plated concrete. Eng Struct 171:56–71
4. Hussein G, Sayed SH, Nasr NE, Mostafa AM (2018) Effect of loading and supporting area on shear strength and size effect of concrete deep beams. Ain Shams Eng J
5. Colajanni P, Recupero A, Spinella N (2017) Increasing the shear capacity of reinforced concrete beams using pretensioned stainless steel ribbons. Struct Concr, 1–10. International Federation for Structural Concrete
6. Aarabzadeh A, Hizaji R (2017) Failure mechanism in fixed-ended reinforced concrete deep beams under cyclic load. World Acad Sci Eng Technol Int J Civ Environ Eng 11(4)
7. Leon Raj J, Appa Rao G, Failure modes of RC deep beam panels. Innov Mater Struct Eng Pract

Experimental Investigation on Lightweight Concrete Using EPS Beads and Metakaolin

Viswanath Gopika and S. P. Akshara

Abstract Expanded polystyrene (EPS) is a lightweight material commonly used in engineering applications since 1950s. Lightweight concrete can be produced by replacing the normal aggregate with lightweight aggregate, either partially or fully, depending upon the requirements of density and strength. Expanded polystyrene waste replaces coarse aggregate to produce light weight non-structural concrete with unit weight varying from 950 to 2000 kg/m^3. Metakaolin is a valuable supplementary cementitious material for concrete and cement applications. Usually 8–20% (by weight) of Ordinary Portland Cement is replaced by Metakaolin. In this study, various mixtures were produced by partially replacing varying percentages of coarse aggregates volume with EPS beads and substituting cement with varying percentage of Metakaolin and their properties were estimated through laboratory experimental study. Such a concrete exhibits appreciable engineering properties. The engineering properties like compressive strength, split tensile strength, density and water absorption of the expanded polystyrene-based lightweight concrete is compared with M30 grade conventional concrete. The obtained results were compared with standard ranges of lightweight concrete as specified in ACI Committee 213.

Keywords Lightweight concrete · EPS beads · Supplementary cementitious · Metakaolin

1 Introduction

There is heavy need for building materials in the construction industry, which is becoming scarce day by day. In concrete construction, dead load represents a very large portion of the load on the structure, and there are considerable advantages

V. Gopika (✉) · S. P. Akshara
Department of Civil Engineering, Federal Institute of Science and Technology, Angamaly, Kochi 683 577, India
e-mail: gopikaviswanath1995@gmail.com

S. P. Akshara
e-mail: akshara027@gmail.com

K. Dasgupta et al. (eds.), *Proceedings of SECON'19*,
Lecture Notes in Civil Engineering 46,
https://doi.org/10.1007/978-3-030-26365-2_71

in reducing the density of concrete. Lightweight concrete is generally accepted as concrete having a density of about 2000 kg/m^3 or less. Using lightweight concrete, improves construction and handling techniques; transportation and on-site handling becomes more economical. Polystyrene beads are a widely used aggregate which can be easily incorporated into concrete to produce light-weight concrete with a wide range of densities. Expanded polystyrene (EPS) is a lightweight cellular plastics material consisting of spherical shape particles which comprised of about 98% air and 2% polystyrene [1]. EPS beads are shown in Fig. 1. It has a good sound and thermal insulation characteristics as well as impact resistance. Using waste polystyrene in concrete production not only solves the issue of disposing this ultra-light solid waste but also helps preserve natural resources. There are many advantages to be gained from the use of lightweight concrete. These include lighter loads during construction, reduced self-weight in structures, and increased thermal resistance [2, 3]. EPS lightweight concrete can be also used in varied structural elements such as cladding panels, curtain walls, composite flooring systems, load-bearing concrete blocks, the sub base material for a pavement, floating marine structures, etc [4, 5].

Cement concrete is the most extensively used construction material. Cement is responsible for about 3% of the greenhouse gas emission and for 5% of the CO_2 emission. Thus it is very important to find a replacement material for cement. Usually 8–20% (by weight) of Portland cement replaced by metakaolin [6–9]. The Metakaolin is shown in Fig. 2. Such a concrete exhibits favourable engineering properties.

Fig. 1 EPS beads

Fig. 2 Metakaolin

1.1 Objectives

i. To find the optimum percentage of metakaolin in expanded polystyrene lightweight concrete.
ii. To test the properties of metakaolin based expanded polystyrene lightweight concrete.
iii. To compare these properties with normal concrete.

2 Methodology and Materials

2.1 Materials Used

The various materials used in the experimental program are:

2.1.1 Cement

Cement is a binder material in concrete which sets and solidifies when it dries furthermore effortlessly ties with alternate materials. The chemical composition of cement is shown in Table 1. Ordinary Portland Cement (OPC) of 53 grade confirming to IS 12269-1987 was used in the experiment. The properties of cement are shown in Table 2.

Table 1 Chemical composition of cement and metakaolin

Chemical composition	Cement %	Metakaolin %
SiO_2	34	54.3
Al_2O_3	5.5	38.3
Fe_2O_3	4.4	4.28
CaO	63	0.39
MgO	1.26	0.08
SO_4	1.92	0.22
K_2O	0.48	0.5

Table 2 Properties of cement and metakaolin

Properties	Cement	Metakaolin
Standard consistency (%)	30.5	34.5
Specific gravity	3.14	2.5
Initial setting time (min)	90	70
Colour	Grey	White

2.1.2 Metakaolin (MK)

Metakaolin is not a by-product; it is obtained as calcination of pure Kaolinite clay at temperature of 650–850 °C, followed by grinding to achieve fineness of 700–900 m^2/kg. The chemical composition of metakaolin is shown in Table 1 and the properties are shown in Table 2.

2.1.3 Fine Aggregate

Locally available M-Sand confirming to zone II of IS 383: 1970 with specific gravity 2.49 was used.

2.1.4 Coarse Aggregate

Coarse aggregates confirming to IS 383: 1970 are used. The aggregate having size more than 4.75 mm are coarse aggregates. The nominal size of 20 mm and specific gravity 2.63 were used.

2.1.5 EPS Beads

The Expanded Polystyrene beads used in this study spherical in shape and size varying in between 4.75 and 8 mm in diameter. The EPS bead density is 20 kg/m^3.

Table 3 Properties of superplasticiser

Relative density	1.24 ± 0.02 at 25 °C
pH	≥6
Chloride ion content	<0.2%
Aspect	Dark brown free flowing liquid

2.1.6 Superplasticizer (SP)

Superplasticiser used is MasterRheobuild 1125 and its dosage is 0.6% by weight of cement. The properties of superplasticiser are shown in Table 3.

3 Concrete Mix Design

Mix design carried out for M30 grade of concrete and various trials were carried out on general guide lines given as per IS 10262: 2009. The complete details of concrete mixes which contain 5, 10, 15, 20, 25, 30% coarse aggregate replaced with EPS beads in normal concrete are summarized in Table 4. For obtaining optimum percentage of metakaolin, 5, 10, 15, 20 and 25% cement is replaced with metakaolin and the compressive strength test was conducted on concrete cubes. The optimum percentage of metakaolin was obtained at 10% of cement replacement. The results are shown in Table 5 and Fig. 3. The details of concrete mixes which contains 5, 10, 15, 20, 25, 30% coarse aggregate replaced with EPS beads in 10% metakaolin based concrete are summarized in Table 6.

Table 4 Mix proportion of different mixes which contain EPS beads as replacement of coarse aggregate in normal concrete

Mix	Notation	Cement (kg/m^3)	Fine aggregate (kg/m^3)	Coarse aggregate (kg/m^3)	EPS beads (kg/m^3)	SP (l/m^3)	Water (l/m^3)
Control mix	CM00	370	692.63	1203.86	0.000	2.22	147.75
5% EPS	NE05	370	692.63	1142.42	0.453	2.22	147.75
10% EPS	NE10	370	692.63	1062.29	0.907	2.22	147.75
15% EPS	NE15	370	692.63	1022.17	1.361	2.22	147.75
20% EPS	NE20	370	692.63	962.04	1.815	2.22	147.75
25% EPS	NE25	370	692.63	901.91	2.269	2.22	147.75
30% EPS	NE30	370	692.63	841.78	2.723	2.22	147.75

Table 5 Compressive strength of concrete cube with varying percentage of metakaolin

Percentage replacement of cement with metakaolin (%)	28th day compressive strength (MPa)
0	38.43
5	39.78
10	42.95
15	37.48
20	34.86
25	31.33

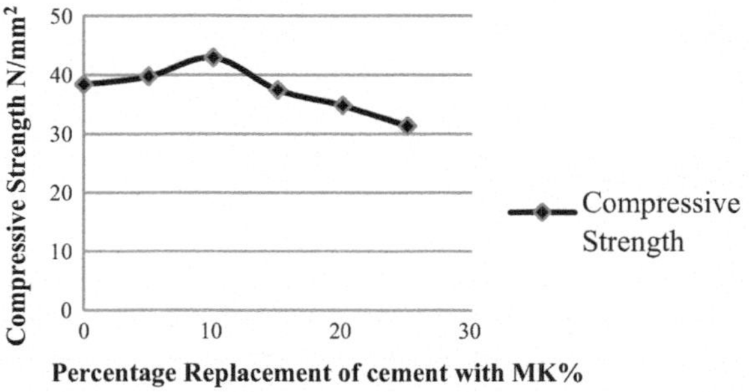

Fig. 3 Compressive strength of concrete cube with varying percentage of metakaolin

Table 6 Mix proportion of different mixes which contain EPS beads as replacement of coarse aggregate in 10% Metakaolin based concrete

Mix	Notation	Cement (kg/m^3)	MK (kg/m^3)	Fine aggregate (kg/m^3)	Coarse aggregate (kg/m^3)	EPS beads (kg/m^3)	SP (l/m^3)	Water (l/m^3)
10% MK + NC	ME00	333	37	692.63	1203.86	0.000	2.22	147.75
5% EPS	ME05	333	37	692.63	1142.42	0.453	2.22	147.75
10% EPS	ME10	333	37	692.63	1062.29	0.907	2.22	147.75
15% EPS	ME15	333	37	692.63	1022.17	1.361	2.22	147.75
20% EPS	ME20	333	37	692.63	962.04	1.815	2.22	147.75
25% EPS	ME25	333	37	692.63	901.91	2.269	2.22	147.75
30% EPS	ME30	333	37	692.63	841.78	2.723	2.22	147.75

Table 7 Variation of slump with EPS percentage

EPS replacement (%)	Slump of concrete without MK (mm)	Slump of concrete with 10% MK (mm)
0	90	87
5	95	91
10	98	93
15	101	98
20	106	102
25	113	109
30	121	116

4 Mixing of EPS Based Lightweight Concrete

The mix was prepared manually. First all dry materials like cement, metakaolin, sand and coarse aggregate were mixed thoroughly, then properly mixed these materials by adding water and superplasticiser. After that, to that wet mix EPS beads are added and again mixed thoroughly to achieve a homogenous mix.

5 Test Results

5.1 Slump

The slump test was conducted on each concrete mix according to IS 1199: 1959.

In general, it was observed that the slump value of concrete increases with increase in partial replacement of coarse aggregate with EPS beads. The slump values obtained are tabulated in Table 7. When cement is replaced with Metakaolin (MK), it was observed that there is slight decrease in workability.

5.2 Density

The density result shows that normal mix has higher density than mix containing EPS beads. When EPS beads replace coarse aggregate partially in normal concrete without any cement replacement (Table 8) the density is lesser than with cement replacement by 10% metakaolin (Table 9). This is because when MK is incorporated in the cement pastes resulted in a very dense microstructure of the paste, with a lower total porosity and finer pore size distribution.

Table 8 Density variation when EPS beads partially replace coarse aggregate in normal concrete

Specimen	Density (kg/m^3)
CM00	2468.15
NE05	1943.7
NE10	1890.37
NE15	1804.4
NE20	1728
NE25	1663.2
NE30	1534.02

Table 9 Density variation when EPS beads replace coarse aggregates in MK based concrete

Specimen	Density (kg/m^3)
ME00	2494.12
ME05	1998.9
ME10	1922.96
ME15	1853.62
ME20	1780.74
ME25	1762.96
ME30	1622.22

5.3 Compressive Strength

Compressive strength tests were carried out using cubical specimen of size 150 mm × 150 mm × 150 mm as per IS 516: 1959. The compressive strength setup was shown in Fig. 4. 7 and 28 day compressive strength test for different replacement of coarse aggregate with EPS beads in normal concrete and MK based concrete were carried out. The results are shown in Figs. 5 and 6.

From the results it was observed that most concrete mixes confirmed to moderate strength lightweight concrete (compressive strength in range 7–17 MPa). Only

Fig. 4 Compression test setup

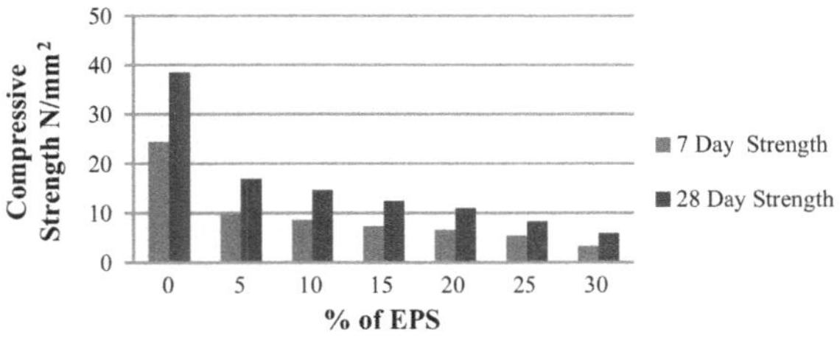

Fig. 5 Compressive strength of cube with varying percentage of EPS beads in normal concrete

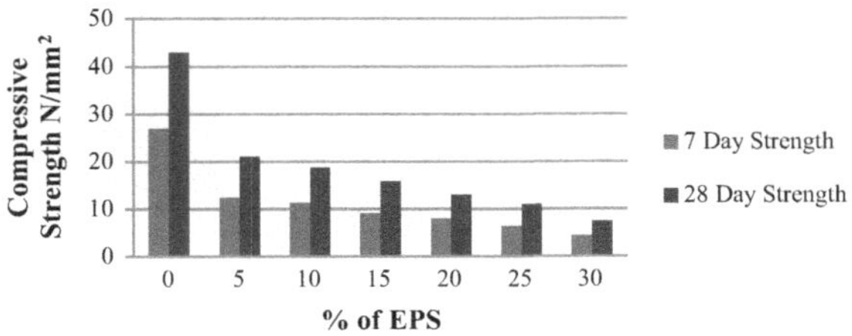

Fig. 6 Compressive strength of cube with varying percentage of EPS beads in MK based concrete

ME05 mix confirmed to structural lightweight concrete (compressive strength greater than 21 MPa and density less than 2000 kg/m^3).NE30 mix is low density concrete (compressive strength less than 7 MPa). Also observed that the 7th day compressive strength of cubes with EPS beads in normal concrete is lesser than cubes with EPS beads in metakaolin based concrete. This is because metakaolin possess high reactivity with calcium hydroxide and had the ability to accelerate the cement hydration. By incorporating metakaolin the early age strength enhances than any other supplementary materials.

5.4 Split Tensile Strength

Split tensile strength of lightweight concrete using EPS beads in normal concrete and metakaolin based concrete were checked on 7th and 28th day. The split tensile strength setup was shown in Fig. 7. The test results were noted as shown in Figs. 8 and 9. When EPS beads replace coarse aggregate in normal concrete the split tensile strength decreases as the EPS percentage increases. When EPS beads replace coarse aggregate in metakaolin based concrete there is approximately 30% increase in split tensile strength than in normal concrete. Also the 7th day split tensile strength obtained for cylinders contain EPS beads in normal concrete is lesser than EPS beads in metakaolin based concrete. This is because of the high reactivity of metakaolin with calcium hydroxide.

Fig. 7 Split tensile strength setup

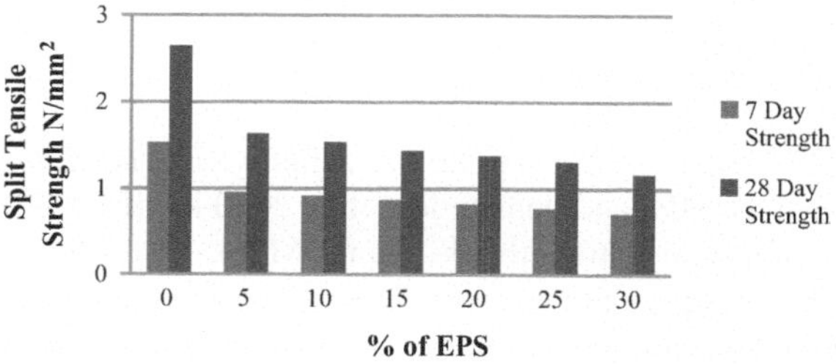

Fig. 8 Split tensile strength of cylinder with varying percentage of EPS beads in normal concrete

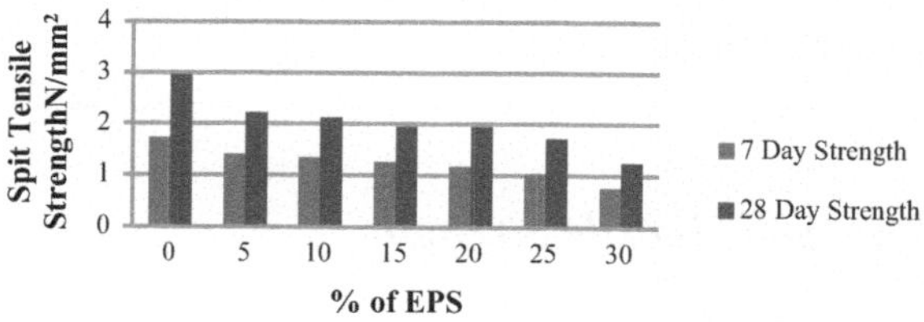

Fig. 9 Split tensile strength of cylinder with varying percentage of EPS beads in MK based concrete

5.5 *Water Absorption*

Water absorption test was done on cubes made by partial replacement of coarse aggregate in normal concrete and metakaolin based concrete. The lowest water absorption was noticed for control mix. When the percentage of EPS beads increases the water absorption also increases. The increase in water absorption of concrete which contains high percentage of EPS is due to the increased amount of voids. It was also observed that by replacing cement by metakaolin, it fills the voids and there by the water absorption reduces. By using metakaolin as partial replacement of cement the water absorption can be reduced by approximately 10% when compared to normal concrete mixes. The results are shown in Fig. 10.

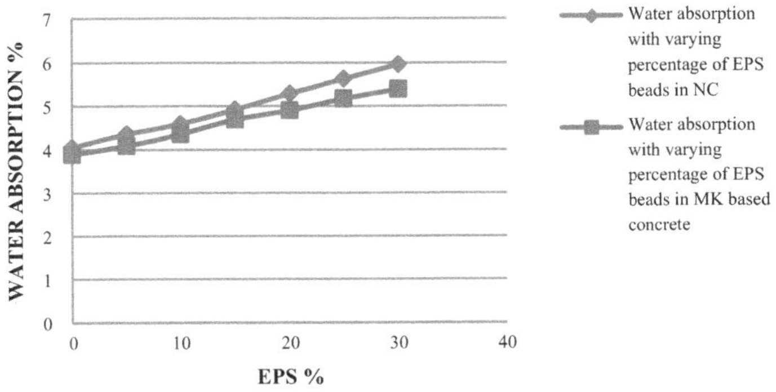

Fig. 10 Water absorption with varying percentage of EPS beads

6 Conclusions

1. From the compressive strength results carried out in normal concrete, by replacing 0–25% cement with metakaolin, the optimum percentage of metakaolin was obtained as 10%. Further tests were carried on 10% metakaolin based concrete.
2. The workability of concrete increases with increase in EPS bead percentage, but slightly decreases for corresponding metakaolin based concrete.
3. The compressive strength and split tensile strength decreases with increase in EPS bead percentage. This reduction rate can be reduced by using metakaolin. The compressive strength and split tensile strength can be increased by approximately 25% and 35% respectively by using metakaolin.
4. Density of MK based EPS concrete lies in the range of light weight concrete (<2000 kg/m^3).
5. The water absorption of concrete increases with increase in EPS bead percentage due to increased amount of voids. This can be reduced by adding metakaolin as partial replacement of cement.
6. Obtained results suggest that expanded polystyrene concrete has scope for applications like wall panels, partition walls, load bearing blocks etc. and it also serves as a solution for the disposal of EPS beads.

References

1. Liu N, Chen B (2014) Experimental study of the influence of EPS particle size on the mechanical properties of EPS lightweight concrete. Constr Build Mater
2. Chen B, Liu J (2007) Properties of lightweight expanded polystyrene concrete reinforced with steel fiber. Cem Concr Res 34:1259–1263
3. Abd SM, Gh D, Hattem M (2016) Effective replacement of fine aggregates by expanded polystyrenebeads in concrete. IJERST 05

4. Sri Ravindrarajah R, Tuck AJ (1994) Properties of hardened concrete containing treated expanded polystyrene beads. Cement Concr Compos 16:273–277
5. Ayse KAYA, Filiz KAR (2016) Properties of concrete containing waste expanded polystyrene and natural resin. Constr Build Mater 105:572–578
6. Al-Akhras NM (2006) Durability of metakaolin concrete to sulfate attack. Cem Concr Res 36:1727–1734
7. Lagier F, Kurtis KE (2007) Influence of Portland cement composition on early age reactions with metakaolin. Cem Concr Res 37:1411–1417
8. Guneyisi E, Gesoglu M, Karaoglu S (2012) Strength, permeability and shrinkage cracking of silica fume and metakaolin concretes. Constr Build Mater 34:120–130
9. Menhosh AA, Wanga Yu, Wang Y (2018) Long term durability properties of concrete modified with metakaolin and polymer admixture. Constr Build Mater 172:41–51

Experimental Study and Optimisation of Best Performance Self Compacting Recycled Aggregate Concrete

M. B. Amrutha Balan and C. A. Abin Thomas

Abstract Construction and demolition waste represents both the largest waste stream as well as an increasingly utilized supply of material to the construction industry. On the other hand, it is now widely recognized and steadily accepted that there is a significant potential for reclaiming and recycling demolished debris for use in value-added applications to maximize environmental and possible economic benefits. Indeed, it is widely acknowledged that recycling of CDW for reuse as fine aggregate in new concrete production is technically viable and under certain circumstances environmentally sustainable and economically feasible route to convert this material in a valuable resource. To synthesis best performance Self Compacting Concrete (SCC), factors such as Super Plasticizer (SP), fly ash and fine recycled concrete aggregate (FRCA) are studied experimentally based on Taguchi analysis, keeping SCC efficiency as the response in MiniTab software. The fresh and hardened properties of SCC incorporating FRCA are investigated by the experimental study to evaluate the efficiency of SCC. Factors for optimum level were fixed by SN analysis using larger the better quality criteria. The analysis was evaluated for a confidence level of 95%. The most influencing factor and optimal level of factors for maximum efficiency are determined. Confirmation experiments were done for the optimal levels of factors and the predicted values were compared with the experimental values.

Keywords Fine Recycled Concrete Aggregate (FRCA) · Self Compacting Concrete (SCC)

1 Introduction

Self-consolidating concrete or self-compacting concrete (commonly abbreviated to SCC) is a concrete mix which has a low yield stress, high deformability, good segregation resistance, moderate viscosity and is able to consolidate under its own weight.

M. B. Amrutha Balan (✉) · C. A. Abin Thomas
Department of Civil Engineering, Federal Institute of Science and Technology, Angamali, Kochi 683 577, India
e-mail: Amritabalan44@yahoo.com

K. Dasgupta et al. (eds.), *Proceedings of SECON'19*,
Lecture Notes in Civil Engineering 46,
https://doi.org/10.1007/978-3-030-26365-2_72

The highly fluid nature of SCC makes it suitable for placing in difficult conditions and in sections with congested reinforcement. There is no standard method for SCC mix design, and many academic institutions as well as admixture, ready-mixed, precast and contracting companies have developed their own mix proportioning methods but EFNARC Guidelines are strictly followed in the development of SCC mixes [1]. In order to achieve desired SCC characteristics, this composite material needs to be designed with a large amount of fines that reduces the coarse aggregate interaction. Thus, SCC seems an interesting opportunity to host and recycle the fine fraction of the recycled aggregates. The increasing scarcity of raw materials and the urgent need to protect the environment against pollution has accentuated the significance of developing new building materials based on industrial waste such as fly ash, Fine Recycled Concrete Aggregates (FRCA). In this study the behaviour of SCC under various percentage replacements for natural aggregate (NA) (fine) with FRCA was examined for its structural property. SCC gains its fluid properties from an unusually high proportion of fine aggregate, such as sand (typically 50%), combined with super plasticizers (additives that ensure particles disperse and do not settle in the fluid mix) and viscosity-enhancing admixtures (VEA). To investigate the three parameters, it is necessary to conduct a large number of experiments in laboratory which is time consuming and tedious. In order to reduce the number of experiments, experiments are designed and analyzed based on Taguchi analysis using MiniTab software. Design of Experiment (DOE) is a structured, organized method to define the relationship between factors that affect a process and its responses. It involves designing a set of experiments, in which all relevant factors are varied systematically and these experiments are conducted in laboratory. The results of these experiments are analysed to help identify optimal conditions, the factors that most influence the results and those that do not. In this study, Taguchi method is used because of its robustness, design for quality and reduction in number of experiments.

2 Experimental Program

2.1 Materials

The cement used for this study is Ordinary Portland Cement (OPC—53 Grade) conforming to IS 12269-1987. Additionally, fly ash class F conforming to ASTM 618 was used as mineral admixture. A polycarboxylate-based super plasticizer conforming to ASTM C 494 Type F and with a specific gravity of 1.09 was used to increase the workability of concrete. In addition, Viscosity modifying agent (VMA) with a specific gravity of 1.0 was used to prevent bleeding and segregation of concrete mix.

Natural river sand and Fine Recycled Concrete Aggregate (FRCA) were used as fine aggregate with SSD condition and had a specific gravity of 2.59 and 2.51 respectively. The gradation of the aggregates were determined by sieve analysis as per IS-383:1970. Coarse aggregate had specific gravity of 2.66 with maximum size of 12 mm.

2.2 Design of Experiments

The design-Expert software Minitab Version 17 was used to design the number of experiments to be performed, calculate the experimental data and evaluate the experimental results. In order to investigate the effects of significant factors and to obtain the optimum condition, the Taguchi method was used. Taguchi's orthogonal array analysis is used to obtain the best parameters for the optimum process design with the least number of experiments. In the Taguchi method, the S/N ratio is used to measure the quality characteristics from the desired value. Steps followed in Taguchi method is shown in Fig. 1.

For conducting experiment three variable parameters influencing SCC mix were selected. Three parameters include fly ash (FA), fine recycled concrete aggregate (FRCA) and Super Plasticizer (SP). Based on the previous studies, three levels of

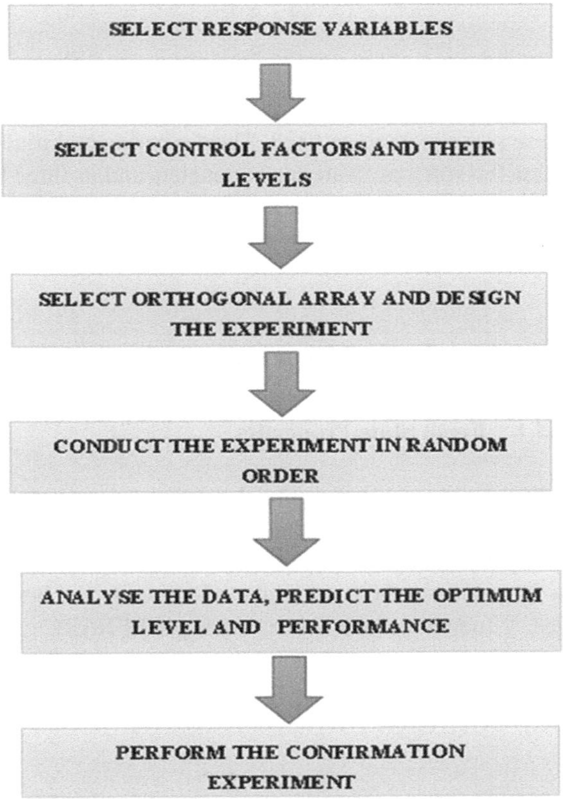

Fig. 1 Steps in Taguchi design

Table 1 Variable parameters and its levels

Variable parameters	Factor notation	Levels		
		1	2	3
Fly ash (% by weight of cement)	A	10	20	30
Fine recycled Concrete aggregate (% by weight of fine aggregate) (FRCA)	B	25	50	75
Super Plasticizer (% by weight of cement)	C	0.15	0.20	0.25

Table 2 Experimental trials after design of experiment

Mix	Fly ash (%)	FRCA (%)	SP (%)
Reference mix	30	0	0.25
1	10	25	0.15
2	10	50	0.20
3	10	75	0.25
4	20	25	0.20
5	20	50	0.25
6	20	75	0.15
7	30	25	0.25
8	30	50	0.15
9	30	75	0.20

these parameters were fixed. Then experimental trials were designed in Taguchi in Mini Tab software. Variable parameters and its three levels are shown in Table 1. In this experiment we obtained 9 different trials of experiments. Table 2 shows the 9 variable experimental trials obtained in Taguchi design.

2.3 *Conducting Experiments*

2.3.1 Fresh State Properties

Filling ability, passing ability, flowing ability and segregation resistance are the characterising properties of self compacting concrete. To characterise a concrete as a Self Compacting Concrete (SCC), it should have the above properties. Slump flow test is for assessing the flowability and viscosity can be assessed by T_{500mm} slump flow test, V funnel test and L-box test as per EFNARC guidelines [2].

2.3.2 Preparation of Specimens

The preparation of specimen includes casting of cubes. Specimens were successfully casted with the concrete after performing the Slump flow, V funnel, L-box and V-funnel test. The concrete is transferred to the lubricated mould and it was seen that the concrete get levelled and compacted under its own weight. Then the specimens are unmoulded after 1 day and taken for water curing after setting as shown in Fig. 2. After required number of days of curing, specimen is taken out of tank and air dried before testing. The compressive strength of casted cubes are determined on 7th, 14th, 21st and 28th day.

2.4 Analysis of Response (Based on S/N Ratio)

Analysis of obtained result data were done by using Taguchi method in Mini Tab software for both fresh properties and compressive strength of SCC mix. Compressive strength at 7th, 14th, 21st and 28th day was taken as response for analysis. Therefore, experimental results after conducting tests in laboratory in each run were entered into the worksheet corresponding to each trials. Analysis of datas were done by signal-to noise ratio (S/N). Finally confirmation experiments were also conducted with the trial of combination of optimum parameters in the laboratory condition to compare with the predicted results.

Fig. 2 Samples of casted of each SCC mix

3 Results and Discussions

The aim of study was to investigate the effect of Fly ash, Fine Recycled Concrete Aggregate (FRCA), Super Plasticizer (SP) on the fresh and compressive strength of SCC. Experimental design was done in Taguchi method using MiniTab 17 software and the developed set of experiments were conducted in laboratory. The experimental results were analysed in the MiniTab software. The effect of various parameters were studied by the analysis of results. For the optimisation of best performance SCC, the effect of the factors such as Fly Ash, FRCA, SP on the fresh and compressive strength of concrete were studied. The parameters were optimized using Taguchi's method using Mini Tab 17 software. The SCC reference mix was developed at 30% Fly ash and 0.25% super plasticizer [3].

3.1 Fresh Properties of SCC Mixes

In order to investigate the significance of variable factors [Fly ash, Fine Recycled Concrete Aggregate (FRCA), Super Plasticizer (SP)], 9 sets of experiments were conducted in random order based on the design of experiments generated in Taguchi. To check whether the generated mixes satisfy the requirements of SCC fresh properties of the mixes were analyzed by conducting slump flow test. Results obtained while conducting tests on fresh properties of all the 10 SCC mixes developed are listed in Table 3 and Fig. 3.

The results show that all the mixes satisfy SCC requirements. FRCA tends to have decreased specific gravity and increased water absorption since they contain the mortar paste from original concrete [4]. Workability decreases with the increase in FRCA [5]. The spread diameter is higher for low replacement of FRCA. From previous studies, it is impossible to develop SCC mix with 100% replacement of FRCA due to the loss of workability [6].

Table 3 Design of experimental runs and their results on fresh properties of SCC

Mix	Fly ash (%)	FRCA (%)	SP (%)	Slump flow (mm)
Reference mix	30	0	0.25	710
1	10	25	0.15	720
2	10	50	0.20	710
3	10	75	0.25	690
4	20	25	0.20	700
5	20	50	0.25	690
6	20	75	0.15	680
7	30	25	0.25	690
8	30	50	0.15	680
9	30	75	0.20	670

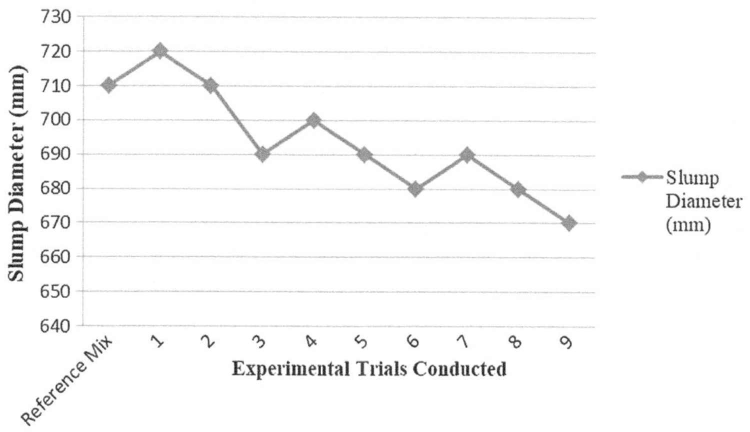

Fig. 3 Slump flow diameter versus experiments conducted in SCC

3.2 Hardened Properties of SCC Mixes

In order to investigate the significance of variable factors (fly ash, FRCA, SP), 9 sets of experiments were conducted based on Taguchi Design generated in MiniTab software. The total 9 runs of experimental data were designed by Taguchi Method and its results, (i.e. compressive strength) are shown in Table 4 and Fig. 4. For the densities hardened concrete, there was a small weight loss by increasing the incorporation of recycled aggregate.

Table 4 Design of experimental runs and their results on compressive strength of SCC

Exptl. trial No.	Fly ash (%)	FRCA (%)	SP (%)	Compressive strength of M30 grade of concrete at			
				7th day	14th day	21th day	28th day
1	10	25	0.15	15.7	21.8	22.9	24.2
2	10	50	0.20	17	23.6	24.8	26.2
3	10	75	0.25	18.1	25.1	26.4	27.9
4	20	25	0.20	18.9	26.1	27.4	29
5	20	50	0.25	18.5	25.7	26.9	28.5
6	20	75	0.15	17.9	24.8	26.1	27.6
7	30	25	0.25	17.8	24.6	25.9	27.4
8	30	50	0.15	17.2	23.9	25.1	26.5
9	30	75	0.20	17.6	24.5	25.7	27.2
Reference mix	30	0	0.25	17.2	23.2	24	26

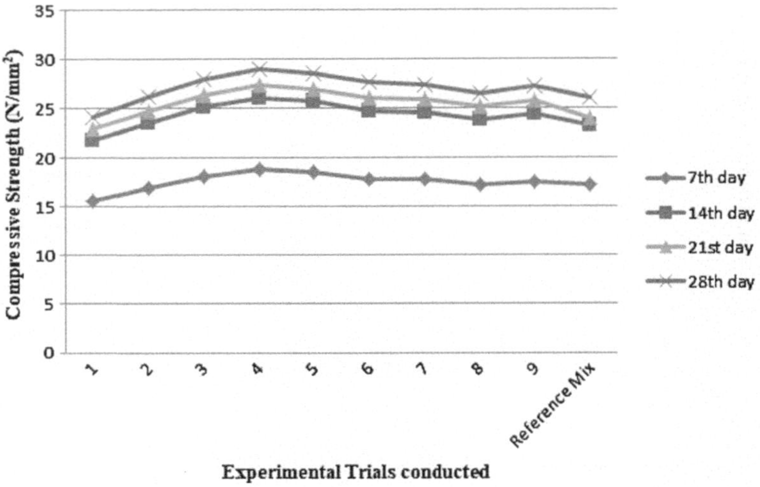

Fig. 4 Compressive strength versus experiments conducted in SCC

The compressive strength of concrete mixes with recycled aggregate were higher than that of SCC reference mix. FRCA contains calcium silicate hydrates (CSH) and calcium hydroxide (CH) [7]. The addition of CH or CSH will accelerate the hydration reactions as it acts as a nuclei for further CSH formation. Crystalline CH acts as sink for calcium and silicate ions in the solution, enabling further dissolution of tricalcium silicate ions (C_3S) and renewed CSH formation [8]. FRCA contains finer particles than natural fine aggregates [9]. Hence, added FRCA particle will densify the microstructure through enhancing particle packing and consequently reducing the amount of space needed to be filled by hydration products [10]. Furthermore, FRCA particle acts as a contact points between hydration products, eventually solidifying the micro-structure. As expected SCC with FRCA exhibited a slightly higher compressive strength than that of reference mix.

3.2.1 Determination of Optimum Parametric Levels for Compressive Strength

The influence of the parameter on the compressive strength of SCC has been evaluated using S/N ratios. The response table for Signal to Noise ratios for compressive strength of SCC is shown in the Table 5. Ranking of predominant parameters influencing the removal % using S/N ratios obtained for different factor levels are also listed in the Table 5.

Out of the three factors, the most influencing factor is the fly ash (Highest S/N ratio). The optimal level of factors for maximum compressive strength are 20% of fly ash, 75% of FRCA replacement in natural aggregate and 0.25% of SP as shown in Fig. 5. When the line is not horizontal (parallel to the x-axis), then there is a

Table 5 Response table for signal to noise ratios (Larger is better)

Level	Fly ash	FRCA	SP
1	26.76	27.01	2676
2	27.50	27.09	27.21
3	2708	2724	2736
Delta	0.74	0.24	0.6
Rank	1	3	2

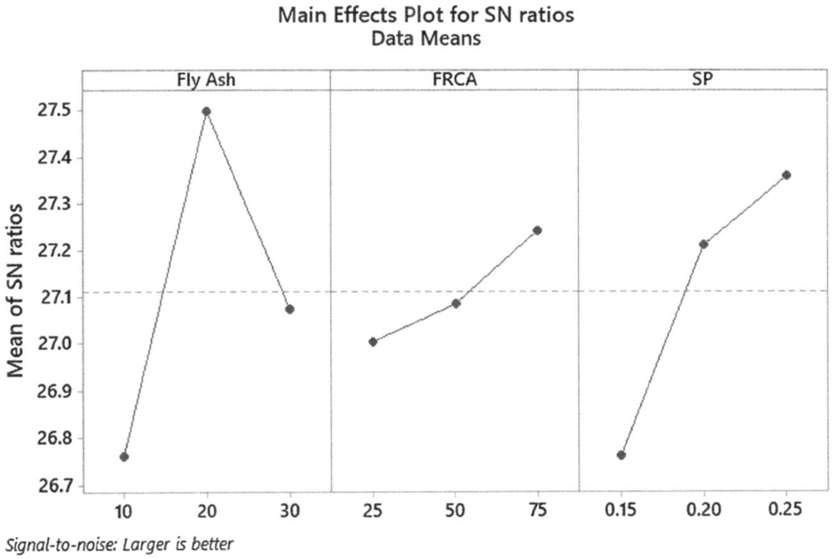

Fig. 5 The main effect plot for S/N ratio of factors

main effect present. Different levels of the factor affect the characteristic differently. The greater the difference in the vertical position of the plotted points (the more the line is not parallel to the X-axis), the greater the magnitude of the main effect. The compressive strength increases with the increases in fly ash content up to 20% and it decreases when fly ash content is 30%. The main effect in increase of compressive strength is for fly ash up to 20% and followed by SP and FRCA. The main effect is determined by the higher delta value and ranking number. Since the rank of fly ash is 1, fly ash has the highest effect when compared to the factors as shown in Table 5.

References

1. Eurocode 2: Design of concrete structures - Part 1–1: General rules and rules for buildings
2. Aggarwal P, Siddique R, Aggarwal Y, Gupta SM (2008 January–June) Self-compacting concrete—procedure for mix design. Leonardo Electron J Pract Technol 12:15–24. ISSN 1583-1078
3. Habibi A, Ghomashi J (2018) Development of an optimum mix design method for self-compacting concrete based on experimental results. Constr Build Mater 168:113–123
4. Fan C-C, Huang R, Hwang H, Chao S-J (2016) Properties of concrete incorporating fine recycled aggregates from crushed concrete wastes. Constr Build Mater 112:708–715 (Science direct)
5. Omrane M, Kenai S, Kadri E-H, Aït-Mokhtar A (2017) Performance and durability of self compacting concrete using recycled concrete aggregates and natural pozzolan. J Clean Prod
6. Vinay Kumar BM, Ananthan H, Balaji KVA (2016) Experimental studies on utilization of coarse and finer fractions of recycled concrete aggregates in self compacting concrete mixes. J Build Eng
7. Yaprak H, Aruntaş HY, Simsek O (2011 May) Effects of the fine recycled concrete aggregates on the concrete properties. Int J Phys Sci
8. Fan C-C, Huang R, Hwang H, Chao S-J (2015) The effects of different fine recycled concrete aggregates on the properties of mortar. Materials 8:2658–2672 (Institute of Materials Engineering, National Taiwan Ocean University)
9. Khoshkenari AG, Shafigh P, Moghimi M, Mahmudb HB (2014) The role of 0–2 mm fine recycled concrete aggregate on the compressive and splitting tensile strengths of recycled concrete aggregate concrete. Mater Des 64:345–354 (Science Direct)
10. Velay-Lizancos M, Martinez-Lage I, Azenha M, Granja J, Vazquez-Burgo P (2018) Concrete with fine and coarse recycled aggregates: e-modulus evolution, compressive strength and non-destructive testing at early ages. Constr Build Mater 193:323–331 (Science direct)

Mechanical and Durability Properties of Microbially Induced Calcite Precipitated Polypropylene Fibre Reinforced Concrete

Riya Tomy and K. N. Resmi

Abstract In fibre reinforced cementitious composites (FRCC) bonding between the fibres and matrix governs many important properties, which includes strengths, fracture energy, ductility, and energy absorption capacities. The study explores the application of a microbially induced calcite precipitation (MICP) to pre-treating surface of polypropylene (PP) fibres for enhancing the interfacial boning strength. The technique utilizes MICP process to produce calcium carbonate that binds onto the fibre surface, leading to increased interfacial bond area and strength. Laboratory tests indicate that MICP modification could increase the mechanical and durability properties of concrete. Tests on hardened concrete to analyse the mechanical properties was done as per IS 516-1959. A durability study was also done as per ASTM standards to analyse the behavior of the developed concrete under adverse weather conditions. Results acknowledged an important role of MICP pre-treatment in enhancing both the mechanical and durability performance of concrete.

Keywords Polypropylene fibre · MICP treatment · Calcite precipitation · Mechanical properties · Durability behaviour

1 Introduction

Polypropylene (PP) fibres are popular for reinforcing cementitious composites due to their unique properties such as high melting point, chemical stability, and relatively low cost compared to other types of fibres [1]. However, polypropylene is typically hydrophobic, resulting in poor bonds with cementitious matrices. Furthermore, PP fibres have low surface free energy and very smooth surface structures that may be susceptible to slipping or de-bonding from the cementitious matrix [2].

R. Tomy (✉) · K. N. Resmi
Department of Civil Engineering, Federal Institute of Science and Technology, Angamaly, Kochi 683 577, India
e-mail: riyatomy@gmail.com

K. N. Resmi
e-mail: reshmikn@gmail.com

K. Dasgupta et al. (eds.), *Proceedings of SECON'19*,
Lecture Notes in Civil Engineering 46,
https://doi.org/10.1007/978-3-030-26365-2_73

Table 1 Steps involved in MICP process

Chemical reactions	Results
$CO(NH_2)_2 + H_2O \rightarrow NH_2COOH + NH_3$	Hydrolyzation of urea
$NH_2COOH + H_2O \rightarrow NH_3 + H_2CO_3$	Release of ammonia
$H_2CO_3 \rightarrow 2H^+ + 2CO_3{}^{2-}$	Formation of bicarbonate
$NH_3 + H_2O \rightarrow NH_4{}^- + OH^-$	Release of hydroxide

It is known that calcium carbonate ($CaCO_3$) crystals are well compatible with cement matrix [3]. By depositing $CaCO_3$ crystals onto the surface of PP fibres, the mechanical properties of FRCC may be improved due to the increased roughness of PP fibres surface and the enhanced interfacial bonding between fibres and matrix. Microbially induced calcite precipitation (MICP) through urea hydrolysis pathway, an emerging research subject in recent years, has been used to enhance the durability of structural materials by forming a protective calcite layer on the surface or acting as healing material in the surface cracks.

Microbially induced carbonate precipitation (MICP) is a natural process that has shaped the earth from ancient time. Table 1 shows the steps involved in MICP Process. In this process, calcium carbonate minerals are formed from calcium and carbonate ions. Because calcium carbonate minerals are a homogenous material compatible with concrete and stone and is environmentally friendly, MICP has been studied as a method for stone crack repair. The technology based on *Bacillus cereus*-induced $CaCO_3$ formation has been commercialized for repairing cracked surface of ornamental stones. Also by incorporating live bacteria in concrete, MICP has been shown to improve mechanical properties and self-healing of concrete.

Bacterial cell walls, which comprise the cell surface, are known to be central to MICP [4]. Bacterial cell walls are negatively charged under environment of neutral or alkaline pH, attract the calcium ions in the extracellular environment to react with the carbonate ions and form calcium carbonate minerals on the cell surface, which serve as nucleation centers for further mineralization. Currently, despite the central role of cell walls in MICP, there is no research studying the effects of directly applying cells walls in concrete. One recent research applying cell walls of *Bacillus subtilis* on cracked stone surface showed increased formation of $CaCO_3$ crystal which significantly decreased water permeability of the stone.

Calcium carbonate can be formed on the surface of the concrete by reacting with CO_2 present in the calcium hydroxide by following reactions

$$CO_2 + Ca(OH)_2 \rightarrow CaCO_3 + H_2O \tag{1}$$

$$Ca(C_3H_5O_2)_2 + 7O_2 \rightarrow CaCO_3 + 5CO_2 + 5H_2O \tag{2}$$

In the current study, the feasibility of using MICP for the surface modification of the PP fibres is investigated, and the corresponding mechanical and durability properties of FRC using MICP modified PP fibres are examined through a series of laboratory tests.

2 Materials and Mix Proportions

The different types of materials that are used in this study are, OPC 53 grade cement conforming to IS 12269-2013, Coarse Aggregate conforming to IS 383-2016, Fine Aggregate conforming to IS 383-2016, Water, Super Plasticizer (Masterglenium 8233 SKY), Silica Fume, *B. subtilis* bacteria, Polypropylene fibres, Calcium chloride and Urea.

Polypropylene fibres (PP) of 36 μm diameter and 12 mm length have been used throughout the study. Modulus of elasticity and tensile strength of fibres were 8 GPa and 500 MPa respectively. Various tests were conducted on the materials and they were checked to conform that they meet the IS standards. Table 2 shows the material test results.

Mix proportioning was done by conducting trials as per IS 10262:2009, and IS 456:2000. Final mix was fixed based on the required conditions for M40 concrete with workability 100 mm as shown in Table 3. Silica fume was added as a partial replacement of cement by 8% whose optimum was found through compressive strength test on cubes. Bacteria used in the current study were *Bacillus subtilis*. The culture was stored in a refrigerator for not more than 1 week prior to use. The cementation solution used consisted of 1 M of calcium chloride (111 g/L) and 1 M of urea (60 g/L).

Table 2 Material properties

Material	Property	Obtained value
Cement	Specific gravity	3.14
	Standard consistency	30.5%
	Initial setting time	90 min
	Fineness	9.6%
Silica fume	Specific gravity	2.6
Fine aggregate	Specific gravity	2.64
	Fineness modulus	2.72
Course aggregate	Specific gravity	2.65
	Fineness modulus	7.21

Table 3 Mix proportion of 1 m^3 of developed concrete

Ingredient	Quantity
Water	157.67 l
Cement	398.73 kg
Silica fume	34.672 kg
Fine aggregate (M-sand)	655.20 kg
Course aggregate (20 mm)	1187.20 kg
Superplasticizer	0.5%
w/c ratio	0.36

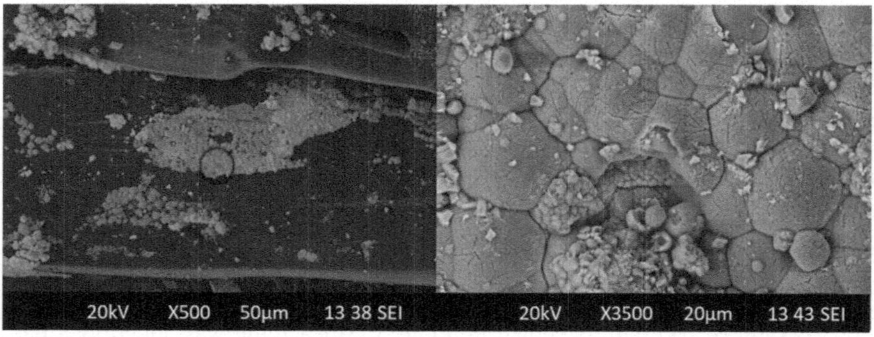

Fig. 1 SEM image of MICP modified PP fibre

2.1 MICP Pre-treatment of PP Fibres

The process of MICP pre-treatment of PP fibres was conducted by submerging the fibres into a biological reagent solution, a mixture of bacterial culture and cementation solution, using a volume ratio of 1:1, for 24 h to allow for the MICP to deposit and coat the fibre surface. The volume ratio between the added fibres and biological reagent solution was about 1:4. The fibres with different amount of coating was calculated using the net dry weight increase of the fibres after the coating divided by the initial weight of the untreated fibres. Figure 1 shows SEM image of MICP modified PP fibre.

3 Experimental Investigation

3.1 Fresh Stage Properties

Slump test have been conducted to assess the workability of the developed concrete.

3.2 *Hardened Stage Properties—Mechanical Properties*

Compressive strength test is conducted on 150 mm × 150 mm × 150 mm cubes cured for 28 days in a compression testing machine of capacity 2000 kN. The load has been applied perpendicular to casting face of cubes. Split tensile strength test is conducted on cylinders with 150 mm diameter and 300 mm diameter in Universal testing machine of capacity 2000 kN and load applied diametrically. Flexural strength test is conducted on beams of size 100 mm × 100 mm × 500 mm by four point loading.

3.3 *Durability Studies*

3.3.1 Water Absorption Test

The test for determining the water absorption characteristics of various mixtures were conducted on three number of 150 mm size concrete cubes for each proportion. After 28 and 56 days of casting, the specimens immersed in curing tank were taken out to measure the saturated weight. Then the specimens were dried in an oven at a temperature of 105 °C and weights of the specimens were found out. Thus the weight of the oven dried specimens after cooled to room temperature were found out.

3.3.2 Acid Resistance—HCl and H_2SO_4

Acid attack test was conducted as per ASTM C267-01 (2012). For acid resistance, the cube specimens were immersed in 5% of 1 M HCl solution by volume added to water after water curing the specimens for 28 days. Similarly another set of cube specimens were immersed in 5% of 1 M H_2SO_4 solution by volume added to water. The percentage of loss in weight and strength was then determined after 28 and 56 days.

3.3.3 Sulphate Resistance (Na_2SO_4)

Sulphate resistance test was conducted accordance with the ASTM C1012. 10% Na_2SO_4 by weight was added to distilled water to make sodium sulphate solution. Cube specimens which was cured for 28 days was kept immersed in the prepared solution for 28 and 56 days. The percentage of loss in weight and strength was determined.

Table 4 Slump test results

W/c ratio	0.30	0.32	0.34	0.36
Slump value (mm)	No slump	70	85	100

4 Results and Analysis

4.1 Fresh Stage Properties

It was observed that the workability increased as the w/c ratio increased. At w/c 0.36 the desired slump of 100 mm was obtained. Table 4 shows the slump test results.

4.2 Hardened Stage Properties

4.2.1 Cube Compressive Strength Test

The cube compressive strength of the mixes was observed after 7 and 28 days. Results showed that fibre dosage with 0.5% of the concrete had the greatest strength. The results was similar in case of concrete with both treated (PP) and untreated fibres (PPT). Figure 2 shows the compressive strength results. The slight decrease in compressive strength may be due to the addition of fibres which can be neglected due to the increased flexural strength and split tensile strength.

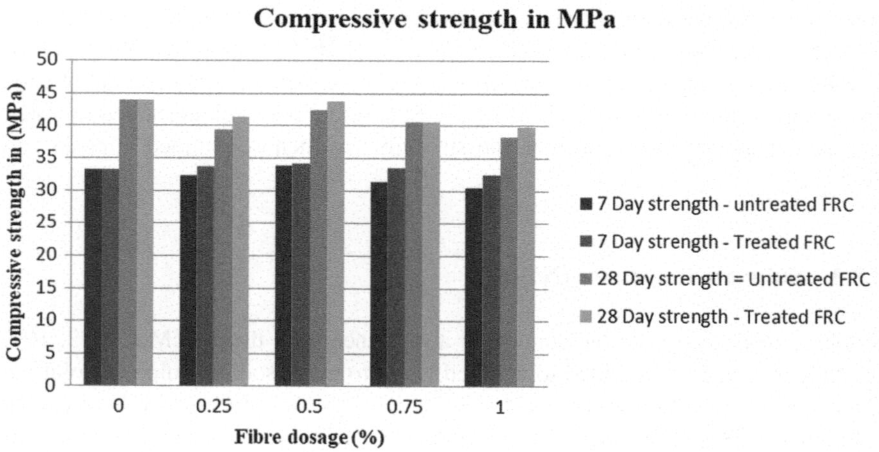

Fig. 2 Compressive strength of concrete with both treated and untreated varying fibre dosage

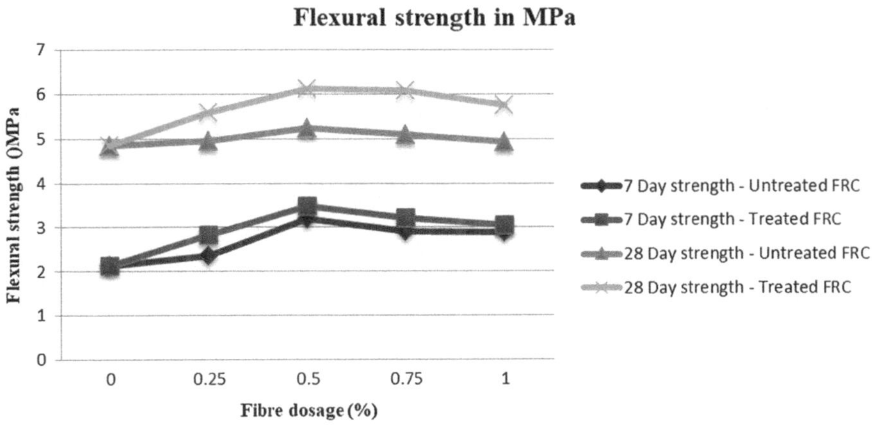

Fig. 3 Flexural strength of concrete with both treated and untreated varying fibre dosage

4.2.2 Flexural Strength Test

The flexural strength of the specimens with and without fibre was determined after 7 and 28 days. It can be seen that addition of fibres increased the flexural strength of concrete by 25–30%. Concrete with 0.5% fibre dosage showed the maximum results. Figure 3 shows the results. Results clearly indicates that the added fibre in the matrix act as a crack arresting mechanism and improve the tensile strength of concrete.

4.2.3 Split Tensile Strength Test

The split tensile strength of the specimens with and without fibre was determined after 7 and 28 days. It can be seen that addition of fibres increased the split tensile strength of concrete by 15–20%. Concrete with 0.5% fibre dosage (treated and untreated) showed the maximum results. Figure 4 shows the results. As a result of the inclusion of fibres into plain concrete, the brittle mode of failure is changed to a ductile mode thus after occurrence of cracks, the specimen did not fail suddenly. Instead, crack propagates due to the bonding of the fibres into the concrete matrix and offers pulling force in the lateral direction to restrain cracking, thereby prolonging the failure state of the specimens under splitting tension [5].

4.3 Durability Test

4.3.1 Water Absorption Test

The results of water absorption test at the age of 28 and 56 days is shown in Fig. 5.

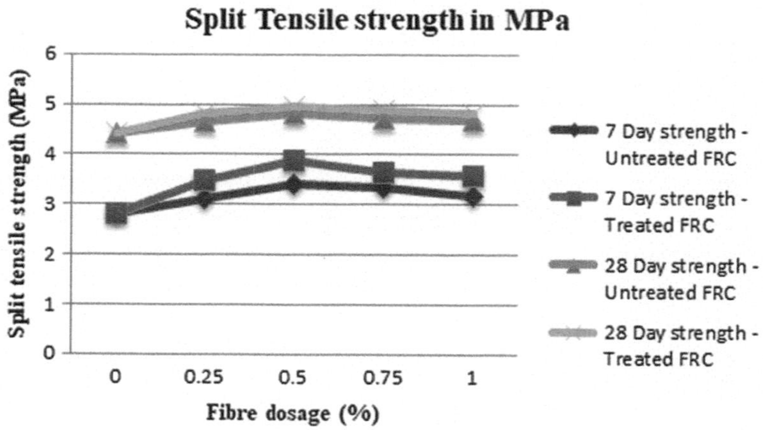

Fig. 4 Split tensile strength of concrete with both treated and untreated varying fibre

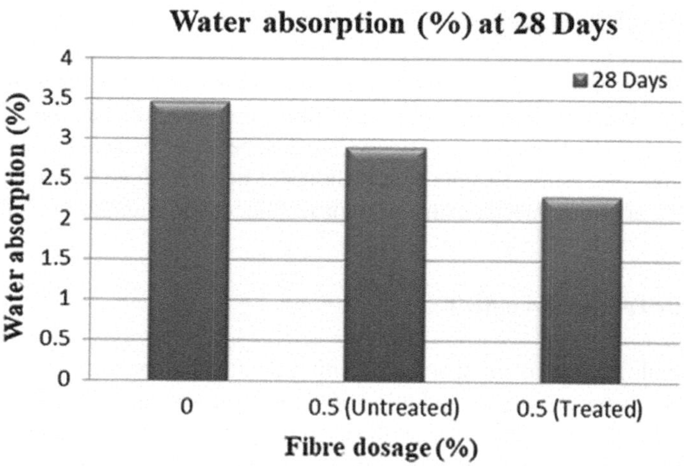

Fig. 5 Water absorption results

It is noted that with addition of fibres, the water absorption gets decreased when compared with Control Concrete. The water absorption percentage varied between 2.3 and 3.47% at 28 days. As per ASTM C 642, the water absorption percentage lesser than 5% is excellent.

4.3.2 Acid Resistance—HCl and H_2SO_4

From the test results, it is found that percentage of weight loss is higher in Control Concrete compared to the concrete with polypropylene fibre. This shows that the fibre reinforced concrete has better resistance to acid attack. Along with percentage of weight loss, the percentage loss of compressive strength also was found out and results are shown in Figs. 6 and 7. It has been reported in various journals that the acid attack and leaching away of the calcium compounds of the cement paste formed in concrete during the hydration process as well as calcium in calcareous aggregate reduced the strength of structural concrete and affected the durability [6].

In case of HCl acid environment the leaching is minimum or low, the percentage loss of weight and compressive strength is less compared to H_2SO_4 acid environment. The percentage of weight and strength loss is shown in Figs. 8 and 9. The maximum

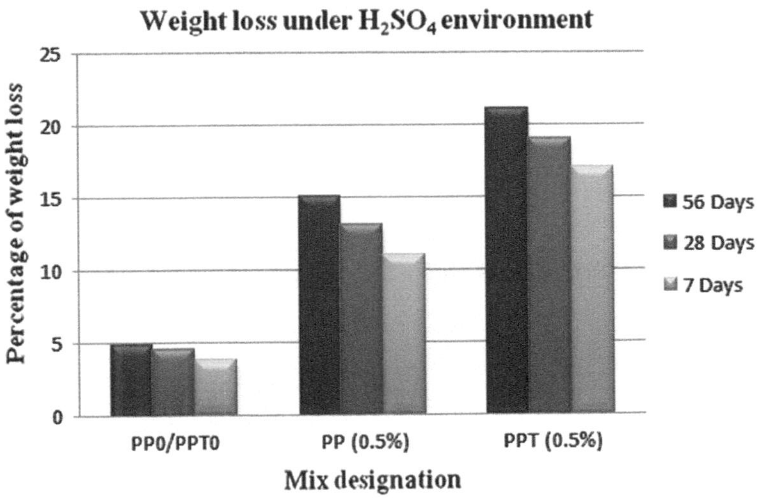

Fig. 6 Weight loss under H_2SO_4 environment

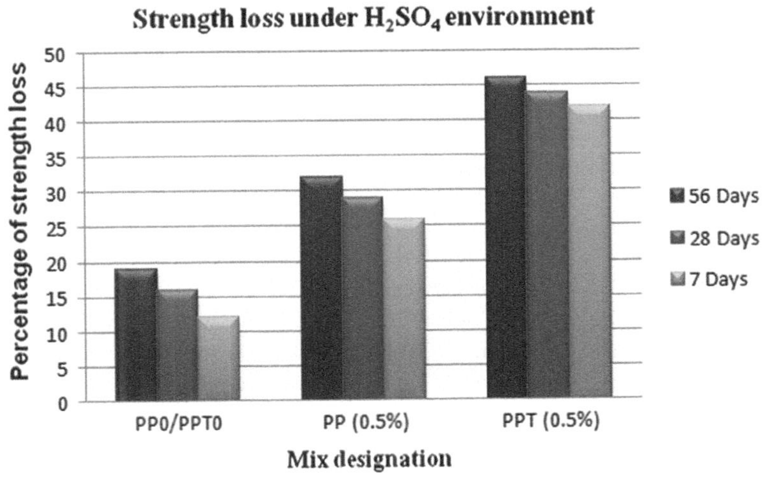

Fig. 7 Strength loss under H_2SO_4 environment

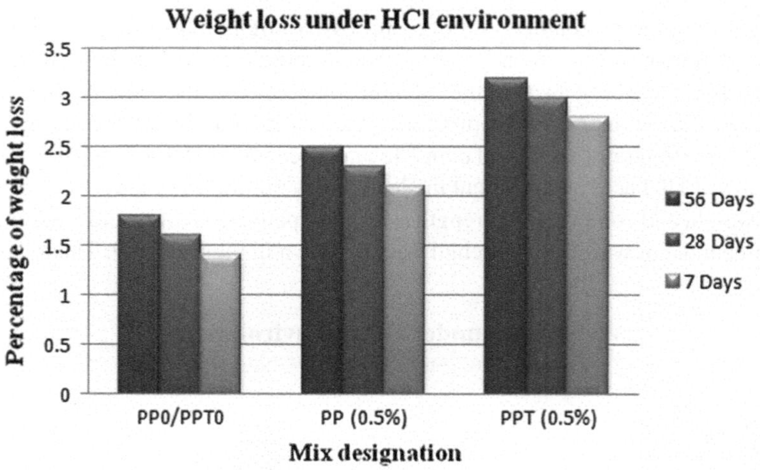

Fig. 8 Weight loss under HCl environment

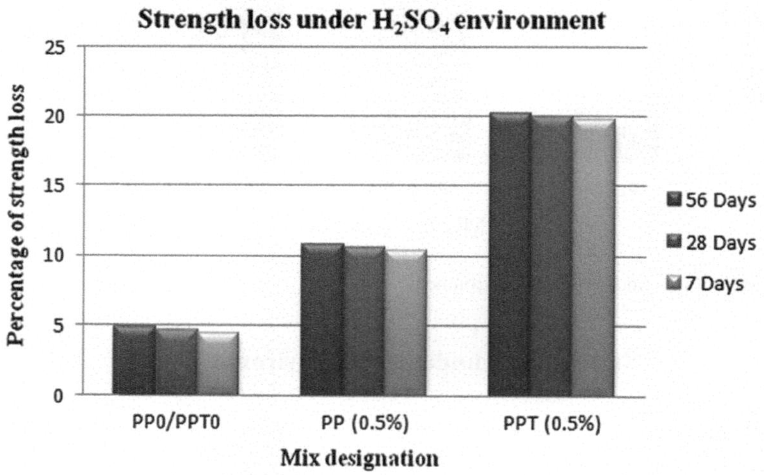

Fig. 9 Strength loss under HCl environment

percentage weight loss is 4.54% at 56 days for Control Concrete which is higher compared to 3.92% for PP (0.5) and 3.63% for PPT (0.5%) at 56 days.

Similarly the strength loss of Control Concrete is 32.3% which is higher compared to PP (0.5) and PPT (0.5%) which is 23.54 and 20.46% at 56 days. The compressive strength of concrete mix exposed to HCl environment is significantly reduced because the acid attack completely converts the hardened cement paste destroying the entire pore system.

4.3.3 Sulphate Resistance (Na_2SO_4)

The percentage loss in weight as well as compressive strength due to sulphate attack is shown in Figs. 10 and 11 respectively. The percentage weight loss of Control Concrete is 0.10% at 56 days. It is observed that the percentage weight loss is decreased with addition of fibres. The percentage of weight loss of PP (0.5%) is 0.5% and for PPT (0.5%) is 0.18% at 56 days, respectively which is lower than control concrete. The higher C3A content in the cement paste makes concrete more vulnerable to

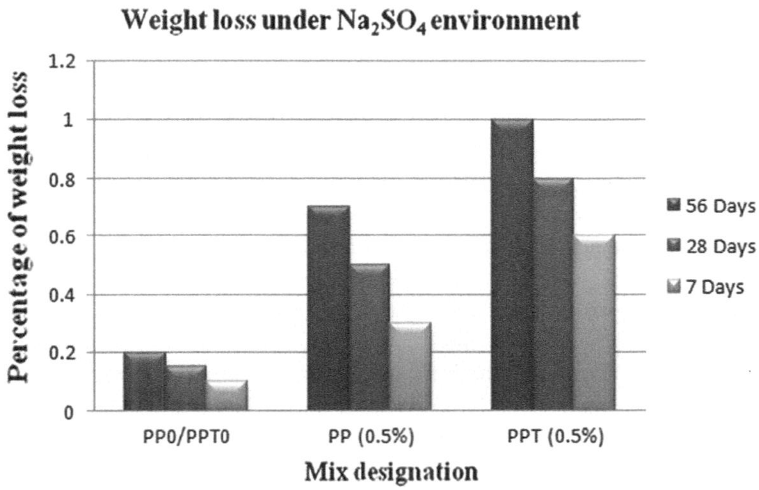

Fig. 10 Weight loss under Na_2SO_4 environment

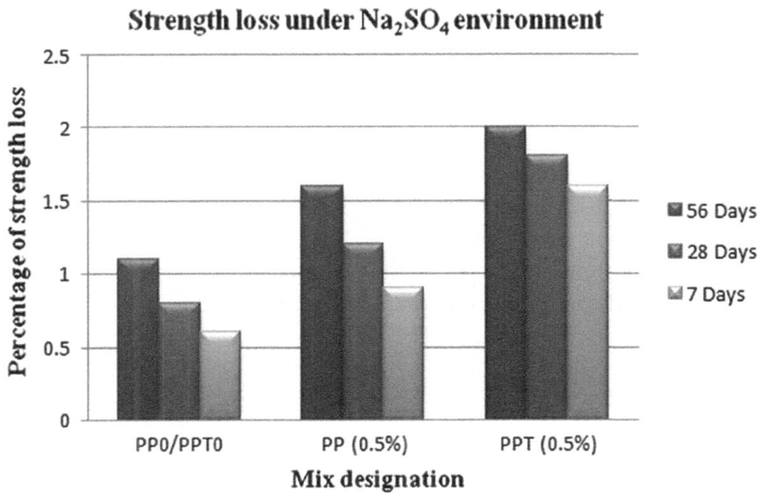

Fig. 11 Strength loss under Na_2SO_4 environment

sulphate attack. The loss of weight in sulphate environment is less than 1%; hence it is negligible. The specimens exposed to Na_2SO_4 environment not show much difference compared to water exposed specimens. The maximum percentage loss in compressive strength was observed for Control Concrete as 0.92% at 56 days and for PP (0.5%) and PPT (0.5%) it was 0.64 and 0.34% at 56 days. It is clear that incorporation of fibre in concrete enhances the sulphate resistance of concrete.

5 Conclusions

The conclusion from the present investigation is based on the limited observations made during the study period.

- Results showed that fibre dosage with 0.5% of the concrete had the greatest strength. The results was similar in case of concrete with both treated and untreated fibres. The slight decrease in compressive strength may be due to the addition of fibres which can be neglected due to the increased flexural strength and split tensile strength.
- The flexural strength of the concrete showed improvement with fibre content than without fibre. Mix with 0.5% fibre content, both in treated and untreated case, showed maximum strength which means the added fibre in the matrix acted as a crack arresting mechanism and improved the tensile strength of concrete. Beyond the optimum dosage (0.5%) the flexural strength was dropped, which can be justified as the extra amount of fibre reduced the free flow of concrete and the compaction achieved.
- The split tensile strength was more for the mix with 0.5% fibre content, both in treated and untreated case, as compared to the mix without fibre. This may be due to the inclusion of fibres into plain concrete resulted in the change of brittle mode of failure into the ductile mode. Also fibres increased the energy absorption capacity and enhanced the lateral stiffness of the concrete.
- The water absorption was found decreasing as the fibre dosage increased. This may be due to the reason that the interfacial zones between fibres and the cement paste provides the additional connected routes for chloride transport in concrete.
- The mix with 0.5% fibre showed good resistance against the acid attack. The strength loss of the mixes PP (0.5%) and PPT (05%) exposed to sulphate environment was found minimum. Therefore the presence of polypropylene fibres, silica fume enhanced sulphate attack resistance to a certain extent.

References

1. Zheng Z, Feldman D (2016) Synthetic fibre-reinforced concrete. J Constr Build Mater 20:185–210

2. Wei QF (2004) Surface characterization of plasma-treated polypropylene fibres. J Constr Build Mater 52:231–235
3. Pei M, Wang D, Zhao Y, Hu X, Xu Y, Wu J, Xu D (2004) Surface treatments of subdenier monofilament polypropylene fibres to optimize their reinforcing efficiency in cementitious composites. J Constr Build Mater 92:2637–2641
4. Wang JY, Snoeck D, Van Vlierberghe S, Verstraete W, De Belie N (2014) Application of hydrogel encapsulated carbonate precipitating bacteria for approaching a realistic self healing concrete. J Constr Build Mater 68:110–119
5. Felekoglu B, Tosun K, Baradan B (2009) A comparative study on the flexural performance of plasma treated polypropylene fibre reinforced cementitious composites. J Constr Build Mater 209:5133–5144
6. Lovata NL, Fahmy MF (2007) Interfacial bond study of a chemically treated polypropylene fibre-reinforced concrete. J Constr Build Mater 1:83–87

An Experimental Study on Addition of Carbon Nanotubes to Improve the Performance of UHPC

A. Aayisha and Regi P. Mohan

Abstract Concrete plays an important role in construction industry. Over the last twenty five years, amazing progress were gained in the research and use of Ultra high performance concrete, as it can provide a durable and cost effective structures. High compressive and tensile strengths, high ductility, and toughness, high flow ability and durability are the outstanding mechanical properties of UHPC. Also, it resists all kinds of corrosion, resulting in maintenance-free constructions. UHPC shows very high brittleness due to its exceptionally high compressive strength and is considered as a weakness. However, UHPC has its drawbacks in terms of tensile strength and brittleness. Hence, these encourage the development of high strength and better performance concrete and create the need to research on the use of Carbon Nanotube (CNT) in UHPC. This is because CNT was believed to be able to further improve concrete's mechanical properties due to its impressive characteristics. In this study, carbon nanotubes is added in different percentages (0.03, 0.06, 0.09, 0.12 and 0.15%) and finding optimum percentage which contributes maximum strength to the UHPC.

Keywords Ultra high performance concrete (UHPC) · Carbon nanotubes (CNTs) · Steel fibre · Fire resistance · Ductility

1 Introduction

Amazing progress were gained in the research and use of Ultra high performance concrete over the last 25 years. UHPC shows excellent mechanical properties and durability. From the sustainability point of view, considerable potential has been shown by UHPC for the future construction industry. It is suitable for lightweight structures that are usually pre-stressed, or for structures and elements exposed to the outdoor environment where it offers excellent durability.

A. Aayisha · R. P. Mohan (✉)
Sree Buddha College of Engineering, Pattoor, Padanilam, India
e-mail: regimohan2007@gmail.com

K. Dasgupta et al. (eds.), *Proceedings of SECON'19*,
Lecture Notes in Civil Engineering 46,
https://doi.org/10.1007/978-3-030-26365-2_74

French Association Francaise de Genie Civil (AFGC) recommended that UHPC have the potential to attain the compressive strength from 150 to 250 MPa. The above range of high strength is attained by replacing the components of conventional concrete mixes. UHPC contains fine aggregate such as quartz sand, silica sand etc., but is devoid of coarse aggregate. In addition, cement is replaced with flyash or silica fume. UHPC has very low water cement ratio compared to conventional concrete. UHPC shows very high brittleness due to its exceptionally high compressive strength. Many studies were carried out to find a technique for increasing the ductility of UHPC. Fibers are commonly used to overcome this weakness and also to enhance the properties of concrete such as flexural strength, ductility or fire resistance. In this study, steel fibres were added at 0.5, 1, 1.5, 2 and 2.5% of volume of concrete and the optimum percentage of steel fibres which contributes to maximum strength is found out. It is found that when steel fibres are added to concrete, the tensile strength of the composite is increased from 10 to 25% depending on the proportion of fibres added and the mix design.

This work mainly focuses on studying the strength parameters of UHPC containing steel fibers and carbon nanotubes and to compare the same with steel fibre only.

2 Objectives of the Study

- To obtain the optimum mix proportion design for the ultra high performance concrete
- To determine the optimum percentage of steel fibre in UHPC
- To study the strength characteristics of UHPC
- To determine the optimum concentration of carbon nanotubes that contributes maximum strength
- To compare the strength properties of UHPC with carbon nanotubes and without carbon nanotubes.

3 Methodology

- A detailed literature survey was carried out on studies on Ultra High Performance Concrete, steel fibre addition etc.
- Materials like cement, fine aggregate, quartz sand, steel fibres, silica fume, carbon nanotubes and super plasticizer were collected.
- Material characterisation was carried out and found out that all are conforming to IS standards.
- Fixing the mix proportion of UHPC by trial and error method. Finding the fresh properties of UHPC mixes to check the self compactibility.

- Cubes, cylindrical and beam specimens with control mix and UHPC with steel fibres addition (0, 0.5, 1, 1.5, 2.0 and 2.5%) of volume of concrete were casted.
- Curing of specimens for 7 and 28 days. Testing to find 7 and 28 day strength properties.
- Determining the optimum percentage of steel fibres from the test results which gives maximum strength values.
- Specimen were casted with optimum percentage of steel fibre and by varying the carbon nanotubes at 0, 0.03, 0.06, 0.09, 0.12 and 0.15% by volume of concrete.
- Compressive strength test, split tensile strength test and flexural strength test were done after 28 days of curing.
- Determine the optimum percentage of carbon nanotubes in UHPC containing steel fibers.
- Conclusions and recommendations were finally made based on the findings and observations.

4 Materials

4.1 Cement

Ordinary Portland cement of 53 grade conforming to IS 12269–1987 is used in the study. Physical properties of cement were tested using Le- Chatlier apparatus and Vicat apparatus. The standard consistency of cement was obtained as 39%. The initial setting time and final setting time was 145 and 350 min respectively. The specific gravity and fineness of cement was obtained as 3.15 and 7.33% respectively.

4.2 Fine Aggregate (FA)

Fine aggregate are soil particles passing through 4.75 mm IS sieve. Generally river sand, crushed stone, crushed gravel, M sand etc. are used as fine aggregate. In this study, M sand conforming to Zone II was used.

4.3 Quartz Sand

Quartz sand reduces porosity of the mixture and thereby increases the final strength and it is an alternative for river sand. Quartz sand with specific gravity 2.67 is used for this study.

4.4 Steel Fibre

Steel fibres mixed into the concrete can provide an alternative to the provision of conventional steel bars or welded fabric in some applications. In this study, steel fibre of 0.5 mm diameter is used.

4.5 Silica Fume (SF)

It is a by-product of silicon and ferrosilicon alloy production and consists of spherical particles with an average particle diameter of 150 μm.

4.6 Super Plasticizer (SP)

Super plasticizer is a chemical used as an admixture to improve the properties of concrete. High Range Water reducing admixture (HRWRA) or Super plasticizer is an important constituent in the production of UHPC to achieve self-compatibility and flow ability. The HRWRA helps in achieving excellent flow at low water content. Conplast SP 430 is used for in this study as super plasticizer.

5 Mix Design

Mix design of ultra high performance concrete has no standard method. The trial and error method was used to obtain the initial mix composition. The trial mix which satisfies the fresh concrete properties and target strength is selected as control mix with a w/c ratio of 0.22%. The mix proportion of concrete is shown in Table 1.

Table 1 Mix proportion of concrete

Sl. No	Cement (kg/m^3)	Fine aggregate (kg/m^3)	Silica fume (kg/m^3)	Quartz sand (kg/m^3)	w/c (%)	Super plasticizer (kg/m^3)
1.	226.03	537.73	113.2	79.6	0.22	2.71

6 Experimental Procedure

The experimental program contains measuring compressive strength, tensile strength and flexural strength of specimens at 7 and 28 day. Compressive strength is measured using cubes of 150 × 150 × 150 mm, the tensile strength is measured using 150 × 150 × 300 mm cylinders and flexural strength is tested by using 500 × 100 × 100 mm beams. 51 cubes, 17 beams and 33 cylinders were casted.

7 Results and Discussion

The compressive strength of the cubical specimens after 7 days and 28 days, tensile strength of cylindrical specimens and flexural strength of beam specimens after 28 days of curing are evaluated as per IS specification. Results are tabulated in Tables 2 and 3 (Figs. 1, 2 and 3).

Table 2 Hardened properties of UHPC mix with different steel fibre volume

Sl. No	% of steel fibre (%)	Avg. compressive strength, MPa		Avg. tensile strength, MPa	Avg. flexural strength, MPa
		7 day	28 day		
1	0	118	139	11.5	16
2	0.5	120	145	12.7	17.1
3	1	128	151	15.4	18.4
4	1.5	136	160	15.59	19.8
5	2	139	178	17.45	21.6
6	2.5	127	148	16.8	19.2

Table 3 Hardened properties of CNT added UHPC

Sl.No.	% of CNTs (%)	Avg. compressive strength, MPa		Avg. tensile strength, MPa	Avg. flexural strength, MPa
		7 day	28 day		
1	0	139	178	17.45	21.6
2	0.03	142.8	179.2	19.2	23.3
3	0.06	143.3	180.5	21	25
4	0.09	145.24	182	22.6	26
5	0.12	146.8	181.3	21.8	24.7
6	0.15	150	179	20	23.2

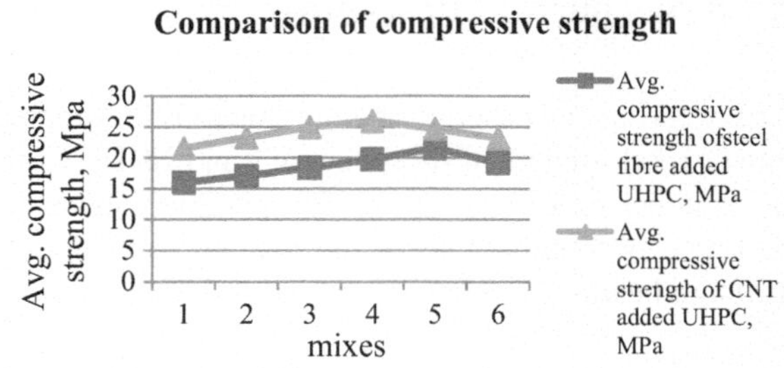

Fig. 1 Comparison of compressive strengths of UHPC

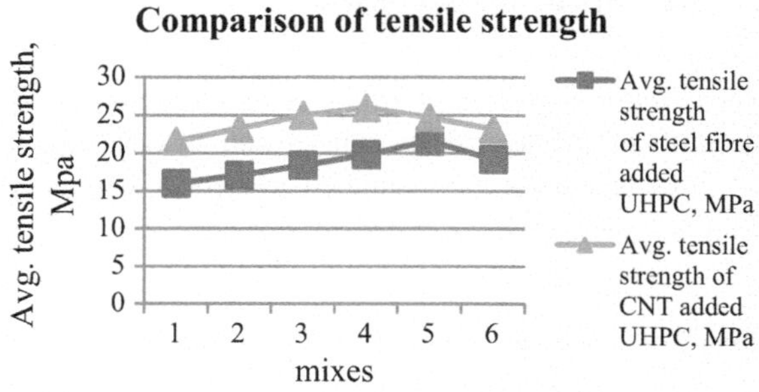

Fig. 2 Comparison of tensile strengths of UHPC

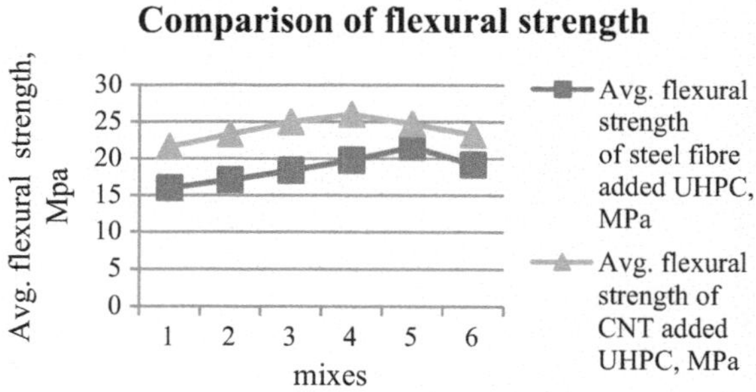

Fig. 3 Comparison of flexural strengths of UHPC

8 Conclusion

From the studies conducted on UHPC samples, it can be concluded that:

- The compressive and tensile strength of UHPC were notably higher than as specified by IS code for normal strength concrete.
- It is found that the optimum quantity of steel fibre to produce UHPC with high mechanical strength at all ages is 2% steel fibers by volume of concrete.
- As expected, UHPC specimen without steel fibers resulted in low tensile strength but with increasing steel fibers to 2% by concrete volume resulted in a higher tensile strength of more than 50% increase at 28 day.

- The compressive strength of UHPC with steel fibers is greater than that of plain UHPC by 11.25% at 2% addition of steel fibers
- The tensile strength of UHPC with steel fibers is greater than that of plain UHPC by 11.93% at 2% addition of steel fibers.
- The incorporation of nanomaterials increased the compressive, tensile and flexural strength of UHPC.
- As the content of CNT was increased from 0 to 0.09%, the compressive strength increased from 178 to 182 MPa.
- 0.09% of carbon nanotubes which gave the maximum strength properties is considered as the optimum percentage of CNT in UHPC.

Mechanical and Durability Characteristics of Bacterial Concrete

Hari Ashwin, V. Abhirami, Ameen Noushad Anzil, Joseph Danty Jerry and Asha Joseph

Abstract Concrete is the most widely used construction material. Micro cracks which occur on the concrete structures can reduce the strength of the structure. Prolonged exposure of these cracks to the stresses and external environment can increase the damage and can even be lethal. Bacterial concrete is a solution for the micro cracks. Bacterial concrete has a self healing property by which it can repair cracks upto 0.5 mm width. An investigation is conducted to study changes in the compressive strength, split tensile strength and flexural strength of concrete by the addition of bacteria. Concrete specimens with bacterial concentrations of 0, 15, 30 and 45 ml mixed with equal amount of calcium chloride are used. When crack occurs, and water passes through the cracks, the bacteria present in the concrete matrix starts to feed on the calcium chloride by consuming oxygen and converts the calcium chloride into the insoluble calcium carbonate (limestone). And this limestone heals the cracks. Durability of the bacterial concrete is also studied and compared with normal concrete by analyzing the water absorption and reaction to sulphate and chloride ions.

Keywords Bacterial concrete · Self healing · Compressive strength · Split tensile strength · Flexural strength · Durability

1 Introduction

Concrete is one of the most consumed materials by humans. Currently, it is the dominant element in the field of construction. Its main feature is its resistance against compressive forces. However cracks can occur over the course of its lifetime which can adversely affect its strength and durability. Cracks can provide pathway for the

H. Ashwin (✉) · V. Abhirami · A. N. Anzil · J. D. Jerry · A. Joseph
Department of Civil Engineerimg, Federal Institute of Science and Technology, Angamaly, Kochi 683577, India
e-mail: ashwinhari2941997@gmail.com

A. Joseph
e-mail: ashameledath@gmail.com

K. Dasgupta et al. (eds.), *Proceedings of SECON'19*,
Lecture Notes in Civil Engineering 46,
https://doi.org/10.1007/978-3-030-26365-2_75

entry of chlorides, sulphates, water, etc which can harm the concrete by affecting its durability. Entry of such substances can cause the reinforcements to corrode. Bacterial concrete [1–3] or self healing concrete incorporates bacteria which reacts with water and metabolizes crystals which fill the cracks and protect the steel within it. The crystals are also formed in the pores of the concrete making it more compact and strong [4]. Bacterial concrete is made by adding calcite precipitating bacteria along with suitable calcium based nutrient in the concrete mix. Bacillus subtilis is the bacteria used to remediate cracks in concrete. It is also known as Hay Bacillus or Grass Bacillus, is a gram-positive bacterium found in gastrointestinal tract of ruminants and humans and can form a tough, end spores, allowing it to tolerate extreme environmental conditions [5]. Subtilis reacts with water and converts Calcium chloride [6] added along with it in the concrete mix to obtain Calcium carbonate crystals. This phenomenon is called Microbiologically Induced Calcite Precipitation [7, 8]. The mineral precipitation is a natural and pollution free process. Hence this method is better when compared to alternative crack remediation techniques.

The present study aims to develop bacterial concrete of M30 grade using Bacillus subtilis bacteria, study and compare the compressive strength, split tensile strength and flexural strength of bacterial concrete with that of the control mix, study and compare the durability of bacterial concrete against water, chloride and acid with the control mix.

2 Methodology

2.1 Selection of Materials

Ordinary Portland Cement of grade 53, conforming to IS 12269 (1987) [9] manufactured by Dalmia Bharat Group is selected. Coarse aggregates conforming to the specifications of IS 10262:2009-Zone II [10] and specific gravity of 2.6 and nominal size 20 mm is used. M sand of specific gravity 2.6 and effective size of 1.3 mm and conforming to the specifications of IS 10262:2009-Zone II is used as fine aggregate. Natural potable water which satisfies the provisions of IS 456:2000 [11] is used. Bacillus Subtilis bacteria of concentration 2×10^8 cells/mL acquired from AbtecAgro Bio Tech Research Centre Ltd. Kottayam, is used. Calcium based nutrient selected for the bacteria is Calcium Chloride.

Table 1 Concrete mix proportions

Materials	Quantity (kg/m^3)
Cement	437.77
Fine aggregate	778
Coarse aggregate	947.79
Water	197
Cement:FA:CA	1:1.777:2.165

2.2 *Mix Proportioning*

The mix design is carried out as per IS 10262:2009 [10]. The grade of concrete adopted for this study is M30. Maximum size of aggregate taken is 20 mm. Water-cement ratio is taken as 0.4. Specific gravity of both fine aggregate and coarse aggregate is 2.6. Workability of 100 mm slump is considered. Mix design is done using these data and the obtained data is as given in Table 1.

2.3 *Preparation of Specimens*

Specimens of three bacterial concentrations and the control mix which is of zero bacteria are prepared. Concentrations of bacterial concrete selected are 15, 30 and 45 mL [12]. Calcium chloride is added as the mineral nutrient for the bacteria. Calcium chloride is added along with bacteria in a 1:1 ratio. That is equal amount of calcium chloride is added for a specific amount of bacterial solution. Concrete mix is prepared as per the mix proportion and desired amount of bacterial solution and equal amount of calcium chloride is mixed and dissolved in the water required for the mix.

Specimens required for compressive strength test and durability tests are concrete cubes of size 150 mm × 150 mm × 150 mm [13]. Three specimens of each concentration are prepared for compression test and two specimens each for the three durability tests. Cylindrical specimens of height 300 mm and diameter 150 mm [13] are prepared for the split tensile strength. Three samples of each concentration are made. Beam specimens of length 500 mm and height and width of 100 mm are made for the flexural strength test [13]. Three samples of each concentration are made. The specimens are designated as per Table 2.

The total number of specimens cast is 96 including cubes, cylinders and beams, Specimens are casted for 7 day test and 28 day test for the three strength tests.

Table 2 Specimen Designation

Mix designation	Bacterial solution added per specimen (mL)	$CaCl_2$ added per specimen (g)
BC0	0	0
BC15	15	15
BC30	30	30
BC45	45	45

2.4 Curing and Testing

Samples are prepared as per Table 2. All the specimens required for the strength tests are cured in water. After 7 days of curing, three specimens of each bacterial concentration are taken out and dried. Compressive strength test are done on the concrete cubes in the Compression Testing Machine (CTM), Split tensile strength test is done on the concrete cylinders in the Universal Testing Machine (UTM) and the Flexural strength test is carried out on the concrete beams in the Flexural Testing Machine. The obtained load and stress values are noted. After 28 days, rest of the specimens for the strength tests are taken out of the curing tank and are tested and the values obtained are noted. Graphs are plotted with the obtained values.

The concrete cubes required for the durability tests are weighed after they are casted and demoulded. The specimens for water absorption test are cured in water. These specimens are taken out after 7 days and are wiped well with a dry cloth to remove the excess water and are weighed. They are put back into the curing tank. The percentage increase in weight is calculated. The same procedure is repeated after 14 and 28 days. Graphs are plotted with the observed values.

The concrete cubes for chloride test are cured in 5% HCl solution and for acid test are cured in 5% H_2SO_4 solution [14]. Weight is taken at 7, 14 and 28 days intervals. The percentage decrease in their weights is calculated for the chloride and acid tests. Graphs are plotted with the observed values.

3 Observations and Results

3.1 Compressive Strength Test

This test gives an idea about the characteristics of concrete. The compressive strength of concrete depends on many factors such as water cement ratio, cement strength, quality of concrete material, quality control during production of concrete etc. A cube sample tested in the CTM is shown in Fig. 1. The compression test results obtained are as shown in Fig. 2.

Fig. 1 Cubes tested in CTM

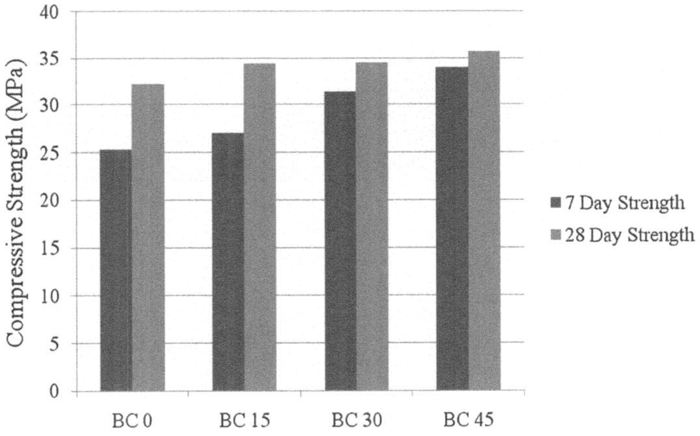

Fig. 2 Result of compressive strength test for control mix and bacterial concrete at 7 days and 28 days

3.2 Split Tensile Strength Test

Split tensile strength test in concrete cylinder is a method to determine the tensile strength of concrete, which is one of its basic and important properties. Concrete due to its brittle nature is not expected to resist the direct tension and hence develop cracks when subjected to tensile forces. Thus this test directly indicates the load at which concrete members tend to crack. A cylinder sample tested in the UTM is shown in Fig. 3. The results of the split tensile strength test obtained are as shown in Fig. 4.

Fig. 3 Cylindrical specimens tested in UTM

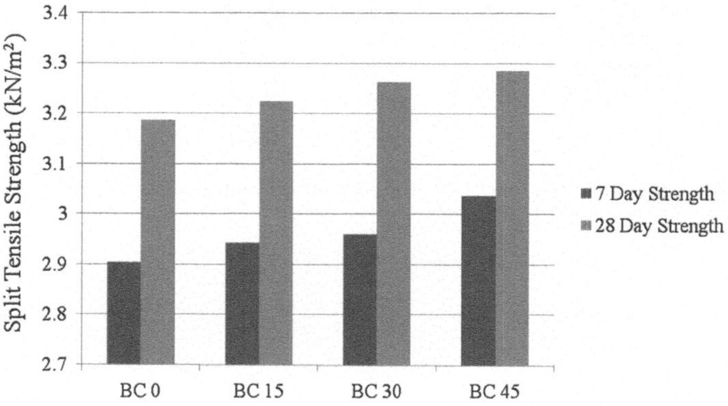

Fig. 4 Result of split tensile strength test for control mix and bacterial concrete at 7 days and 28 days

3.3 Flexural Strength Test

Flexural strength is one measure of the tensile strength of concrete. It is a measure of an unreinforced concrete beam to resist failure in bending. When the flexural strength of concrete is increased the concrete can withstand greater forces without any major defects. A beam sample tested in the flexural testing machine is shown in Fig. 5. The flexural strength test results obtained are as shown as in Fig. 6.

Fig. 5 Beam tested in flexural testing machine

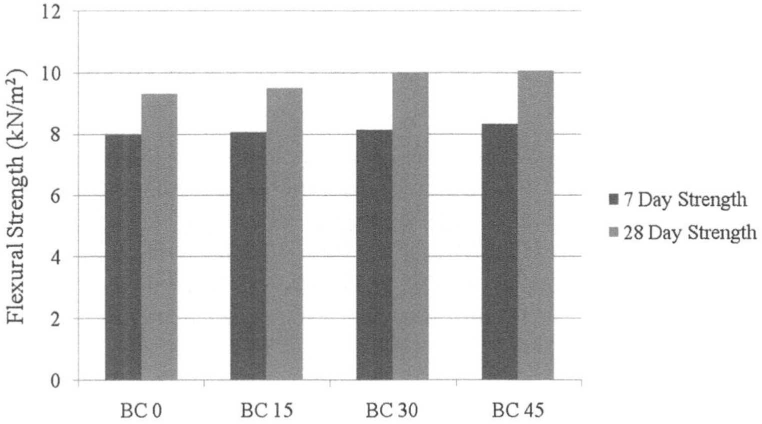

Fig. 6 Result of flexural strength test for control mix and bacterial concrete at 7 days and 28 days

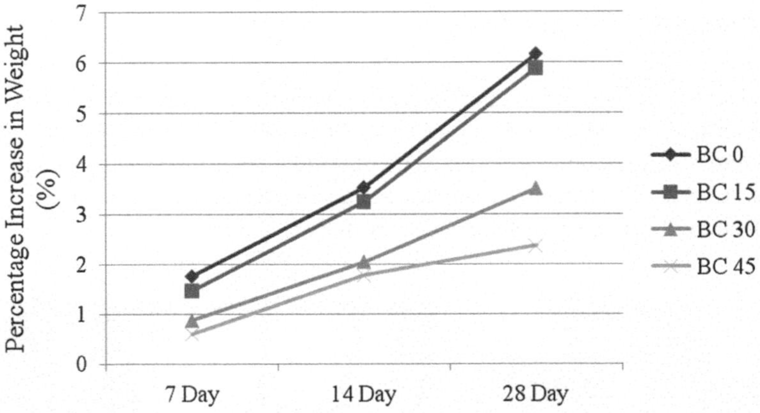

Fig. 7 Result of water absorption test for control mix and bacterial concrete at 7, 14 and 28 days

3.4 *Water Absorption Test*

The water absorption test is used to analyse the water absorption by concrete cube specimen. Initial weight of the specimen after casting is compared with the weights after 7, 14 and 28 days of curing and percentage increase in the weight of the specimens are calculated. Low water absorption is desirable for the durability of concrete. The water absorption results obtained are as shown in Fig. 7.

3.5 *Acid Test*

Acid test is conducted to study the reaction of concrete specimen in an acidic medium. Concrete cube specimens of each bacterial concentration are casted and are cured in 5% Sulphuric acid solution. Specimens are weighed at 7, 14 and 28 day intervals and are compared with their initial weight. Their percentage decrease in weight is calculated. Low weight decrease values are desirable. Specimen cured in sulphuric acid is shown in Fig. 8. The results of the acid test conducted are given in Fig. 9.

3.6 *Chloride Test*

Chloride test is conducted to study the action of chloride ions towards the concrete cube specimen. Concrete cube specimens of each bacterial concentration are casted and are cured in 5% Hydrochloric acid solution. Specimens are weighed at 7, 14 and 28 day intervals and are compared with their initial weight. Their percentage decrease in weight is calculated. Low weight decrease values are desirable. Specimen cured

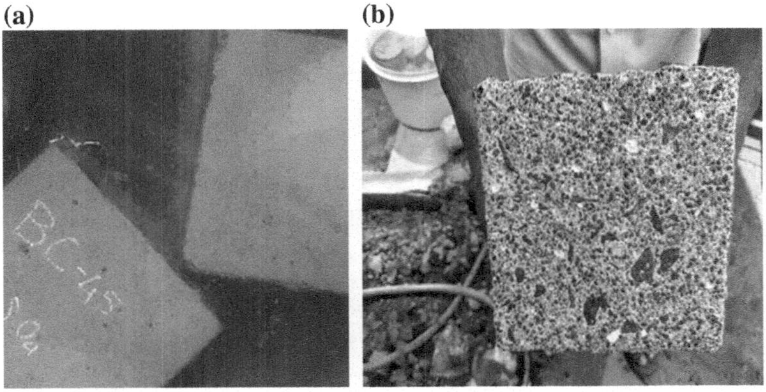

Fig. 8 **a**. Cube in Sulphuric acid after 1 day. **b**. Cube after 7 days of curing in sulphuric acid

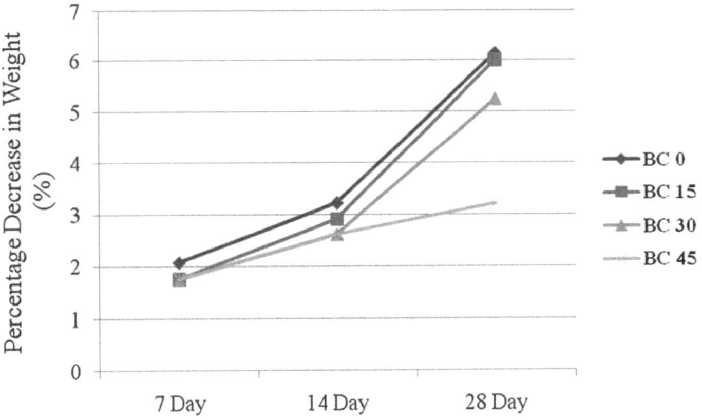

Fig. 9 Effect of acid on control mix and bacterial concrete

in hydrochloric acid is shown in Fig. 10. The results of the chloride test conducted are as shown in Fig. 11.

4 Conclusion

Compressive strength, split tensile strength and flexural strength tests were carried out on specimens designated as BC 0, BC 15, BC 30 and BC 45 of four bacterial concentrations 0 mL, 15 mL, 30 mL and 45 mL respectively. Compression test was done on cubes of size 150 mm × 150 mm × 150 mm, split tensile strength test on cylinders of height 300 mm and diameter 150 mm and flexural strength test was carried out on beams of 500 mm length and 100 mm × 100 mm cross section.

Fig. 10 **a**. Specimen immersed in HCl Solution for curing. **b**. Specimen after 7 days of curing in HCl Solution

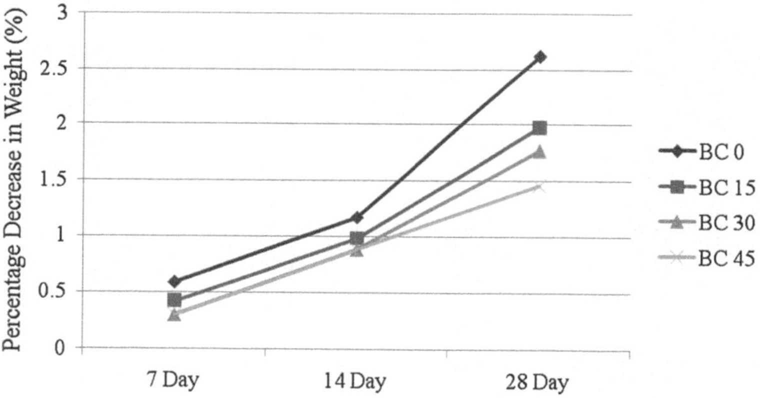

Fig. 11 Effect of chloride on control mix and bacterial concrete

Durability was studied by conducting water absorption test, chloride test and acid test on cube specimen of all four bacterial concentrations.

The following observations were made from the result of the tests conducted.

- The compressive strength increases with the bacterial concentration and maximum is observed for BC 45 specimens.
- The split tensile strength increases with the bacterial concentration and maximum is observed for BC 45 specimens.
- The flexural strength increases with the bacterial concentration and maximum is observed for BC 45 specimens.
- For specimens cured in water, the percentage weight increase is lesser for higher bacterial concentration and is least for BC 45.

- For specimens cured in sulphuric acid solution, the percentage weight decrease is lesser for higher bacterial concentration and is least for BC 45.
- For specimens cured in HCl solution, the percentage weight increase is lesser for higher bacterial concentration and is least for BC 45.

The conclusion that can be drawn from these observations is that incorporating bacteria in concrete can improve strength as well as durability of the concrete.

References

1. Kunamineni V, Meena M, Shirish VD (2017) Bacteria based self healing concrete—a review. Constr Build Mater
2. Meera CM, Subha V (2017) Strength and durability assessment of bacteria based self healing concrete. Constr Build Mater 152:1008–1014
3. Sandip M, Palash D, Arun KC (2017) Application of bacteria in concrete. Mater Today: Proc 4:9833–9836
4. Mondal S, Ghosh AD (2018) Investigation to optimal bacterial concentration for compressive strength enhancement of microbial concrete. Constr Build Mater 183:202–214
5. Joshi KA, Kumthekar MB, Godake VP (2016) Bacillus subtilis bacteria impregnation in concrete for enhancement in compressive strength. Int J Sci Technol Eng 03(05)
6. Hsu C-M, Huang Y-H, Nimje VR, Lee W-C, Chen H-J, Kuo Y-H, Huang C-H, Chen C-C, Chen C-Y (2018) Comparative study on the sand bioconsolidation through calcium carbonate precipitation by sporosarcina pasteurii and bacillus subtilis. Multidisciplinary Digital Publishing Institute
7. Antonio F, Brunella P, Giorgio M, Milvia LR (2010) Study of bacillus subtilis genes involved in calcium carbonate biomineralization. University of Florence
8. Savitha J, Reshma GV, Alina R, Jisbin J, Devika V (2018) A study on strength characteristics of bacterial concrete. Int Res J Eng Technol
9. IS 12269 (1987) Specification for 53 grade ordinary Portland cement [CED 2: Cement and Concrete], Bureau of Indian Standards
10. IS 10262:2009 - Concrete mix proportioning—guidelines, Bureau of Indian Standards
11. IS 456 (2000) Plain and reinforced concrete—code of practice [CED 2: Cement and Concrete], Bureau of Indian Standards
12. Dileep Kumar RB, Salma S, Krishna Chaitanya PVS, Mounika K (2017) Experimental study on bacterial concrete. Int J Sci Technol Eng 3(11)
13. IS 516 (1959) Method of tests for strength of concrete [CED 2: Cement and Concrete], Bureau of Indian Standards
14. Hiren P, Piyush J, Kaizad E, Vasim MK (2017) The experimental investigation of durability test on concrete cubes. Int J Adv Eng Res Dev

Experimental Investigation on Fresh and Hardened Properties of Hybrid Fibre-Reinforced Self-Compacting Concrete

Ashly Joseph and S. Sreerath

Abstract Self-Compacting Concrete (SCC) is a concrete mix having high flowability and resistance to segregation. It fills the framework and consolidates under its own weight without external vibration. SCC has a brittle nature, therefore to improve its tensile properties as well as the behavior under the impact, impregnation of different kinds of fibres can be adopted [1]. Reinforced SCC with polypropylene fibre has shown promisable result in both fresh and hardened state, addition of Hooked end steel fibres along with polypropylene fibre could enhance flexural residual strength, producing a hybrid fibre-reinforced SCC, which is well suited for structural use. Three volume ratios of steel fibres (0.5, 1.0, 1.5%) were mixed with three amounts (0.3, 0.6, 0.9%) of polypropylene fibres. Studies are carried out on the fresh and hardened properties of SCC for all the mixes. The fresh property tests such as Slump flow test, V-Funnel test, U-Box test and L-box test was conducted on the developed SCC to check the compatibility, filling ability, passing ability and segregation resistance. The hardened properties such as compressive strength, split tensile strength, flexural strength was determined by conducting suitable tests as per the European guidelines of SCC (EFNARC) and compared with that of standard specimen. From the experimental test results compressive strength of concretes did not change considerably, but the addition of fibres had noteworthy effects on splitting tensile and flexural strengths. As expected, the increase in fibre content led to an increase in these strengths. Fresh properties of SCC were found to decrease with an increase in fibre content.

Keywords Self -Compacting concrete · Polypropylene fibre · Hooked end steel fibre

A. Joseph (✉) · S. Sreerath
Department of Civil Engineering, Federal Institute of Science and Technology, Angamaly, Kochi 683577, India
e-mail: ashlyjoseph35@gmail.com

S. Sreerath
e-mail: iamsreerath@gmail.com

K. Dasgupta et al. (eds.), *Proceedings of SECON'19*, Lecture Notes in Civil Engineering 46, https://doi.org/10.1007/978-3-030-26365-2_76

1 Introduction

Self-compacting concrete (SCC) can be considered as a concrete which has little resistance to flow so that it can be placed compacted under its own weight with little or no vibration effort, even it possesses enough viscosity to be handled with segregation or bleeding. SCC was developed in Japan in the late 1980 as a solution to attain durable concrete structure independent of the quality of construction work. The concept of using fibres as a reinforcement in the concrete mixture is not a new study. The use of fibres has been carried out from ancient times. There are different types of fibre reinforced concrete that are categorized based on the fibre that is employed. If steel fibre is used we get steel fibre reinforced concrete. Similarly, nylon reinforced concrete, glass fibre reinforced concrete, carbon fibre reinforced concrete etc. are some of the types. A composite can be stated as a hybrid when two or more type of fibres is used in a combined matrix to produce a composite that will reflect the benefit of each of the individual fibre used. This will finally furnish a synergetic response to the whole structure. Such a composite of concrete is termed as the Hybrid Fibre Reinforced Concrete (HFC) [2]. The mechanical properties of concrete are enhanced appreciably using fibres. This will reduce the chances of brittleness and hence small crack formation, as small cracks are the main factors behind propagation and larger cracks formation. Fibre debonding or chances of pull out are less as this cause requires large energy absorption. This is the reason that gives fracture resistance and toughness to HFC during dynamic as well as cyclic loads.

2 Experimental Program

2.1 Materials

The cement used for this study is Ordinary Portland cement (OPC-53 grade) conforming to IS: 12269-1987 Additionally, fly ash class F is designated in ASTM C 618 was used as mineral admixture [3]. The shape of the steel fibre and polypropylene fibres are provided in Figs. 1 and 2 [4]. Also, Table 2 presents the properties

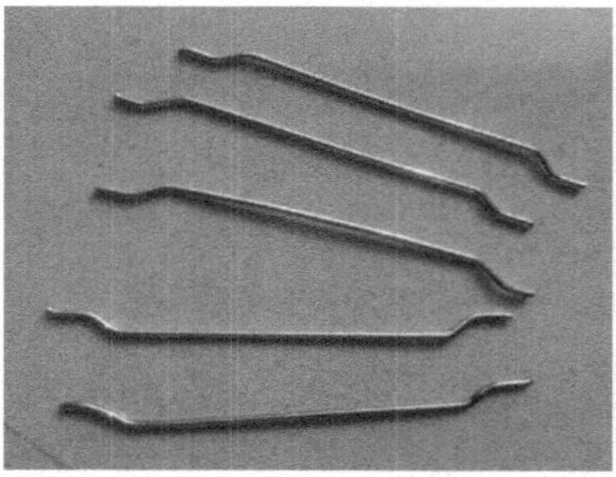

Fig. 1 Hooked end steel fibre

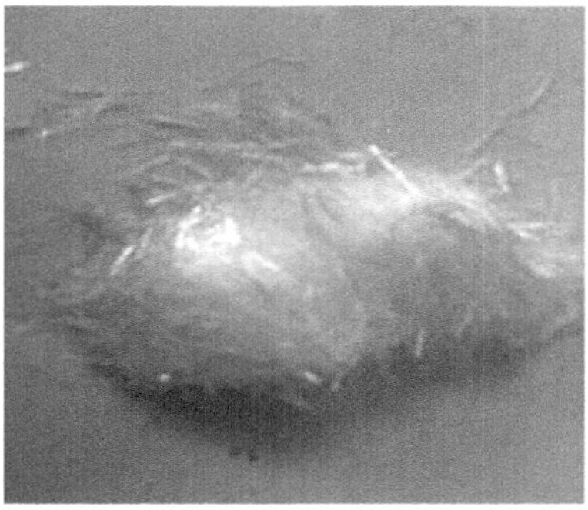

Fig. 2 Polypropylene fibres

of hooked-end steel and polypropylene fibres that were used in concrete mixtures. A polycarboxylate-based superplasticizer conforming to ASTM C494 Type-F and with a specific gravity of 1.09 was used to increase the workability of concrete. In addition, the viscosity modifying agent (VMA) with a specific gravity of 1.0 was used to prevent from bleeding and segregation of aggregates from cement paste. Fine and coarse aggregates with SSD condition and maximum size of 12 mm were used in all mixtures. The sand and gravel had a specific gravity of 2.64 and 2.68 respectively. The sand had a fineness modulus of 2.71.

2.2 Mixing Design and Testing Procedures

For self-compacting concrete, there is no specific method for mix design. Mix design is done as per IS 10262:2019 and EFNARC 2005. The composition of the mixes is presented in Table 1. In the present paper the hybrid mixes were prepared with the use of steel and polypropylene fibres. Three volume ratios of steel fibres 0.5, 1.0 and 1.5%, were tested. The examined steel fibres were 35 mm long with a sector of the circle cross-section and hooked end shape (Table 2). The considered polypropylene fibres of lengths 12 mm (Table 2) were added in the amounts of 0.3, 0.6 and 0.9%, by volume of concrete [5]. The different combinations of steel and polypropylene fibres can be found in Table 3, where the rheological parameters of hybrid fibre-reinforced self-compacting concrete (HFR-SCC) including slump-flow and L-box tests are also summarized. The mechanical properties of the HFR-SCC mixes were tested in compressive, split tensile and flexural strength at the age of 28 days. All the test was conducted as per the European guidelines of SCC (EFNARC). The compressive tests have been carried out in 2000 kN hydraulic compression testing machine on cubes with dimensions of 150 × 150 × 150 mm. For every mix, 6 specimens have been tested with a constant loading rate. The flexural tensile parameters were tested in two-point bending tests on three beams for each mix with dimensions of 100 × 100 × 500 mm. The split tensile strength test was tested in 1000 kN universal testing machine on cylinders with dimensions of 150 × 300 mm.

Table 1 Composition of HFR-SCC mix

Cement (kg/m^3)	Fine aggregate (kg/m^3)	Coarse aggregate (kg/m^3)	Water (kg/m^3)	Fly ash (kg/m^3)	Steel fibres (%) by volume	Polypropylene fibres (%) by volume	Superplasticizer (kg/m^3)	Viscosity modifying agent (kg/m^3)
508	438	564	395	353	0.5, 1.0, 1.5	0.3, 0.6, 0.9	7.48	1.6

Table 2 Properties of fibres

Designation	SF	PP
Length [mm]	35	12 mm
Tensile strength [MPa]	1100 MPa	400
Effective diameter	0.65 mm	0.022 mm
Aspect ratio (length/effective diameter)	53.85	545
Fibre shape	Hooked end steel fibre	Straight
Acid/Alkali resistance	–	Excellent

Table 3 Properties of fresh HFR-SCC mix

Mix	Slump flow (mm)	T_{500} slump flow (s)	V-funnel flow time (s)	L-box (PA)	U-box (mm)
SCC control mix	755	3	6	0.86	1
0.5SF	750	3	6.44	0.85	2
0.5SF+0.3PP	740	3.3	7.51	0.84	4
0.5SF+0.6PP	670	3.8	8.46	0.82	4
0.5SF+0.9PP	610	4.1	9.86	0.8	4.1
1SF	740	4.3	6.77	0.85	5
1SF+0.3PP	733	4.6	7.84	0.84	5.2
1SF+0.6PP	660	4.8	8.91	0.83	5.2
1SF+0.9PP	600	5.1	9.98	0.82	5.3
1.5SF	715	5	7	0.9	6
1.5SF+0.3PP	710	5.2	7.94	0.92	6.4
1.5SF+0.6PP	650	5.6	9.24	0.93	6.5
1.5SF+0.9PP	555	6	10.1	0.98	6.6

3 Test Results and Discussion

3.1 Properties of Mixes in the Fresh State

In general, fresh parameters decrease with an increase in the amount of fibres, which is a well-set knowledge [6]. The mixes prepared with the use of only steel fibres satisfied the requirements for the SCC in the fresh state. All the hybrid mixes conformed to the demanded range of slump flow test (650–800 mm), except for those containing 0.9% of the PP fibres. The passing ability of all hybrid mixes were investigated in L-box tests and U-box test, where 0.9% of PP fibres was applied, gives higher value than control specimen. Considering the summary amount of fibres it can be concluded that for the $V_f > 2\%$ the hybrid mixes did not satisfy the requirements for the SCC.

This trend was observed without regard for superplasticizer amount. In all hybrid mixes the increase in porosity in comparison to plain SCC was observed.

3.2 Compressive Strength

As expected, the application of hybrid fibres did not have any significant influence on the compressive strength of the SCC matrix (Fig. 3), which can be attributed to the fact that the fibres influence mostly the post-cracking behaviour of the matrix [7]. An increase in the compressive strength was noted when the steel fibres were added up to 1.5%. Increase in compressive strength is due to the effective role of steel fibres as agents preventing propagation of cracks, reducing their development, providing higher contact surface area and denser matrix. A slight increase in the compressive strength was noted for the mix containing 0.3% of PP fibres and all other specimens containing polypropylene fibres showed lower compressive strength than of control mix at the age of 28 days and the value of f_c varied in the range of 0.5–16%. The compressive strength reduction can be generally explained by the fact that the fibres are some perturbation which causes higher amount of voids in the matrix. The highest observed decrease of f_c in case of application of the PP fibres can be also attributed to the instability of the mix in the fresh state [8].

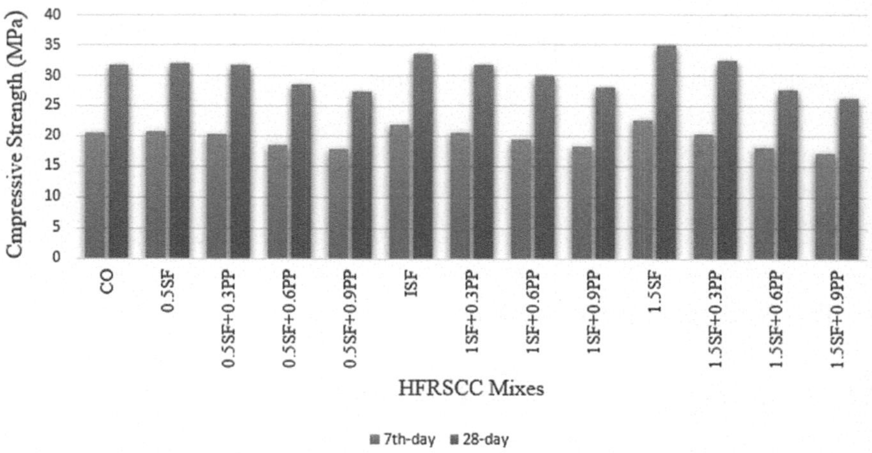

Fig. 3 Graphical representation for compressive strength of HFRSCC mixes

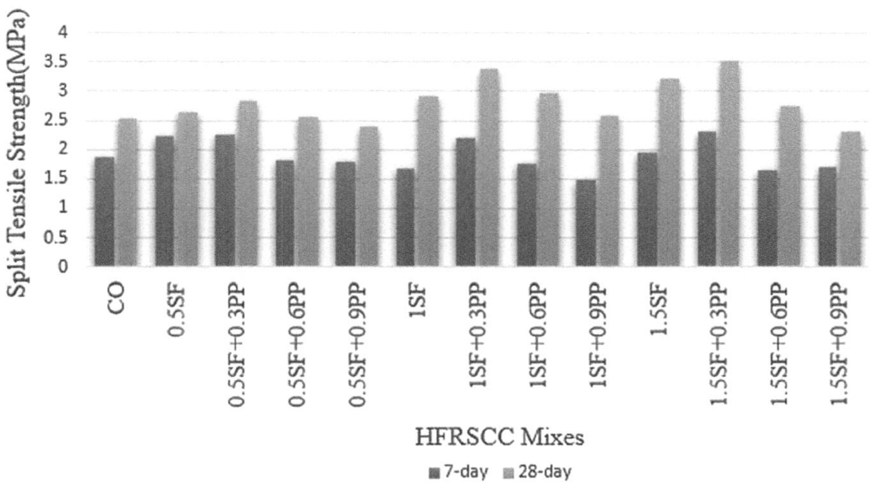

Fig. 4 Graphical representation for split tensile strength of HFRSCC mixes

3.3 *Split Tensile Strength*

The results show that while the compressive strength of concretes did not change considerably, the addition of fibres had noteworthy effects on splitting tensile strength. As expected, the increase in fibre content led to an increase in the strength. The addition of 0.5, 1.0, and 1.5% of steel fibres into concrete led to an increase of splitting tensile strength at the age of 28 days by 0.4, 15 and 27% compared to control mix. However, the hybrid mix 1.5SF+0.3PP specimen shows higher splitting tensile strength than the control specimen and percentage in increase were 39%. As a matter of fact, due to the longer length, more suitable shape, greater modulus of elasticity and higher tensile stress of steel fibres than polypropylene fibres, steel fibres adequately bridge between the internal micro cracks of concrete and improve the splitting tensile strength. Exceptionally, for the hybrid mixes containing 0.6% of polypropylene fibre shows a relative increase in split tensile strength when compared with control specimen. Lower splitting tensile strength was found in hybrid mixes containing 0.9% polypropylene fibre. The reduction in the splitting tensile strength may be because of their different characteristics such as length, shape, modulus of elasticity and tensile stress. The comparison between the splitting tensile strength of the control specimen and specimens containing fibres is presented in Fig. 4 [9].

3.4 *Flexural Strength*

The results of the flexural tests generally reaffirmed the role of the long fibres in attracting the macro-cracks and short fibres in arresting the micro-cracks. The results

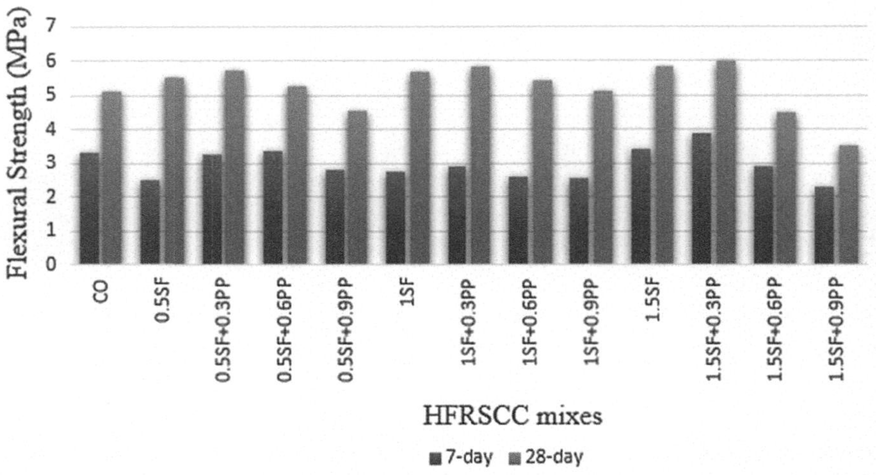

Fig. 5 Graphical representation for flexural strength of HFRSCC mixes

showed that the flexural parameters of HFR-SCC were pronouncedly enhanced due to hybrid fibre addition. By adding 0.5, 1.0 and 1.5% of steel fibres into SCC, the flexural strength increased by 7, 10 and 14% at the age of 28 days [10]. The hybrid mix 1.5SF+0.3PP specimen shows higher flexural strength than the control specimen and percentage in increase were 17%. The specimens containing 0.9% polypropylene fibres showed lower flexural strength. For the hybrid mixes containing 0.6% of polypropylene fibre shows a relative increase in flexural strength when compared with control specimen. On the contrary, the reduction in the flexural strength may be because of their different characteristics such as length, shape, modulus of elasticity and tensile stress. The increase of the total amount of fibres for 2% even to 3% can be ineffective in enhancing the flexural parameters of hybrid fibre reinforced mixes. The comparison between the splitting tensile strength of the control specimen and specimens containing fibres is presented in Fig. 5.

4 Summary and Conclusions

The following main conclusions can be drawn from the laboratory investigations:

- Analyzing the results of the hybrid mixes in the fresh state, Fresh properties of SCC were found to decrease with an increase in fibre content, considering the summary amount of the fibres, it can be concluded that for the fibres volume ratio higher that 2% the mixes did not satisfy requirement for the SCC. The mixes containing 0.9% PP fibres should be analyzed as conventional concrete.

- The steel fibres decreased the slump flow in SCC due to the presence of fibres which prevent cement paste from flowing. Polypropylene fibres (0.6 and 0.9%) drop the slump flow of hybrid fibre-reinforced mixtures considerably
- In comparison with steel fibres, polypropylene fibres showed greater effects on increasing the viscosity and V-funnel time flow. Addition of steel fibres with polypropylene fibres makes the V-funnel flow time to increase up to 66%. All the V-funnel flow times were within the allowable range (6–12 s) for SCCs
- The compressive strength of SCC slightly increased with the addition of steel fibres at the 28 days. The reason is attributed to the effective role of steel fibres as agents preventing propagation of cracks, reducing their development, and providing higher contact surface area
- Hybrid mix containing 0.6 and 0.9% of polypropylene gives a reduction in compressive strength compared to the control sample up to 38%. The reason is attributed to polypropylene fibres increased the porosity and rate of trapped air in the concrete but for the hybrid mix containing 0.3% polypropylene fibre shows a slight influence on compressive strength of the SCC matrix.
- The hybrid mixes 1.5SF+0.3PP specimen shows higher splitting tensile strength than the control specimen and percentage in increase was 39%. Hybrid mix containing 0.9%, polypropylene decreased the splitting tensile strength gradually and even became lower than the control sample in some hybrid fibre-reinforced specimens.
- Flexural parameters of HFR-SCC were pronouncedly enhanced due to hybrid fibre addition. Steel fibres play the most important role in enhancement of the flexural parameters of hybrid fibre-reinforced SCC mixes. The hybrid mix 1.5SF+0.3PP shows higher flexural strength than the control specimen and percentage in increase 17%.
- Among all the hybrid combination, 1.5SF+0.3PP mix performed better in all respects compared to other hybrid combination.

Acknowledgements The authors gratefully acknowledge the contribution of the faculties for thier guidance regarding the technical aspects of the work. The author would also like to thank the co-author Sreerath S. for contributions to this project. We thank FISAT engineering college for providing us the platform for working on the paper. All contributions are greatly acknowledged for generation of the paper.

References

1. Ahmad S (2018) Rheological and mechanical properties of self-compacting concrete with glass and polyvinyl alcohol fibres. Build Eng 49:69–77
2. Sai Nitesh KJN (2016) An experimental investigation on effect of hybrid fibre on high strength self compacting concrete and vibrated concrete. Earth Sci Eng 9:400–403
3. Yaman O, Sahmaran M (2007) Hybrid fibre reinforced self-compacting concrete with a high volume coarse fly ash. Constr Build Mater 21:150–156
4. Sivakumar A (2007) Experimental investigation on high strength concrete reinforced with hybrid fibres (combination of hooked steel and a non-metallic fibre). Cem Concr Res 35:27–30

5. Aydin AC (2017) Self compactability of high-volume hybrid fibre reinforced concrete. Constr Build Mater 21:1149–1154
6. Sahmaran M, Yurtseven A, Yaman IO (2005) Workability of hybrid fibre reinforced self-compacting concrete. Build Environ 40:1672–1677
7. Akcay B, Tasdemir MA (2012) Mechanical behaviour and fibre dispersion of hybrid steel fibre reinforced self-compacting concrete. Constr Build Mater 28:287–293
8. Singh AK (2010) Effect of inclusion of polypropylene and steel fibres on the compressive and flexure properties of fibre reinforced concrete. Build Environ 40:72–77
9. Hameed R (2010) Experimental investigation on flexural properties of metallic-hybrid-fibre-reinforced concrete. Cem Concr Res 25:55–62
10. Pajak M, Ponikiewskib T (2017) Experimental investigation on hybrid steel fibres reinforced self-compacting concrete under flexure. Procedia Eng 193:218–225
11. Shetty MS (2012) Concrete technology. S. Chand and Publications Pvt Ltd

Experimental Investigation on PCE and SNF Type Admixture on Early Age Strength of M40 Grade Green Concrete

Basil Baby, A. V. Daniel, Emmanuel Jose, P. P. Gokul, Naveen John and Sachu Saju

Abstract Super plasticizers are chemical admixtures used where well-dispersed particle suspension is required. Its application in concrete facilitates the reduction of the water with respect to the cement ratio without adversely influencing the workability of the concrete sample. Nowadays, due to rapid constructions there is a huge demand for the use of chemical admixtures. Among these admixtures, super plasticizer poly carboxylic ether (PCE) and naphthalene formaldehyde sulphonate (SNF) has great market relevance. The optimum dosage of the admixture is to be determined and the percentage reduction of water content is also noted. Various tests are conducted on the sample at different mix proportion and at different age as per code specifications. Perlite, which is a type of amorphous volcanic glass, is replace with fine aggregate at optimum dosage of the super plasticizer and strength is determined [1]. Perlite addition will results in green concrete of superior strength, fire resistance, acoustic properties, thermal insulation, HVAC conditions along with the replacement of conventional materials used in concrete. This research investigates on the determination of optimum dosage of super plasticizer and possible replacement of fine aggregate with perlite for sustainable concrete.

B. Baby (✉) · A. V. Daniel · E. Jose · P. P. Gokul · N. John · S. Saju
Department of Civil Engineering, Viswajyothi College of Engineering and Technology, Vazhakulam, India
e-mail: basilpisharath@vjcet.org

A. V. Daniel
e-mail: daniel.av85@gmail.com

E. Jose
e-mail: emmanueljose0107@gmail.com

P. P. Gokul
e-mail: gokulpramod999@gmail.com

N. John
e-mail: naveenathickal@gmail.com

S. Saju
e-mail: sachuoonnanal@gmail.com

K. Dasgupta et al. (eds.), *Proceedings of SECON'19*,
Lecture Notes in Civil Engineering 46,
https://doi.org/10.1007/978-3-030-26365-2_77

Keywords Super plasticizer · Green concrete · Water reduction

1 Introduction

Super plasticizers, also known as high range water reducers, are chemical admixtures used where well-dispersed particle suspension is required. Super plasticizers produce a homogeneous, cohesive concrete generally without any tendency for segregation and bleeding. But the optimum dosage value of super plasticizer is not available for various mixes. The optimum dosage suggested by the manufactures is only a range value. But by providing a value other than optimum or correct value it will leads to reduction in strength parameters. This study aims on determining the optimum dosage need for selected grade of concrete [1]. Super plasticizers, Sikament 2002 NS and C Flow 175 M is selected for the analysis and testing, after conducting market studies and survey among construction companies around Thodupuzha, Muvattupuzha and Ernakulam. M40 mix was selected for sample preparation as per market feedback. Strength tests and material properties were determined before designing the mix proportion for testing and verified as per code specifications.

Perlite is an amorphous volcanic glass which can be used for the replacement fine aggregate. Their addition will not affect any reaction procedure. The addition of perlite will improve the properties like fire resistance thermal insulation compatibility with cement, resistance to spalling, acoustic properties and HVAC condition in a building. The replacement of fine aggregate fraction with perlite will results in a more eco-friendly concrete [3]. The necessary investigation on the effect of perlite on strength of concrete is further investigated, which results in a sustainable concrete.

2 Methodology

2.1 Material Selection

2.1.1 Cement

OPC 43 cement should conform to IS: 8112-1989 and the designed strength of 28 day curing should be minimum 43 MPa or 430 kg/cm^2. The specific gravity of the cement used is 3.12.

2.1.2 Coarse Aggregate

Locally available crushed aggregates were used for the mix proportion. 20 and 12.5 mm aggregates were taken for the sample at a percentage ratio of 60 and 40%.

The specific gravity of 20 and 12.5 mm aggregate are 2.85 and 2.83 respectively. The fineness modulus of 20 and 12.5 mm are 8.25 and 7.5 respectively

2.1.3 Fine Aggregate

Manufactured sand has been used as fine aggregate which has the following properties. The specific gravity and water absorption was found to be 2.59 and 1.63% respectively. The fineness modulus of fine aggregate was found to be 2.94.

2.1.4 Super Plasticizer

Super plasticizer Sikament 2002 NS (SNF) imparts high workability to concrete. 0.5–2.0% by weight of cement is the dosage value recommended by the manufactures. The specific gravity of Sikament 2002 NS is 1.17. C FLOW 175 M is a PCE (Poly carboxylic ether) super plasticizer formulated to reduce the water content of a concrete mixture with a specific gravity of 1.12. The recommended dosage is between 0.4 and 0.8% by weight of cement [5]. These super plasticizers works as a dispersant by preventing the flocculation of fine particles of cement. The hydrophilic tip formed by plasticizers is able to reduce the surface tension of water and the adsorbed polymer keeps the cement particles apart by electrostatic repulsion and thereby increasing the workability.

2.1.5 Perlite

Perlite is a type of amorphous volcanic glass which is formed by the hydration of obsidian. It's an ultra-lightweight thermal resisting & insulating material. The specific gravity of the selected perlite sample is 2.22. The chemical composition of Perlite are: 70–75% SiO_2, 12–15% Al_2O_3, 3–4% Na_2O, 3–5% K_2O, 0.5–2% Fe_2O_3, 0.2–0.7% MgO and 0.5–1.5% CaO. It has a well-sealed tough structure which prevents bead damage during mixing and facilitates low water absorption, hence provides proper curing of cement.

2.2 *Mix Design*

Mix design was mainly based on the dosage of super plasticizer and water reduction. The M40 mix design was carried out as per the suggestions in IS 10262:2009, IS 456:2000 and IS 383:1997. The values are (Tables 1 and 2).

Table 1 Quantity for 1 m^3 concrete with SNF type super plasticizer

Materials for 1 m^3 of concrete						
Using SNF type super plasticizer						
% of super plasticizer by weight of cement	% of water reduction	Cement (kg)	Water (kg)	Fine aggregate (kg)	Coarse aggregate (kg)	Admixture (kg)
0.5	20	350.51	157.73	699.35	1196.70	1.75
0.5	25	320.60	147.87	715.52	1224.28	1.64
0.75	25	320.60	147.87	715.52	1224.28	2.46
1.0	30	306.39	138.01	731.67	1252.05	3.07

Table 2 Quantity for 1 m^3 concrete with PCE type super plasticizer

Materials for 1 m^3 of concrete						
Using PCE type super plasticizer						
% of super plasticizer by weight of cement	% of water reduction	Cement (kg)	Water (kg)	Fine aggregate (kg)	Coarse aggregate (kg)	Admixture (kg)
0.5	20	350.51	157.73	699.35	1309.93	1.75
0.5	25	328.60	147.87	715.52	1340.22	1.64
0.6	25	328.60	147.87	715.52	1340.22	1.97
0.7	30	306.69	138.01	731.69	1370.52	2.15

2.3 *Tests Conducted on Concrete*

2.3.1 Slump Test

Slump test is conducted before the setting of cement to determine the workability and consistency of the concrete prepared. The trial carried out helped to determine the dosage of admixture and water reduction in the mix preparation. The test is conducted on a conical frustum of height 30 cm.

2.3.2 Strength Test

Compressive strength test is conducted for cubes and cylinders. The compressive strength of 7 and 28 days curing is checked for cubes and peak values are selected. The cube specimens are of size 15 × 15 × 15 cm and tested on compression testing machine. Split tensile strength is conducted on cylinders to determine the tensile strength of specimen. The specimen is loaded to make the cylinder split across the vertical diameter.

Flexure test is conducted to evaluate the tensile strength of the specimen. It measure the ability of concrete beam to withstand the failure in bending. The result is expressed in terms of modulus of rupture (f_b).

3 Results and Analysis

3.1 Specimens of Naphthalene Formaldehyde Sulfonate (SNF) Type

Specimens of various dosages of plasticizer and water reduction were considered based on the slump value.

The obtained values are satisfying with medium workability (At a range of 75–100 mm slump) which is suitable [2] for the specimen preparation. It satisfies the requirements for the concrete in the construction of members like beams, columns etc. as per IS specifications (Table 3).

The optimum dosage of SNF Super Plasticizer is 0.5 and 25% water reduction from the analysis and it is used to determine the strength results of the Perlite added concrete at a replacement percentage of 10, 15, 20 and 25% [4] (from Table 4; Fig. 1).

The percentage of perlite addition along with SNF SP which results maximum strength is determined from the analysis. The obtained strength is more than that of conventional concrete mix (from Table 6; Fig. 2).

Table 3 Slump values-SNF specimens

Mix proportion (SP dosage/water reduction)	Slump value (mm)
0.5/20	82
0.5/25	90
0.75/25	98
1.0/30	100

Table 4 Strength results of SNF super plasticizer specimens

Specimen	7 days compressive strength of cubes (N/mm^2)	28 days compressive strength of cubes (N/mm^2)	Split tensile strength of cylinder after 28 days (N/mm^2)	Compressive strength of cylinders after 28 days (N/mm^2)	Flexure strength of beam after 28 days (N/mm^2)
0.5/20	35.6	40.6	2.43	19.72	7.86
0.5/25	34.9	46.3	4.52	32.38	8.24
0.75/25	35.7	43.8	3.32	25.64	7.92
1.0/30	13.2	23.6	1.31	10.42	4.64

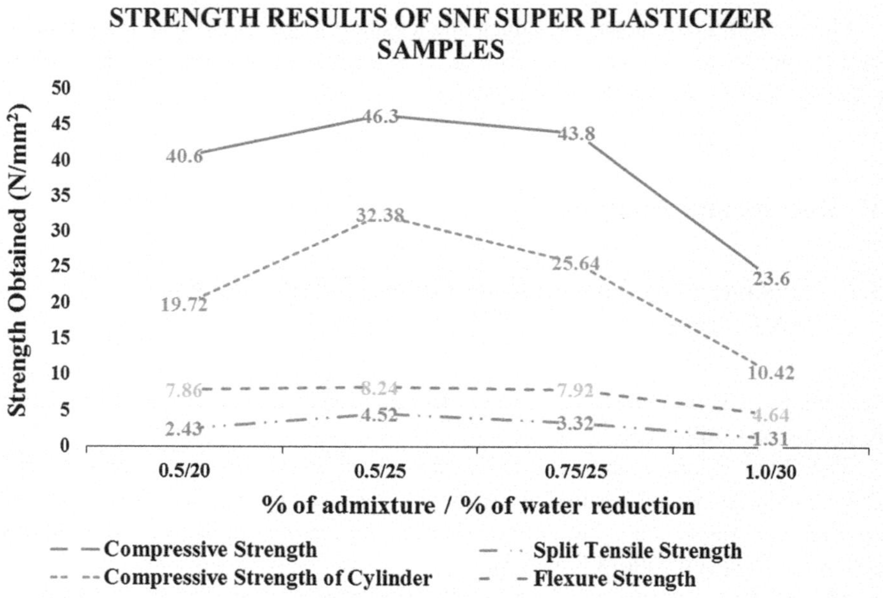

Fig. 1 Analysis of SNF super plasticizer specimens

Table 5 Materials for 1 m^3 of concrete SNF SP-perlite specimens

Perlite replacement and SNF super plasticizer dosage at optimum amount							
% of perlite addition by total % fine aggregate	% of m sand by total % fine aggregate	Weight of cement (kg)	Weight of water (kg)	Weight of m sand (kg)	Weight of perlite (kg)	Weight of coarse aggregate (kg)	Weight of SNF SP (kg)
10	90	328.6	147.8	643.9	60.7	1340.2	1.64
15	85	328.6	147.8	608.1	91.1	1340.2	1.64
20	80	328.6	147.8	572.4	121.5	1340.2	1.64
25	75	328.6	147.8	536.6	151.9	1340.2	1.64

3.2 *Specimens of Poly Carboxylic Ether (PCE) SP*

The trial dosage values of PCE type Super Plasticizer can be determined from slump values as given, Table 7.

The obtained values are satisfying with medium workability (At a range of 75–100 mm slump) which is suitable [2] for the specimen preparation.

Table 6 Strength results of SNF SP-perlite specimens

SNF super plasticizer and perlite addition					
Specimens perlite dosage using SNF SP	7 day compressive strength of cubes (N/mm^2)	28 day compressive strength of cubes (N/mm^2)	Split tensile strength of cylinder after 28 days (N/mm^2)	Compressive strength of cylinders after 28 days (N/mm^2)	Flexure strength of beam after 28 days (N/mm^2)
10	27.56	46.32	4.25	36.82	8.86
15	28.92	48.62	4.62	39.52	9.66
20	26.32	45.24	4.02	38.51	9.32
25	25.34	40.24	3.62	33.62	8.24

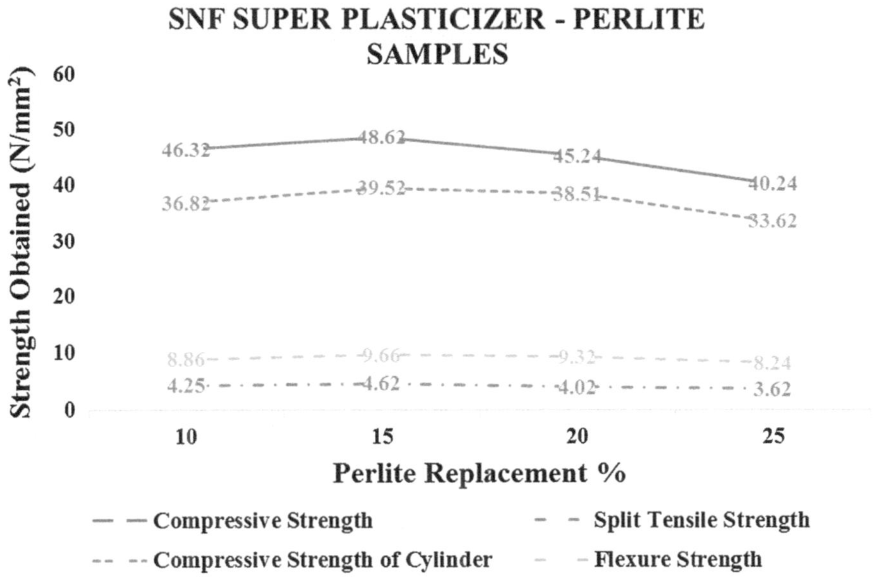

Fig. 2 Analysis of SNF super plasticizer-perlite specimens

Table 7 Slump values of PCE SP-specimens

Mix proportion (SP dosage/water reduction)	Slump value (mm)
0.5/20	94
0.5/25	92
0.6/25	96
0.7/30	99

Table 8 Strength results PCE SP specimens

Specimen	7 days compressive strength of cubes (N/mm^2)	28 days compressive strength of cubes (N/mm^2)	Split tensile strength of cylinder after 28 days (N/mm^2)	Compressive strength of cylinders after 28 days (N/mm^2)	Flexure strength of beam after 28 days (N/mm^2)
0.5/20	23.91	39.85	3.86	32.88	7.52
0.5/25	24.6	41.48	4.12	33.18	8.46
0.6/25	28.88	48.14	4.56	38.51	9.65
0.7/30	16.50	27.50	2.65	22.46	5.35

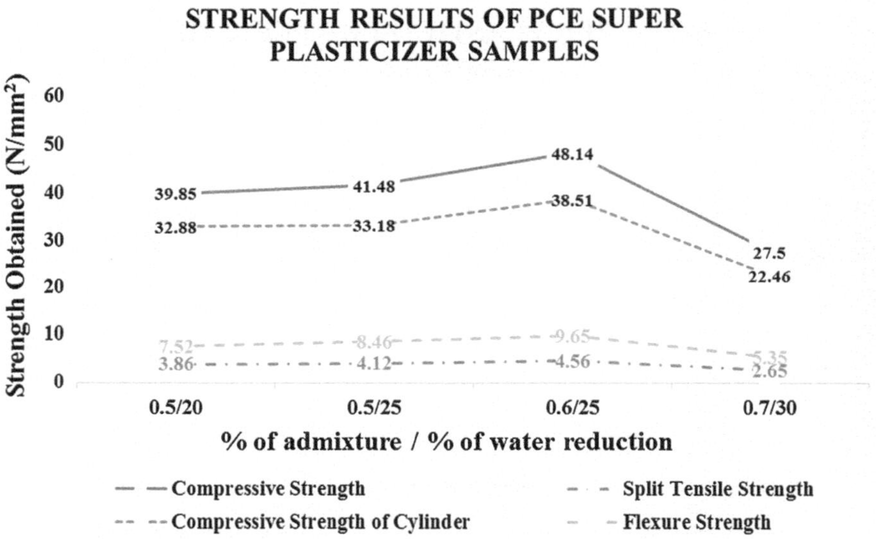

Fig. 3 Analysis of PCE SP specimens

The optimum dosage of PCE super plasticizer is 0.6 and 25% water reduction from the analysis and it is used to determine the strength results of the Perlite added concrete at a replacement percentage of 10, 15, 20 and 25% [3] (Table 8; Fig. 3).

The percentage of perlite addition along with PCE Super Plasticizer which results maximum strength is determined from the analysis (from Table 10; Fig. 4).

4 Conclusion

Based on the study and analysis performed on the strength characteristics of the Specimens with respect to the workability constraints and code parameters the following inferences can be formulated:

Table 9 Materials For 1 m^3 of concrete PCE SP-perlite specimens

Perlite replacement and PCE SP dosage at optimum amount							
% of perlite addition by total % fine aggregate	% of m sand by total % fine aggregate	Weight of cement (kg)	Weight of water (kg)	Weight of m sand (kg)	Weight of perlite (kg)	Weight of coarse aggregate (kg)	Weight of C flow 175 M (kg)
10	90	328.6	147.8	643.9	60.8	1340.2	1.97
15	85	328.6	147.8	608.2	91.2	1340.2	1.97
20	80	328.6	147.8	572.4	121.2	1340.2	1.97
25	75	328.6	147.8	536.6	151.9	1340.2	1.97

Table 10 Strength results perlite and PCE SP specimens

Specimen	7 days compressive strength of cubes (N/mm^2)	28 days compressive strength of cubes (N/mm^2)	Split tensile strength of cylinder after 28 days (N/mm^2)	Compressive strength of cylinders after 28 days (N/mm^2)	Flexure strength of beam after 28 days (N/mm^2)
10	26.81	44.68	4.62	35.88	8.63
15	30.56	50.52	4.96	39.56	9.86
20	29.24	48.48	4.78	38.82	9.54
25	24.28	42.56	3.58	31.24	8.22

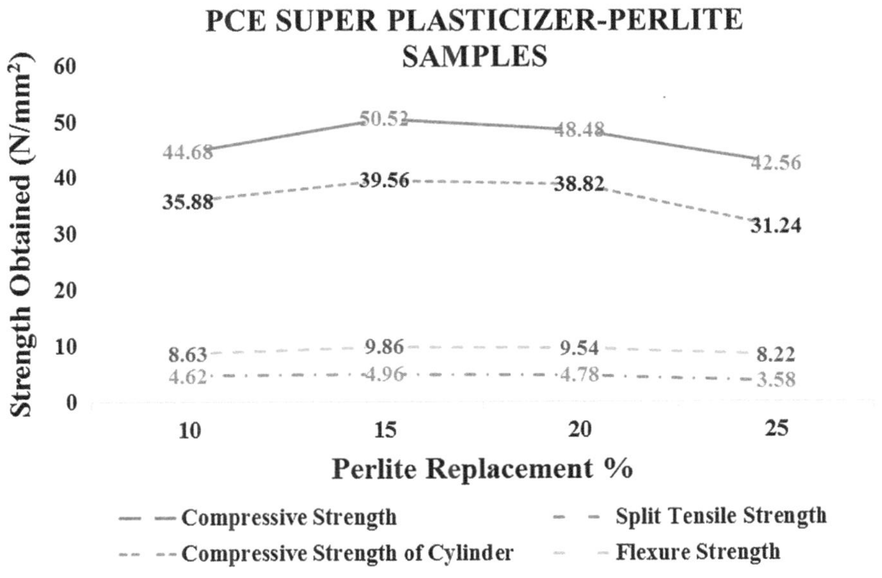

Fig. 4 Analysis of PCE SP-perlite specimens

- Optimum dosage of super plasticizer improves the workability of concrete without adding excess water.
- At a dosage of 0.5% of SNF super plasticizer (Sikament 2002 NS) and 25% water reduction, maximum strength achieved.
- At 0.6% of PCE super plasticizer (C Flow 175 M) and 25% water reduction all the strength parameters reached peak value.
- Excess dosage leads to strength reduction in the specimen after aging and less dosage results in stiff concrete production.
- At 15% perlite addition the strength values increased while using SNF super plasticizer and PCE super plasticizer.
- Sustainable concrete mix can be provided for future design and construction with the use of perlite and super plasticizers.

References

1. Tamrakar R, Mishra SP (2013) Experimental studies on property of concrete due to different ingredient based super plasticizer. Int J Sci Eng Technol Res (IJSETR) 2(5):1036
2. Alsadey S (2015) Effect of super plasticizer on fresh and hardened properties of concrete. J Agric Sci Eng 1(2):70–74
3. Plank J, Gruber M (2009) Effectiveness of poly carboxylate super plasticizers in ultra-high strength concrete: the importance of PCE compatibility with silica fume. J Adv Concr Technol 7(1):5–12
4. Munoz A, Cifuentes S, Colorado HA (2018) Admixture optimization in concrete using super plasticizer. In: The minerals, metals & materials society
5. Technical data guide of concrete admixture-C flow 175 M by CBS chemicals

Studies on the Utilization of Alternative Fine Aggregate in Geopolymer Concrete

T. Saranya, P. S. Ambily and Bharati Raj

Abstract Development of a nation is directly dependent on the available infrastructure. Due to the tremendous growth of infrastructure, demand for building materials such as river sand and cement is also increasing day by day. As the demand for river sand increases; large volume of sand is extracted from the rivers resulting in many adverse environmental effects as well as increase in cost. The cement production is also to be increased to meet these demands and this production has a very bad impact on environment due to CO_2 emission. In order to reduce both these problems, an eco-friendly solution using sea sand and crushed sand as alternative fine aggregate in geopolymer concrete (GPC) is proposed. In this study, an M40 grade GPC with fly ash and Ground Granulated Blast Furnace Slag as binder and a combination of hydroxide and silicates of sodium is used as reaction liquid. Various tests like compression, flexure, split tension, water absorption, sorptivity and Rapid Chloride Permeability were carried out. It is found that, performance of GPC with sea sand and crushed sand are comparable or marginally higher than that of GPC made of river sand in terms of mechanical and durable properties.

Keywords Geopolymer concrete · Sea sand · Strength · Alternative fine aggregate · Crushed sand · Durability

1 Introduction

Portland cement is commonly used as the main binder to produce concrete, which is the key building material used for construction activities. Even though it is a widely used material, it is not eco-friendly because of the CO_2 emission caused by its production. The production of 1 tonne of cement produces about 600–800 kg of CO_2 [1]. Cement production is expected to increase in the future to meet the growing

T. Saranya (✉) · B. Raj
Department of Civil Engineering, NSSCE, Palakkad, India
e-mail: saranyathulasidharan3@gmail.com

P. S. Ambily
CSIR-Structural Engineering Research Centre, Chennai, India

K. Dasgupta et al. (eds.), *Proceedings of SECON'19*,
Lecture Notes in Civil Engineering 46,
https://doi.org/10.1007/978-3-030-26365-2_78

population and construction demands. Thus it is necessary to use an eco-friendly alternative to Ordinary Portland Cement (OPC). Such an alternative is geopolymer, it utilizes industrial aluminosilicate based byproduct materials such as fly ash, rice husk ash, granulated blast furnace slag (GGBS), etc. to produce a low cost, eco-friendly alternative to OPC. These materials with the presence of activator solution, i.e., a mix of hydroxides and silicates of sodium/potassium reacts, dissociates and the gets polymerized to form a strong material, called geopolymer [2]. In addition to OPC, fine aggregate is another main component of concrete. Since river sand is typically used fine aggregate in concrete, its natural sources gets depleted with rising demands for concrete. This also raises the need of an alternate material for river sand. Dredged marine sand, which is a nuisance to docking of ships in the harbor, can be a possible solution [3]. Crushed sand, can also be used as an alternative to river sand. It is being extensively used as fine aggregate in many places.

The main problem with the usage of sea sand is the probable effects of its chloride and sulphate salt contents in the matrix [4]. However, GPC has a different hardening process than OPC, i.e., geopolymerisation instead of hydration, and thus the final products in the concrete matrix will be different. Thus sea sand as fine aggregate in GPC might be a potentially advantageous method [5].

This study aims at arriving at a mix design for GPC with sea sand and crushed sand as fine aggregate and comparing its mechanical and durability properties with river sand. Compression, split tension, flexural, sorptivity and water absorption tests were carried out after 28 days of casting.

2 Experimental Program

In this study, 3 types of concrete is cast with same proportion by volume using different types of sand i.e., sea sand, river sand and crushed sand. The binder used was a mixture of fly ash and GGBS with alkaline activator solution (AAS) containing sodium hydroxide flakes, sodium silicate solution and water. Class F fly ash conforming to IS 3812-1: 2013 [6]; with a specific gravity of 2.2 and GGBS conforming to IS 12089: 1987 (R2008) [7] with a specific gravity of 2.69 is used. The physical properties of the aggregates used are given in Table 1. Sodium hydroxide flakes were dissolved in water and allowed to cool for 24 h, after which sodium silicate solution was added into it. The Fly ash, GGBS, fine aggregate and coarse aggregate are dry mixed in the mixer machine and solution was added after the uniform blending of dry materials. This mixing was continued for 3–5 min.

Mix design was carried out for M40 grade mix. Details about mix are given in Table 2. The binder consisted of fly ash and GGBS in equal quantity. Sodium hydroxide Solution of 3.4 M and silicate to hydroxide ratio of 2 was used for preparation of AAS.

The fresh concrete was tested for workability by slump test. Compaction of concrete was carried out using a vibration table. Cubes of size 100 × 100 × 100 mm, cylinders of size 100 × 200 mm and prisms of size 100 × 100 × 500 mm were cast.

Table 1 Physical properties of aggregates

Property	River sand	Crushed sand	Sea sand	10 mm coarse aggregates
Fineness modulus	2.85	2.34	1.62	7.22
Specific gravity	2.6	2.63	2.41	2.54
Apparent specific gravity	2.66	2.66	2.43	2.69
Water absorption	0.81%	0.4%	0.2%	2.27%
Maximum bulking	35.71%	50%	35.48%	–
% of water content at maximum bulking	4%	4%	6%	–
% of water content when bulking is zero	18%	20%	30%	–

Table 2 Proportion of concrete mix (in weight proportion)

Type of concrete	Binder	Fine aggregate	Coarse aggregate	AS
River sand concrete (GFRS)	1	1.77	2.77	0.63
	379.25 kg/m^3	671.28 kg/m^3	1050.5 kg/m^3	238.93 kg/m^3
Crushed sand concrete (GFCS)	1	1.79	2.77	0.63
	379.25 kg/m^3	678.86 kg/m^3	1050.5 kg/m^3	238.93 kg/m^3
Sea sand concrete (GFSS)	1	1.72	2.77	0.63
	379.25 kg/m^3	652.32 kg/m^3	1050.5 kg/m^3	238.93 kg/m^3

Table 3 Fresh concrete properties

Property	GFRS	GFCS	GFSS
Slump (mm)	55	53	50
Fresh concrete density (kg/m^3)	2360	2380	2360

The cast specimens were demoulded on the next day and kept for curing under ambient temperature conditions. The specimens were tested for compressive strength on 7th, 14th and 28th days; all other tests were performed after 28 days of age. Fresh concrete properties such as density and slump were determined and are given in Table 3.

3 Results and Discussions

3.1 Density of Hardened Concrete

Density of the hardened concrete was calculated by dividing the weight of specimen, by their volume. Three samples from each mix were taken after 28 days of casting

Table 4 Properties of hardened concrete

Properties	GFRS	GFCS	GFSS
Density (kg/m^3)	2310	2336	2324
Split tensile strength (MPa)	3.5	4.4	4.5
Flexural strength (MPa)	4.0	5.1	4.3

and average value was noted as density. The results are given in Table 4. It was found that crushed sand GPC had maximum density.

3.2 Compressive Strength

Compression test was carried out on the samples after 7, 14 and 28 days of casting as per IS 516: 1959 [8] on 100 × 100 × 100 mm cubes. The compression testing machine (CTM) used was of 2000 kN capacity. A loading rate of 2.3 kN/s was applied gradually. The specimens were kept on the platens of CTM in such a way that the load was not applied on the casting face.

The compression test was performed on 5 samples from each mix on 7th and 14th days and 10 samples from each mix after 28th days and the average was calculated as the compressive strength and plotted in Fig. 1. GFCS and GFSS were found to have more compressive strength than GFRS at all the ages of 7, 14 and 28 days. GFSS was found to have 9.67% more compressive strength than GFRS. A faster

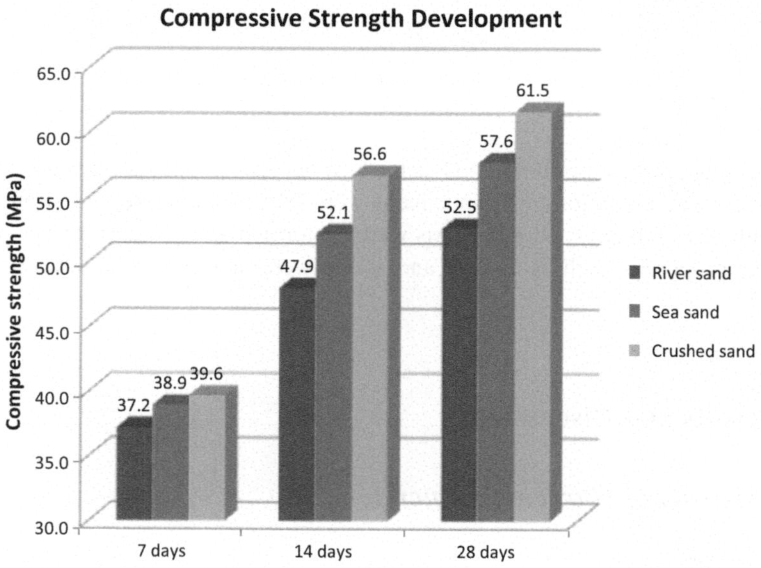

Fig. 1 Compressive strength development

strength development i.e., more than 90% of 28 days strength, was achieved for all the specimens at the age of 14 days itself.

3.3 Split Tensile Strength

Split tensile test was carried out on cylinders of height 200 mm and diameter 100 mm, after 28 days of casting as per IS 5816 (1999) [9]. A CTM was used to apply gradual load at a rate of 1.2 kN/s. The cylinders were placed with the long surface touching the platens and circular surface perpendicular to the plane of platens. Two small plywood strips were kept on top and bottom of the specimen in order to apply line load on the cylinders so that it gets split into two parts.

Five specimens from each mix were tested and average value is given in Table 4. The failure of all the specimens was through a straight line, which split the specimen to equal halves. The maximum load (kN) reading from the CTM was divided by the area of the specimen to get split tensile strength (MPa). GFSS was found to have more splitting tensile strength than GFRS and GFCS. GFSS had 28.57% more split tensile strength than GFRS.

3.4 Flexural Test

Flexural test is used to evaluate the tensile strength of concrete in an indirect way. It tests the ability of unreinforced concrete beam or slab to withstand failure in bending. The results are expressed as modulus of rupture (MR) in MPa. The flexural test on concrete can be conducted using either three point load test or four point load test as per IS 516 (1959). Test method used here is four point load test. The test was carried out on beams of breadth 100 mm, height 100 mm and length 500 mm. Test was carried out after 28 days of casting. A compression testing machine was used to apply gradual load. The beams were placed above 2 rollers which were 400 mm apart, symmetrically placed below the specimen to be tested. Two similar rollers were kept above the specimen in such a way that the portion of specimen between the supporting roller is divided into three equal parts, i.e., two rollers were kept at a distance of 133 mm from adjacent supporting roller. A block was placed on top of the loading rollers to distribute the load equally. The test results are given in Table 4. The failure of all specimens except one was in the central span. GFCS was found to have more flexural strength than GFSS and GFRS. GFSS had 7.5% more flexural strength than GFRS.

Fig. 2 Water absorption

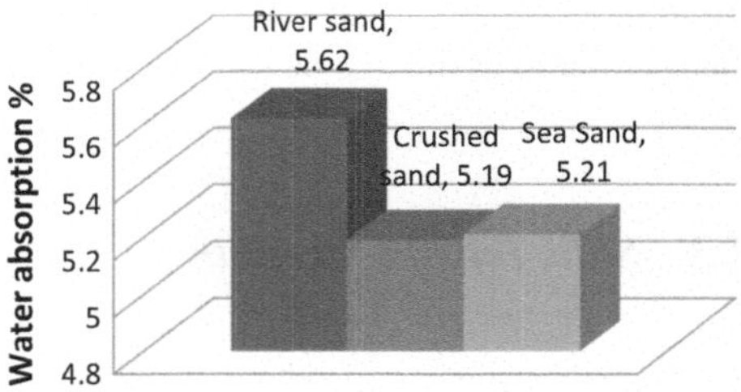

3.5 Water Absorption

Water absorption test was carried out (as per ASTM C642-13 [10]) after 28 days for all the disc specimens; which are cut from 100 × 200 mm cylinders. The disc specimens were of height 50 mm and diameter 100 mm. They were oven dried for 24 h at 100 °C and dry weight was noted. Then they were fully immersed in a tub of water and saturated weight was noted after 24 h. The relative increase in weight is the water absorption percentage for the sample.

The results are given in Fig. 2. GFRS was found to have more water absorption percentage than GFCS and GFSS.

3.6 Rapid Chloride Permeability Test (RCPT)

It was carried out on disc specimens of size 100 × 50 mm, as per ASTM C1202-18 [11], to find the electrical conductance of concrete, which indicates its resistance to chloride ion precipitation. Specimen was placed between two RCPT cells, which contain 0.3 N NaOH solution in one cell and 0.3% by mass NaCl solution filled in the other one. 60 V voltage was made to pass through this setup and current flow readings were noted every 30 min for 6 h. Total charge passed is a measure of electrical conductance of concrete during the test period. It is calculated using the formula,

$$Q = 900(I_0 + 2I_{30} + 2I_{60} + \cdots + 2I_{300} + 2I_{330} + I_{360})$$

As per ASTM C1202, the concrete samples whose Q (total charge passed during the test, i.e., 6 h) value is within the range of 1000–2000 are under the category of 'low chloride ion permeability'. Results are plotted in Fig. 3. It can be seen that GFRS was found to have more chloride ion permeability than GFCS and GFSS.

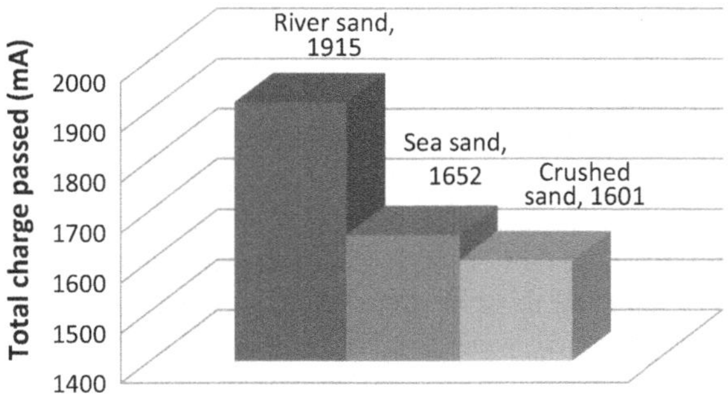

Fig. 3 RCPT results

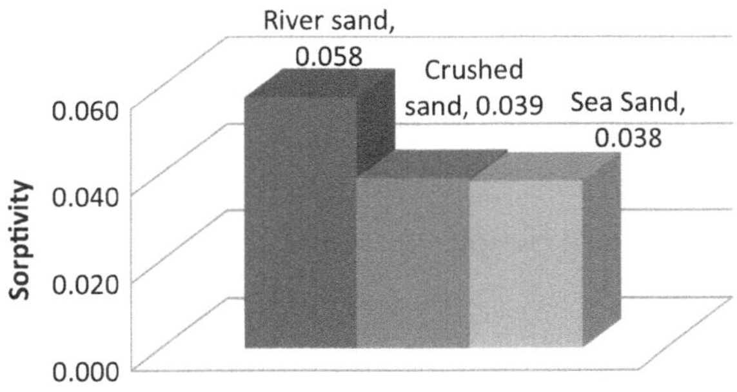

Fig. 4 Sorptivity

3.7 *Sorptivity*

Sorptivity or rate of absorption of water is determined by measuring the increase in the mass of a specimen when only one surface of the specimen is exposed to water represented as a function of time (ASTM C1585-13 [12]). Concrete discs of 100 × 50 mm were cut from cylinder specimens and sealant was applied on top surface and curved surface, exposing the bottom surface. This specimen is placed in water such that the exposed surface comes in contact with water and a height of 1–3 mm of water is kept. Weights were noted at specific intervals and graph is plotted between absorption (I) in mm and square root of time in seconds. The slope of this graph is known as sorptivity.

It was found that sorptivity was also more for GFRS compared to GFCS and GFSS as can be seen from Fig. 4.

4 Conclusions

Mechanical and durability properties of geopolymer concretes, with sea sand, river sand and crushed sand as fine aggregates were studied. From the study the following conclusions are arrived.

- Density of crushed sand based geopolymer concrete (GFCS) was found to be more compared to that of river sand based GPC (GFRS) and sea sand based GPC (GFSS).
- Compared to conventional concrete, an early strength development is observed in GPC i.e., more than 90% of 28 days strength was achieved for all the samples during 14 days itself.
- GFSS and GFCS was found to have marginally higher strength (compression, split tensile and flexural strength) than GFRS, after 28 days.
- As expected, water absorption, sorptivity and chloride ion permeability was found to be less for GFSS and GFCS than GFRS.

Thus it may be concluded that sea sand and crushed sand can be used as alternatives to river sand in geopolymer concrete, without any loss in strength and durability related properties, in fact, they have equal or marginally better properties than river sand based GPC.

5 Future Scope

Durability studies and corrosion studies on Reinforced GPC using sea sand and crushed sand as fine aggregate need to be conducted. Further studies with various mix proportions could also be conducted.

References

1. Gunasekara C, Law D, Bhuiyan S, Setunge S, Ward L (2019) Chloride induced corrosion in different fly ash based geopolymer concretes. Constr Build Mat 200:502–513
2. Liu W, Cui H, Dong Z, Xing F, Zhang H, Lo TY (2016) Carbonation of concrete made with dredged marine sand and its effect on chloride binding. Constr Build Mat 120:1–9
3. Babaee M, Castel A (2016) Chloride-induced corrosion of reinforcement in low-calcium fly ash-based geopolymer concrete. Cem Concr Res 88:96–107
4. Liu W, Huang R, Fu J, Tang W, Dong Z, Cui H (2018) Discussion and experiments on the limits of chloride, sulphate and shell content in marine fine aggregates for concrete. Constr Build Mat 159:725–733
5. Nguyen KT, Le TA, Lee K (2018) Evaluation of the mechanical properties of sea sand-based geopolymer concrete and the corrosion of embedded steel bar. Constr Build Mat 169:462–472
6. BIS IS 3812-1 (2013) Indian standard specifications for pulverized fuel ash part I, for use as pozzolana in cement, cement mortar and concrete. Bureau of Indian Standards, New Delhi, India
7. IS 12089:1999 (R2008) Indian standard specifications for granulated slag for the manufacture of Portland slag cement. Bureau of Indian Standards, New Delhi, India
8. IS 516:1959 (2005) Method of tests for strength of concrete, Bureau of Indian Standards, New Delhi
9. IS 5816:1999 Splitting tensile strength of concrete (First revision), Bureau of Indian Standards, New Delhi

10. ASTM C (2013) Standard test method for density, absorption, and voids in hardened concrete. C642–13
11. ASTM C (2012) Standard test method for electrical indication of concrete's ability to resist chloride ion penetration. C1202–18
12. ASTM C (2004) Standard test method for measurement of rate of absorption of water by hydraulic cement concretes. C1585–13

Green Cell Reinforcement in Soil

Alinda Jose, Anina P. Jose, P. Chethana Unni, Dhanya Rose Babu and V. Vasudha

Abstract Development in construction field is a matter of great importance in the present day scenario. To achieve all the infrastructural needs of present ever growing population, a large number of residential, commercial, industrial and transportation structures are to be constructed. The structural load acting on a building has to be safely transferred to the ground. But due to space constrain and lack of hard stratum, structures have to be constructed on the weak soil. This is a great challenge to civil engineers as there is a chance of settlement of foundation or failure of structured. Hence different ground improvement techniques have to be implemented to improve the property of soil. A variety of ground improvement techniques are available now a days, among them most widely used technique of reinforcing soil is with geosynthetic materials. Screw pine-cell is the most advanced form of Geocell. It is a three dimensional, natural, honeycomb like structure of cells interconnected at joints. The soil particles can be trapped inside these cells providing an overall confinement to the soil layer and improves its properties regarding support to civil engineering structures. In our study we focus on how to increase the load carrying capacity and reduce the settlement of weak soil using Screw pine-cells.

Keywords Green cell · Screw pine cells · Settlement of weak soil · Load carrying capacity · Ground improvement

1 Introduction

Nowadays most of the countries are giving so much importance to the infrastructural growth. But the soil around us are not of good quality, thus designing and construction of various works on these weak and compressible soil lead to many problems such as losing of shear strength and excessive settlement etc. Hence the stabilization of the soil has been an area of major concern in the field of construction. Soil stabilization means to improve the engineering properties of the soil and make it more stable. So,

A. Jose · A. P. Jose · P. Chethana Unni · D. R. Babu · V. Vasudha (✉)
Civil Engineering Department, Sahrdaya College of Engineering and Technology, Thrissur, India
e-mail: vasudha@sahrdaya.ac.in

K. Dasgupta et al. (eds.), *Proceedings of SECON'19*,
Lecture Notes in Civil Engineering 46,
https://doi.org/10.1007/978-3-030-26365-2_79

here we use the concept of cellular confinement system. This method creates a new composite form that possesses mechanical and geotechnical properties. Geocells are three dimensional honeycombed synthetic cellular structures. These 3 dimensional zones of confinement infilled with sand reduces the lateral movement of soil particles while vertical loading act on it, which results in high lateral stress and resistance on the cell-soil interfaces. We introduce the Green cells which are prepared by naturally and locally available screw pines, since synthetic material are not economical and eco-friendly. The screw pines are cut into thin layer of strips and are woven together using galvanized wires to form a grid as shape resembling geocell. These cells possess more lateral and vertical confinement. This proves to be a very good improvement in the footing performance especially when provided as a mattress. Since these are obtained naturally, it is eco-friendly, it possess no negative impact to the environment.

2 Methodology

The prime objective of our study is to reduce settlement of foundation soil by reinforcing it with green cell. Thus the reinforcing property of green cell is checked by applying load on model test sample and noting settlement which is then compared with unreinforced soil.

2.1 Materials Required

For experiment the materials required are weak soil, sand for infilling green cell, screw pine for making green cell. Various tests are to be done on soil and sand to understand the properties and to check whether soil is weak or not.

2.2 Weak Soil

The foundation soil was collected from field. In order to prove the selected soil is weak various tests were conducted like Atterberg's limit, Specific gravity test for soil, Unconfined compression test, Particle size distribution, Standard proctor test and Direct shear test.

2.3 Green Cell

Green cell is made with screw pine in the form of three dimensional geocell. The screw pine are easily available in Kerala and for our study screw pine strips with 10 mm width were collected and waved to form grids which were then tied at required interval with galvanized steel wire of 1 mm to form cells. The size of the green cell has decided based on similitude with commercially available geocell. The sizes of green cell has been made as 125 mm × 105 mm for inside diagonals and height as 75 mm which is half of the geocell with cell aperture size 250 mm × 210 mm × 150 mm.

In our study we compared the property of green cell reinforced soil and unreinforced soil under three cases with different cell sizes of 125 mm × 85 mm, 125 mm × 105 mm and 145 mm × 105 mm with same height of 75 mm (Fig. 1).

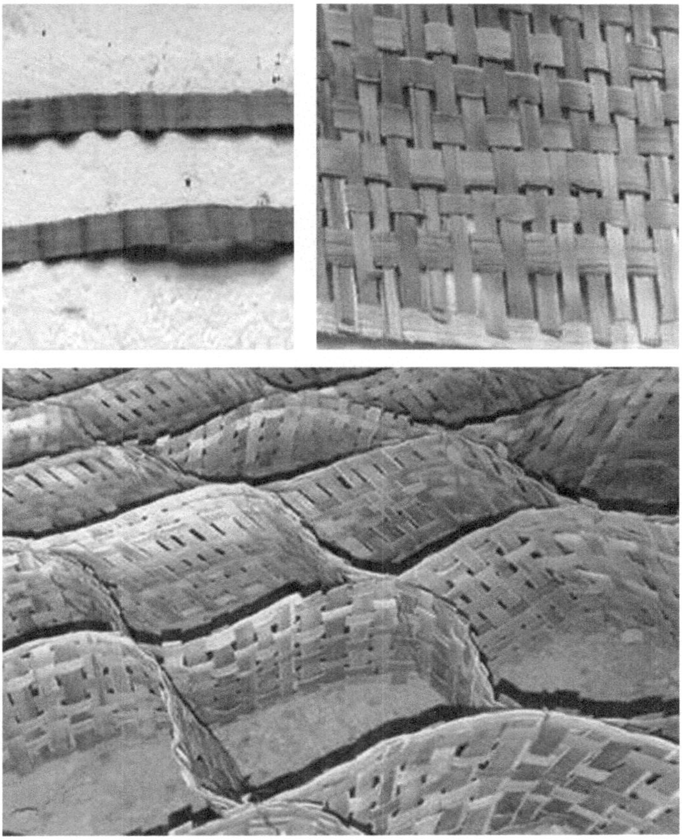

Fig. 1 Screw pine strips, grid and green cell

3 Tests on Soil

3.1 Results Obtained

See Table 1.

The liquid limit of soil is obtained as 75% which is greater than 50% hence the soil is highly compressible. The type of soil having a plastic limit greater than 30% is highly plastic in nature and here we have a plastic limit of 66.6% which indicate that soil is highly plastic. The specific gravity value we obtained is 2.584 which fall under organic soil. For Unconfined compression test we have obtained a value of 4.668 KN/m^2 as unconfined compressive strength. The unconfined compressive strength between 0 and 25 indicates that the soil is having very soft consistency. The graph obtained for particle size distribution shows that the soil is poorly distributed. In case of direct shear test it is seen that the sand used as infill is loose as it has an

Table 1 Test results on weak soil

SI. No	Description	Values
Atterberg's limit		
1	Liquid limit	75%
2	Plastic limit	66.67%
3	Flow index	55.14%
4	Toughness index	0.3010
J	Plasticity index	16.67
6	Shrinkage index	44.1483
7	Shrinkage limit	22.5217
8	Shrinkage ratio	1.6381
9	Volumetric shrinkage	69.943
Specific gravity		
1	Specific gravity of soil	2.584
Unconfined compression test		
1	Unconfined compressive strength q_u	4.668 kN/m^2
2	Un drained shear strength	2.334 kN/m^2
Particle size distribution		
1	Uniformity coefficient C_u	2.714
2	Coefficient of curvature C_o	1.806
Standard proctor test		
1	Maximum dry density	0.927
2	Optimum moisture content	11.11%
3	Zero air void line (100% saturation)	1.898
Direct shear test		
1	Angle of internal friction	9.349°

angle of friction of 9.349°. Standard proctor test was conducted and the optimum moisture content was obtained as 11.11% with maximum dry density of 0.927 (Table 1). This test is important for the preparation of foundation bed.

4 Laboratory Load Carrying Test

Differential settlement between different parts of structure can endanger the entire structure. Providing reinforcement in soil can avoid such situation in construction fields. The effect of settlement depends upon many factors such as magnitude, type of structure etc. Load bearing test conducted for green cell reinforcement helps to find out how much settlement can be reduced in soil by using this particular method. Load bearing test is conducted for both plain soil and soil placed with green cell of three different sizes. The size of the cells were selected by comparing with the size of commercial geocell. For conducting load bearing text we have constructed a steel tank of size 450 mm × 450 mm × 30 mm. This steel tank represents the land area where there is a chance for settlement.

First we determined the deformation values for plain soil which is very weak. Plain soil is set as the standard for conducting other tests because based on settlement in plain soil we will compare the settlement values for soil were green cell is provided as reinforcement. Foundation bed is prepared using natural clayey soil which is very weak. Filling of soil is done in three layers by providing optimum moisture content at which it obtains maximum dry density and 25 blows for each soil layer. Constant fall of height for each blow should be maintained throughout the filling inside steel tank. The sides of steel tank should be covered with polyethylene sheet in order to avoid side friction. For plain soil test we fill the weak soil up to a height of 300 mm. As a representation of footing we are placing plywood of size 75 mm × 75 mm × 15 mm. Load is applied to the soil using universal testing machine. Steel tank is placed inside universal testing machine and load is applied over center of plywood without having any eccentricity. Load values are noted from dial clock and corresponding settlement values are noted from deformation scale. Load application is done till the reinforcement fails. Second part of test is to conduct the experiment after placing green cell reinforcement in soil. Here we fill soil up to height of 200 mm and optimum moisture content is added to soil at which it obtains maximum dry density. Green cell which is having a cell size of 125 mm × 85 mm × 75 mm is placed over weak soil. The placing of green cell should be always in fully stretched manner inside the steel tank. Cell pockets of green cell are infilled with sand which gives green cell more stability. Above the green cell layer we will place the weak soil with same water content and plywood of same size as a representation of footing. Loading and corresponding settlement values are noted. Results show that settlement values are lesser for reinforced soil when compared with that of plain soil. This cell is the one which is having maximum number of cell pockets with minimum cell area.

After successfully completing first green cell reinforced test, second green cell of size 125 mm × 105 mm × 75 mm is reinforced in soil inside steel tank with similar arrangement which is given for first green cell and load settlement values are noted from deformation scale. Last test is conducted for green cell of size 145 mm × 105 mm × 75 mm. This is the largest cell size among other three cell sizes hence more amount of sand is required as infill. Similar to other two cells, load application arrangement remains same for this cell size too. Settlement values are noted according to the applied loads. Load application for all three cell sizes remains the same. The reason behind load bearing tests conducted for three cell size is in order to make a comparison among various cell sizes and finally to find which cell size is having least settlement. From the values collected it is understood that performance of green cell reinforcement stabilizes the soil.

5 Results and Discussions

5.1 Load Settlement Curve

In this chapter, the results obtained from Load carrying tests on green-cell reinforced soil along with the unreinforced one are presented. In order to evaluate the performance improvement due to reinforcements, reference tests were carried out with unreinforced soils as well. The obtained results are plotted as load settlement curve (Table 2).

Combined load settlement curve was used to represent the load-settlement response of the three different cell size cases along with unreinforced soil case. The post-test exhumation of the green cell shows the deformation on the ribs. Out of the three different cell sizes used in the tests, the performance of the cell size 125 mm × 85 mm was found to be better compared to the other two. The maximum load taken by unreinforced soil was observed as 40 KN and the settlement of footing

Table 2 Load carrying test observations

Load (KN)	Settlement (mm)			
	Plain soil	Cell aperture size 125 mm × 105 mm	Cell aperture size 125 mm × 85 mm	Cell aperture size 145 mm × 105 mm
0	0	0	0	0
5	10	8	6	7
10	15	12	9	11
15	18	14	11	13
20	20	16	12	15
25	24	18	14	17
30	26	20	15	18
35	30	22	17	20
40	35	24	18	22

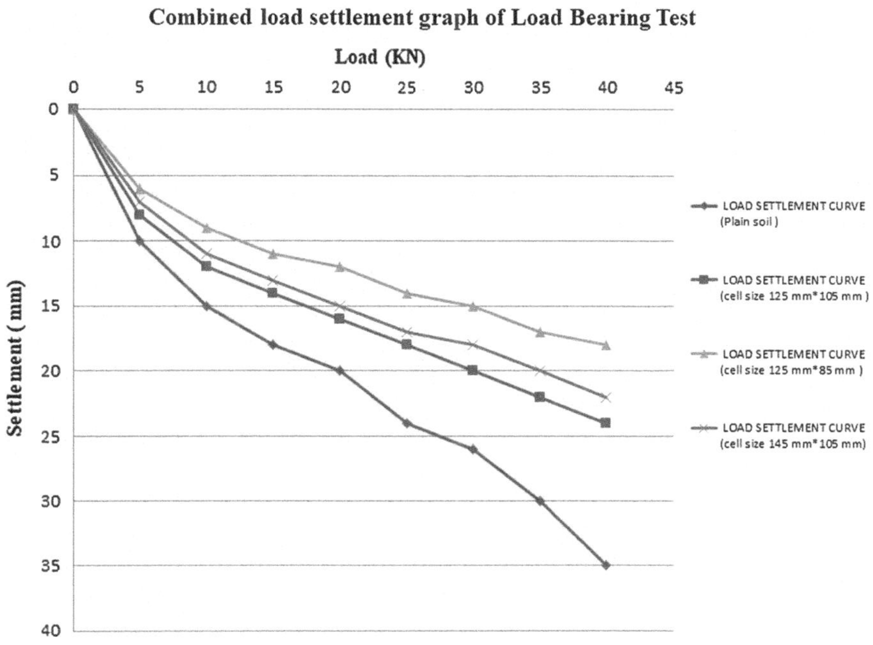

Fig. 2 Combined load settlement graph

at that load was 35 mm. For the green cell reinforced soil, the settlement has reduced much more at the same level of load application. Thus it was found from the studies that increase in load carrying capacity and reduction in the settlement of the soil both takes place because of green cell reinforcement. The settlement was least for the cell with least single cell area of 125 mm × 85 mm (Fig. 2).

5.2 *Load Carrying Capacity Improvement Factor (IF)*

The extent of improvement in the load carrying capacity due to the different cell sizes was quantified more precisely using load carrying capacity improvement factor. The increase in the load carrying capacity due to the provision of the reinforcement can be found through a non-dimensional parameter called load carrying capacity improvement factor (IF), which is defined as,

$$IF = q_r/q_o$$

where q_r is the load carried by the reinforced bed at a particular settlement and q_o is the load carried by the unreinforced bed at the same settlement [1]. The improvement factor graph was drawn with reference to the load carrying capacity of the unreinforced soil. The Improvement factor calculation is shown in the following table and its graph is plotted below it (Table 3).

Figure 3 represents the variation of the load carrying improvement factors with the footing settlement for green cell with different cell sizes. It was found that the IF values increase with the increase in the settlement. For the green cell with cell size 125 mm × 850 mm, the maximum IF = 2.85 was observed at settlement =

Table 3 Preliminary patch specifications

Settlement (mm)	Load taken				Improvement factor for load carrying capacity IF = q_r/q_0		
	By plain soil (q_0)	By reinforced soil with (q_r)					
		Cell size 125 mm × 105 mm	Cell size 125 mm × 85 mm	Cell size 145 mm × 105 mm	Cell size 125 mm × 105 mm	Cell size 125 mm × 85 mm	Cell size 145 mm × 105 mm
6	3	3.75	5	4.28	1.25	1.66	1.42
8	4	5	8.33	6.25	1.25	2.08	1.56
10	5	7.5	12.5	8.75	1.5	2.5	1.75
12	7	10	20	12.5	1.42	2.85	1.78
14	9	15	25	17.5	1.67	2.77	1.94
16	11.67	20	32.5	22.5	1.71	2.78	1.92
18	16	25	40	30	1.56	2.5	1.87

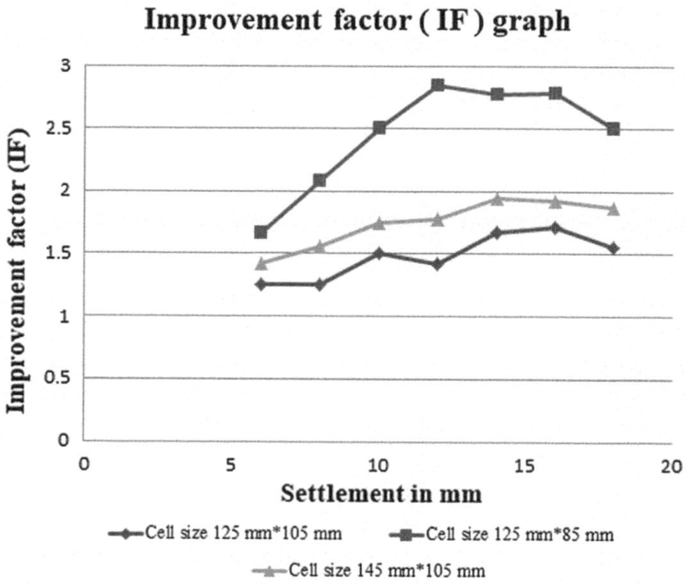

Fig. 3 Improvement factor (IF) graph

12 mm. The IF curve had a parabolic shape with IF value increasing initially and later decreasing slowly. IF values vary from minimum to maximum in a range between 1.25 and 1.71 for the cell size 125 mm × 105 mm, between 1.66 and 2.85 for the cell size 125 mm × 850 mm and between 1.42 and 1.94 for cell size 145 mm × 105 mm. Green cell with cell size 125 mm × 850 mm were proved to be the better reinforcement as the improvement factors observed were higher as compared to other two green cells.

6 Discussion and Conclusions

From the results of laboratory investigation performed on a clay bed reinforced with natural screw pines, we can conclude that this method is the best one in terms of cost effectiveness, bearing capacity improvement and reduction of differential settlement. Also the green cell confinement provides all round confinement to the materials and hence prevents the lateral spreading of soil on the application of load. Sizes of cell pockets are found to be the significant parameters for compressive strength because when tested, it is observed that cells with minimum cell area and maximum number of perfect cell pockets possesses least settlement. While considering the improvement factor it is maximum for the above type of cell again.

Hence from our study we concluded that the green cell of size 125 mm × 85 mm × 75 mm was the best among the three green cells we have considered. So these results will be very helpful in providing guidelines for design and construction of various foundation reinforcement. The study has its own limitations as it is a model test and only three different sized cells are considered, thus overall behaviour of the green cell reinforced soil is of interest for future research.

Acknowledgements The authors would like to thank Vasudha V, Assistant Professor Sahrdaya College of Engineering and Technology, for guidance and support.

References

1. Hegde A, Sitharam TG (2015) Experimental and analytical studies on soft clay beds reinforced with bamboo cells and geocells. Int J Geosynth Gr Eng 1(2):13

Effect of Metakaolin in the Mechanical Properties of Ambient Cured Flyash Based Geopolymer Concrete

N. N. Aswathy and R. Ritzy

Abstract Cement is one of the most versatile construction material used worldwide. But cement industry is responsible for several environmental issues like global warming. To address these issues there is a need for alternative binders to substitute cement. Geopolymer concrete (cement less concrete) is found to be a better alternative to conventional concrete. Geopolymer concrete (GPC) utilizes the cementitious properties of industrial byproducts like flyash, silicafume, GGBS etc. Flyash based geopolymer concrete is the common type of GPC. Flyash based GPC requires heat curing and is suitable only for precast members. Development of GPC which does not require heat curing will widen its application. This work aims to study the effect of metakaolin on the properties of ambient cured flyash based GPC. Experimental investigation is carried out by replacing flyash by metakaolin in varying percentages of 20, 30, 40, 50 and 60%. It was observed that ambient cured flyash-metakaolin GPC possess characteristics suitable for structural applications. The mechanical properties of GPC increased with increase in metakaolin content while workability properties showed a slight reduction.

Keywords Geopolymer concrete · Precast · Flyash · Metakaolin · Ambient cured

1 Introduction

Concrete has through the last hundred years established itself as one of the major construction material. The combination of excellent compressive strength, durability, readily available and affordable subcomponents has made concrete a highly demanded construction material and the backbone of our society's infrastructure. Ordinary Portland Cement (OPC) is the main component in concrete which provides the versatility to concrete. With rapid expansion of construction industry, there is

N. N. Aswathy · R. Ritzy (✉)
Sree Buddha College of Engineering, Pattoor, Alappuzha, India
e-mail: ritzyrose@gmail.com

N. N. Aswathy
e-mail: aswathynn8@gmail.com

K. Dasgupta et al. (eds.), *Proceedings of SECON'19*,
Lecture Notes in Civil Engineering 46,
https://doi.org/10.1007/978-3-030-26365-2_80

an increase in demand and consumption of cement. Massive production of cement causes various impacts on health and environment. Cement industry is one of the primary producer of CO_2, a major greenhouse gas, SO_2 and NO_2. A single industry accounts for 5% of global CO_2 emissions [1]. The production of one ton of Portland clinker produces approximately one ton of carbon dioxide, which results in global warming. Large scale production of cement leads to the depletion of various natural resources. Also cement production is highly energy intensive. Due to these adverse effects of cement, there is a need of alternate binders that are ecofriendly and contribute towards waste management.

The name geopolymer was coined by a French Professor Davidovits in 1978 to represent a broad range of materials characterized by networks of inorganic molecules [1]. The primary difference between geopolymer concrete and Portland cement concrete is the binder. It is an alkali-activated binder, produced by a polymeric reaction of alkaline liquids with the silicon and the aluminium oxides in source materials like flyash, GGBS, silicafume, redmud etc. These silicon and aluminium is dissolved in an alkaline activating solution and subsequently polymerizes into molecular chains and become the binder Geopolymers are environmental friendly materials that do not emit greenhouse gases during polymerisation process. The polymerization requires a considerably quick reaction of silica (Si)-alumina (Al) under alkaline condition which subsequently creates three-dimensional polymeric chain of Si-O-Al-O bonds [2]. Dissimilar to OPC or pozzolanic cements, geopolymer utilizes the polycondensation of silica and alumina and a high alkali content to attain compressive strength [2].

Most of the earlier studies were conducted on temperature-cured geopolymer concrete that is considered to be ideal for precast concrete members. Very limited researches have been carried out on flyash based geopolymer concrete at ambient temperature condition because the geopolymerization reaction of flyash based GPC is slow at ambient temperature, and also flyash based GPC mix is highly workable. This study focuses on the effect of metakaolin on the workability of fresh flyash based geopolymer concrete and the strength development with age when cured at ambient temperature.

2 Objectives of the Study

- To obtain the mix proportion of flyash based geopolymer concrete.
- To obtain the optimum binder composition of flyash and metakaolin.
- To investigate the workability and mechanical behaviour (compressive strength, split tensile strength and flexural strength) of FA-MK based GPC.

3 Methodology

- A detailed literature survey was carried out on previous studies related to geopolymer concrete.
- Physical properties of materials were tested and found that all properties are conforming to IS standards.
- Fixing the mix proportion of GPC by trial and error method as per "Modified guidelines for geopolymer concrete mix design using Indian standard".
- Casting of standard test specimens by replacing flyash with metakaolin by 20, 30, 40, 50 and 60% by volume of flyash.
- Testing of specimens on 7th day and 28th day.
- Determining the optimum percentage of FA-MK from the test results.
- Conclusions and recommendations are finally made based on the findings and observations.

4 Materials

4.1 Cement

Ordinary Portland cement of 53 grades conforming to IS 12269-1987 was used in the study for casting control specimens of M30 grade.

4.2 Fine Aggregate

Fine aggregate are soil particles passing through 4.75 mm IS sieve. Generally river sand, crushed stone, crushed gravel, M sand etc. are used as fine aggregate. In this study, M sand conforming to Zone II is used.

4.3 Coarse Aggregate

Coarse aggregates passing through 20 mm IS sieve and retained on 4.75 mm IS sieve is used for the study.

4.4 Fly Ash (FA)

Fly ash is a by-product from burning pulverized coal in electric power generating plants. Class F fly ash is used and it exhibit pozzolonic properties. Tests are conducted on fly ash as per IS 2386 (Part III)-1963.

4.5 Metakaolin (MK)

Metakaolin used in this study is the byproduct from Kerala Ceramics Limited, Kundara. It is baby pink in colour. It is a highly reactive material which includes a higher contribution of silica and alumina.

4.6 Alkaline Solution

The alkaline solution used in this study was a combination of Sodium Hydroxide (NaOH) and Sodium Silicate (Na_2SiO_3). The sodium hydroxide and sodium silicate of laboratory grade were purchased from Haseena Chemicals, Trivandrum. Sodium hydroxide is available in the form of flakes and sodium silicate is available in the form of liquid. Sodium hydroxide solution was prepared by dissolving flakes in distilled water. 12 M concentration of sodium hydroxide was used for the entire study.

4.7 Super Plasticizer (SP)

To improve the workability of Geopolymer concrete, a commercially available naphthalene-based super plasticizer Conplast SP430 was used. In this study super plasticizer was taken as 1.1% by amount of binding material.

4.8 Water

Potable water is considered as being acceptable for casting. Hence clean drinking water available in college water supply system was used for casting.

Table 1 Mix proportion of control mix

Material	Cement	Fine aggregate	Coarse aggregate	Water
Weight	437.78	664.22	1139.29	197
Ratio	1	1.52	2.6	0.45

Table 2 Mix proportion of geopolymer concrete

Flyash (kg/m^3)	Fine aggregate (kg/m^3)	Coarse aggregate (kg/m^3)	Na_2SiO_3 (kg/m^3)	NaOH (kg/m^3)	Extra water (kg/m^3)
450	539.94	1010.29	217.85	87.15	9
1	1.07	2.02	0.61		0.02

5 Experimental Investigation

The experimental investigation is done on the fresh concrete (slump test) and hardened concrete (compressive strength, split tensile strength and flexural strength) of flyash-metakaolin based geopolymer concrete.

5.1 Mix Design

In this study M30 grade conventional concrete was used as control mix. As per IS: 10262-2009, a mix proportion was suitably designed for M30 grade concrete. The mix proportioning was done based on the properties of various materials like cement, fine aggregate and coarse aggregate. The mix proportion of control mix is given in Table 1.

There is no standard method for the mix design for geopolymer concrete. The mix design for geopolymer concrete of M30 grade was done based on "Modified guidelines for geopolymer concrete mix design using Indian standard" by Rangan et al. [3] and by trial and error method. The mix proportion for flyash based geopolymer concrete is given in Table 2.

5.2 Mix Proportioning

To study the effect of metakaolin on the mechanical properties of flyash based geopolymer concrete, different proportions are prepared with different percentages of metakaolin (20, 30, 40, 50 and 60%) by volume of flyash. For each mix at least 3 specimens are tested for compressive strength, split tensilestrength, flexural strength at 7th day and 28th day.

Table 3 Slump variation of FA-MK GPC

Concrete	OPCC M30	F80M20	F70M30	F60M40	F50M50	F40M60
Slump value (mm) (cm)	73	71.3	69	66.1	62.4	60.5

5.3 Specimen Details

Mechanical properties include compressive strength test, split tensile strength test and flexural strength test. The compressive strength test is done by casting concrete cube of size 150 mm × 150 mm × 150 mm. Concrete cylinder of height 300 mm and diameter 150 mm is used to test the split tensile strength and flexural test is conducted by casting plain concrete beam of size 500 mm × 100 mm × 100 mm. The concrete is poured in the mould with no compaction. After 24 h these moulds are removed and test specimens are placed for ambient curing. The compression test is conducted on 7th day and 28th day. The split tensile and flexural strength test is conducted on 28th day.

6 Results and Discussions

6.1 Workability

In this study, the workability of various mixes were assessed by determining slump values as per IS 1199:1959 specification and the results are tabulated in Table 3. It is observed that slump values of all the mixes lies in the range of 50–100 mm, represents medium workability. True slump was obtained and it was observed that by increasing the metakaolin content slump value is decreasing. It may be due to higher fineness and angular shape of metakaolin particles compared to spherical shaped flyash particles.

6.2 Hardened Concrete Properties

Hardened concrete properties are tested as per IS standards and the results are tabulated in Table 4. Strength properties of ambient cured flyash based GPC is significantly lesser than conventional concrete. But by replacing flyash with metakaolin compressive strength, split tensile strength and flexural strength keeps on increasing up to 50% replacement. Flyash based geopolymer have low strength due to incomplete reaction of flyash particles at ambient temperature. Enhancement of mechanical

Table 4 Hardened properties of GPC

Mixes	Compressive strength (MPa)		Split tensile strength (MPa)	Flexural strength (MPa)
	7 day	28 day	28 day	28 day
OPCC M30	23.9	42.5	4.03	4.28
F80M20	18.33	23.66	2.26	2.62
F70M30	24.66	33.33	2.85	2.84
F60M40	26.4	37.8	3.82	3.92
F50M50	33.33	43.2	4.17	4.35
F40M60	30.6	32.3	3.02	3.82

properties due to the incorporation of metakaolin is may be due to complete polymerisation reaction of metakaolin particles at ambient curing condition. Metakaolin particles do not require elevated temperature curing for geopolymerisation process.

7 Conclusions

Geopolymer concrete (zero cement concrete) technology is a sustainable solution for solving environmental problems caused by cement industry and problems related to disposal of industrial wastes like flyash, metakaolin etc. Flyash based geopolymer concrete requires heat curing inorder to attain sufficient mechanical properties. Metakaolin blended FA based GPC mixes attained enhanced mechanical properties at ambient room temperature curing itself without the need of heat curing as in the case of only FA based GPC mixes. Metakaolin particles do not require elevated temperature curing for polymerisation reaction. Elimination of the elevated temperature for curing will reduce the cost and energy required for heat curing process. So incorporation of metakaolin to FA based GPC will widen the application of flyash based geopolymer concrete beyond precast concrete industry. Keeping in view of environmental protection, sustainability, savings in natural resources and production cost, metakaolin blended FA based geopolymer concrete can be considered as a sustainable concrete for future (Figs. 1, 2 and 3).

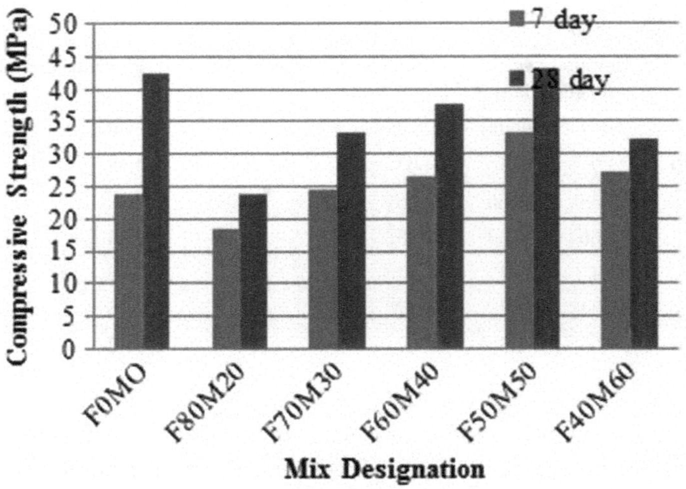

Fig. 1 Effect of metakaloin on compressive strength

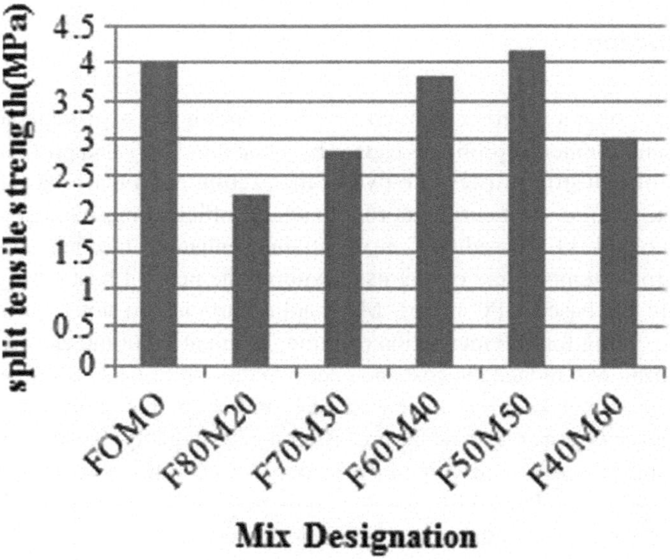

Fig. 2 Effect of metakaloin on split tensile strength

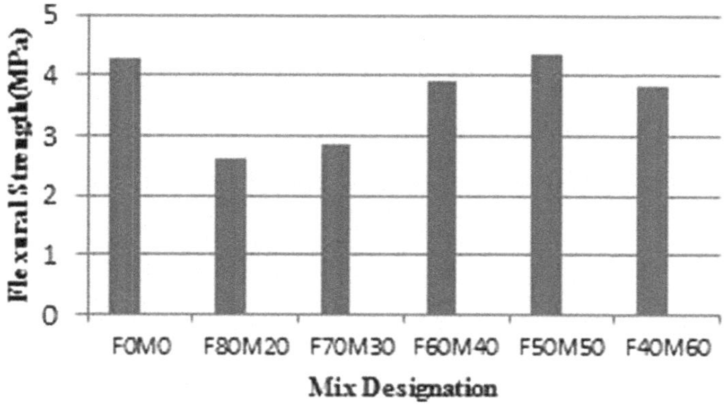

Fig. 3 Effect of metakaloin on flexural strength

References

1. Roy DM, Jiang W, Silsbee M (2000) Chloride diffusion in ordinary, blended, and alkali-activated cement pastes and its relation to other properties. Cem Concr Res 30(12):1879–1884
2. Rashad A, Bai Y, Basheer P, Collier N, Milestone N (2012) Chemi-cal and mechanical stability of sodium sulfate activated slag after exposure to elevated temperature. Cem Concr Res 42(2):333–343
3. Anuradha R, Sreevidya V, Venkatasubraman R, Rangan BV (2011) Modified guidelines for geopolymer concrete mix design using Indian standard. Asian J Civil Eng 13:353–364
4. Habert G, De Lacaillerie JDE, Roussel N (2011) An environmental evaluation of geopolymer based concrete production: reviewing current research trends. J Cleaner Prod 19(11):1229–1238
5. McLellan BC, Williams RP, Lay J, Van Riessen A, Corder GD (2011) Costs and carbon emissions for geopolymer pastes in comparison to ordinary portland cement. J Cleaner Prod 19(9):1080–1090
6. Bakharev T, Sanjayan JG, Cheng Y-B (2003) Resistance of alkali-activated slag concrete to acid attack. Cem Concr Res 33(10):1607–1611
7. Part WK, Mahyuddin R, Cheah CB (2015) An overview on the influence of various factors on the properties of geopolymer concrete derived from industrial by-products. Constr Build Mater 77:370–395

Effect of Acid Environment on Bond Durability of Basalt Fiber Reinforced Polymer Bars and Steel Bars Embedded in Concrete

Aiswarya Sreehari and P. E. Kavitha

Abstract Bond durability of ribbed basalt fiber reinforced polymer bars and steel bars exposed to sulphuric acid environment for different durations was investigated. Seventy-two pull out specimens were tested under direct tensile load to study the bond performance in terms of their bond strength and bond failure modes. The test parameters included the bar material such as basalt and steel, the duration of exposure such as 28 and 56 days, grade of concrete such as M25, M30 and M35, and bar diameter such as 10, 12, 20 mm. Experimental results showed that the bond strength of BFRP bars to concrete decreased by 19% of that of steel bars. All the specimens failed by concrete splitting mode of failure. 28 days of acid conditioning increased the bond strength of BFRP specimens by 5.45–53.9% and for steel specimens by 1.93–28.73% of their initial strength. 56 days of acid conditioning decreased the bond strength of BFRP specimens by 1.33–24.86% and for steel specimens by 8.78–18.44% of their initial strength.

Keywords Bond durability · BFRP · Pull out · Bond failure modes · Bond strength

1 Introduction

In certain aggressive environments, deterioration of concrete occurs resulting in the corrosion of steel bars thus eliminating the bond strength between the two materials. This trend paved way for using alternate reinforcing bars such as Fiber Reinforced Polymer (FRP) bars instead of steel bars. Several studies have recognized the mechanical characteristics of basalt composites but the lack of knowledge and

A. Sreehari (✉)
Department of Civil Engineering, SCMS School of Engineering and Technology, Ernakulam, Kerala, India
e-mail: aiswaryasreehari@yahoo.com

P. E. Kavitha
Department of Civil Engineering, Federal Institute of Science and Technology, Ernakulam, Kerala, India
e-mail: pe_kavitha@yahoo.com

K. Dasgupta et al. (eds.), *Proceedings of SECON'19*,
Lecture Notes in Civil Engineering 46,
https://doi.org/10.1007/978-3-030-26365-2_81

research studies on basalt products as reinforcing materials has resulted in BFRP bars not being listed amongst the recognized FRP reinforcing bars in most design codes. Thus, the research on the practicability of using BFRP products in reinforcing concrete structures has been very constricted. Therefore, the current study aims at narrowing the gap in the knowledge on bond behaviour and bond durability of BFRP bars to concrete. Most of the previous researches indicate that, the bond strength is influenced by the type of resin used for manufacturing the bar. Epoxy resin made BFRP bars shows superior performance compared to vinyl ester made BFRP bars [1, 2]. As the frictional force increases, the bond strength also increases, which is confirmed when sand coated bars were used [1, 3–5]. As the bar diameter and bar embedment length increases, the bond strength decreases. This is because as the bar diameter increases, on account of poisons effect there will be more reduction in the bar diameter when subjected to longitudinal stresses and as the embedment length increases, there will be a nonlinear distribution of bond stress along the surrounded portion of the bar [2, 5]. The corrosion due to acid attack increases the bond stress, the increase is remarked due to the friction caused by the swelling of bar [3, 6]. In an alkaline environment, the bond stress reduces due to the formation of reaction products which results in peeling of the outer layer surface of bars, therefore reducing the frictional force [3, 6]. High temperature seems to be corrosive for BFRP bars, as it reduces the bond strength. At high temperature the elastic modulus of bar diminishes resulting in poor bond strength [4].

Bond characteristics of bars in concrete are the most crucial parameter for implementation of the material to the concrete structures. Thus, one of the fundamental aspects of structural behaviour is bond development, as bond governs the serviceability, ductility, flexural, shear and torsion load bearing capacities of reinforced concrete members. Durability is a superior concern for concrete structures exposed to aggressive environments. The various corrosive agents that attacks RC structures are acids such as sulphuric acid, organic acids and nitrous oxide compounds, alkalis such as potash, washing soda and lime, salts and various atmospheric agents such as oxygen and airborne moisture. Among the common corrosive agents mentioned above, acids prove to be more dangerous. Hence a study was conducted to assess whether the BFRP bars assures the minimum bond strength required as per IS 456:2000 by exposing BFRP embedded specimens to 5% sulphuric acid solution for 28 days and 56 days in comparison with steel embedded specimens.

2 Experimental Program

2.1 Parameters Included

The test parameters included were bar material such as basalt and steel as shown in Figs. 1 and 2, the duration of exposure such as 28 and 56 days, grade of concrete such as M25, M30 and M35, and bar diameter such as 10, 12 and 20 mm.

Fig. 1 Ribbed BFRP bars used in the study

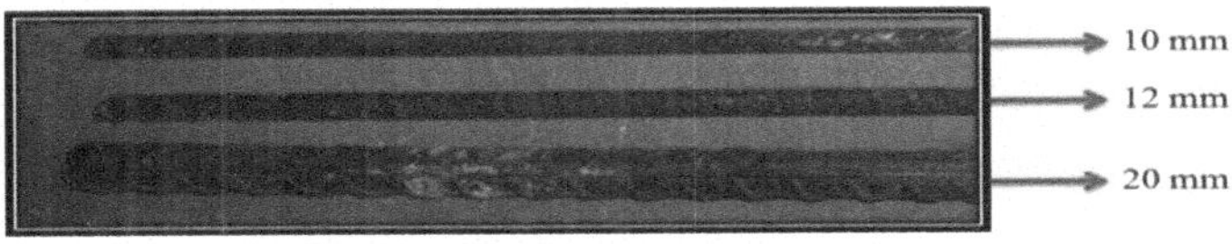

Fig. 2 Ribbed steel bars used in the study

Table 1 Properties of materials

Sl. No	Material	Test	Value obtained	Standard value
1	Cement (OPC 53) {As per IS: 12269-1987}	Specific gravity	3.2	2.9–3.2
		Fineness	7%	Not more than 10%
		Consistency limit	30%	25–35%
2	Fine aggregate (M sand) {As per IS: 383-1970}	Specific gravity	2.66	2.5–2.8
		Sieve analysis	Zone I	Zone I–Zone IV
3	Coarse aggregate (20 mm and downgraded) {As per IS: 2386-1963}	Specific gravity	2.78	2.6–2.9

2.2 Concrete Mix

Concrete mixes with a compressive strength of 25, 30 and 35 MPa were used for this study. The concrete mix design was done as per IS 456:2000 and IS 10262:2009. Admixture of type MASTER GLENIUM SKY 8433 produced by BASF Incorporation, of specific gravity 1.08, was added to increase the workability of concrete for obtaining a desired slump range of 75–125 mm. Materials were tested for its various properties which is required for the mix design as shown in Tables 1 and 2.

2.3 Properties of BFRP & Steel Bars

As per the manufacturer's specification, the guaranteed modulus of elasticity of BFRP bars was 50 GPa and for steel was 210 GPa. The physical properties and mechanical properties are shown in Tables 3 and 4.

Table 2 Concrete mix proportions (For 1 m^3 of concrete)

Sl. No	Mix	Mix proportion	Cemen (kg)	Fine aggregate (kg)	Coarse aggregate (kg)	w/c ratio	Super-plasticizer (%)
1	M25	1:1.84:3.14	394.00	726.00	1238.00	0.40	0.20
2	M30	1:1.78:3.09	403.00	720.00	1248.00	0.38	0.25
3	M35	1:1.67:2.96	422.00	707.00	1258.00	0.36	0.30

Table 3 Physical properties of bars

Bar material	Surface treatment	Diameter (mm)	Weight (g/m)
BFRP bars	Ribbed	10.19	166.00
		13.06	246.00
		20.26	570.00
Steel bars	Ribbed	10.36	547.00
		12.05	739.00
		20.00	2261.00

Table 4 Mechanical properties of bars

Bar material	Nominal yield stress (MPa)	Ultimate load (kN)	Ultimate tensile strength (MPa)	
			Guarante	Tested
BFRP—10 mm	N/A	60.0	950.0	735.7
BFRP—12 mm	N/A	78.0	900.0	582.3
BFRP—20 mm	N/A	120.0	800.0	372.2
Steel—10 mm	512.5	51.6	–	612.2
Steel—12 mm	514.7	53.0	–	517.4
Steel—20 mm	518.1	198.0	–	505.8

2.4 Preparation of Test Specimen

The specimens for the pull out tests as shown in Fig. 3, were prepared according to the recommendations given in IS: 2770 (Part 1)-1967.

2.5 Curing Conditions

After 24 h of casting, the specimens were taken off from the moulds and water cured in crates for 28 days. After curing the specimens for a period of 28 days in normal water as per ASTM C267-01 Clause 9.4, the specimens were exposed to

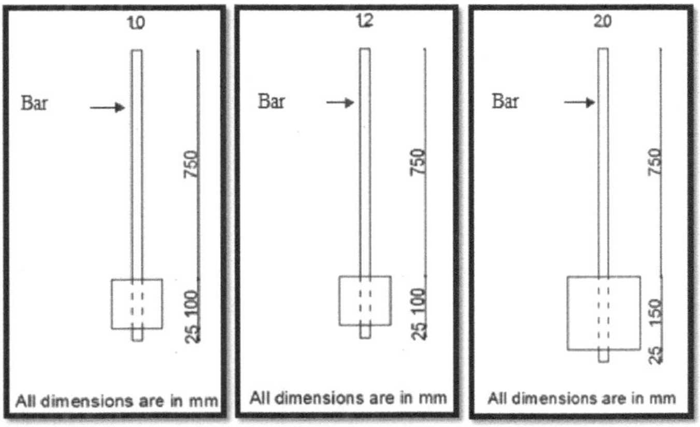

Fig. 3 Schematic of the pull out specimens

Fig. 4 Acid conditioning of specimens in crates

natural environmental condition for two days before moving them to the investigated environmental exposure. After capping, the specimens were then conditioned in 5% sulphuric acid solution for a period of 28 days and 56 days as per ASTM C267-01 Clause 11.2, as shown in Fig. 4.

2.6 *Pull Out Test*

Pull out tests were carried out as per IS: 2770 (Part 1)-1967. After the curing period and corroding period, the bar- concrete bond strength was determined through the pull out tests, conducted using Universal Testing Machine of capacity 1000 kN [7] as shown in Fig. 5. For all the specimens, maximum load at failure was recorded. Nominal bond stress was obtained from the equation below

Fig. 5 Pull out test set up of specimens

Fig. 6 Brown traces in BFRP embedded specimens

Fig. 7 Peeling of outer layer of BFRP bar

$$\tau = P/\pi DL \quad (1)$$

where, τ is the nominal bond stress in N/mm^2, P is the load at failure in N, D is the diameter of the bar in mm and L is the embedment length in mm.

3 Results and Discussion

3.1 Bond Strength of Acid Conditioned Specimens

The results of bond behaviour between the BFRP bars and steel bars with the surrounding concrete after acid immersion for a period of 28 days and 56 days of grades M25, M30 and M35 are shown in Tables 5, 6 and 7 respectively. It was observed that there was an increase in the bond strength for both steel and BFRP embedded specimens after 28 days of acid immersion and decrease in the bond strength after 56 days of acid immersion. This trend was noticed in all the three grades of concrete for all the diameters. The increase in the bond strength after 28 days is attributed to two main aspects such as (a) Due to the sulphate attack blocking the micro cracks in the concrete; thereby increasing its strength as reported by Wu et al. [8] and (b) Due to the curing of concrete in its early stage as notified by Altalmas et al. [3]. There was a decrease in the bond strength after 56 days of acid immersion. The decrease in the bond strength is attributed to two main aspects such as (a) As the deterioration increased, the micro cracks expanded further due to the sulphate, which resulted in the cracking of the concrete surface as reported by Wu et al. [8] and (b) There was a weak adhesion between the layers of the bar which can be due to the poor manufacturing quality of the bars as failure was governed by the shear stresses between the bar layers rather than between bar and concrete, this observable fact was reported by Altalmas et al. [3], Ei Refai et al. [1, 4].

Table 5 Variation of bond stress with immersion in acid for a period of 28 and 56 days in M25 grade concrete

Sl.no	Diameter	Control specimens	28 days acid conditioned specimens		56 days acid conditioned specimens	
		Bond stress (N/mm^2)	Bond stress (N/mm^2)	% increase in bond stress (%)	Bond stress (N/mm^2)	% decrease in bond stress (%)
1	Steel—10	11.97	15.41	28.73	10.89	9.04
2	BFRP—10	8.15	12.54	53.90	7.58	6.95
3	Steel—12	8.97	10.82	20.70	7.86	12.28
4	BFRP—12	7.27	9.18	26.28	6.02	17.12
5	Steel—20	7.75	9.44	21.92	6.38	17.50
6	BFRP—20	6.24	8.06	29.27	4.99	20.07

Table 6 Variation of bond stress with immersion in acid for a period of 28 and 56 days in M30 grade concrete

Sl.no	Diameter	Control specimens	28 days acid conditioned specimens		56 days acid conditioned specimens	
		Bond stress (N/mm^2)	Bond stress (N/mm^2)	% increase in bond stress (%)	Bond stress (N/mm^2)	% decrease in bond stress (%)
1	Steel—10	13.37	16.8	25.60	11.33	15.25
2	BFRP—10	9.49	14.07	48.33	9.36	1.33
3	Steel—12	10.03	12.84	28.05	8.97	10.60
4	BFRP—12	8.59	11.35	32.10	7.32	14.80
5	Steel—20	8.38	9.85	17.50	7.64	8.85
6	BFRP—20	6.79	8.70	28.13	5.94	12.50

Table 7 Variation of bond stress with immersion in acid for a period of 28 and 56 days in M35 grade concrete

Sl.no	Diameter	Control specimens	28 days acid conditioned specimens		56 days acid conditioned specimens	
		Bond stress (N/mm^2)	Bond stress (N/mm^2)	% increase in bond stress (%)	Bond stress (N/mm^2)	% decrease in bond stress (%)
1	Steel—10	14.52	17.58	21.11	13.24	8.78
2	BFRP—10	12.29	16.50	34.28	11.46	6.73
3	Steel—12	11.62	13.85	17.21	10.19	12.33
4	BFRP—12	10.61	12.51	18.00	8.12	23.50
5	Steel—20	10.93	11.14	1.93	8.91	18.44
6	BFRP—20	8.75	9.23	5.45	6.58	24.86

3.2 *Influence of Bar Diameter on Bond Strength*

The magnitudes of bond stress for smaller diameter bars are comparatively larger than the bars with larger diameter [3, 5, 9–13]. As the diameter of bar increases, the surface area of the bar in contact with the concrete also increases. According to the Eq. (1), the increase in the bar diameter ultimately causes reduction in the bond stress. the bond stress decreases with increase in the surface area of bar. In M25, M30 and M35 grade concrete, for 12 mm BFRP bar the bond stress decreased by 10.8%, 9.4% and 13.7% respectively compared to 10 mm diameter bar. In M25, M30 and M35 grade concrete, for 20 mm BFRP bar the bond stress decreased by 23.4%, 28.4% and 28.8% respectively compared to 10 mm diameter bar. The increase in the bond strength for small diameter bars is attributed to two main aspects such as (a) Smaller diameter bars shows better adherence to concrete than larger diameter bars thereby increasing the bond strength, and (b) The bond strength of 20 mm diameter BFRP bars is less as it possesses a different surface texture which is smooth when compared to 10 and 12 mm diameter bars. BFRP specimens possess highest percentage increase in the bond strength compared to steel specimens after acid conditioning. This is attributed to the friction that was developed between the BFRP bars and the concrete, due to the existence of products (brownish traces) that resulted from the reaction between the bars and the acid.

3.3 Influence of Grade of Concrete on Bond Strength

Altalmas et al. [3] reported that the concrete compressive strength plays a major role in the development of the bond stress of FRP bars to concrete. By increasing the concrete strength, a result occurs in increasing the bond strength between the bars and the surrounding concrete. M35 grade concrete cube possess a dense structure resulting in higher compressive strength thereby increasing the bond performance. In M30 and M35 grade concrete, for 10 mm BFRP bar the bond stress increased by 16.4% and 50.8% respectively compared to M25 grade. In M30 and M35 grade concrete, for 12 mm BFRP bar the bond stress increased by 18.2% and 50% respectively compared to M25 grade. In M30 and M35 grade concrete, for 20 mm BFRP bar the bond stress increased by 8.8% and 40.3% respectively compared to M25 grade.

3.4 Bond Failure Modes

Splitting of concrete specimens in a brittle mode of failure as shown in Fig. 8, was observed for all the tested BFRP and steel except in one specimen the bar ruptured before attaining its bond strength which is the steel rupture failure mode. This can be due to that the bond strength between the concrete and steel was greater than the tensile strength of the steel. In some of the specimens, on account of the reaction

Fig. 8 Concrete splitting

between the acid and concrete, reaction products were formed on the inner concrete surface which is indicated by brownish colour as shown in Fig. 6. Colored spots (brownish traces) were observed on the BFRP bar indicating the existence of reaction products on the bar surface which is responsible for the increase in bond strength due to increase in frictional force, was also observed by Altalmas et al. [3] and Ei Refai et al. [4], where the brownish traces were formed due to the presence of iron ions in BFRP bars. In some BFRP bars, the outer layer was entirely peeled off from the subsequent bar layers as shown in Fig. 7, which indicates that the failure is controlled by the shear strength of the bar ayers and not the shear strength of concrete.

4 Conclusion

The major conclusions derived from the experimental study are as follows:

- BFRP bars possess a high strength to weight ratio compared to steel bars in which the strength to weight ratio of BFRP bars is greater than steel bars by 74.7, 70.5 and 66.2% for 10, 12 and 20 mm diameter bars.
- 28 days of acid conditioning increases the bond strength of BFRP bars by 5.45–53.9% and for steel bars by 1.93–28.73% of their initial strength. 56 days of acid conditioning decreases the bond strength of BFRP bars by 1.33–24.86% and for steel bars by 8.78–18.44% of their initial strength.
- All the BFRP and steel acid conditioned specimens possess a bond stress greater than the minimum bond stress given in IS 456:2000 Clause 26.2.1.1 for all the three grades of concrete such as 2.24 N/mm^2 for M25, 2.4 N/mm^2 for M30 and 2.72 N/mm^2 for M35.
- BFRP specimens possess highest percentage increase in the bond strength compared to steel specimens after acid conditioning. This is attributed to the friction that was developed between the BFRP bars and the concrete, due to the existence of products (brownish traces) that resulted from the reaction between the bars and the acid.
- The bond stress decreases with increase in the surface area of bar i.e., in M25, M30 and M35 grade concrete, for 20 mm BFRP bar the bond stress decreased by 35.7%, 38.16% and 44.05% respectively compared to 10 mm diameter bar.
- The bond stress increases with the increase in the compressive strength of concrete cube i.e., in M30 and M35 grade concrete, for 10 mm BFRP bar the bond stress increased by 12.2% and 31.5% respectively compared to M25 grade.
- The mode of bond failure observed for all the acid conditioned BFRP and steel specimens were concrete splitting mode of failure, except one which failed by steel rupture failure mode.

The study confirmed the great potential of using BFRP bars as reinforcing materials. Even though BFRP bars showed lower bond strength than that of their steel counterparts, the current findings have demonstrated the promise of BFRP bars to be used as reinforcing materials for concrete elements exposed to an acid environment.

References

1. Ei Refai A, Abed F, Altalmas A (2015) Bond durability of BFRP bars embedded in concrete under direct pull out conditions. J Composites Constr ASCE 19(5):04014078(1)–04014078(11)
2. Tighiouart B, Benmokrane B, Gao D (1998) Investigation of bond in concrete member with fiber reinforced polymer bars. J Constr Build Mater 12:453–462 Elsevier
3. Altalmas A, Ei Refai A, Abed F (2015) Bond degradation of basalt fiber reinforced polymer bars exposed to accelerated aging conditions. J Constr Build Mater 81:162–171 Elsevier
4. Ei Refai A, Ammar MA, Masmoudi R (2015) Bond performance of BFRP bars to concrete. J Composites Constr ASCE 19(3):04014050(1)–04014050(12)
5. Benmokrane B, Elgabbas F, Ahmed EA, Cousin P (2015) Characterization and comparitive durability study of glass/vinylester, basalt/vinylester and basalt/epoxy FRP bars. J Composites Constr ASCE 19(6):04015008(1)–04015008(12)
6. Serbescu A, Guadagini M, Pilakoutas K (2015) Mechanical characterization of basalt FRP rebars and long term strength predictive model. J Composites Constr ASCE 19(2):04014037(1)–04014037(13)
7. Aiswarya S, Kavitha PE (2017), Influence of exposure to acid environment on the bond strength of basalt fiber reinforced polymer bars and steel bars embedded concrete blocks. Int Res J Eng Technol 04(04)
8. Fang Z, Wu Y, Lin H, Zhou S (2016) Degradation model of bond performance between deteriorated concrete and corroded deformed steel bars. J Constr Build Mater 119:89–95 Elsevier
9. Wu G, Dong ZQ, Wang X, Zhu Y (2015) Prediction of long term performance and durability of BFRP bars under the combined effect of sustained load and corrosive solutions. J Composites Constr ASCE 19(3):04014058(1)–04014058(9)
10. Appa Rao G (2014) Parameters influencing bond strength of rebars in reinforced concrete. Int J Appl Eng Technol 4(1):72–81
11. Antonietta Aiello M, Leone M, Pecce M (2007) Bond performance of FRP rebars reinforced concrete. J Mater Civil Eng ASCE 19(3):205–213
12. Okelo R (2007) Realistic bond strength of FRP rebars in NSC from beam specimens. J Aerosp Eng ASCE 20(3):133–140
13. Gu X, Dong Q (2012) Laboratory test and numerical simulation of bond performance between basalt fiber reinforced polymer and concrete. J Test Eval ASTM Int 40(7):1–8

Experimental Study on Strengthening of Steel Beam Using Carbon Fiber Reinforced Polymer Sheet

Samithamol Salim and P. E. Kavitha

Abstract The number of civil infrastructures which have deteriorated and no longer fulfill the requirements of safety standards is continuously increasing day by day. Over the past few decades, there has been increasing interest in applying adhesively bonded composites to repair existing and/or strengthen new civil engineering structures. Extensive research has been conducted on Fiber Reinforced Polymer (FRP) strengthening of concrete structures, whereas relatively less work has been done on FRP strengthening of steel structures. Among the FRP materials available, Carbon Fiber Reinforced Polymers (CFRP) were found to be the most suitable for strengthening steel structures because of their well known high mechanical properties and high strength to weight ratio. This paper presents an experimental investigation on the effect of CFRP strengthening in the load carrying capacity, stiffness and deformation characteristics of undamaged and damaged steel beams. The effect of end anchorage on flexural strengthening of steel beam is also investigated.

Keywords Steel beam · CFRP · Strengthening · Flexure · Debonding

1 Introduction

Steel structures need to be rehabilitated for different reasons such as material degradation, design fault, and change in the load acting on the structures. Conventional rehabilitation techniques make use of welded or bolted steel plates to strengthen the damaged elements but the problem of steel corrosion still remains and difficulties in fitting complex profiles can arise. Fiber reinforced polymer materials, in specific

S. Salim (✉)
Computer Aided Structural Engineering, Department of Civil Engineering, Sree Narayana Gurukulam College of Engineering, Kadayiruppu P.O., Kolenchery 682311, India
e-mail: samithamolsalim@gmail.com

P. E. Kavitha
Department of Civil Engineering, Federal Institute of Science and Technology, Angamaly, India
e-mail: pe_kavitha@yahoo.com

K. Dasgupta et al. (eds.), *Proceedings of SECON'19*,
Lecture Notes in Civil Engineering 46,
https://doi.org/10.1007/978-3-030-26365-2_82

carbon fiber reinforced polymers display outstanding mechanical properties with typical tensile strength and modulus of elasticity of more than 1200 MPa and 140 GPa, respectively [1, 2]. In addition, the CFRP laminates weigh less than one fifth of the steel and are corrosion resistant [2]. By addition of the CFRP layers, the stress level in the original member will decrease, that in turn results in a longer fatigue life [2]. The strength and stiffness of steel beams can be enhanced by epoxy bonding CFRP plates or sheets to its tension face [2, 3].

The application of different length, thickness and type of CFRP plates in strengthening steel I beams will cause changes in the CFRP failure modes and load bearing capacity [4–7]. Strengthening using shorter and thicker CFRP strips cause premature end-debonding where as longer CFRP plates increase the resistance against end-debonding and will enhance the flexural behavior of steel beams [4, 8, 9]. Using CFRP plates with higher modulus will further improve the load carrying capacity and stiffness of steel beam [5–7]. But this higher load carrying capacity was found for steel beams with thicker CFRP plates [10, 11]. It is advocated to use thin CFRP plates with higher modulus as thicker CFRP plates results in premature debonding [10]. Steel sections require CFRP plates of a matching stiffness (i.e., MOE) or higher in order to get a reasonable stiffness or strength gain [12–14]. Parametric study conducted by Deng et al. [15] showed that the maximum shear and normal stresses decreased with the increase in thickness of the adhesive and with the decrease in thickness of CFRP plate [15]. Damage to the bottom flange degrades both the flexural stiffness as well as the strength of the beam [10, 12]. The elastic flexural stiffness of damaged beams could be partially restored (up to 50%) with the use of CFRP plates. Strength of damaged beams could be fully restored to its original, undamaged state with the use of CFRP plates [12]. Adding CFRP plate to the steel web not only increase the flexural strength but also reduce the longitudinal strain and shear stresses in the bottom CFRP plate [13]. The flexural strength of steel beams can also be increased by increasing the reinforcement ratio of CFRP [5]. The ultimate load carrying capacity, yield load and the stiffness after yielding of steel beams could be enhanced, also by increasing the number of layers of epoxy bonded CFRP sheets [2, 16]. Provision of end anchoring of CFRP could further enhance the load bearing capacity of steel beams and could decrease the effects of end-peeling and debonding of CFRP [17, 18].

The main aim of this paper is to study the effectiveness of CFRP on flexural strengthening of steel beam by varying the CFRP strengthening scheme, providing end anchorages and by inducing damage on steel beam to simulate corrosion and then repairing with CFRP.

2 Materials and Methods

2.1 Materials

Steel beam, CFRP sheet and adhesive were the three important materials used for the study. The steel beam used was ISMB 100 with overall height 100 mm and flange width 50 mm and of length 1 m. The grade of steel was E250 with tensile strength 410 MPa. The CFRP sheet used for strengthening steel beams was of thickness 0.41 mm. For strengthening, 3 layers of CFRP sheets were used in each portion so as to get an overall thickness of 1.2 mm and the length adopted for strengthening was 600 mm which was 2/3rd the effective span of the steel beam [4, 5]. The width of CFRP adopted to strengthen different portions of steel beam was 50, 40 and 30 mm. According to manufacturer's data sheet, the tensile strength of the CFRP sheet was 4000 Mpa with modulus of elasticity 240 GPa. The adhesive used was Araldite AW 106 epoxy resinwith hardener HV 953 IN and is mixed in equal quantities by volume so as to get an adequate bonding property. According to manufacturer's data sheet, the ultimate tensile strength of the epoxy adhesive was 33 MPa with shear strength of 17.6 Mpa for 5 days curing.

2.2 Specimen Details

There were 8 specimens designated as UB, SBS1, SBS2, SBS3, SBS4, SBS4E, DUB and DSBS3 respectively. UB denotes the unstrengthened beam which serves as control beam (Fig. 1). SBS1, SBS2, SBS3 and SBS4 were the beams strengthened using scheme1, scheme 2, scheme 3 and scheme 4. In scheme 1 strengthening, CFRP was bonded only to the lower portion of bottom flange (Fig. 2) where as in scheme 2, CFRP was applied to the lower and upper portions of bottom flange (Fig. 3). CFRP was bonded to the lower portion of bottom flange as well as in bottom web portion in the beam strengthened using scheme 3 (Fig. 4) while CFRP was also bonded to the upper portion of bottom flange in the beam strengthened using scheme 4 (Fig. 5). The beam SBS4E was strengthened using scheme 4 with end anchorages at the ends

Fig. 1 UB

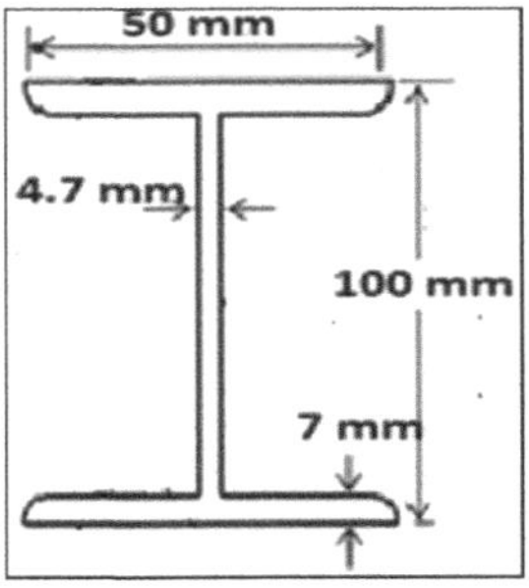

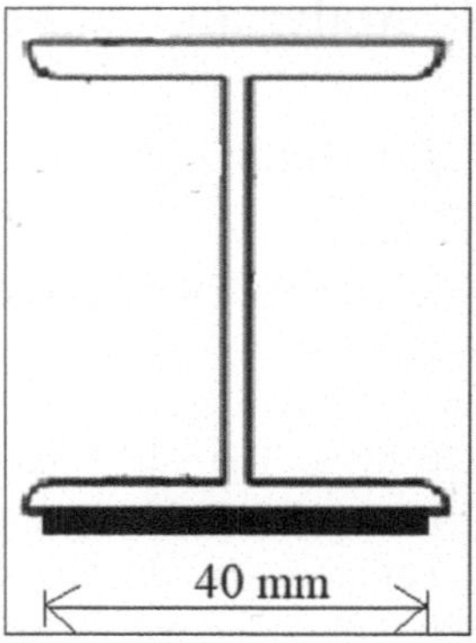

Fig. 2 SBS1

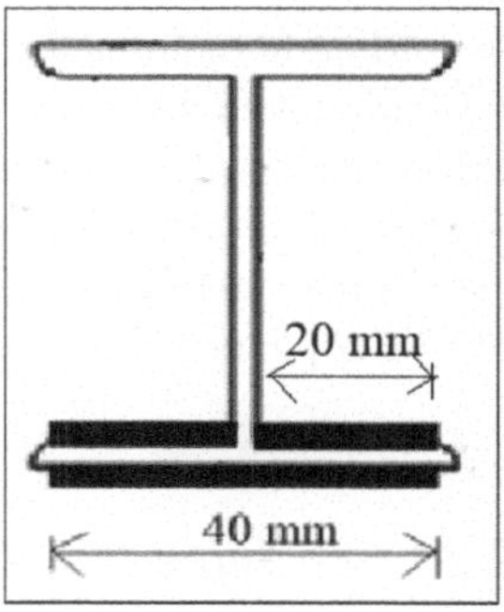

Fig. 3 SBS2

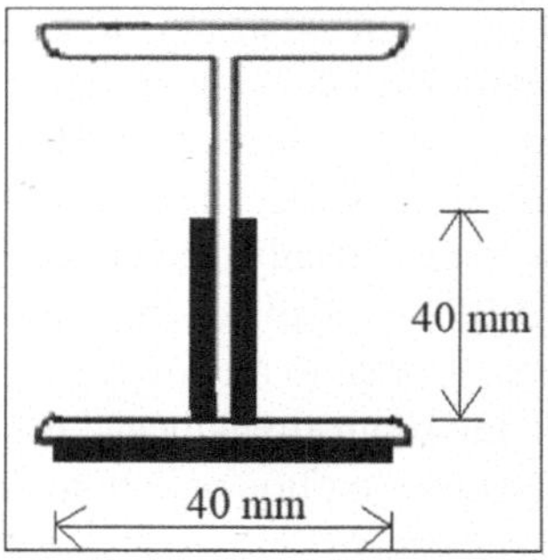

Fig. 4 SBS3

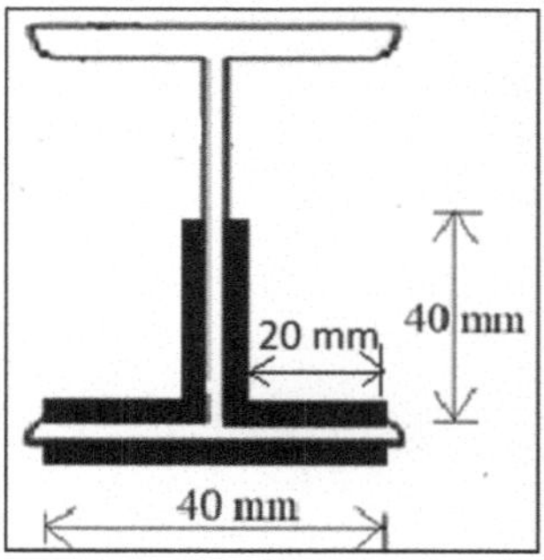

Fig. 5 SBS4

Fig. 6 SBS4E

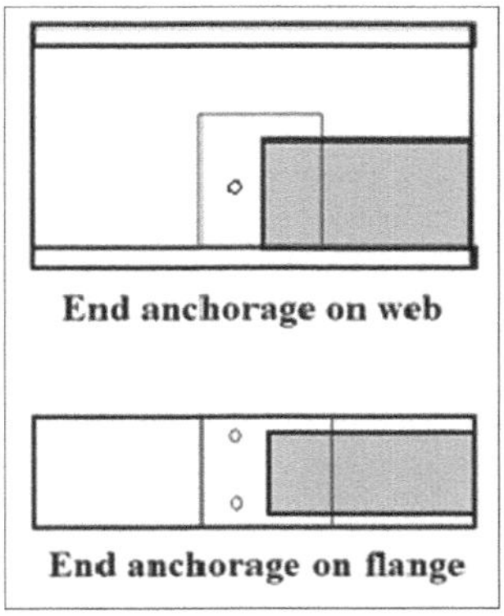

Fig. 7 DUB

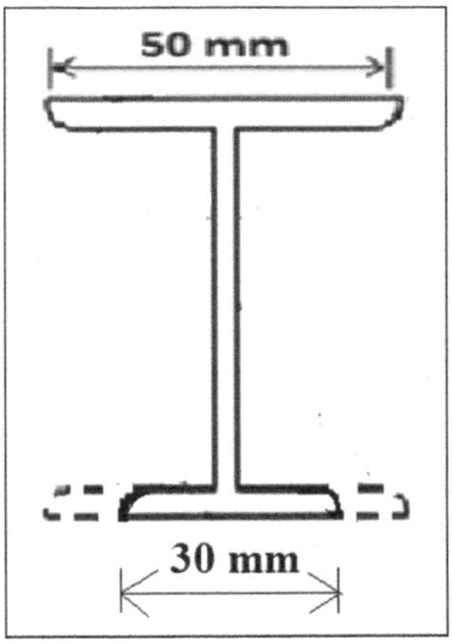

of CFRP (Fig. 6). End anchorage was provided by using steel plates and bolts. DUB was the damaged unstrengthened beam and the damage was induced by removing 10 mm of bottom flange from both sides to simulate corrosion damage (Fig. 7). DSBS3 was the damaged beam strengthened using scheme 3 (Fig. 8).

2.3 Specimen Preparation and Procedure

The surface of specimens were treated with sandpaper of grit size 120 to remove the rust and imperfections on the surface if any and then cleaned with a brush. Epoxy adhesive is then applied on the steel beam and on CFRP. CFRP is then placed over the prepared beam surface and then pressed tightly using a roller. Similarly, each portion is strengthened using 3 layers of CFRP. The strengthened specimens were tested after 1 week of curing at room temperature.

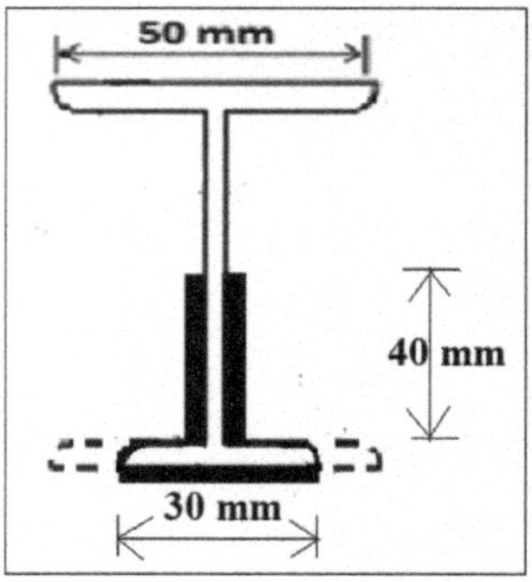

Fig. 8 DBS3

2.4 *Test Setup*

All the beams were tested under four point bending static loading and is performed on Universal Testing Machine (UTM) of 1000 kN capacity. Midspan deflections were measured using a dial guage of accuracy 0.01 mm. Figure 9 shows the loading scheme of the test.

3 Results and Discussions

The load versus midspan deflection curves of tested beams were shown in Fig. 10. From the figure, it can be seen that the load carrying capacity and stiffness of strengthened beams were enhanced w.r.t the control beam. Damage considerably reduced the flexural strength of steel beam where as on strengthening with CFRP, the lost strength and stiffness is regained to some extent.

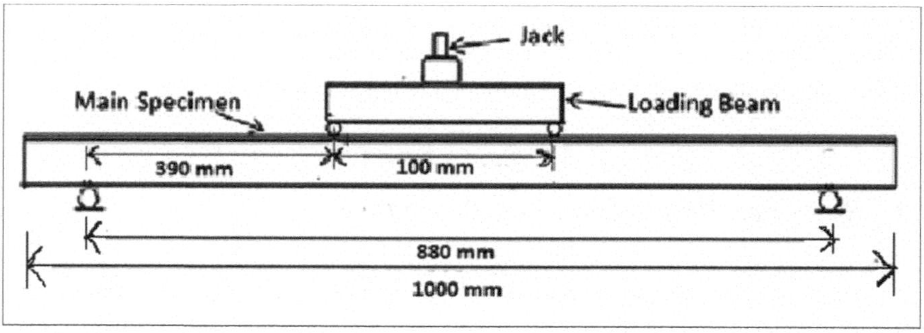

Fig. 9 Loading scheme

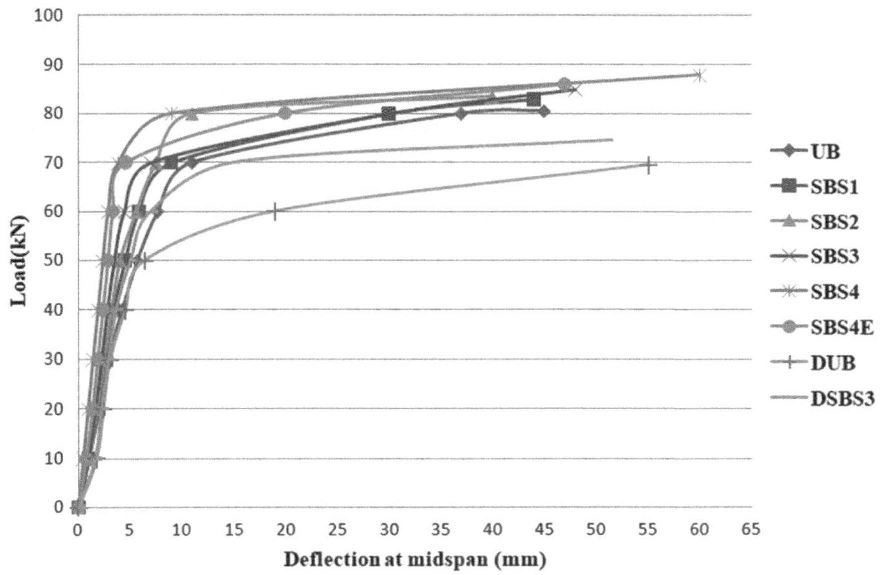

Fig. 10 Load versus midspan deflection curves of tested beams

3.1 Load Carrying Capacity

The most important parameter required in the strengthening of structures is the increment in the load carrying capacity of the strengthened specimens compared to the unstrengthened specimens. Effectiveness of CFRP strengthening on both yield load and ultimate load of the steel beam is studied. Table 5.1 shows the yield load, ultimate load and the percentage increment/decrement in yield load and ultimate load of the tested specimens.

3.1.1 Yield Load and Ultimate Load

Yield load is the load corresponding to the beginning of yielding in steel whereas the load corresponding to the failure of the beam is the ultimate load. The control beam yielded at 35 kN and failed at 80.5 kN while the beam strengthened using scheme 1 yielded at 45 kN and failed at 83 kN i.e., 28.57% increment in yield load and 3.11% increment is obtained when the steel beam was strengthened using CFRP bonded only to the lower portion of tension flange. The higher increment in yield load (82.26%) and ultimate load (9.32%) was observed for the beam strengthened using scheme 4. The percentage increments in yield load and ultimate load for the beams SBS3 and SBS4 is respectively double and triple the increments observed for beams SBS1 and SBS2 and the reason for this higher increments is the presence of CFRP bonded to the web portion. Thus CFRP bonded to the web portion enhances the load carrying capacity of strengthened beams. When damage is induced in the steel beam, the yield load and ultimate load w.r.t the control beam is reduced by 14.28 and 13.66% respectively. Strengthening using CFRP incremented the yield load of the damaged beam by 28.57% w.r.t the control beam and the yield load observed was 45 kN which was the yield load of the beam SBS1 which means that, for the damaged beam to act as a strengthened undamaged beam in the service life, strengthening has to be done using CFRP bonded to the lower portion of bottom flange as well as in bottom web portion. Through strengthening, the decrement in ultimate load carrying capacity was reduced to 7.45%.

3.2 *Stiffness*

Stiffness is an important parameter in strengthening of structures. Variations in elastic stiffness and post elastic stiffness of tested beams were studied.

3.2.1 Elastic Stiffness and Post Elastic Stiffness

Elastic stiffness is the stiffness in the elastic stage i.e., prior to the yielding of the beam. A deflection of 3 mm corresponds to the elastic stage of the tested beams as none of the beams has yielded at this deflection level. The percentage increment/decrement in stiffness in the elastic range with respect to the control beam can be found from the loads taken by each beams at this deflection level. Post elastic stiffness is the stiffness in the post yielding stage of the steel beam. A deflection of 9 mm corresponds to the post elastic stage of the tested beams as all the tested beams has already yielded at this deflection level. The percentage increment/decrement in stiffness in the post elastic stage with respect to the control beam can be found from the loads taken by each beams at this deflection level. Table 2 gives the percentage increment/decrement in elastic and post elastic stiffness for the tested beams.

It can be seen from the above table that both elastic and post elastic stiffness of the steel beam is improved upon strengthening with CFRP w.r.t the control beam and the higher increments was observed for the beam strengthened using scheme 4. The elastic stiffness of the beam SBS4 was doubled w.r.t UB. The higher increment in the post elastic stiffness of SBS2 comparing with SBS3 is due to the presence of CFRP bonded to the steel beam in the upper portion of bottom flange. It is much more clear from the percentage increments of SBS3 and SBS4. The only difference between the two beams is the presence of CFRP in the upper portion of bottom flange in SBS4 which is responsible for the higher increment of stiffness in post elastic range for beams SBS2 and SBS4.

The load taken by the undamaged unstrengthened beam at 3 mm deflection was 31 kN which was reduced to 29 kN when damage is induced. Upon strengthening with CFRP, the lost stiffness was restored as DSBS3 took 31 kN load at 3 mm deflection same as that of UB. The percentage increment in elastic stiffness w.r.t DUB was 6.9%.

The load taken by UB at 9 mm deflection was 64 kN whereas the damaged beam has taken only 52 kN load at same deflection. There was a decrement of 18.75% in stiffness at the post elastic stage due to damage on beam. When strengthening using CFRP is done, the beam took 63 kN load at 9 mm deflection which was near to the load taken by UB. The decrement in post elastic stiffness observed was only 1.56% after strengthening w.r.t UB. 21.15% increment in post elastic stiffness is observed for the strengthened beam DSBS3 when compared with DUB.

3.3 *Deformation Characteristics*

One of the most important parameters in the flexural strengthening of structures is the decrement of the vertical deflection for the strengthened specimen compared to the unstrengthened specimen. The deflection of the beams corresponding to the yield load and ultimate load of the unstrengthened beam is found and analysed to study the effectiveness of CFRP strengthening in the deformation characteristics of tested specimens.

3.3.1 At the Yield Load and Ultimate Load of Control Beam

The change in deformation observed for the tested beams at the yield load and ultimate load of control beam is shown in Table 3.

It can be seen from the table that the deflection in the undamaged strengthened beams at the yield load and ultimate load of unstrengthened beam is reduced up to 40 and 72.89% by CFRP strengthening. Higher reduction in deformation is observed for the beam strengthened using scheme 4. The reason behind the higher reduction in deformation for beams SBS2 and SBS4 compared to SBS1 and SBS3 at the ultimate load of UB is the presence of CFRP bonded to the upper portion of bottom flange. 8.57% increment in deformation was occurred in the damaged beam at the yield load of undamaged beam whereas 5.71% reduction in deformation was observed upon strengthening with CFRP. The percentage reduction in deformation for DSBS3 w.r.t DUB was 13.16%. DUB failed at 69.5 kN and the corresponding deflection was 55 mm. A higher decrement in deformation was observed for the damaged strengthened beam at the ultimate load of unstrengthened damaged beam. The observed decrement in deformation was 73.45% which shows the effectiveness of strengthening using CFRP on the damaged beam.

3.4 *Influence of End Anchorage on Flexural Strength*

It can be observed from the Tables 1, 2 and 3 that provision of end anchorage at the ends of CFRP enhanced the load carrying capacity and stiffness of the steel beam

Table 1 Load carrying capacity of tested specimens

Beam designation	Yield load (kN)	Increment/Decrement in yield load w.r.t UB (%)	Ultimate load (kN)	Increment/Decrement in ultimate load w.r.t UB (%)
UB	35	–	80.5	–
SBS1	45	28.57	83	3.11
SBS2	50	42.86	83.5	3.73
SBS3	55	57.14	85	5.59
SBS4	64	82.86	88	9.32
SBS4E	60	71.43	86	6.83
DUB	30	−14.28	69.5	−13.66
DSBS3	45	28.57	74.5	−7.45

Table 2 Elastic stiffness and post elastic stiffness of tested beams

Beam designation	Load taken at 3 mm deflection (kN)	Increment/Decrement in elastic stiffness w.r.t UB (%)	Load taken at 9 mm deflection (kN)	Increment/Decrement in post elastic stiffness w.r.t UB (%)
UB	31	–	64	–
SBS1	35	12.9	70	9.38
SBS2	37	19.35	74	15.63
SBS3	43	38.71	71	10.94
SBS4	62	100	80	25
SBS4E	53	70.97	73	14.06
DUB	29	−6.45	52	−18.75
DSBS3	31	0	63	−1.56

Table 3 Deformation characteristics of the tested specimens

Beam designation	Deflection at middle span corresponding to the yield load of UB (35 kN) (mm)	Change in deformation w.r.t UB at the yield load of UB (%)	Deflection at middle span corresponding to the ultimate load of UB (80.5 kN) (mm)	Change in deformation w.r.t UB at the ultimate load of UB (%)
UB	3.5	–	45	–
SBS1	3	−14.29	32.3	−28.22
SBS2	2.9	−17.14	15.1	−66.44
SBS3	2.5	−28.57	31.8	−29.33
SBS4	1.6	−54.29	12.2	−72.89
SBS4E	2.1	−40	22.3	−50.44
DUB	3.8	8.57	–	–
DSBS3	3.3	−5.71	–	–

and reduced the deformations when compared with the unstrengthened beam. No improvements in flexural strength is observed when the beam provided with end anchorage is compared with the beam without end anchorage. In fact, a decrement is observed. The presence of holes on the smaller I section brought a damaging effect to it. Thus for smaller sections, end anchorage is not recommended. But the debonding characteristics can be improved by providing end anchorages. In the beam SBS4, partial debonding was observed on CFRP sheets at the web portion and on both sides of bottom flange whereas in SBS4E, only a small portion of CFRP bonded on the lower portion of bottom flange detached from the steel beam. End debonding was not prominent.

3.5 Failure Modes

All the tested specimens failed in flexure showing good ductile behaviour before failure. The unstrengthend beam has undergone flexural failure after the yielding of steel (Fig. 11). The failure mode observed for CFRP in the strengthened specimens were partial debonding from one end. No complete debonding of CFRP was occurred in any of the specimen. End debonding was occurred because of high stress and strain intensity on adhesive at the CFRP tips. For the beam strengthened using scheme 1 (SBS1), debonding initiated from one end and progressed up to middle portion of CFRP length (Fig. 12). In the beam strengthened using scheme 2 (SBS2), the debonding of CFRP from the lower portion was similar to that of SBS1 while

Fig. 11 Flexural failure of UB

Fig. 12 Debonding of CFRP in SBS1

only end debonding was observed for CFRP in the upper portion of bottom flange (Fig. 13). The debonding of CFRP from one end till the middle portion of CFRP length was observed in the flange portion as well as in the web portion of beam SBS3 (Fig. 14) where as in beam SBS4, mid debonding of CFRP was observed in web and debonding of CFRP from one end till middle portion of CFRP length is observed in the flange portion (Fig. 15). In the case of the beam with end anchorage (SBS4E), only a small portion of CFRP from quarter portion of CFRP length detached from steel beam (Fig. 16). No end debonding was observed which shows the adequacy of steel plates and bolts as end anchorage to prevent premature debonding of CFRP. The damaged unstrengthened beam (DUB) has shown large ductility prior to the failure in flexure (Fig. 17). The damaged beam strengthened using scheme 3 (DSBS3) has

Fig. 13 Debonding of CFRP in SBS2

Fig. 14 Debonding of CFRP in SBS3

Fig. 15 Debonding of CFRP in SBS4

Fig. 16 Debonding of CFRP in SBS4E

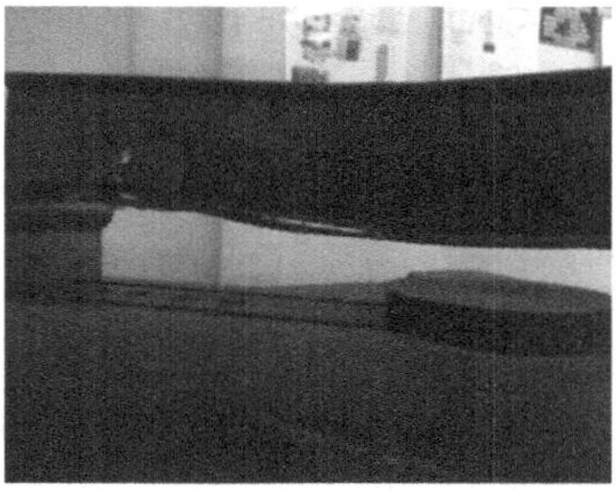

Fig. 17 Flexural failure of DUB

Fig. 18 Debonding of CFRP in DSBS3

also undergone flexural failure after yielding of steel with the partial debonding of CFRP from bottom flange as well as from both sides of web (Fig. 18).

4 Conclusions

Based on the experimental investigations carried on control and CFRP strengthened beam specimens, the following conclusions are drawn:

- CFRP strengthening was found effective in enhancing the flexural strength of steel beam.

- An increase of 28.57–82.86% and 3.11–9.32% were observed in yield load and ultimate load for the strengthened specimens respectively.
- The increase of ultimate load was relatively less than the increase of the yield load due to the early debonding of CFRP from the steel beam.
- Elastic stiffness of the steel beam was incremented from 12.9 to 100% and post elastic stiffness was incremented from 9.38 to 25% through CFRP strengthening.
- Of the strengthening schemes investigated, scheme 4 in which CFRP sheets were attached to the lower and upper portions of the bottom flange as well as in the web portion was found more effective in enhancing the load carrying capacity and stiffness of the steel beam.
- It was observed that bonding of CFRP in the web portion has a major role in incrementing the yield load and ultimate load of steel beam. Also, CFRP bonded to the upper portion of bottom flange increases the post elastic stiffness and reduces the deformation characteristics of the steel beam.
- The modulus of elasticity of the CFRP sheet has to be similar or higher than that of the steel to achieve reasonable gains in strength and stiffness.
- The higher increments in stiffness and the higher reduction in deformation characteristics of strengthened specimens were attributed to the high elastic modulus of CFRP sheet used.
- All the tested beams failed in flexure after the yielding of steel. The failure mode of CFRP in the strengthened specimens was partial debonding from one end.
- Provision of end anchorage reduced the chances for premature end debonding of CFRP but it couldn't enhance the load carrying capacity and stiffness of the steel beam w.r.t the strengthened beam without end anchorage. Therefore strengthening without end anchorage is sufficient for strength gain in smaller steel I sections.
- Damage to the bottom flange degraded both the flexural stiffness as well as the strength of the steel beam.
- CFRP repair was found effective in improving the lost strength and stiffness of the damaged beam.
- The elastic and post elastic stiffness of damaged beams can be restored almost fully with the use of CFRP sheets for the system investigated.
- The yield strength of the damaged beams can be fully restored to its original undamaged state upon strengthening with CFRP. 7.19% gain in ultimate strength was observed for the strengthened beam w.r.t the damaged control beam for the system investigated.
- A good reduction in deformation was observed for the damaged strengthened specimen at the yield load of unstrengthened specimen and at the ultimate load of damaged specimen.

References

1. Tavakkolizadeh M, Saadatmanesh H (2003) Fatigue strength of steel girders strengthened with carbon fiber reinforced polymer patch. J Struct Eng, ASCE 129:186–196
2. Tavakkolizadeh M, Saadatmanesh H (2003) Strengthening of steel-concrete composite girders using carbon fiber reinforced polymer sheets. J Struct Eng, ASCE 129:30–40
3. Mosavi SM, Sadeghi Nik A (2014) Strengthening of steel-concrete composite girders using carbon fibre reinforced polymer (CFRP) plates. Sadhana 40:249–261
4. Lenwari A, Thepchatriand T, Albrecht P (2005) Flexural response of steel beams strengthened with partial-length CFRP plate. J Compos Constr ASCE 9:296–303
5. Fam A, MacDougalland C, Shaat A (2009) Upgrading steel-concrete composite girders and repair of damaged steel beams using bonded CFRP laminates. Thin-Walled Struct 47:1122–1135 Elsevier
6. Narmashiri K, Ramli Sulong NH, Jumaat MZ (2011) Failure analysis and structural behaviour of CFRP strengthened steel I-beams. Constr Build Mater 30:1–9 Elsevier
7. Narmashiri K, Ramli Sulong NH, Jumaat MZ (2011) Flexural strengthening of steel I-beams by using CFRP strips, Int J Phys Sci, Academic J 6(7):1620–1627
8. Deng J, Lee MMK (2007) Behaviour under static loading of metallic beams reinforced with a bonded CFRP plate. Compos Struct 78:232–242 Elsevier
9. Yousefi O, Narmashiri K, Ghods A (2014) Investigation of flexural deficient steel beams strengthened by CFRP. Indian J Fundam Appl Life Sci, CIBTech 4:372–380
10. Al-Saidy AH, Klaiber FW, Wipf TJ, Al-Jabriand KS, Al-Nuaimi AS (2008) Parametric study on the behaviour of short span composite bridge girders strengthened with carbon fibre reinforced polymer plates. Constr BuildMater 22:729–737 Elsevier
11. Ellobody E (2011) Performance of composite girders strengthened using carbon fibre reinforced polymer laminates. Thin-Walled Struct 49:1429–1441 Elsevier
12. Al-Saidy AH, Klaiber FW, Wipf TJ (2004) Repair of steel composite beams with carbon fiber-reinforced polymer plates. J Compos Constr ASCE 8:163–172
13. Al-Saidy AH, Klaiberand FW, Wipf TJ (2007) Strengthening of steel-concrete composite girders using carbon fibre reinforced polymer plates. Constr Build Mater 21:295–302 Elsevier
14. Linghoff D, Al-Emrani M, Kliger R (2010) Performance of steel beams strengthened with CFRP laminate—Part 1: Laboratory tests. Compos Part B 41:509–515 Elsevier
15. Deng J, Lee MMK, Moy SSJ (2004) Stress analysis of steel beams reinforced with a bonded CFRP plate. Compos Struct 65:205–215 Elsevier
16. Colombi P, Poggi C (2006) An experimental, analytical and numerical study of the static behaviour of steel beams reinforced by pultruded CFRP strips. Compos Part B 3:64–73
17. Narmashiri K, Jumaat MZ, Ramli Sulong NH (2010) Investigation on end anchoring of CFRP strengthened steel I-beams. Int J Phys Sci, Acad J 5(9):1360–1371
18. Sen R, Liby L, Mullens G (2001) Strengthening steel bridge sections using CFRP laminates. Compos Part B 32:309–322

A Review on Nano TiO_2—A Repellent in Paint

Aparna Varma, Amala Rose James and Sunitha A. Daniel

Abstract Many technologies had origin through the continuous and careful observation of unnoticed features of the species present in our nature. The self cleaning property of lotus leaves is a remarkable one among them, known in the name "lotus effect". It has found a huge place in the civil industry. The exterior house paints are always subjected to wide fluctuation in atmospheric temperature conditions such as exposure to winters, hot summers, raining, etc. which reduces its durability and causes algal growth. Thus by effective incorporation of nanotechnology with this lotus effect, it has become able to provide self-cleaning ability along with air purification and antibacterial performance to construction materials like paint, tiles, etc. Nanomaterials like silicon, Titanium Dioxide etc. help to improve functionalities like water or stain resistance, UV protection, scratch resistance etc. When nano TiO_2 was added to paint it degraded the polluting compounds at material surface by photocatalysis. Addition of it with cementitious materials displayed multiple photocatalytic functions and self cleaning properties. This review paper throws light into the repellent property of nano TiO_2 when used in paint thereby enhancing aesthetic appearance of surfaces and effectively reducing the cost of routine maintenance. Rhodamine B removal ability of nano TiO_2 is reviewed.

Keywords Nanotechnology · Self cleaning · Nano TiO_2 · Paint · Photocatalysis · Rhodamine B

1 Introduction

Nanotechnology deals with particles having their size in nano range. This size reduction helps to enhance their material properties. These materials have unique characteristics compared to that of their macro and micro counterparts. This feature of nanomaterials has greatly influenced scientists and engineers which paved the way to the usage of nanoparticles in various industries including civil industry. The widely

A. Varma (✉) · A. R. James · S. A. Daniel
Department of Civil Engineering, Saintgits College of Engineering, Kottayam, India
e-mail: aparna.varma1620@saintgits.org; varmaaparna98@gmail.com

K. Dasgupta et al. (eds.), *Proceedings of SECON'19*,
Lecture Notes in Civil Engineering 46,
https://doi.org/10.1007/978-3-030-26365-2_83

used materials in construction industry are concrete, steel, timber, glass etc. When we compare the properties like Young's Modulus, tensile strength etc. of these materials with that of a nanomaterial, say carbon nanotube, the results are astonishing. Young's Modulus &tensile strength of carbon nanotubes(CNT) are higher than that of all construction materials stated (proved to have 150 times value of Young's Modulus that of steel and 6 times more lighter than steel). This reveals the relevance of nanomaterials in this industry.

This field when utilized in civil industry can produce much better products. Various nanomaterials like nano TiO_2, CNT, nano-silica etc. are added to cementitious materials so that the properties and strength are enhanced [1]. Dirt absorption by paint coating is a problem in this industry which can raise the maintenance cost and reduce the aesthetics of the surface. By mixing nano TiO_2 with paint, bio mimicking and nanotechnology are joined together, giving a solution for this problem. It can act as a hydrophilic material and by the process of photocatalysis in nano TiO_2, impurities are broken down under the action of sunlight and thus can remove dirt.

2 Significance

Paints improve the aesthetic appearance of the buildings and also protect them from external attacks. But paint may have so many defects while acting as a coating which may be due to poor workmanship or low quality of the paint. Poor dirt resistance of paint is one such defect which reduces the aesthetics of building and increase the routine maintenance cost. This defect is due to the use of low quality binders in paints. Binders give adhesion to substrates and provide cohesion between pigments and fillers, weather resistance, wear resistance, chemical and solvent resistance which are dependent on binder quality.

Studies were undergone to find out a solution for stain removal on paint surfaces. By continuous observation of our surroundings (biomimicking) it was identified that the self cleaning properties are exhibited by many surfaces present in nature. It includes leaves of lotus and cabbage, the wings of butterflies, geckofeet and shark skin etc. [2]. It was found that self cleaning coatings if provided, it could resist the walls being stain filled. Self cleaning coatings fall under two types, hydrophilic and hydrophobic, both work by the operation of water. These properties are seen in nature itself but relying on the action of water alone could not make paint a dirt-repellant one. Thus evolved the use of nano TiO_2 which works by hydrophilic characters and photocatalysis. But TiO_2 also has a negative impact on paint. It can degrade the binders used in paints thus reducing its life. This problem can be reduced by incorporating TiO_2 into the paint as a thin coating on glass microspheres which act as carrier particles [3]. These glass particles actually provide a surface for the photocatalysis reaction to take place and thus does not affect the binder.

Apart from self-cleaning, TiO_2 can act as an air purifier, fire retardant, thermal insulator etc. It is also used in other fields too due to its photocatalytic effects. It is used in paper industry to improve opacity of papers. It is used in medical field to kill bacteria on surface in photodynamic therapy to kill cancer cells. Also used in cosmetics like sunscreen creams [4], skin milks etc. due to its ability to filter UV radiations [5].

3 Titanium Dioxide

TiO_2 in the pigment form is not obtained directly from the nature. TiO_2 is mostly found as mineral ilmenite. TiO_2 exists in different forms [6] mainly as anatase (Fig. 1a), rutile (Fig. 1b) and then as brookite (Fig. 1c). Titanium bearing mineral deposits are widely found on earth which contains Titanium precursors like Titanium Tetrachloride and Titanium sulphate. These precursors are extracted from deposits by chloride or sulphate processes which can be used for the production of TiO_2 pigments as raw material.

TiO_2 can be synthesized by various methods [8]. Sol-gel method, ultrasonic-assisted sol-gel method, micro emulsion method, colloidal synthesis, hydrolysis method, thermos hydrolysis method, thermal plasma process, solvo-thermal method, reactive plasma processing, induction plasma torch, plasma electrolytic oxidation, co precipitation method [8] etc. Experimental preparation by Sol-gel method [9] is discussed below.

Sol-gel method involves two processes, hydrolysis and polymerization. In the preparation process 10 ml of Tri-Tetra-Iso- Propoxide(TTIP) was blended with 40 ml of 2-propanol in a flask with round bottom having a Ph- meter mounted on it, which was placed above a hotplate or magnetic stirrer. Water was then added to crystallizer dish having the flask and it was heated at a temperature of 80 °C. At this same temperature 2-propanol and Tri-Tetra-Iso-Propoxide precursor were mixed by continuously

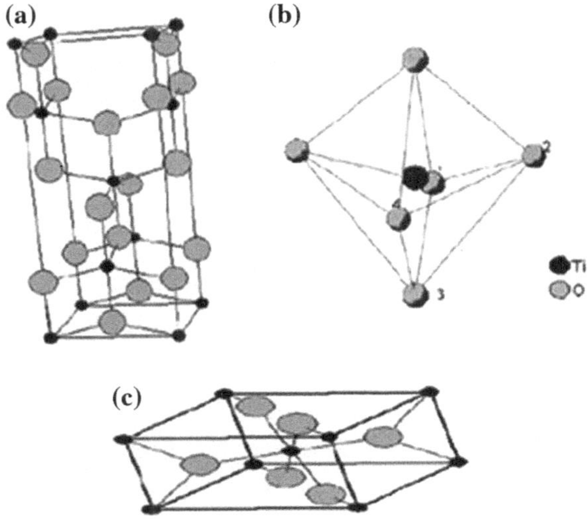

Fig. 1 Different forms of TiO_2 [7]. **a.** Anatase, **b.** Rutile, **c.** Brookite

stirring for one hour. Then acidified & deionized water was added to the alcoholic solution with the titanium precursor whose pH value was adjusted to 2 with the help of nitric acid to initialize hydrolysis. The stirring was continued for 6 hours and a colloidal liquid with 5 g/L TiO_2 was obtained. By continuously stirring on hotplate at 80 °C, all the excess solvent was removed by evaporation. The reaction involved was [9]:

$$Ti(OCH(CH_3)_2)_4 + 2H_2O \rightarrow TiO_2 + 4(CH_3)_2CHOH$$

4 Self-Cleaning Property

Selfcleaning properties can be imparted to building materials by developing superhydrophobic or superhydrophilic surfaces. The hydrophobic and hydrophilic properties are based on the ratio of interfacial tension between liquid and surface which physically indicates the contact angle [10, 11] between water drops and the surface. When contact angle equals 180°, it corresponds to non-wetting and when the same equals 0°, it corresponds to complete wetting.

Hydrophobic behavior of the surface is based on Lotus Effect [10]. The top and bottom surface of the lotus leaves has a special and characteristic roughness. It also has nano-size wax crystals which have high water repellent chemical properties. The surface roughness and water-repellent wax crystals, when combined make the lotus leaves extremely non-wettable. This micro rough surfaces (Fig. 2) has contact angles greater than 130–140°. As a result, when water comes in contact with such surfaces water takes spherical shapes and roll down the surfaces taking away dirt with them. Thus these hydrophobic properties impart self cleaning.

Hydrophilic behavior of the surfaces can be obtained by using soaps and detergents. They decrease the surface tension of water and reduce the contact angle. To get low contact angles without detergents we can use active thin films on the surface and this can be achieved by TiO_2. When TiO_2 coated materials is exposed to light contact

Fig. 2 Structure of surface of lotus leaves

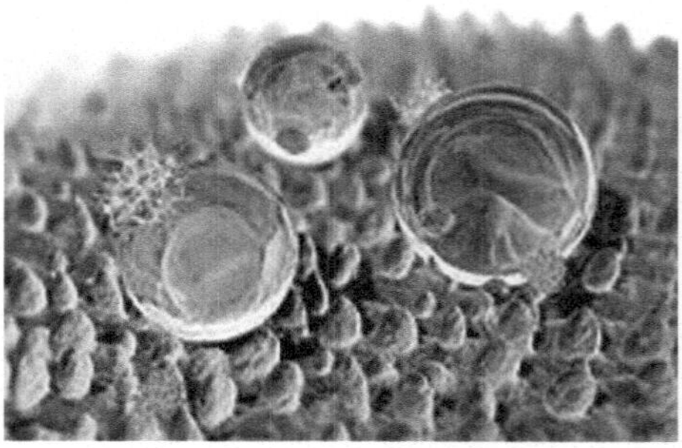

angles less than one degree is obtained and as a result instead of becoming droplets water form sheets. The presence of water resists dirt to get stick on the surface.

5 Photocatalysis Mechanism

Photocatalysis is the process which accelerates a photoreaction with the help of a catalyst. In this process an adsorbed substrate absorbs light. In photo accelerated catalysis, the photocatalytic activity (PCA) relies on the ability of the catalyst to create electron-hole pairs which generate free radicles (e.g. hydroxyl radicles) which can undergo secondary reactions [11]. The photocatalysts are semiconductor particles that act as sensitizers for light-induced redox processes. These semiconductors have filled valance band and an empty conduction band. The minimum light energy required to make the material electrically conductive equals band gap energy E_g which is the energy difference between the lowest energy level of the conduction band and the highest energy level of the valence band (Fig. 3).

When a photon with an energy of hν exceeds the band gap energy, an electron (e^-) jumps to the conduction band from the valence band leaving a hole (h^+) in it. These electron-hole pairs get trapped at the surface of the particle in semiconductors and involve in different chemical reactions [11]. They are:

- Oxidation of donor molecules by holes

$$D + h^+ \rightarrow D^{\bullet +}$$

- Reduction of acceptor molecules conduction band electrons.

$$A + e^- \rightarrow A^{\bullet -}$$

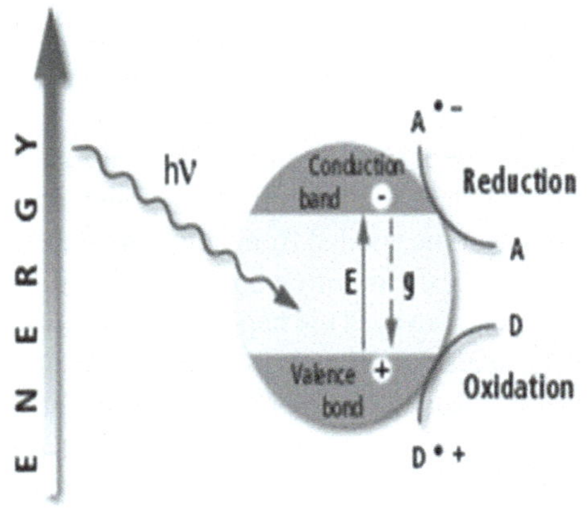

Fig. 3 Operation of photocatalyst

- Reaction of holes h^+ with water to produce highly reactive hydroxyl radicle (•OH). These hydroxyl radicles and holes both can be utilized to oxidize organic contaminants.

$$H_2O + h^+ \rightarrow {}^{\bullet}OH + H^+$$

- Oxygen in air plays the role as electron acceptor giving super-oxide ion ($O_2^{\bullet -}$) which are highly reactive particles that helps to oxidize the organic materials.

$$O_2 + e^- \rightarrow O_2^{\bullet -}$$

6 Role of TiO_2 in Photo Catalysis

Titanium dioxide is a semiconducting material which acts as a photocatalyst. It is chemically activated by light; under the influence of light the material decomposes organic materials. TiO_2 is one of the best materials that shows semiconducting properties to be used in the areas of chemical conversion and storage of solar energy. Among the different forms of TiO_2, anatase performs highest photoactivity [11].

The photocatalytic splitting of water on TiO_2 electrodes was discovered by Fujishima and Honda in 1972 [12]. This was a milestone in the era of heterogenous photocatalysis. The semiconductor TiO_2 has a band gap energy of 3.2 eV. When photons having an energy higher than the band gap energy of TiO_2 irradiates this material, an electron from the valence band gets promoted to the conduction band [11]. As a result charge-carrier generation takes place which is indicated by the following equation: $TiO_2 + hv \rightarrow h^+ + e^-$ [11]. The redox potentials of the adsorbates and the band energy positions of semiconductor governs the energy induced e^- transferring ability of the semiconductor. Either the acceptor species should have a potential level much below the conduction band of the semiconductor or the donor should have a potential level much above the valence band of the semiconductor so that the e- could jump to the empty hole.

Titanium dioxide photocatalysis has wide range of uses. Some of them are self-cleaning effect, air-cleaning effect, anti-bacterial effect, water treatment, anti-fogging effect [11]. When we consider an outdoor application of TiO_2, the external building walls become spoiled from automobile exhaust gases containing oil components. So by coating them with a super-hydrophilic photocatalyst, the dirt on the walls can be washed away. This happens because super-hydrophilic surface shows affinity to water than the oil and also generation of potential cleaning agents on the surface by the formation of a photogenerated hole-electron pair that reacts with oxygen and water in the environment. The agents •OH and •OOH decompose large organic molecules to smaller ones [11]. Thus grease and dirt can be swept away with water by the combination of photocatalysis and super-hydrophilicity.

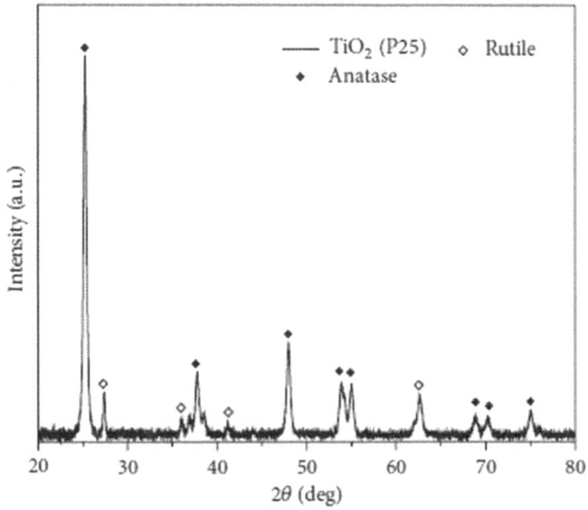

Fig. 4 XRD pattern of TiO_2 (P25) nanoparticles [13]

7 Coating Characterization

7.1 *Phase Analysis*

Microstructural characterization of the coating was investigated by X-ray diffraction (XRD). It is a mix of anatase and rutile phases. This study used commercial bicrystalline TiO_2 with 80% anatase, and 20% rutile. (Figure 4) It is clear from the XRD pattern that the addition of TiO_2 nanoparticles makes a change in the surface morphology [13].

7.2 *Morphology Study*

Scanning electron microscope operated at a voltage of 10 kV and probe current of 10 mA were used to observe the surface and cross-section morphologies of the coating. Elemental composition of the coatings was analyzed using energy dispersive X-ray spectroscopy (EDX).

The SEM images [13] showed a uniform morphology a dense surface with no signs of major cracking (Fig. 5).

A study on the dirt removing efficiency of nano TiO_2 on paint coatings was made. Rhodamine B is a dye which plays the role of dirt that sticks on the paint surface. TiO_2 was added to paint by coating and intermixing methods [14].

The samples were then exposed to UV-A and visible radiations to study the dirt removing efficiency of TiO_2 coatings.

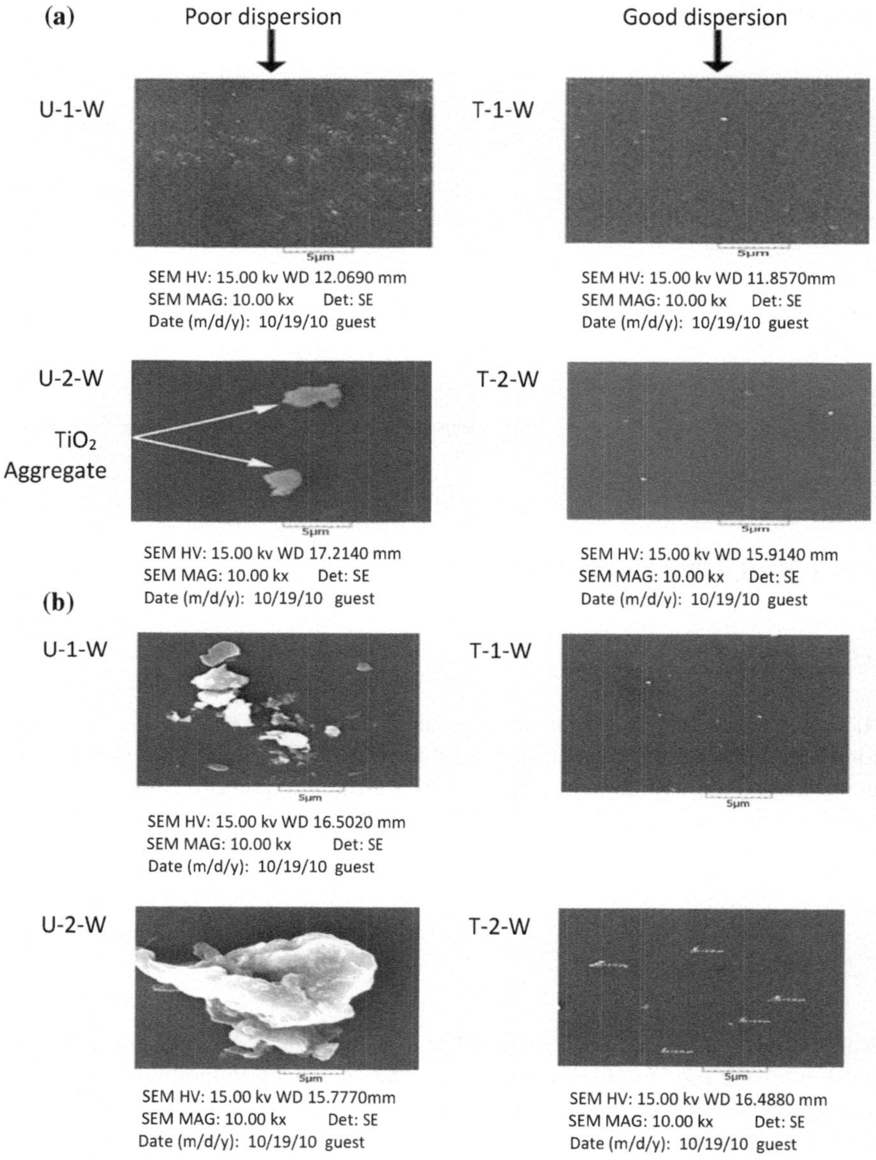

Fig. 5 SEM micrographs of **(a)** cross section and **(b)** surface of acrylic coatings containing 1 and 2 wt% of untreated (P25) and silica treated TiO_2 (12TT) nanoparticles [13]

RhB removal under visible light irradiation shows different result as in the case of UV irradiation before and after weathering conditions are applied. RhB removal efficiencies under visible light irradiations were lower than UV-A irradiations. The difference lies in the fact that the degradation under UV irradiation is by ordinary photocatalytic photoreaction but under visible light irradiations it degrades by dye sensitized photoreaction i.e., RhB molecules absorbs photon of visible light and forms RhB^+ radicles on TiO_2 surface so that TiO_2 is not directly activated [15].

8 Conclusion

Surface damage to structures is a vital problem in the civil industry. Nano particles like nano TiO_2 when mixed with paint improves the durability of the surface. The size, shape and greater surface area of these materials improve the existing properties of paint coated surfaces due to their specific structural characteristics. This review paper has analyzed the self- cleaning effect of nano TiO_2 due to its photo catalytic nature along with hydrophilic properties.

The overall results of the review made proves that the TiO_2 coated surfaces are water and dirt repellent and hence proven good to be used as additives in paints.

9 Recommendations

Treated nanoparticles have a better dispersion in the coating than their untreated counterparts. If this finding is made useful in nano TiO_2 it will create better paints and more durable surfaces.

References

1. Khitab A, Anwar W, Mansouri I, Tariq MK, Mehmood I (2015) Future of civil engineering materials. A Rev Recent Dev
2. Ganesh VA, Raut HK, Nair AS, Ramakrishna S (2011) A review on self-cleaning coatings. J Mater Chem
3. Gunnarsson SG (2012) Self cleaning paint: Introduction of photocatalytic particles into paint system. Technical University of Denmark
4. Jafari SM, Sadeghi Hashtjin G, Koohi MK (2019) Titanium Dioxide nanoparticles as a common component of sunscreens, An experimental study of dermal or occular safety assessment. Iran J Vet Med
5. Varner K (2010) State of the science literature review: nano titanium dioxide environmental matters. United States Environmental Protection Agency
6. Carp O, Huisman CL, Reller A (2004) Photoinduced reactivity of Titanium Dioxide. Progress in solid state. Scientific Research, An Academic Publisher
7. Malekshahy Byranvand M, Nemati Kharat A, Fatholahi L, Malekshahi Beiranvand Z (2013) A review on synthesis of nano titanium dioxide via. different methods. J Nanostruct

8. Macwan DP, Dave PN, Chaturvedi S. A Review on nano titanium dioxide sol-gel type syntheses and its applications. J Mater Sci
9. Giampiccolo A, Tobaldi DM, Leonardi SG, Murdoch BJ, Seabra MP, Ansell MP, Neri G, Ball RJ (2018) Sol gel/graphene TiO_2 nanoparticles for the photocatalytic—assisted sensing and abatement of NO_2. Environ Appl Catal B
10. GreBler S, Fiedeler U, Simko M, Gazso A, Nentwich M (2000) Self cleaning, dirt and water-repellent coatings on the basis of nanotechnology
11. Benedix R, Dehn F, Quass J, Orgass M (2000) Application of titanium dioxide photocatalysis to create self-cleaning building materials. Lacer 5
12. Fujishima A, Honda K (1972) Electrochemical photolysis of water at a semiconductor electrode. Nature, Int J Sci
13. Pazokofard S, Esfandeh M, Mirabedini SM, Mohseni M, Ranjbar Z (2012) Investigating the role of surface treated TiO_2 nanoparticles on self-cleaning behaviour of an acrylic facade coating. Springer, Berlin
14. Guo M-Z, Maury-Ramirez A, Poon CS (2015) Self-cleaning ability of titanium dioxide clear paint coated architectural mortar and its potential in field application. J Cleaner Prod
15. Graniani L, Quagliarini E, Bondioli F, D′Orazio M (2014) Durability of self cleaning TiO_2 coatings on fired clay brick facades: effects of UV exposure and wet and dry cycles. Build Environ

Shear Performance of Embedded Through Section (ETS) Over Near Surface Mounting (NSM) Method of RC Beams

Preetha Prabhakaran and Glory Joseph

Abstract Reinforced concrete structures often have to face modifications in performance during their service life mainly because of ageing and exposure to aggressive environment. Near Surface Mounting (NSM) and Embedded Through-Section (ETS) are two effective techniques for the shear strengthening of reinforced concrete (RC) beams. NSM technique involves cutting of a series of shallow grooves in the concrete cover of the structures and inserting steel/FRP bars into them and ETS technique consists of drilling holes across the full depth of the beam cross-section and inserting steel/FRP bars into them, thereby bonding the bars with the surrounding concrete using appropriate bonding material. This paper presents the results of experimental investigations on RC beams strengthened in shear using CFRP bars in Embedded through section (ETS) and Near Surface Mounting (NSM) methods. Three point bending tests are conducted to find the effect of ETS/NSM bar spacing and ETS/NSM bar orientation on the load carrying capacity of RC beams. The experimental results confirms the effectiveness of ETS and NSM methods and a comparison is made in terms of shear strengthening ratio, load carrying capacity of RC beams.

Keywords Near surface mounting (NSM) · Embedded through-section (ETS) · CFRP bars · Shear strengthening ratio

1 Introduction

Reinforced concrete structures often have to face modification and improvement of their performance during their service life mainly because of ageing or inability of existing structures to meet the current standards or requirements because of their out dated standards. At present, the technique of strengthening reinforced concrete

P. Prabhakaran (✉)
Cochin University of Science and Technology, Cochin, India
e-mail: preethanimesh@yahooo.co.in

G. Joseph
Division of Civil Engineering, Cochin University of Science and Technology, Kalamassery, India

K. Dasgupta et al. (eds.), *Proceedings of SECON'19*,
Lecture Notes in Civil Engineering 46,
https://doi.org/10.1007/978-3-030-26365-2_84

structures with Fibre Reinforced Polymer (FRP) bars has attracted much attention. Due to their light weight, high tensile strength and ease to install on irregular surfaces, the use of FRP materials for strengthening of RC elements has become an accepted practice within civil engineering community. Different strengthening techniques, like externally bonded reinforcement (EBR), near surfaces mounted (NSM) and Embedded Through Section (ETS) are being investigated and applied for the shear strengthening of RC beams [1].

In Near Surface Mounting (NSM) method, grooves are first cut into the concrete structures and the FRP bars or strips are then embedded and bonded therein with epoxy adhesives [2]. The embedded through-section (ETS) is also an effective technique for the shear strengthening of reinforced concrete (RC) elements. This technique is based on the execution of holes drilled through the element cross section, in which steel or fibre reinforced polymer (FRP) bars are inserted and bonded to the surrounding concrete with an epoxy adhesive [3, 4] .

This paper aims to investigate the influence of percentage of CFRP bars in ETS and NSM techniques to improve the shear strength of RC beams. The experimental results are compared for both strengthening methods in various aspects such as shear strengthening ratio, bar spacing and bar orientation.

2 Experimental Programme

Experimental investigation is done on the structural performance of reinforced concrete (RC) beams strengthened in shear using NSM as well as ETS FRP bars of varying configurations. In this section, the details of specimen, material properties, strengthening techniques and loading test setup of this experimental program is presented. The test program consists of two series of strengthening (i.e. NSM and ETS) which includes casting, instrumentation, and testing twenty six RC beams having rectangular cross-section. The ultimate load for the specimen is determined using three point bending test in the Universal Testing Machine (UTM) and the increase in ultimate load is determined with respect to the control specimen to find out the effectiveness of each NSM/ETS shear strengthened beams.

The RC beams have length of 1000 mm with cross section of 150 mm width, 200 mm depth and effective span of 700 mm. The design of reinforcements for the beam is done according to IS: 456 codes. 2 numbers of 10 mm diameter bars are used as tension reinforcement and 2 numbers of 10 mm diameter bars are used as compression reinforcement. 3 numbers of 8 mm diameter bars are provided as shear reinforcement, one each at the two support points and also at the point of loading. There are thirteen NSM configurations and thirteen ETS configurations in this study with parameters being FRP bar spacing and bar orientation (vertical and inclined). For this work, the diameters of FRP bars used are 8 mm, 10 mm and 12 mm respectively. Here, the spacing of FRP bars studied are 220 mm and 146.7 mm, i.e., one-fourth and one-sixth of the effective span, L of the beam. Orientation of bars selected are

Table 1 Material specifications

Material	Parameters	Value
CFRP bar	Tensile strength	2172 MPa
	Modulus of Elasticity	124 GPa
	% elongation	1.75%
	Density	0.1116 kg/m
Epoxy (polyester resin)	Compressive strength	87 MPa
	Tensile strength	11 MPa
	Flexural strength	19 MPa
	Compressive modulus	13 GPa
	Density	1880 kg/m^3
Araldite epoxy resin	Specific gravity	1.17
	Viscosity at 25 °C (Pa.s)	50
	Colour	Creamy viscous
Polyester resin Lokfix-S	Compressive strength	83 MPa
	Tensile strength	11 MPa
	Flexural strength	20 MPa
	Compressive modulus	13 GPa
	Density	1880 kg/m^3

vertical and inclined i.e., 90° and 45° respectively. The properties of CFRP bar, epoxy etc. are presented in Table 1.

In this paper an attempt is made to evaluate the shear performance of RC beams using NSM as well as ETS techniques with same percentage of CFRP reinforcement. The experimental results are compared to evaluate the efficiency of each shear strengthening method.

2.1 Shear Strengthening Setup

A schematic diagram consisting of cross section and longitudinal section of the control specimen (CB) showing their geometry and reinforcement detail is shown in Fig. 1. Two numbers of 10 mm diameter bars are used as tension reinforcement and two numbers of 10 mm diameter bars are used as compression reinforcement. Three numbers of 8 mm diameter bars are provided as shear reinforcement, one each at the two support points and also at the point of loading.

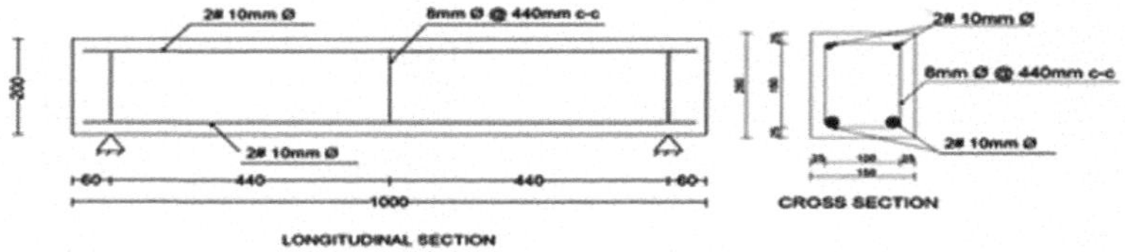

Fig. 1 Geometry and reinforcement detailing of control beam

2.2 Near Surface Mounting (NSM) Method

Thirteen beam specimens are cast in moulds of dimension 1000 mm × 150 mm × 200 mm. For the purpose of introducing NSM FRP bars in the beams for strengthening, grooves of dimension equal to 1.5 times the diameter of NSM FRP bars is adopted as per ACI 440.2R-08(13.3, Page 39), i.e., 12 mm × 12 mm, 15 mm × 15 mm and 18 mm × 18 mm for NSM bar diameters of 8 mm, 10 mm and 12 mm respectively are made in them.

The designation CB stands for control beam, and NSM strengthened beam is designated as NB-8d-sL/4-90, where NB stands for NSM beam, d stands for the diameter i.e. 8 mm, s stands for spacing of NSM bars i.e. L/4, and 90° vertical. Table 2 includes the cross section of the beams for NSM shear strengthening. The study is focused on the effect of bar spacing and bar orientation (90° and 45°) which can influence the bond performance and hence the structural behaviour of NSM strengthened beams. The different NSM strengthened beam configurations with 10 mm dia bars are diagrammatically represented in Table 2.

2.3 Embedded Through Section (ETS) Method

There are thirteen ETS configurations in this study with parameters being ETS bar spacing and ETS bar orientation. For the purpose of introducing ETS steel bars in the beams for strengthening, holes with full depth i.e., 200 mm are made. The diameter of holes is such that it is 8 mm more than the diameter of ETS FRP bar to be inserted. The diameters of ETS bars used are 8 mm, 10 mm and 12 mm respectively. Also the spacing of ETS bars studied are 220 mm and 146.7 mm, i.e., one-fourth and one-sixth of the effective span, L of the beam.

The ETS strengthened beam is designated as EsL/4d08-90, where E stands for ETS beam, s stands for spacing of ETS bars i.e. L/4, d stands for the diameter i.e. 8 mm. The orientation of bars selected are vertical and inclined i.e., 90° and 45° respectively. The various ETS strengthened beam specimens with 10 mm dia ETS bar and their configuration characteristics are diagrammatically represented in Table 3. (ETS strengthening technique is shown in Fig. 2).

Table 2 NSM shear strengthened RC beams (using 10 mm dia CFRP bars)

Beam Crossection	Longitudinal Section	Beam specimen
NB-10d-sL/4-90		
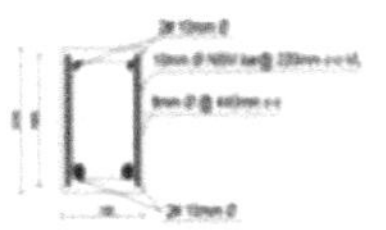	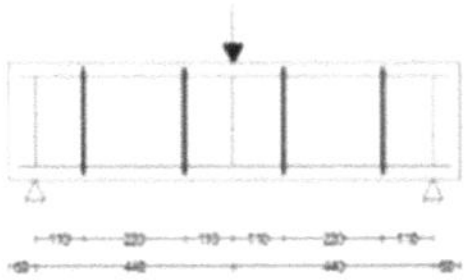	
NB-10d-sL/6-90		
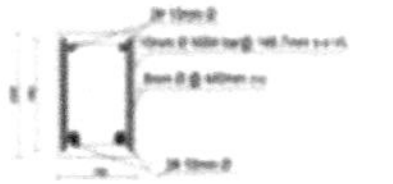	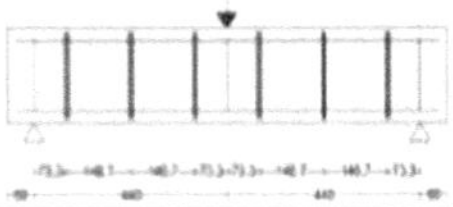	
NB-10d-sL/4-45		
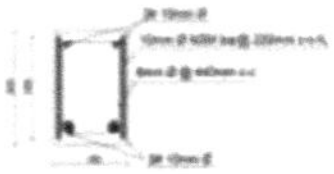	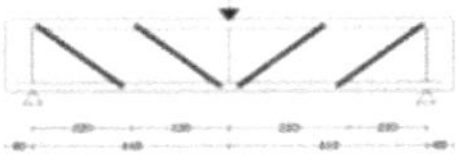	
NB-10d-sL/6-45		
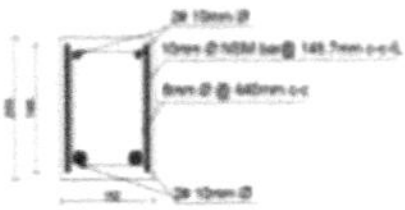	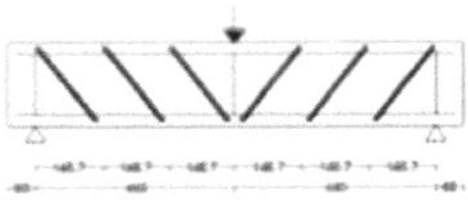	

The NSM (or ETS) shear strengthening ratio can be calculated as follows.

$$NSM(or\ ETS) shear\ strengthening\ ratio = \frac{Cross\ sectional\ area\ of\ NSM(or\ ETS) bar}{Breadth\ of\ RC\ beam\ X Spacing\ between\ NSM(or\ ETS) bars}$$

A summary of test beam specifications is presented in Table 4.

3 Results and Discussion

3.1 Failure Modes

Reinforced concrete elements strengthened with FRP systems are observed with additional failure modes compared with reinforced concrete elements that are not strengthened with FRP systems. The control beams NCB, ECB and the strengthened beam specimens all failed in shear as designed. From the crack patterns seen in

Table 3 ETS beam configurations

Beam Cros section	Longitudinal Section
EsL/4d10-90	
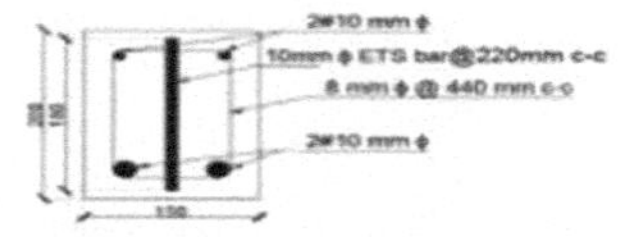	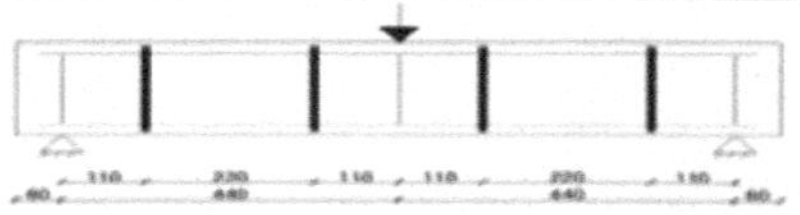
EsL/6d10-90	
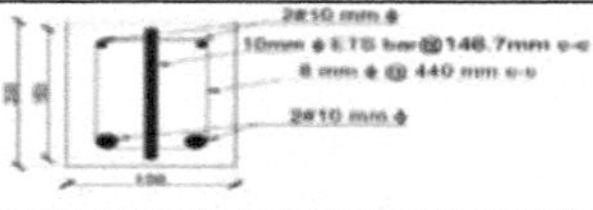	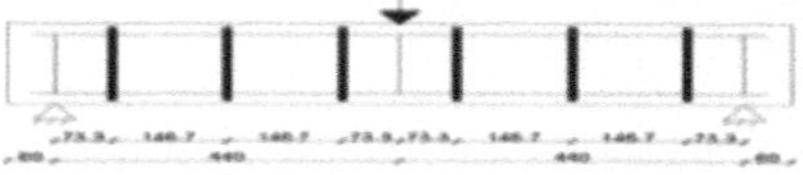
EsL/4 d10-45	
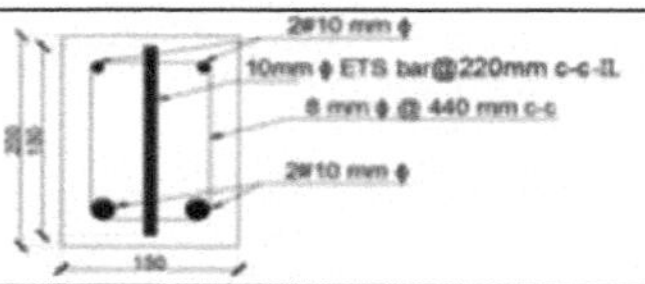	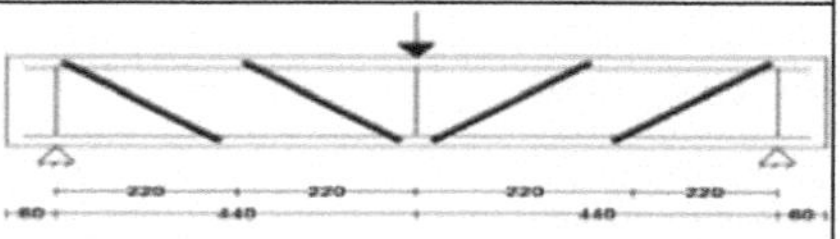
EsL/6 d10-45	
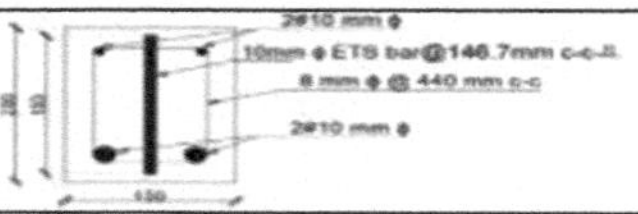	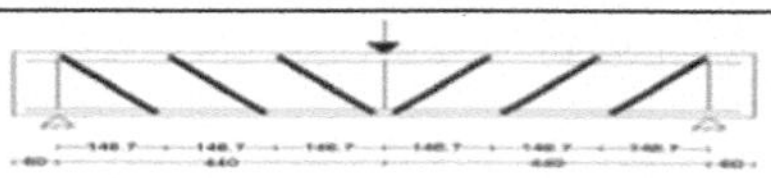

a) ETS beam with holes drilled at L/6 spacing

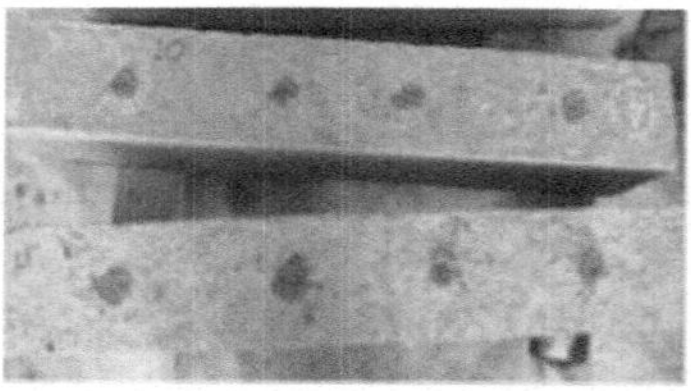

b) ETS beam strengthened beam at L/4 spacing

Fig. 2 ETS strengthening technique

Fig. 3, the diagonal shear cracks were clearly visible on the sides of the beam. All the specimens showed a similarity in their patterns. Also, from the failure patterns it is clear that no de-bonding of NSM FRP bars from the epoxy or the epoxy from the concrete substrate occurred for the NSM shear strengthened specimens Fig. 3 shows the various failure patterns obtained for the 10 mm dia FRP strengthened tested beams.

The crack patterns clearly indicates the efficient crack arresting mechanism of the strengthened specimens with NSM bars at inclined (45°) orientation by showing lesser intensity and also an irregularity in cracks for specimens, NB-8d-sL/6-45, NB-10d-sL/4-45, NB-10d-sL/6-45, NB-12d-sL/4-45 and NB-12d-sL/6-45 than the

Table 4 Specifications of tested beams

Beam ID	Method	Description	Shear strengthening ratio
NCB	–	NSM control beam	–
ECB	–	ETS control beam	–
NB-8d-sL/4-90 and NB-8d-sL/4-45	NSM	Strengthened beam with 8 mm dia NSM/ETS bar spaced at L/4 distance and orientation 90° and 45° respectively	0.15
EsL/4d08-90 and EsL/4d08-45	ETS		
NB-10d-sL/4-90 and NB-10d-sL/4-45	NSM	Strengthened beam with 10 mm dia NSM/ETS bar spaced at L/4 distance and orientation 90° and 45° respectively	0.23
EsL/4d10-90 and EsL/4d10-45	ETS		
NB-12d-sL/4-90 and NB-12d-sL/4-45	NSM	Strengthened beam with 12 mm dia NSM/ETS bar spaced at L/4 distance and orientation 90° and 45° respectively	0.34
EsL/4d12-90 and EsL/4d12-45	ETS		
NB-8d-sL/6-90 and NB-8d-sL/6-45	NSM	Strengthened beam with 8 mm dia NSM/ETS bar spaced at L/6 distance and orientation 90° and 45° respectively	0.23
EsL/6d08-90 and EsL/6d08-45	ETS		
NB-10d-sL/6-90 and NB-10d-sL/6-45	NSM	Strengthened beam with 10 mm dia NSM/ETS bar spaced at L/6 distance and orientation 90° and 45° respectively	0.35
EsL/6d10-90 and EsL/6d10-45	ETS		
NB-12d-sL/6-90 and NB-12d-sL/6-45	NSM	Strengthened beam with 12 mm dia NSM/ETS bar spaced at L/6 distance and orientation 90° and 45° respectively	0.513
EsL/6d12-90 and EsL/6d12-45	ETS		

specimens with NSM bar orientation at 90°. It is because when the NSM bars are arranged in a direction opposite to the shear cracks, these bars resist the shear forces acting in the shear span of the specimens in a greater efficiency, thus resulting in an irregular crack patterns. Whereas, crack lines are almost straight in NCB and most of the beams strengthened with NSM bars at vertical orientation i.e., NB-8d-sL/6-90, NB-10d-sL/6-90, NB-12d-sL/4-90 and NB-12dsL/6-90.

Carefully observing the crack patterns in the Fig. 3 we can see that, the beams strengthened with ETS bars of various diameters with L/4 spacing have lesser intensity of cracks than those beams strengthened with L/6 spacing. Also, crack lines are almost straight in ECB and beams strengthened with ETS spacing L/4; but there is an irregular pattern for the cracks in the strengthened beams with L/6 ETS spacing.

NSM shear strengthening

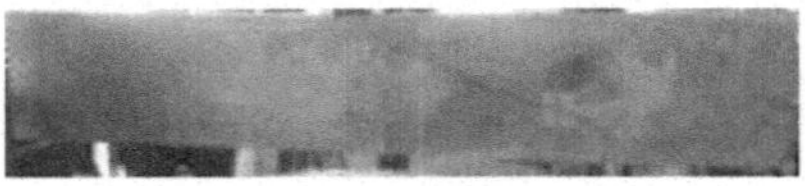

NCB (108kN)

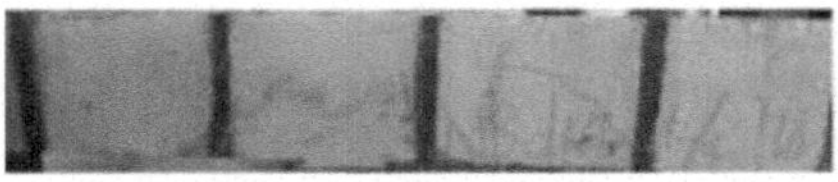

NB-10d-sL/4-90(126kN)

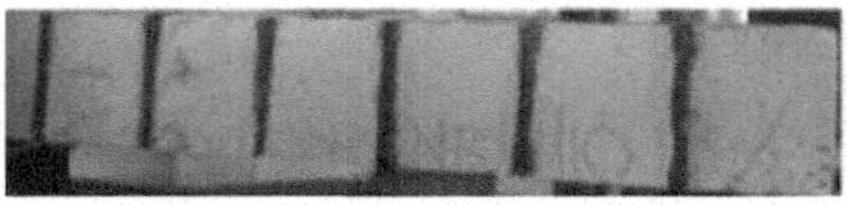

NB-10d-sL/6-90 (132kN)

NB-10d-sL/4-45(134kN)

NB-10d-sL/6-45(166kN)

ETS shear strengthening

ECB(98kN)

EsL/4d10-90(120kN)

EsL/6d10-90(125kN)

EsL/4d10-45(127kN)

EsL/6d10-45(131kN)

Fig. 3 Failure modes of NSM and ETS strengthened beams

It is because when the ETS bar spacing is L/6, there are more ETS bars than in those with spacing of L/4; thus more number of ETS bars resist the shear forces acting in the shear span thus resulting in an irregular pattern for the shear crack.

3.2 Load Deflection Characteristics

In this study, all the specimens strengthened as well as non-strengthened failed in shear and for NSM no de-bonding failure at epoxy-concrete interface or at bar-epoxy interface observed. It is clear from Table 5 that the load-deflection characteristics of all the specimens tested follows a uniform straight pattern till its yield point and shows an abrupt increase in the deflection followed by a sudden shear failure.

Table 5 Load deflection characteristics

Method of strengthening	Spacing	Beam ID	Pu (kN)	ΔPu (%)	δy (mm)	δu (mm)
Near surface mounting (NSM)	L/4	NCB	108	–	2.15	3
		NB-8d-sL/4-90	120	11.11	3.25	5
		NB-8d-sL/4-45	126	16.66	2.95	5.15
		NB-10d-sL/4-90	126	16.66	3.20	5
		NB-10d-sL/4-45	134	24.1	2.90	5.35
		NB-12d-sL/4-90	138	27.8	3.5	5.5
		NB-12d-sL/4-45	146	35.2	2.86	5.15
	L/6	NB-8d-sL/6-90	124	14.8	2.85	4.5
		NB-8d-sL/6-45	154	42.6	3.45	6.25
		NB-10d-sL/6-90	132	22.22	3.65	6.0
		NB-10d-sL/6-45	166	53.7	3.95	7.0
		NB-12d-sL/6-90	142	31.5	3.35	5.25
		NB-12d-sL/6-45	178	64.8	3.55	6.5
Embedded through section (ETS)	L/4	ECB	98	–	2.47	3.2
		EsL/4d08-90	119	21.4	2.93	4.3
		EsL/4d08-45	125	27.55	2.86	4.38
		EsL/4d10-90	120	22.45	2.9	4.55
		EsL/4d10-45	127	29.6	2.8	4.63
		EsL/4d12-90	132	34.7	3.05	4.58
		EsL/4d12-45	158	61.2	2.98	5.12
	L/6	EsL/6d08-90	125	27.55	2.7	3.98
		EsL/6d08-45	133	35.7	2.96	4.32
		EsL/6d10-90	131	33.7	2.78	4.0
		EsL/6d10-45	145	47.9	2.85	4.8
		EsL/6d12-90	140	42.8	3.37	5.25
		EsL/6d12-45	167	70.4	3.52	6.48

Pu—Ultimate Load; δu—Deflection at ultimate load ΔPu—% ultimate load increase; δy—Deflection at yield load

3.3 *Influence of CFRP Bar Spacing*

The influence of CFRP bar spacing on the load carrying capacity of the strengthened beams is one of the significant parameters considered in this study. Two different spacing are chosen for the study. They are L/4 and L/6 where L is the effective span of the beam. All the specimens tested in this study follow a similar fashion for its load-deflection curve till its yield point and then there is an abrupt increase in the deflection followed by a sudden shear failure with minimal warnings.

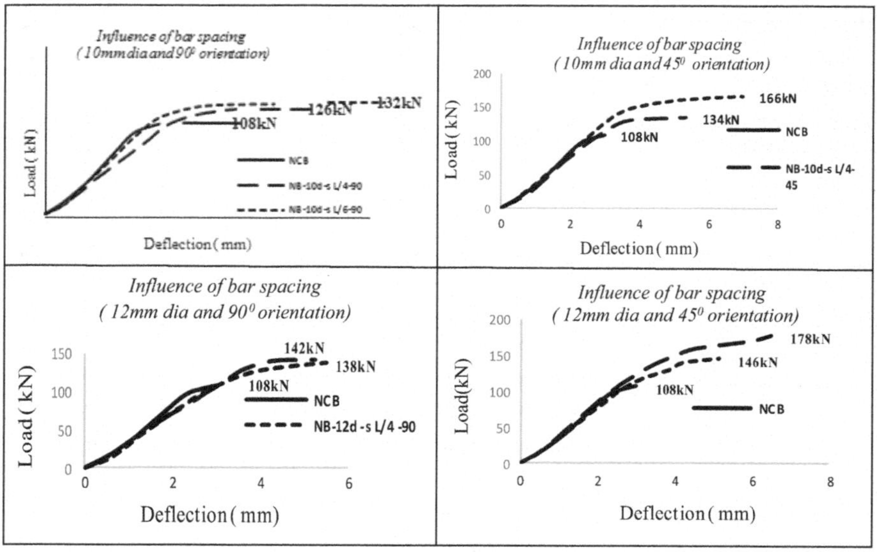

Fig. 4 Influence of bar spacing for two different bar orientations (90° and 45°) in NSM

NSM strengthened beams with L/4 bar spacing, and 8 mm, 10 mm and 12 mm CFRP bar diameters, have shown an average percentage increase of 13.9%, 20.38% and 31.5% in ultimate load capacity. Whereas for beams with L/6 bar spacing, the average percentage increase in ultimate load capacity increased to 36.1%, 37.96% and 48.15% respectively. From it, we can clearly see that, the ultimate load is greater for the strengthened beam with NSM bar spacing of L/6 in both the NSM bar orientations. It can be noticed that the ultimate load carrying capacity is positively affected by the decreased spacing of the NSM bars for various NSM bar diameters at both vertical and inclined orientations; it is because more number of NSM bars are present in those specimens, which thereby effectively takes part in the crack arresting mechanism [5]. Figure 4 shows the graphical representation of influence of bar spacing for different bar orientations.

In ETS strengthened beams, a very similar result is obtained for the set of strengthened beams. There is an average percentage increase of 24.48%, 26.03% and 47.95% respectively in ultimate load carrying capacity for 8 mm, 10 mm and 12 mm dia CFRP bar strengthened beams with L/4 bar spacing. At the same time, for bars with L/6 spacing, the average percentage increase in ultimate load carrying capacity is 31.63%, 40.8% and 56.6% respectively for 8 mm, 10 mm and 12 mm dia CFRP bars. The ultimate load is clearly greater for the strengthened beam with ETS bar spacing of L/6. The three sets of results in it is a solid indication of the fact that the ultimate load carrying capacity is positively affected by the decreased spacing of the ETS bars. As the spacing of the ETS bars decreases, the shear strengthening ratio increases. This causes the increase in the shear strengthening efficiency (The influence of spacing for two different bar orientation is shown in Fig. 5).

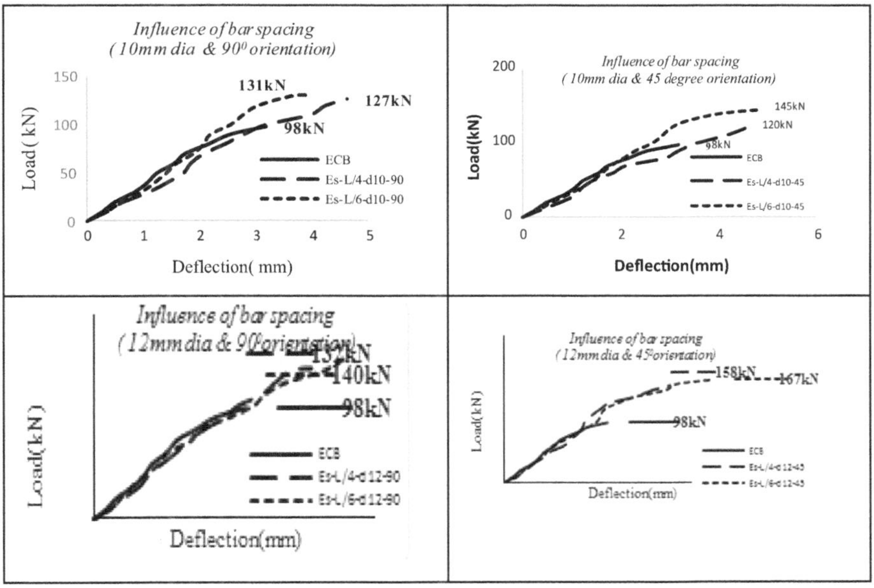

Fig. 5 Influence of bar spacing for two different bar orientations (90° and 45°) in ETS

3.4 Influence of NSM Bar Orientation

The effect of NSM bar orientation on shear strength is the second parameter considered in this study. The different NSM bars are at two different spacing, i.e., L/4 and L/6, where L is the effective span of the beam as mentioned in the previous session. Also, the different NSM bar orientation chosen for the study are vertical (90°) and inclined (45°) orientations.

We can see that, the ultimate load is greater for the strengthened beams: NB-8d-sL/4-45, NB-10d-sL/4-45 and NB-12d-sL/4-45 with NSM bars oriented at 45° with an increase in ultimate load of 16.66%, 24.07% and 35.18% respectively compared to the 11.11%, 16.66%, and 27.77% increase for specimens NB-8d-sL/4-90, NB-10d-sL/4-90 and NB-12d-sL/4-90 respectively. The failure mode shown by these beams are diagonal shear failure and no de-bonding of NSM bars observed. When the NSM bars are oriented at an inclined pattern, that is perpendicular to these diagonal shear cracks showed much better result than vertical orientation [6].

Now, similar set of results as aforementioned is obtained for the remaining strengthened beams of varying diameters and orientations in which NSM bars are placed at spacing of L/6 as shown in Fig. 6, which is of course an increase in the ultimate load of RC beams as a result of the better performance of orienting NSM bars inclined, specimens NB-8d-sL/6-45, NB-10d-sL/6-45 and NB-12d-sL/6-45 has an increase in ultimate load of 42.59%, 53.7% and 64.81% respectively compared to the 14.81%, 22.22% and 31.48% increase for the specimens NB-8d-sL/6-90, NB-10d-sL/6-90 and NB-12d-sL/6-90 respectively.

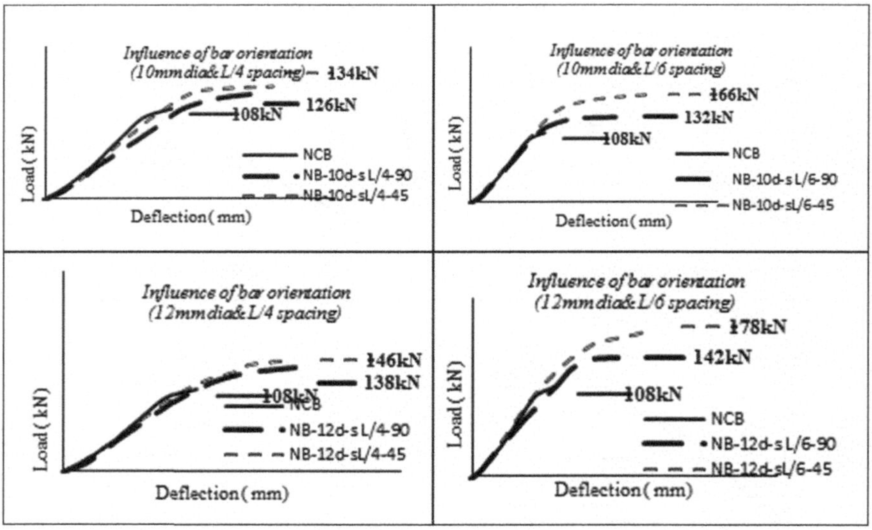

Fig. 6 Influence of bar orientation for two different spacing (L/4 and L/6) in NSM

In ETS strengthened beams, a very similar result is obtained for the set of strengthened beams in 90° and 45° orientations. Es-L/4-d08-45, Es-L/4-d10-45 and Es-L/4-d12-45 with ETS bars oriented at 45° with an increase in ultimate load of 27.55%, 29.6% and 61.2% respectively compared to the 21.4%, 22.45%, and 34.7% increase for specimens Es-L/4-d08-90, Es-L/4-d10-90 and Es-L/4-d12-90 respectively. At the same time, from Fig. 6, it can be seen that for bars with L/6 spacing, the average percentage increase in ultimate load carrying capacity is 35.7%, 47.9% and 70.4% respectively for 8 mm, 10 mm and 12 mm dia CFRP bars at 45° orientation. It is clear that the ultimate load is clearly greater for the strengthened beam with ETS bar spacing of L/6 and 45° orientation. The three sets of results in it is a solid indication of the fact that the ultimate load carrying capacity is positively affected by the decreased spacing and inclined orientation of the ETS bars (Fig. 7 shows the influence of bar orientation for two different spacing).

Hence these results clearly presents the significance of ETS/NSM bar orientation in strengthening purpose, for its contribution towards the load carrying capacity of RC beams. The ultimate load carrying capacity for specimens strengthened with NSM bars at inclined orientation (45°) is more compared to those with vertical orientation (90°) because of its better crack arresting property, as bars are placed perpendicular or in a direction opposite to the diagonal shear cracks.

The inclination of cracks is close to 45° in the RC beams that fail in shear. The efficiency of the ETS shear strengthening technique can be increased by placing the ETS bars in the direction perpendicular to that of the direction in which the shear cracks occur, thus slowing down the propagation of the shear cracks [7]. This increase in efficiency can also be attributed to the large available resisting bond length. Vertical ETS bars provided an increase of load carrying capacity in the interval of

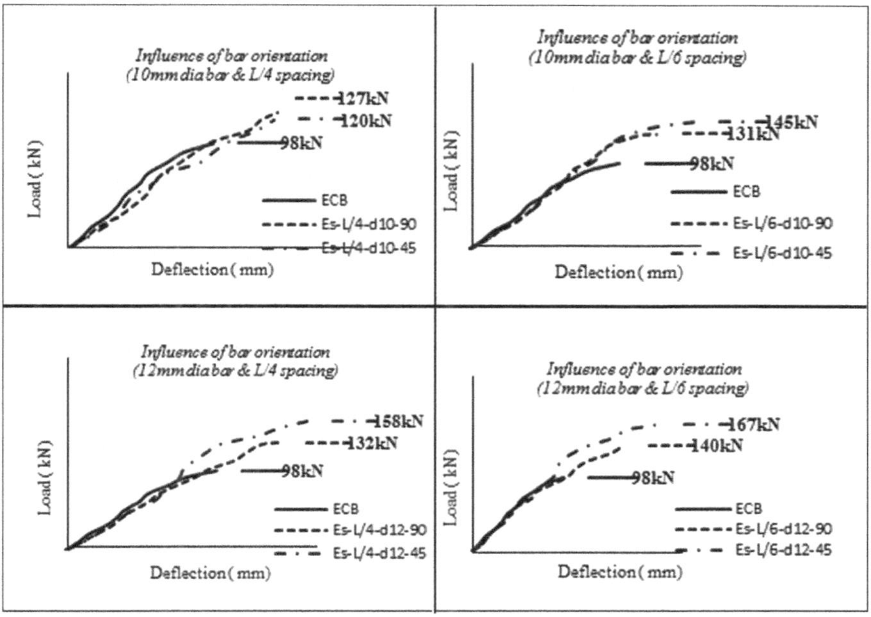

Fig. 7 Influence of bar orientation for two different spacing (L/4 and L/6) in ETS

5–68% while the inclined ETS bars assured a higher strengthening effectiveness with an increase in load carrying capacity from 53 to 136%. Hence, providing 45° inclination in ETS bars increases the efficiency of shear strengthening.

4 NSM Versus ETS—A Comparison

The effectiveness in terms of shear strengthening of near surface mounting (NSM) method is compared with embedded Through Section (ETS) method with the usage of same cross sectional area of CFRP and similar shear strengthening ratio. The effectiveness of these strengthening methods was measured by comparing the gain achieved by each strengthening method for an equivalent amount of FRP to carry shear loads.

The NSM technique consisting of epoxy bonding of bars or strips into grooves created on the sides of the beam, limits the deboning failure to an extent. They usually fail by separation of the side concrete covers at the internal steel stirrups. Strengthening with NSM requires less man hours to install while the ETS method offers more protection against loss of strengthening by integrating the additional reinforcement within the member, which proved effective in preventing splitting or separation. The concrete cover separation failure usually associated with NSM is completely absent in ETS technique. It is because the ETS technique relies on concrete core to transfer the stresses between concrete and ETS bar. The concrete core provides better confinement and better bond performance to overcome debonding failure. Thus the ETS technique is the most efficient technique for developing FRP tensile strength potential before the final failure occurs

In this study, the average increase in shear capacity reached 22.04% for the beam strengthened with NSM bars and 32.81% for the beams strengthened with ETS FRP rods with L/4 spacing. Whereas with L/6 spacing, the average increase in shear capacities are 38.27% and 43% respectively. At 45° CFRP bar orientation, the maximum deflection of the beam NB-10d-sL/6-45 at the ultimate load was 1.39 times that of the EsL/6d10-45 beam at the ultimate load (3.95 mm at load 166 kN versus 2.85 mm at 145 kN) and at 90° orientation, the maximum deflection of the beam NB-10d-sL/6-90 was 1.31 times that of the EsL/6d10-90 beam at the ultimate load (3.65 mm at load 132 kN vs. 2.78 mm at 131 kN), whereas the control beams NCB and ECB achieved the an average smallest deflection of 2.31 mm at the ultimate load 103 kN.

5 Conclusions

Based on the experimental investigations carried out on the non-strengthened and shear strengthened specimens using NSM as well as ETS bars of varying configurations, the following conclusions are drawn:

- Shear strengthening of RC beams by NSM method is an efficient method in increasing its load carrying capacity. The NSM shear strengthened beams have 11.11–64.81% increase in ultimate load compared to the non-strengthened beam (Control beam).whereas the ETS strengthened beams are having 21.4–70.4% increase in ultimate load capacity compared to control beam. It proves that ETS Technique is more efficient than NSM Technique.
- The load carrying capacity of NSM as well as ETS strengthened beams increases by decreasing the spacing of the strengthening bars. An average increase of 18.8% in ultimate load is obtained when the spacing of NSM bars decreased from L/4 to L/6. But in ETS strengthened beams the average increase in ultimate load is only 10.2% with decrease in spacing from L/4 to L/6.
- The load carrying capacity of NSM and ETS strengthened beams increases for those having strengthening bars placed at inclined (450) orientation. Up to 33.3% increase in ultimate load is found for the NSM bars oriented at 45° compared to the NSM bars oriented at 90° and for ETS strengthened beams there is an increase up to 27.6%.

References

1. Adhikary BB, Mutsuyoshi H (2006) Shear strengthening of reinforced concrete beams using various techniques. Constr Build Mater 20:366–373
2. Bilotta A, Ceroni F, Nigro E, Pecce M (2015) Efficiency of CFRP NSM strips and EBR plates for flexural strengthening of RC beams and loading pattern influence. Compos Struct 124:163–175
3. El-Ghandour AA (2011) Experimental and analytical investigation of CFRP flexural and shear strengthening efficiencies of RC beams. Constr Build Mater 25:1419–1429
4. Mofidi A, Challal O, Benmokrane B, Neale K (2012) Experimental tests and design model for RC beams strengthened in shear using the embedded through section FRP method. J Compos Constr 16:540–550
5. Barros JAO, Dias SJE (2006) Near surface mounted CFRP laminates for strengthening of concrete beams. Cement Concr Compos 28:276–292
6. Rahal KN, Rumaih HA (2011) Tests on reinforced concrete beams strengthened in shear using near surface mounted CFRP and steel bars. Eng Struct 33:53–62
7. Barros JAO, Dalfre GM (2012) Assessment of the effectiveness of the embedded through-section technique for the shear strengthening of reinforced concrete beams. Strain-An Int J Exp Mech 49:75–93

Study on Mechanical Properties of Self Healing Self Curing Concrete

Chinnu Susan Jacob and Vidya Jose

Abstract Concrete is one of the most widely used construction material in industry. It has high compressive strength but low tensile strength. The low tensile strength may result in crack formation which seriously affects the life of concrete. A solution to this is the introduction of self-healing properties in concrete. Also the properties of hardened concrete will mainly depend on its extent of hydration. Proper curing methods can be used to achieve proper hydration. Bacterially induced self-healing by calcium carbonate precipitation has been proposed as an eco-friendly method in recent years. The bacterium selected for the study is Bacillus Subtilis JC3. The present study concentrates on the effect of bacteria as self -healing agent in self-curing concrete and its optimum percentage in strength aspects of hardened concrete.

Keywords Concrete · Self healing · Bacillus Subtilis bacteria

1 Introduction

Concrete is a composite material which has high compressive strength but low in tensile strength. Therefore cracking is a common phenomenon due to low tensile strength. The sources of tensile stresses can be expansive reactions, external loads or imposed deformations. The presence and propagation of cracks may eventually affect the strength and durability of concrete. The strength in concrete is attributed by the process of hydration of cement. Curing allows the continuous hydration of cement. In conventional practices, curing is performed externally after construction of concrete structures. The method of self-curing is a technique in which the moisture within concrete is preserved and used for its effective hydration. There can be many external factors which hinders the curing process of conventional concrete by evaporating the moisture in concrete. Polyethylene glycols (PEGs) are family of water soluble linear

C. S. Jacob (✉) · V. Jose
Department of Civil Engineering, Toc H Institute of Science and Technology, Arakkunnam, India
e-mail: chinnusj93@gmail.com

V. Jose
e-mail: vidyajose@tistcochin.edu.in

K. Dasgupta et al. (eds.), *Proceedings of SECON'19*,
Lecture Notes in Civil Engineering 46,
https://doi.org/10.1007/978-3-030-26365-2_85

polymers formed by the additional reaction of Ethylene oxide (EO) with Mono ethylene glycol (MEG) or Diethylene glycol. For this experimental study, PEG 400 is used which consists of a distribution of polymers of varying molecular weights with an average of 400. One common feature of PEG appears to be its water-soluble nature. The polymers added in the mix mainly form hydrogen bonds with water molecules and reduce the chemical potential of the molecules which in turn reduces the vapour pressure, thus reducing the rate of evaporation from the surface. The properties such as curing and healing are collaborated especially to meet the needs of concrete structures experiencing water inadequacy and inaccessibility. The bacteria used in the study is inducing self-healing is Bacillus Subtilis JC3. The same is varied in different concentrations and optimum concentration for strength is found out based on its action in concrete specimens. The self-healing self-curing system can achieve a tremendous cost reduction in terms of damage detection and maintenance of concrete structures, assuring a safe service life of the structure.

2 Materials and Methods

2.1 Materials Used

The different materials used in this investigation are

2.1.1 Cement

53grade ordinary Portland cement confirming IS: 12269: 1987.

2.1.2 Fine Aggregate

M sand with specific gravity 2.67. The fine aggregate conforming to zone II according to IS: 383-1970 was used.

2.1.3 Coarse Aggregate

Crushed granite was used as coarse aggregate. The coarse aggregate according to IS: 383-1970 was used. Maximum coarse aggregate size used was 20 mm, having a specific gravity of 2.78.

Table 1 Biochemical characteristics Bacillus Subtilis JC3

Properties	Specifications
Shape, size, gram strain	Long rods, 0.6–0.8 μm in width and 2.0–3.0 μm in length, gram positive
Colony morphology(on nutrient agar plate)	Irregular, dry, white, opaque colonies
Fermentation	
Lactose	No acid, no gas
Dextrose	No acid, no gas
Sucrose	Acid and gas
H2S production	Absent
Nitrate reduction	Absent
Indole production	Absent
Citrate utilization	Absent
Catalase activity	Present
Gelatine liquefaction	Present
Starch hydrolysis	Present
Lipid hydrolysis	Present

2.1.4 Polyethylene Glycol-400

Polyethylene glycol is a condensation polymer of ethylene oxide and water which have the general formula $H(OCH_2CH_2)nOH$, where n is the average number of repeating oxyethylenegroups typically from 4 to about 180. The abbreviation (PEG) is termed in combination with anumeric suffix which indicates the average molecular weight. One common feature of PEG appears to be its water-soluble nature. The PEG-400 use in the study had Molecular Weight 400, Clear liquid appearance, pH 5-7 and Specific Gravity 1.126

2.1.5 Bacillus Subtilis JC3

Bacillus Subtilis JC3, a laboratory cultured bacterium collected from Kerala Agricultural University Mannuthy, was used. Bacteria in suspension with a concentration of 10^8 cells/ml were collected and 10^3,10^5 and 10^7 cells/ml of bacterial concentration was made from the obtained sample (Table 1).

2.1.6 Water

Potable water was used in the experimental work for both mixing and curing purposes.

Table 2 Properties of calcium lactate

S. No.	Properties	Calcium lactate
1	Appearance	White powder
2	Assay	98.0–101.0%
3	Calcium assay	13.4–14.5%
4	Loss on drying	22–30%
5	Heavy metals	20 ppm max
6	Mesophilic bacteria	1000/g
7	Mould	100/g
8	Yeast	100/g
9	Formula	$C_6H_{10}O_6Ca . 5H_2O$
10	Molecular wt.	308.3
11	Solubility in water	9 g/100 ml in water
12	Toxicity	Non toxic
13	Synonyms	Lactic acid
14	Melting point	240 °C
15	Density	1.49 g/cm^3

2.1.7 Calcium Lactate

Calcium lactate is a black or white crystalline salt which is made by the action of lactic acid on calcium carbonate. Healing is a result of the bacterial conversion of calcium lactate into calcium carbonate. In this experimental study, the amount of calcium lactate is 3% of binder content. As the binder content is kept constant, the calcium lactate content also remains constant (Table 2).

2.2 Preparation of Self-Curing Concrete

Concrete mix with PEG was considered as the control mix of M40 grade. Based on literatures, the amount of PEG was fixed to 1% of total binder. The materials were weighed using a digital balance and mixed in a pan mixer. The slump of the fresh concrete was determined to study the workability of concrete. The specimens were demoulded after 24 h and air cured, then tested in saturated surface dry condition at the required age.

Table 3 Nomenclature

SCC	M 40—SELF CURING CONCRETE (Control specimen)
SHSCC1	M 40—SELF HEALING SELF CURING CONCRETE (with bacteria of concentration 10^3 cells/ml)
SHSCC2	M 40—SELF HEALING SELF CURING CONCRETE (with bacteria of concentration 10^5 cells/ml)
SHSCC3	M 40—SELF HEALING SELF CURING CONCRETE (with bacteria of concentration 10^7 cells/ml)

2.3 *Preparation of Self-Healing Self-Curing Concrete*

Bacteria were added to self-curing concrete mix in suspension state at different concentrations (10^3,10^5 and 10^7 cells/ml). Concrete specimens were designated as SHSCC1, SHSCC2 and SHSCC2 respectively. Bacteria used as self-healing agent should have the capability to survive high alkaline environment of concrete for long durations and should be able to form spores (highly resistant structures) withstanding mechanical forces during concrete mixing. A bacterial concrete mix was prepared using Bacillus subtilis JC3 along with nutrients from which the bacteria could potentially produce calcite based bio-minerals. The bacteria genus Bacillus has been found to sustain the high-alkaline environment of concrete due to its extremely thick outer cell membrane that enables them to remain viable until a suitable environment is available to grow. They would become active when the cracks form on concrete surface allowing water or CO_2 to enter into the structure. This phenomenon will reduce the pH of the concrete environment thereby incorporated bacteria will become activated. Calcium lactate nutrients supplied along with bacteria content in suspension helps to produce calcite crystals (Table 3).

3 Results and Discussions

The results of compressive strength and Split tensile strength of concrete specimens for self-curing concrete and bacteria enriched self-curing concrete are as follows.

3.1 *Compressive Strength*

Compressive strength tests were conducted on 150 mm × 150 mm × 150 mm cubes in compression testing machine. Load at the failure divided by area of specimen gives the compressive strength of concrete. Tests were carried out after 7 days and 28 days respectively (Fig. 1).

Fig. 1 Cube under compression test

It was observed that there were slight variations in strength as the bacterial concentration was varied. The compressive strength was highest in concrete specimen having bacterial concentration 10^5 cells/ml (Fig. 2; Table 4).

3.2 *Split Tensile Strength*

Split tensile strength tests were conducted on 300 mm × 150 mm cylinders as per IS 5816-1999 by placing the cylindrical specimen horizontally between the loading surfaces of compression testing machine and then load was applied until the failure of the cylinder along the vertical diameter. Tests were carried out after 7 days and 28 days respectively (Fig. 3).

The strength observed was higher than the control specimen and it increased gradually up to specimen having bacterial concentration 10^5 cells/ml and then decreased. Therefore the optimum strength was obtained for SHSCC2 (Fig. 4; Table 5).

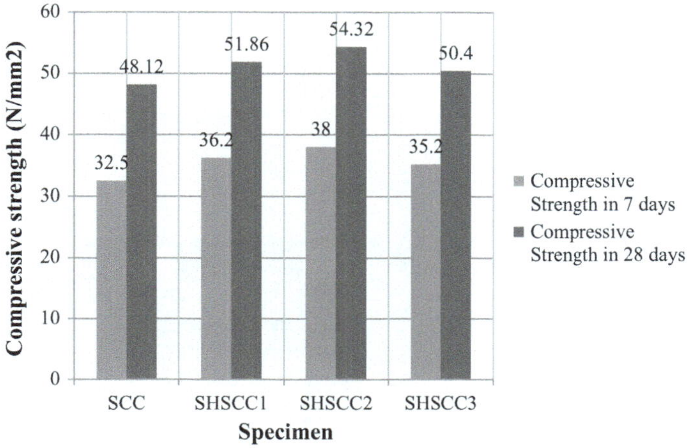

Fig. 2 Compressive strength test of various mixes

Table 4 Compressive strength of concrete specimens

Cell concentrations/ml of mixing water	Average compressive strength of concrete in N/mm^2			
	7 days	% increase	28 days	% increase
SCC	32.5	–	48.12	–
SHSCC1	36.2	11.38	51.86	7.23
SHSCC2	38	16.92	54.32	12.88
SHSCC3	35.2	8.3	50.4	4.7

4 Conclusion

The experimental study reveals that the addition of bacteria Bacillus Subtilis JC3 in concrete improves the properties of concrete in terms of compressive strength and split tensile strength. The amount of PEG was fixed in each case. As bacteria can be produced in the laboratory, it could be proved to be a safe and very cost effective way. Bacterial concrete with a concentration of 105 cells/ml bacteria was found to give best results out of the samples used. Hence it could be concluded that this particular concentration give optimum results which is proven by 12.88% increase in compressive strength and 17.9% increase in split tensile strength when compared to conventional concrete. The study accomplishes that the use of bacteria (Bacillus Subtilis) in self-curing concrete enhances its strength and hence using this type of bacteria for self-healing mechanism in concrete can produce cost effective and strong structures.

Fig. 3 Cylinder under split tensile test

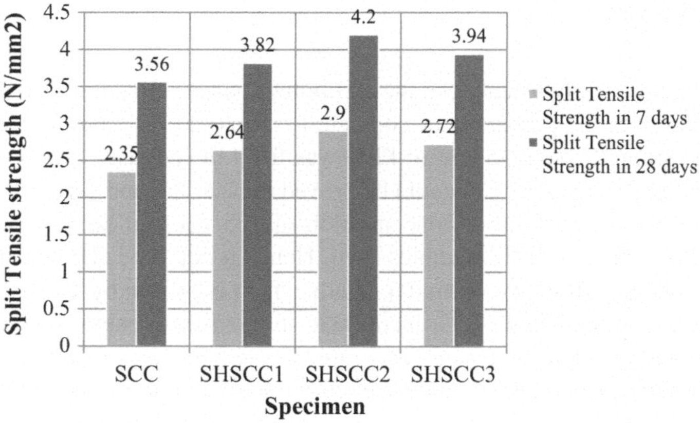

Fig. 4 Split tensile strength

Table 5 Split tensile strength of concrete specimens

Cell concentrations/ml of mixing water	Average split tensile strength of concrete in N/mm^2			
	7 days	% increase	28 days	% increase
SCC	2.35	–	3.56	–
SHSCC1	2.64	12.34	3.82	7.30
SHSCC2	2.90	23.4	4.2	17.9
SHSCC3	2.72	15.7	3.94	10.67

References

1. Ananthi A, Ranjith R, Swarna Latha S, Vimal Raj R (2017) Experimental study on the properties of self-curing concrete. Int J Concr Technol 3:8–13
2. Babitha Rani H, Shivakumar S, Veerendra Babu D, Kothari K, Sindhu S, Harish J (2017) Studies on self-curing concrete with the addition of polyethylene glycol – 400. Int J Eng Technol Sci Res 4:2394–3386
3. Joseph BM (2016) Studies on properties of self-curing concrete using polyethylene Glycol. IOSR J Mech Civil Eng 12–17
4. Jonkers HM, Thijssen A, Muyzer G, Copuroglu O, Schlagen E (2010) Application of bacteria as self-healing agent for the development of sustainable concrete. J Ecol Eng 36:230–235
5. Kumar J, Srikanth M, Rao KJ (2012) Strength characteristics of self-curing concrete. Int J Res Eng Technol 1:51–56
6. Xu J, Wang X (2018) Self-healing of concrete cracks by use of bacteria-containing low alkali cementitious material. J Constr Build Mater 167:1–14
7. Van Tittelboom K, De Belie N, De Muynck W, Verstraete W (2010) Use of bacteria to repair cracks in concrete. J Cem Concr Res 40:157–166
8. Mousa MI, Mahdy MG, Abdel-Reheem AH, Yehia AZ (2014) Mechanical properties of self-curing concrete (SCUC). J Hous Build Res Centre 1687–4048
9. Luo M, Qian C (2016) Influeneces of bacteria-based self-healing agents on cementitious materials hydration kinetics and compressive strength. J Constr Build Mater 121:659–663
10. Luo M, Qian C, Li R (2015) Factors affecting crack repairing capacity of bacteria-based self-healing concrete. J Constr Build Mater 87:1–7
11. Mondal S, Ghosh AD (2018) Investigation into the optimal bacterial concentration for compressive strength enhancement of microbial concrete. J Constr Build Mater 183:202–214
12. Thiru Chelve SR, Sivakumar S, Raj Kumar M, Shanmugaraja G, Nallathambi Vishnu Priyan M (2017) Effect of polyethylene glycol as internal curing agent in concrete. Int J Innovative Res Sci Eng Technol 6

Model Analysis of Rhombic Grid Hyperboloid Latticed Shell Structures

G. S. Amritha and Jency Sara Kurian

Abstract The RGHLS is a type of diagrid structure that has been developed in recent years. It consist of two columns. The major columns are designed as inclined members in major direction from bottom to top of RGHLS. They sustain the major percentage of vertical loads. The secondary columns have to be segmented in a helical direction in order to form the hyperboloid latticed shells. Secondary columns sustain much less vertical load than major columns. They are more commonly used in purpose-driven structures, such as water towers (to support a large mass), cooling towers, and aesthetic features. The objective of this study is to compare the result of the Model analysis by changing the shape of major and secondary column and also the performance of the structure when the position of the central axis of the structure is varied with respect to its height. The software used for this study is ANSYS. Height on the either side of the horizontal axis is varied and optimum height that exhibit best result is determined.

Keywords Rhombic grid hyperboloid latticed structure (RGHLS) · Model analysis

1 Introduction

Hyperboloid structures are architectural structures designed using a hyperboloid in one sheet. Often these are tall structures such as towers where the hyperboloid geometry's structural strength is used to support an object high off the ground, but hyperboloid geometry is also often used for decorative effect as well as structural economy.

Hyperboloid structures are more commonly used in purpose-driven structures, such as water towers (to support a large mass), cooling towers, and aesthetic features. At the bottom of cooling tower, the widening provides a large area for installation of fill to promote thin film evaporative cooling of the circulated water. As the water first evaporates and rises, the narrowing effect helps accelerate the laminar

---

G. S. Amritha (✉) · J. S. Kurian
Amal Jyothi College of Engineering, Koovapally P.O, Kanjirapaly, Kottayam 686518, India
e-mail: amrithags@ce.ajce.in

K. Dasgupta et al. (eds.), *Proceedings of SECON'19*, Lecture Notes in Civil Engineering 46,
https://doi.org/10.1007/978-3-030-26365-2_86

flow, and then as it widens out, contact between the heated air and atmospheric air supports turbulent mixing.

Rhombic grid hyperboloid structures act as a ground floor and they sustain enormous vertical and horizontal loads induced from upper building structures on top of them. The major columns are designed as inclined members in major direction without any stops from bottom to top of RGHLS. They sustain the major percentage of vertical loads. The secondary columns have to be segmented in a helical direction in order to form the hyperboloid latticed shells. Secondary columns sustain much less vertical load than major columns (Fig. 1).

The RGHLS is a type of diagrid structures that has been developed in recent years. It is favoured by architects due to its concise and transparent design as well as its artistic building forms, and it has been widely adopted in high-rise building and TV tower structures. When the diagrid structure is adopted in high-rise buildings with a core inside, it bears mainly horizontal loads induced from wind or earthquake and possesses excellent lateral stability because the slab connects the diagrid structures and the core. Hence, most civil engineering researchers have focused on investigations of the lateral stiffness and resistance, shear lag effect, aseismic performance, and progressive collapse resistance of diagrid structures. RGHLS sustain vertical loads from Dead Loads and Live Loads of overall building structures and horizontal load from seismic /wind actions. In addition to direct vertical load, RGHLS sustains

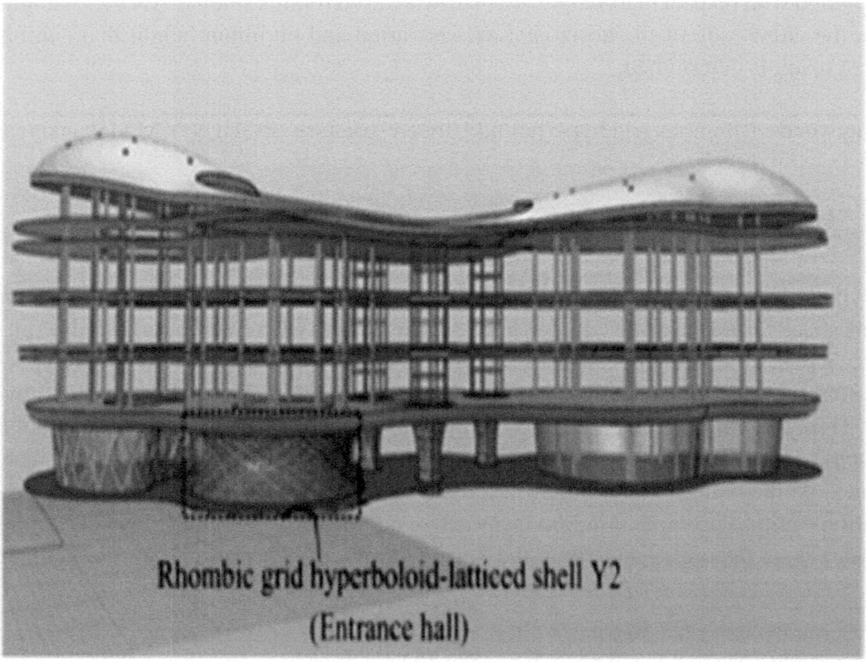

Fig. 1 Structural system of the china comic and animation museum [3]

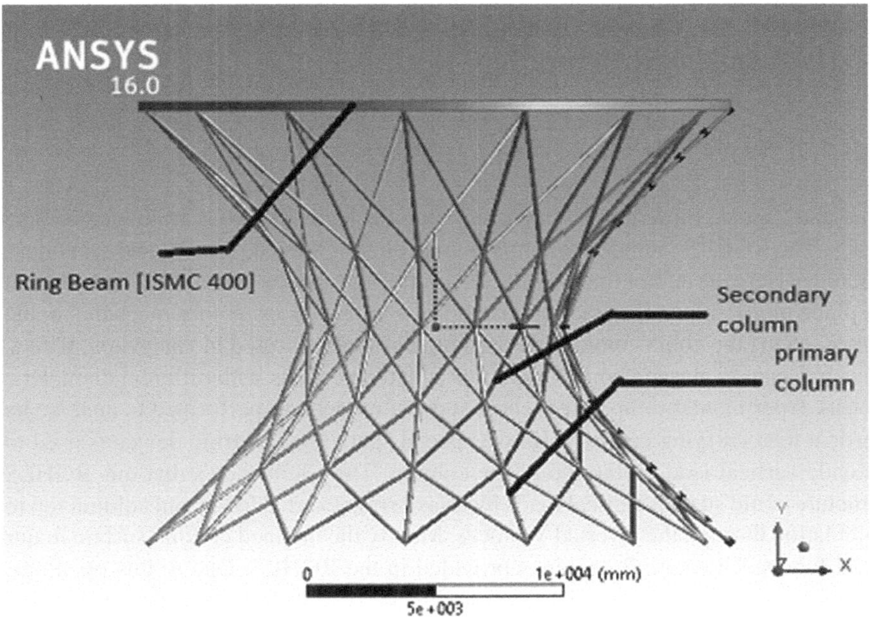

Fig. 2 Rhombic grid hyperboloid structure from model analysis

a combined in-plane and out-plane bending moment actions on its surface at the waist portion, making RGHLS prone to fail by Multi-column interaction instability (Fig. 2).

1.1 Objective

1. To analysis the RGHLS by changing the position of the central axis of the structure with respect to its height.
2. To study the effect of circular, square and triangular shapes of major and secondary column during its performance.

1.2 Scope

1 The effect of circular, square, triangular shapes of major and secondary columns are analysed in this project.
2. Only dead load and live load are to be considered.

2 Review on Various Studies on Rhombic Grid Hyperboloid Structure

2.1 *Grid Structures*

Guo and Zhu [1] present about the instability mechanism of RGHLS under vertical loads. The RGHLS includes two directional columns named primary and secondary columns. Here the author find outs the load carrying capacity corresponding to the out of plane multi-column interaction instability. For testing instability mechanism, the paper selects the china comic and Animation museum located in Hangzhou, China. This museum is placed above 6 hyperboloid latticed shells with different diameters. For its experimental setup, a reduced-scale test model was performed to analyze its vertical load carrying capacity. Here a special spatial beam-string device is used to provide vertical load on the top of the column. The result shows that this RGHLS structure could sustain vertical loads by transferring exerted load from column top to foundation through the inverted V-braces. Mainly the inclined column sustain major axial forces. There are 2 openings provided in the RGHLS. Due to this openings, un-uniform vertical compressive deformation occur on the top of the RGHLS during loading. The ultimate load carrying capacity obtained from the test is 3.37 times its design load. So, we can conclude that it has sufficient load carrying capacity and is in the safe margin.

Izadi and Bargi [2] conducted studies in Optimization and performance evaluation Natural draft steel hyperbolic cooling towers. Natural draft hyperbolic cooling towers that are doubly curved thin shell concrete structures that contribute to protection of environment and to efficient power generation. These structures are beneficial for cooling the thermal power plants by minimizing the need of water and thus avoiding thermal pollution of water bodies. Natural draft hyperbolic Cooling towers are important structures of thermal power plants, providing cooling across a broad range of applications. They are the structures which liberates heat to the atmosphere by the process known as evaporation, and are widely used in industries such as power plants, oil refining, steel mills, chemical processing and for manufacturing processes. Here the behaviour of this cooling tower is estimated by giving dead, wind and earthquake loads. Use of stiffening rings at the height of the tower gives a less weight structure but it may increase the cost of structure. Also observed that steel tower remain strong under earthquakes loads. Concrete tower does not show a good performance during strong earthquake.

Guo and Zhu [3] conducts analysis on load resistance and hysteretic response of rhombic grid hyperboloid-latticed shell. RGHLS is a structure consist of numerous bidirectional major and secondary columns interwine with each other to form the x-shaped joint with rhombic grid. The application of RGHLS are used as ground floor and they bear enormous amount of vertical and horizontal loads induced from upper building structures on top of them. The failure of this structure is in the form of in-plane and out-of plane multi-column interaction stability due to the twist deformation under vertical loads. Here the author selects RGHLS in the china comic and animation museum as a prototype. This prototype is then converted into reduced scale for the ease of experimental done in the laboratory. Out of two columns, major column sustain major percentage of vertical load. So compressive stiffness of major column is more as compared to secondary column. here the structure undergoes a cyclic displacement of 5, 10, 15, 20 mm etc. the conclusion of this analysis is that the reduced scale model failed due to strong in-plane and out-plane interaction instability between major and secondary columns under combined vertical and horizontal load actions.

Maden and Korkmaz [4] explains about the development of deployable doubly ruled hyperboloid surfaces. Geometric principles of hyperboloids are generated by ruled surface generation method and the morphology study had been discussed in detail. From this analysis, we can finalize that the behavior of hyperboloid is related to joint types at the intersection point of the bars. Result obtained from proposed deployable hyperboloid is that can be used as self-standing tower structures, so a no. of hyperboloid section can be joined together to form new geometric sections which serves as a shelter structures.

Morozov and Lopatin [5] conducted studies about Finite-element modelling and buckling analysis of anisogrid composite lattice cylindrical shells anisotropic grid composite lattice cylindrical shells are made of composite materials having high specific strength and stiffness. These structures are composed of curvilinear helical and circumferential ribs. The main load bearing members of these structures are ribs. Ribs can withstand both membrane and bending stiffnesses of the shell. These anisogrid structures are mainly used in rocket interstages, payload adapters for spacecraft launchers etc. One of the major failure mode of anisogrid are buckling under various loading. For analysing buckling, there are two approaches are to be used. First one is continuum model, structures fully taken as a shell and analysed. Second approach is discrete modelling. Discrete modelling is done through finite element modelling. In the case of axial loading, three different buckling modes have been identified. From these analysis, we can conclude that the impact of critical load decreases with the increase of shell length.

Loukaides and Seffen [6] conducted studies about Multistable grid and honeycomb shells. Multistable shell is a combination of pre-stressed and composite materials. These shell have more applications since they offer a construction of morphing structures. Multistable shell have the ability to support loads in distinct geometries, without any elaborate mechanisms. Now a days, unsymmetrical laminates plays an important role in engineering applications of bistable shells. Aim of this study is to prove these honeycomb and grid shells can be made to be bistable, when initially free from stresses and is of singly curved one. For lower the dependence on more expensive and harder to manufacture composite materials, the attractive approach is to use local geometry of a material. Both shells are of connected networks of beam-like elements, but their elastic responses vary due to the difference in their local geometric proportions. The cross-section of grid shell beams have comparable width and depth, so they act as classical euler-bernoulie beams. But in honeycomb beams, depth to width ratio is more than 10 and it behave like slender webs. Therefore it bears more local distortions and instabilities when bending and torsion are applied.

3 Analysis

The FEM analysis consist of three stages: Modelling, Solving and Post processing. Modelling phase include geometric modelling of the structure, material definition, meshing and applying boundary conditions and mechanical forces. Solving phase consists of applying loads under specific load sequence and solving through analysis. Post processing stage includes the interpretation and analysis of FEM results in order to find the structural response of the model under applied loadings.

3.1 Modelling

Modelling is carried out by changing the position of the central axis of the structure with respect to its height. Here three models are carried out.

1. **Model 1**: Equal height on either side of the central axis.
2. **Model 2**: One-fifth height of the model above central axis and four-fifth below central axis.
3. **Model 3**: One-fifth height of the model below central axis and four-fifth height above central axis.

3.1.1 General Equation of Hyperbola

$$\frac{X^2}{a^2} - \frac{Y^2}{b^2} = 1,\ \text{where } \mathrm{a}^2 = 25 \text{ and } \mathrm{b}^2 = 20 \tag{1}$$

See Table 1.

Table 1 Values obtained from the general equation of Hyperbola

$\frac{X^2}{a^2}$	$\frac{Y^2}{b^2}$	X	y
1	0	5	0
1.4999	0.4999	6.12352	3.16196
2	1	7.0716	4.472
2.8	1.8	8.3666	6
3.6	2.6	9.4868	7.2111
4.4	3.4	10.4880	8.2462
5.2	4.2	11.4017	9.16515

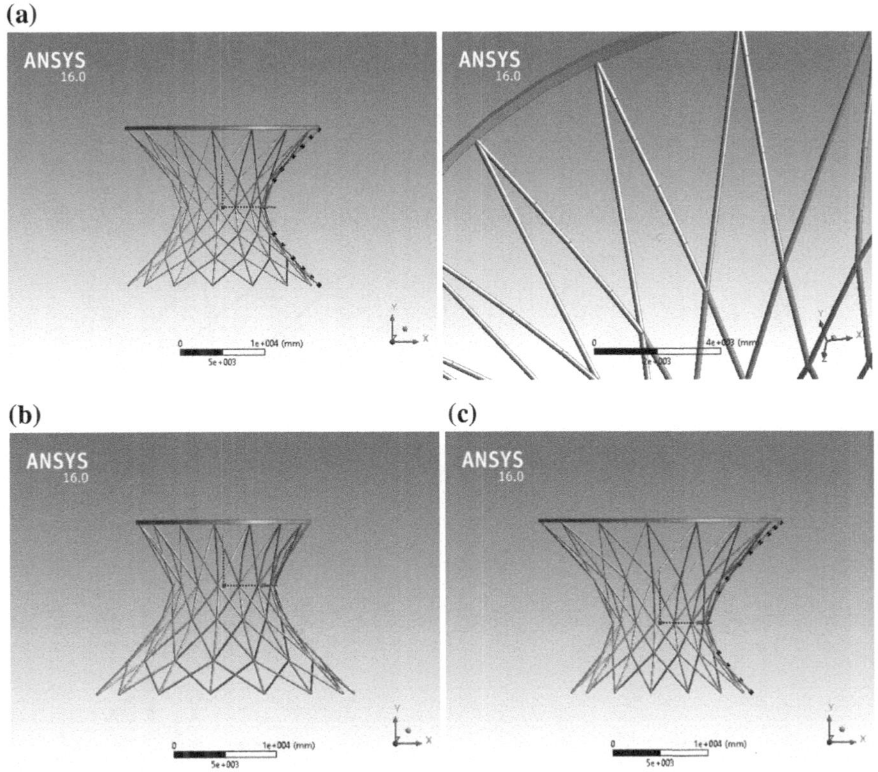

Fig. 3 Models of circular grid shape: **a** Model 1. **b** Model 2. **c** Model 3

3.1.2 Circular Grid Shape

See Fig. 3.

3.1.3 Square Grid Shape

See Fig. 4.

3.1.4 Triangular Grid Shape

Analysis is carried out by applying load on the top of the structure. Provide a yield strength of 446 Mpa. Fixed support is provided at bottom of each columns. The result obtained from the analysis are total deformation, equivalent stress and strain (Fig. 5).

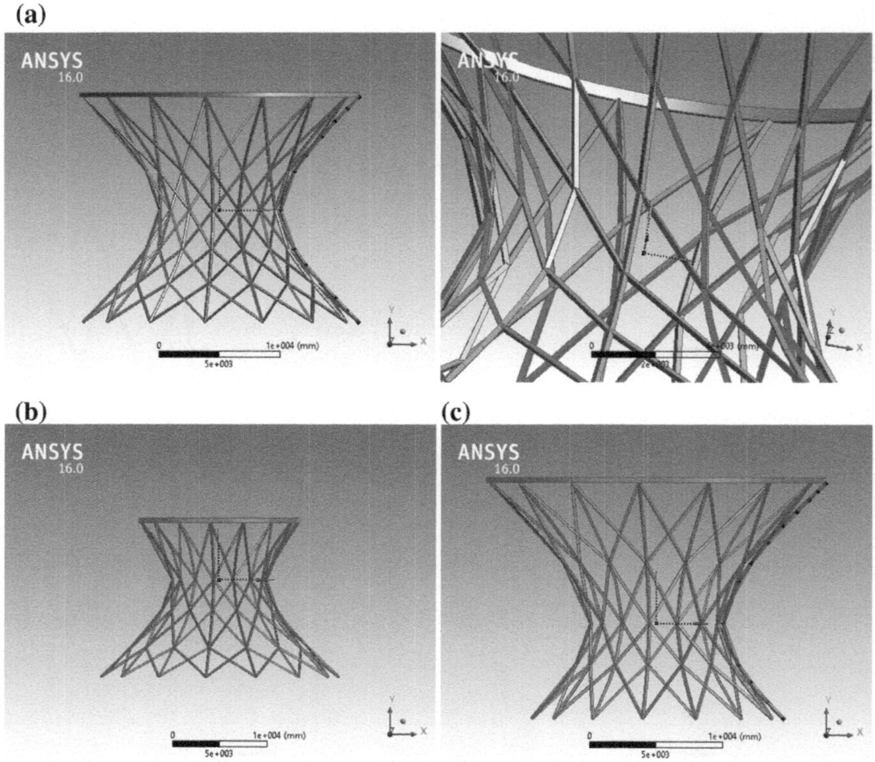

Fig. 4 Models of square grid shape: **a** Model 1. **b** Model 2. **c** Model 3

4 Result and Discussion

In circular shaped structure (Table 2; Figs 6, 7 and 8).

Model 1 shows a better performance up to the load of 2600 KN, Model 2 shows a better performance up to 3100 KN and Model 3 up to the load 2350 KN, after that stress value become reducing.

In square shape,

Model 1—load up to 2150 KN
Model 2—load up to 2400 KN
Model 3—load up to 1700 KN

In Triangular cross-sectional structure,

Model 1—load up to 3450 KN
Model 2—load up to 3800 KN
Model 3—load up to 3700 KN

From the three above shapes, Model 2 shows better result.

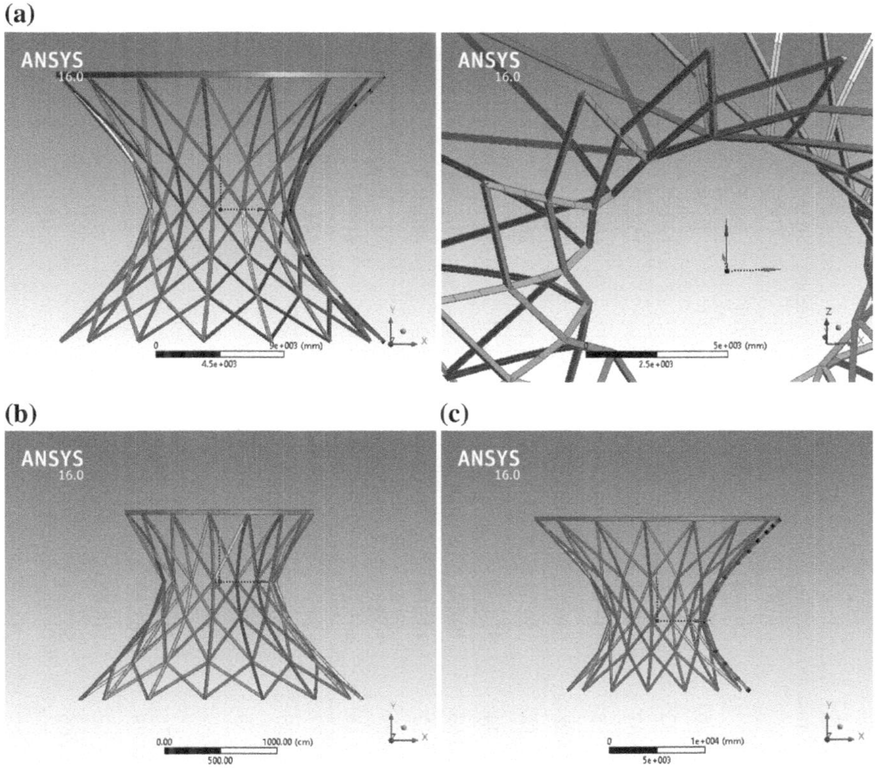

Fig. 5 Models of triangular grid shape: **a** Model 1. **b** Model 2. **c** Model 3

From the above showing deformation and stress diagrams, Maximum deformation is occur at top of the ring beam and maximum stress occurs at middle portion of the structures. Here, the structure acts like a cantilever beam as the load is applied at top of the structure and fixed support is provided at the bottom.

From the three above models, Model 2 shows better result. Among the three cross-sectional shape models, triangular cross-sectional shape shows maximum load carrying capacity. Mostly the model 2 structures are used as Cooling towers, water towers etc. At the bottom of cooling tower, the widening provides a large area for installation of fill to promote thin film evaporative cooling of the circulated water. As the water first evaporates and rises, the narrowing effect helps accelerate the laminar flow, and then as it widens out, contact between the heated air and atmospheric air supports turbulent mixing. So model 2 is best for this type of structures.

Table 2 Result obtained from the analysis

	Load (kN)	Model 1			Model 2			Model 3		
		Deformation (mm)	Stress (MPa)	Strain	Deformation (mm)	Stress (MPa)	Strain	Defromation (mm)	Sress (MPa)	Strain
Circular	200	31.684	37.365	0.00024	30.631	150.46	0.000761	43.976	38.088	0.00030425
	400	63.369	74.731	0.00048	61.263	253.29	0.0013	87.951	38.088	0.00030425
	600	95.053	112.1	0.000723	91.925	279.32	0.0017	131.93	114.26	0.00091274
	800	126.74	149.46	0.000964	122.67	320.69	0.003	175.9	152.35	0.001217
	1000	158.42	186.83	0.00121	153.65	343.93	0.002445	219.88	190.44	0.0015212
	1200	190.11	224.19	0.00145	184.98	346.57	0.00259	263.85	228.53	0.0018255
	1400	221.79	261.56	0.0015	216.58	359.37	0.00262	307.83	266.62	0.0020163
	1600	253.48	298.92	0.00193	248.44	376.48	0.0026	351.81	304.7	0.0021698
	1800	285.16	336.26	0.00213	280.63	404.31	0.00268	395.8	342.78	0.0024411
	2000	316.85	373.65	0.00232	313.12	433.69	0.00272	439.8	380.86	0.0026062
	2200	348.56	411.02	0.0025049	345.89	456.22	0.0026057	483.81	397.56	0.0024922
	2300	351.55	422.66	0.0025051	357.65	465.64	0.0026061	505.84	415.52	0.0024672
	2350	365.67	435.87	0.0026216	365.77	486.23	0.0027244	516.85	424.53	0.0024207
	2400	380.29	448.38	0.0027237	378.97	490.39	0.0027252	527.97	420.02	0.0024481
	2600	412.06	463.67	0.0028823	412.39	520.22	0.0026187			
	2650	420.01	459.59	0.0029379						
	2800				446.42	545.62	0.0028307			
	3000				481.78	563.91	0.0031732			
	3100				500.02	578.85	0.0029058			
	3200				518.66	567.54	0.0028485			

(continued)

Table 2 (continued)

	Load (kN)	Model 1			Model 2			Model 3		
		Deformation (mm)	Stress (MPa)	Strain	Deformation (mm)	Stress (MPa)	Strain	Defromation (mm)	Sress (MPa)	Strain
Square	200	34.067	59.851	0.00033864	19.902	63.651	0.00042588	31.204	54.022	0.0003035
	400	68.137	59.851	0.00067728	39.803	89.311	0.00065195	62.408	108.04	0.000607
	600	102.2	179.55	0.0010159	59.705	114.04	0.00089	93.612	162.07	0.0009105
	800	136.27	239.4	0.001355	79.608	147.04	0.00112	124.82	216.09	0.001214
	1000	170.34	299.25	0.000169	99.511	181.94	0.001344	156.02	270.11	0.0015175
	1200	204.4	359.11	0.002032	119.41	218.33	0.00156	187.22	324.14	0.001821
	1400	238.47	345.85	0.001971	139.32	254.72	0.00155	218.43	385.74	0.0021403
	1600	272.54	333.52	0.001865	159.22	291.11	0.00171	234.03	385.74	0.0021403
	1700	271.43	323.33	0.001855	165.88	301.77	0.00172	265.24	409.86	0.0022741
	1800	306.62	375.2	0.001982	179.13	327.49	0.00193	280.84	405.51	0.0022745
	2000	340.71	452.98	0.002284	199.03	363.88	0.00214			
	2150	366.28	448.15	0.002262						
	2160	367.98	444.75	0.002243						
	2200	374.81	452.98	0.002284	218.94	359.76	0.0021304			
	2400				238.84	393.11	0.0022035			
	2600				258.75	367.94	0.00219			
Triangular	200	28.685	42.923	0.00025396	13.79	20.562	0.000195	31.221	34.014	0.000174
	400	57.37	85.847	0.00050792	27.579	41.123	0.000389	62.442	68.028	0.00035
	600	86.054	128.77	0.00076188	41.369	61.685	0.000584	93.663	102.04	0.000523

(continued)

Table 2 (continued)

Load (kN)	Model 1			Model 2			Model 3		
	Deformation (mm)	Stress (MPa)	Strain	Deformation (mm)	Stress (MPa)	Strain	Defromation (mm)	Sress (MPa)	Strain
800	114.74	171.69	0.00010158	55.158	82.246	0.000778	124.88	136.06	0.000698
1000	143.42	214.62	0.0012698	68.948	102.81	0.000972	156.1	170.07	0.0008722
1200	172.11	257.54	0.0015238	82.737	123.37	0.001167	187.33	204.09	0.0010466
1400	200.83	300.46	0.0017777	96.527	143.93	0.001361	218.55	238.1	0.001221
1600	229.59	343.38	0.0020316	110.32	164.49	0.001556	249.77	272.11	0.0013955
1800	258.35	386.3	0.0022855	124.11	185.05	0.001459	280.09	306.13	0.0015699
2000	287.18	421.19	0.0021112	137.92	205.63	0.0016167	312.21	340.14	0.0017443
2200	316.15	463.52	0.0023215	151.73	216.45	0.0017782	343.43	374.16	0.0019188
2400	345.16	505.88	0.0025331	165.55	227.37	0.0019556	374.65	408.17	0.0020932
2600	374.27	548.24	0.0027458	179.37	250.25	0.0021395	405.93	442.22	0.0022678
2800	403.45	590.63	0.0029581	193.2	272.32	0.0023132	437.23	476.27	0.0024424
3000	432.66	633.04	0.0031705	207.02	295	0.002228	468.61	510.33	0.0026171
3100	447.28	654.25	0.00322768						
3200	476.59	640.15	0.0033117	220.85	317.83	0.0023492	500.01	544.41	0.0027919
3400	491.27	682.34	0.0034178	234.39	340.76	0.0024713	531.45	563.18	0.0028841
3450	498.63	692.93	0.003471						
3500	506.01	652.7	0.0032677				547.21	570.24	0.0028844
3600	504.11	650.44	0.0032544	248.52	361.78	0.0026115	563.09	563.26	0.0028859
3700							579.03	577.99	0.00296

(continued)

Table 2 (continued)

	Load (kN)	Model 1			Model 2			Model 3		
		Deformation (mm)	Stress (MPa)	Strain	Deformation (mm)	Stress (MPa)	Strain	Defromation (mm)	Sress (MPa)	Strain
	3800	500.99	624.65	0.0030145	262.39	368.2	0.0027177	595.09	571.19	0.0029284
	3850	489.66	600.11	0.00299101	265.86	356.12	0.0027487	590.01	569.11	0.0029275

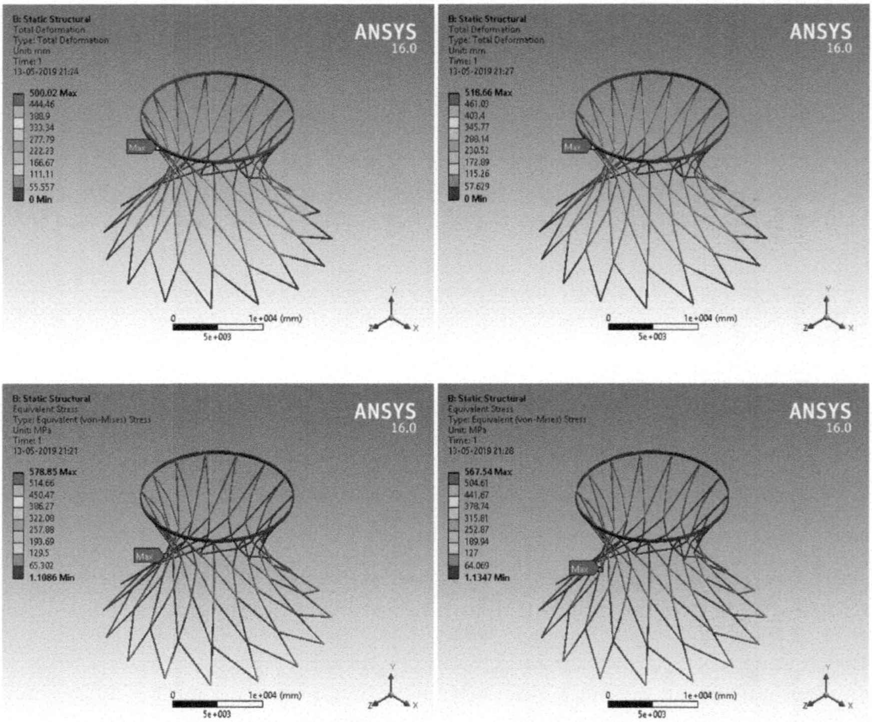

Fig. 6 Figure showing the values of deformations and stress of Model 2 under the loads 3100 and 3200 KN (Circular)

5 Conclusions

Main application of RGHLS is act as a ground floor and they sustain enormous vertical and horizontal loads induced from upper building structures on top of them. Rhombic grid provide more strength as compared to other grid shapes. From the result of the analysis, Model 2 (one-fifth height above Central axis and four-fifth height below Central axis) shows more load carrying capacity as compared to other two models in three different cross-sectional shapes. So, the structure with one-fifth height above central axis and four-fifth height below central axis shows better result compared to others. Model 2 shows better result because of its larger diameter at bottom surface provide wider area and can withstand more load as compared to other two models. Also, from the three shapes of the cross-section of the structures, triangular shape shows maximum load carrying capacity. So it can concluded that Model 2 with triangular cross-section shows best performance compared to others.

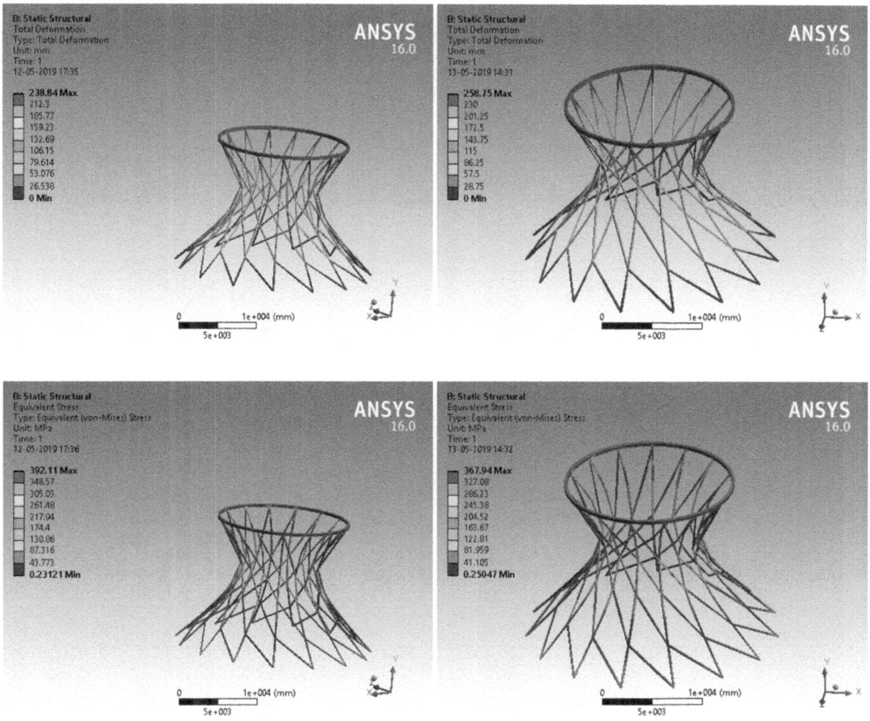

Fig. 7 Values of deformations and stress of Model 2 under the loads 2400 and 2600 KN (Square)

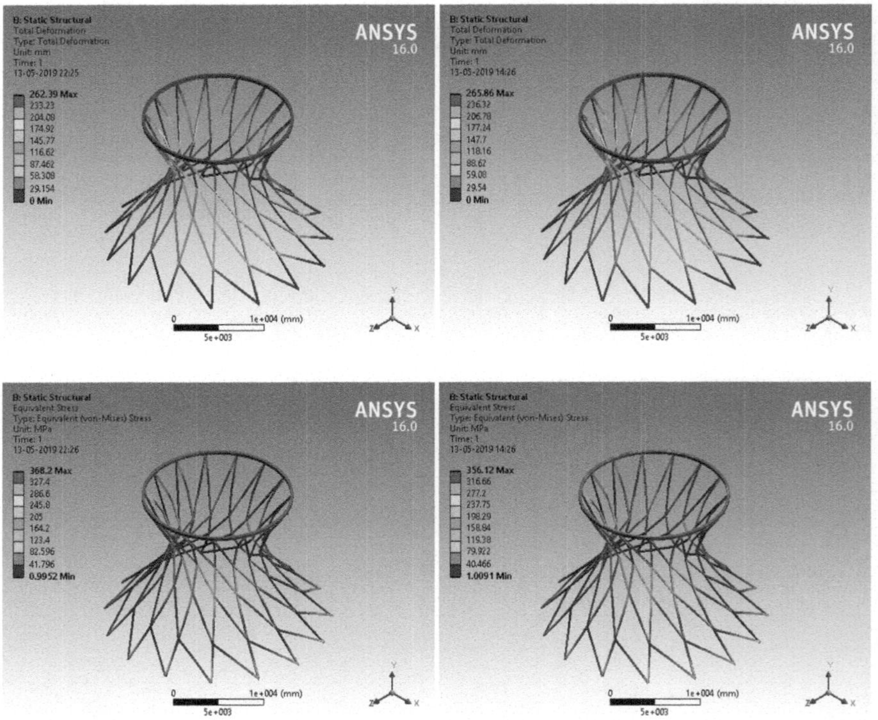

Fig. 8 Values of deformations and stress of Model 2 under the loads 3800 and 3850 KN (Triangular)

References

1. Guo Y-L, Zhanga Y-H, Zhu B-L (2018) Experimental and numerical studies of instability mechanism and load resistance of rhombic grid hyperboloid- latticed shells under vertical load. J Eng Struct 166(2018):167–186
2. Izadi M, Bargi K (2013) Natural draft steel hyperbolic cooling towers: optimization and performance evaluation. J Struct Design Tall Spec Build. https://doi.org/10.1002/tal.1081
3. Guo Y-L, Zhu B-L, Zhou P (2017) Experimental and numerical investigation into the load resistance and hysteretic response of rhombic grid hyperboloid-latticed shells. J Eng Struct 153:700–716
4. Maden F, Korkmaz K (2017) Geometric and kinematic analysis of deployable doubly ruled hyperboloids. Megaron 12(3):343–354
5. Morozov EV, Lopatin AV (2011) Finite-element modelling and buckling analysis of anisogrid composite lattice cylindrical shells. J Compos Struct 93(2011):308–323
6. Loukaides EG, Seffen KA (2015) Multistable grid and honeycomb shells. J Solids Struct 59(2015):46–57

The manufacturer's authorised representative in the EU is Springer Nature Customer Service Centre GmbH, Europaplatz 3, 69115 Heidelberg, Germany. If you have any concerns regarding our products, please contact ProductSafety@springernature.com

Printed and bound by CPI Group (UK) Ltd, Croydon, CR0 4YY

30/09/2026

02225463-0001